T0131435

Modelling Transport

Modelling Transport

Fifth Edition

Juan de Dios Ortúzar
Department of Transport Engineering and Logistics,
Instituto Sistemas Complejos de Ingenieria (ISCI).
Pontificia Universidad Catolica de Chile

Luis G. Willumsen
Nommon Solutions and Technologies

Registered Offices

John Wiley & Sons, Inc., 111 River Street, Hoboken, NJ 07030, USA

John Wiley & Sons Ltd, The Atrium, Southern Gate, Chichester, West Sussex, PO19 8SQ, UK

For details of our global editorial offices, customer services, and more information about Wiley products visit us at www.wiley.com.

Wiley also publishes its books in a variety of electronic formats and by print-on-demand. Some content that appears in standard print versions of this book may not be available in other formats.

Library of Congress Cataloging-in-Publication Data applied for:

Hardback ISBN 9781119282358

Cover Design: Wiley

Cover Image: © Scharfsinn/Shutterstock

Set in 9.5/12.5pt STIXTwoText by Straive, Pondicherry, India

SKY10079021_070424

Contents

Preface

This book results from over 50 years of collaboration, often at a distance and sometimes working together in Britain and Chile. During this half-century, we have often discussed the more substantial and weaker aspects of transport modelling and planning. We have also speculated, researched, and tested some new and not-so-new ideas in practice. We have agreed and disagreed on topics such as the level of detail required for modelling or the value of disaggregate and activity-based models in forecasting; some 33 years ago, we took advantage of a period when our views converged to put them in writing, but they have evolved and continue to do so.

At that time, we decided to present the most important (in our view) transport modelling techniques in a form accessible to students and practitioners alike. We attempted this giving particular emphasis to key topics in contemporary modelling and planning:

- the practical importance of theoretical consistency in transport modelling;
- the issues of data and specification errors in modelling, their relative importance and methods to handle them;
- the key role played by the decision-making context in the choice of the most appropriate modelling tool;
- how uncertainty and risk influence the choice of the most appropriate modelling tool;
- the advantages of variable resolution modelling; a simplified background model coupled with a much more detailed one addressing the decision questions in hand;
- the need for a monitoring function relying on regular data collection and updating of forecasts and models so that courses of action can be adapted to a changing environment.

However, since we wrote that first edition, a lot has changed. The clear distinction between private and schedule-based public transport has been disrupted by new forms of mobility: vehicle sharing, electric bikes, and, sometime in the future, autonomous vehicles. Also, transport planning objectives have evolved: today, the reduction of emissions and the provision of equitable access to all sections of the community are at the top of the agenda; they were secondary, if at all, in the twentieth century. More complex problems call for better tools to deal with them and develop a broader understanding and more considered judgment. Luckily, the profession has responded to this challenge, and this fifth edition of *Modelling Transport* attempts to bring these better tools and understanding to our audience.

In writing this book, we aimed to create both a text for a diploma or Master's course in transport and a reference volume for practitioners. However, we present the material in such a way as to be useful for undergraduate courses in civil engineering, geography, and town planning. We approached the subject from the point of view of a modelling exercise, discussing the role of theory, data, model specification in its broadest sense, model estimation, validation, and forecasting. Initially, we based the book on our lecture notes, prepared and improved over several years of teaching at undergraduate and graduate levels; we have also used them to teach practitioners through in-house training programmes and short skills-updating courses. We have extended and enhanced our lecture notes to cover additional material and to help the reader tackle the book without the support of a supervisor.

Chapters 2–9, 13, and 18 provide all the elements necessary to run a good 30 sessions course on transport demand modelling; in fact, such a course – with different emphases on certain subjects – has been taught by us at undergraduate level in Chile and at postgraduate level in Australia, Britain, China, Colombia, Germany, Italy, Mexico, Portugal, and Spain; the addition of material from Chapters 10–12 would make it a transport modelling course. Chapters 4–6, 10–12, and 14 provide the basic core for a course on network modelling and equilibrium in transport; a course on transport supply modelling would require more material, particularly relating to certain aspects of public transport supply which we do not discuss in enough detail. Chapters 15–17, 19, and 20 cover material which is getting more important as time goes by, in particular as the shift in interest in the profession is moving from passenger issues to freight and logistics and to the role that models play not only in social evaluation but also in the analysis of private projects. Chapter 1 provides an introduction to transport planning issues and outlines our view on the relationship between planning and modelling.

During our professional life, we have been fortunate to combine teaching with research and consultancy practice. We have learnt from the literature, our research and experimentation, and our mistakes. The latter has not been too expensive in terms of inaccurate advice. However, this is not just luck; a conscientious analyst pays for mistakes by working harder and longer to sort out alternative ways of dealing with a problematic modelling task. We have also learnt the importance of choosing appropriate techniques and technologies for each task; the ability to tailor modelling approaches to decision problems is an essential skill in our profession. Throughout the book, we examine the practical constraints to transport modelling for planning and policy-making in general, particularly given the limitations of current formal analytical techniques and the nature and quality of the data likely to be available.

We have avoided the intricate mathematical detail of every model to concentrate instead on their basic principles, identifying their strengths and limitations, and discussing their use. The level of theory supplied by this book is sufficient to select and use the models in practice. We have tried to bridge the gap between the more theoretical publications and a simplistic 'how to' book offering a blueprint to each modelling problem. In this latest edition, we have also marked, with a shaded box, material which is more advanced or still under development but essential enough to be mentioned. There are no single solutions to transport modelling and planning issues. A recurring theme in the book is the dependence of modelling on context and theory. We aim to provide enough information and guidance so that readers can go and use each technique in the field; to this end, we have striven to look into practical questions about the application of each methodology. Wherever the subject area is still under development, we have striven to extensively reference more theoretical papers and books, which the interested reader can consult as necessary. Concerning other, more settled modelling approaches, we have kept the references to those essential for understanding the evolution of the topic or serving as entry points to further research.

Nobody can aspire to become a qualified practitioner in any area without working in a laboratory or field. Therefore, we have gone beyond the sole description of the techniques and have accompanied them with various application examples. These illustrate some of the theoretical or practical issues related to particular models. We provide a few exercises at the end of key chapters; these can be solved with the help of a scientific pocket (or better still, a spreadsheet) calculator and should assist in the understanding of the models discussed.

Although the book is ambitious in covering many themes, it must be made clear from the outset that we do not intend (nor believe it possible) to be up-to-the-minute on every topic. The book is a good reflection of the state-of-the-art, but for leading-edge research, the reader should use the references provided as signposts for further investigation.

We wrote most of the first edition during a sabbatical visit by the first of us to University College London in 1988–89. This was possible owing to the support provided by the *UK Science and Engineering Research Council*, *The Royal Society*, *Fundación Andes* (Chile), *The British Council*, and *The Chartered Institute of Transport*. We thank them for their support as we acknowledge the funding provided for our research by many institutions and agencies over the past 50 years. The third and fourth editions benefited greatly from further sabbatical stays at University College London in 1998–99 and 2009; these were possible owing to the support provided by the *UK Engineering and Physical Sciences Research Council*. We also wish to acknowledge the support for our research provided by several *FONDECYT* projects in Chile, by *Instituto Sistemas Complejos de Ingeniería* (ISCI), through grant ANID PIA/PUENTE AFB230002; the *Centre for Sustainable Urban Development* (CEDEUS), through grant CEDEUS/FONDAP/15110020; and the *BRT+ Centre of Excellence* funded by the Volvo Research and Educational Foundations.

We have managed to maintain a fairly even intellectual contribution to the contents of this book, but in writing and researching material for it, we have also benefited from numerous discussions with friends and colleagues. Richard Allsop taught us a good deal about methodology and rigour. Huw Williams's ideas are behind many of the theoretical contributions in Chapter 7; Andrew Daly and Hugh Gunn helped to clarify many issues in Chapters 3, 7–9, and 18. Dirck Van Vliet's emphasis in explaining assignment and equilibrium in simple but rigorous terms inspired Chapters 10–12. Tony Fowkes made valuable comments on car ownership forecasting and stated-preference methods. Jim Steer provided a constant reference to practical issues and the need to develop improved approaches to address them.

Many parts of the first edition of the book also benefited from a free, and sometimes very enthusiastic, exchange of ideas with our colleagues J. Enrique Fernández and Joaquín de Cea at Pontificia Universidad Católica de Chile, Sergio Jara-Díaz and Jaime Gibson at Universidad de Chile, Marc Gaudry then at Université de Montréal, Roger Mackett at University College London, and Dennis Gilbert and Mike Bell then at Imperial College. Many others also contributed, without knowing, to our thoughts.

Subsequent editions of the book have benefited from comments by a number of friends and readers, apart from those above, who have helped to identify errors and areas for improvement. Among them we should mention Francisco Bahamonde-Birke at Tilburg University; Michel Bierlaire at Ecole Polytechnique Fédérale de Lausanne; Patrick Bonnel at the French Laboratoire d'Economie des Transports; David Boyce at University of Illinois; Victor Cantillo at Universidad del Norte, Barranquilla; Elisabetta Cherchi at Newcastle University; Michael Florian at Université de Montréal; Rodrigo Garrido, Ricardo Hurtubia, Luis I. Rizzi, and Francisca Yáñez from Pontificia Universidad Católica de Chile; David Hensher at ITLS, Sydney University; Ben Heydecker at University College London; Frank Koppelman at Northwestern University; Mariëtte Kraan at University of Twente; C. Angelo Guevara, Francisco J. Martínez, and Marcela Munizaga at Universidad de Chile; Piotr Olszewski at Warsaw University of Technology; Alejandro Tudela at Universidad de Concepción; Joan L. Walker at University of California at Berkeley; and Sofia Athanassiou, Neil Chadwick, Yaron Hollander, Gloria Hutt, Serbjeet Kohli, and John Swanson while they were at Steer. Special thanks are due to John M. Rose at ITLS, University of Sydney, for his contributions to Chapter 2;

to Stephane Hess at Leeds University and Camila Balbontín at Pontificia Universidad Católica de Chile, for their contributions to Chapters 7 and 8; and to Jose Holguín-Veras at Rensselaer Polytechnic Institute for his contribution to Chapter 15.

We have not taken on board all suggestions, as we felt some required changing the approach and style of the text; we are satisfied that future books will continue to clarify issues and provide greater rigour to many of the topics discussed here; transport is indeed a very dynamic subject.

We are grateful to the skilled editors at Wiley, whose efforts have helped us to express our ideas more clearly with their attention to detail and excellent work.

Our final thanks go to our graduate and undergraduate students in many countries; they are always sharp critics and provided the challenge to put our money (time) where our mouth was. We are also grateful to Tomás Ramírez for having carefully re-drawn most of the figures in the book.

Despite all this generous assistance, we are, as usual, solely responsible for any errors remaining in this latest edition of the book. We genuinely value the opportunity to learn from our mistakes.

Juan de Dios Ortúzar and Luis G. Willumsen

About the Companion Website

This book is accompanied by a companion website.

www.wiley.com/go/ortuzar5e

This website includes:

- Solution Manual

1

Introduction

1.1 Background

We cannot predict the future with certainty, but we can prepare for it by designing policies and projects that we expect will improve the welfare and quality of life of a community. We can adopt an experimental approach to the design of some simple schemes, for example, changing the allocation of lanes to different types of users, using markers that can be removed later; we can then check whether the results from the experiment are positive and make them permanent, or return the roadway to its original use. We can test interventions like these in an emergency; for example, during the 2020 pandemic some of the road lanes were repurposed for cyclists, pedestrians, or pavement cafes on a temporary basis; as most people liked them, many have become permanent features. Nevertheless, this is not feasible for more significant interventions like a new road or tram link because the political, economic, and environmental costs would be too large. Moreover, most investments in transportation take several years to plan, implement, and mature, and, therefore, their impacts will happen mostly in the future and probably under different conditions, so no experiment would be truly valid; we need a different approach.

A good alternative is to develop a sufficiently realistic transport model where various interventions, policies, and projects can be tested for their performance against given objectives. This phrase hides two difficult issues: (i) how realistic the model should be, and (ii) what are the most relevant indicators of the performance of an intervention against objectives. This is the central topic of this introductory chapter. We look, first, at transport problems and the objectives to tackle them; indeed, they are two interconnected problems. Then, we discuss the nature of the transport models required to address these problems and use a simple example to illustrate their character and scope. Afterwards, we consider key issues in modelling and model design, concluding with comments on the apparent conflict between theory and practice.

The first part of the twenty-first century has seen two powerful agents of change affecting most aspects of life and welfare, and of course, transport modelling. The first is *technological progress*, in particular cheap and fast telecommunications facilitating new forms of requesting, using, and paying for mobility. A consequence of this progress is the possibility of working and procuring goods and services remotely: telework and e-commerce continuously change travel patterns. The second agent is the requirement that *equity* and *fairness* take a central role in designing transport interventions; this refers to both intra- and inter-generational fairness. The former deals with the importance of a fair distribution of access to opportunities for the community rather than just the provision of time savings, for example. Inter-generational equity, on the other hand, demands that protecting the environment for future generations be central to any policy and programme of investment. Notwithstanding, these two key agents of change, technology and equity, must be addressed in the context of a third important element, *high uncertainty,* which has a profound effect on modelling and decision-making.

When we think about these issues, we cannot forget the role of transport infrastructure in enhancing the economic competitiveness of nations, the role of telecommunications in reducing the need to travel, the transformational impact of new technologies and the overarching objectives of reducing emissions and providing better access to opportunities to all sections of the community.

1.2   Models and Their Role

A *model* is a simplified representation of a part of the real world – the system of interest – which focuses on certain elements considered important from a particular point of view. Models are, therefore, problem- and viewpoint-specific. Such a broad definition allows us to incorporate both physical and abstract models. In the first category, we find, for example, those used in architecture or fluid mechanics, which are basically aimed at design. In the latter, the range spans from the mental models all of us use in our daily interactions with the world to formal and abstract (typically analytical) representations of some theory about the system of interest and how it works. Mental models play an important role in understanding and interpreting the real world and our analytical models. They are enhanced through discussions, training, observation and, above all, experience. Mental models are, however, difficult to communicate and to discuss.

In this book, we are concerned mainly with an important class of abstract models: mathematical models. These models attempt to replicate the system of interest and its behaviour by means of mathematical equations based on certain theoretical statements about it. Although they are still simplified representations, these models may be very complex and often require large amounts of data to be used. However, they are invaluable in offering a 'common ground' for discussing policy and examining the inevitable compromises required in practice with a certain level of objectivity. Another important advantage of mathematical models is that during their formulation, calibration, and use, the planner can learn much, through experimentation, about the behaviour and internal workings of the system under scrutiny. In this way, we also enrich our mental models, thus permitting more intelligent management of the transport system.

A model is only realistic from a particular perspective or point of view. It may be reasonable to use a knife and fork on a table to model the position of cars before a collision, but not to represent their mechanical features or their route choice patterns. The same is true of analytical models: their value is limited to a range of problems under specific conditions. The appropriateness of a model is, as discussed in the rest of this chapter, dependent on the context where it will be used. The ability to choose and adapt models for particular contexts is one of the most important elements in the complete planner's toolkit.

This book is concerned with the contribution transport modelling can make to improved decision-making and planning in the transport field. We argue that the use of models is inevitable and that formal models are highly desirable. However, transport modelling is only one element in transport planning: administrative practices, an institutional framework, skilled professionals, and good levels of communication with decision-makers, the media, and stakeholders are some of the other requisites for an effective planning system. Moreover, transport modelling and decision making can be combined in different ways depending on local experience, traditions, and expertise.

Transport modelling has a long trajectory from the early beginnings in Detroit in the 1950s. Boyce and Williams (2015) provide us with an excellent non-mathematical presentation of the evolution of theories, methods, and models underpinning transport forecasts and policy analysis. Their book is worth reading to gain an in-depth understanding of the evolution and role of transport modelling and planning.

Before we discuss how to choose a modelling and planning approach, it is worth outlining some of the main characteristics of transport systems and their associated problems. We will also discuss some important modelling issues, which will find application in other chapters of this book.

1.3 Characteristics of Transport Problems

Our understanding of transport problems has evolved over time. Traditionally, we have seen a general increase in road traffic and transport demand that has resulted in congestion, delays, accidents, and environmental problems well beyond what has been considered acceptable so far. These problems have not been restricted to roads and car traffic alone. Economic growth seems to have generated levels of demand exceeding the capacity of many transport facilities. In this sense, transport problems have been perceived as a mismatch between demand expectations and the supply of transport services, including road space. This is consistent with conventional economic thinking.

1.3.1 Characteristics of Transport Demand

The demand for transport is *derived*; it is not an end in itself. With the possible exception of sightseeing, people travel in order to satisfy a need (work, leisure, or health) undertaking an *activity* at particular locations. This is equally significant for goods movements. In order to understand the demand for transport, we must understand the way in which these activities are distributed over space, in both urban and regional contexts. A good transport system should provide equitable opportunities to satisfy these needs; a heavily congested or poorly connected system restricts options and may restrain economic and social development.

The demand for transport services is highly *qualitative* and *differentiated*. There is a whole range of specific demands for transport that are differentiated by time of day, day of the week, journey purpose, type of cargo, importance of speed and frequency, and so on. A transport service without the attributes matching this differentiated demand may well be useless. This characteristic makes it more difficult to analyse and forecast the demand for transport services: tonne and passenger kilometres are extremely coarse units of performance hiding an immense range of requirements and services.

Transport demand takes place over *space*. This seems a trivial statement, but it is the distribution of activities over space, which creates transport demand. There are a few transport problems that may be treated, albeit at a very aggregate level, without explicitly considering space. However, in the vast majority of cases, the explicit treatment of space is unavoidable and highly desirable. The most common approach to treat space is to divide study areas into zones and code them, together with transport networks, in a form suitable for processing with the aid of computer programs. In some cases, study areas can be simplified by assuming that the zones of interest form a corridor, which can be collapsed into a linear form. However, different methods for treating distance and for allocating origins and destinations (and their attributes) over space are an essential element in transport analysis.

Finally, transport demand and supply have very strong *dynamic* elements. A good deal of the demand for transport is concentrated on a few hours of a day, in particular in urban areas where most of the congestion takes place during peak periods. In effect, the peak of demand usually happens at different times and in different locations in large cities. This time-variable character of transport demand makes it more difficult – and interesting – to analyse and forecast. It may well be that a transport system could cope well with the *average* demand for travel in an area but that it breaks down during peak periods. A number of techniques exist to try to spread the peak and average the load on the system: flexible working hours, staggering working times, premium pricing, and so on. However, peak and off-peak variations in demand remain a central, and fascinating, problem in transport modelling and planning.

1.3.2 Characteristics of Transport Supply

The first distinctive characteristic of transport supply is that it is a *service* and not a good. Therefore, it is not possible to stock it, for example, to use it in times of higher demand. A transport service must be consumed when and where it is produced; otherwise, its benefit is lost. For this reason, it is very important to estimate demand with as much accuracy as possible to save resources by tailoring the supply of transport services to it.

Many characteristics of transport systems derive from their nature as a service. In very broad terms, a transport system requires certain fixed assets, the *infrastructure*, and a number of mobile units, the *vehicles*. The combination of these, together with a set of rules for their operation, makes possible the movement of people and goods.

The infrastructure and vehicles are not often owned or operated by the same group or company. This is certainly the case for most transport modes, with the notable exception of some rail and ferry systems. This separation between supplier of infrastructure and provider of the final transport service generates a complex set of interactions between government authorities (central or local), stakeholders, construction companies, developers, transport operators, travellers and shippers, and the general public. The latter plays several roles in the supply of transport services: it represents the residents affected by a new scheme or the unemployed in an area seeking improved accessibility to foster economic growth; it may even be car owners wishing to travel unhindered through somebody else's residential area.

The provision of transport infrastructure is particularly important from a supply point of view. Transport infrastructure is 'lumpy' one cannot provide half a runway or one-third of a railway station. In certain cases, there may be scope for providing a gradual build-up of infrastructure to match growing demand. For example, one can start providing an unpaved road, upgrade it later to one or two lanes with surface treatment; at a later stage a well-constructed single and dual carriageway road can be built, to culminate perhaps with motorway standards. In this way, the provision of infrastructure can be adjusted to demand and avoid unnecessary early investment in expensive facilities. This is more difficult in other areas such as airports, metro lines, and ports. Technology innovation is making the future increasingly fluid and difficult to predict; with greater uncertainty, the importance of building flexibility into the design of transport infrastructure will become even more critical.

Investments in transport infrastructure are not only lumpy but also take a long time to complete. These are usually large projects. The construction of a major facility may take from 5 to 15 years from planning to full implementation. This is even more critical in urban areas where a good deal of disruption is also required to build them. This disruption involves additional costs to users and non-users alike.

Moreover, transport investment has an important political role. For example, politicians in developing countries often consider a road project a safe bet: it shows they care and is difficult to prove wrong or uneconomic by the popular press. In emerging and advanced nations, transport projects usually carry the risk of alienating large numbers of residents affected by them or travellers suffering from congestion and delay in overcrowded facilities. Political judgement as well as forethought and planning are essential in making choices of this kind.

The separation of providers of infrastructure and suppliers of services introduces economic complexities too. For a start, it is not always clear that all travellers and shippers perceive the total costs incurred in providing the services they use. These can be wide-ranging. The resources used when travelling include much more than the direct vehicle operating costs or fares. Most roads are provided for free to most users. Charging for road space, for example, is seldom carried out directly, and when it happens, the price does not include congestion costs or other external effects; perhaps the nearest approximation to this is toll roads and modern road-pricing schemes. The use of taxes on vehicles and fuels is only a poor approximation to charging for the provision of infrastructure.

Transport is an essential element in the welfare of nations and the well-being of urban and rural dwellers. If those who make use of transport facilities do not perceive the resource implications of their choices, they are likely to generate a balance between supply and demand that is inherently inefficient. Under-priced limited resources will be overused whilst other desirable but priced resources may be underused. Car owners probably see depreciation, insurance, and annual taxes as fixed, *sunk*, costs, which at most affect the decision to buy a car but not that of using it. The marginal perceived cost of using a car is quite low.

An additional element of distortion is provided by a number of external costs borne by non-users – associated with the production of transport services: accidents, delays to others, pollution, and environmental degradation in general. These *externalities* are seldom *internalised*; the user of a transport service rarely perceives or pays for the costs of cleaning the environment or the delays they cause to other travellers. Internalising these costs could also help everyone to make better decisions and improve the allocation of demand to alternative modes. Moreover, the costs of global warming and a poor environment will be borne by future generations, and the young have become more vocal about this unfairness.

A prominent feature of transport demand and supply is congestion, which is difficult to define as we all believe we know its exact meaning. However, most practitioners are aware that what is considered congestion in Leeds or Lampang is often accepted as normal flow in London or Lagos. Congestion arises when demand levels approach the capacity of a facility and the time required to use it (travel through it) increases well above the average under low demand conditions. In the case of transport infrastructure, the inclusion of an additional vehicle generates a supplementary delay for each of the other users as well, as shown in Figure 1.1. The social (or marginal) cost includes the external effects of congestion, which are perceived by others but not by the driver, who only perceives his/her private (or average) cost. Authorities consider these external costs and may design schemes such as *Congestion Charging* to internalise them, and help more reasoned decision-making by individuals. The shape of these curves, monotonically increasing, infuses most problems, arguments, and transport policies.

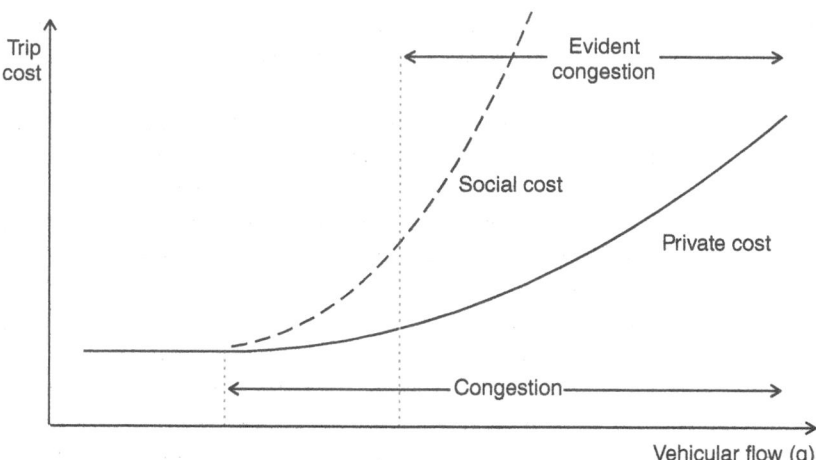

Figure 1.1 Congestion and its external effects

1.3.3 A View of Transport Problems

Our perception of transport problems evolves over time. It is now clearer that the overarching objective should be something like 'providing equitable access to opportunities to satisfy needs' and that there are three broad approaches to address this objective. The first is to plan and provide better transport infrastructure and services so that anybody can access suitable work, education, and services at a reasonable cost in time and money. This is the traditional view of transport planning, characterised by a process of forecasting future transport needs and selecting the best projects to serve them; this is an approach often characterised as *Predict and Provide*. It must be said, though, that the provision of new links and services seems to generate additional demand.

The second approach is that there can be alternatives to travel provided by remote working, education, and procurement of goods and services; this has become abundantly clear after 2020. This virtual access reduces the demand for travel and often widens further the reach of opportunities; of course, it may also generate some additional travel for deliveries or occasional trips. The third approach is to encourage or provide better access to essential activities through dense mixed land use, where residents can easily access work and services; this is a longer-term approach that may be implemented through land use planning and regulations. Typically, cities in Europe and places in Asia have higher densities where it is easier to access work, education, and leisure at shorter distances. Despite declaring strong feelings in favour of suburban or inner city living, most people seem to be equally happy in very different density settings. This general view of the issues, as one of accessibility to opportunities to satisfy need rather than mobility by itself, is more helpful under conditions of fast technological changes.

Of course, this poses the additional question of what are the needs that must be satisfied and whether they are permanent or evolve over time. The most common view of this issue considers the hierarchy of needs proposed by Maslow (1943) and illustrated in Figure 1.2.

Satisfying basic needs usually requires access to work and education to secure resources to attend to other needs. Work and education generate the most recurrent, and therefore modellable, trips. After the basic physiological needs of shelter, health, and safety are met, people seek to satisfy other needs of a more psychological nature like friendship and accomplishment. Indeed, this is not strictly a hierarchical process, as friendship and human contact are also seen as essential to health and well-being. But the manner in which these needs are expressed and satisfied evolves over time.

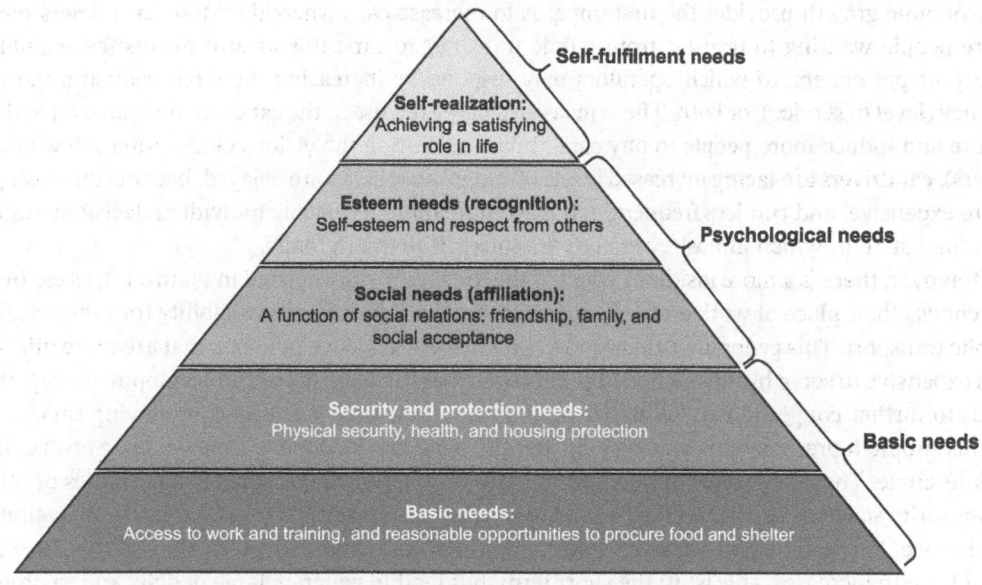

Figure 1.2 Maslow hierarchy of needs

An interesting application of these ideas to the topical issue of public transport satisfaction was done by Allen et al. (2019), where they confirmed the existence of a hierarchy of transit needs using a benchmark study applied in four Latin American metropolises; this included three macro-domains: *basic* (functional), *safety* (security, protection), and *hedonic* (customer services, comfort).

1.3.4 A Simple Model

Sometimes very simple cause-and-effect relationships can be depicted graphically to help understand the nature of some transport problems. A typical example is the car/public-transport vicious circle depicted in Figure 1.3.

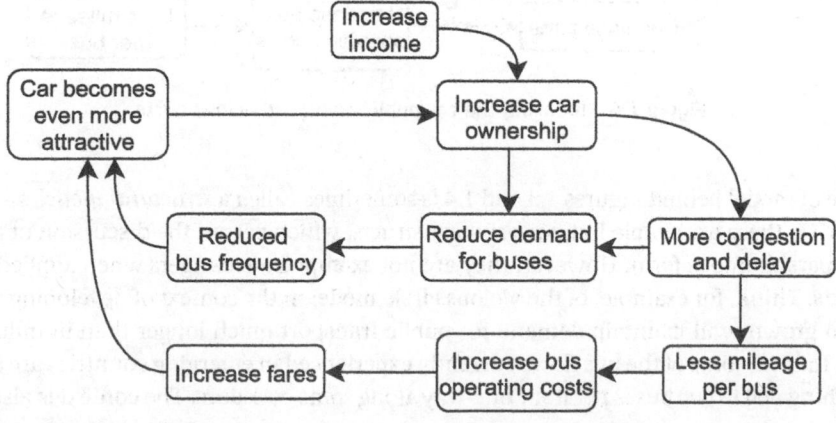

Figure 1.3 Car and public-transport vicious circle

Economic growth provides the first impetus to increase car ownership. More car owners mean more people wanting to transfer from public transport to cars; this in turn means fewer public-transport passengers, to which operators may respond by increasing the fares, reducing the frequency (level of service), or both. These measures make the use of the car even more attractive than before and induce more people to buy cars, thus accelerating the vicious circle. After a few cycles (years), car drivers are facing increased levels of congestion; buses are delayed, become increasingly more expensive, and run less frequently; the accumulation of sensible individual decisions results in a final state in which almost everybody is worse off than originally.

Moreover, there is a more insidious effect in the long term, not depicted in Figure 1.3, as car owners choose their place of work and residence without considering the availability (or otherwise) of public transport. This generates urban sprawl and low-density developments that are more difficult and expensive to serve by more efficient public transport modes. This is the 'development trap' that leads to further congestion and a higher proportion of our time spent in slow-moving cars.

This simple representation can also help identify what can be done to slow down or reverse this vicious circle. These ideas are summarised in Figure 1.4. Physical measures like bus lanes or other bus-priority schemes are particularly attractive as they also result in a more efficient allocation of road space. Public transport subsidies have strong advocates and detractors; they may reduce the need for fare increases, at least in the short term, but tend to generate large deficits and to protect poor management from the consequences of their own inefficiency. Car restraints, and in particular congestion charging, can help to internalise externalities and generate a revenue stream that can be distributed to other areas of need in transportation.

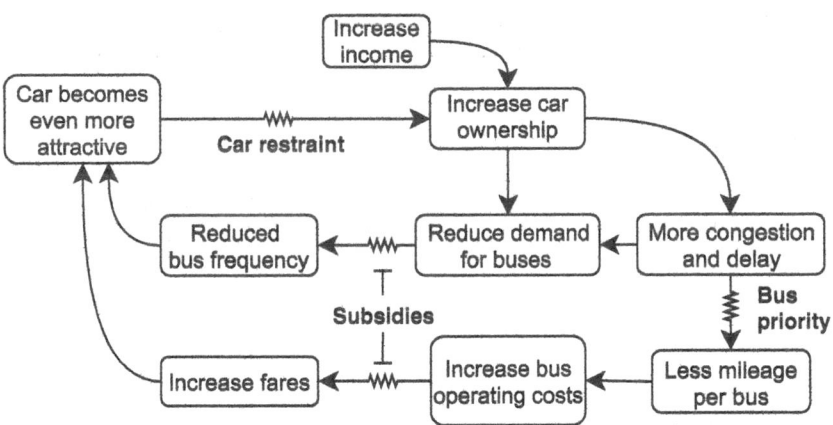

Figure 1.4 Breaking the car/public-transport vicious circle

The type of model behind Figures 1.3 and 1.4 is sometimes called a *structural model*, as discussed in Chapter 14; these are simple but powerful constructs, which permit the discussion of key issues in a fairly parsimonious form. However, they are not exempt from dangers when applied to different contexts. Think, for example, of the vicious circle model in the context of developing countries. Population growth will maintain demand for public transport much longer than in industrialised countries. Indeed, some of the bus flows currently experienced in emerging countries are extremely high, reaching 400 to 600 buses per hour one-way along some corridors. The context is also relevant when looking for solutions; it has been argued that one of the main objectives of introducing bus-priority schemes in emerging countries is not to protect buses from car-generated congestion but to organise bus movements (Gibson et al. 1989). High bus volumes often implement a de facto priority, and interference between buses may even become a greater source of delay than car-generated congestion. To be of value, the vicious circle model must be revised in this new context.

It should be clear that it is not possible to characterise all transport problems in a unique, universal form. Transport problems are context dependent, and so should be the ways of tackling them. Models can offer a contribution in terms of making the identification of problems and selection of ways of addressing them more solidly based.

1.3.5 Classic and New Modes of Transport

Throughout this book, we tend to use the term 'public transport' in its classic meaning of buses, trams, metros, and rail-based modes. This is somewhat restrictive, as other forms of transportation are also public in the sense of being available without having to own them: car-hire per day or minute, other vehicle-hire by-the-minute (motorcycles, bicycles, e-Scooters, and the like), taxis, and similar vehicle and ride-sharing services. The classic public transport modes are characterised by running more or less on a fixed route and schedule; the rest are on-demand services whose route is partially or fully determined by the user.

Some of the proposed new modes will run scheduled services, for example, the so-called Hyperloop, which can be modelled using existing techniques. Demand-responsive public transport poses a different modelling challenge: the level of service offered depends on the size of the fleet and passenger capacity, the location of units over space and time, and how these can be combined to deliver a service to a particular demand.

Note also that the models implicit in Figures 1.3 and 1.4 ignore taxis, ride-sharing, on-demand mobility, and future autonomous vehicles. These new alternatives make modelling considerably more complex, something we can only ignore when they are expected to represent a small fraction of the demand. A useful thought experiment is to modify both these figures to consider the impact of demand-responsive services, bearing in mind that some of these reduce the need to own a vehicle in order to use it.

Not all public transport systems, in their wider conception, are equally efficient in dealing with congestion and emissions. To compare them, one must consider the number of passengers moved per passenger car unit (pcu) or equivalent (pce), in other words, the density of passengers per unit equivalent to a car (a bus is generally considered to be equivalent to two or three cars in terms of use of road space). Therefore, the use of taxis and car sharing services may not reduce congestion but rather have the opposite effect, as there will be some empty vehicle kilometres spent to relocate the unit where it is most needed (only passengers count, not taxi drivers).

1.4 Modelling and Decision-Making

The main role of transport models is to support better decision-making with respect to access and mobility. This book assumes that the decision style adopted involves the use of models, but it does not advocate a single (i.e. a normative) decision-making approach. The acceptability of modelling, or a particular modelling approach, within a decision style is very important. Models that end up being ignored by decision makers not only represent wasted resources and effort but also result in frustrated analysts and planners. For an excellent introduction of transport modelling to modellers and non-modellers it is useful to consult Hollander (2016); this deceptively simple book is profound in its analysis of the basis for such models.

There are several features of transport problems and models, which must be considered when specifying an analytical approach:

1) **Precision and accuracy required**. These concepts are sometimes confused. *Accuracy* is the degree to which a measurement or model results match true or accepted values. Accuracy is an issue pertaining to the quality of data and models. The level of accuracy required for particular applications varies greatly. It is often the case that the accuracy required is just that necessary to discriminate between a good scheme and a less good one. In some cases, the best scheme may be quite obvious, requiring less accurate modelling. Remember, however, that common sense has been blamed for some very poor transport decisions in the past.

   Precision refers to the level or units of measurement used to collect data and deliver model outputs. One may measure travel times between two points in fractions of a second, but individuals may estimate and state the same much less precisely, in five-minute intervals. Precision is not accuracy, and it is often misleading. Reporting estimates with high precision is often interpreted as indicating confidence in their accuracy, whereas transport modellers often use precise numbers to report uncertain estimates. There is a difference between stating that ... 'traffic on link X was measured as 2347 vehicles between 8:00 and 9:00 AM yesterday' and saying that ... 'traffic on link X between 8:00 and 9:00 AM will be 3148 vehicles in five years' time': the first statement may be both precise and accurate, while the second is equally precise but certainly inaccurate. It is less misleading to report the second figure as 3150. As in the quote attributed to John Maynard Keynes ... 'it is much better to be roughly right than precisely wrong'. Moreover, there are inherent reasons why it is not possible to aim for high accuracy when forecasting beyond the next five years, as discussed later in this book.

2) **The decision-making context**. This involves the adoption of a particular *perspective* and choice of a *scope* or coverage of the system of interest. The choice of perspective defines the types of decisions that will be considered: *strategic* issues or schemes, *tactical* (transport management) schemes, or even specific operational problems. The choice of scope involves specifying the level of analysis: is it just transport or does it involve activity location too? In terms of the transport system, are we interested in just demand or also on the supply side at different levels: system or suppliers' performance, cost minimisation issues within suppliers, and so on? The question of how many options need to be considered to satisfy different interest groups or to develop a single best scheme is also crucial. Therefore, the decision-making context will also help define the requirements for the models to use and decide which variables to include in them or consider given or exogenous.

3) **Level of detail, or granularity, required**. The level of resolution of a model system can be described along four main dimensions: geography, unit of analysis, behavioural responses, and handling of time.

   Space is vital; we can handle it aggregately, as a few zones with area-wide speed flow curves or at the level of individuals' addresses for trips with links described in detail. There is a wide range of options, and the choice will depend on the application. If the issue is a detailed design for traffic in a small area, highly disaggregated zones with an accurate account of the physical characteristics of links would be appropriate in a micro-simulation model. However, strategic planning may call for a more aggregate zoning system with links described in terms of their speed-flow relationships alone.

The *unit of analysis* for modelling may be the same zone with trips emanating and ending there or, at the other end of the spectrum, sampled or synthesised individuals negotiating at home who does what activity over the next week; somewhere in between, there will be different household or person strata as representative of the travelling population and the trips or tours they undertake.

The *behavioural responses* included may vary from fairly simple route choice actions in a traffic model to changes in time of travel, mode, destination, tour frequency, and even land use and economic activity impacts in an integrated land use/transport model.

Time, in turn, can be treated either as a discrete or a continuous variable. In the first case, the model may cover a full day (as in many national models), a peak period, or a smaller time interval; all relevant responses will take place in that period, although there may be interactions with other periods. Alternatively, we can consider time as a continuous variable, which allows for more dynamic handling of traffic and behavioural responses like the choice of time of travel. Considering discrete time slices is a standard option, as treating time as a continuous variable is much more demanding.

4) **The availability of suitable data**, their stability, and the difficulties involved in forecasting their future values. In some cases, very little data may be available; in others, there may be reasons to suspect the information quality, or to have less confidence in future forecasts for key planning variables as the system is not sufficiently stable. In many cases, the data available will be the key factor in deciding the modelling approach.

5) **The state of the art in modelling** for particular types of interventions in the transport system. This in turn can be subdivided into:
   - behavioural richness;
   - mathematical and computer tractability;
   - availability of good solution algorithms.

It must be borne in mind that, in practice, all models assume that some variables are exogenous. Moreover, many other variables are omitted from the modelling framework on the grounds of being not relevant to the task at hand, too challenging to forecast, or expected to change little and not influence the system of interest. An explicit consideration of what is left out of the model may help decide its appropriateness for a given problem.

6) **Resources available for the study**. These include money, data, computer hardware and software, technical and people skills, and so on. However, two types of resources are worth highlighting here: time and level of communication with decision-makers and the general public. Time is probably the most crucial: if little time is available to choose between schemes, we need shortcuts to provide timely advice. Decision-makers are prone to setting up absurdly short timescales to assess projects that will take years to process through multiple decision instances, years to implement, and many more years to confirm as right or wrong. On the other hand, a good *level of communication* with decision-makers and stakeholders will alleviate this problem: fewer unrealistic expectations about our ability to accurately model transport schemes will arise, and a better understanding of the advantages and limitations of modelling will moderate the extremes of blind acceptance or total rejection of study recommendations.

7) **Levels of training and skills of the analysts**. Training costs are relatively high, so much so that it is sometimes better to use an existing model or software that is well understood rather

than embark on acquiring and learning to use a more advanced one. Although this looks like a recipe for stifling innovation and progress, it should always be possible to build up strengths in new advanced techniques without rejecting the experience gained with earlier models.

1.5 Issues in Transport Modelling

We have identified the interactions between transport problems, decision-making styles, and modelling approaches. We need to discuss now some of the critical modelling issues relevant to the choice of model. These issues cover some general points like the roles of theory and data, model specification, and calibration. Nevertheless, perhaps even more critical choices are those between aggregate or disaggregate approaches, cross-section or time-series models, and revealed or stated preference techniques.

1.5.1 General Modelling Issues

Wilson (1974) provides an interesting list of questions to be answered by any would-be modeller; they range from broad issues such as the *purpose* behind the model-building exercise to detailed aspects such as *what techniques* are available for building the model. We will discuss some of these below, together with other modelling issues, which are particularly relevant to the development of this book.

1.5.1.1 The Roles of Theory and Data

Many people tend to associate the word 'theory' with an endless series of formulas and algebraic manipulations. In the urban transport modelling field, this association has been largely correct: it is difficult to understand and replicate the complex interactions between human beings, which are an inevitable feature of transport systems.

Some theoretical developments attempting to overcome these difficulties have resulted in models lacking adequate data and/or computational software for their practical implementation. This has led to the view, held strongly by some practitioners, that the gap between theory and practice is continually widening; this is something we have tried to redress in this book.

An important consideration on judging the contribution of a new theory is whether it places any meaningful restrictions on, for example, the form of a demand function. There is at least one documented case of a 'practical' transport planning study, lasting several years and costing several million dollars, which relied on 'pragmatic' demand models with a faulty structure (i.e. some of its elasticities had a wrong sign; see Williams and Senior 1977). Although this could have been diagnosed *ex ante* by the pragmatic practitioners had they not despised theory, it was only discovered *post hoc* by theoreticians.

Unfortunately (or perhaps fortunately, as a pragmatist would say), it is sometimes possible to derive similar functional forms from different theoretical perspectives (this, the *equifinality issue*, is considered in more detail in Chapter 8). The interpretation of the model output, however, is heavily dependent on the theoretical framework adopted. For example, the same functional form of the Gravity Model can be derived from an analogy with physics, from entropy maximisation and from maximum utility formalisms, but the interpretation of the output may depend on the theory adopted. If one is just interested in flows on links, it may not matter which theoretical framework underpins the analytical model function. However, if an evaluation measure is required, then an economically based theory of human behaviour, even if flawed, will be more helpful in this case. In other cases, phrases like 'the gravitational pull of this destination will increase', or 'this is the most probable arrangement of trips' or 'the most likely trip matrix consistent with our information about the system' will be used; these provide no help in devising evaluation measures but assist in the interpretation of the nature of the solution found. The theoretical framework will also lend some credence to the ability of the model to forecast future behaviour. In this sense, it is interesting to reflect on the influence that practice and theory may have on each other. For example, it has been noted that models or analytical forms used in practice, have traditionally had a guiding influence on the assumptions employed in the development of subsequent theoretical frameworks. It is also well-known that widely implemented forms, like the Gravity-Logit model we will discuss in Chapters 6 and 7, have been the subject of strong *post hoc* rationalisation (Williams and Ortúzar 1982b):

> ... *theoretical advances are especially welcome when they fortify existing practice which might be deemed to lack a particularly convincing rationale.*

The two classical approaches to the development of theory are known as *deductive* (building a model and testing its predictions against observations) and *inductive* (starting with data and attempting to infer general laws). The deductive approach has been found more productive in the pure sciences, and the inductive approach has been preferred in the analytical social sciences. It is interesting to note that data are central to both; in fact, it is well-known that data availability usually leaves little room for negotiation and compromise in the trade-off between modelling *relevance* and modelling *complexity*. Indeed, in very many cases, the nature of the data restricts the choice of model to a single option.

The question of data closely connects with issues such as the type of variables represented in the model, which is, of course, closely linked again to questions about theory. Models predict some dependent (or endogenous) variables given other independent (or explanatory) variables. To test a model, we would typically need data about each variable. Of particular interest are the *policy variables*, which are those assumed to be under the decision-makers control, e.g. those the analyst may vary to test the value of alternative policies or schemes.

Another important issue in this context is that of aggregation:

- How many population strata or types of people do we need to achieve a good representation and understanding of a problem?
- In how much detail do we need to measure certain variables to replicate a given phenomenon?
- Space is crucial in transport; at what level of detail do we need to code the origin and destination of travellers to model their trip-making behaviour?

1.5.1.2 Model Assumptions

All models need assumptions. Some of them are explicit, and others are implicit. Good and complex models usually involve a long list of assumptions; there is an implicit recognition that if one or more of these assumptions fails to happen, for example, that economic growth will not continue uninterrupted, then any forecasts may also fail to materialise. Most of the explicit assumptions are about the future and, in particular, about external inputs to the model, for example population or economic forecasts, future technologies, evolution of fuel prices, and so on. There may be some assumptions about parameters that have not been fully estimated for the model; for example, they may have been transferred from other contexts.

However, there are some basic assumptions that are seldom made explicit, and these relate more to the nature of the model. We mention here four of these and discuss their validity:

a) Travel behaviour is assumed to be essentially recurrent. We may observe only a small sample of users over one or a few days, and we extrapolate their behaviour for the rest of the population and, essentially, for the future. In fact, there has been plenty of evidence that travel behaviour is quite variable. Trip making is less recurrent than one would imagine.

b) Activities are assumed to be stable, at least within family groups of similar characteristics. They will evolve with the changes in the family, and although they may be undertaken by different members at different times, they remain essentially stable. Technology is quickly weakening this assumption, which was never entirely true. We no longer rent videos and thirty years ago we would not have thought about going to a gym. We do not really know how generative AI and Virtual Reality will change our activities in the future.

c) Travel behaviour can be modelled as choices. Choices of the times to undertake activities, where to carry them out, the mode of travel to reach that location, and the route to follow. But in reality, we do not make this kind of choice every time we travel. We mostly do what we did last time, or apply a simple heuristic like ... 'its Saturday so there will be less/more traffic therefore I will ...'. However, most of our models do not consider *habit* or *inertia* and assume that we make these choices, at least every time that something changes.

d) When making these choices, we seek to maximise our benefit (utility), and in doing so, we act fairly rationally; essentially, we ignore anybody else's needs. See our thoughts on rationality further down this chapter.

e) The system will eventually converge to equilibrium. Although we acknowledge that transport system equilibrium will never happen, we seek to run our models to something close to equilibrium to compare alternative interventions. Equilibrium offers a 'level playing field' at some cost in realism.

1.5.1.3 Model Specification

In its widest and more interesting sense, this issue considers the following themes.

Model structure. Is it possible to replicate the system to be modelled with a simple structure that assumes, for example, that all alternatives are independent? Or is it necessary to build more complex models which proceed, for example, to calculate probabilities of choice conditional on previous selections? Disaggregate models, such as those discussed in Chapters 7–9, usually have parameters, which represent aspects of model structure, and the extensions to methodology achieved since the mid-1980s have allowed the estimation of more and more general model forms. However, as Daly (1982b) has remarked, although it might be supposed that ultimately all issues concerned with model form could be resolved by empirical testing, such resolution is neither possible nor appropriate.

Functional form. Is it possible to use linear forms, or does the problem require postulating more complex non-linear functions? The latter may represent the system of interest more accurately, but certainly will be more demanding in terms of resources and techniques for model calibration and use. Although theoretical considerations may play a big role in settling this question, it is also possible to examine it in an inductive fashion by means of 'laboratory simulations', for example, in stated preference experiments.

Variable specification. This is the more usual meaning attached to the specification issue: which variables to use and how (in what form) they should enter a given model. For example, if income is assumed to influence individual choice, should the variable enter the model as such or by deflating a cost variable? Methods to advance on this question range from the deductive ('constructive') use of theory, to the inductive statistical analysis of the data using transformations.

1.5.1.4 Model Calibration, Validation, and Use

A model can be simply represented as a mathematical function of variables $\mathbf{X}$ and parameters θ, such as:

$$Y = f(\mathbf{X}, \theta) \tag{1.1}$$

It is interesting to mention that the twin concepts of *model calibration* and *model estimation* have traditionally had a different meaning in the transport field. Calibrating a model requires choosing its parameters, assumed to have a non-null value, to optimise one or more *goodness-of-fit* measures, which are a function of the observed data. This procedure has been associated with the physicists and engineers responsible for aggregate transport models, who did not worry unduly about the statistical properties of these indices (e.g. how large any calibration errors could be).

Estimation involves finding the values of the parameters, which make the observed data more likely under the model specification; in this case, one or more parameters can be judged *non-significant* and left out of the model. Estimation also considers the possibility of examining empirically certain specification issues; for example, structural and/or functional form parameters may be estimated. This procedure has tended to be associated with the engineers and econometricians responsible for disaggregate models, who placed much importance on the statistical testing possibilities offered by their methods. However, in essence, both procedures are the same because the way to decide which parameter values are best is by examining certain previously defined goodness-of-fit measures. The difference is that these measures generally have well-known statistical properties, which in turn allow confidence limits to be built around the estimated values and model predictions.

Because the large majority of transport models have been built on the basis of *cross-sectional* data, there has been a tendency to interpret model *validation* exclusively in terms of the goodness-of-fit achieved between observed behaviour and base-year predictions. Although this is a *necessary*, it is by no means a *sufficient* condition for model validation; this has been demonstrated by a number of cases which have been able to compare model predictions with observed results in *before-and-after* studies (see the discussion in Williams and Ortúzar 1982a). Validation requires comparing the model predictions with information *not used* during the process of model estimation. This obviously puts a more stringent test on the model and requires further information or more resources.

One of the first tasks a modeller faces is to decide which variables will be predicted by the model and which are possibly required as inputs to it. Some will not be included at all, either because the modeller lacks control over them or simply because the theory behind the model ignores them (see Figure 1.5). This implies immediately a certain degree of error and uncertainty (we will come back to this problem in Chapter 2), which of course gets compounded by other errors which are also inherent to modelling; for example, sampling errors and, more importantly, errors due to the unavoidable simplifications of reality the model demands in order to be practical.

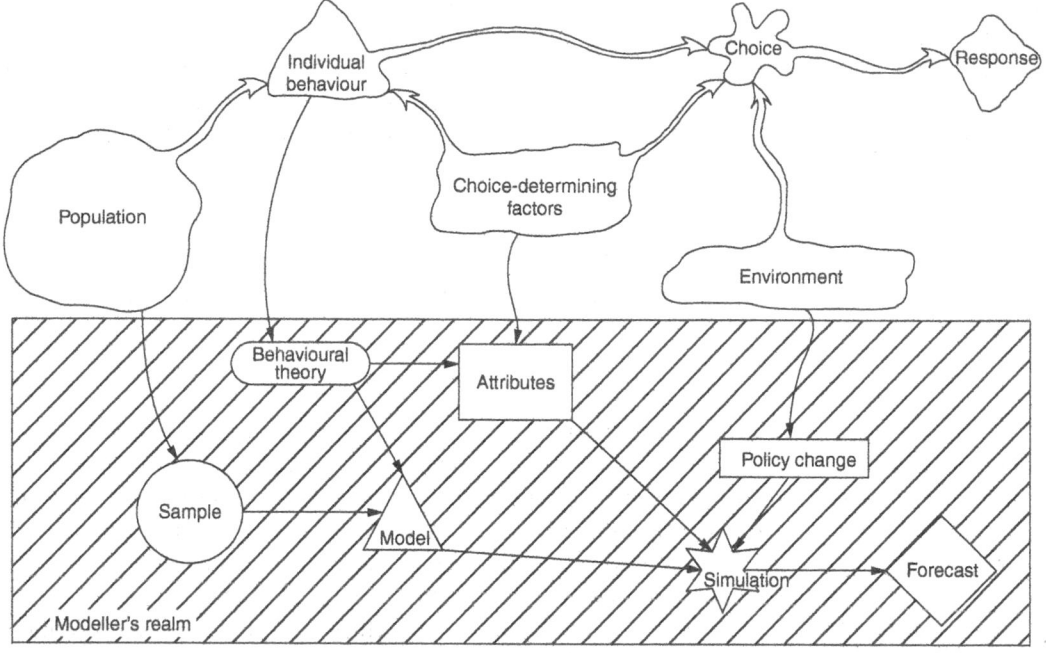

Figure 1.5 Modelling and sampling

Thus, the main use of models in practice is for *conditional forecasting*: The model will produce estimates of the dependent variables given a set of independent variables. In fact, typical forecasts are conditional in two ways (Wilson 1974):

- in relation to the values assigned to the policy variables in the plan, the impact of which is being tested with the model;
- in relation to the assumed values of other variables.

A model is normally used to test a range of alternative plans for a range of possible assumptions about the future value of the other variables (e.g. low- and high-income scenarios). This means that it might be 'run' many times in the context of examining a particular problem. For this reason, it may be of importance that its specification allows for quick turn-around time in a computer; this is not an easy task in the case of a full-scale transportation model, which involves complex processes of equilibration between supply and demand, as we will discuss in Chapter 12.

1.5.1.5 Modelling, Forecasting, and Judgement

There is a subtle difference between modelling and forecasting. Modelling focuses on building and applying appropriate tools that are sensitive to the choices of interest and respond logically to changes in key policy instruments. The successful modeller will provide useful and timely advice to the decision-making process, even if the data and timescales are limited. In this case, it is important that the model produces consistent results for all expected interventions, policies, and projects, such that they can be ranked fairly even if the correspondence to reality is not perfect.

Forecasting is an attempt to envision and quantify future conditions. It typically involves estimating future travel demand and the resulting multimodal flows and costs over time. In the case of private sector projects (see Chapter 19), these projections are usually accompanied by revenue forecasts, and investors will take considered risks based on them. Forecasting is usually based on formal models, but they alone cannot provide the full picture; it is necessary to incorporate other analyses and assumptions. Given the uncertainty about the future, several complementary approaches might be used in forecasting. For example, a formal model may be supported by considering the main economic drivers of future travel activity in a region; in that way, it is made clear how forecasts are dependent on the future of these activities. The success of forecasts can only be objectively measured through before-and-after studies.

The importance of formal models increases as the interventions considered diverge further from what is on the ground and known today. For example, when introducing a mode not currently available in a city, the model will often have to rely on stated preference data, information from other regions, or rational decision-making theory. The same is true when evaluating any sort of policy not currently in existence (congestion charging) or when considering fuel prices or congestion conditions radically different than those experienced at present. In general, we cannot provide sound advice on these issues only based on good modelling, even if the model is excellent. Good advice requires intelligent consideration of other factors and assumptions, particularly the limitations of the selected modelling approach.

Given the nature of analytical models, interpretation of their output is essential. Interpretation requires good judgement, and this is only acquired with experience and a thorough understanding of the theories underpinning models and their limitations. For instance, most models described in this text are supported by random utility theory (see Chapter 7) that in turn assumes rational decision-making on the part of travellers. However, there is an increasingly solid body of evidence, provided mostly by Behavioural Economics and Psychology, that humans are neither entirely rational nor consistent in their choices. This evidence (see Ariely 2009) punctures the theory underpinning our models – even the most advanced activity-based approaches – and makes it even more important the application of judgement in the interpretation of model outputs.

1.5.2 Aggregate and Disaggregate Modelling

The level of aggregation selected for measuring the data is an important issue in the general design of a transportation planning study. Of central interest is the aggregation of exogenous data, that is, information about items other than the behaviour of travellers, which is assumed to be endogenous (i.e. the model attempts to replicate it). For example, throughout the years, it has been a cause for concern whether a given data item represents an average over a group of travellers rather than being collected specifically for a single individual. When a model aims at representing the behaviour of more than one individual (e.g. a population segment like car owners living in a zone), such as in the case of the *aggregate* models we will examine in Chapters 5 and 6, a certain degree of aggregation of the exogenous data is inevitable. But when a model attempts to represent the behaviour of individuals, such as in the case of the *disaggregate* models we will study in Chapters 7–9, it is conceivable that exogenous information can be obtained and used separately for each traveller. An important issue is then whether, as often, it might be preferable on cost or other grounds to use less detailed data (see Daly and Ortúzar 1990).

Forecasting future demand is a crucial element of the majority of transport planning studies. Being able to predict the likely usage of new facilities is an essential precursor to make a rational decision about the advantages or otherwise of providing such facilities. It may also be important to have an idea about the sensitivities of demand for important variables under the control of the analyst (e.g. the price charged for their use). In most cases, the forecasts and sensitivity estimates must be provided at the aggregate level, that is, they must represent the behaviour of an entire population of interest. Therefore, the analyst using disaggregated models must find a sound method for aggregating model results to provide these indicators.

Aggregate models were used almost without exception in transport studies up to the late 1970s; they became familiar, demanded relatively few skills on the part of the analyst (but required arcane computer knowledge), and had the property of offering a 'recipe' for the complete modelling process, from data collection through the provision of forecasts at the level of links in a network. The output of these models, perhaps because they were generated by obscure computer programs, was often considered more accurate than intended, for example in predicting turning movement flows 15 years in the future. However, aggregate models have been severely (and sometimes justifiably) criticised for their inflexibility, inaccuracy, and cost. Unfortunately, many disaggregate approaches, which have adopted sophisticated treatments of the choices and constraints faced by individual travellers, have failed to take the process through to the production of forecasts, sometimes because they require data which cannot reasonably be forecast.

Disaggregate models, which became increasingly popular during the 1980s, offer substantial advantages over the traditional methods while remaining practical in many application studies. However, one important problem in practice is that they demand a high level of statistical and econometric skills for their use (in particular for the interpretation of results), certainly much higher than in the case of aggregate models. Moreover, the differences between aggregate and disaggregate model systems have often been overstated. For example, the disaggregate models were first marketed as a radical departure from classical methods, a 'revolution' in the field, while eventually it became clear that an 'evolutionary' view was more adequate (see Williams and Ortúzar 1982b). In fact, in many cases, there is complete equivalence between the forms of the forecasting models (Daly 1982a). The essential difference lies in the treatment of the description of behaviour, particularly during the model development process; in many instances, the disaggregate approach is clearly superior to the grouping of behaviour by zones and by predefined segments.

Attempts to clarify the issue of whether disaggregate or aggregate approaches should be preferred, and in what circumstances, have basically concluded that there is no such thing as a definitive approach appropriate to all situations (see Daly and Ortúzar 1990). These attempts have also established the need for guidelines to help the despairing practitioner to choose the most appropriate model tools to apply in a particular context. We have striven to answer that call in this book.

1.5.3 *Homo Sapiens* and *Homo Economicus*

The large majority of practical transport-modelling approaches to date rely on a simplified assumption of how human beings behave, in particular how we make choices between alternatives. In common with many economic theories, the assumption is that individual travellers are rational and capable of comparing alternatives and always choose the one that is best for them, the one that maximises their own utility. This assumed behaviour has been characterized as *Homo economicus*, a being that, in contrast with *Homo sapiens* (an imperfect but real human being), is rational, has consistent tastes and preferences, and almost invariably has full and perfect information about all alternatives.

The *homo economicus* assumption is convenient as it lends itself to useful mathematical formulations that are easier to interpret. However, Behavioural Economics and Social Psychology have thoroughly demonstrated that *Homo sapiens* is far more complex and occasionally rational; see for example, the works of Ariely (2009) and Kahneman (2013). McFadden (2013), one of the key originators of *Random Utility Theory* underpinning some of the most useful *homo economicus*-based transport models, has recognized this for quite some time.

There are several relevant departures from the perfectly rational human being that influence our interpretation of model results based on the *homo economicus* assumptions. In our view the most important are:

a) There is a difference between the *experiencing self* and the *remembering self*. The first one experiences the congested and free-flowing elements of a journey, but the second may only remember the salient frustration of five minutes queueing at a standstill. Decisions on future travel are made by the remembering self. This may also explain why we are so poor at estimating waiting times and distances.

b) We sometimes overvalue what we *own*, and other times we overvalue what we want but do not yet have; we may not be willing to contribute €500 to create a garden in an empty plot nearby, but once the garden exists, we ask for much more in compensation for its use as a parking lot. A consequence of this is our *loss aversion*. We are more averse to a loss than favourable to a gain of equivalent value, and it has been shown that people tend to value more attributes that they do not possess (Greene and Ortúzar 2002). This failure is also related to Maslow's hierarchy, as discussed in Figure 1.2.

c) We care more about changes from the *status quo* than about absolute values; we tend to compare ourselves to our neighbours rather than assess our objective level of welfare.

d) We may not react to *small changes*. We may tolerate small increases in travel time over time but react quickly to the introduction of a congestion charge. In fact, it has been shown – and we discuss this later in Chapters 7 and 8 – that we tend to have associated thresholds of perception, and some behavioural theories base choices on these concepts (Simon 1957; Tversky 1972). The concept of *search costs* is associated with this: if a small improvement is expected relative to the 'cost' (effort, time, hassle) of searching for a better choice, then it is probably not worthwhile to seek improvements. Also related to this is the reality that travel decisions are always made under conditions of *imperfect* and *asymmetric knowledge*. Note that marketing and good customer information may play a role in reducing this search cost.

e) There are *lags* in our responses to changes in the level of service. We cannot change jobs or residence quickly, but we can take advantage of a new metro line for our shopping trips almost straight away.

f) We have a *diminishing sensitivity to changes in utility*. The value of a $100 gain when we have $1000 in our bank account is greater than the same amount when we have $10 000 in savings. Saving 5-minutes in a 30-minute trip is worth more than in a 3-hour journey.

g) We cannot rationally *cope with too many alternatives*; we only consider a few route options, not all possible ones, and may use different heuristics when choices are too complex. We will discuss this in some detail in Chapter 7.

h) We seem to have *two modes of thinking*. Most of the time we use intuition and inertia to make fast decisions (**System 1** thinking in Kahnemann's terms); whenever we face a significant new problem or change, we engage a more rational and reflective mode (**System 2**). A small increase in fuel prices does not change our commuter journey, but the introduction of an equivalent parking charge at work (to promote carpooling) makes us think afresh. In fact, some people may apply different choice heuristics in a given choice situation (González-Valdés et al. 2022) or change heuristics depending on the complexity of the choice situation, context, and time available to make the choice.

i) Our behaviour is influenced directly by what other people do or say they do, not just indirectly through supply, demand, and price relationships. Terms such as 'social norm' and 'herding or bandwagon effect', and their measurement have been incorporated into contemporary models, for example in the context of fare evasion behaviour in public transport.

j) Cultural preconceptions and self-image matter. 'Winners drive their cars; losers use public transport' and/or 'I am not going to pay for something that is my right: the freedom of the road' may evolve with time into more rational and sensitive attitudes. Linked to this is the perception of public transport in some places as an inferior mode where a reduction in price does not lead to an increase in demand. This is partially related to the previous point, but also to 'latent variables' associated with intimate attitudes and perceptions of individuals (see Bahamonde-Birke et al. 2017a) that can be incorporated in models, as we will see in Chapter 7.

These issues influence how we perceive attributes, such as time, in a choice context and how we value prices, as will be discussed in subsequent sections. The acceptance that our models do not reflect human nature well should help in interpreting model outputs and delivering better advice to decision-makers.

1.5.4 Cross-Section and Time Series

The vast majority of transport planning studies up to the late 1980s relied on information about trip patterns revealed by a cross-section of individuals at a single point in time. Indeed, the traditional use of the cross-sectional approach transcended the differences between aggregate and disaggregate models discussed above.

A fundamental assumption of the cross-sectional approach is that a measure of the response to an incremental change may simply be found by computing the derivatives of a demand function with respect to the policy variables in question. This makes explicit the assumption that a realistic *stimulus–response* relation may be derived from model parameters estimated from observations at one point in time. This would be reasonable if there were always enough people changing their choices, say of mode or destination, in *both* directions and without habit or time-lag effects.

However, the cross-sectional assumption has two potentially serious drawbacks. First, a given cross-sectional data set may correspond to a particular 'history' of changes in the values of certain key variables influencing choice. For example, changes in mode or location in time may have been triggered by a series of different stimuli (petrol prices, life-cycle effects, etc.), and the extent to which a system is considered to be in *disequilibrium* (because of, say, inertia) will depend on these. The trouble is that it can be shown (see Chapter 7) that the responses of groups with exactly the same current characteristics but having undergone a different path of change may be very different indeed. Second, data collected at only one point in time will usually fail to discriminate between alternative model formulations, even between some arising from totally different theoretical postulates. It is always possible to find 'best-fit' parameters from base-year data even if the model suffers severe mis-specification problems; the trouble is, of course, that these do not guarantee good response properties for a future situation. As we saw in Section 1.5.1.4, a good base-year fit is not a sufficient condition for model validation.

Thus, in general, it is not possible to discriminate between the large variety of possible sources of dispersion within a cross-sectional data set (i.e. preference dispersion, habit effects, constraints, and so on). Real progress in understanding and assessing the effectiveness of forecasting models, however, can only be made if information is available on response over time. From a theoretical point of view, it is also desirable that appropriate frameworks for analysis are designed, which allow the eventual refutation of hypotheses relating to response. Until this is achieved, a general problem of potential misrepresentation will continue to cast doubt on the validity of cross-sectional studies.

The discussion above has led many people to believe that, where possible, longitudinal or time-series data should be used to construct more dependable forecasting models. This type of data incorporates information on response by design. Thus, in principle, it may offer the means to directly test and even perhaps reject hypotheses relating to response.

Longitudinal data can take the form of *panels* or more simply *before-and-after* information. Unfortunately, models built on this type of data have severe technical problems of their own; in fact, progress in this area has been limited (Cherchi et al. 2017). We will discuss some of the issues involved in the collection and use of this type of information in Chapters 2 and 7.

1.5.5 Revealed and Stated Preferences

The development of good and robust models is quite difficult if analysts cannot set up experiments to observe the behaviour of the system under a wide range of conditions. Experimentation of this kind is neither practical nor viable in transport, and analysts are restricted, like astronomers, to making observations on events and choices they do not control. Up until the mid-1980s, it was almost axiomatic that modelling transport demand should be based on information about observed choices and decisions, that is, *revealed-preference* data. Within this approach, project evaluation requires expressing policies in terms of changes in attributes, which 'map onto' those considered to influence current behaviour. However, this has practical limitations, basically associated with survey costs and the difficulty of distinguishing the effects of attributes, which are not easy to observe (e.g. those related to notions such as quality or convenience). Another practical embarrassment has been traditionally the 'new option' problem, whereby it is required to forecast the likely usage of a facility not available at present and perhaps even radically different from all existing ones.

Stated-preference (SP) techniques, borrowed from the field of market research, were put forward by the end of the 1970s as offering a way of experimenting with transport-related choices, thus solving some of the problems outlined above. SP techniques base demand estimates on an analysis of the response to *hypothetical choices*; these, of course, can cover a wider range of attributes and conditions than the real system. However, these techniques were severely discredited in their beginning because it was not known how to discount for the over-enthusiasm of certain respondents, e.g. not even half of the individuals stating they would take a given course of action actually did so when the opportunity eventually arose.

It took a whole decade for the situation to change, but by the end of the 1980s, SP methods were perceived by many to offer a real chance to solve the above-mentioned difficulties. Moreover, it has been found that revealed-and stated-preference data and methods may be employed in complementary senses, with the strengths of both approaches recognised and combined. In particular, they are considered to offer an invaluable tool for assisting in the modelling of completely new alternatives. Despite this advantage, it is necessary to recognise that the SP approach has limits. In particular, for its results to be valid, respondents must be able to imagine the hypothetical futures presented as realistically as possible. Sometimes, this is very difficult; for example, it is difficult to imagine future autonomous vehicles and how they will change mobility, just as it would have been difficult to conceive of smartphones and their impact five years before their introduction in 2007.

We will examine data-collection aspects of stated-preference methods in Chapter 2 and modelling issues in Chapter 8.

1.6   The Structure of the Classic Transport Model

1.6.1   The Classic 4/5 Stage Model

Years of experimentation and development have resulted in a general structure, which has been called the *classic transport model*. This structure is, in effect, a result of practice in the 1960s but has remained relatively unaltered despite major improvements in modelling techniques since then.

The general form of this model is depicted in Figure 1.6. The approach starts by considering a zoning and network system, and the collection and coding of planning, calibration, and validation data. These data should include base-year levels for population of different types in each zone of the study area, as well as levels of economic activity including employment, shopping space, educational, and recreational facilities. These data are then used to estimate a model of the total number of trips generated and attracted by each zone of the study area (*trip generation*). The next step is the allocation of these trips to particular destinations in other words, their *distribution* over space (or *destination choice*), thus producing a trip matrix. The following stage normally involves *mode choice* modelling, that is, the allocation of trips in the matrix to different modes. But the traveller also has the opportunity to adjust the time of the trip, either to avoid the worst of congestion or to benefit from a lower off-peak fare (*time of travel choice*). Finally, the last stage in the classic model requires the *assignment* of trips by each mode to their corresponding networks: typically, private and public transport. The results from the assignment stage are flows on the road and public transport networks, and travel times and costs between each origin and destination.

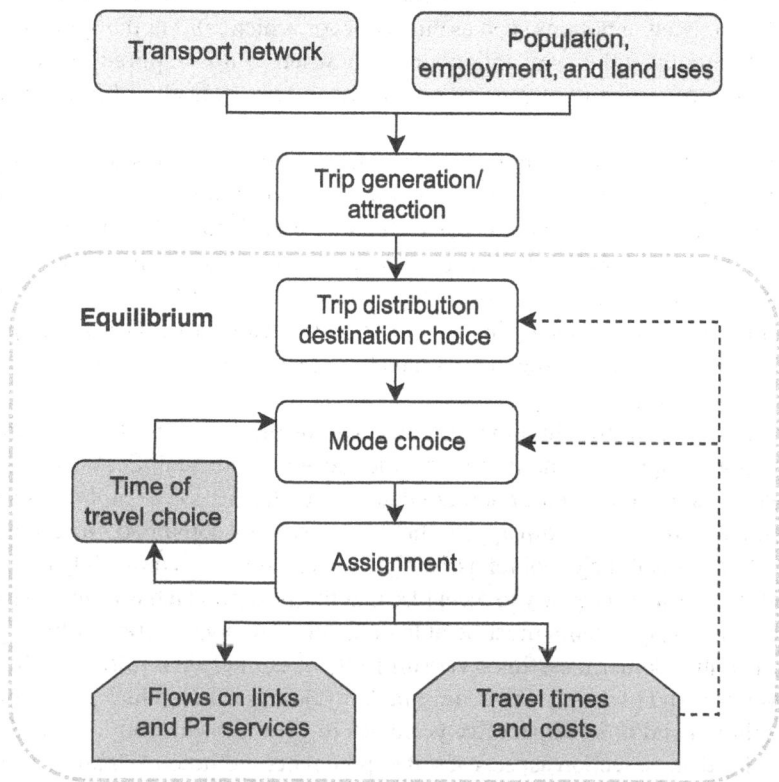

Figure 1.6 The classic 4/5-stage transport model

The original classic model was presented as a sequence of only four sub-models: trip generation, distribution, modal split, and assignment; this was later extended to consider the time of travel choice as well. It is generally recognised that travel decisions are not actually taken in this type of sequence; a contemporary view is that the 'location' of each sub-model depends on the form of the utility function assumed to govern all these travel choices (see Williams 1977). Moreover, the 4/5 stage model is seen as concentrating attention on only a limited range of travellers' responses. For example, when faced with a new toll road or public transport service, a trip maker can have a range of behavioural responses:

- Change their route to benefit from the shorter travel time using a new road or public transport service.
- Change the time of travel now that the journey can be made shorter.
- Change the destination of some trips in the short term, for example shopping trips, to take advantage of the new accessibility.
- Change jobs or residence in the medium term, for the same reason.
- Change mode of travel, perhaps from rail to car or even from car to Bus Rapid Transit (BRT) for example;
- Change the make-up of the linked trips or tours they perform. For example, it may be possible to go back home and then play squash rather than going directly to the club from work.
- Change the number of people that travel together. It may be advantageous to combine car trips with family members or neighbours to share the cost of travel every day.
- Change some of the activities that are undertaken during the week. It may now be worth going to a more distant supermarket to get better food but only once a week, rather than buying food more often on the way back from work.
- In the case of an interurban toll road, it may now be possible to go to a destination on business and return the same day, rather than having to stay there overnight to get the work done.
- Decide to acquire a car to take advantage of the new toll road (travelled by rail beforehand).

Despite these comments, the 4/5-stage sequential model provides a point of reference to contrast alternative methods. For example, some approaches attempt to treat simultaneously the choices of trip frequency (trips per week), destination, and mode of travel, thus collapsing trip generation, distribution, and mode choice into one single model. Other approaches emphasise the role of household activities and the travel choices they entail; concepts like sojourns, circuits, and time and money budgets are used in this context to model travel decisions and constraints. These modelling strategies are more difficult to cast in terms of the four main decisions or sub-models above. However, the improved understanding of travel behaviour these activity-based models provide is likely to enhance more conventional modelling approaches, as we will see in Chapter 16.

The trip generation–distribution–modal split and assignment sequence is the most common but not the only possible one. Some past studies have put modal split before trip distribution and immediately after (or with) trip generation. This permits a greater emphasis on decision variables depending on the trip generation unit, the individual, or the household. However, forcing modal split before the destination requires 'averaging' the attributes of the journey and modes in the model. This detracts policy relevance from the modal-split model. Another approach is to perform distribution and mode choice simultaneously, as discussed in Chapter 6. Note also that the classic model tends to make trip generation inelastic, that is, independent of the level of service provided in the transport system. This is probably unrealistic, but only recently have techniques been developed, which can take systematic account of these effects.

Once the model has been calibrated and validated for base-year conditions, it must be applied to one or more planning horizons. To do this, it is necessary to develop *scenarios* and plans describing the relevant characteristics of the transport system and planning variables under alternative futures. The preparation of realistic and consistent scenarios is not a simple task as it is easy to fall into the trap of constructing futures, which are neither financially viable nor realistic in the context of the likely evolution of land use and activities in the study area. In fact, scenario writing is still more of an art than a rigorous technique and requires a good deal of engineering expertise combined with sound political judgement; unfortunately, these are scarce resources not always found together in planning teams.

Having prepared realistic scenarios and plans for testing, the same sequence of models is run again to simulate their performance. A comparison is then made between the costs and benefits, however measured, of different schemes under different scenarios; the idea is to choose the most attractive programme of investment and transport policies, which satisfies the demand for movement in the study area.

An important issue in the classic model is the consistent use of variables affecting demand. For example, at the end of the traffic assignment stage, new flow levels, and therefore new travel times, will be obtained. These are unlikely to be the same travel times assumed when the distribution and mode choice models were run, at least when the models are used in the forecasting mode. This seems to call for a re-run of the distribution and modal-split models based on the new travel times. The subsequent application of the assignment model may well result in a new set of travel times; it will be seen that in general, the naive feedback of the model does not lead to a stable set of distribution, modal split, and assignment models with consistent travel times (i.e. equilibrium). This problem will be treated in some detail in Chapter 12; its particular relevance lies in the risk of choosing the wrong plan depending on how many cycles one is prepared to undertake.

1.6.2 Granularity

As stated above, the level of detail, or granularity, of a model can be seen in different dimensions. We distinguish here the most relevant in the case of passenger transport, as outlined in Table 1.1.

Take, for example, the level of detail in describing infrastructure. It may provide detailed information about all junctions, stations, and bus stops in a large network, or only some information in a local area network (part of a city), or even less detail on the full multi-modal network of the study area. Finally, infrastructure may be represented very simply with a network displaying only the connectivity of different places, what is sometimes called a *skeleton network*.

The attributes of each element of these networks may in turn be a simple link representation, perhaps with an associated flow-delay relationship and junctions represented as nodes. At the other extreme, each link in the network may have a detailed representation of the lane widths, gradient, any lateral friction from pedestrians and vendors, on-street parking availability, facilities for buses (bus stops and bus-only lanes) and for bicycles (cycle lanes, bike parking), signal timings at junctions, curve radii, etc. In the case of public transport, the description of the network may include the type and capacity of the buses or metro carriages, special facilities at stops/stations, timetable details, an indication of travel time reliability, etc. Careful observation and measurement can provide most of this information today, and good engineering judgement may be able to estimate conditions in the future where, for example, autonomous vehicles offer mobility as a service.

Table 1.1 Model granularity in different dimensions

	High		← Granularity →		Low
Infrastructure	Junction/station/stop	Detailed local area network	City multimodal network		Skeleton network
Infrastructure attributes	Lane width, signal timings, merge length, curve radii	Link length, capacities, flow delay functions			Total length by mode
Space/location	Address coordinates	Zones	Large zones/regions		A-spatial representation
Unit of analysis	Each individual	Representative individual	Number of individuals in zones		Population
Individual characteristics	Age, income, employment, vehicle availability, attitudes, membership		Car ownership, socio-economic group		Population
Reason to travel	Activities agreed in households	Fixed activities	Many trip purposes	Few trip purposes	Trend extrapolation
Movement analysed	Many multiple tours	Few standard tours	Trips	Flows on links	Vehicle-km travelled
Modes considered	All including car, bus, metro, rail, taxi, motorcycle, bicycle, walk, paratransit, mixed mode (P&R), etc.		Car, bus, metro, rail, active modes		Car or public transport only
Time	Year, month, day, period, hour, fifteen minute intervals	Summer and Winter peak periods	Average day peak periods		Average day
Behavioural response	All: choice of trip frequency, time of travel, destination, mode, route, people travelling together, etc.		Few responses: e.g. mode-route		Only one: e.g. route choice only

Locations in space may be described very precisely with the coordinates of each address for origins and destinations, but more common is the use of zones of different sizes to identify locations. It is also possible to use very aggregate descriptions of the level of mobility using non-spatial indicators like vehicle kilometres travelled (VKT).

The unit of analysis was, for decades, trips by groups of people based in zones, one-way movements from a zone of origin to a destination zone. Trips were then used in the 4/5 stages of the model. Of course, as not all trips are the same, distinguishing them by their regularity or other characteristics helped to estimate them; trip purpose is, therefore, a natural distinction to use. Another is to identify different types of individuals by car ownership, income, and perhaps the type of job they hold. This led to recognise that trips by a traveller were not entirely independent; if the journey to work was by bus, it is very unlikely that the trip back home would be by car. The unit of analysis was changed to linked trips or *tours* as a better way of explaining choices of mode and perhaps time of travel as well. In practical terms, many different tours are possible, so most models characterise only some combinations of purposes, focussing on those more likely to be recursive and relatively simple.

Another important improvement was to replace aggregate groups of people in zones with individuals based at specific addresses. This has two benefits: (i) the specific characteristics of each individual (e.g. gender, income, car availability, etc.) can be considered instead of using the proportion of individuals in a zone with them; and (ii) the particular components of a journey (e.g. precise walking distance to station, parking availability) are considered in the choice models rather than the averages from zone centroid to zone centroid. Of course, it is neither possible, nor practical, to consider and model each individual in a city; therefore, a sample of 'representative individuals' is chosen for modelling purposes. The next step was to consider activities directly as the main reason for travel and, potentially at least, the fact that which individual does what and when in a household is negotiable. In most cases, the use of activities for modelling is simplified to a smaller set of linked ones – in essence – a set of tours by representative individuals connecting fixed activities.

Some models deal with only one or two modes of transport, say cars and buses. Others deal with the full set of modes available in a study area. These would include walking and cycling (*active modes*), motorcycles, perhaps paratransit, and so on. Taxis are difficult to model as they can also travel without passengers; for this reason, they are often ignored unless they constitute an important component of demand. A particular modelling issue is the treatment of mixed-mode journeys, for example, Park and Ride. These may come in different combinations of access, mainline, and distribution modes; even these concepts do not fully describe all possible mixed-mode journeys. The model can treat each mode-mix as a separate mode or handle the choice of mode combinations as a 'route choice' problem.

The treatment of time is another granularity dimension. In principle, we would like to understand the performance of the transport system for a full year, including holidays and week-ends. In effect, in the case of transport concessions at least, an aggregate measure of performance, namely revenue, has to be provided for the full 365 days, year after year, for the duration of the contract. In practice, however, it is customary to model only a few time periods and expand results to cover the full year. The usual minimum is a peak period, often the AM peak, but including a PM peak and some off-peak periods, including week-ends, is also desirable to get a better representation of performance. The modelled time period may be an hour or, in some cases, a few hours. Some models will go down to simulate periods of 15 minutes or less to better capture the local behaviour of traffic. From the perspective of models based on activities, the ideal unit of time would be a week, as it is always possible to re-allocate some household activities to individuals over time, for example shopping. However, most current models are based on a simple set of activities and tours during a representative day that is never fully defined.

Finally, the set of behavioural travel responses to an intervention in the transport system can be very wide, as discussed above and in the rest of this text. However, very few transport models aim to incorporate all of them. For example, some models that treat in great detail the traffic interactions of individual vehicles, say at a signal-controlled junction, treat the number of trips by car every 15 minutes as a fixed input without aiming to represent the potential impact of signal timings on time of travel, mode, and destination choice. The focus is on traffic delays, and the assumption is made that the impacts on time of travel, mode, and destination choice would be minimal; including them would require additional data collection and, more importantly, introduce a new source of error in the estimates of impacts. Therefore, the range of behavioural responses considered relevant to study a particular intervention is also another dimension of the granularity of a model.

1.6.3 Macro, Meso, and Micro Models

Most transport models are developed on specific software platforms, most of them commercial packages that tend to dominate the market. The four or five most important software platforms can perform many modelling requirements to handle contemporary and future problems; they offer different packages to deal with different levels of granularity or detail. This has led to the definition of basically three levels of modelling: macro, meso, and micro simulation packages.

Macro-simulation models are usually employed for cities or regions with zones numbering 200 and above. The networks are reasonably simplified with road links, each one with a flow-delay relationships. The main assumption behind this type of model is that all trips clear the network during the modelled period. Flows are handled as rates, vehicles per hour, and no physical queues are modelled. These models are sometimes called 'strategic', as they are used for transport planning and policy development in the long term.

Meso-simulation models recognise and model the production of physical queues and allow for demand to be variable during a modelled period; they represent delays in urban areas better but are more complex; vehicles are sometimes treated as platoons. The networks contain more details, like signal timings and queue storage capacities.

Micro-simulation models treat each vehicle, and sometimes pedestrians and cyclists, individually following known physical rules; they require a much more detailed specification of the networks and parameters specifying normal driving practices: acceleration, deceleration, and so on. These are often called 'tactical' models and are typically used for short-term analyses. Agent and activity based models are often implemented at this micro-simulation level as well.

1.7 Transport Planning and Uncertainty

Transport planning models on their own do not solve transport problems. To be useful, they must be utilised within a decision process adapted to the chosen decision-making style. The classic transport model was originally developed for an idealised normative decision-making approach. Its role in transport planning can be presented as contributing to the key steps in a 'rational' decision-making framework, as in Figure 1.7:

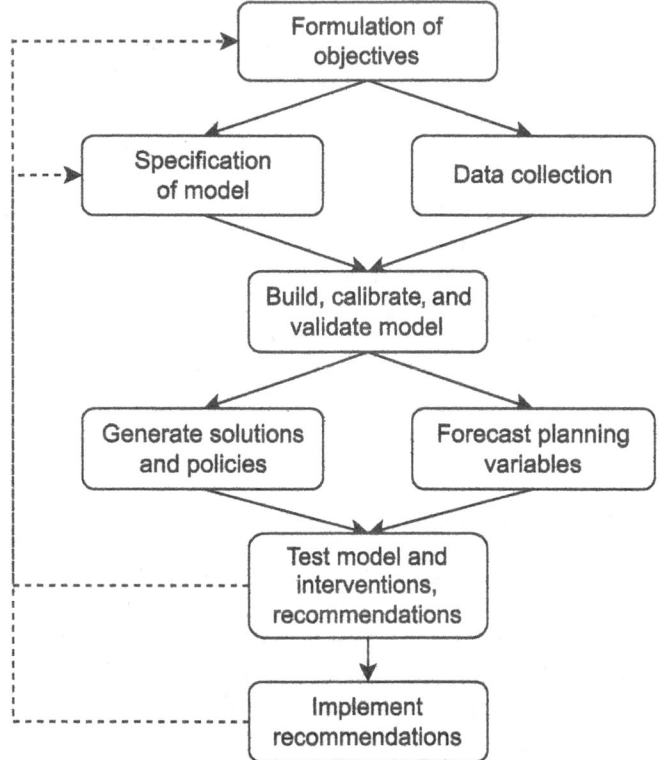

Figure 1.7 A framework for rational decision-making with models

1) **Formulation of the problem and objectives**. A problem can be defined as a mismatch between expectations and perceived reality. The formal definition of a transport problem requires reference to objectives, standards, and constraints. The first reflects the values implicit in the decision-making process, a definition of an ideal but achievable future state. Standards are provided to compare, at any one time, whether minimum performance is being achieved at different levels of interest. For example, the fact that many signalised junctions in a city operate at more than 90% degree of saturation can be taken to indicate an overloaded network. Constraints can be of many types: financial, temporal, geographical, technical, or simply certain areas or types of buildings that should not be threatened by new proposals.

2) **Collection of data** about the present state of the system of interest to support the development of the analytical model. Of course, data collection is not independent from model development, as the latter defines which types of data are needed: data collection and model development are closely interrelated.

3) **Construction of an analytical model** of the system of interests. The tool set provided in this book can be used to build transport models, including demand and system performance procedures, from a tactical and strategic perspective. In general, one would select the simplest modelling approach, which makes possible a choice between schemes on a sound basis. The construction of an analytical model involves specifying it, estimating or calibrating its parameters, and validating its performance with data not used during calibration.

4) **Generation of solutions** for testing. This can be achieved in a number of ways, from tapping the experience and creativity of local transport planners and interested parties to the construction of a large-scale design model, perhaps using optimisation techniques. This involves supply- and cost-minimisation procedures falling outside the scope of this book. To test the solutions or schemes proposed, it is necessary to *forecast the future values of the planning variables*, which are used as inputs to the model. This requires the preparation of consistent quantified descriptions, or scenarios, about the future of the area of interest, normally using forecasts from other sectors and planning units. We will come back to this issue in Chapter 18.

5) **Testing the model and solution**. The performance of the model is tested under different scenarios to confirm its reasonableness; the model is also used to simulate different solutions and estimate their performance in terms of a range of suitable indicators. These must be consistent with the identification of objectives and problem definition above and combined with the evaluation of solutions and recommendations. This involves the operational, economic, financial, and social assessment of alternative courses of action on the basis of the indicators produced by the models. A combination of skills is required here, from economic analysis to political judgement.

6) **Implementation of the solution** and search for another problem to tackle; this requires recycling through this framework, starting again at point (1).

There are two problems with this description of the process. First, real transport systems do not obey the restrictions above: objective functions and constraints are often difficult to define. With hindsight, these definitions often turn out to be blinkered: by narrowing a transport problem, we may gain the illusion of being able to solve it; however, transport problems have the habit of 'biting back', of reappearing in different places and under different guises; new features and perspectives are added as our understanding of the transport system progresses; changes in the external factors and planning variables throw our detailed transport plans off course.

The second issue is that future uncertainty is unavoidable, and this figure ignores it. The future is certainly not going to be what we expect. Therefore, we are planning to solve not just current problems but also those likely to crop up in the future. This uncertainty results from a number of sources: new technologies, evolution of energy costs, resource constraints, and health crisis. Moreover, people's preferences and values change over time as well.

How can we improve this general approach to cope with an ever-changing world? It seems essential to recognise that the future is much more fluid than our forecasting models would lead us to believe. If this is the case, master plans need revising at regular intervals and other decision-making strategies need supporting with the inclusion of fresh information regularly collected to check progress and correct course where necessary. Adaptive or mixed-mode decision-making styles seem more flexible and appropriate to the characteristics of transport problems. They recognise the need to continually redefine problems, arenas, and goals as we understand them better, identify new solution strategies, respond to political and technological changes, and enhance our modelling capabilities through training, research, and experience.

The introduction of a *monitoring function* is an important addition to the scheme in Figure 1.7. A monitoring system is not restricted to regular data collection; it should also facilitate all other stages in the decision-making framework, as highlighted in Figure 1.8. There are two key roles for a monitoring system. First, it should provide data to identify departures from the estimated behaviour of the transport system and of exogenous key variables such as population and economic growth. Second, the data collected should be valuable in further validating and enhancing the modelling approach followed in preparing the plans.

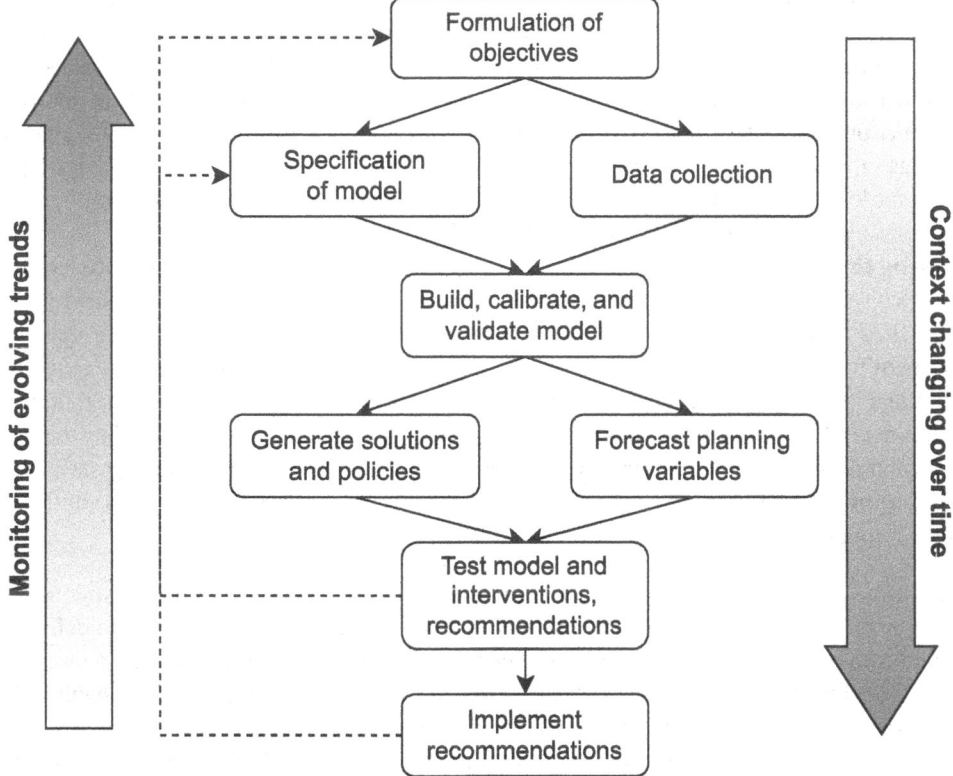

Figure 1.8 Planning and monitoring with the help of models

A good monitoring system should also facilitate learning by the planning team and provide ideas on how to improve and modify models. In this sense, major disruptions to the transport system, like public-transport strikes, short-term fuel shortages, or major roadworks, which may temporarily change the network structure and its characteristics, should provide a major source of information on transport behaviour to contrast with model predictions. These unplanned experiments should enable analysts to test and enhance their models. A monitoring system fits very well with the idea of a regular or *continuous planning approach* in transport. If the monitoring system is not in place, it should be established as part of any transportation study.

Monitoring the performance of a transport system and plans is such an important function that it deserves to influence the choice of transport models used to support planning and policy making. The use of models, which can be re-run and updated using low-cost and easy-to-collect data, seems particularly appropriate to this task. As we shall see in subsequent chapters, these simpler models cannot provide all the behavioural richness of other more detailed approaches. However, there is scope for combining the two techniques by applying the tool with the highest resolution to the critical parts of the problem and using coarser tools that are easier to update to monitor progress and identify where and when a new detailed modelling effort is needed. We have made an attempt to identify the scope for trade-offs of this kind in the remainder of this book.

There is a plethora of new data sources to support this monitoring function. These include all the digital traces that the devices we use leave behind: mobile phones, smart cards, applications, and navigation units. These are complemented by an abundance of sensors from induction and video counters, passenger counters, WiFi, and Bluetooth sensors. The adoption of a monitoring function enables the implementation of a continuous planning process. This is in contrast to the conventional approach of spending considerable resources over a period of one or two years to undertake a large-scale transport study. This burst of activity may be followed by a much longer period of limited effort in planning and updating plans. Soon the reports and master plans become obsolete or simply forgotten, and nobody capable of running the models again is left in the planning unit. Some years later, a new major planning and modelling effort is embarked upon, and the cycle is repeated. This style of planning with the help of models in fits and starts is wasteful of resources, does not encourage learning and adaptation as a planning skill, and alienates analysts from real problems. This approach is particularly painful in developing countries: they do not have resources to waste, and the rapid change experienced there speeds up plan and data obsolescence. The use of models that are simpler and easier to update is advocated in Chapter 14 to help with the implementation of a sound but low-cost monitoring function.

1.8 Theoretical Basis Versus Expedience

One of the recurring themes of transport modelling practice is the distance, and some would say mistrust, between theoreticians and practitioners. The practitioner would often refer to the need to choose between a theoretically sound but difficult to implement set of models and a more pragmatic modelling approach reflecting the limitations of the data, time, and resources available for a study. The implication is that the 'pragmatic' method can deliver the answers needed in the time available for the study, even if shortcuts must be taken.

We have nothing against pragmatic approaches provided they deliver the answers needed to make sound decisions. There is no point in using sophisticated and expensive (but presumably theoretically sound) models for the sake of winning some credit in the academic fraternity. However, there are several reasons to prefer a model based on a sound theoretical background:

1) To guarantee stable results. The recommendations from a study should not depend on how many iterations of a model were run. Prescriptions like 'always start from free flow costs' or 'iterate twice only' are not good enough reasons to assume stable results: next time somebody will suggest running a couple more iterations or a different, and quite justifiable, starting point; this should not be able to change the recommendations for or against a particular scheme.
2) To guarantee consistency. One should be careful about using a particular model of travellers' choice in one part of a model system and a different one in another. Pragmatic models sometimes fail to pass this test. Model consistency is necessary to pass the test of 'reasonableness' and public scrutiny.
3) To give confidence in forecasting. It is almost always possible to fit a model to an existing situation. However, there are plenty of examples of well-fitting models that make no sense, perhaps because they are based on correlated variables. Variables which are correlated today may not be so tomorrow; for example, a strong correlation between banana exports and car ownership in a particular country may disappear once oil is discovered there. Therefore, models should be backed by some theory of travel behaviour so that one can interpret them consistently and have some confidence that they will remain valid in the future.
4) To understand model properties and develop improved algorithms for their solution. When one is able to cast a problem in mathematical programming or maximum likelihood terms, to mention two popular approaches to model generation, one has a wealth of technical tools to assist in the development of good solution algorithms. These have been developed over the years by researchers working in many areas besides transport.

5) To understand better what can be assumed to be constant and what must be accepted as variable for a particular decision context and level of analysis. The identification of exogenous and endogenous variables and those that may be assumed to remain constant is a key issue in modelling economics. For example, for some short-term tactical studies, the assumption of a fixed trip matrix may be reasonable, as in many traffic-management schemes. However, even in the short term, if the policies to be tested involve significant price changes or changes to accessibility, this assumption no longer holds valid.

On the other hand, practitioners have sometimes abandoned the effort to use theoretically better models; some of the reasons for this are as follows:

1) They are too complex. This implies that heuristic approaches, rules of thumb, and *ad hoc* procedures are easier to understand and therefore preferable. This is a reasonable point; we do not advocate the use of models as 'black boxes'; quite the contrary. Model output needs interpretation, and this is only possible if a reasonable understanding of the basis for such a model is available. Without ignoring the important role of academic literature in advancing the state of the art, there is a case for more publications explaining the basis of models without recourse to difficult notation or obscure (to the practitioner) concepts. Most models are not that complex, even if some of the statistics and computer implementations needed may be quite sophisticated. Good publications bridging the gap between the practitioner and the academic are an urgent need.

2) Theoretical models require data, which are not available and are expensive to collect. This is often not entirely correct; many advanced models make much better use of small-sample data than some of the most pragmatic approaches. Improvements in data collection methods have also reduced these costs and improved the accuracy of the data.

3) It is better to work with 'real' matrices than with models of trip-making behaviour. This is equivalent to saying that it is better to work with fixed trip matrices, even if they have to be grossed up for the planning horizon. We will see that sampling and other data-collection errors cast doubts on the accuracy of such 'real' matrices; moreover, they cannot possibly respond to most policies (e.g. improvements in accessibility, new services, and price changes) nor be reasonable for oversaturated do-minimum future conditions. Use of observations alone may lead to 'blinkered' decision-making, a false sense of accuracy, and an underestimation of the scope for change.

4) Theoretical models cannot be calibrated to the level of detail needed to analyse some schemes. There may be some truth in this statement, at least in some cases where the limitations of the data and time available make it necessary to compromise in detail if one wishes to use a better model. However, it may be preferable to err in this way than to work with the illusion of sufficient detail but undermined by potentially pathological (predictions of the wrong sign or direction) or insensitive results from *ad hoc* procedures.

5) It is better to use the same model (or software) for most problems because this ensures consistency in the evaluation methods. This is, in principle, correct, provided the model remains appropriate to these problems. It has the advantage of a consistent approach, ease of use and interpretation, and reduced training costs. However, this strategy breaks down when the problems are not of the same nature. Assumptions of fixed trip matrices, or insensitivity to mode choice or pricing policies, may be reasonable in some cases but fail to be acceptable in others. The use of the same model with the same assumptions may be appropriate in one case and completely misleading in another.

The importance of these criteria depends, of course, on the context of the decision and the levels of analysis involved in the study. We argue in this book for using the appropriate level of resolution to the problem at hand. We prefer to strive to use good sound models as much as possible, even if we have to sacrifice some level of detail. One must find the best balance between theoretical consistency and expedience in each case and decision-making context. We have striven to provide material to assist in this choice.

1.9 Becoming a Better Modeller

Earlier editions of this book had a chapter on certain minimum mathematical and statistical prerequisites for interacting successfully with contemporary transport models; we have omitted this chapter now as there are many books and courses on the internet that provide this kind of material. Notwithstanding, key elements of that chapter that were a bit more specialised (particularly in statistics) were moved to where they are needed in the text. We outline here some requirements to become a better modeller.

The need for a good mathematical background in this field is natural because a good level of *technical skills* is essential. No one should use a technically flawed model. There is enough knowledge and good practice to avoid models that do not converge or use inconsistent assumptions in different sub-models. This technical knowledge must be complemented with experience in the practical implementation of models. However, our understanding of human behaviour and travel is imperfect, and, therefore, we need additional skills to improve model results and their interpretation.

Curiosity is essential to gain a deeper understanding of travel demand and choice. Young professionals need an inquisitive mind to learn from experience. We make forecasts all the time, such as when we choose a university or a partner for life, and the rationality for these choices often departs from that of *homo economicus*. We learn when we can identify flaws in our own decision-making, in particular about travel-related choices: new car and choosing places to live and work influenced by an expected lifestyle.

Listening skills should also be near the top, as without them there is little point in having curiosity or a wide range of interests. This skill should be developed as prejudices creep in all the time and we tend to incorporate them as part of our identity.

A good modeller must also cultivate a *questioning mind*. It is far too easy to become enamoured of the complex and demanding model one has just completed validating. Good analysts should be the first to question their results. The professionals who can say, 'These results cannot be right' before sharing them with others are on their way to success.

Communication skills are also key in our field. A set of numbers is a limited outcome from a modelling effort. Interpretation and communication of results in a manner that is understood by the listeners and stakeholders is paramount.

Finally, modellers tend to have a poor reputation for sticking to timescales. One must develop a realistic view of how long it takes to deliver reasonable results. Note that it is often much longer than the original estimate of an inexperienced analyst. Few things are more damaging to a reputation than the late delivery of results, especially if they are no more than a set of numbers without a suitable interpretation.

Exercises

1.1 The following objectives have been proposed for transport planning and modelling: (a) to reduce congestion, (b) to reduce travel time, (c) to provide more equitable access to opportunities, (d) to strengthen public transport and active modes, and (e) to protect the environment for future generations. Take each one of these objectives and think how it would shape the identification of problems and solutions to them. Select only one as the most important and discuss why you would defend it.

1.2 Take the example depicted in Figure 1.3. Build a simple model in a spreadsheet depicting this case with yearly changes and covering a horizon of 10 years. Search, adopt, and document the assumptions you will need, for example, about elasticities of demand, car ownership, population growth, cost impacts, and so on. Contrast results over ten years.

1.3 Now consider Figure 1.4 and use the assumptions above. Now add the impact of car restraint, for example, road user charges, and public transport support (e.g. subsidies). Repeat the process of Exercise 1.2.

1.4 Modify the simple model depicted in Figure 1.3 to consider the role of low cost, widely available, app procured taxi-like services.

1.5 Modify the simple model depicted in Figure 1.4 to consider the role of low-cost, demand responsive ride-sharing services. Discuss the policy implications.

1.6 Modify the simple model in Figure 1.3 for a country like Vietnam or Colombia, where motorcycle ownership may be more important and higher than car ownership. Your model should include both cars and motorcycles and any interaction between them.

2

Data

This chapter is devoted to data collection issues and their representation for use in transport modelling. Although we present a wide range of data collection methods this is by no means complete. The nature of the data to be collected depends, of course, on the models chosen for a particular study. Moreover, advances in telecommunications are changing the travel data collection process, with new ways of processing the digital traces from devices we use, for example, GPS navigation or mobile phones.

As we are trying to develop models that represent mobility accurately, we need precise information about reality. Data may come from two main sources: (i) specific data collected to develop and validate the models and (ii) general, or secondary data collected for different purposes but useful for model development. A typical example of this secondary source is Census data; this is an essential description of the population residing at each location and is particularly useful for the expansion and correction of most sample data. Other key secondary sources are Employment and Education Enrolment data.

There are two basic ways to collect data: *active* and *passive*. In an active data collection exercise, a subject is asked to respond to specific questions like 'where are you coming from and what is your destination'. The accuracy of this data relies on two assumptions: (i) there is a perfect understanding of the question with no ambiguity, and (ii) the respondent is able and willing to provide a truthful answer. Unfortunately, there are many reasons why these two assumptions are seldom fully valid: words mean different things to different people, the person may not accurately recall an item such as the exact fare paid or the real waiting time for a bus; the respondent may be reluctant to disclose some data or may even wish to simplify answers to reduce the length of the survey.

Passive data collection methods rely either on direct observation, for example, person counts at bus stops, or require extracting information from a valid source, for example, automatic image recognition of a video taken at the same bus stop. Traffic and person counts are typical examples of this data but new data sources like video image processing, Bluetooth sensors, and digital traces from mobile phone data, can also be processed to extract useful information about movements.

All data have errors. Some of these stem from the time when the data was collected; this is more likely to happen with secondary data sources such as Census data collected years before a specific study. Other errors are due to the instrument used (for example, reporting errors or inaccurate location of a device); the impossibility or impracticality of collecting a particular type of data for the whole population during the full period of interest, leads to sampling errors; these could be random or have a bias that requires correction.

Modelling Transport, Fifth Edition. Juan de Dios Ortúzar and Luis G. Willumsen.
© 2024 John Wiley & Sons Ltd. Published 2024 by John Wiley & Sons Ltd.
Companion website: www.wiley.com/go/ortuzar5e

The treatment in this chapter is general. We will consider four issues that are a prerequisite for other subjects treated in the rest of the book. Firstly, we will provide a brief introduction to statistical sampling theory. Interested readers are advised that there is a complete book on the subject (Stopher and Meyburg 1979), which may be consulted for more details. In Section 2, we will discuss the nature and importance of errors, which can arise both during model estimation and when forecasting with the aid of models; the interesting question of data accuracy versus model complexity and cost is also addressed.

In Section 3, we will consider various types of surveys used in applied transport planning; we will be particularly interested in problems such as the correction, expansion, and validation of survey data, and we will also discuss issues involved in the collection of longitudinal (e.g. panel) data, and travel time data. Finally, Section 4 gives a fairly complete treatment of the most important issues involved in the experimental design and collection of stated preference (SP) data.

2.1 Basic Sampling Theory

2.1.1 Statistical Considerations

Statistics may be defined as the science concerned with gathering, analysing, and interpreting data to obtain the maximum quantity of useful information. It may also be described as one of the disciplines concerned with decision-making under uncertainty; its goal would be, in this case, to help determine the level of uncertainty associated with measured data in order to support better decisions.

Data usually consist of a sample of observations taken from a certain population of interest that is not economically (or perhaps even technically) feasible to observe in its entirety. These observations are made about one or more attributes (say income) of each member of the population. Inferences can be made then about the mean value of these attributes, often called *population parameters*. Sample design aims at ensuring that the data to be examined provide the greatest amount of useful information about the population of interest at the lowest possible cost; the problem remains of how to use the data (i.e. expand the values in the sample) to make correct inferences about this population. Thus, two difficulties exist:

- how to ensure a *representative* sample; and
- how to extract valid conclusions from a sample satisfying the above condition.

Neither of these would constitute a problem if there was no variability in the population. To solve the second difficulty, a well-established procedure exists, which does not present major problems if certain conditions and assumptions hold. The identification of a representative sample, however, may be a more delicate task in certain cases, as we shall see below.

2.1.1.1 Basic Definitions

Sample. The sample is defined as a collection of units that has been especially selected to represent a larger population with certain attributes of interest (i.e. height, age, income). Three aspects of this definition have particular importance: first, which population the sample seeks to represent; second, how large the sample should be; and third, what is meant by 'especially selected'.

Population of interest. This is the complete group about which information is sought; in many cases, its definition stems directly from the study objectives. The population of interest is composed of individual elements; however, the sample is usually selected on the basis of sampling units, which may not be equivalent to these individual elements, as aggregation of the latter is often deemed necessary. For example, a frequently used sampling unit is the household while the elements of interest are the individuals residing in it. There is also a time dimension associated with defining the population of interest: we may focus on different time periods, for example, a 'neutral' working day, week-ends, summer/winter, or a full week of movements.

Sampling method. Most acceptable methods are based on a form of *random sampling*. The key issue here is that the selection of each unit is carried out independently, with each unit having the same probability of being included in the sample. The more interesting methods are:

1) **Simple random sampling**, which is not only the simplest method but constitutes the basis of all the rest. It consists in first associating an identifier (number) to each unit in the population and then selecting these numbers at random to obtain the sample; the problem is that far too large samples may be required to ensure sufficient data about minority options of particular interest. For example, it may well be that sampling households at random in a developing country would provide little information on multiple car ownership.
2) **Stratified random sampling**, where *a priori* information is first used to subdivide the population into homogeneous strata (with respect to the stratifying variable) and then simple random sampling is conducted inside each stratum using the same sampling rate. The method allows the correct proportions of each stratum in the sample to be obtained; thus, it may be important in those cases where there are relatively small subgroups in the population as they could lack representation in a simple random sample.

It is also possible to stratify with respect to more than one variable, thus creating an n-dimensional matrix of group cells. However, care must be taken with the number of cells created as it increases geometrically with the number of strata; large figures may imply that the average number of sampling units per cell is too small. Nevertheless, even stratified sampling does not help when data are needed about options with a low probability of choice in the population; in these cases, a third method called *choice-based sampling*, actually a subset of the previous one, is required. The method consists in stratifying the population based on the result of the choice process under consideration. This method is fairly common in transport studies, as we will see in Section 3. A major advantage is that data may be produced at a much lower cost than with the other sampling methods; its main drawback is that the sample thus formed may not be random and, therefore, the risk of bias in the expanded values is greater.

Sampling error and sampling bias. These are the two types of error that might occur when taking a sample; combined, they contribute to the measurement error of the data. The first is simply due to the fact that we are dealing with a sample and not with the total population (i.e. it will always be present due to random effects). This type of error does not affect the expected values of the means of the estimated parameters; it only affects the variability around them, thus determining the degree of confidence that may be associated with the means; it is basically a function of sample size and of the inherent variability of the parameter under investigation.

The sampling bias, on the other hand, is caused by mistakes made either when defining the population of interest, or when selecting the sampling method, the data collection technique, or any other part of the process. It differs from the sampling error in two important respects:

- it can affect not only the variability around the mean of the estimated parameters but the values themselves; therefore, it implies a more severe distortion of the survey results;
- while the sampling error may not be avoided (it can only be reduced by increasing sample size), the sampling bias may be virtually eliminated by taking extra care during the various stages of sampling design and data collection.

Sample size. Unfortunately, there are no straightforward and objective answers to the calculation of sample size in every situation. This happens, although sample size calculations are based on precise statistical formulae, because many of their inputs are relatively subjective and uncertain; therefore, they must be produced by the analyst after careful consideration of the problem in hand. Determining sample size is a problem of trade-offs, as:

- a much too large sample may imply a data-collection and analysis process, which is too expensive given the study objective and its required degree of accuracy; but
- a far too small sample may imply results that are subject to an unacceptably high degree of variability reducing the value of the whole exercise.

Somewhere between these two extremes lies the most efficient (in cost terms) sample size given the study objective. In what follows it will be assumed that this consists in estimating certain population parameters by means of statistics calculated from sample data. As any sample statistics are subject to sampling error, it is also necessary to include an estimate of the accuracy that may be associated with its value.

2.1.1.2 Sample Size to Estimate Population Parameters

This depends on three main factors: variability of the parameters in the population under study, degree of accuracy required for each, and population size. Without doubt the first two are the most important; this may appear surprising at first sight because, to many, it seems intuitively necessary to take bigger samples in bigger populations in order to maintain the accuracy of the estimates. However, as will be shown below, the size of the population does not significantly affect sample size except in the case of very small populations.

The *Central Limit Theorem*, which is at the heart of the sample size estimation problem, postulates that the estimates of the mean from a sample tend to become distributed Normal as the sample size (n) increases. This holds for any population distribution if n is greater than or equal to 30; the theorem holds even in the case of smaller samples if the original population has a Normal-like distribution.

Consider a population of size N and a specific property that is distributed with mean μ and variance σ^2. The Central Limit Theorem states that the means ($\bar{x}$) from successive samples distribute Normal with mean μ and standard deviation se ($\bar{x}$), known as the *standard error* of the mean, and given by:

$$\text{se}\,(\bar{x}) = \sqrt{(N-n)\sigma^2/[n(N-1)]} \tag{2.1}$$

If only one sample is considered, the best estimate of μ is $\bar{x}$ and the best estimate of σ^2 is s^2 (the sample variance); in this case, the standard error of the mean can be estimated as:

$$\text{se}\,(\bar{x}) = \sqrt{(N-n)s^2/nN} \tag{2.2}$$

and, as mentioned above, it is a function of three factors: the parameter variability (s^2), the sample size (n), and the size of the population (N). However, for large populations and small sample sizes (the most frequent case) the factor $(N-n)/N$ is very close to 1 and Eq. (2.2) reduces to:

$$se\,(\bar{x}) = \frac{s}{\sqrt{n}} \tag{2.3}$$

Thus, for example, quadrupling sample size will only halve the standard error (i.e. it is a typical case of diminishing returns of scale). The required sample size may be estimated by solving Eq. (2.2) for n and this is usually simpler to do in two stages, first calculating n from Eq. (2.3) such that:

$$n' = \frac{s^2}{se\,(\bar{x})^2} \tag{2.4}$$

and then correcting for finite population size, if necessary, by:

$$n = \frac{n'}{1 + \dfrac{n'}{N}} \tag{2.5}$$

Although the above procedure appears to be both objective and relatively trivial, it has two important problems that impair its application: estimating the sample variance s^2 and choosing an acceptable standard error for the mean. The first one is obvious: s^2 can only be calculated once the sample has been taken, so it has to be estimated from other sources. The second one is related to the desired degree of confidence to be associated with the use of the sample mean as an estimate of the population mean; normal practice does not specify a single standard error value, but an interval around the mean for a given confidence level. Thus, two judgements are needed to calculate an acceptable standard error:

1) A confidence level for the interval must be chosen; this expresses how frequently the analyst is prepared to make a mistake by accepting the sample mean as a measure of the true mean (e.g. the typical 95% level implies an acceptance to err in 5% of cases).
2) It is necessary to specify the limits of the confidence interval around the mean, either in absolute or relative terms; as the interval is expressed as a proportion of the mean in the latter case, an estimate of this is required to calculate the absolute values of the interval. A useful option considers expressing sample size as a function of the expected *coefficient of variation* (CV = σ/μ) of the data.

For example, if a Normal distribution is assumed and a 95% confidence level is specified, this means that a maximum value of 1.96 se $(\bar{x})$ would be accepted for the confidence interval (i.e. $\mu \pm 1.96\sigma$ contains 95% of the Normal probability distribution); if a 10% error is specified we would get a $(\mu \pm 0.1\mu)$ interval and it may be seen that:

$$se\,(\bar{x}) = 0.1\mu/1.96 = 0.051\mu$$

and replacing this value in (2.4) we get:

$$n' = (s/0.051\mu)^2 = 384\text{CV}^2 \tag{2.6}$$

Note that if the interval is specified as $(\mu \pm 0.05\mu)$, that is, with half the error, n' would increase fourfold to 1536 CV^2.

To complete this point, it is important to emphasise that the above exercise is relatively subjective; thus, more important parameters may be assigned smaller confidence intervals and/or higher levels of confidence. However, each of these actions will result in smaller acceptable standard errors and, consequently, bigger samples and costs. If multiple parameters need to be estimated, the sample may be chosen based on that which requires the largest sample size.

2.1.1.3 Obtaining the Sample

The last stage of the sampling process is the extraction of the sample itself. In some cases, the procedure may be easily automated, either on site or at the desk (in which case care must be taken that it is actually followed on the field), but it must always be conducted with reference to a random process. Although the only truly random processes are those of a physical nature (i.e. roll of a dice or flip of a coin), they are generally too time-consuming to be useful in sample selection. For this reason, pseudo-random processes, capable of generating easily and quickly a set of suitable random-like numbers, are usually employed in sampling.

Example 2.1 Consider a certain area the population of which may be classified in groups according to: automobile ownership (with and without a car); and household size (up to four and more than four residents).

Let us assume that m observations are required by cell in order to guarantee a 95% confidence level in the estimation of, say, trip rates; assume also that the population can be considered to have approximately the distribution in Table 2.1 (i.e. from historic data).

Table 2.1 Population distribution for Example 2.1

Car ownership	Household size	% of population
With car	Four or less	9
	More than four	16
Without car	Four or less	25
	More than four	50

There are two possible ways to proceed:

1) Achieve a sample with m observations by cell by means of a random sample. In this case, it is necessary to select a sample size that guarantees this for each cell, including that with the smallest proportion of the population. Therefore, the sample size would be:

$$n = 100m/9 = 11.1m$$

2) Alternatively, one can undertake first a preliminary random survey of $11.1m$ households where only cell membership is asked for; this low-cost survey can be used to obtain the addresses of m households even in the smallest group. Subsequently, as only m observations are needed by cell, it would suffice to randomly select a (stratified) sample of $3m$ households from the other groups to be interviewed in detail (together with the m already detected for the most restrictive cell).

As can be seen, a much higher sample is obtained in the first case; its cost (approximately three times more interviews) must be weighed against the cost of the preliminary survey.

Example 2.2 Assume that for the purposes of a transport study, the population of a certain area has been classified according to two income categories and that there are only two transport modes available (car and bus) for the journey to work. Let us also assume that the population distribution is as in Table 2.2:

Table 2.2 Population distribution for Example 2.2

	Low income	High income	Total
Bus user	0.45	0.15	0.60
Car user	0.20	0.20	0.40
Total	0.65	0.35	1.00

1) **Random sample**. If a random sample is taken, it is clear that the same population distribution would be obtained.
2) **Stratified or exogenous sample**. Consider a sample with 75% low-income (LI) and 25% high-income (HI) travellers. From the previous table, it is possible to calculate the probability of a low-income traveller using bus, as:

$$P(\text{Bus/LI}) = \frac{P(\text{LI and Bus})}{P(\text{LI and Bus}) + P(\text{LI and Car})} = \frac{0.45}{0.45 + 0.20} = 0.692$$

Now, given the fact that the exogenous sample has 75% of individuals with low income, the probability of finding a bus user with low income in the sample is $0.75 \times 0.692 = 0.519$. Doing this for the rest of the cells, the table of probabilities for the stratified sample shown in Table 2.3 may be built:

Table 2.3 Probabilities for the stratified sample

	Low income	High income	Total
Bus user	0.519	0.107	0.626
Car user	0.231	0.143	0.374
Total	0.750	0.250	1.000

3) **Choice-based sample**. Let us assume now that we take a sample of 75% bus users and 25% car users. In this case, the probability of a bus user having low income may be calculated as:

$$P(\text{LI/Bus}) = \frac{P(\text{LI and Bus})}{P(\text{LI and Bus}) + P(\text{HI and Car})} = \frac{0.45}{0.45 + 0.15} = 0.75$$

Therefore, the probability of finding a low-income traveller choosing bus in the sample is 0.75 times 0.75, or 0.563. Proceeding analogously, the probabilities for the choice-based sample may be built as in Table 2.4:

Table 2.4 Probabilities for the choice-based sample

	Low income	High income	Total
Bus user	0.563	0.187	0.750
Car user	0.125	0.125	0.250
Total	0.688	0.312	1.000

As expected, each sampling method produces in general a different distribution in the sample. The importance of this example grows if we consider what is involved in the estimation of models using the various samples. For this, it is necessary to acquire an intuitive understanding of what calibration programs do; they simply search for the 'best' values of the model coefficients associated with a set of explanatory variables; in this case, *best* consists in replicating the observed choices more accurately.

For the population as a whole, the probability of actually observing a given data set may be found, conceptually, simply by calculating the probabilities of choosing the observed option by different types of travellers (with given attributes and choice sets). For example, in Table 2.2 (simple random sample) the probability that a high-income traveller selects car is given by the ratio between the probability of her having high income and using car, and the probability of her having high income, that is:

$$\frac{0.20}{0.15 + 0.20} = 0.572$$

If we consider Table 2.3 (stratified sample), the same probability is now given by:

$$\frac{0.143}{0.107 + 0.143} = 0.572$$

This is no coincidence; in fact, it was one important finding of an interesting piece of research by Lerman et al. (1976) in the USA. In practice, it means that standard software may be used to estimate models with data obtained from an exogenous sample.

It is also important to note that this is not the case for choice-based samples. To prove this, consider calculating the same probability but using information from Table 2.4:

$$\frac{0.125}{0.187 + 0.125} = 0.400$$

As can be seen, the result is completely different. To end this theme, it is interesting to mention that Lerman et al. (1976) proposed a method to use data from choice-based samples in model estimation avoiding bias at the expense only of requiring knowledge of the actual market shares. This involves weighting the observations of the choice-based sample by factors calculated as:

$$\frac{\text{Prob (select the option in a random sample)}}{\text{Prob (select the option in a choice based sample)}}$$

Thus, in our example, the weighting factor for bus-based observations should be:

$$\frac{0.45 + 0.15}{0.563 + 0.187} = 0.8$$

and for car users:

$$\frac{0.20 + 0.20}{0.125 + 0.125} = 1.6$$

Note that it is necessary to have data about choices on each alternative (i.e. it would not be possible to calibrate a model for car and bus, based on data for the latter mode only). We will come back to this problem in Section 8.4.2.

2.1.2 Practical Considerations in Sampling

2.1.2.1 The Implementation Problem

Stratified (and choice-based) sampling requires random sampling inside each stratum; but it may be difficult to do so in certain cases. Consider, for example, a situation where the population of interest are all potential travellers in a city. Thus, if we stratify by area of residence, it may be relatively simple to isolate the subpopulation of residents inside the city (e.g. using data from a previous survey); the problem is that it is extremely difficult to isolate and take a sample of the rest (i.e. those living outside the city).

An additional problem is that in certain cases even if it is possible to isolate all subpopulations and conforming strata, it may still be difficult to ensure a random sample inside each stratum. For example, if we are interested in taking a mode choice-based sample of travellers in a city, we will need to interview bus users and for this, it is first necessary to decide which routes will be included in the sample. The problem is that certain routes might have, say, higher than average proportions of students and/or old-age pensioners, and this would introduce bias (see Lerman and Manski 1979).

2.1.2.2 Finding the Size of Each Subpopulation

This is a key element in determining how many people will be surveyed. Given a certain stratification, there are several methods available to find out the size of each subpopulation, such as:

1) **Direct measurement**. This is possible in certain cases. Consider a mode choice-based sample of journey-to-work trips; the number of bus and metro tickets sold, plus traffic counts during the peak hour in an urban corridor, may yield an adequate measure (although imperfect as not all trips during the peak are to work) of the number of people choosing each mode. If, on the other hand, we have a geographical (i.e. zonal) stratification, the last census may be used to estimate the number of inhabitants in each zone.
2) **Estimation from a random sample**. If a random sample is taken, the proportion of observations corresponding to each stratum is a consistent estimator of the fraction of the total corresponding to each subpopulation. It is important to note that the cost of this method is low as we only seek to establish the stratum to which the respondent belongs.
3) **Solving a system of simultaneous equations**. Assume we are interested in stratifying by chosen mode and that we have data about certain population characteristics (e.g. mean income and car ownership). Taking a small on-mode sample we can obtain modal average values of these variables and postulate a system of equations, which has the subpopulation fractions as unknowns.

Finally, the 'failure rate' of different types of surveys must be considered when designing sampling frameworks. The sample size discussed above corresponds to the number of successful and valid responses to the data-collection effort. Some survey procedures are known to generate low valid response rates (e.g. some postal surveys), but they may still be used because of their low cost (Richardson et al. 1995).

Example 2.3 Assume the following information is available:

$$\text{Average income of population (I): } 33\,600\,\$/\text{year}$$

$$\text{Average car ownership (CO): } 0.44\,\text{cars/household}$$

Assume also that small on-mode surveys yield the data in Table 2.5:

Table 2.5 Strata information for Example 2.3

Mode	I ($/year)	CO (cars/household)
Car	78 000	1.15
Bus	14 400	0.05
Metro	38 400	0.85

If F_i denotes the subpopulation fraction of the total, the following system of simultaneous equations holds:

$$33\,600 = 78\,000F_1 + 14\,400F_2 + 38\,400F_3$$
$$0.44 = 1.15F_1 + 0.05F_2 + 0.85F_3$$
$$1 = F_1 + F_2 + F_3$$

the solution of which is:

$$F_1 = 0.2451$$
$$F_2 = 0.6044$$
$$F_3 = 0.1505$$

This means that if the total population of the area was 180 000 inhabitants, there would be approximately 44 100 car users 108 800 bus users and 27 100 metro users.

2.2 Errors in Modelling and Forecasting

The statistical procedures normally used in (travel demand) modelling assume not only that the correct functional specification of the model is known *a priori*, but also that the data used to estimate model parameters have no errors. In practice, however, these conditions are often violated; furthermore, even if they were satisfied, model forecasts are usually subject to errors due to inaccuracies in the values assumed for the explanatory variables in the design year.

The ultimate goal of modelling is often forecasting (i.e. the number of people choosing given options); an important problem model designers face is to find which combination of model complexity and data accuracy fits best the required forecasting precision and study budget. To this end, it is important to distinguish between different types of errors, in particular:

- those that could cause even correct models to yield incorrect forecasts, that is, errors in the prediction of the explanatory variables, transference and aggregation errors; and
- those that actually cause incorrect models to be estimated, for example, measurement, sampling, and specification errors.

In the next section, consideration is given first to the types of errors that may arise with the broad effects they may cause; then the trade-off between model complexity and data accuracy is examined with particular emphasis on the role of simplified models in certain contexts.

2.2.1 Different Types of Error

Consider the following list of errors that may arise during the processes of building, calibrating, and forecasting with models.

2.2.1.1 Measurement Errors

These occur due to the inaccuracies inherent in the process of measuring the data in the base year, such as questions badly registered by the respondent, answers badly interpreted by the interviewer, network measurement errors, coding, and digitising errors, and so on. These errors tend to be higher in less developed countries but they can always be reduced by improving the data-collection effort (e.g. by appropriate use of computerised interview support) or simply by allocating more resources to data quality control; however, both of these cost money.

Measurement error, as defined here, should be distinguished from the difficulty of defining the variables that ought to be measured. The complexity that may arise in this area is indicated in Figure 2.1. Regretfully, modeller and travellers use different 'units' to express variables like time and distance. The latter may find it difficult to convert their experience into precise minutes and kilometres. Modellers just hope that measurements reflect, with some unknown degree of error, the travellers' perceptions that influence their choices.

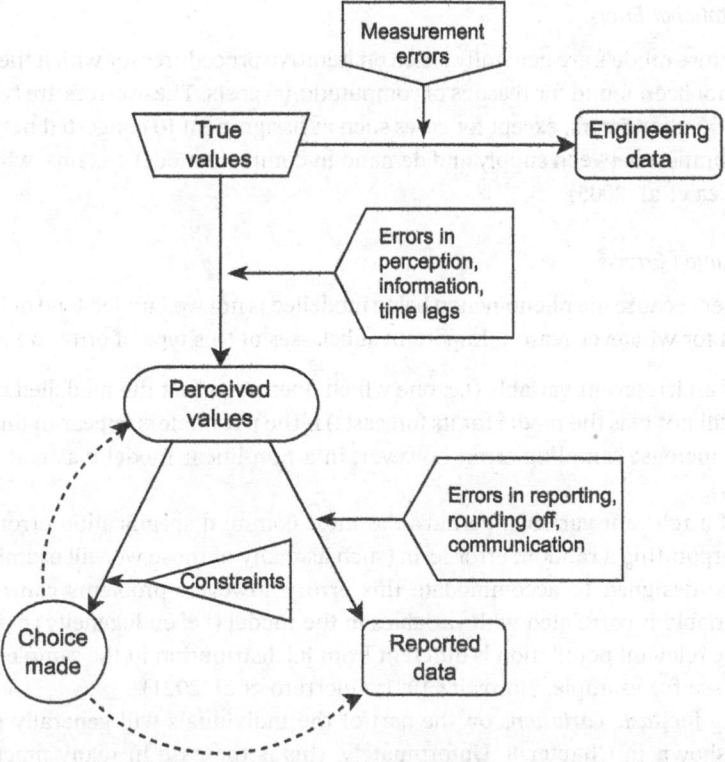

Figure 2.1 Attribute measurement and choice

Ideally, modelling should be based on the information perceived by individual travellers but whilst reported data may give some insight into perception, its use raises the difficult question of how to forecast what users are going to perceive in the future. So, it appears inevitable that models will be endowed with perception errors which tend to be greater for non-chosen alternatives due to the existence of *self-selectivity* bias (i.e. the attributes of the chosen option are perceived as better and those of the rejected option as worse than they are, to reinforce the rationality of the choice made).

Most models are used to forecast future conditions. This poses an interesting problem in the choice of variables to be included. Let us assume that we can fit a model well to a set of variables with current data but that it may be difficult to forecast the future value of at least some of them. If this is the case, even if we can get accurate values for these independent variables during the survey, their future values will only be known with great uncertainty (a wide confidence interval). This problem may take a particularly complex form if the issue is to estimate the number of individuals with specific characteristics of age, gender, income, employment type, marriage status, and number of children that will reside in a particular zone or location ten years from now.

2.2.1.2 Sampling Errors

These arise because the models must be estimated using finite data sets. Sampling errors are approximately inversely proportional to the square root of the number of observations (i.e. to halve them it is necessary to quadruple sample size as we saw above); thus, reducing them may be costly. Daganzo (1980) has examined the problem of defining optimal sampling strategies in the sense of refining estimation accuracy.

2.2.1.3 Computational Errors

These arise because models are generally based on iterative procedures for which the exact solution, if it exists, has not been found for reasons of computational costs. These errors are typically small in comparison with other errors, except for cases such as assignment to congested networks and problems of equilibration between supply and demand in complete model systems, where they can be large (see De Cea et al. 2005).

2.2.1.4 Specification Errors

These arise either because the phenomenon being modelled is not well understood or because it needs to be simplified for whatever reason. Important subclasses of this type of error are the following:

1) Inclusion of an irrelevant variable (i.e. one which does not affect the modelled choice process). This error will not bias the model (or its forecasts) if the parameters appear in linear form, but it will tend to increase sampling error; however, in a non-linear model bias may be caused (see Tardiff 1979).
2) Omission of a relevant variable, perhaps the most common specification error. Interestingly, models incorporating a random error term (such as many of those we will examine in Chapters 4 and 7) are designed to accommodate this error; however, problems can arise when the excluded variable is correlated with variables in the model (i.e. endogeneity) or when its distribution in the relevant population is different from its distribution in the sample used for model estimation (see for example, Horowitz 1981; Guerrero et al. 2021).
3) Not allowing for *taste variations* on the part of the individuals will generally produce biased models, as shown in Chapter 8. Unfortunately, this is the case in many practical models of choice; exceptions are the less-yielding Multinomial Probit and Mixed Logit models, which we discuss in Chapters 7 and 8.

4) Other specification errors, in particular the use of model forms that are not appropriate, such as linear functions to represent non-linear effects, *compensatory* models to represent behaviour that might be *non-compensatory* (see the discussion in Chapter 8), or the omission of effects such as *habit* or *inertia* (see Cantillo et al. 2007; Gutierrez et al. 2020). A full discussion of these forms of error is given by Williams and Ortúzar (1982a). The underlying assumption that human beings act as *homo economicus* is a key reason for many errors of this type and one that is very difficult to eliminate.

All specification errors can be reduced, in principle, simply by increasing model complexity; however, the total costs of doing this are not easy to estimate as they relate to model operation, but may induce other types of errors which might be costly or impossible to eliminate (e.g. when forecasting more variables and at a higher level of disaggregation). Moreover, removal of some specification errors may require extensive behavioural research and it must simply be conceded that such errors will be present in all feasible models. The modeller should then interpret outputs based on the knowledge of the potential mismatch between model and reality.

2.2.1.5 Transfer Errors

These occur when a model developed in one context (time and/or place) is applied in a different one. Although adjustments may be made to compensate for the transfer, ultimately the fact must be faced that behaviour might just be different in different contexts. In the case of spatial transfers, the errors can be reduced or eliminated by partial or complete re-estimation of the model to the new context (although the latter would imply discarding the substantial cost savings obtainable from transfer). However, in the case of temporal transfer (i.e. forecasting), this re-estimation is not possible and any potential errors must just be accepted (see the discussions in Chapter 9). This type of error will be greater for long-range planning applications as time will reduce the validity of the model, and perhaps more importantly, the accuracy of the planning variables used as input.

2.2.1.6 Aggregation Errors

These arise basically out of the need to make forecasts for groups of people, while modelling often needs to be done at the level of the individual to capture behaviour better. The following are important subclasses of aggregation errors:

1) **Data aggregation**. In most practical studies the data used to define the choice situation of individual travellers is aggregated in some form or another. Even when travellers are asked to report the characteristics of their available options, they can only have based their choice on the expected values of these characteristics. When network models are used, there is aggregation over routes, departure times, and even zones; this means that the values thus obtained for the explanatory variables are, at best, averages for groups of travellers rather than exact values for any particular individual. Models estimated with aggregate data will suffer from some form of specification error (see Daly and Ortúzar 1990). Reducing this type of aggregation error implies making measurements under more sets of circumstances: more zones, more departure times, more routes, more socioeconomic categories; this costs time and money and increases model complexity.

2) **Aggregation of alternatives**. Again, due to practical considerations, it may just not be feasible to attempt to consider the whole range of options available to each traveller; even in relatively simpler cases such as the choice of mode, aggregation is present as the large variety of services encompassing a bus option, say (e.g. one-man operated single-decker, two-man operated double-decker, mini-buses, express services), are seldom treated as separate choices.

3) **Model aggregation**. This can cause severe difficulties for the analyst except in the case of linear models where it is a trivial problem. Aggregate quantities such as flows on links are a basic modelling result in transportation planning, but methods to obtain them are subject to aggregation

errors which are often impossible to eliminate. We will examine this problem in some detail in Chapters 4 and 9.

2.2.2 The Model Complexity/Data Accuracy Trade-off

Given the difficulties discussed above, it is reasonable to consider the dual problem of how to optimise the return of investing in increasing data accuracy, given a fixed study budget and a certain level of model complexity, to achieve a reasonable level of precision in forecasts. To tackle this problem, we must understand first how errors in the input variables influence the accuracy of the model we use.

Consider a set of observed variables $\mathbf{x}$ with associated errors $\mathbf{e_x}$ (i.e. standard deviations); to find the output error derived from the propagation of input errors in a function such as:

$$z = f(x_1, x_2, ..., x_n)$$

the following formula may be used (Alonso 1968):

$$e_z^2 = \sum_i \left(\frac{\partial f}{\partial x_i}\right)^2 e_{x_i}^2 + \sum_i \sum_{j \neq i} \frac{\partial f}{\partial x_i} \frac{\partial f}{\partial x_j} e_{x_i} e_{x_j} r_{ij} \tag{2.7}$$

where r_{ij} is the coefficient of correlation between x_i and x_j; the formula is exact for linear functions and a reasonable approximation in other cases. Alonso (1968) used it to derive some simple rules to be followed during model building to prevent large output errors; for example, an obvious one is to avoid using correlated variables, thus reducing the second term of the right-hand side of Eq. (2.7) to zero.

If we take the partial derivative of e_z with respect to e_{x_i} and ignore the correlation term, we get:

$$\frac{\partial e_z}{\partial e_{x_i}} = \left(\frac{\partial f}{\partial x_i}\right)^2 \frac{e_{x_i}}{e_z} \tag{2.8}$$

Using these marginal improvement rates and an estimation of the marginal costs of enhancing data accuracy it could be possible, in principle, to determine an optimum improvement budget; this is not an easy problem in practice though, not least because the law of diminishing returns (i.e. each further percentage reduction in the error of a variable will tend to cost proportionately more), that should lead to a complex iterative procedure. However, Eq. (2.8) serves to deduce two logical rules (Alonso 1968):

- concentrate the improvement effort on those variables with a large error; and
- concentrate the effort on the most relevant variables, that is, those with the largest value of $(\partial f / \partial x_i)$ as they have the largest effect on the dependent variable.

Example 2.4 Consider the model $z = xy + w$, and the following measurement of the independent variables:

$$x = 100 \pm 10; \quad y = 50 \pm 5; \quad w = 200 \pm 50$$

Assume also that the marginal cost of improving each measurement is the following:

Marginal cost of improving x (to 100 ± 9) = $5.00
Marginal cost of improving y (to 50 ± 4) = $6.00
Marginal cost of improving w (to 200 ± 49) = $0.02

Applying Eq. (2.7) we get:

$$e_z^2 = y^2 e_x^2 + x^2 e_y^2 + e_w^2 = 502\,500$$

then $e_z = 708.87$; proceeding analogously, values of improved e_z in the cases of improving x, y, or w may be found to be 674.54, 642.26, and 708.08, respectively. Also, from (2.8) we get:

$$\frac{\partial e_z}{\partial e_x} = \frac{10y^2}{708.87} = 35.2; \quad \frac{\partial e_z}{\partial e_y} = 70.5; \quad \frac{\partial e_z}{\partial e_w} = 0.0705$$

These last three values are the marginal improvement rates corresponding to each variable. To work out the cost of the marginal improvements in e_z we must divide the marginal costs of improving each variable by their respective marginal rates of improvement. Therefore, we get the following marginal costs of improving e_z arising from the various variable improvements:

Marginal improvement in $x = 5/35.2 = \$0.142$

Marginal improvement in $y = 6/70.05 = \$0.085$

Marginal improvement in $w = 0.02/0.705 = \$0.284$

In this case, then, we would decide to improve the measurement accuracy of variable y if the marginal reduction in e_z was worth at least \$0.085.

Let us now define *complexity* as an increase in the number of variables of a model and/or an increase in the number of algebraic operations with the variables (Alonso 1968). It is obvious that in order to reduce specification error (e_s) complexity must be increased; however, it is also clear that as there are more variables to be measured and/or greater problems for their measurement, data measurement error (e_m) will probably increase as well.

If total modelling error is defined as $E = \sqrt{(e_s^2 + e_m^2)}$, it is easy to see that the minimum of E does not necessarily lie at the point of maximum complexity (i.e. maximum realism). Figure 2.2 shows not only that this is intuitively true, but also that as measurement error increases, the optimum value can only be attained at decreasing levels of model complexity.

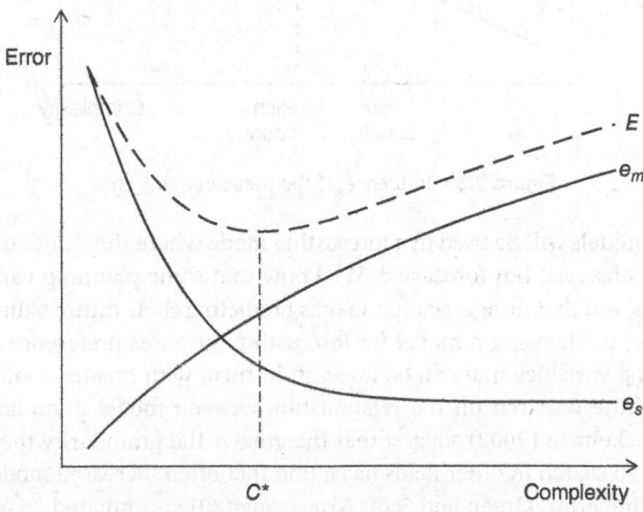

Figure 2.2 Variation of error with complexity

Example 2.5 Imagine that we have to make a choice between an extremely simple model, which is known to produce a total error of 30% in forecasts, and a new model which has a perfect specification (i.e. $e_s = 0$) given by:

$$z = x_1 x_2 x_3 x_4 x_5$$

where x_i are independent variables measured with a 10% error (i.e. $e_m = 0.1\ x_i$). To decide which model is more convenient we will apply Eq. (2.7):

$$e_z^2 = 0.01\left[x_1^2(x_2 x_3 x_4 x_5)^2 + x_2^2(x_1 x_3 x_4 x_5)^2 + \dots + x_5^2(x_1 x_2 x_3 x_4)^2\right]$$
$$e_z^2 = 0.05[x_1 x_2 x_3 x_4 x_5]^2 = 0.05\ z^2$$

that is, $e_z = 0.22z$ or a 22% error, in which case we would select the second model.

The interested reader can check that if the x_i variables could only be measured with 20% error, the total error of the second model would come out as 44.5% (i.e. we would now select the first model even if its total error increased up to 44%).

Figure 2.3 serves to illustrate this point, which may be summarized as follows: if the data are not of a very good quality, it might be safer to predict with simpler and more robust models (Alonso 1968). However, to learn about and understand the phenomenon, a better-specified model will always be preferable.

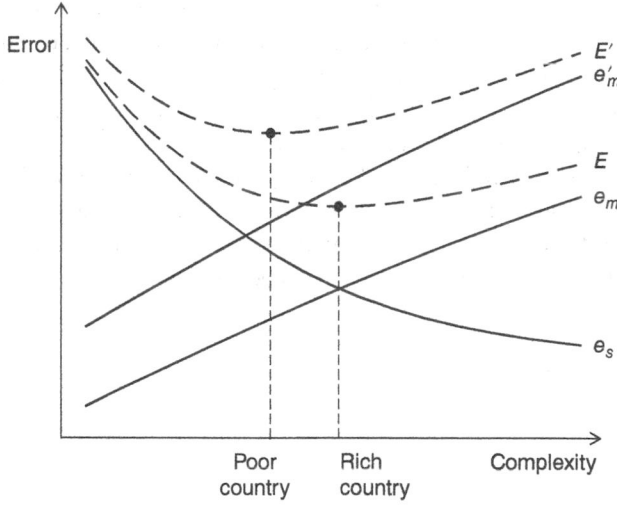

Figure 2.3 Influence of the measurement error

Moreover, most models will be used in a forecasting mode where the values of the planning variables x_i will not be observed but forecasted. We know that some planning variables are easier to forecast than others and that disaggregation makes predicting their future values an even less certain task. Therefore, in choosing a model for forecasting purposes preference should be given to those using planning variables that can be forecast, in turn, with greater confidence.

There has been little research on the relationship between model granularity and modelling error. Zhao and Kockelman (2002) suggest that the greater the granularity the greater the uncertainty propagation. Research in other fields has found that often increased model complexity leads to increase forecasting error. Green and Scott Armstrong (2015) compared 25 business forecasting models and found that added complexity increased forecasting error in them by 27%.

Consider, for example, that x_1 in Example 2.5 is fuel price. An accuracy of 10% can be expected in gasoline costs over the year in the areas where data were collected. However, the accuracy of the estimate of fuel prices in ten years from now will decrease, probably to something like 40%. If the errors in other variables remain as stated, it will be enough for the error in x_1 to increase to 25% to make the total error 32% and the simpler model preferable.

2.2.3 Forecasting Errors

Forecasting travel demand is central to most modelling work and, therefore, the accuracy that can be expected for these forecasts is critical. Several studies have questioned the accuracy that is generally achieved in travel demand forecasts.

In particular, there is sufficient evidence that the profession has not been very good at forecasting traffic and revenue for private and public sector projects. Standard & Poor's (S&P), the rating agency, collected data for over 100 toll road and public transport projects, and compared traffic and revenue projections with the actual outturns in each case. For toll roads, S&P concluded that there was, on average, a 23% overestimation of traffic and revenue (Bain and Polakovic 2005). It also observed that traffic and revenue projections for financial institutions suffered less from an overestimation of future traffic. S&P concluded there was an *optimism bias* inherent in most forecasts and that a key element in auditing traffic and revenue projections was to try and compensate for that bias.

S&P found similar results when dealing with public transport projects, albeit from a smaller and more local sample (Bain and Plantagie 2003). Moreover, the work of Flyvbjerg et al. (2005) covered over 200 road and rail projects in 14 countries and found that errors in forecasting rail projects were, in general, greater than for road schemes. They concluded that over 70% of rail projects over predicted demand by two-thirds. For about 50% of the road projects, the difference between forecasts and actual traffic was greater than 20%.

There have been several other papers contrasting projected against actual flows and revenues, and all reached similar conclusions. Hartgen (2013) has summarised most results and offers some suggestions to tackle these issues.

Faced with this evidence it is difficult to claim that there is no bias present in our traffic and revenue projections. It can be argued that unless one has an optimistic view of the future no project would ever get implemented. Nevertheless, there is at least one observation regarding this reported over-prediction bias. Sample bias is inevitable in studies of the nature considered by Flyvbjerg et al. (2005) and S&P. The forecasts made available to them were those that resulted in implemented projects or successful bids. More pessimistic views of the future led to lost bids or to projects that remained postponed indefinitely. Therefore, the results do not reflect the whole range of traffic projections but only those that were based on a more upbeat expectation of economic growth and demand.

To get a better understanding of the possible accuracy of demand forecasts, it is worth considering four main sources of inaccuracy in any model:

1) **Difficulty in establishing an accurate base year**. This is often more difficult than it sounds because of the limitations of current data collection methods. Travel matrices are a poorly defined 'average' of current conditions; parameters like the 'value of time' are sensitive to changes in the economy and the mood of residents, perceptions of operating costs are fuzzy, and so on.

2) **Uncertainty due to model quality**. Any model leaves some aspects out and includes others; the assumption of rational behaviour may be too unrealistic, the choices may be imperfect, the assumptions of consistent future behaviour (no change in tastes) may be unwarranted, the level of disaggregation for willingness to pay too coarse, network coding too naïve.

3) **Uncertainty about future data inputs**. This refers to uncertainty about future growth drivers such as population, income levels, economic activity, and their materialisation in space and time. These will be all projections and, as such, subject to much greater errors than observed data for the base year.

4) **Uncertainty about future scenarios**. These are unexpected or semi-expected changes in the future context of our models. Take for example the huge effects on transport that the COVID pandemic had worldwide. They also includes changes in technologies and attitudes, and how they affect travel (i.e. self-driving cars, remote presence, eCommerce, etc.). We know that they will happen but cannot say, yet, exactly how they will impact travel demand.

The first type of error is, in principle, avoidable; however, in practice, we can reduce, but never quite eliminate, it with better data. Models will always be simplifications and therefore an imperfect representation of reality. A better model may reduce this imperfection. Moreover, as models get more complex and realistic there is greater opportunity to incur modelling mistakes: wrongly coded links, erroneous instructions to the software in macros and scripts, and errors in parameters and functions. These avoidable errors can be reduced with good peer reviews and auditing of models, but will never be fully eliminated.

The third source of error, future data, is not avoidable and can only be reduced with better forecasts for items like population and the economy. However, the track record for these data items is not good. Finally, the twenty-first century seems to have augmented the source of scenario uncertainty. Technology innovation, changes in tastes, practices and values, political instability, pandemics, and changes in the world of work, all point towards greater uncertainty on how exactly the future will pan out. Figure 2.4 illustrates how these sources of uncertainty interact to limit the accuracy of any model forecast.

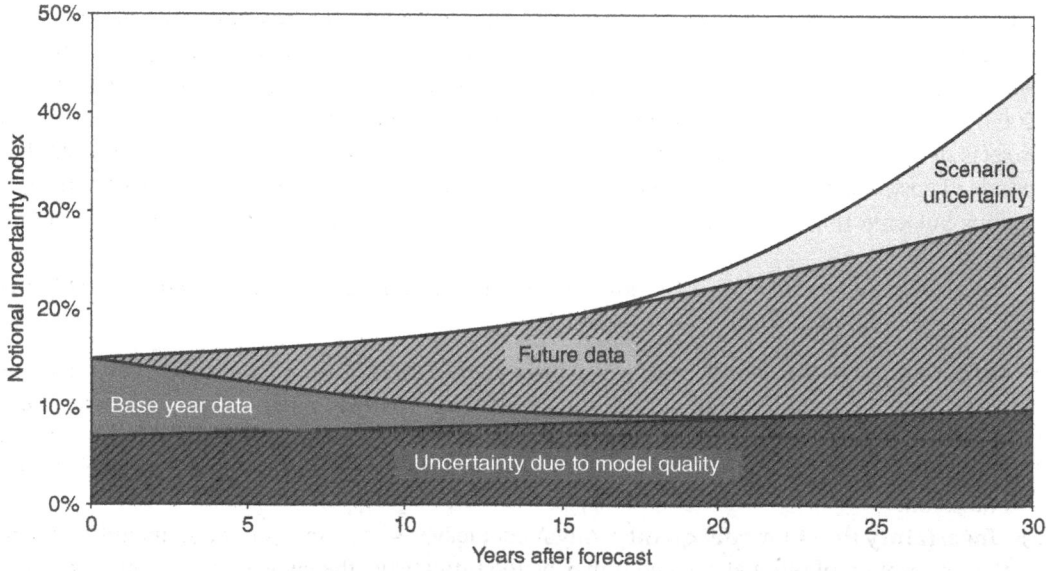

Figure 2.4 Contributors to future uncertainty

2.3 Basic Data-collection Methods

2.3.1 Practical Considerations

The selection of the most appropriate data collection methods will depend significantly on the type of models used in a study; they will define what type of data is needed and therefore what data collection methods are more appropriate. However, practical limitations will also have a strong influence in determining the most appropriate type of survey for a given situation. The specification of the desired model system and the design of a survey plan are not simple matters and require considerable skill and experience.

For example, many models adopt the objective to represent an 'average' or 'representative' day and consider specifically some time periods. This is a strong assumption that presupposes travel patterns are reasonably recurrent. We may call this the *Activity Regularity Assumption*, but there is little evidence about it. In fact, we can observe significant day-to-day variations in the level of traffic on any link in a network (Cherrett and McDonald 2002). One would typically expect some 10% variation in flow levels on similar days and on the same day of the week over similar weeks (i.e. excluding seasonal variations). Moreover, when we look at the make-up of this traffic, we find that only about half of it is truly recurrent. One can record vehicle number plates at the same location on several days to identify how many are 'unique' (appear only once) and how many reappear on other days. Cherrett and McDonald (2002) surveyed four different locations during peak periods around Southampton and found that only between 25% and 49% of vehicles reappeared on subsequent days. Del Mistro and Behrens (2008) undertook a survey over three weeks for several roads in Cape Town; they found that in arterial roads only 34–43% of vehicles would reappear on subsequent days; this percentage increased to around 54% for residential streets. Despite this evidence, limitations in data collection make it extremely difficult to abandon the Activity Regularity Assumption.

For basic information on recruiting, training, questionnaire design, supervision, and quality control, the reader is referred to the classical book by Moser and Kalton (1985). Information about survey procedures with a particular transport planning flavour may also be found in Stopher and Meyburg (1979) and Richardson et al. (1995). In what follows we briefly discuss some of the most typical practical constraints in transport studies.

2.3.1.1 Length of the Study

This obviously has great importance because it determines indirectly how much time and effort it is possible to devote to the data-collection stage. It is important to achieve a balanced study (in terms of its various stages) avoiding the all too frequent problem of eventually finding that the largest part of the study budget (and time) was spent in data collection, analysis, and validation (see Boyce et al. 1970).

2.3.1.2 Study Horizon

There are two types of situations worth considering in this respect:

- if the design year is too close, as in a tactical transport study, there will not be much time to conduct the study; this will probably imply the need to use a particular type of analysis tool, perhaps requiring data of a special kind;
- in strategic transport studies, on the other hand, the usual study horizon is 20 or more years into the future; although in principle this allows time to employ almost any type of analytical tool (with their associated surveys), it also means that errors in forecasting will only be known in 20 or more years' time; this calls for flexibility and adaptation if a successful process of monitoring and re-evaluation is to be achieved.

2.3.1.3   Limits of the Study Area

Here it is important to ignore formal political boundaries (i.e. of county or district) and concentrate on the whole area of interest. It is also necessary to distinguish between this and the study area as defined in the project brief; the former is normally larger as we would expect the latter to develop in a period of, say 20 years. The definition of the area of interest depends again on the type of policies examined and decisions to be made; we will come back to this issue below.

2.3.1.4   Study Resources

It is necessary to know, as clearly and in as much detail as possible, how many personnel and of what level will be available for the study; it is also important to know what type of computing facilities will be available and what restrictions to their use will exist. In general, the time available and study resources should be commensurate with the importance of the decisions to be taken as a result. The greater the cost of a wrong decision, the more resources should be devoted to get it right.

There are many other possible restrictions, ranging from physical (i.e. sheer size and topography of the locality) to social and environmental (e.g. known reluctance of the population to answer certain types of questions), which need to be considered and will influence sample design.

A general practical consideration is that travellers are often reluctant to answer 'yet another' questionnaire. Responding to questions takes time and may sometimes be seen as a violation of privacy. This may result in either flatly refusing to answer or in the provision of simplistic but credible responses, which is actually worse. In many countries, it is necessary to obtain permission from the authorities before embarking on any traffic survey involving disruptions to travellers.

Modern technology offers a number of methods to collect information about trips, tours, destinations, modes, or route choices, without requiring the active participation of the traveller, for example tracking mobile phone locations. However, these present privacy issues which need careful handling, with sensitivity and, of course, according to the law which at the time of writing is not particularly clear in most places. Moreover, these tracking methods do not offer much insight into the underlying behavioural intentions although some may be inferable from locations and timings. These methods are discussed in Section 2.3.6.

2.3.2   Types of Surveys

Up to the mid-1970s a large number of household origin–destination (O–D) surveys, using a simple random sample technique, were undertaken in urban areas of industrialised countries and also in many important cities in developing countries. This large effort was very expensive and demanded enormous quantities of time (a problem with collecting too much information is that a lot of time and money must also be spent analysing it); in fact, as we have commented already, the data-collection effort has traditionally absorbed a vital part of the resources available to conduct these large studies leaving, in many cases, little time and money for the crucial tasks of preparing and evaluating plans.

In many urban areas, particularly in large metropolitan areas, there is an important role for travel survey data. In some situations, this kind of data is used almost entirely for its richness in portraying the existing situation and thus helping the analyst to identify problems related to the transport system. In others, and such is our main interest, data is collected primarily for use in strategic transport modelling and hence forecasting, but it still may be used for both purposes (Battellino and Peachman 2003).

Understanding the use of the data is one of the key steps in determining the survey methodology for any travel survey. For example, *activity models* (Beckman et al. 1995) require large amounts of data not only about the activities people perform, but also on the activity 'infrastructure' (e.g. opening times of shops). However, the usual needs of mobility survey data are to provide the basis for accurate predictions, typically by a strategic transport planning model. In this case, the key data elements are trips between origins and destinations, rather than the underlying behavioural determinants, hence the term, 'origin–destination' study.

Stopher and Jones (2003) provide a rigorous, complete, and useful guide to the aspects a state-of-the-art survey should consider. Here we will only concentrate on certain key elements required to enhance the usefulness of the data as an aid to calibrating a contemporary supply–demand equilibration strategic transport planning model. In that case, current best practice suggests that the data set would be likely to have the following characteristics (Ampt and Ortúzar 2004):

- consideration of stage-based trip data, ensuring that analyses can relate specific modes to specific locations/times of day/trip lengths, etc.;
- inclusion of all modes of travel, including non-motorised trips;
- measurements of highly disaggregated levels of trip purposes;
- coverage of the broadest possible time period, e.g. 24 hours a day, seven days a week, and perhaps 365 days a year (to cover all seasons);
- achieve data from *all* members of the household;
- obtain high-quality information robust enough to be used even at a disaggregate level (Daly and Ortúzar 1990);
- be part of an integrated data collection system incorporating household interviews as well as O–D data from other sources such as cordon surveys.

Unfortunately, collecting data at this level of precision is not an easy task and is often precluded by the sheer difficulty of convincing a sufficiently large sample of individuals to participate in such a strenuous effort. Then, depending on the modelling objectives (i.e. strategic analysis versus detailed tactical studies), the analyst may need to ease the burden on the respondents and settle for less detailed information.

2.3.2.1 *Survey Scope*

Figure 2.5 is useful to describe the scope of a study to capture all trips affecting a metropolitan area. It is first necessary to define the study's area of interest. Its external boundary is known as the *external cordon*. Once this is defined, the area is divided into zones (we will look at some basic zoning rules in Section 3.1) to have a clear and spatially disaggregated idea of the origin and destination of trips, and so we can spatially quantify variables such as population and employment.

The area outside the external cordon is also divided into zones but at a lesser level of detail (larger zones). Inside the study area, there can also be other *internal* cordons, as well as *screen lines* (i.e. an artificial divide following a natural or artificial boundary with few crossings, such as a river or a railway line), the purposes of which are discussed below. There are no hard and fast rules for deciding the location of the external cordon and hence which areas will be considered external to the study; it depends on the scope and decision levels adopted for the study (i.e. it is a very contextual problem).

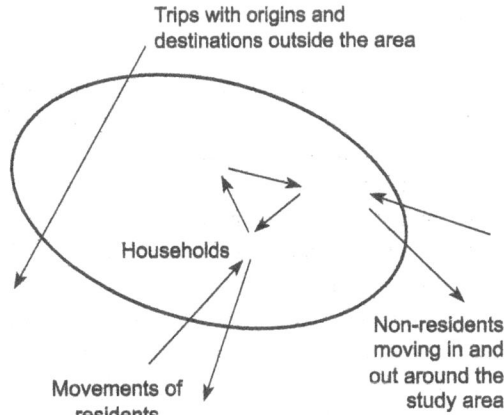

Trips with origins and
destinations outside the area

Households

Non-residents
moving in and
out around the
study area

Movements of
residents

Figure 2.5 Scope of data collection needed for a metropolitan O–D survey

Figure 2.5 implies that the following data are needed:

1) **Household survey**. Trips made by all household members by all modes of transport both within the study area and leaving/arriving at the area during the survey period; this survey should include socio-economic information (income, car ownership, family size structure, etc.), which is very efficient to generate data to estimate trip generation and mode split models; furthermore, data on household travel provides good information on the distribution of trip lengths in the city, an important element in the estimation of trip distribution models.

2) **Intercept surveys, external cordon**. Data on people crossing the study area border, particularly non-residents of the study area. This data can also be used to check and amplify the household data on study area crossings, since there is usually only a small amount of data collected, even in a very large survey. These are shorter surveys, carried out at points that intercept trips arriving and departing the study area: off-kerb surveys, on-board public transport vehicles, or at mode interchange points (i.e. airports).

3) **Intercept surveys, internal cordons, and screen lines**. These are required to measure trips by non-residents, and again to verify household data to some extent. They are important inputs to other models.

4) **Traffic and person counts**. They are low cost and are required for calibration, validation, and for further checks to other surveys. The integration of this data into the survey methodology is discussed below.

5) **Travel time surveys**. These are required to calibrate and validate most models and may be needed for both car and public transport travel.

6) **Other related data**. To create robust forecasting models as needed in large metropolitan areas, it is also important to have a survey methodology that allows integration of related data items that influence travel behaviour (Richardson et al. 1995). Here we include:

   - land-use inventory; residential zones (housing density), commercial and industrial zones (by type of establishment), parking spaces, etc.; these are particularly useful for trip generation models;
   - infrastructure and existing services inventories (public and private transport networks, fares, frequency, etc.; traffic signal location and timings); these are essential for model calibration, especially distribution and assignment models;
   - information from special surveys on attitudes and elasticity of demand (e.g. stated preference and other methods).

Each of the above survey components requires a detailed design together with a carefully selected sampling strategy. In what follows we give insights into the total methodological design and clues to the necessary measures to integrate the diverse survey components.

2.3.2.2 Home Interview Travel Surveys

Home Interview or *Household Travel Surveys* are the most expensive and difficult type of survey but offer a rich and useful data set. However, on many occasions interest will not be centred on gathering data for the complete model system, but only for parts of it: the most typical case is that of mode choice and assignment in short-term studies.

An interesting method, particularly suitable for corridor-based journey-to-work studies and which has proved very efficient in practice, is the use of workplace interviews (see Dunphy 1979; Ortúzar and Donoso 1983). These involve the local authority asking a sample of institutions (employers) in, for example, the Central Business District (CBD), permission to interview a sample of their employees; in certain cases, it has been found efficient to ask for the sample to be distributed by residence of the employee (e.g. those living in a certain corridor). It must be noted, however, that contrary to the case of a random household survey, the data obtained in this case are choice based in terms of destination; nevertheless, it is mostly random with respect to mode.

Although we will be referring mainly to household surveys, most aspects of the general discussion and indeed those about the design of measurement instrument are equally applicable to any other type of O–D survey.

General considerations. Both the procedures and measurement instruments used to collect information on site have a direct and profound influence on the results derived from any data-collection effort. This is why it has been recommended to include the measurement procedure as yet another element to be considered explicitly in the design of any project requiring empirical data for its development. Wermuth (1981), for example, has even proposed a categorisation of all the stages comprising a measurement procedure. In this part, we will refer to only two of these categories: the *development* and the *use* of measurement instruments designed to measure activity patterns outside the household.

We have already mentioned that the empirical measurement of travel behaviour is one of the main inputs to the decision-making process in urban transport planning; in fact, it provides the basis for the formulation and estimation of models to explain and predict future travel activities. For this reason, methodological deficiencies at this stage will have direct repercussions in all subsequent stages of the transport planning process.

Frequent criticisms about household or workplace travel surveys have included:

- the surveys only measured average rather than actual travel behaviour of individuals;
- only part of the individual's movements could be investigated;
- level-of-service information (for example about travel times) is poorly estimated by the respondent.

In fact, it has been found that variable measurements derived from traditional O–D surveys – for example, related to times, distances, and costs of travel – have proved inadequate when compared with values measured objectively for the same variables. That is, the reported characteristics have tended to differ substantially from reality in spite of the fact that the individuals responding to the survey experience the actual values of these level-of-service variables twice per day. It has also been concluded that the bias has a systematic nature and is apparently related to user attitudes with respect to each mode; for example, in the case of public transport, access, waiting, and transfer times (which are rather bothersome) tend to be severely overestimated. It is interesting to note that from a conceptual point of view these results would indicate that the subjective perception of level-of-service variables constitutes an important determinant in modal choice (see the discussion in Ortúzar et al. 1983).

A methodological analysis of these criticisms leads to two conclusions (Brög and Ampt 1982). First, travel behaviour information should not be sought in general terms (i.e. average values) but with reference to a concrete temporal point of reference (e.g. a pre-assigned travel day). Second, it is not recommended to examine the various activities in isolation, but rather to take the complete activity pattern as the basis for analysis; for example, it can be shown that asking for starting and ending times of a trip yields more accurate results than asking for its total duration. Thus, contemporary travel surveys employ an *activity recall framework* (Ampt and Ortúzar 2004).

An ongoing data collection process. The best approach to household travel data collection postulates that information should be gathered for *each day* of the week *throughout the year* and over several years (Richardson et al. 1995). In order to allow the use of data at any level of aggregation and to move away from the need for a standard zoning system, the methodology also recommends geocoding all origin and destination information.

Collecting data for each day of a given year allows capturing seasonal variations, as well as week-end/weekday differences. Therefore, the approach has numerous advantages:

- it permits measuring changes in demand over time, and in particular, it allows the correlation of these changes to changes in the supply system;
- since respondents only report data for one or two days, it makes their task easy and reliable, at the same time giving data over a longer period;
- the spread of the study over a year also results in lower operational costs;
- it allows for better quality control.

On the other hand, in this new approach, there are several issues that need to be addressed in each specific circumstance:

- it is necessary to wait for up to a year before there is sufficient data to meet the purposes for which the study was designed (typically calibrate a full-scale model);
- if interviewers are used, it is necessary to keep them motivated over a longer period;
- it is necessary to develop weighting processes that take account of seasonal variations;
- it is necessary to develop special methods for post-weighting annual data if it is combined with ongoing survey data (as we discuss below).

Periodic update of matrices and models. Matrices and models to match the ongoing data collection system should be updated periodically to maximise the benefit of the continuous information. Notwithstanding, although it is possible to consider the preparation of partial trip matrices given specific requirements, trip tables and models for the whole study area should not be updated more frequently than every 12 18 months, depending on the type of city under study.

Implications for data collection. Periodic updating of models and matrices is likely to have an effect on the data collected. For example, which information is most sensitive to updating? In this context there may be several elements that are worth periodic updating:

- trip generation and attraction models;
- travel matrices, reflecting the differential growth in different parts of the study area;

- modal split, including non-motorised modes, reflecting the possible impacts of different transport policies;
- traffic levels in different parts of the network, allowing identification of differential growth in the primary, secondary, access, and local networks;
- car ownership and household formation trends in various city boroughs.

The priority to apply in the case of each of these indicators will depend on the type of transport policies being considered, the need to monitor their performance, and the general modelling needs. It will also depend on their expected rates of change, their importance, and the costs associated with collecting data for updating, including the social cost of bothering users of the transport system (see DICTUC 1998 for a more detailed discussion).

It is also important to note the convenience of allowing estimation of other types of models likely to be needed in the future, such as time-of-day choice models and dynamic models. On the other hand, the availability of data collected on a continual basis allows monitoring user behaviour with respect to radical interventions in the transport system. Examples are environmental emergencies where CBD car-entry restrictions are increased, main road works, bus strikes, or changes in petrol prices, bus fares, or parking charges. The response to such policies (predictable or otherwise) provides basic information about users' behavioural thresholds and creates a temporal database that should facilitate the development of more sophisticated models.

Questionnaire format and design. Since one of the aims of a survey is to achieve the highest possible response rate to minimise non-response bias, it is recommended that mixed methods (i.e. based on self-completion and personal interviews) are used to collect the data (Goldenberg 1996). In particular, a self-completion system seems more appropriate in districts where people are used to 'filling in forms' (with personal interview validation follow-up) or where households cannot be accessed other than by remote security bell systems, where attempts at personal interviews result in low response rates. This combination capitalises on the cost-effectiveness and efficiency of high-quality self-completion designs and ensures minimum response burden for all participants (Richardson et al. 1995).

Telephone-based surveys, although widely used in North America for their relative cost-effectiveness, are not recommended for several reasons (Ampt and Ortúzar 2004):

1) Even where prompts are sent to households in advance (e.g. in the form of mini-diaries), they tend to suffer from extensive proxy reporting (i.e. one person reporting on behalf of others) leading to significant under-reporting of trips.
2) Although phone ownership is very high in countries where phone lists are used as sampling frame, there are often up to 40% of unlisted households (Ampt 2003); this means that at least 40% (plus those households without phones) of the population cannot give data if a single method approach was used.
3) Where random digit dialling is used to overcome this difficulty, the problem can be exacerbated by the ire of people with unlisted numbers receiving calls.
4) An increasing number of people in many countries have now only mobile phones; this means that, even if mobile phone numbers are available, there is a mixture of household-based and person-based sampling, which would need considerable effort if the weighting stage is to be effective.

In terms of layout, the order of the questions normally seeks to minimise the respondent's resistance to answering them, so difficult questions (e.g. relating to income) are usually put at the end. The survey instrument (and any personal interviews) should try to satisfy the following criteria:

1) The questions should be simple and direct.
2) Make sure each question serves a specific purpose; an excessive number of questions degrades the response rate and increases trip omissions.
3) The number of open questions should be minimised.
4) Travel information must include the purpose of the trip. It is interesting to acquire *stage-based trip data* (i.e. all movements on a public street) to ensure that analyses can relate specific modes to specific locations, times of day, etc.
5) Collect information so that complete tours can be re-constructed during analysis.
6) Seek information about *all modes of travel*, including non-motorised travel.
7) All people in the household should be included in the survey, including non-family members, like maids in developing countries.
8) To facilitate the respondent's task of recording all travel, an activity-recall framework is recommended, whereby people record travel in the context of activities they have undertaken rather than simply trips they have made; this has been shown to result in much more accurate travel measurement (Stopher 1998).
9) Since people have difficulty recalling infrequent and discretionary activities, even when they are recent, a travel day or days should be assigned to each household in advance. Respondents should be given a brief diary in advance of these days; the information in the diary may then be transferred to the self-completion form or reported to the interviewer at the end of the day (or as soon as possible).
10) Finally, all data should be collected at the maximum level of disaggregation (x–y co-ordinate level) based on a geographical information system (GIS).

The survey instrument needs to be designed for minimum respondent burden (Ampt 2003), maximum response rate (CASRO 1982), and hence greatest robustness of the data; for these reasons:

- self-completion designs need to focus on overall layout since they are the researchers' only contact with the respondent; the layout needs to be clear and concise, and in general, it should lead respondents onto the next question; layouts should usually be designed to encourage every respondent to reply, whether or not they are used to filling out forms (i.e. be user friendly, nicely presented and using simple language).
- the strength of personal interviews lies in the ease of response for survey participants, so the focus needs to be on training interviewers to understand the context of the survey and making sure the survey designs are easy for them to administer.

For either type of household survey, it is recommended that the survey must be divided into two parts: (i) personal and household characteristics and identification and (ii) trip data. We will briefly review the information sought in each part:

1) **Personal and household characteristics and identification**. This part includes questions designed to classify the household members according to their relation to the head of the household (e.g. wife, son), gender, age, possession of a driving licence, educational level, and occupation. To reduce the possibility of a subjective classification, it is important to define a complete set of occupations (non-household surveys are usually concerned only with the person being interviewed; however, the relevant questions are the same or very similar). This part also includes questions designed to obtain socio-economic data about the household, such as characteristics of the house, identification of household vehicles (including a code to identify their usual user), house ownership, and family income.

2) **Trip data**. This part of the survey aims at detecting and characterising all trips made by the household members identified in the first part. A trip is now defined as any movement outside a building or premises with a given purpose; but the information sought considers trips by stages, where a stage is defined by a change of mode (including walking). Each stage is characterised on the basis of variables such as origin and destination (normally expressed by their nearest road junction or full postcode, if known), purpose, start and ending times, mode used, amount of money paid for the trip, and so on. Ideally, analysis should be able to link trips in a logical way to re-construct tours and to generate trip *productions* and *attractions* by household (we will come back to these in Chapter 4).

Definition of the sampling framework. The scope of mobility surveys usually includes all travellers in an area (Figure 2.5). Thus, it not only includes residents but also visitors to households, in hotels, other people in non-private dwellings (such as hospitals), and travellers that pass through the area on the survey days.

Once the scope has been defined, the sampling frame needs to be determined. In other words, what type of list will provide information on all residents, visitors, and people who pass through the area, to choose a sample of those people and trips. Although there are various options, the household sample frame, while complex, is usually the most straightforward. If a Census has been conducted recently and information on all dwellings is available, this can be ideal. Alternatively, a block list of the whole region (prepared for any reason, e.g. for a utility company or for a previous survey) could be used, but a key issue is that it should be very up-to-date. Although Census data is only available every ten years in most places, in certain countries the government possesses a list of all dwellings officially registered for paying property taxes and this can be a useful starting point. If such lists are not available, several other methods can be used, the most typical one in industrialised nations being telephone listings (Stopher and Metcalf 1996) complemented by other methods if telephone ownership or listings are not universal. If no 'official' frame is available, it is always possible to simply sample blocks at random, enumerate the households in the block, and randomly sample from these.

Choosing the sampling frame for travel by non-residents is far more complicated. We recommend to do this in the following way:

- obtain a list of all non-private dwellings and select a sample (possibly stratified by size or by type of visitor);
- obtain a list of public transport interchanges where people are likely to arrive and leave the metropolitan region (e.g. airports, train stations, and long-distance bus stations); ideally, this will produce a sampling of travellers at each intercept point, although in some cases it will be necessary to sample sites;
- obtain a list of all road crossing points of the area's external cordon; as with public transport interchanges, ideally, all cross-points should be included, although in some cases they will need to be sampled.

However, the above procedure does not guarantee a perfect sampling frame. Fortunately, with some clear exceptions, the importance of trips made by visitors is generally much smaller than that of residents in any given study area.

Sample size. Travel surveys are always based on some type of sampling. Even if it were possible to survey all travellers on a specific service on a given day, this would only be a sample of travellers making trips in a given week, month, or year. The challenge in sampling design is to identify sampling strategies and sizes that allow reasonable conclusions, and reliable and unbiased models, without spending excessive resources on data collection. Often there is more than one way of obtaining the relevant information. For some data needs it may be possible to gather the information either through household surveys or through intercept surveys. In these situations, it is best to use the method that delivers the most precise data at the lowest cost (DICTUC 1998).

There are well-documented procedures for estimating the sample size of household surveys so that it is possible to satisfy different objectives; for example, estimation of trip rates, and trip generation by categories, levels of car ownership, and even of mode choice variables for different income strata (Stopher 1982). Given reasonable budget limitations, the analyst faces the question of whether it is possible to achieve all these objectives with a given sample of households in a certain metropolis (see for example, Purvis 1989). In general, these methods require knowledge about the variable to be estimated, its coefficient of variation, and the desired accuracy of measurement together with the level of significance associated with it.

The first requirement, although both obvious and fundamental, has been ignored many times in the past. The majority of household O–D surveys have been designed on the basis of vague objectives, such as 'to reproduce the travel patterns in the area'. What is the meaning of this? Is it the elements of the O–D matrix which are required, and if this is the case, are they required by purpose, mode, and time of day, or is it just the flow trends between large zones that are of interest?

The second element (coefficient of variation of the variable to be measured) was unknown in the past, but now it may be estimated using information from the large number of household O–D surveys that have been conducted since the 1970s. Finally, the accuracy level (percentage error acceptable to the analyst) and its confidence level are context-dependent matters to be decided by the analyst on the basis of personal experience. Any sample may become too large if the level of accuracy required is too strict. It can be said that this aspect is where the 'art' of sample size determination lies.

Once these three factors are known, the sample size (n) may be computed using the following formula (M.E. Smith 1979):

$$n = \frac{CV^2 Z_\alpha^2}{E^2} \tag{2.9}$$

where CV is the coefficient of variation, E is the level of accuracy (expressed as a proportion) and Z_α is the standard Normal value for the confidence level (α) required.

Example 2.6 Assume that we want to measure the number of trips per household in a certain area in the USA, and that we have data about the coefficient of variation of this variable for various locations as in Table 2.6:

Table 2.6 Coefficient of variation of trip rates

Area	CV
Average for U.S.A. (1969)	0.87
Pennsylvania (1967)	0.86
New Hampshire (1964)	1.07
Baltimore (1962)	1.05

As all the values are near to one, we can choose this figure for convenience. As mentioned above, the decision about accuracy and confidence level is the most difficult; Eq. (2.9) shows that if we postulate levels, which are too strict, sample size increases exponentially. On the other hand, it is convenient to fix strict levels in this case because the number of trips per household is a crucial variable (i.e. if this number is badly wrong, the accuracy of subsequent models will be severely compromised). In this example, we will ask for 0.05 level of accuracy at a 95% level.

For $\alpha = 95\%$ the value of Z_α is 1.645 in a one-sided test, therefore we get:

$$n = 1.0(1.645)^2/(0.05)^2 = 1084$$

that is, it would suffice to take a sample of approximately 1100 observations to ensure trip rates with a 5% tolerance 95% of the time. The interested reader may consult M.E. Smith (1979) for other examples of this approach.

The situation changes, however, if it is necessary to estimate O–D matrices. For example, M.E. Smith (1979) argues that a sample size of 4% of all trips in a given study area would be needed to estimate levels higher than 1100 trips between O–D pairs at the 90% confidence level with a standard error of 25%. This effectively means that if there are less than 1100 trips between two zones, a sample size of less than 4% would not be sufficient to detect them.

Example 2.7 Trips by O–D cell in Santiago de Chile, at the municipality level (e.g. just 34 zones), were analysed using data from the 1991 Household Mobility survey, which interviewed over 33 000 households (Ortúzar et al. 1993). It was observed that only 58% of the O–D cells contained more than 1000 trips. Thus, it would seem necessary to postulate a sample size of 4% of trips (and by deduction, 4% of households) to estimate an O–D matrix at the municipality level with a 25% standard error and 90% confidence limits. However, if the effect of response rates was considered (even if they were as high as 75%), as there were about 1.4 million households in the city, this would imply an initial sample size of nearly 75 000 households. It is doubtful that such a large sample size (and the associated costs and levels of effectiveness) would be justified to accomplish such a meagre objective.

Clearly, the driving force behind large sample sizes is the need to obtain trip matrices at the zone level. It has also been shown that it is very difficult to reduce the measurement error to an acceptable level in areas with more than say 100 zones since the sample size required is close to that of the population (M.E. Smith 1979). Hence, if the objective of the study includes estimating an O–D matrix, it is necessary to use a combination of survey methods, including both household and intercept surveys, to take advantage of their greater efficiencies for different data objectives.

Optimisation strategies for sample design. To achieve a sampling design that yields a smaller sample size, it is necessary to devise strategies that allow to estimate, say, trip generation rates by socio-economic status. One approach is to use a multi-stage stratified random sampling heuristic that produces better results than the classic method devised by M.E. Smith (1979); unfortunately, it requires a lot of effort by the analyst and does not guarantee a unique solution (DICTUC 1998). The heuristic begins by ordering the socio-demographic classes according to the degree to which they are represented in the population. Next the zones in the study area are allocated a class based on the most frequently occurring socio-economic group in them. Then a random sample of zones of each socio-economic type is selected (i.e. of the order of 1% of all households). After that, the remaining zones are categorised in priority order and are chosen as necessary to reach the difference between the sample already selected and the minimum required for each new class. The procedure is repeated until all classes have the minimum sample size required (say 30 or 50 observations).

This procedure was applied in Santiago for the 264-zone system defined in the 1991 O–D survey (Ortúzar et al. 1993) and for a stratification of 14 classes based on income and car ownership. The final solution achieved was a sample size of 1312 households located in only 15 zones, which guaranteed a minimum of 30 observations per class. However, in certain zones, notably, those containing high-income people, the solution implied somewhat unreasonable sample sizes (i.e. around 20% of the zonal population). A better solution was found by solving the following optimisation problem (Ampt and Ortúzar 2004):

Minimise

$$\sum_{i\in\{\text{classes}\}} \sum_{j\in\{\text{zones}\}} \alpha_j \eta_{ij}$$

subject to

$$0 \leq \alpha_j \leq \delta$$

$$\sum_{j\in\{\text{zones}\}} \alpha_j \eta_{ij} \geq \mu_i$$

where α_j is the proportion of households to interview in zone j and δ a reasonable threshold (e.g. a maximum of 5%), η_{ij} is the number of households of class i in zone j, and μ_i is the minimum acceptable sample size for each class i (i.e. 30 or 50 observations).

Using the same information as in the previous case, it was found that the problem could be optimally solved yielding a sample of just 482 households. More interestingly, for a stratification with 26 classes (i.e. adding household size as stratifying variable), it was found that an optimum sample size of 1372 households, guaranteeing a minimum of 30 observations in each of the specified classes, would be possible by collecting data in only 17 of the 264 zones.

However, as no limit was enforced for δ, some values of α_j were again near 20% of the zone population. So, by applying the restriction of δ being less than 5%, a sample of 1683 households in only 27 zones was finally obtained. Note that the method permits segmentations other than by socio-economic criteria. For example, it is also possible to identify spatial differences in terms of physical area (i.e. distance from the CBD) or access to the public transport network, and to increase the number of classes considered for the optimisation (Ortúzar et al. 1998).

Finally, remember that the design can also be improved by allowing for different response rates between different groups. In principle it is possible to estimate the number of households required in a gross sample (μ_i) to achieve a given minimum number of responses for each class, thereby ensuring a design that will yield even higher-quality trip generation data.

Sample size for a continuous survey. A final challenge consists of designing a sampling strategy for a continuous survey. If a sample of say 15000 households is required in year 1 to fulfil the initial modelling requirements of a metropolitan area, an ongoing survey would probably have the following form or something similar:

Year 1	Year 2	Year 3	Year 4	Year 5
15 000	5000	5000	5000	5000

This method requires smaller ongoing input after the first year, which offers several advantages:

- a smaller well-trained field force and administrative procedures which are likely to ensure very high-quality data with minimal effort in subsequent years;
- the appropriate authorities make a financial commitment for four years in year 1, reducing the risk of difficulties over repeat funding in say year 4.

But it does require the development of an annual weighting and integration system to ensure the data is readily usable for modelling, and this system needs to be robust and easy to use. We need to ensure that *all* the data at the end of year 2 is representative of year 2, that *all* the data at the end of year 3 is representative of that year, and so on. Such a procedure provides an up-to-date representation of existing travel behaviour for modelling and other purposes. In developing cities (i.e. where rapid changes occur in car ownership, land-use spread, and distribution), this would mean a more accurate modelling capability than has ever been possible in the past. It would also provide a larger sample size for use in the second and subsequent years, enabling more detailed questions to be asked of the data in them. Furthermore, if it is assumed that the data will be used for other purposes as well as modelling, the annual data collection method will provide essential time series data. Here are some examples:

- changes in travel patterns (by mode) related to changes in car ownership levels and distribution, pollution levels, or land-use patterns;
- changes in choice of mode related to changes in supply patterns, for example, improvements for pedestrians, expansion of the public transport network.

The way in which data from the second and subsequent years should be integrated and combined with data from the first year has to occur at four levels: household, vehicle, person, and trip. In this sense it is important to consider three things:

- careful sample selection and high response rates to ensure the 15 000 households in year 1 are representative of the city; then weighting and expansion procedures should be applied as described below;
- make sure that the 5000 households in year 2 are representative of the city (i.e. spatially and on all other parameters used for the first year of the sample selection); again, weighting procedures need to be applied;
- at the end of year 2, the databank will consist of 20 000 households but it will contain the raw data and the weighting factors only.

In smaller-sized cities, or in areas where there is little change in size and structure, it may not be necessary to have such a complicated sampling strategy, but it still depends on the uses of the data. For example, an equal sample for each of the years in a five-year period could be appropriate.

2.3.2.3 Other Important Types of Surveys

Roadside interviews. These provide useful information about trips not registered in household surveys (i.e. external–external trips in a cordon survey). They are often a better method for estimating trip matrices than home interviews as larger samples are possible. For this reason, the data collected are also useful in validating and extending the household-based information.

Roadside interviews involve asking a sample of drivers and passengers of vehicles (e.g. cars, public transport, goods vehicles) crossing a roadside station, a limited set of questions; these include at least origin, destination, and trip purpose. Other information, such as age, sex, and income, is also desirable but seldom asked due to time limitations; however, well-trained interviewers can easily add at least part of these data from simple observation of the vehicle and occupants (with obvious difficulties in the case of public transport).

Conducting these interviews requires a good deal of organisation and planning to avoid unnecessary delays, ensure safety and deliver quality results. The identification of suitable sites, coordination with the police, and arrangements for lighting and supervision are important elements in the success of these surveys. We shall concentrate here on issues of sample size and accuracy.

Example 2.8 Let us assume a control point where N cars cross and we wish to survey a sample of n vehicles. Let us also assume that of these n, X_1 cars travel between the origin–destination pair $O–D_1$. In this case, it can be shown that X_1 has a hyper geometric distribution $H(N, N_1, n)$, where N_1 is the total number of travellers between pair $O–D_1$, and that its expected value and variance are given by:

$$E(X_1) = np \text{ with } p = N_1/N$$
$$V(X_1) = np(1-p)(1-n/N)$$

Using a Normal approximation (based on the Central Limit theorem) we get that the distribution of X_1 is:

$$X_1 \sim N(np, np(1-p)(1-n/N))$$

and an estimator for p is:

$$\hat{p} = \frac{X_1}{n}$$

Therefore

$$\hat{p} \sim N\left(p, \frac{p(1-p)(1-n/N)}{n}\right)$$

and an approximate $100(1-\alpha)\%$ confidence interval for p is given by:

$$\left[\hat{p} - z\sqrt{\frac{p(1-p)(1-n/N)}{n}}, \hat{p} + z\sqrt{\frac{p(1-p)(1-n/N)}{n}}\right]$$

where z is the standard Normal value for the required confidence level (1.96 for the 95% level). We typically require that the absolute error e associated with $\hat{p}$ does not exceed a pre-specified value (usually 0.1), that is:

$$E = z\sqrt{\frac{p(1-p)(1-n/N)}{n}} \leq e$$

Working algebraically on this expression we get:

$$n \geq \frac{p(1-p)(1-n/N)}{(e/z)^2}$$

or equivalently:

$$n \geq \frac{p(1-p)}{(e/z)^2 + p(1-p)/N} \tag{2.10}$$

It can be seen that for a given N, e, and z, the value $p = 0.5$ yields the highest (i.e. most conservative) value for n in (2.10). Taking this value and considering $e = 0.1$ (i.e. a maximum error of 10%) and $z = 1.96$, we obtain the values in Table 2.7.

Table 2.7 Variation of sample size with observed flow

N (passengers/period)	n (passengers/period)	100 n/N (%)
100	49	49.0
200	65	32.5
300	73	24.3
500	81	16.2
700	85	12.1
900	87	9.7
1100	89	8.1

Example 2.9 An examination of historical data during preparatory work for a roadside interview revealed that flows across the survey station varied greatly throughout the day. Given this, it was considered too complex to try to implement the strategy of Table 2.7 in the field. Therefore, the following simplified Table 2.8 was developed:

Table 2.8 Simpler variation of sample size with flow

Estimated observed flow (passengers/period)	Sample size (%)
900 or more	10.0 (1 in 10)
700 to 899	12.5 (1 in 8)
500 to 699	16.6 (1 in 6)
300 to 499	25.0 (1 in 4)
200 to 299	33.3 (1 in 3)
1 to 199	50.0 (1 in 2)

The fieldwork procedure requires stopping at random the corresponding number of vehicles, interviewing all their passengers and asking origin, destination, and trip purpose. In the case of public-transport trips, given the practical difficulties associated with stopping vehicles for the time required to interview all passengers, the survey may be conducted with the vehicles in motion. For this it is necessary to define road sections rather than stations and the number of interviewers to be used depends on the observed vehicle-occupancy factors at the section. However, even this approach may be unworkable if the vehicles are overloaded.

Cordon surveys. These provide useful information about external–external and external–internal trips. Their objective is to determine the number of trips that enter, leave, and/or cross the cordoned area, thus helping to complete the information coming from the household O–D survey. The main one is taken at the external cordon, although surveys may be conducted at internal cordons as well. In order to reduce delay, they sometimes involve stopping a sample of the vehicles passing a control station (usually with police help), to which a short mail-return questionnaire is given. In some Dutch studies, a sample of licence plates had been registered at the control station and the questionnaires were sent later to the corresponding addresses stored in the Incomes and Excise computer. An important problem here is that return-mail surveys may produce biased results: this is because less than 50% of questionnaires are usually returned and the type of person who returns them may be different to those that do not (see Brög and Meyburg 1980). This is why in many countries roadside surveys often ask a rather limited number of questions (i.e. occupation, purpose, origin, destination, and modes available) to encourage better response rates.

Screen-line surveys. Screen lines divide the area into large natural zones (e.g. at both sides of a river or motorway), with few crossing points between them. The procedure is analogous to that of cordon surveys and the data also serve to fill gaps in and validate (see Figure 2.6) the information coming from the household and cordon surveys, including those of heavy vehicles. Care has to be taken when aiming to correct the household survey data in this way because it might not be easy to conduct the comparison without introducing bias.

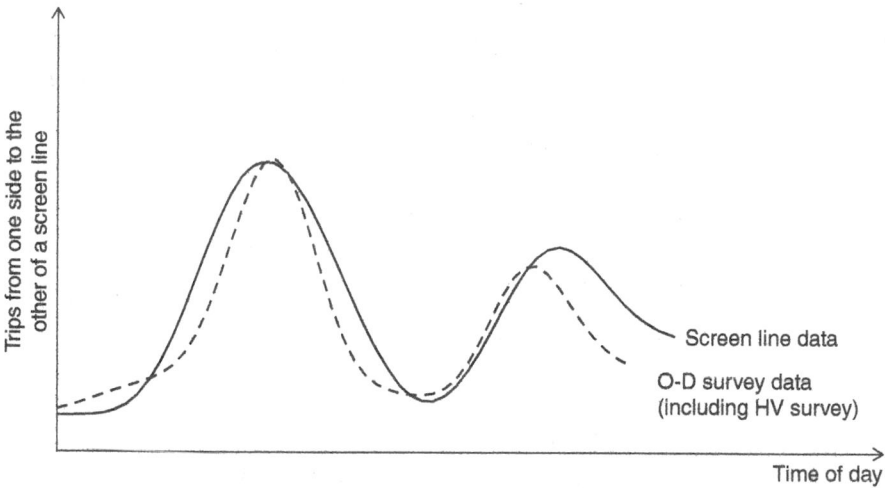

Figure 2.6 Household survey data consistency check

2.3.3 Data Correction, Expansion, and Validation

Correction and weighting are essential in any travel survey (Stopher and Jones 2003); the following sections discuss an approach deemed appropriate for the contemporary surveys described above, which are conducted over a period of several years.

2.3.3.1 Data Correction

The need to correct the O–D survey data to achieve results that are not only representative of the whole population but also reliable and valid, has been discussed at length (Brög and Erl 1982; Wermuth 1981). It is now accepted that simply expanding the sample is not appropriate, although for many years it was the most commonly practised method. Brög and Ampt (1982) identify a series of correction steps as follows.

Corrections by household size and socio-demographic characteristics. To make corrections that guarantee that the household size, age and sex, housing type, and vehicle ownership distributions of the sampled data represent that in the population (based on Census data), an iterative approach is needed since more simplistic methods do not guarantee correct results (see the discussion by Deville et al. 1993). Multi-proportional fitting (see Sections 5.2.3 and 5.6.2), also known as 'raking ratio' (Armoogum and Madre 1998), is probably the best approach in this case since it guarantees convergence in few iterations. Furthermore, its application has the additional advantage of not requiring the subsequent calculation of expansion factors. Stopher and Stecher (1993) give an almost pedagogical example of this approach.

The method is particularly valid if the secondary population data has been gathered close to the time of the travel survey. However, it may not be appropriate if the travel survey is done several years after the Census as the urban population may change rather quickly, particularly in less-developed countries. In this case, it would be necessary to calculate proportions of households in each group and to compute expansion factors in the more traditional form (Ortúzar et al. 1993; Richardson et al. 1995).

The multi-proportional method does not guarantee that each cell value will be identical in the Census and in the travel survey since in any matrix there is an important degree of indeterminacy (i.e. many combinations of cell values can give rise to the same totals of rows and columns). In particular, due to its multiplicative characteristics, a cell with a zero will always end up with a zero value. Furthermore, certain special matrix structures (that contain zeroes in some key positions) can lead to non-convergence of the method (see Section 5.6.1 for an example).

To avoid bias in the multi-proportional correction, because we are correcting by items as diverse as, say, household size (number of persons) on the one hand, and personal characteristics (gender and age) on the other, it is better to define unique categories, thus avoiding classes that consider – for example – two to four persons, six or more persons, etc. Nevertheless, it is easy to imagine occasions on which it would be necessary to group some category because it is not represented in the sample for a given zone. In that case, it is convenient to check if it is possible to group similar zones instead of making the correction at such a disaggregate level (Stopher and Stecher 1993).

Additional corrections in household surveys. In addition to the corrections by household size, vehicle ownership, and socio-demographics, there are two other correction procedures necessary – depending on whether it is a personal interview or self-completion survey (Richardson et al. 1995). These procedures are noted below:

Corrections for non-reported data These are needed when certain elements of the survey have not been answered (item non-response). In self-completion surveys, interviewing a validation sample of people using personal interviews and then weighting the data accordingly (Richardson et al. 1995) is used to address this. This type of correction is not usually needed when personal interviews are used because interviewers must be well trained and supervised, thereby decreasing the incidence of item non-response (but see the discussion in Stopher and Jones 2003).

Corrections for non-response. These are needed when a household or individual does not respond, that is, does not return the survey instrument or refuses verbally or by mail to respond to the survey (Zimowski et al. 1997). This can be attributed to a variety of causes, and it is important to differentiate between genuine sample loss (e.g. vacant dwellings that do not generate travel should be ineligible), and refusals (where the person could be travelling but not responding, clearly eligible). In the case of personal interviews, it has been recommended that corrections should be based on the number of visits necessary to achieve a response since it has been shown that this is associated with strong differences in travel behaviour (Kam and Morris 1999; Keeter et al. 2000); however, there is also evidence suggesting that these differences might be small (Kurth et al. 2001).

In self-completion surveys, on the other hand, it was originally believed that corrections could be done based on the number of follow-up reminders needed to generate a household response (Richardson et al. 1995) but the problem is likely to be more complex than for personal interviews (see the discussions by Polak 2002, and Richardson and Meyburg 2003). Related to this, it is interesting to mention that reductions in non-response bias due to the inadequate representation of certain population strata (i.e. by income) have been reported using special factoring techniques that take into account the differences in return rates by different types of households by zones (Kim et al. 1993).

A final related point is how to decide when the response by a household is considered complete. The US National Travel Survey uses the 'fifty-percent' rule (at least 50% of adults over 18 years of age completed the survey), after arguments that excluding households where not everybody responded may exaggerate bias, and data are weighted to mitigate the person-level non-response in sampled households. Research on this subject allowed detecting the most likely types of households and the most likely non-respondent (DRCOG 2000). Interestingly, trip rates by the sample including 50% households have been found to be not statistically different from a sample including only households with 100% of members responding (see Ampt and Ortúzar 2004).

Integration weighting for a continuous survey. Integration weighting is required to unite each wave of a continuous survey; in this case, it is recommended to proceed as follows (Ampt and Ortúzar 2004):

1) *Household weighting* should occur for each 'important' variable (as chosen in prior consultation), for example, household size, car ownership, or household income.
2) *Vehicle weighting* should be done in the same way. A variable of particular importance here is the age of the vehicle since without correct weighting it would appear as if the fleet was not ageing.
3) *Person weighting*. Here factors of importance are likely to be, for example, income and education.
4) *Trip weighting*. Number of trips and mode are likely to be the key variables in this case – all done according to the same general principles described above.

In this way, the data will be representative of the population in every year of the survey. Of course, this is not perfect, but with a good sampling scheme, it should be very robust. For example, in year 2 the sample will actually reflect real changes in household size (say) that may be occurring. Hence if one wanted to use years 1 and 2 to reflect the situation in year 2 (which is exactly what a government agency would like to do), it would be necessary to weigh the year 1 data set to have the proper household size that is actually observed in year 2. Clearly, if a given year coincides with a Census year, the weighting process can take on a whole new meaning, although this is likely to occur only about once a decade.

Example 2.10   Table 2.9 presents the number of samples gathered in the first three years of a continuous survey, stratified according to household size. If we consider households of size 1 say, we can see that they constitute 13.33% of the sample in year 1 (i.e. 2000/15 000), 17% of the sample of year 2, and if added without reweighting 14.25% of the sample for both years. However, this would be akin to the proverbial mixing of apples and pears.

To integrate the data properly we need first to calculate (appropriate) weights for year 1 to ensure that both sets have the same proportions as measured in the latest year (based on the assumption that the new sample drawn each year represents the characteristics of that year's population). These weights are calculated in the next part of the table and are equal to the ratio between the percentages (for each stratum) of years 2 and 1 (i.e. 24/20 = 1.2 in the case of households of size 2). The final part of the table shows the result of adding the weighted year-1 data to the year-2 data, to achieve a final sample of 20 000 households that have the same distribution according to household size as it occurs in year 2.

Table 2.9   Weighting procedures for integration

	Household size									
	\multicolumn 1		\multicolumn 2		\multicolumn 3		\multicolumn 4		\multicolumn 5		Total
Year 1	2000	13.33%	3000	20.00%	4000	26.67%	5000	33.33%	1000	6.67%	15 000
Year 2	850	17.00%	1200	24.00%	1000	20.00%	1500	30.00%	450	9.00%	5000
Total	2850	14.25%	4200	21.00%	5000	25.00%	6500	32.50%	1450	7.25%	20 000

Weighting values for year 1

	1	2	3	4	5
Year 1	17/13.33 = 1.275	1.20	0.75	0.90	1.35

Reweighting procedure

	1		2		3		4		5		Total
Year 1	2550		3600		3000		4500		1350		15 000
Year 2	850		1200		1000		1500		450		5000
Total	3400	17%	4800	24%	4000	20%	6000	30%	1800	9%	20 000

2.3.3.2   *Imputation Methods*

Survey non-response makes identification of population parameters problematic and, normally, identification of non-response data is only possible if certain assumptions (frequently not testable) are made about the distribution of missing data. Non-response does not, however, necessarily preclude identification of the bounds on parameters. There are several state-of-practice imputation methods ranging from *deductive imputation*, to use of overall or class means, to *hot* and *cold-deck imputation*, and so on (Armoogum and Madre 1998). In fact, the organizations conducting major surveys usually release data files that provide non-response weights or imputations to be used for estimating population parameters. Stopher and Jones (2003) recommend distinguishing between imputation and inference. The latter can be used initially and is particularly useful for certain types of variables (i.e. if a person does not indicate s/he is a worker but reports trips to work). However, there are some variables that may not be safe to infer due, for example, to changing social structures.

Imputation is defined as the substitution of values for missing data, based on certain rules or procedures. It is worth noting, however, that most imputation methods do not preserve the variance of the imputation variable (for example income), and therefore, they can produce inconsistent estimates when the variable that contains imputations is included in a model. For this reason, some researchers even believe that imputing values increases the bias in some instances, and is simply translated in makeshift data. Thus, it is recommended that the changes produced upon imputing values are registered, and if it is possible, to have their effects evaluated. Horowitz and Manski (1998) show how to bind the asymptotic bias of estimates using typical weights and imputations. They provide a thorough mathematical treatment of the subject and illustrate it with empirical examples using real data.

Another approach to solving this and other problems consists of making multiple imputations and thereafter combining the estimators of the resulting models in each case to obtain consistent values that include a consideration of the errors associated with the imputation process (Brownstone 1998).

In the Santiago 2001 O–D Survey (DICTUC 2003), 543 households out of 15 537 did not answer the family income question. Due to the strong asymmetry of the income distribution, a logarithmic transformation of the data was used to centre the distribution and achieve a better resemblance to a Normal distribution. Multiple imputations were successfully produced using a linear model based on the Student t-distribution with five degrees of freedom (Lange et al. 1989), estimated using Gibbs sampling (Geman and Geman 1984). Outliers were detected and removed from the estimation process; as it turned out, they were found to be wrongly coded, meaning that the process had the secondary advantage of allowing for further checks on the quality of the data.

2.3.3.3 Sample Expansion

Once the data have been corrected it is necessary to expand them in order to represent the total population; to achieve this expansion factors are defined for each study zone as the ratio between the total number of addresses in the zone (A) and the number obtained as the final sample. However, often data on A are outdated leading to problems in the field. The following expression is fairly general in this sense:

$$F_i = \frac{A - A(C + CD/B)/B}{B - C - D}$$

where F_i is the expansion factor for zone i, A is the total number of addresses in the original population list, B is the total number of addresses selected as the original sample, C is the number of sampled addresses that were non-eligible in practice (e.g. demolished, non-residential), and D is the number of sampled addresses where no response was obtained. As can be seen, if A was perfect (i.e. $C = 0$) the factor would simply be $A/(B - D)$ as defined above. On the other hand, if $D = 0$ it can be seen that the formula takes care of subtracting from A the proportion of non-eligible cases, in order to avoid a bias in F_i.

2.3.3.4 Validation of Results

Data obtained from O–D surveys are normally submitted to three validation processes. The first simply consider on site checks of the completeness and coherence of the data; this is usually followed by their coding and digitising in the office. The second is a computational check of valid ranges for most variables and in general of the internal consistency of the data. Once these processes are completed, the data is assumed to be free of obvious errors.

In mobility studies the most important validation is done within the survey data itself and not with secondary data such as traffic counts at screen lines and cordons in the study area. The reason is that each method has its own particular biases which confound this task. For example, gross comparisons, such as number of trips crossing a cordon or number of trips by mode, often give relatively poor comparisons.

Although state-of-the-art survey techniques minimise these problems, the use of independent data to check figures from all components of a metropolitan O–D travel survey is still recommended (see the discussion in Stopher and Jones 2003). Objective comparisons of these figures, taking into account the strengths and weaknesses of each survey method allow us to detect potential biases and to take steps to amend them. Furthermore, if matrices are to be adjusted (see Section 14.4), it is essential to reserve independent data to validate the final results. This requires good judgement and experience since if insufficient care is given to the task, it is easy to produce corrections to the O–D matrices that do not correspond to reality.

2.3.4 Longitudinal Data Collection

Most of the discussion so far has been conducted under the implicit assumption that we are dealing with cross-sectional (snap-shot) data. Usually, the main source of longitudinal data is traffic counts collected on several years at the same location. They provide valuable information on traffic growth but little understanding of changes in travel behaviour. As we saw in Chapter 1, travel behaviour researchers are becoming increasingly convinced that empirical cross-sectional models have suffered from lack of recognition of the inter-temporality of most travel choices. Panel data are a good alternative to incorporate temporal effects because in this data structure, a given group of individuals is interviewed at different points in time.

In this part, we will attempt to provide a brief sketch of longitudinal or time-series data-collection methods and problems; we will first define various approaches and then we will concentrate on the apparently preferred one: panel data. In Chapter 8, we will consider the added problems of modelling discrete choices in this case.

We will finally examine some evidence about the likely costs of a panel data-collection exercise in comparison with the more typical cross-sectional approach.

2.3.4.1 Basic Definitions

1) **Repeated cross-sectional survey**. This is one which makes similar measurements on samples from an equivalent population at different points in time, without ensuring that any respondent is included in more than one round of data collection. This kind of survey provides a series of snapshots of the population at several points in time; however, inferences about the population using longitudinal models may be biased with this type of data and it may be preferable to treat observations as if they were obtained from a single cross-sectional survey (see Duncan et al. 1987).

2) **Panel survey**. Here, similar measurements (i.e. the panel *waves*), are made on the same sample at different points in time. There are several types of panel surveys, for example:
   - **rotating panel survey**. This is a panel survey in which some elements are kept in the panel for only a portion of the survey duration;
   - **split panel survey**. Which is a combination of panel and rotating panel survey;
   - **cohort study**. This is a panel survey based on elements from population sub-groups that have shared a similar experience (e.g. birth during a given year).

Although the use of panel data has increased in many areas, especially since the pioneering work of Heckman (1981), in transport there are only a few examples, which can be classified into two groups:

1) **Long survey panels**. These consist of repeating the same survey (i.e. with the same methodology and design) at 'separate' times, for example, once or twice a year for a certain number of years or before-and-after an important event. Some famous examples are the Dutch Panel (Van Wissen and Meurs 1989) and the Puget Sound Transportation Panel (PSTP) in the United States (Murakami and Watterson 1990). The main problem of this kind of panel is attrition (i.e. loosing respondents) between successive surveys (known as waves).

2) **Short survey panels**. These are multi-day data where repeated measurements on the same sample of units are gathered over a 'continuous' period of time (e.g. two or more successive days), but the survey is not necessarily repeated in subsequent years. Some recent examples of this type of panel are the two-day time-use diary for the US National Panel Study of Income Dynamics and the six-week travel and activity diary data panels collected in Germany (Axhausen et al. 2002) and Switzerland (Axhausen et al. 2007). In this case, attrition is not a problem, but the infrequent changes in mode choice and low data variability (both the attributes of each mode and the respondents' socioe-conomic characteristics are practically fixed) are, as this may cause difficulties in estimating models, as discussed by Cherchi and Ortúzar (2008b).

If a substantive intervention is planned for a system, panels have even more significant advantages for evaluating changes (Kitamura 1990). Indeed, Van Wissen and Meurs (1989), based on the Puget Sound Panel, described how the effects of policies could change trends; also, it is easier to capture these changes using observations of the same individuals, as part of their current behaviour may be explained by previous experiences. Although the advantages of panels seem clear, there are precious few panels built around a substantial system change that would allow modelling changes in mode choice; notable exceptions are the before and after study developed in Amsterdam around an extension of its urban motorway system (Kroes et al. 1996) and the *Santiago Panel* (Yáñez et al. 2010b) built around the introduction of Transantiago, a radical change to the public transport system of Santiago, Chile (Muñoz et al. 2009).

It is important to distinguish between longitudinal surveys and panel data. The former consists of periodic measurements of certain variables of interest. Finally, although in principle it is possible to obtain panel data from a cross-sectional survey, measurement considerations argue for the use of a panel survey design rather than retrospective questioning to obtain reliable panel data.

2.3.4.2 Representative Sampling

Panel designs are often criticised because they may become unrepresentative of the initial population as their samples necessarily age over time. However, this is only strictly true in cohort study designs considering an unrepresentative sample to start with; for example, if the sample consists of people with a common birth year, individuals joining the population either by birth or immigration will not be represented in the design.

In general, a panel design should attempt to maintain a representative sample of the entire population over time. So, it must cope not only with the problems of birth, immigration, or individual entry by other means but also be able to handle the incorporation of whole new families into the population (e.g. children leaving the parental home, couples getting divorced). A mechanism is needed to maintain a representative sample that allows families and individuals to enter the sample with known probabilities, but this is not simple (for details see Duncan et al. 1987).

2.3.4.3 Sources of Error in Panel Data

A panel design may add to (or detract from if it is not done with care) the quality of the data. Although repeated contact and interviewing are generally accepted to lead to better-quality information, panels have typically higher rates of non-response than cross-sectional methods and run the risk of introducing *contamination* as we discuss below.

Effects on response error. Respondents in long survey panels have repeated contact with interviewers and questionnaires at relatively long-time intervals; this may improve the quality of the data for the following reasons:

- repeated interviewing over time reduces the amount of time between event and interview, thus tending to improve the quality of the recalled information;
- repeated contact increases the chances that respondents will understand the purpose of the study; also, they may become more motivated to do the work required to produce more accurate answers;
- it has been found that data quality tends to improve in later waves of a panel, probably because of learning, by respondents, interviewers, or both.

However, in the case of short survey panels, the quality of responses tends to decrease with the number of days considered (i.e. less trips are reported and travel by slow modes is omitted) due to fatigue.

Non-response issues. The generic 'non-response' label includes several important issues which have two basic sources: the loss of a unit of information (attrition) and/or the loss of an item of information. Hensher (1987) discusses in detail how to test and correct this type of error.

The non-response problems associated with the initial wave of a panel are not different from those of cross-sectional surveys, so very little can be done to adjust for their possible effects. In contrast, plenty of data have been gathered about non- respondents in subsequent waves; this can be used to determine their main characteristics, enabling non-response to be modelled as part of the more general behaviour of interest (see Kitamura and Bovy 1987).

Typical large panel designs spend a great amount of effort attending to the 'care and feeding' of respondents: this involves instructing interviewers to contact respondents many times and writing letters of encouragement specifically tailored to the source of respondents' reluctance. These 'maintenance policies' are often considered important by panel administrators, as is the use of incentives to encourage cooperation (Yáñez et al. 2010b).

Response contamination. Evidence has been reported that initial-wave responses in panel studies may differ from those of subsequent waves; for this reason, in some panel surveys the initial interviews are not used for comparative purposes. A crucial question is whether behaviour itself or just its reporting, is being affected by panel membership. Evidence about this is not conclusive, but it seems to depend on the type of behaviour measured. For example, Traugott and Katosh (1979) found that participants in a panel about voting behaviour increased their voting (i.e. changed behaviour) as time went by; however, it was also found that this was caused partly by greater awareness of the political process and partly by the fact that individuals who were less politically motivated tended to drop out of the panel.

Treatment of repeated observations. Another problem, which is more specific to short survey panels, relates to the presence of repeated observations. It is normal to expect that certain individuals, in different days, may repeat exactly the same trips (typical cases are the systematic trips to work that are often made every day with the same characteristics: time, cost, purpose, mode, and so on). So, especially when these data are used for model estimation, a crucial question here is which should be the optimum length of the short survey panel as the way in which repeated information is treated may affect the estimation results (Cherchi et al. 2017; Yáñez et al. 2011)

2.3.4.4 Relative Costs of Longitudinal Surveys

Questions about the relative costs of longitudinal studies cannot be answered without reference to the alternatives to them. One obvious comparison is between a single cross-sectional survey, with questions about a previous period, and a two-wave panel. However, if the longitudinal study is designed to keep its basic sample representative each year and if enough resources are devoted to the task, it can also serve as an (annual) source of representative cross-sectional data and thus ought to be compared with a series of such surveys rather than just a single one.

Duncan et al. (1987) made rough calculations on these lines, concluding that in the first case, the longitudinal survey would cost between 20% and 25% more than the cross-sectional survey with retrospective questions. However, they also conclude that in the second case, the field costs of each successive wave of the cross-sectional study would be between 30% and 70% higher than additional waves of the panel survey, depending on the length of the interview.

Other costs are caused by the need to contact and persuade respondents in the case of panels and by the need to sample again with each fresh cross-section in the other case. Finally, there are other data processing costs associated with panels but these must be weighed against the greater opportunity to check for inconsistencies, analysis of non-response, consideration of inertia effects in modelling, and so forth.

2.3.5 Travel Time Surveys

The requirements for detailed and accurate travel time, vehicle speed, and delay data are important for the calibration and validation of model systems. As travel times are a key determinant of travel costs, it is important to ensure that the models correctly represent delays in the network of interest. In principle, one would expect traffic on the road network to be subject to variability in their travel time. Travel times on buses, and even metro, can also be affected by congestion and disruption. The focus here is mostly on vehicle travel times on congested networks but the principles are applicable to other modes.

Travel times can be divided into:

- running times, whilst the vehicle is moving, and
- delays, when the vehicle is stopped because of congestion or traffic control measures (traffic lights, stop signs, etc.)

Note that the measurement of spot speeds, say with a radar gun, provide a poor indication of link travel times as they miss precisely the second source above.

Travel times can be highly variable as a result of traffic control measures and congestion levels. For short links, measurement errors may also be significant. As the variability over a single link would be too high for most models, it is preferable to observe travel times on segments covering several junctions to reduce it and make results more representative of the type of model used. For strategic models, segments should include at least five links or be at least one kilometre long (whatever is the longest). For smaller-scale models one may use shorter segments but observations will have to be repeated more frequently to obtain a reliable estimate despite local variability in travel times.

The most common technique for travel time measurements is known as the *moving observer method*. In this case, a probe car is driven at the average speed of the traffic stream and times are recorded for stretches of road. Maintaining an average speed is difficult and the normal requirement is for the driver to overtake as many vehicles (in the relevant class) as vehicles overtake him. An observer in the car (or a GPS-based instrument), record times at regular intervals or when passing identifiable locations (i.e. key junctions, a particular bridge, or building).

The design of a travel time survey requires:

- specifying the level of accuracy required;
- identifying one or more circuits to be surveyed;
- identifying the road sections of interest;
- selecting a method for data collection: observer, GPS, or other;
- selecting the days and times when the surveys will be conducted, and
- deciding the number of runs that will be needed for each circuit and survey times.

The accuracy required will depend on the objective of the model. For large-scale strategic models, it is desirable to have an accuracy of some 5–8 km/h (i.e. around ± 10%). For operational studies, a better accuracy of 2–5 km/h is desirable. In the case of disaggregate mode choice models, as discussed in Chapters 7–9, it has been found that the level of accuracy should be very high indeed (i.e. average travel times for the peak period will not adequately represent those experienced by travellers within the peak, and it has been recommended to group individuals according to departure time in, at most, 15 minutes intervals, see Daly and Ortúzar 1990).

The circuits to be surveyed should be representative of the study area of interest. They should cover roads and streets of different types and flow levels, with emphasis on those types considered most important. The length of the road sections should be chosen to reduce the variability encountered at junctions, especially if they are signal controlled. For dense urban areas, sections should contain between 7 and 10 signal-controlled junctions.

The sample size and the number of runs to be undertaken will depend also on the variability observed on different types of roads under different conditions. Eq. (2.9) above can be used to estimate more accurately both the number of segments (links between junctions) in a road section and the number of runs. Ideally, the coefficient of variation CV should be estimated from recent observations. Typical values for the CV are between 9 and 15 for roads with low and high variability. If we require 90% confidence to be within a 10% error, this results in three to seven runs for this range. It is often recommended that at least five runs are undertaken to ensure any special circumstance does not unduly affect the results.

2.3.6 Digital Data Sources

Reductions in the cost of technology and the widespread use of devices that leave a digital crumb of information usable for data collection have resulted in the introduction of a number of new travel data sources. These can be classified as:

- techniques that identify movements between points where sensors have been installed, including, for example, number plate and Bluetooth surveys, and
- techniques that exploit, anonymously, the digital traces left by the use of devices or means of payment in the hands of travellers; these include the use of payment smart cards and mobile phone data.

Number plate surveys. It is possible to observe the registration or number plate of vehicles at different locations in the network either using human observers, video recordings, or *Automatic Number Plate Recognition* (ANPR or ALPR for license plates). Observers or video cameras must be located at entry and exit points in an area during the period of observations. ANPR can be used directly or mediated by high-quality digital video recordings. Software will then match number plates and times, construct point-to-point trip matrices and compute travel times between the places where the number plates were recorded. These partial matrices can be contrasted with estimated ones to confirm and potentially improve them.

A limitation of this method is the number of locations where recordings can be organised simultaneously. In the end, ANPR is not 100% accurate, as not all trips are matched (some begin or end without crossing a second point) and, therefore, not all travel times and trip tables are established. Sample rates, however, are generally quite good.

Bluetooth surveys. It is currently possible to locate low-cost Bluetooth sensors in a way analogous to a number plate survey. The sensors identify the MAC (hardware) number of each Bluetooth device, anonymised if required. Travel times and partial (point-to-point) matrices are obtained in this way. Bluetooth provides a smaller sample than number plates and there may be more than one such device in a vehicle. However, this is a low-cost survey method.

Note that both these point-to-point methods require a bit of data cleaning. The raw data cannot differentiate between a long inter-point time, because of congestion, from journeys where the driver stopped to buy a cup of coffee. These exceptional matches must be weeded out of the data before analysis.

Smart card data. The use of smart cards for payment for public transport and other services generates valuable data. It is possible, for example, to estimate stop-to-stop matrices using this source. This is easier if the traveller validates both on boarding and alighting the service. If she only validates on boarding (as in most bus services) then the next validated boarding can be used to estimate the end of the first trip (see Munizaga and Palma 2012; Munizaga et al. 2014). This type of data is seldom available to institutions other than operators and the government. Moreover, processing it requires data mining skills and software not common in our field.

Mobile phone data. The normal operation of mobile phones requires knowledge of where the unit is to route the call or message via the nearest mast/antenna. This information is activated when a call or a message is sent/received and also at some regular intervals or because the phone has moved from one area to another. Phone location, in this sense, is specified in terms of the nearest antenna; many of these are directional and in some cases, the information on the strength of the signal can be used to estimate distance to the mast. This information is always anonymised (strictly speaking pseudonymised), sometimes twice over. It is always possible to establish a relationship between antenna's location and a transport model zoning system.

But processing this type of data to produce useful information is complex and requires good understanding of how a particular mobile phone company operates, what data is transmitted, stored, and at what interval (Willumsen 2021). Different companies offer these services using this data, providing trip matrices and other mobility indicators used in traffic and transport studies.

The first step in this analysis is to identify 'stays' that is locations where the mobile phone has remained for at least some time for the user to perform an activity, typically 30 minutes. A sequence of stays can then be linked as a tour with its trips between activities. This information, processed over several days, helps to establish the place of residence and work/study. Other data sources are needed to expand this data, usually, a combination of Census information and movement counts.

The attraction of this type of data is significant. The sample size is an order of magnitude larger than a Household Travel Survey; observations over several days and weeks enable extracting information on the regularity and frequency of trips and activities useful in analysing changes in working-from-home patterns. The pattern of trip-making can also be used to identify professional drivers of taxis and trucks. On the other hand, only a few trip purposes can be identified with confidence, and associating a trip with a mode of travel in urban areas is difficult because of the limited space granularity of mobile phone data.

Nevertheless, the availability of movement data for different days and times may help to solve the paucity of longitudinal data; further, its capacity to identify changes in behaviour in response to disruptions to travel (road closures, service interruptions, poor weather) or evolving trends like remote working is likely to help gaining a better understanding of the changing mobility patterns.

2.4 Stated Preference Surveys

2.4.1 Introduction

The previous discussion has been conducted under the implicit assumption that any choice data corresponded to *revealed preference* (RP) information; this means data about actual or observed choices made by individuals. It is interesting to note that we are seldom in a position to actually observe choice; normally we just manage to obtain data on what people report they do (or more often, what they have done on the previous day or, better, in the pre-assigned travel day).

In terms of understanding travel behaviour, RP data have limitations:

1) Observations of actual choices may not provide sufficient variability for constructing good models for evaluation and forecasting. For example, the trade-offs between alternatives may be difficult to distinguish so the attribute level combinations may be poor in terms of statistical efficiency.
2) Observed behaviour may be dominated by a few factors making it difficult to detect the relative importance of other variables. This is particularly true with secondary qualitative variables (e.g. public-transport information services, security, décor) which may also cost money and we would like to find out how much travellers value them before allocating resources among them.
3) The difficulties in collecting responses for policies that are entirely new, for example, a completely new mode (perhaps a people mover) or cost-recovery system (e.g. electronic road pricing).

These limitations would be surmounted if we could undertake real-life controlled experiments within cities or transport systems, but the opportunities for doing this in practice are very limited. Thus, where data from real markets are not available for predicting behaviour or eliciting reliable preference functions, researchers have turned to *stated preference* (SP) methods. These cover a range of techniques, which have in common the collection of data about respondent's intentions in hypothetical settings as opposed to their actual actions as observed in real markets. The three most common SP methods have been *contingent valuation* (CV), *conjoint analysis* (CA), and *stated choice* (SC) techniques. In transport, SC techniques have tended to dominate and for this reason, we will focus on this method providing only a brief description of the CV and CA survey approaches. Note also that in the transport arena, the SP label has not embraced contingent valuation, as in fields such as marketing or environmental economics; further, in transport practice the SP label has generally referred to either CA or SC without a formal distinction.

2.4.1.1 Contingent Valuation and Conjoint Analysis

As a coherent technique, CV primarily deals solely with eliciting *willingness-to-pay* (WTP) information for various policy or product options (Mitchell and Carson 1989). In this case, the policy (e.g. a way to reduce accident risk) is presented to respondents who are then asked how much they are willing to pay for having it. Four types of CV questions are typically used in practice; direct questioning, biding games, payment options, and referendum choices. In CV studies, the policy or product is kept static and the outcome, in the form of WTP, is for the entire product or policy. As such, CV questions cannot be used to disentangle the WTP for individual characteristics or attributes of the product or policy under study. We will come back to this technique in section 18.5.2.2.

Unlike CV, traditional conjoint analysis allows the researcher to examine the preferences, and even WTP if a price or cost attribute is included, not only for the entire policy or product but also for individual characteristics of the object(s) under study. In CA, respondents are presented with a number of alternative policies or products and are asked to either rate or rank them (see Figure 2.7). The levels of the characteristics or attributes of the various policies or products are systematically varied and become the independent variables that are regressed against the ratings or rankings data. The parameter weights for each attribute reflect the marginal preference or 'part-worth' for that attribute. Thus, if a cost or price attribute is included as part of the product or policy presented to respondents, then the ratio of any non-price parameter to the price or cost parameter reflects the marginal WTP for the associated non-price attribute (Gaudry et al. 1989). The special difficulties associated with estimating WTP when flexible discrete choice functions, such as those we will discuss in Section 8.6 are used to model the situation in hand, are discussed by Sillano and Ortúzar (2005).

Fare	Interchange	Time on bus	Walk time
70 p	No change	15 mins	10 mins

Fare	Interchange	Time on bus	Walk time
70 p	No change	20 mins	8 mins

Fare	Interchange	Time on bus	Walk time
85 p	No change	15 mins	10 mins

Fare	Interchange	Time on bus	Walk time
85 p	1 change	15 mins	8 mins

Figure 2.7 Example of stated-preference ranking exercise

Traditional CA has had limited acceptance in transport studies due to a number of criticisms that have been levelled against the method over the years (Louviere and Lancsar 2009). Firstly, it has been argued that the statistical methods primarily used to analyse CA data are inappropriate, in that the dependent variable of a linear regression model should be, at a minimum, interval scaled. As such, using ranking data as a dependent variable certainly violates this assumption, although some argue that even ratings data also is not interval-level data, given how respondents psychologically use the ratings metric. A second criticism lies not in how the data is analysed but with the very use of ratings or rankings data as measurement metric. Respondents in real life do not rate or rank alternatives and even if they did different people would approach such scales in psychologically different manners. As such, it has been argued that outputs of CA surveys have no psychologically meaningful interpretation (Louviere and Lancsar 2009). So, SC methods have tended to dominate transport studies.

2.4.1.2 Stated Choice Methods

SC studies are similar to CA methods insofar as respondents are presented with a number of hypothetical alternatives; however, the two methods differ in terms of the response metric. Whereas CA asks respondents to rank or rate the alternatives (with all alternatives shown to respondents at the same time), respondents undertaking a SC survey are asked to choose their preferred alternative from amongst a subset of the total number of hypothetical alternatives constructed by the analyst. In asking respondents to make a choice, rather than a rating or ranking, the two criticisms levelled at CA are avoided. Firstly, the analysis of discrete choice data requires a different set of econometric models specifically developed to analyse such data; thus, the choice metric is consistent with the statistical model applied to it. Secondly, the selection of the single preferred alternative is psychologically consistent across respondents and a task that is common to individuals in real markets. A further distinction between the two methods is that CA tasks typically present respondents with a relatively large number of alternatives, simultaneously, to rate or rank, whereas SC methods typically present only a few alternatives at a time (and in most cases only two), changing them and having respondents repeat the choice task.

On the other hand, the primary distinction between RP and SC surveys is that in the latter case, individuals are asked about what they would choose to do (or how would they rank/rate certain options) in one or more hypothetical situations. The degree of artificiality of these situations may vary, according to the needs and rigour of the exercise:

- the *decision context* may be a hypothetical or a real one; in other words, the respondent may be asked to consider an actual journey or one that she might consider undertaking in the future;
- some of the *alternatives* offered may be hypothetical although it is recommended that at least one of them be an existing one, for example, the mode just chosen by the respondent including all its attributes.

A crucial problem with SP data collection in general, is how much faith we can put on individuals actually doing what they stated they would do when the case arises (for example, after introducing a new option). In fact, experience in the 1970s was not good in this sense, with large differences between predicted and actual choice (e.g. only half the people doing what they said they would) found in many studies (see Ortúzar 1980).

The situation improved considerably in the 1980s and good agreement with reality was reported from models estimated using SC data (Louviere 1988a). This occurred because data-collection methods improved enormously and became very demanding, not only in terms of survey design expertise but also in their requirements for trained survey staff and quality-assurance procedures. The interested reader can consult the excellent book by Louviere et al. (2000).

The main features of an SC survey may be summarised as follows:

a) It is based on the elicitation of respondents' statements of how they would respond to different hypothetical (travel) alternatives.
b) Each option is represented as a 'package' of different attributes like travel time, price, headway, reliability, and so on.
c) The analyst constructs these hypothetical alternatives so that the individual effect of each attribute can be estimated; this is achieved using *experimental design* techniques that ensure the parameters of the chosen attributes are estimated with the smallest standard errors. In reality, an experimental design is nothing more than a matrix of numbers used to assign values to the attributes of each

alternative. By using experimental design theory, the assignment of these values occurs in some non-random manner, and by systematically varying the design attributes, the analyst is able to control as many factors as possible influencing the observed choices. In creating the design in a specific and precise manner, the analyst seeks to ensure the ability to obtain reliable parameter estimates with minimal *confoundment* with the other parameter estimates.

d) The researcher has to make sure that respondents are given hypothetical alternatives they can understand, appear plausible and realistic, and relate to their current level of experience;

e) The responses given by individuals are analysed to provide quantitative measures of the relative importance of each attribute; for this choice, models are estimated as discussed in detail in Chapter 8.

However, the process of constructing effective SP surveys is far from simple and quite time-consuming if done correctly. Extensive qualitative and secondary research is advised to determine the relevant set of alternatives, attributes, and attribute levels that will be used to make up the hypothetical alternatives. In what follows we give advice based on useful discussions and comments by Prof. John M. Rose at the University of Sydney, one of the leading experts in this subject.

In preparing an SP survey, the analyst will need to address at least the following issues:

1) Will the experiment be *labelled* (i.e. the names of the alternatives have substantive meaning beyond their ordering; see Figure 2.8) or *unlabelled* (see Figure 2.9), and will a *non-purchase* or *status quo* alternative be presented (see Figure 2.8b)? We will come back to this last issue in Section 2.4.2.6.

	Train	Bus		Car
Fare	$3.00	$4.00	Petrol cost	$1.00
			Toll cost	$3.00
			Parking cost	$8.00
Access time	5 mins	10 mins		
In-vehicle time	35 mins	25 mins	In-vehicle time	15 mins
Egress time	15 mins	10 mins	Egress time	5 mins
I would choose	●	●	or	●

(a) Standard design

	Train	Bus		Car	None
Fare	$3.00	$4.00	Petrol cost	$1.00	
			Toll cost	$3.00	
			Parking cost	$8.00	
Access time	5 mins	10 mins			
In-vehicle time	35 mins	25 mins	In-vehicle time	15 mins	
Egress time	15 mins	10 mins	Egress time	5 mins	
I would choose	●	●	or	●	●

(b) Design including a non-purchase option

Figure 2.8 Example of labelled mode SC tasks

	Route A	Route B	Route C
Petrol costs	$1.50	$2.00	$1.00
Toll cost	$2.00	$4.00	$0.00
Prob. of arriving late	0.30	0.50	0.10
Prob. of arriving early	0.10	0.20	0.30
Free flow time	15 mins	10 mins	20 mins
Congested time	10 mins	15 mins	20 mins
Egress time	15 mins	10 mins	5 mins
Please rank in order of preference (1 = best)	☐	☐	☐

Figure 2.9 Example of unlabelled route SC task

2) In deciding what attributes to use, we need to determine what factors best represent those influencing choices between the various alternatives. Note that other external criteria may also influence this task; for example, if the outputs from the study will be used as inputs into, say a network model, the attributes should accommodate the constraints or needs of the latter (e.g. if a network model does not allow for a comfort attribute, the analyst will need to determine whether it is worthwhile including comfort in the SC study); we will come back to this issue below.

3) The analyst also needs to define values for the levels of each attribute, including specific quantitative values (e.g. $5, $10, and $20) or qualitative labels ('low', 'medium', and 'high'). Once the above have been defined, further pre-testing and piloting are also recommended. This may result in further refinements of the survey instrument. Only once the analyst is satisfied with the survey, should the SC study be put out to field.

2.4.2 The Survey Process

In setting up an SP survey, analysts should aim to follow the five stages illustrated in Figure 2.10. The first requires that the study objectives are clearly defined and clarified. This involves identifying the population of interest as well as refining the experimental objects, or alternatives that will be studied. Definitions and descriptions of new alternatives should also be defined and tested.

The second stage requires outlining the set of assumptions reflecting our overall beliefs as to what qualities are important for an experimental design to display. These assumptions will dictate the statistical properties of the design generated in Stage 3 of the process. As there exist many different possible experimental designs for any given problem (each with different statistical properties), specifying the assumptions and outlining the properties that the analyst deems important is critical to generate the design. Unfortunately, in the vast majority of SP studies, this second stage is generally ignored and researchers generate designs without fully appreciating what assumptions led to them, or whether the generated designs are appropriate for meeting the needs of the study.

The actual method for constructing a design in Stage 3 is dependent on the assumptions made in Stage 2, with different assumptions requiring different design generation methods. Thus, even if the analyst skips Stage 2, implicit assumptions are still being made in generating the design.

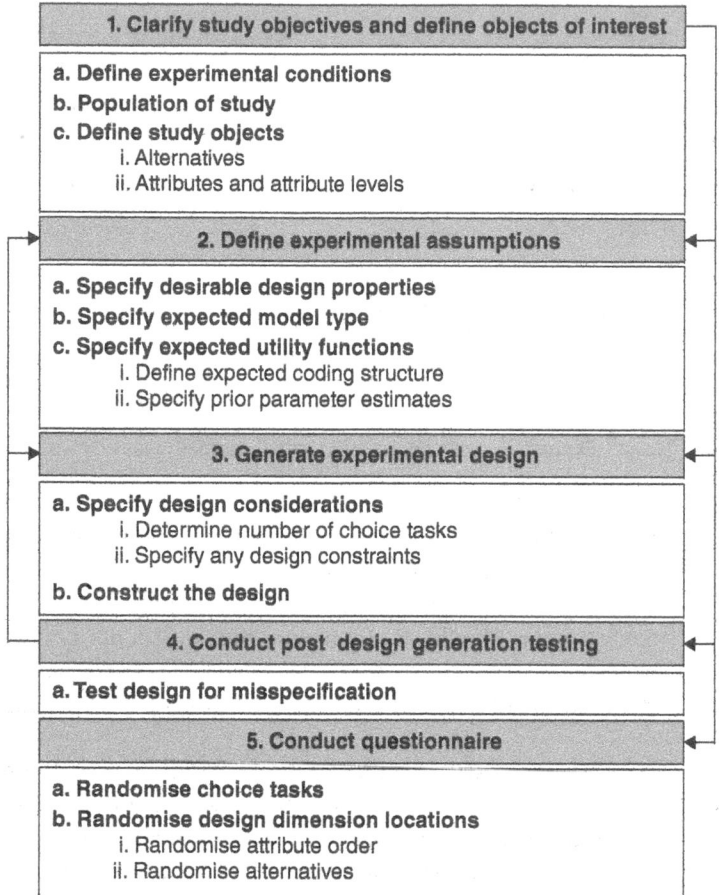

Figure 2.10 Steps in designing a stated preference experiment

Stage 4 represents an ideal stage in the process rather than a necessary one; unfortunately, as with Stage 2, it is often ignored in practice. In this stage, the analyst performs tests, usually in the form of simulations, in order to determine how the design is likely to perform in practice. This type of test may allow the analyst to correct any issues with the design before going to field.

The final stage of the design generation process involves taking the design and using it to construct the questionnaire that will be given to respondents.

2.4.2.1 Clarifying Study Objectives and Defining Objects of Interest

This stage involves the analyst gaining an understanding of the specific context or problem under study, the population of interest, as well as the types of choices that sampled respondents will be asked to make.

Typically, the choice context (experimental conditions) is an input that is not under the analyst's control, being supplied by an external client or determined by the study objectives. Nevertheless, understanding the context of the study is crucial to the success of SP studies and in special circumstances in-depth interviews and seeking specialist knowledge may be vital in this task (Ortúzar and Palma 1992). Armed with this knowledge, the analyst will then need to determine what behavioural outputs are of direct interest, such as determining the subjective value of time (SVT), WTP for risk reductions, or just estimating a generalised cost of travel formulation.

After gaining a full understanding of the problem under study, the analyst is next required to identify and understand the population of interest. This involves determining who the sample respondents are, where they are likely to be located, how they will be sampled and how will they be surveyed. Understanding such questions at this stage is important, as they will influence the type of questionnaire that will be used and this will likely influence the type of experimental design generated.

For example, if respondents are located in a geographically dispersed pattern and have limited access to internet, mail-back paper and pencil surveys may be the only option. In such cases, the experimental design will be more difficult to adapt to individual specific circumstances. Where respondents may be surveyed using a computer or over the internet, the experiment may adapt to each individual's reported circumstances (e.g. if a respondent does not have access to a car, then the car alternative may be removed from the survey for that respondent). The type of survey used will also have implications in terms of the data collected, which will determine whether the assumptions made in Stage 2 of the survey design process transfer from the design over to the data finally gathered. As well as having an impact upon survey design, understanding the population of interest will also provide further insights in the sampling required for the study (see Stage 3). Finally, in understanding the population of interest, the analyst may determine for example, whether different segments should be sampled, and hence whether more than one experimental design or survey questionnaire is necessary.

Understanding the population of interest as well as the overall study objectives will provide insights into the number and diversity of alternatives applicable to various sampled individuals when making decisions within the study context. Such knowledge will assist in constructing the choice tasks that will be used in the SP survey. Figure 2.8 showed two different choice tasks that might be considered for a mode choice study. Figure 2.8a requires respondents to choose between train, bus, and car alternatives, whereas the choice task in Figure 2.8b allows the respondent to also select none of these alternatives. If the objective of the study was to model commuter choice, then the initial choice task can be applied without problems as respondents cannot choose not to travel to work; however, it might not be adequate for respondents who can telecommute (i.e. work from home), and in that case, the second-choice task should be preferable. Similarly, for non-commuting trips the second-choice task might be more appropriate given that most non-commuting trips can be considered discretionary in nature (at least for non-teenagers).

Attributes and alternatives. The construction of realistic, or technologically feasible alternatives, requires the following four distinct tasks:

a) The range of options is usually given by the objective of the exercise; however, one should not omit realistic alternatives a user might consider in practice. For example, in studying potential responses of car drivers to new road-pricing initiatives, say, it may not be sensible to consider only alternative modes of travel; changes to departure time or to alternative destinations (to avoid the most expensive road charges) may be relevant responses. By ignoring them one places the respondent in a more artificial (less realistic) context, perhaps triggering inappropriate or unrealistic responses (we will discuss this type of issues further in Section 2.4.2.6).

b) The set and nature of the attributes should also be chosen to ensure realistic responses. The most important attributes must be present and they should be sufficient to describe the technologically feasible alternatives. Care must be applied here as particular combinations of attributes (e.g. a high-quality, high-frequency, low-cost alternative) may not be seen as realistic by respondents, reducing the value of the whole exercise. Care must be taken also if the number of attributes is deemed excessive (say higher than six); Carson et al. (1994) found that fatigue effects make respondents simplify their choices by focusing on a smaller number of attributes or simply answering at random or in lexicographic fashion (Sælensminde 2001). In this sense, it has been shown that there may be limits, which are culturally affected, on the number of choice tasks, alternatives, attributes, and even their range of variation, that are acceptable in a given study (Caussade et al. 2005; Rose et al. 2009a).

c) To ensure that the right attributes are included and that the options are described in an easy-to-understand manner, it is advantageous to undertake a small number of group discussions (e.g. focus groups) with a representative sample of individuals. A trained moderator will make sure all relevant questions regarding perception of alternatives, identification of key attributes and the way in which they are described and perceived by subjects, and the key elements establishing the context of the exercise are all discussed and reported. Focus groups are expensive and in many cases, the researcher will be tempted to skip them believing a good understanding of the problem and context already exists. In that case, it will be even more essential to undertake a carefully monitored pilot survey where any issues of attribute description and alternative presentation can be explored.

d) The selection of the metric for most attributes is relatively straightforward. However, there are some situations that may require more careful consideration, in particular with respect to qualitative attributes like 'comfort' or 'reliability'. For example, travel time reliability can be presented as a distribution of journey times on different days of a working week, or as the probability of being delayed by more of a certain time. For more on this issue see the discussion of stimulus presentation below.

e) Finally, in relation to the number of levels that each attribute can take, it is important to bear in mind that Wittink et al. (1982) found evidence that variables with more levels could be perceived as more important by respondents; we will come back to this issue in relation to another topic below.

2.4.2.2 Defining Experimental Assumptions

For any SC study, several potential experimental designs can be constructed. The analyst's aim is to choose a particular design construction method and generate the design. This will depend upon different considerations, most of which reflect the personal beliefs of the analyst as to what are important properties the design must possess. However, some decisions do not reflect the personal biases or beliefs of the analyst but, rather, are influenced by the problem being studied.

Labelled or unlabelled experiments. In many instances, the decision to treat an experiment as either *labelled* or *unlabelled* will depend upon the problem under study. In particular, mode choice studies will generally require a labelled experiment, whereas route choice problems are in general amenable to unlabelled SC experiments. Nevertheless, the decision as to whether either type of experiment is used is crucial, as it typically impacts upon the number and type of parameters that will be estimated as part of the study.

Generally, unlabelled experiments require only the estimation of generic parameters whereas labelled experiments may require the estimation of either alternative specific or generic parameters or combinations of both. Prior knowledge of the number of likely (design-related) parameter estimates is important as each one represents an additional degree of freedom required for estimation purposes. General experimental design theory posits that the Fischer Information (or Hessian) matrix

(**I**) will be singular if the number of choice observations (each one equivalent to a choice task) is smaller than the number of parameters (see Goos 2002). As such, the minimum number of choice tasks required for an experimental design is equal to or greater than the number of (design-related) parameters to be estimated. Note that the inclusion of a *status quo* alternative does not impact upon the minimum number of choice tasks required for a design, as it does not require the estimation of any attribute-related parameter estimates. The decision to use a labelled rather than an unlabelled choice experiment may also impact upon the generation of *orthogonal designs*, as discussed in Section 2.4.2.3.

Imposing attribute level balance. This is another consideration in generating designs. Attribute level balance occurs when each attribute level appears an equal number of times, within each attribute, over the entire design. This is generally considered a desirable property, although it may impact upon the statistical efficiency of the design (see Section 2.4.2.3). If present, it ensures that each point in preference space is equally represented, so that parameters can be estimated equally well on the whole range of levels, instead of having more or less data points at only some of the attribute levels (which may affect how the design performs in practice). Nevertheless, it is worth noting that attribute level balance may require larger designs than dictated by the number of parameter's estimates requirement.

Example 2.11   Consider a design with four attributes, where two have two levels, one has three levels and the last has four levels. In the classical jargon in this field, we would refer to this as a $2^2 3^1 4^1$ factorial design; note that the product of levels to the power of attributes (48 in this case) represents the total number of choice tasks needed to recover all effects (i.e. main or linear effects and all interactions), that is, a full factorial design (more about this below).

Assuming each attribute will produce a unique parameter estimate (i.e. main effects only), the smallest design would require just four choice tasks based on the number of parameters criterion; however, to maintain attribute level balance, the smallest possible design would require 12 choice tasks (12 being divisible without remainder by 2, 3, and 4).

Number of attribute levels. This should reflect the researchers' belief as to the relationship each level has to the overall contribution to utility and whether the relationship is expected to be linear or nonlinear from one level to the next. If nonlinear effects are expected for a certain attribute and the analyst suspects that the attribute will be, say, *dummy coded* (see Example 2.13) prior to analysis, then more than two levels will be needed to model appropriately the suspected nonlinearities. Where dummy-coded (or *effects* and/or *orthogonal*-coded) attributes are included, the number of levels for these attributes is predetermined. However, the more levels used, the higher the potential number of choice tasks required due to additional parameters being estimated. Also, mixing the number of levels for different attributes may yield a higher number of choice tasks (due to attribute level balance).

Varying the range of attributes. Research suggests that using a wide range (e.g. $0–$30) is statistically preferable to using a narrow range (e.g. $0–$10) as this will theoretically lead to parameter estimates with a smaller standard error; however, using too wide a range may also be problematic (see Bliemer and Rose 2010). In fact, having too wide an attribute level range may result in choice tasks with dominated alternatives; whereas having too narrower a range may result in alternatives for which the respondent will have trouble distinguishing between (see Cantillo et al. 2006). However, such considerations are purely statistical in nature and analysts should also consider practical limitations upon the possible range that the attribute levels can take; that is, the attribute levels shown to respondents must make sense to them (must be realistic). Hence there will be often a trade-off between the statistical preference for a wider attribute level range and practical considerations that may limit this range.

Inclusion of interaction effects. Interactions may be important when the effects of two variables are not additive (see Figure 2.11); including interactions will impact upon the number of choice tasks required of a design. This is because each interaction effect will have a corresponding parameter estimate and hence it requires an additional degree of freedom, and in turn, an additional choice task. As such, Rose and Bliemer (2009) suggest starting the design generation process by specifying the 'worst case' utility specification (i.e. in terms of all the effects that might be tested, along with any non-linear parameterization that may be estimated). Generating a design with too few choice tasks will likely preclude the estimation of potentially valid utility specifications at a later stage, whilst generating a design with more than the minimum number of choice tasks does not preclude the estimation of simpler model forms.

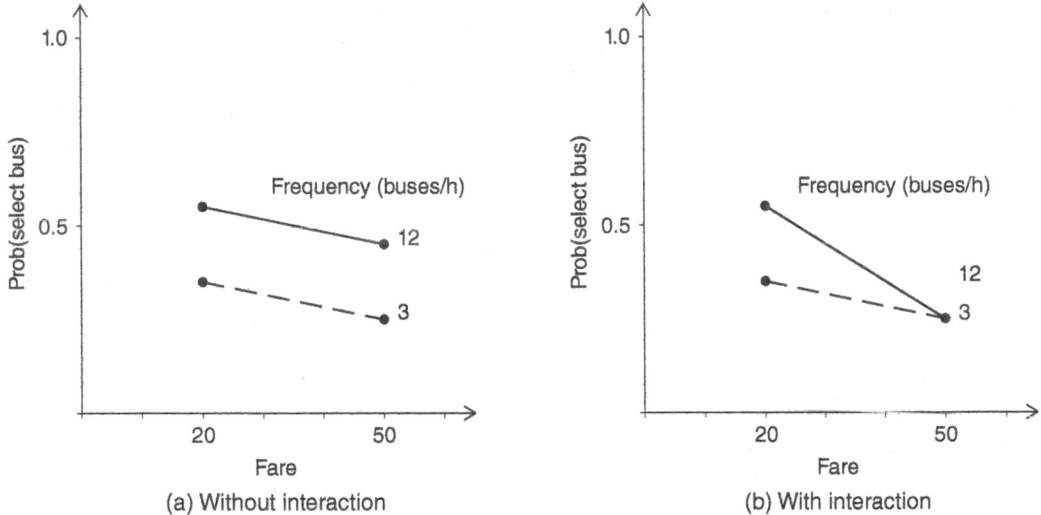

(a) Without interaction   (b) With interaction

Figure 2.11 Presence and absence of attribute interaction

Once decisions for each of the above have been made, several different experimental *design generation procedures* can be considered. The easiest method is to employ a *full factorial* (FF) design, that is, one consisting of all possible choice tasks. One benefit of using a FF design is that all main effects and interaction effects will be orthogonal. Unfortunately, the number of choice tasks in a FF design will typically be too large and many of the choice tasks will have dominated or unrealistic alternatives.

Fractional factorial designs. Due to the practical impossibility of dealing with FF designs, many analysts rely on the so-called *fractional factorial* designs, which consist of a subset of choice tasks from the full factorial. To construct a fractional factorial design, one could randomly select choice tasks from the FF; however, this may lead to bias and more intelligent strategies are possible. Numerous methods have been explored within the literature as to how to select choice tasks in a structured manner so that the best possible data from the SC experiment will be produced for estimating models. The most widely known fractional factorial design type is the *orthogonal design*, which is produced so as to have zero correlations between the attributes within the SC experiment (and thus it is excellent for estimating linear models, see Rose and Bliemer 2009). Although there are several types of orthogonal designs, we will consider here only the most popular one, consisting of constructing a simple orthogonal array.

Example 2.12 Consider a situation with five attributes, two at two levels and the rest at three levels (i.e. a $2^2\ 3^3$ design). In this case, depending on the number of interactions to be tested, the number of options required would vary as follows in a classical orthogonal design:

- 108 to consider all effects (i.e. a full factorial design);
- 81 to consider principal effects and all interactions between pairs of attributes, ignoring effects of a higher order;
- 27 to consider principal effects and interactions between one attribute and all the rest;
- 16 only if no interactions are considered.

More recently, several researchers have suggested other types of fractional factorial designs such as *D-optimal* or *D-efficient* designs (Rose and Bliemer 2008). In generating these types of designs, researchers define its efficiency in terms of variances and covariances of the parameter estimates; the lower these (co)variances, the more efficient the experimental design. As such, the objective in generating this type of design is to choose attribute-level combinations that will result in the smallest possible parameter (co)variances. To do so, the analyst must make a number of assumptions about the model to be estimated as well as the parameter estimates that will be obtained. This enables us to calculate the expected asymptotic covariance matrix ($\mathbf{S}^2$) of the design, from which the (co)variances are derived; note that it is calculated as the negative inverse of $\mathbf{I}$ (see Section 8.4.1), the Fisher information or Hessian matrix. Understanding what model will be estimated is important, as $\mathbf{S}^2$ for a given design will be different for different econometric model specifications.

Nevertheless, two competing schools of thought have emerged within the literature as to what parameter priors are appropriate to use in generating experimental designs for SC studies. The first creates designs under the so-called *null hypothesis*, namely zero-valued parameter priors (Street et al. 2005), whilst the competing school assumes non-zero-valued parameter priors. Clearly, in the latter case, one has to decide what these non-zero valued parameter priors are, typically leading to more efficient designs if the population parameter estimates are truly non-zero. However, this comes at the expense of the effort to obtain parameter priors (Rose and Bliemer 2008).

Within the first school of thought, whilst not necessary, further assumptions are typically made in generating SC designs. Firstly, it is generally assumed that designs that are orthogonal within alternatives and which maximise the differences in attribute levels between alternatives will be optimal (see Burgess and Street 2003, 2005; Street and Burgess 2004, 2007). This is because under the assumption that the parameters are all zero, any discrete choice model will collapse to a linear model and therefore an orthogonal design will be optimal. This also acknowledges the fact that discrete choice models are really difference in the utilities' models (see Section 7.3.1).

A second assumption which has been less well communicated for this class of designs is that $\mathbf{S}^2$ is usually constructed under the assumption that the analyst will be applying orthogonal codes to models estimated using data collected with the design. As such, the efficiency of the design may be reduced if a different coding system is used in practice, which is typically the case. Nevertheless, the appeal of this approach is two-fold:

- respondents are forced to make trade-offs on each and every attribute of the design, as no two attributes in any given choice situation will, where possible, take the same value;
- the approach does not require *a priori* knowledge of the parameter estimates, and this may be particularly suitable for designs that have mainly qualitative attributes.

The second design school of thought utilises non-zero valued parameter priors, assuming some prior knowledge about their values. Original research into generating *efficient* designs assumed that researchers had *exact* knowledge about the expected parameter estimates (e.g. $\theta_1 = -0.2$; $\theta_2 = 1.0$).

Designs generated under such an assumption are known as *locally optimal designs* as their (co)var-iances will be minimised only at the precise values assumed for the prior parameters (see Rose and Bliemer 2009; Scarpa and Rose 2008).

Later on, researchers produced *Bayesian efficient designs* that do not assume precise knowledge of the parameter estimates. Such designs allow for the true population to fall within some distribution of possible parameter estimates, such that the analyst optimises over the distribution of possible priors, for example, $\theta_1 \sim N(-0.8, 0.2)$. In this approach, we generally let go of the principle of orthogonality and generate designs in a manner that can be expected to minimise the elements of $\mathbf{S}^2$ associated with the (nonlinear) discrete choice model estimated on the data (Bliemer and Rose 2010; Carlsson and Martinsson 2002; Ferrini and Scarpa 2007; Fowkes 2000; Huber and Zwerina 1996; Kanninen 2002; Kessels et al. 2006; Rose and Bliemer 2008; Sándor and Wedel 2001, 2002, 2005; Scarpa and Rose 2008; Toner et al. 1998; Watson et al. 2000).

The main advantage of this approach is that the generated design is directly related to the expected outcome of the modelling process. Besides, it can be optimised for any type of model, not just MNL, or for a range of model types (Rose et al. 2009b). Further, the approach can assume any data structure (i.e. it is not limited to assuming orthogonal coding). However, while the first school can prove optimality of their designs (under the null hypothesis), the second school generally cannot (under the non-null hypothesis). Therefore, these designs are typically called *efficient* and not *optimal* designs.

A Note on Dummy, Effects and Orthogonal Coding

If the marginal impact upon utility is believed to be non-linear from one attribute level to another, the analyst may wish to test this by transforming the data using dummy, effects, or orthogonal coding. Within the literature, the former remains the preferred method, although effects and orthogonal coding offer a number of advantages over it.

In all cases, the analyst creates $D = L-1$ new variables in the data, where L is the total number of levels for the attribute being transformed. For *dummy coding* transformations, the researcher uses a series of zeros and ones to map the original levels to the newly created D variables. To create the mapping, each new dummy variable corresponds to the first $L-1$ levels of the original attribute. To create the dummy codes, every time level l appears for the original attribute, the corresponding newly created dummy variable takes the value 1; otherwise, it takes the value zero. This occurs for all but the last attribute level (the base level), which does not have a corresponding dummy variable. In that case, the last attribute level will take the value zero for all dummy coded variables (see Table 2.10). *Effects coding* uses the same pattern of mapping creating $E = L-1$ effects coded variables; however, the base level takes the value -1 for all the E effects coded variables (see Table 2.10).

Finally, like dummy and effects coding, *orthogonal coding* requires the creation of $O = L-1$ new variables. However, unlike dummy and effects coding, here we use orthogonal polynomial contrasts to populate the O new variables (see Table 2.10). Thus, each successive orthogonal coded variable corresponds to a higher order polynomial effect for the pre-transformed attribute (i.e. $o_{k1} \rightarrow x_k$ (linear effect), $o_{k2} \rightarrow x_k^2$ (quadratic effect), $o_{k3} \rightarrow x_k^3$ (cubic effect), etc.). Rather than directly taking the polynomial power of the original variable and using these directly (which will induce correlation in the data), orthogonal coding uses orthogonal polynomial contrasts which retain orthogonality within each attribute (see Chihara 1978).

Table 2.10 Example of dummy, effects and orthogonal coding

Attribute levels		Dummy coding					Effects coding					Orthogonal coding				
		D1	D2	D3	D4	D5	E1	E2	E3	E4	E.5	O1	O2	O3	O4	O5
2	1	1	—	—	—	—	1	—	—	—	—	1	—	—	—	—
	2	0	—	—	—	—	−1	—	—	—	—	−1	—	—	—	—
3	1	1	0	—	—	—	1	0	—	—	—	−1	1	—	—	—
	2	0	1	—	—	—	0	1	—	—	—	0	−2	—	—	—
	3	0	0	—	—	—	−1	−1	—	—	—	1	1	—	—	—
4	1	1	0	0	—	—	1	0	0	—	—	−3	1	−1	—	—
	2	0	1	0	—	—	0	1	0	—	—	−1	−1	3	—	—
	3	0	0	1	—	—	0	0	1	—	—	1	−1	−3	—	—
	4	0	0	0	—	—	−1	−1	−1	—	—	3	1	1	—	—
5	1	1	0	0	0	—	1	0	0	0	—	−2	2	−1	1	—
	2	0	1	0	0	—	0	1	0	0	—	−1	−1	2	−4	—
	3	0	0	1	0	—	0	0	1	0	—	0	−2	0	6	—
	4	0	0	0	1	—	0	0	0	1	—	1	−1	−2	−4	—
	5	0	0	0	0	—	−1	−1	−1	−1	—	2	2	1	1	—
6	1	1	0	0	0	0	1	0	0	0	0	−5	5	−5	1	−1
	2	0	1	0	0	0	0	1	0	0	0	−3	−1	7	−3	5
	3	0	0	1	0	0	0	0	1	0	0	−1	−4	4	2	−10
	4	0	0	0	1	0	0	0	0	1	0	1	−4	−4	2	10
	5	0	0	0	0	1	0	0	0	0	1	3	−1	−7	−3	−5
	6	0	0	0	0	0	−1	−1	−1	−1	−1	5	5	5	1	1

Example 2.13 Consider the orthogonal design for two attributes, A and B, and four choice tasks shown in Table 2.11. The rows represent choice tasks and the columns have the attribute level values that would be shown in each choice task. In this case, the correlation structure (i.e. the correlation coefficients for the variables in each column) is orthogonal by construction (see Wonnacott and Wonnacott 1990).

Table 2.11 Original orthogonal design

Choice task	A	B
1	1	3
2	2	1
3	3	4
4	4	2
Correlation structure		
	A	B
A	1	0
B	0	1

Table 2.12 demonstrates dummy, effects, and orthogonal coding transformations for this design. The correlation structure is also given at the base of the table for each coding type (it can be simply computed using the *data analysis tool* in Excel). As can be seen, both dummy and effects coding induce correlation within the data, even if the original design from which they were created was orthogonal (uncorrelated). Orthogonal coding, despite being largely ignored within the literature, avoids this problem.

Table 2.12 Dummy, effect, and orthogonal code comparison

Choice task	Dummy codes						Effects codes						Orthogonal codes					
	A_{D1}	A_{D2}	A_{D3}	B_{D1}	B_{D2}	B_{D3}	A_{E1}	A_{E2}	A_{E3}	B_{E1}	B_{E2}	B_{E3}	A_{O1}	A_{O2}	A_{O3}	B_{O1}	B_{O2}	B_{O3}
1	1	0	0	0	0	1	1	0	0	0	0	1	−3	1	−1	1	−1	−3
2	0	1	0	1	0	0	0	1	0	1	0	0	−1	−1	3	−3	1	−1
3	0	0	1	0	0	0	0	0	1	−1	−1	−1	1	−1	−3	3	1	1
4	0	0	0	0	1	0	−1	−1	−1	0	1	0	3	1	1	−1	−1	3
Correlation structure																		
	A_{I1}	A_{I2}	A_{I3}	B_{I1}	B_{I2}	B_{I3}	A_{I1}	A_{I2}	A_{I3}	B_{I1}	B_{I2}	B_{I3}	A_{I1}	A_{I2}	A_{I3}	B_{I1}	B_{I2}	B_{I3}
A_{I1}	1	0	−0.3	−0.3	−0.3	−0.3	1	0.5	0.5	0	−0.5	0.5	1	0	0	0	0	1
A_{I2}	−0.3	1	−0.3	1	−0.3	−0.3	0.5	1	0.5	0.5	−0.5	0	0	1	0	0	−1	0
A_{I3}	−0.3	−0.3	1	−0.3	−0.3	−0.3	0.5	0.5	1	−0.5	−1	−0.5	0	0	1	−1	0	0
B_{I1}	−0.3	1	−0.3	1	−0.3	−0.3	0	0.5	−0.5	1	0.5	0.5	0	0	−1	1	0	0
B_{I2}	−0.3	−0.3	−0.3	−0.3	1	−0.3	−0.5	−0.5	−1	0.5	1	0.5	0	−1	0	0	1	0
B_{I3}	1	−0.3	−0.3	−0.3	−0.3	1	0.5	0	−0.5	0.5	0.5	1	1	0	0	0	0	1

Aside from the issue of correlation, a further reason for preferring to use effects and orthogonal codes over dummy codes, is that the base level of dummy coded variables will be perfectly confounded with the model constants and hence indistinguishable from each other. By using non-zero-base level codes, effects, and orthogonal coding avoid such confoundment and allow for independent estimates of the base level.

2.4.2.3 Generating the Experimental Design

In practice, there are several different approaches that one might employ to generate a workable experimental design, all of which reflect the analyst own beliefs about what are the most important properties for a SC experimental design to display.

Traditional orthogonal designs methods. These have been historically the most common experimental design types. Orthogonality is related to the correlation structure of the design attributes. By forcing them to have zero correlations, each attribute is independent of all others. Several methods for constructing orthogonal designs exist in practice, including but not limited to methods such as generating balanced incomplete blocked designs (BIBD), Latin Squares designs, orthogonal in the differences' fractional factorial designs, and fold-over designs; all these have been discussed extensively elsewhere (Fowkes and Wardman 1988; Louviere et al. 2000; Rose and Bliemer 2008).

General

We will only consider the most widely applied orthogonal design type: the L^{KJ} orthogonal fractional factorial design (where L is the number of levels, K the number of attributes, and J the number of alternatives); two types of L^{KJ} designs have been described in the past. The first involves attributes that are uncorrelated both within and between alternatives; such designs are termed *simultaneous* orthogonal designs, as all alternatives are generated at the same time. The second type involves first locating an orthogonal design for the first alternative, and using the same design to construct subsequent alternatives by re-arranging the rows of the design (Louviere et al. 2000); such designs are known as *sequentially generated L^{KJ}* orthogonal fractional factorial designs.

In generating an orthogonal design sequentially, the analyst needs only locate an orthogonal design for a single alternative, whereas the simultaneous design approach requires the generation of an orthogonal design considering the correlation structure of all attributes, irrespective of which alternative they belong to. For this reason, the sequential design approach will generally result in designs with smaller numbers of choice tasks, as the theoretical minimum number of choice tasks required for a design does not necessarily guarantee that an orthogonal design may be located.

Example 2.14 Consider a design with three alternatives, each described by seven attributes with three attribute levels. The smallest simultaneous fractional factorial orthogonal design that can be constructed with 21 design attributes (7 attributes across 3 alternatives) has 72 choice tasks (see the web sites mentioned below).

In comparison, the smallest sequential orthogonal design (where it is only necessary to locate an orthogonal design that is uncorrelated for 7 attributes) has only 12 choice tasks.

One limitation of the sequential design process is that such designs are appropriate only for unlabelled SC experiments. Note also that designs generated under the null hypothesis of zero prior parameters, as described above, are sequentially generated L^{KJ} orthogonal fractional factorial designs.

Independent of the actual process used, a number of useful websites and software are available for obtaining orthogonal designs. As part of the book by Hedayat et al. (1999), Neil Sloane maintains online a library of over 200 orthogonal arrays (http://neilsloane.com/oadir/). Further, several software packages such as SPSS (www.spss.com), SAS (www.sas.com) and Ngene 1.0 (www.choice-metrics.com) are also able to generate a range of orthogonal designs.

D-optimal design method under the null hypothesis. Traditionally, an analyst would construct a sequential orthogonal design by simply assigning choice tasks randomly from the first alternative to make up the second and subsequent alternatives. More recently, a new optimality criterion has been developed to construct optimal orthogonal SC designs specifically generated for MNL models using orthogonal codes. These designs maintain (within alternatives) orthogonality, whilst also minimizing S^2 under the assumption that the parameters will be zero and the attributes will be orthogonal coded.

In practice, this typically results in the attribute levels across alternatives being made as different as possible. As such, these designs will generally increase the trade-offs that respondents are forced to make across all attributes maximising the information obtained in terms of the importance that each attribute plays on choice (Burgess and Street 2005; Street and Burgess 2004). Street and Burgess (2007), Street et al. (2005), and Rose and Bliemer (2008) provide detailed discussions of the exact procedures used in generating this class of design. Finally, Ngene 1.0 and Burgess (http://crsu.science.uts.edu.au/choice/choice.html) provide computing capabilities for generating such designs.

D-efficient design methods under the non-null hypothesis. An alternative approach to generating SC experiments involves selecting a design that is likely to provide an S^2 matrix containing values that are as small as possible, under the assumption that the parameters will not be zero. Given that their asymptotic standard errors obtained from discrete choice models are simply the square roots of the leading diagonal of this matrix, the smaller the matrix elements (or at a minimum, its diagonal elements), the smaller the asymptotic standard errors for each parameter. However, these *efficient* designs are unlikely to be orthogonal.

Efficient designs constructed under the non-null hypothesis differ from those generated under the null hypothesis in that they attempt to mimic the performance of the model to be estimated post data collection. If after extensive pre-design research, including focus groups, in-depth interviews, and pilot studies, one expects that the selected attributes have no impact upon choice (equivalent to assuming that the parameters will be statistically equal to zero), then one could question why the study is being conducted at all. In this way, the objective function defining optimality in generating a non-null prior parameter efficient design may be considered a practical one, as the design seeks to minimise the standard errors one is expecting to obtain in practice.

Nevertheless, efficient designs constructed under the non-null parameter prior hypothesis require two strong assumptions:

1) The analyst must first decide what model type is likely to be estimated once the data has been collected, in order to decide what matrix will be specifically used in generating the design. This is because S^2 for one discrete choice model will differ from that of any other discrete choice model; for example, that corresponding to the MNL model is different to that of a Nested Logit (NL) or Mixed Logit (ML) models (see Bliemer and Rose 2010; Rose et al. 2009b).
2) The analyst must also assume what the population parameter estimates will be in order to predict S^2 for a design. The reason is that for any Logit model S^2 is analytically equal to the negative inverse of the second derivatives of the model's log-likelihood function (as we will see in Section 8.4) and these are, in turn, a function of the model probabilities. But the model probabilities are a function of the utilities, which are in turn a function of the design attributes and the parameter estimates.

As discussed above, the analyst can assume prior parameter estimates in a Bayesian-like fashion when constructing a design. The assumption of prior parameters does not need to be too restrictive. Precise prior parameter values need not be provided (though such designs have been generated in the past; see for example, Carlsson and Martinsson 2002). Rather, prior parameter distributions that (hopefully) contain the true population parameter values can be used. Such designs are then optimised over a range of possible parameter values, without the analyst having to know the precise population values in advance (see Sándor and Wedel 2001; Kessels et al. 2006). This, however, increases the computing time required to generate the design. Rose and Bliemer (2008) outline the precise steps used to generate this type of design, whilst Bliemer and Rose (2010), Bliemer et al. (2009) and Rose et al. (2009b) provide details of the analytical second derivatives (needed to compute $\mathbf{S}^2$) for a range of different Logit models.

Notwithstanding, a simple approach to solve this issue in practice is as follows:

- use priors when they are available, and for those attributes without known priors either first build and orthogonal design (i.e. no priors required) or specify them very loosely – as discussed above;
- with this preliminary design, go to the field and apply the survey to a pilot sample; then, use the data to estimate a model;
- use the estimated parameters as new priors to build a better *D-efficient* design;
- apply the survey in stages, for example, only interview, at random, 10% of the total sample size first, and use that data to estimate new priors; use them to build a new design if their values were significantly different from the priors used at the previous stage;
- this sequential approach should converge in a couple of iterations.

Measuring statistical efficiency. Rather than attempting to minimise the elements of $\mathbf{S}^2$ of a given design directly, a number of measures of the statistical efficiency of a design have been proposed and can be used instead. The most common is the *D-error*, which uses the scaled determinant (i.e. raised to the power $1/K$ to account for the number of parameters to be estimated) of $\mathbf{S}^2$ to measure efficiency. Another, less popular measure, is *A-error* that is based on the trace of $\mathbf{S}^2$.

The determinant of a matrix is a single summary statistic of the magnitude of the elements contained within the matrix. The smaller the determinant, the smaller, on average, the values contained within the matrix will be. In the case of designs generated under the null hypothesis, assuming orthogonal coding and a MNL model structure, the *D-error measure* is converted to a *D-optimality* statistic, which is a percentage value of the design's overall efficiency. Typically, designs with values of around 90% or higher, are said to represent desirable designs of this class. For all other efficient designs, the objective is to minimise the *D-error*.

Another measure of statistical efficiency, proposed by Bliemer and Rose (2009), is *S-error*. McFadden (1974) described a direct relationship between the matrix $\mathbf{S}^2$ of Logit models and the sample size required to locate statistically significant parameter estimates. Bliemer and Rose (2009) proposed exploiting this relationship to calculate the sample size requirements for SC experiments. The *S-error* of a design provides the theoretically minimum sample size required to obtain asymptotically significant parameter estimates from it. As with *D-error*, the objective is to find a design that minimises *S-error*. In order to calculate the *S-error* of a design, the analyst must also construct its matrix $\mathbf{S}^2$.

Independent of the precise efficiency measure used, minimizing the elements of $\mathbf{S}^2$ for a design also reduces the expected asymptotic standard errors (i.e. the square roots of the diagonals of the matrix). As such, for any given sample size, smaller asymptotic standard errors mean smaller confidence intervals around the parameters estimates, as well as larger asymptotic *t*-ratios for each parameter. Hence, efficient designs are constructed specifically for the purpose of producing more reliable study results.

Alternatively, efficient designs may produce the same asymptotic standard errors as other designs given smaller sample sizes. This is because the $\mathbf{S}^2$ of all discrete choice models are divisible by N, the sample size, and as such, the asymptotic standard errors are also divisible by the square root of N. The result of this is that there exists an inescapable diminishing return in terms of the statistical significance of the parameter estimates obtained from each additional respondent added to a survey.

Example 2.15 Let the Fisher information matrix with N respondents be denoted by $\mathbf{I}_N(\boldsymbol{\theta})$. Since $\mathbf{I}_N(\boldsymbol{\theta}) = N \cdot \mathbf{I}_1(\boldsymbol{\theta})$, it holds that $\mathbf{S}_N^2 = (\mathbf{I}_N(\boldsymbol{\theta}))^{-1} = \frac{1}{N}(\mathbf{I}_1(\boldsymbol{\theta}))^{-1} = \frac{1}{N}\mathbf{S}_1^2$, such that:

$$se_N(\boldsymbol{\theta}) = \frac{se_1(\boldsymbol{\theta})}{\sqrt{N}}. \tag{2.11}$$

Hence, it is clear that the asymptotic standard errors provide diminishing improvements (decreases) for larger sample sizes.

Methods for Generating Designs Under the Non-null Hypothesis

A number of algorithms have been implemented for generating efficient designs under the non-null hypothesis; these tend to be either row or column-based algorithms. *Row-based algorithms* (e.g. Modified Federov algorithm, see Cook and Nachtsheim 1980) typically start by creating a set of candidate choice tasks (either generating an FF or a fractional factorial design) and then select choice tasks either randomly or based on some form of efficiency criterion.

Column-based algorithms start with a random design and change the attribute levels within each attribute of the design. Row-based algorithms have the benefit of being able to remove dominated choice tasks from the candidate set; however, such algorithms typically struggle with maintaining attribute level balance. Column-based algorithms, on the other hand, typically have little difficulty in maintaining attribute level balance but can often result in dominated choice tasks.

The most popular algorithm appears to be the RSC (relabelling, swapping, and cycling) algorithm (Huber and Zwerina 1996; Sándor and Wedel 2001). It has three separate operations; however, some may be omitted if required. The algorithm begins with a randomly constructed design after which the columns and rows are changed using *relabelling*, *swapping*, and *cycling* techniques, or combinations thereof. Relabelling occurs when all attribute levels of an attribute are switched (e.g. the combination {1, 2, 1, 3, 2, 3} might be relabelled {3, 2, 3, 1, 2, 1}). The *swapping* operation involves switching the levels of only a few attribute levels at a time rather than all attribute levels (e.g. the attribute combination {**1**, 2, 1, **3**, 2, 3} might become {**3**, 2, 1, **1**, 2, 3}). Finally, *cycling* replaces all attribute levels in each choice task simultaneously, by replacing the first attribute level with the second level, the second level with the third, etc. As such, cycling can only be performed if all attributes have exactly the same set of feasible levels.

A note on blocking of designs. Often the total number of choice tasks generated from a design may be too large for any one respondent to handle. In such cases, the researcher may turn to *blocking* the design. One way of doing this is by means of modular algebra selecting one effect to be confounded (see Galilea and Ortúzar, 2005). For orthogonal designs, blocking involves finding an additional 'blocking' column that may be used to allocate subsets of the generated choice tasks to different respondents. By using an orthogonal blocking column, the allocation of the choice tasks to respondents will be independent of the attribute levels shown to each (i.e. one respondent will not view choice tasks with only high prices and another choice task with low prices). For non-orthogonal efficient designs, it is unlikely that an orthogonal blocking column may be located. In such a case, the analyst may produce a near orthogonal blocking column by minimizing the largest correlation between the blocking column and the design attributes. This approach will be demonstrated in Section 2.4.3.

A note on the need for prior information in generating designs. One of the main criticisms of generating SC experiments is the requirement for prior knowledge about parameter priors and model structure. With regards to the first issue, researchers have shown that where no prior information about likely parameter values is known, a traditional orthogonal design will most likely generate good results as it is actually generated under the assumption of no priors. However, where prior information is available, it is generally possible to obtain greater statistical efficiency by relaxing the orthogonality constraint (see Rose and Bliemer 2009 for a review of this literature). Thus, even if the only information that the analyst has is that a parameter will take a particular sign (e.g. a cost parameter will be negative), a Bayesian uniform distribution may be used to generate a D-efficient design under the non-null hypothesis with advantage.

The issue of requiring advanced knowledge about the model structure is far more complicated. Different discrete choice models have different S^2 matrices as each one produces more or fewer parameter estimates. As such, optimizing a design for a MNL model, say, does not ensure that it will be optimal for other model forms.

Equations for constructing the S^2 matrix for other model structures have been reported; for example, Bliemer et al. (2009) compare designs optimized for the Nested Logit (NL) model with those generated for the MNL. Sándor and Wedel (2002) examined the cross-sectional formulation of the random parameters Mixed Logit (ML) model, and Bliemer and Rose (2010) explored the panel formulation of this same model. It is worth noting, however, that these researchers have found that designs optimized for the MNL model typically perform well when analysed using other model forms, with the exception of the cross-sectional version of the random parameters ML model. In any case and to overcome this criticism, Rose et al. (2009b) introduced a form of model averaging, where the S^2 matrix for different model structures can be computed and a weighted efficiency measure generated.

A note on interaction effects and SC designs. Quite often, analysts are interested in estimating interaction effects in addition to main effects (see Figure 2.11). Using the traditional design method of constructing orthogonal designs, this meant that the attributes of the design were allocated to particular columns so that not only were the effects orthogonal with each other, but so too were some or all of the interaction columns (formed by multiplying two or more main effects columns together). Recall that orthogonal designs minimise the elements contained within the S^2 matrix of linear models. Indeed, they tend to produce zero covariances suggesting that the parameter estimates of the effects of interest are independent of one another.

Given that the experimental design literature originated with the examination of linear models (i.e. ANOVA and regression models), such designs were deemed important. Nevertheless, discrete choice models are not linear models (although they do collapse to linear models when all parameter estimates are simultaneously zero). As such, the same problems that exist for main effects exist for interaction terms when it comes to estimating discrete choice models. In fact, a design that is capable of detecting independent interaction effects under the null hypothesis will produce non-zero covariances when the parameters are no longer zero, suggesting that there exists correlation between the parameters of interest. Thus, to minimize the standard errors and covariances of any interaction effects of interest, prior parameter estimates are required for these also.

2.4.2.4 Conduct Post Design Generation Testing

Once a design has been generated, it is possible to test how it might be expected to perform in practice. When conducted, such tests have typically taken one of two forms. Given a design, fix it and change the parameter priors to test its efficiency under the new set of assumed parameters (see Rose and Bliemer 2008). In taking this approach, the researcher is able to determine the robustness of the design to misspecification of the prior parameters. Some analysts have also employed Monte Carlo simulation to test whether the correct data generation process has been used in generating the design as well as how accurate the parameter estimates will be for various sample sizes (see Kessels et al. 2006; Ferrini and Scarpa 2007).

A number of different statistical measures have been used to compare the prior parameters to those obtained from the Monte Carlo simulation process. The two most popular are the mean square error (MSE) and the relative absolute error (RAE), defined as follows:

$$\text{MSE} = \frac{1}{R} \sum_{r=1}^{R} \left(\theta_k^{(r)} - \bar{\theta} \right)^2 \tag{2.12}$$

$$\text{RAE} = \frac{1}{R} \sum_{r=1}^{R} \left(\theta_k^{(r)} - \bar{\theta} \right)^2 / \bar{\theta} \tag{2.13}$$

where $\theta_k^{(r)}$ is the parameter estimate for attribute k obtained at sample iteration r, and $\bar{\theta}$ is the known prior parameter estimate used in running the Monte Carlo simulation.

Another popular statistic often used in these tasks is the expected mean square error of the parameter estimates (EMSE). Unlike the MSE and RAE, this measure provides a single summary statistic of the overall bias and variance across all parameter estimates, rather than for individual parameter estimates:

$$\text{EMSE} = \frac{1}{R} \sum_{r=1}^{R} \left(\theta_k^{(r)} - \bar{\theta} \right)^T \left(\theta_k^{(r)} - \bar{\theta} \right) \tag{2.14}$$

2.4.2.5 Conduct Questionnaire

Once the experimental design has been generated, the next stage is to construct the questionnaire. Given that the design is simply nothing more than a matrix of values, the analyst needs to convert this matrix into something that respondents can meaningfully interpret and respond to. The task for the analyst is therefore to convert each row of the design into a choice similar to that shown in Figures 2.8 and 2.9. This may call, for example, for the use of high-quality graphic material to convey an impression of what new rolling stock might be like. The researcher must be careful to avoid any implicit bias in the illustrative material used. Graphic illustrations are often preferred to photographs because of the higher control afforded in respect of the details included in them (Iglesias et al. 2013).

Example 2.16 In a study of the role of train frequency over demand for intercity travel (Steer and Willumsen 1983) it was found that although different people perceived the key variable (frequency) in different ways, almost nobody thought about it in terms of trains per hour or per day. Therefore, the SC survey started by ascertaining how was frequency (i.e. the analyst's concept) viewed by the traveller, for instance:

- I took the last train that puts me in Newcastle before 11 a.m.; it was the 7:50 from Kings Cross, or
- I just turned up at the station and found that the next train to Newcastle was due in 15 minutes.

The interviewer then converted the different frequency attributes of the experimental design into the same terms, for example: 'To get to Newcastle before 11 a.m. you must now take the 7:30 train' in a low-frequency option, or '...the 8:00 train', in a high-frequency one. Alternatively, '...the next train to Newcastle was in 30...' or '...10 min', for each option. Travellers were then asked to choose among alternatives described in terms they were familiar with and which affected their current journey choices thus increasing realism and relevance.

In constructing the questionnaire, it is advised that the analyst randomise the order of the choice tasks shown to different respondents to minimise possible order effects. That is, respondents may use the first few choice tasks to learn what it is they are being asked to do, whereas they might suffer fatigue for the last few choice tasks. Randomising the choice tasks over respondents should reduce any interaction of these biases with the specific choice tasks of the design that might otherwise occur. Although less common, it is also advised that the order of the alternatives and attributes be randomised between respondents.

2.4.2.6 Nothing is Important

It has been recommended to add a null option to the experimental design, also known as a *non-purchase option*. The reason is that if two or more options are presented to an individual who finds them all unacceptable, and has no opportunity to reject the lot, it is possible that this will trigger a secondary decision-making mechanism that could bias the results of the experiment. This important problem has been ignored far too often in practice.

Example 2.17   Olsen and Swait (1998) studied the veracity of the following propositions for the case of buying a product the sale of which was subject to strict prerequisites (e.g. concentrated orange juice, where it is expected that many consumers would require it to be unsweetened):

- if a non-purchase option (NPO) is not present, the attribute weights will differ from those observed when an NPO is offered in the design;
- if an NPO is included in the experimental design, the analyst should be able to identify more non-linear preference structures than if the NPO were missing;
- models based on data without an NPO may show low predictive capacity for choice situations including an NPO, whereas models based on data including an NPO will present good predictive capacity in any scenario.

Olsen and Swait (1998) used an experimental design with three brands, two levels for orange quality, two levels of sweetness, two types of packing (single and in lots of four) and two levels of price per unit. They also added a cheap option (with the supermarket brand name), consisting of a sweet orange juice made from low-quality oranges. They postulated a factorial design allowing them to estimate the main effects and all interactions between pairs of attributes.

Equal-sized samples (70 individuals) were presented with 16 situations involving three options, for the designs with and without NPO. They also asked if consumers would veto the sale of a product if one of its attributes had an unacceptable level.

They found that the parameters of the estimated models not only differed in magnitude but, as expected, the model with NPO presented significant non-linear (interaction) effects. These results were confirmed by an analysis of the responses to the question about vetoing a product depending on its characteristics (i.e. 63% of the sample found the sweetened juice unacceptable; 57% of it found unacceptable the requirement to buy packages of four units). Table 2.13 shows the percentage error in the predictions for each data set using the parameters estimated with the other set. There is no doubt that their initial hypotheses were confirmed; so, it may be concluded that nothing *is* indeed important.

Table 2.13 Cross errors of prediction in market shares

Alternative	Prediction error (%)	
	with NPO → without NPO	without NPO → with NPO
1	3.8	24.8
2	−1.9	21.6
3	−2.8	37.9
NPO	–	−47.8

2.4.2.7 Realism and Complexity

A key element in the success of SP surveys is the degree of realism achieved in the responses. Realism must be preserved in the *context* of the exercise, the *options* that are presented, and the *responses* that are allowed. This can be achieved in a number of ways:

1) Focusing on *specific* rather than general behaviour; for example, respondents should be asked how they would respond to an alternative on a given occasion, rather than in general; the more abstract the question the less reliable the response.
2) Using a realistic choice context, in particular, one the respondents have had recent personal experience of (i.e. a *pivot* design, see Train and Wilson 2008).
3) Retaining the constraints on choice required to make the context realistic; this usually means asking respondents to express preferences in relation to a very recent journey without relaxing any of its constraints: for example, 'if today you would prefer to use the car to visit your dentist in the evening directly from work, then retain this restriction in your choices'. Easing these constraints will just produce unrealistically elastic responses.
4) Using existing (perceived) levels of attributes so that the options are built around existing experience.
5) Using respondents' perceptions of what is possible to limit the attribute values in the exercise. For example, in considering improved rail services, do not offer options where the station is closer to home than feasible.
6) Ensuring that all relevant attributes are included in the presentation; this is especially important if developing travel choice models and not just measuring the relative importance of different attributes.
7) Keeping the choice experiments as simple as possible, without overloading the respondent. Remember we respond to very complex choices in practice but we do so over a long period of time, acquiring experience about alternatives at our own pace and selecting the best for us. In an SP exercise, these choices are compressed on a very short period of time and must, therefore, be suitably simplified.
8) Allowing respondents to opt for a response outside the set of experimental alternatives. For example, in a mode choice exercise if all options become too unattractive the respondent may decide to change destination, time of travel or not to travel at all; allow her a 'will do something else' alternative. If a computer-based interview is used it could be programmed to branch then to another exercise exploring precisely these other options.
9) Making sure that all the options are clearly and unambiguously defined. This could be quite difficult when dealing with qualitative attributes like security or comfort; for example, do not express alternatives as 'poor' or 'improved' as this is too vague and prone to different

interpretations by respondents. Describe instead what measures or facilities are involved in improving security or ride comfort (for example, closed circuit TV in all stations/attendants present at all times, air-conditioning in all coaches as in InterCity trains, etc.).

2.4.2.8 Use of Computers in SP Surveys

Computers have been used now for several years in the conduct of surveys of many kinds, including SC surveys. Computers offer significant advantages over 'paper and pen' methods but they have, given present technology, a few limitations. Let us consider them first.

In the case of SC surveys, one is most likely to use portable, preferably notebook-size computers. In the past, their main limitations were battery life and weight but modern machines have practically overcome these problems. The second restriction is screen size and quality. Contemporary portable computers offer a reasonable screen size and high-resolution colour screens permit the display of more information, at a price. Also, a computer screen is perfectly suited to paired choices in their pure and generalised form (i.e. with an associated rating scale, see Ortúzar and Garrido 1994a). It is also possible to display not only the attributes that vary as part of the SC experiment but also other features that remain fixed, such as destination, clock times, or indeed anything else that may be relevant or useful to the respondent.

What makes computer-based interviewing most attractive is, however, the task of tailoring the experiment to the subject. Most stated preference interviews will include a questionnaire in which information about the respondent and a recent journey (or purchase, etc.) is collected and used to build a subsequent experiment (i.e. pivot design). This questionnaire can be reproduced in software with the added advantages of automatic entry validation and automatic routing (see Figure 2.12). With a computerised system, the responses to this initial questionnaire can be used to generate the SC experiments and options automatically for each subject, following a specified design. Automatic routeing can be used to select the appropriate experiment for each individual depending on her circumstances. Furthermore, range and logic checks on the responses and pop-up help screens or look-up information windows (e.g. for timetables) can be incorporated to improve the quality of the interview.

	Option A	Option B	Option C	Alternative mode
Departure time to work	7:21	8:11	9:06	8:15
Usual travel time to work (Usual arrival time to work)	44 (8:05)	54 (9:05)	49 (9:55)	40 (8:55)
Travel time to work once a week (Usual arrival time to work)	49 (8:10)	62 (9:13)	56 (10:02)	48 (9:03)
Comfort	Crowded vehicle, sitting	Crowded vehicle, standing, usually have to wait next for boarding	Crowded vehicle, sitting	
Adittional cost ($)	$5.27	$5.61	$4.93	$15.00
Which option would you choose?	○	○	○	
Departure time from work	17:00	18:00	18:45	18:10
Usual travel time after work (Usual arrival time after work)	40 (17:40)	56 (18:56)	49 (19:34)	40 (18:50)
Travel time at destination after work once a week (Usual arrival time at destination after work)	55 (17:55)	65 (19:05)	54 (19:39)	51 (19:01)
Comfort	Crowded vehicle, sitting	Crowded vehicle, standing	Half crowded vehicle, standing	
Adittional cost ($)	$4.34	$5.27	$5.61	$12.00
Which option would you choose?	○	○	○	○

52%   (⬅ Previous) (➡ Next)

Figure 2.12 Example of computerised questionnaire

Computers also allow for experimenting with adaptive designs, that is, modifying the experimental design in the light of the responses of the subject (Holden et al. 1992; Toubia et al. 2007); although there can be gains by adapting the design in a Bayesian sense, care must be exercised not to lose the desirable properties of the sample and general design. In fact, Bradley and Daly (2000) caution against the use of adaptive designs as they may lead to bias; also, the methods to implement this in the case of efficient designs for more complex discrete choice models are very complex.

The use of computers for SC surveys also allows to design more complex interviews than might be attempted manually, although this complexity may never be apparent to the respondent, or even to the interviewer. Moreover, good software permits randomisation of the order in which the options are offered to each individual thus removing a further potential source of bias in the responses. Finally, as all responses are stored directly on disk there are no data entry costs nor errors and data are available immediately for processing.

A number of software packages offer excellent facilities for designing and coding complex interviews with a minimum of understanding of computing itself. In summary, the practical advantages of computer-based SP interviews are:

- an interesting format that is consistent across interviews and respondents;
- automatic question branching, prompting, and response validation;
- automatic data coding and storage;
- the ease with which the SP exercise can be tailored to each individual;
- the reduction in interview time achieved because the interviewer does not have to calculate and prepare written options;
- reduced training and briefing costs;
- the statistical advantages of randomising the sequence of choices.

On the debit side one has the initial cost of investing in hardware, software, insurance, and the requirement to provide some back-up services (disks, spare battery packs, modems, technical advice to interviewers and supervisors, etc.) on location.

Another, everyday more attractive, possibility is conducting the interview remotely via a Web page survey distributed through the Internet (see for example, Iragüen and Ortúzar 2004; Hojman et al. 2005). In this case, the sampling frame is an important issue as well as the even more careful design of the survey instrument; this has to follow the already noted special recommendations for mail-back surveys.

2.4.2.9 Quality Issues in Stated Preference Surveys

Stated-preference (SP) techniques have proved to be a powerful instrument in research and model development in transport and other fields. Their value depends on the careful application of the guidelines developed so far and discussed in the preceding pages. A key element in this is restricting the artificiality of the exercise to the minimum required. The more the analyst is interested in predicting future behaviour the more important it is to make sure the *decision context* is specific (an actual journey, not a hypothetical one) and the *response space* is behavioural.

But one of the dangers of these techniques is that it is relatively easy to cut corners in order to reduce costs. For example, one can allow the decision context to become less specific and more generic; this makes the sampling easier, the questionnaire simpler and, not surprisingly, the resulting models quite believable as they reflect 'ideal' rather than constrained behaviour. The value of goodness-of-fit indicators in SP surveys is entirely dependent on the quality and realism of the experiment. The problem is that the models resulting from 'cheaper' studies will only be found to be flawed much later.

The same is true of the analysis techniques discussed in Chapter 8. Good analysis will often require combining SP and RP data to make sure the resulting models are well anchored (scaled) in the restrictions and noise of real behaviour.

SP surveys can be a cost-effective way of refining and improving modelling tools but too much emphasis on low cost, at the expense of quality assurance and sound analysis, is likely to lead to disappointments and poor decision support.

2.4.3 Case Study Example

Consider a simple hypothetical transport evaluation study in which respondents will be asked to choose between three different hypothetical routes. The study objective is to determine the role that prices (i.e. petrol and toll costs), and travel times (i.e. travelling in free flow and congested traffic conditions), have upon route choice. The study also requires determining how travel time reliability may influence the choice of route.

Assume that secondary and qualitative research confirmed the above as the relevant set of attributes influencing choice; the same research further identified the attribute levels shown in Table 2.14 as being relevant to the study. Note that each attribute has three levels in this example but this needs not be the case, as different attributes are allowed to take different numbers of attribute levels. This was only done here for the sake of simplicity.

Further, given that we have chosen a route choice problem as our case study, an unlabelled SC experiment is the most appropriate experimental design approach to consider. Recall, however, that the processes and principles in constructing unlabelled SC surveys are a little different to those for generating labelled SC surveys. As such, where differences do exist, we will make a special note.

Table 2.14 Attribute and attribute levels

Travel costs ($)	
Petrol	1.00, 1.50, 2.00
Toll	0.00, 2.00, 4.00
Travel times (min)	
Free flow time	10, 15, 20
Congested time	10, 15, 20
Egress time	5, 10, 15
Travel time reliability	
Probability of arriving early	0.1, 0.2, 0.3
Probability of arriving late	0.1, 0.3, 0.5

Now armed with the appropriate set of alternatives (3), attributes (7), and attribute levels (3 for each attribute), the goal becomes to generate an experimental design that can be used to capture data on the behavioural responses of individuals that will assist in answering the identified study objectives.

The first step is to write down the most likely set of representative utility functions (V) that will be estimated for the study. Eq. (2.15) shows the expected utility functions to be used once data has been collected. In writing out the equations, as shown below, it is easy to see that we anticipate estimating generic parameters for the main effects only (i.e. the parameters are the same across alternatives and there are no interactions) and no alternative specific constants (ASC); note that if we had assumed a labelled choice experiment, the utility functions should reflect the mix of expected alternative specific and generic parameters to be estimated. If ASC or interaction effects were expected, these should be included in the utility specification. Likewise, any dummy, effects, or orthogonal coded variables should also be included.

$$
\begin{aligned}
V(A) = {} & \theta_1 x_{Pet_A}\{1.00,1.50,2.00\} + \theta_2 x_{Toll_A}\{0,2,4\} + \theta_3 x_{FFT_A}\{10,15,20\} + \theta_4 x_{CongT_A}\{10,15,20\} \\
& + \theta_5 x_{EgT_A}\{10,15,20\} + \theta_6 x_{Pr\ earl_A}\{0.1,0.2,0.3\} + \theta_7 x_{Pr\ earl_A}\{0.1,0.3,0.5\}, \\
V(B) = {} & \theta_1 x_{Pet_B}\{1.00,1.50,2.00\} + \theta_2 x_{Toll_B}\{0,2,4\} + \theta_3 x_{FFT_B}\{10,15,20\} \\
& + \theta_4 x_{CongT_B}\{10,15,20\} + \theta_5 x_{EgT_B}\{10,15,20\} + \theta_6 x_{Pr\ earl_B}\{0.1,0.2,0.3\} + \theta_7 x_{Pr\ earl_B}\{0.1,0.3,0.5\}, \\
V(C) = {} & \theta_1 x_{Pet_C}\{1.00,1.50,2.00\} + \theta_2 x_{Toll_C}\{0,2,4\} + \theta_3 x_{FFT_C}\{10,15,20\} \\
& + \theta_4 x_{CongT_C}\{10,15,20\} + \theta_5 x_{EgT_C}\{10,15,20\} + \theta_6 x_{Pr\ earl_C}\{0.1,0.2,0.3\} + \theta_7 x_{Pr\ earl_C}\{0.1,0.3,0.5\}.
\end{aligned}
\tag{2.15}
$$

At the same time as the utility specification is considered, the most likely model structure to be estimated should also be decided. This is because different model structures will have more or less parameter estimates (as we will see in Chapter 7). For example, if a Mixed Logit model with random parameters is to be estimated (see Section 7.6.2), then more than one parameter may be associated with each attribute. Similarly, the Nested Logit model will require the estimation of additional scale-related parameters (see Section 7.4.3) than the simpler MNL.

Given the model structure and expected utility specification, it is then possible to determine the smallest number of choice tasks required in generating the design. For the present case study, assume that we would like to estimate a simple MNL model with utility specification (2.15) on the data; in that case, seven parameters need to be estimated. As such, the design is required to have seven choice tasks at a minimum. As we are uncertain as to whether interaction effects might be present, and we would like to allow for the possibility of a more advanced econometric model being estimated, we may wish to produce a design with more than seven choice tasks.

On the other hand, if attribute level balance is required, the final design should aim to have more than seven choice tasks with their total number being divisible by three (i.e. as all attributes have three levels, the number of choice tasks must be divisible by this number). Given this, we decided to generate a design with 12 choice tasks, but we will block it so that each respondent faces only six of them (i.e. assume that initial qualitative research indicated that respondents could answer only this number of tasks comfortably).

We have constructed three different designs in our example (although in practice, it would be usual to construct only one): an orthogonal design, a *D*-optimal design under the null-hypothesis, and a *D*-efficient design under the non-null hypothesis. In generating the latter, information is required as to the expected values of the parameter estimates. Now, prior parameter values may be established from a number of sources. For example, the analyst may have some prior expectation as to the likely sign, such as a cost parameter should be negative in a utility function. Alternatively, previous research may also provide evidence as to what values the parameter estimates might take, or a pilot study may provide an indication as to reasonably likely values.

For the current case study, assume that a review of the literature established that the set of parameter estimates in Table 2.15 could act as good priors in setting up the experiment. Note that these do not include an *Egress time* attribute, and hence the prior parameter for this attribute was taken as zero.

Table 2.15 Parameter priors and prior standard errors

Travel costs ($)	
Petrol	−0.479 (0.0311)
Toll	−0.426 (0.0362)
Travel times (min)	
Free flow time	−0.098 (0.0087)
Congested time	−0.147 (0.0108)
Egress time	0.0 (0.0)
Travel time reliability	
Probability of arriving early	−0.120 (0.0827)
Probability of arriving late	−0.305 (0.032)

Note that should the exact parameter estimates reported in Table 2.15 be used, with no additional information, then a locally optimal design would result. In fact, the parameter estimates for the current study are unlikely to match exactly those obtained in our reference study; for this reason, we will assume Bayesian prior parameter distributions in generating the design. To do this, we will take draws from Bayesian multivariate Normal distributions using the reported parameter estimates as the means of the distributions and their standard errors as the standard deviations (e.g. the Bayesian prior parameter distribution for the petrol attribute will be $\theta_1 \sim N$ (−0.479, 0.0311)). Although we have chosen a multivariate Normal distribution here, we could have just as easily assumed Uniform distributions or any other distributional assumption in generating the design (Rose and Bliemer 2009).

The three designs generated are shown in Table 2.16 and their correlation structures are reported in Table 2.17. All designs were generated using Ngene. The first one, the traditional orthogonal design, was constructed using the sequential design process. Thus, an orthogonal design was first built for the first alternative; then its choice tasks were randomly re-arranged to build the second and third alternatives (e.g. the first-choice task in alternative 1 was randomly selected to be choice task 11 in alternative 2 and choice task 8 in alternative 3, and so on). As can be seen in Table 2.18, this results in a design where the *within alternative* correlations are zero for the design attributes, but the *between alternative* correlation structure may be non-zero. We will return to discuss the second design after discussing the third.

Design 3 was constructed using the prior parameter estimates mentioned above. Here we require estimating the design's $\mathbf{S}^2$ matrix, and hence the Fisher information matrix $\mathbf{I}_N(\theta)$. Re-arranging the design so that each row represents an alternative, and hence several rows combine to form a choice task, $\mathbf{I}_N(\theta)$ may be calculated using simple matrix algebra by means of Eqs. (2.16) and (2.17):

$$\mathbf{I}_N(\theta) = (\mathbf{Z}^T \mathbf{Z}) = \sum_{c=1}^{C} \sum_{j=1}^{J_c} \mathbf{z}_{jc}^T \mathbf{z}_{jc} \tag{2.16}$$

where

$$\mathbf{z}_{jc} = \left(x_{jkc} - \sum_{i=1}^{J_c} x_{ikc} P_{ic} \right) \sqrt{P_{ic}} \tag{2.17}$$

here j and i are used to denote alternatives, c a choice task, k a particular attribute, and x_{jkc} the attribute level for the kth attribute of alternative j in choice task c. Finally, P_{ic} represents the choice probability of alternative i being chosen in choice task c.

Table 2.16 Experimental designs

Orthogonal design

Choice task	Alternative A							Alternative B							Alternative C							
	Petrol	Toll	Pr. Late	Pr. Early	F.F. Time	Cong. Time	Eg. Time	Petrol	Toll	Pr. Late	Pr. Early	F.F. Time	Cong. Time	Eg. Time	Petrol	Toll	Pr. Late	Pr. Early	F.F. Time	Cong. Time	Eg. Time	Block
1	1.5	2	0.3	0.1	15	10	15	1.5	2	0.5	0.2	10	10	15	2	0	0.1	0.1	20	15	10	1
2	2	0	0.5	0.3	20	15	10	1	0	0.5	0.2	15	20	5	1	4	0.1	0.3	20	15	10	1
3	2	4	0.3	0.1	10	20	5	1	4	0.1	0.3	20	15	10	2	0	0.5	0.3	20	15	10	1
4	1	0	0.1	0.2	10	10	5	1	4	0.5	0.1	20	15	10	2	4	0.3	0.3	15	10	5	1
5	1	4	0.1	0.3	20	15	10	1.5	2	0.1	0.2	15	20	15	1	4	0.5	0.1	20	15	10	1
6	1.5	2	0.1	0.2	15	20	15	1	0	0.1	0.2	10	10	5	1.5	2	0.3	0.3	10	20	15	1
7	2	4	0.3	0.3	15	10	5	2	0	0.1	0.1	20	15	10	1	0	0.5	0.2	15	20	5	2
8	1	0	0.5	0.2	15	20	5	2	4	0.3	0.3	15	10	5	1.5	2	0.3	0.1	15	10	15	2
9	2	0	0.1	0.1	20	15	10	1.5	2	0.3	0.3	10	20	15	1.5	2	0.1	0.2	15	20	15	2
10	1	4	0.5	0.1	20	15	10	2	4	0.3	0.1	10	20	5	1.5	2	0.5	0.2	10	10	15	2
11	1.5	2	0.3	0.3	10	20	15	1.5	2	0.3	0.1	15	10	15	1	0	0.1	0.2	10	10	5	2
12	1.5	2	0.5	0.2	10	10	15	2	0	0.5	0.3	20	15	10	2	4	0.3	0.1	20	20	5	2

Efficient design under null hypothesis assumption

Choice task	Alternative A							Alternative B							Alternative C							
	Petrol	Toll	Pr. Late	Pr. Early	F.F. Time	Cong. Time	Eg. Time	Petrol	Toll	Pr. Late	Pr. Early	F.F. Time	Cong. Time	Eg. Time	Petrol	Toll	Pr. Late	Pr. Early	F.F. Time	Cong. Time	Eg. Time	Block
1	1.5	2	0.3	0.1	15	10	15	2	4	0.5	0.2	10	15	10	1	0	0.1	0.3	20	20	5	1
2	1	0	0.5	0.2	15	20	5	1.5	2	0.1	0.3	10	10	15	2	4	0.3	0.1	20	15	10	1
3	2	0	0.1	0.1	20	15	10	1	2	0.3	0.2	15	20	5	1.5	4	0.5	0.3	10	10	15	1
4	1.5	2	0.5	0.2	10	10	15	2	4	0.1	0.3	20	15	10	1	0	0.3	0.1	15	20	5	1

Choice task	Alternative A							Alternative B							Alternative C							
	Petrol	Toll	Pr. Late	Pr. Early	F.F. Time	Cong. Time	Eg. Time	Petrol	Toll	Pr. Late	Pr. Early	F.F. Time	Cong. Time	Eg. Time	Petrol	Toll	Pr. Late	Pr. Early	F.F. Time	Cong. Time	Eg. Time	Block
5	2	4	0.3	0.1	10	20	5	1	0	0.5	0.2	20	10	15	1.5	2	0.1	0.3	15	15	10	1
6	1	4	0.1	0.3	20	15	10	1.5	2	0.5	0.1	10	20	5	2	2	0.5	0.2	10	10	15	1
7	2	4	0.3	0.3	15	10	5	1	2	0.5	0.1	20	15	5	2	2	0.1	0.2	20	20	10	2
8	2	0	0.5	0.3	20	15	10	1	0	0.1	0.2	10	20	15	2	4	0.3	0.2	10	10	15	2
9	1	4	0.5	0.1	20	15	10	1.5	4	0.1	0.1	15	15	5	1	2	0.3	0.3	10	10	15	2
10	1.5	2	0.3	0.3	10	20	15	2	2	0.5	0.1	15	20	10	1.5	0	0.1	0.1	15	15	5	2
11	1	0	0.1	0.2	10	10	5	1	0	0.5	0.3	20	15	10	1	4	0.5	0.3	15	20	10	2
12	1.5	2	0.1	0.2	15	20	15	2	4	0.3	0.3	20	10	10	1	0	0.3	0.3	20	15	5	2

Efficient design under non-null hypothesis assumption

Choice task	Alternative A							Alternative B							Alternative C							
	Petrol	Toll	Pr. Late	Pr. Early	F.F. Time	Cong. Time	Eg. Time	Petrol	Toll	Pr. Late	Pr. Early	F.F. Time	Cong. Time	Eg. Time	Petrol	Toll	Pr. Late	Pr. Early	F.F. Time	Cong. Time	Eg. Time	Block
1	2	2	0.1	0.2	15	15	10	1.5	2	0.1	0.2	10	10	5	1	0	0.5	0.2	15	15	15	1
2	1	4	0.1	0.1	10	10	15	2	0	0.5	0.3	20	10	15	1.5	4	0.3	0.2	20	15	10	1
3	1.5	2	0.5	0.1	20	10	5	1.5	0	0.1	0.3	10	20	5	1.5	4	0.3	0.1	15	20	10	1
4	1	0	0.3	0.3	20	10	15	1.5	4	0.3	0.1	10	15	15	2	2	0.5	0.2	10	15	10	1
5	1.5	4	0.3	0.2	20	15	5	2	0	0.5	0.1	15	15	5	1	2	0.1	0.3	15	15	5	1
6	1	0	0.5	0.3	10	20	10	1	2	0.3	0.2	15	20	10	2	0	0.1	0.1	15	10	15	1
7	2	4	0.5	0.3	10	10	15	2	2	0.3	0.2	20	20	10	1	0	0.1	0.1	20	20	5	2
8	1	2	0.1	0.2	20	15	5	1.5	4	0.1	0.1	15	15	5	2	0	0.5	0.3	10	20	10	2
9	2	0	0.1	0.1	15	15	5	1	4	0.5	0.3	15	10	15	1.5	2	0.3	0.2	20	20	15	2
10	1.5	4	0.3	0.3	15	20	10	2	0	0.1	0.3	20	15	15	1	4	0.5	0.1	10	10	15	2
11	2	2	0.3	0.2	15	20	10	1	2	0.5	0.1	10	20	5	1.5	2	0.1	0.3	10	10	5	2
12	1.5	0	0.5	0.1	10	20	15	1	4	0.3	0.2	20	10	10	2	4	0.3	0.3	10	10	5	2

Table 2.17 Design correlation structures

	Traditional orthogonal design														
	Petrol	Toll	Pr. Late	Pr. Early	F.F. Time	Cong. Time	Eg. Time	Petrol	Toll	Pr. Late	Pr. Early	F.F. Time	Cong. Time	Eg. Time	Block
Petrol	1														
Toll	0	1													
Pr. late	0	0	1												
Pr. early	0	0	0	1											
F.F. time	0	0	0	0	1										
Cong. time	0	0	0	0	0	1									
Eg. time	0	0	0	0	0	0	1								
Petrol	−0.25	0.25	0.5	0	0.13	−0.25	0	1							
Toll	−0.5	0	0	−0.5	−0.13	0.25	−0.5	0	1						
Pr. late	−0.13	−0.63	0.38	−0.13	−0.13	−0.5	0.25	0	0	1					
Pr. early	0.25	−0.25	0.13	−0.38	0	0.25	0	0	0	0	1				
F.F. time	0.13	0.13	0.13	0.38	−0.63	−0.25	−0.5	0	0	0	0	1			
Cong. time	0.13	0.13	0	0	0.63	−0.25	−0.25	0	0	0	0	0	1		
Eg. time	0.13	−0.13	−0.5	0	−0.13	−0.25	0.25	0	0	0	0	0	0	1	
Block	0	0	0.41	0	0	0	0	0.82	−0.2	−0.2	−0.41	−0.61	0	0	1

D-optimal design under the null hypothesis

Traditional orthogonal design

	Petrol	Toll	Pr. Late	Pr. Early	F.F. Time	Cong. Time	Eg. Time	Petrol	Toll	Pr. Late	Pr. Early	F.F. Time	Cong. Time	Eg. Time	Petrol	Toll	Pr. Late	Pr. Early	F.F. Time	Cong. Time	Eg. Time	Block
Petrol	1																					
Toll	0	1																				
Pr. late	0	0	1																			
Pr. early	0	0	0	1																		
F.F. time	0	0	0	0	1																	
Cong. time	0	0	0	0	0	1																
Eg. time	0	0	0	0	0	0	1															
Petrol	-0.5	0	0	0	-0.38	0	0.75	1														
Toll	0	-0.5	0	0	-0.38	0	0.75	0.75	1													
Pr. late	0.38	0.38	-0.5	0	-0.38	0	0	0.38	0	1												
Pr. early	-0.38	-0.38	0	-0.5	-0.38	0	0	0	0.38	-0.38	1											
F.F. time	0	0	0	0	-0.5	0	0	0	0	0	0	1										
Cong. time	0	0	0	0	0.75	-0.5	0	-0.38	-0.38	-0.38	-0.38	0	1									
Eg. time	0	0	0	0	-0.75	0	-0.5	-0.5	0	0.38	0.38	0	-0.75	1								
Petrol	-0.5	0	0	0	0.38	0	-0.75	-0.75	-0.5	0	0	0	0	-0.5	1							
Toll	0	-0.5	0	0	0.38	0	-0.75	-0.5	-0.5	0	0	0	0	-0.75	0.75	1						
Pr. late	-0.38	-0.38	-0.5	0	0.38	0	0	0	0	-0.5	0	0	0	0	0.38	0.38	1					
Pr. early	0.38	0.38	0	-0.5	0.38	0	0	-0.38	-0.38	0	-0.5	0	0	0	0	0	-0.38	1				
F.F. time	0	0	0	0	-0.5	0	0	0.38	0.38	0.38	0.38	-0.5	-0.5	0	-0.38	-0.38	-0.38	-0.38	1			
Cong. time	0	0	0	0	-0.75	-0.5	0	0.38	0.38	0.38	0.38	0	-0.5	0	-0.38	-0.38	-0.38	-0.38	0.75	1		
Eg. time	0	0	0	0	0.75	0	-0.5	-0.75	-0.75	-0.38	-0.38	0	0	-0.5	0.75	0.75	0.38	0.38	-0.75	-0.75	1	
Block	0	0	0	0.41	0	0	0	0	0	0	0	0	0	0	0	0	0	-0.2	0	0	0	1

(Continued)

Table 2.17 (Continued)

	Petrol	Toll	Pr. Late	Pr. Early	F.F. Time	Cong. Time	Eg. Time	Petrol	Toll	Pr. Late	Pr. Early	F.F. Time	Cong. Time	Eg. Time	Petrol	Toll	Pr. Late	Pr. Early	F.F. Time	Cong. Time	Eg. Time	Block
Petrol	1																					
Toll	0.13	1																				
Pr. late	0	−0.13	1																			
Pr. early	−0.13	0.13	0.25	1																		
F.F. time	−0.13	0	−0.25	0	1																	
Cong. time	0.13	−0.25	0.13	0.13	−0.25	1																
Eg. time	−0.13	0	0.25	0.25	−0.63	−0.13	1															
Petrol	−0.13	0.88	−0.13	0.25	0.13	−0.5	0.13	1														
Toll	0	−0.75	−0.13	0	0	0.13	0.13	−0.63	1													
Pr. late	0.13	0	−0.13	−0.25	−0.25	0	0.13	−0.13	0	1												
Pr. early	0.13	0.13	0	−0.38	−0.38	−0.13	0	0.13	−0.38	−0.13	1											
F.F. time	−0.13	0.38	0.13	0	−0.63	0.13	0.38	0.38	−0.13	0.13	0.38	1										
Cong. time	0	0.13	0.63	0.5	0.13	0	−0.13	0	−0.25	−0.13	−0.25	−0.25	1									
Eg. time	0	0.13	−0.13	0	0.5	0	−0.5	0.25	0.13	−0.13	0	0.25	−0.13	1								
Petrol	−0.63	−0.75	0.13	−0.13	0	0.13	0	0.25	0.13	0	−0.25	−0.13	0	−0.13	1							
Toll	−0.13	0.13	0.13	−0.13	0	0	0.13	−0.63	0.63	0	−0.25	0.25	−0.25	0.13	0	1						
Pr. late	−0.25	−0.13	−0.5	−0.38	0.38	−0.13	0	0.13	−0.38	−0.63	0.5	−0.13	−0.5	0.25	0.13	0.13	1					
Pr. early	0	−0.13	−0.38	−0.25	0.25	0.25	−0.13	−0.25	0.38	0.38	0.13	−0.13	−0.38	0.13	0.25	0	0	1				
F.F. time	0.5	−0.13	−0.13	−0.25	−0.38	−0.25	−0.38	0	−0.25	−0.13	−0.63	0	0.13	−0.38	−0.38	−0.13	−0.63	−0.13	1			
Cong. time	0.13	0.13	−0.25	−0.25	0.38	−0.75	−0.25	0.25	0.13	−0.13	0.13	−0.13	0	0.38	−0.13	−0.25	0.13	−0.13	0.25	1		
Eg. time	0.13	−0.5	−0.38	−0.13	0	0.13	−0.25	−0.63	0.25	0.13	0	−0.63	0	−0.5	0.25	−0.38	0	0	0.38	0	1	
Block	0.41	0	0	0	−0.2	0.41	0	−0.2	0.41	0	0	0.41	0	0.41	0	0	0	0.2	0	0	−0.2	1

Table 2.18 $I_N(\theta)$ and S^2 for the third design

	$I_N(\theta)$						
	Var(θ_1)	Var(θ_2)	Var(θ_3)	Var(θ_4)	Var(θ_5)	Var(θ_6)	Var(θ_7)
Var(θ_1)	**2.088**	−1.974	0.026	0.045	−0.807	−1.525	2.761
Var(θ_2)	−1.974	**27.323**	0.065	−0.077	−20.218	−30.855	−5.141
Var(θ_3)	0.026	0.065	**0.376**	0.002	−2.105	0.608	0.362
Var(θ_4)	0.045	−0.077	0.002	**0.091**	0.374	−0.33	0.507
Var(θ_5)	−0.807	−20.218	−2.105	0.374	**188.628**	−36.823	3.175
Var(θ_6)	−1.525	−30.855	0.608	−0.33	−36.823	**159.857**	−14.522
Var(θ_7)	2.761	−5.141	0.362	0.507	3.175	−14.522	**213.312**
	S^2						
	Var(θ_1)	Var(θ_2)	Var(θ_3)	Var(θ_4)	Var(θ_5)	Var(θ_6)	Var(θ_7)
Var(θ_1)	**0.588**	0.081	0.002	−0.179	0.016	0.024	−0.004
Var(θ_2)	0.081	**0.071**	0.016	0.027	0.011	0.017	0.002
Var(θ_3)	0.002	0.016	**2.852**	−0.161	0.034	−0.001	−0.005
Var(θ_4)	−0.179	0.027	−0.161	**11.441**	−0.018	0.022	−0.022
Var(θ_5)	0.016	0.011	0.034	−0.018	**0.008**	0.004	0
Var(θ_6)	0.024	0.017	−0.001	0.022	0.004	**0.011**	0.001
Var(θ_7)	−0.004	0.002	−0.005	−0.022	0	0.001	**0.005**

For the third design, Figure 2.13 shows the calculations used to construct $I_N(\theta)$ assuming the mean of the Bayesian parameter distributions given in Table 2.16. For known design values $\mathbf{x}$, the first problem is to calculate the expected probabilities for each choice task given the assumed model and utility functions. Once the choice probabilities have been calculated, it is possible to construct the $\mathbf{Z}$ matrix in Eq. (2.16) as well as $I_N(\theta)$. Taking the inverse of $I_N(\theta)$ produces S^2 for the third design, under the parameter priors assumed; these two matrices are shown in Table 2.18. The D-error value for this design is then calculated as:

$$D\text{-error} = \det\left(S^2(\mathbf{x}, \theta)\right)^{1/K} \tag{2.18}$$

and the value computed for the matrix S^2 in Table 2.12 is 0.109.

However, in generating this design, we will not simply assume the parameter estimates shown in the above figure and tables. Rather, we will use simulation methods to take draws from the assumed multivariate Bayesian prior parameter distributions; let us call these Ω in general (i.e. recall they can be Normal or any other distribution), and calculate S^2 for each draw taken. The Bayesian D_b-error can therefore be calculated as:

$$D_b\text{-error} = \int_\theta \det\left(S^2(\mathbf{x}, \theta)\right)^{1/K} \phi(\theta \mid \Omega) d\theta \tag{2.19}$$

By fixing the simulated draws and changing the attribute level combinations of the design using the algorithms discussed in Section 2.4.2.3, the Bayesian D_b-error can be calculated for each newly generated design. That producing the lowest value is retained and used. In particular, the Bayesian D_b-error for Design 3 using 100 Halton draws (see Bhat 2001) in the simulation was 0.10975. Note that the actual D-error or Bayesian D_b-error value is actually meaningless, and can only be used to compare designs constructed under the same set of assumptions.

Input table

e =		-0.48	-0.43	-0.31	-0.12	-0.10	-0.15	0.00
S	J	Petrol	Toll	Late	Early	FF	CongT	EgT
1	1	2	2	0.1	0.2	15	15	10
1	2	1.5	2	0.1	0.2	10	10	5
1	3	1	0	0.5	0.2	15	15	15
2	1	1	4	0.1	0.1	10	10	15
2	2	2	0	0.5	0.3	20	10	5
2	3	1.5	4	0.3	0.2	20	15	10
3	1	1	4	0.1	0.1	10	10	15
3	2	2	0	0.5	0.3	20	10	5
3	3	1.5	4	0.3	0.2	20	15	10
4	1	1.5	2	0.5	0.1	20	10	5
4	2	1.5	0	0.1	0.3	10	20	10
4	3	1.5	4	0.3	0.1	15	20	10
5	1	1.5	2	0.5	0.1	20	10	5
5	2	1.5	0	0.1	0.3	10	20	10
5	3	1.5	4	0.3	0.1	15	20	10
6	1	1.5	2	0.5	0.1	20	10	5
6	2	1.5	0	0.1	0.3	10	20	10
6	3	1.5	4	0.3	0.1	15	20	10
11	1	1.5	4	0.3	0.2	20	15	5
11	2	2	0	0.5	0.1	15	15	15
11	3	1	2	0.1	0.3	15	15	5
12	1	1	0	0.5	0.3	10	20	10
12	2	1	2	0.3	0.2	15	20	5
12	3	2	0	0.1	0.1	15	10	15

V and P

S	J	V	P
1	1	-5.54	0.12
1	2	-4.07	0.50
1	3	-4.33	0.39
2	1	-4.67	0.45
2	2	-5.58	0.49
2	3	-6.70	0.06
3	1	-4.67	0.45
3	2	-4.58	0.49
3	3	-6.70	0.06
4	1	-5.16	0.36
4	2	-4.71	0.57
4	3	-6.93	0.06
5	1	-5.16	0.36
5	2	-4.71	0.57
5	3	-6.93	0.06
6	1	-5.16	0.36
6	2	-4.71	0.57
6	3	-6.93	0.06
11	1	-6.70	0.08
11	2	-4.80	0.52
11	3	-5.07	0.40
12	1	-4.59	0.31
12	2	-5.85	0.09
12	3	-3.94	0.60

Z matrix

S	J	Petrol	Toll	Late	Early	FF	CongT	EgT
1	1	0.22	0.26	-0.05	0.00	0.85	0.85	0.19
1	2	0.10	0.54	-0.11	0.00	-1.77	-1.77	-3.13
1	3	-0.23	-0.76	0.15	0.00	1.55	1.55	3.46
2	1	-0.35	1.32	-0.14	-0.07	-3.69	-0.20	3.50
2	2	0.34	-1.42	0.13	0.07	3.15	-0.21	-3.35
2	3	-0.01	0.48	0.00	0.00	1.09	1.14	0.05
3	1	-0.35	1.32	-0.14	-0.07	-3.69	-0.20	3.50
3	2	0.34	-1.42	0.13	0.07	3.15	-0.21	-3.35
3	3	-0.01	0.48	0.00	0.00	1.09	1.14	0.05
4	1	0.00	0.62	0.15	-0.07	3.65	-3.84	-1.92
4	2	0.00	-0.74	-0.12	0.06	-2.99	2.75	1.38
4	3	0.00	0.75	0.01	-0.03	0.26	0.90	0.45
5	1	0.00	0.62	0.15	-0.07	3.65	-3.84	-1.92
5	2	0.00	-0.74	-0.12	0.06	-2.99	2.75	1.38
5	3	0.00	0.75	0.01	-0.03	0.26	0.90	0.45
6	1	0.00	0.62	0.62	-0.07	3.65	-3.84	-1.92
6	2	0.00	-0.74	-0.74	0.06	-2.99	2.75	1.38
6	3	0.00	0.75	0.75	-0.03	0.26	0.90	0.45
11	1	-0.02	0.81	-0.01	0.00	1.29	0.00	-1.46
11	2	0.32	-0.80	0.13	-0.06	-0.28	0.00	3.45
11	3	-0.36	0.56	-0.14	0.07	-0.25	0.00	-3.30
12	1	-0.33	-0.10	0.14	0.07	-1.92	3.35	-1.43
12	2	-0.18	0.54	0.02	0.01	0.47	1.78	-2.24
12	3	0.31	-0.14	-0.11	-0.06	1.21	-3.11	1.89

Figure 2.13 Calculating the Z matrix

Although we do not show it here, a $\mathbf{S}^2$ matrix can also be calculated for the other two designs. The interested reader can check that under the same assumptions used to generate the *D*-efficient design for the non-null hypothesis (i.e. the same utility functions and prior parameter estimates), the traditional orthogonal design would produce a *D*-error of 0.20037 and a Bayesian D_b-error of 0.20129, whilst the *D*-optimal design under the null hypothesis would produce a *D*-error of 0.18799 and a Bayesian D_b-error of 0.189708.

Coming back to the second design now (i.e. under the null-hypothesis), recall that it was generated assuming a MNL model structure with parameter priors equal to zero and attribute levels recoded into orthogonal codes. We do not show the precise design generation process here and the interested reader is referred to Rose and Bliemer (2008) or Street et al. (2005) for the methods employed to come up with the exact attribute level combinations for generating it. However, the design can also be generated using similar methods as those shown for constructing the *D*-efficient design under the non-null hypothesis.

In taking this approach, the design matrix would need to be transformed to allow for orthogonal codes rather than the actual levels shown to respondents. Next, the parameter priors would be set to zero and $\mathbf{I}_N(\boldsymbol{\theta})$ calculated. Unlike the *D*-efficient design under the non-null hypothesis, Street and Burgess (2004) have derived a set of equations that allow the analyst to determine precisely how efficient this class of design is. Rather than use the *D*-error value, Street and Burgess (2004) maximise the *D*-efficiency of the design. Note that for our example, the *D*-efficiency is 72.07%, which suggests that there could possibly be a better (i.e. nearer to the optimum) design under the assumptions used in its construction.

Examining the design itself (Table 2.16), it is clear that under the assumptions used to generate it, a typical outcome is that the attribute level differences will be maximised between the alternatives. This can be seen, for example, by examining the petrol price attribute which never takes the same level across alternatives. Similar differences exist for all other attributes. So, as we had mentioned, this particular design generation process tends to maximise the trade-offs that respondents are asked to make in choosing amongst their choice alternatives.

Once the design has been generated, various tests of how it might be expected to perform in practice can be conducted. A simple one is to fix the design and systematically vary the parameter priors, observing the impact of each change on the efficiency of the design. Table 2.19 shows the *D*-error calculations when the first parameter prior is varied over a range of values. By examining the impact upon efficiency given such parameter changes, the analyst may gain an understanding of how robust the design will be over varying degrees of prior parameter misspecification.

Even though not explicitly stated, traditional orthogonal designs are generated under the null hypothesis. As such, testing for prior parameter misspecification is not limited to *D*-efficient designs constructed under the non-null hypothesis. Indeed, although rarely done in practice, it can be applied to any generated design, including those generated under the null hypothesis assumption.

Once the design has been generated the survey can be finalised. To construct the choice tasks, we need to translate the generated design into a format that respondents can meaningfully interpret and respond to. An examination of Figure 2.9 will reveal that the choice task shown there corresponds to the first-choice task taken from the *D*-optimal design under the null hypothesis in Table 2.17.

Table 2.19 Prior parameter misspecification test

% variation	θ_1	Design 1	Design 2	Design 3	% variation
−100	0	0.195474	0.177806	0.119139	40.223309
−80	−0.0958	0.196038	0.17958	0.117869	32.2328728
−60	−0.1916	0.196801	0.181486	0.116844	24.2427384
−40	−0.2874	0.197772	0.183523	0.116066	16.2529087
−20	−0.3832	0.198958	0.18569	0.11532	8.2632126
0	**−0.479**	**0.200367**	**0.187988**	**0.115242**	**0.274161**
20	−0.5748	0.202007	0.190418	0.115196	−7.7147528
40	−0.6706	0.203888	0.192982	0.115396	−15.7033546
60	−0.7664	0.20617	0.195684	0.11584	−23.691612
80	−0.8622	0.208405	0.19528	0.11653	−31.681235
100	−0.958	0.21106	0.201519	0.117468	−39.6672541

Sample size and SC experiments. Eq. (2.11) provides clues as to the sample size requirements for SC experiments. Bliemer and Rose (2009) demonstrate how the sample size requirement to obtain asymptotically statistically significant parameter estimates can be derived from this equation. The asymptotic t-ratio value for a given parameter θ_k may be calculated as:

$$t_N(\theta_k) = \theta_k / \left(\frac{se_k(\theta_k)}{\sqrt{N_k}} \right) \qquad (2.20)$$

where N_k is the sample size that would be derived from the calculation for attribute k. Re-arranging it we get:

$$N_k = \frac{t_N(\theta_k)^2 se_k(\theta_k)^2}{\theta_k^2} \qquad (2.21)$$

Thus, for any desirable asymptotic t-ratio value, say that for a 95% confidence level in a two-sided test, 1.96, it is possible to calculate the sample size requirement to achieve that asymptotic t-ratio value for a design under various prior parameter assumptions.

For the case study above, taking the means of the prior parameter distributions as the true population parameters, and obtaining the parameter standard errors from Table 2.11, the sample size requirements according to each parameter, in order, would be: 11.12, 1.74, 125.27, 3096.82, 3.25, and 2.30 (the reader can easily check this). Note that no sample size can be calculated for *Egress time* given that it was assumed with a zero-prior parameter. Taking the largest sample size as the critical one, we could say that the design requires at least 3097 respondents for all parameter to be statistically significant. However, note that in making such calculations, Bliemer and Rose (2009) suggest that these sample sizes represent a theoretical minimum and that a larger sample size should probably be adopted.

2.4.4 Limitations of Stated Preference Methods

When respondents choose one of a set of hypothetical alternatives, they have to imagine how satisfied they will be with that selection. This requires the ability to accurately envisioning future conditions with characteristics that have not actually been experienced: a new Light Rail System, say, when currently only an unreliable bus service is available. We make our current choices on the basis of what we remember of the alternatives. Alas, our short and long-term memories are not that accurate; in the words of Gilbert (2007)

> ...*memory is not a dutiful scribe that keeps a complete transcript of our experiences, but a sophisticated editor that clips and saves key elements of an experience and then uses these elements to rewrite the story each time we ask to reread it.*

These salient elements can be positive or negative and are mostly remembered because they were different from previous experience or expectations. In SP studies, the analyst defines these salient elements and requires respondents to use them to 'rewrite the story', to remember forward, in a realistic manner without omitting elements of the reference journey and its constraints; a very difficult task indeed.

Moreover, in defining the attributes of each alternative, the analyst makes salient some elements that may not be so in practice. For example, vehicle operating costs are often an attribute in stated mode choice models, in spite of the fact that most travellers are only vaguely aware of what they are in normal decision-making. SP cannot be used to determine certain parameters, for example, any perceived cost discount when paying by electronic means (instead of cash) a toll or public transport fare. Any perceived discounting takes place below the conscious level and bringing it to the fore may force an unrealistic response.

Exercises

2.1 We require estimating the population of a certain area for the year 2020 but we only have available reliable census information for 1990 and 2000, as follows:

$$P_{1990} = 240 \pm 5 \text{ and } P_{2000} = 250 \pm 2$$

To estimate the future population, we have available the following model:

$$P_n = P_b t^d$$

where P_n is the population in the forecast year (n), P_b the population in the base year (b), t is the population growth rate and $d = (n - b)/10$, is the number of decades to extrapolate growth.

Assume that the data from both censuses are independent and that the model does not have any specification error; in that case,

(a) Find out with what level of accuracy is it possible to forecast the population in the year 2020;

(b) You are offered the census information for 2010, but you are cautioned that its level of accuracy is worse than that of the previous two censuses:

$$P_{2010} = 265 \pm 8$$

Find out whether it is convenient to use this value.

(c) Repeat the analysis assuming that the specification error of the model is proportional to d, and that it can be estimated as $12d\%$.

2.2 Consider the following modal-split model between two zones i and j (but we will omit the zone indices to alleviate notation):

$$P_1(\Delta t/\theta) = \frac{\exp\,(-\theta t_1)}{\exp\,(-\theta t_1) + \exp\,(-\theta t_2)} = \frac{1}{1 + \exp - \theta(t_2 - t_1)} = \frac{1}{1 + \exp\,(-\theta\,\Delta t)}$$

$$P_2(\Delta t/\theta) = 1 - P_1 = \frac{\exp\,(-\theta\,\Delta t)}{1 + \exp\,(-\theta\,\Delta t)}$$

where t_k is the total travel time in mode k, and θ a parameter to be estimated.

During the development of a study, travel times were calculated as the average of five measurements (observations) for each mode, at a cost of $1 per observation, and the following values were obtained:

$$t_1 = 12 \pm 2 \text{ min} \quad t_2 = 18 \pm 3 \text{ min}$$

(a) If the estimated value for θ is 0.1, compute a confidence interval for P_1.
(b) Assume you would be prepared to pay $3 per each percentage point of reduction in the error of P_1; find out whether in that case, it would be convenient for you to take 10 extra observations in each mode whereby the following values for t_k would be obtained:

$$t_1 = 12 \pm 1 \text{ min} \quad t_2 = 17.5 \pm 1.5 \text{ min}$$

2.3 Consider an urban area where 100 000 people travel to work; assume you possess the following information about them:
i) General information:

Mode	Average number of cars per household	Family income (1000 $/year)
Car	2.40	120
Underground	1.60	60
Bus	0.20	10
Total	0.55	25

ii) Population distribution

	Cars per household			
Family income (1000 $/year)	0	1	2+	Total
Low (<25)	63.6	15.9	0.0	79.5
Medium (25–75)	6.4	3.7	2.4	12.5
High (>75)	0.0	2.4	5.6	8.0
Total	70.0	22.0	8.0	100.0

You are required to collect a sample of travellers to estimate a series of models (with a maximum of eight parameters) which guarantee a negligible specification error if you have available at least 50 observations per parameter. You are also assured that if you take a 20% random sample of all travellers the error will be negligible and there will be no bias.

Your problem is to choose the most convenient sampling method (random, stratified or choice-based), and for this you have available also the following information:

Hourly cost of an interviewer	$2 per hour
Questionnaire processing cost	$0.3 per form
Time required to classify an interviewee	4 min
Time required to complete an interview	10 min

You are also given the following table containing recommended choice-based sample sizes:

Subpopulation size	% to be interviewed
<10 000	25
10 000–15 000	20
15 000–30 000	15
30 000–60 000	10
>60 000	5

2.4 Consider the following results of having collected stratified (based on income I) and choice-based samples of a certain population:

Stratified sample

Mode	Low I	Medium I	High I
Car	3.33	18.00	20.00
Bus	33.34	7.20	4.00
Underground	3.33	4.80	6.00
Total	40.00	30.00	30.00

Choice-based sample

Mode	Low I	Medium I	High I	Total
Car	6.67	20.00	13.33	40.00
Bus	17.24	2.07	0.69	20.00
Underground	16.67	13.33	10.00	40.00

(a) If you know that the income-based proportions in the population are 60%, 25%, and 15%, respectively for low, medium, and high income, find an equivalent table for a random sample. Is it possible to validate your answer?

(b) Compute the weighting factors that would be necessary to apply to the observations in the choice-based sample in order to estimate a model for the choice between car, bus, and underground using standard software (i.e. that developed for random samples).

3

Zones and Networks

One of the most important early choices facing the transport modeller is the level of detail (granularity) to be adopted in a study. This problem has many dimensions: they refer to the schemes to be tested, the type of behavioural variables to be included, the treatment of time, and so on. This chapter concentrates on design guidelines for two of these choices: zoning system and network definition.

We shall see that in these two cases, as in other key elements of transport modelling, the final choices reflect a compromise between three conflicting objectives: accuracy/precision, forecasting ability, and cost.

In principle, greater accuracy could be achieved by using a more detailed and precise zoning and network system; in the limit, this would imply recognising each individual household, its location, distance to access points to the network, and so on. With a large enough sample (100% rate over several days), the representation of the current system could be made very accurate indeed. However, the problem of stability over time weakens this vision of accuracy as one would need to forecast, at the same level of detail, changes at the individual household level that would affect transport demand. It is simply not possible to forecast the locations and characteristics of each person and household 10 or 30 years from now; any such forecasts would be full of errors thus weakening the accuracy of any model projection. On top of this is the issue of costs; very granular data are generally more expensive to collect, check, clean, and validate and, also, more expensive to update when necessary.

It is also possible to set up a synthetic population (households and individuals) based on a recent Census and Household Mobility Survey. However, privacy considerations require that no personal identifiable information exists in this synthetic population. As this would be an extrapolation from, say, a 2% sample in the survey, it would be controlled by more aggregate totals based on census tracts that cannot, again, provide personal identifiable information. This approach provides high precision but not necessarily accuracy. We discuss the generation and use of synthetic populations in greater detail in Chapter 16.

Something similar takes place when considering travel networks. As we know from experience, it is possible to have a full detail of the road and public transport networks of a city, including timetables and real-time information. These details include pedestrian and cycle lanes as well as information of importance to users with lower mobility levels (ramps, stairs, lifts, and gradients). However, using this very detailed network for all stages in a model imposes significant overheads, not just in computer running time (this is a matter of adding more cores) but in terms of cleaning, updating, validating, and interpreting the results.

Modelling Transport, Fifth Edition. Juan de Dios Ortúzar and Luis G. Willumsen.
© 2024 John Wiley & Sons Ltd. Published 2024 by John Wiley & Sons Ltd.
Companion website: www.wiley.com/go/ortuzar5e

An increasingly common approach to these two difficulties is to establish hierarchical zoning and network systems; these are maintained at the highest level of detail but used only at the level commensurate with the modelling and forecasting task in hand; think how uncertain it must be any detailed description of the networks 10–30 years from now.

One can also have a different zoning system for different sub-models. For example, the *demand* models (Generation, Distribution, Time of day travel, and Mode choice) can be run with a coarser zoning system, but the Assignment stage is usually run with a more detailed zoning and network system. For example, in 2017, London had a mesoscopic model with some 5400 zones and 25 500 simulated junctions; its strategic transport model used 1300 demand zones and 4100 assignment zones with some 100 000 highway links and 250 000 public transport links. Zoning and network systems evolve quickly as demands for precision, travel demand management, and computer power increase.

3.1 Zoning Design

A zoning system is used to aggregate the individual households and premises into manageable chunks for modelling purposes. The main two dimensions of a zoning system are the number of zones and their size. The two are, of course, related. The greater the number of zones, the smaller they can be to cover the same study area. It has been common practice in the past to develop a zoning system specifically for each study and decision-making context. This is obviously wasteful if one performs several studies in related areas; moreover, the introduction of different zoning systems makes it difficult to use data from previous studies and compare modelling results over time.

The first choice in establishing a zoning system is to distinguish the study area itself from the rest of the world. The following ideas may help in making this choice:

1) In choosing the study area, one must consider the decision-making context, the schemes to be modelled, and the nature of the trips of interest: mandatory, optional, long or short distance, and so on.
2) For strategic studies, one would like to define the study area so that the majority of trips have their origin and destination inside it; however, this may not be possible for the analysis of transport problems in smaller urban areas where the majority of trips of interest are through-trips and a bypass is considered.
3) Similar problems arise with traffic management studies in local areas where, again, most trips will have their origin, destination, or both, clearly outside the area of interest. What matters in these cases is whether it is possible to model changes to these trips arising as a result of new schemes.
4) The study area should be somewhat bigger than the specific area of interest covering the schemes to be considered. Opportunities for re-routeing, changes in destination, and so on, must be allowed for; we would like to model their effects as part of the study area itself.

The region external to the study area is normally divided into a number of *external* zones. In some cases, it might be enough to consider each of these to represent 'the rest of the world' in a particular direction; the boundaries of these different slices of the rest of the world could represent the natural catchment areas of the transport links feeding into the study area. In other cases, it may be advantageous to consider external zones of increasing size with the distance to the study area. This may help in the assessment of impacts over different types of travellers (e.g. long- and short-distance).

The study area itself is divided into smaller *internal* zones. Their number will depend on a compromise between a series of criteria discussed below. For example, the analysis of traffic management schemes will generally call for smaller zones, often representing car parks or major generators/attractors of trips. Strategic studies, on the other hand, will often be carried out on the basis of much larger zones. Examples of zone numbers chosen for various studies in the past are presented in Table 3.1.

Table 3.1 Typical zone numbers for studies

Location	Population	Number of zones	Comments
Abu Dhabi (2021)	2.8 million	~3200	Strategic and mesoscopic
Bogotá (2015)	7.4 million	~900	Normal zones
Dallas-Forth Worth (2004)	6.5 million	~4900	Including 61 external zones
Dublin (2010)	1.7 million	~650	Including ~10 000 road links
London (2011)	8.9 million	~1300	Demand zones
		~5400	Assignment zones
Melbourne (2009)	5.1 million	~3500	Normal zones
Santiago (2018)	6.5 million	~1000	Normal zones
Sydney (2006)	3.6 million	~2700	Normal zones
Washington DC (2008)	6.5 million	~2200	Normal zones
		~460	Coarse zones

As can be seen, there is a wide variety of models and number of zones per million inhabitants. These vary depending on the nature of the model (tactical and short-term planning vs strategic and long-term), the resources available, and the particular focus or set of problems addressed.

Note also that even if a more aggregate zoning system is used for part of the model system, for example, trip generation and distribution, a finer system of zones is often used for mode choice and assignment: these are often referred to as *Traffic Assignment Zones* or TAZ.

Zones are represented in computer models as if all their attributes and properties were concentrated in a single point called the *zone centroid*. This notional spot is best thought of as floating in space and not physically on any location on a map. Centroids are attached to the network through *centroid connectors*, representing the average costs (time, distance) of joining the transport system for trips with origin or destination in that zone. The nodes selected to connect with each centroid are nearly as important as the costs associated with their centroid connectors. These should be close to natural access/egress points for the zone itself; this means that they should not connect directly to major intersections and additional mid-link nodes may be needed. In the case of public transport networks, the centroid should connect to a station or stop of the service. Locating centroids automatically at the centre of gravity of each zone and measuring their distance to key nodes to produce centroid connectors is a quick fix valid only for the simplest 'first cut' network runs. As zones represent the natural catchment area of the transport networks, their centroid connectors should represent the mean costs to access them.

Centroids and centroid connectors play a key role in the quality of the rest of the models, but their definition and coding do not follow a strict and objective approach; they rely a good deal on the experience of the modeller. The centroid connector influences the route followed to load trips onto both the road and public transport networks and, therefore, affects the total cost of travelling from origin to destination, and all the models that include them. In fact, a key task in applied modelling is the *network calibration* stage, where checks are made as to the reasonableness of the set of connectors to and from each zone, and if they allow reasonable routeing for all trips in an observed O-D trip matrix.

The following is a list of zoning criteria that has been compiled from experience in several practical studies:

1) Zoning size must be such that the aggregation error caused by assuming that all activities are concentrated at the centroid is not too large. It might be convenient to start postulating a system with many small zones, as this may be aggregated in various ways later depending on the nature of the projects to be evaluated.
2) The zoning system must be compatible with other administrative divisions, particularly with Census tracts; this is probably the fundamental criterion and the rest should only be followed if they do not lead to inconsistencies with it.
3) Zones should be as homogeneous as possible in their land use and/or population composition; Census tracts with clear differences in this respect (i.e. residential sectors with vastly different income levels) should not be aggregated, even if they are very small.
4) Zone boundaries must be compatible with cordons and screen lines and with those of previous zoning systems. However, it has been found in practice that the use of main roads as zone boundaries should be avoided, because this increases considerably the difficulty of assigning trips to zones, when these originate or end at the boundary between two or more zones.
5) Zones do not have to be of equal size; if anything, they could be of similar dimensions in travel time units, therefore generating smaller zones in congested than in uncongested areas.
6) Bear in mind that computer execution time for most transport sub-models is proportional to the square of the number of zones.

It is advantageous to develop a hierarchical zoning system, as in the London Transportation Studies, where subzones are aggregated into zones that, in turn, are combined into districts, traffic boroughs, and finally sectors. This facilitates the analysis of different types of decisions at the appropriate level of detail. Hierarchical zoning systems benefit from an appropriate zone-numbering scheme, where the first digit indicates the broad area, the first two the traffic borough, the first three the district, and so on.

3.2 Road Network Representation

The transportation network is deemed to represent a key component of the supply side of the modelling effort, that is, what the transport system offers to satisfy the movement needs of trip makers in the study area. The description of a transport network in a computer model can be undertaken at different levels of detail and requires the specification of its structure, its properties or attributes, and the relationship between those properties and traffic flows. A basic understanding of traffic flow theory helps in deciding the best representation of road networks in a model.

3.2.1 Traffic Flow

Traffic behaves in a complex and nonlinear way, depending on the interactions among vehicles and with traffic control devices, traffic lights, stop signs, etc. It is useful to distinguish two different flow regimes:

- free-flowing traffic, as observed on a motorway operating below capacity, and
- interrupted flow, as observed whenever at-level junctions are present; interruptions may also be due simply to high levels of traffic close to capacity.

Due to the individual reactions of human drivers, vehicles display cluster formations (*platoons*) and shock wave propagation, both forward and backward, depending on vehicle density.

In a free-flowing link, the key variables are speed, flow, and concentration/density. Flow conditions are found to be stable and free-flowing when the density is well below 80 vehicles per km per lane. As density reaches the maximum flow rate (capacity) and exceeds the optimum density (as more vehicles attempt to join the stream), traffic flow becomes unstable, and even a minor incident can result in stop-and-start driving conditions. The maximum or *jam density* refers to extreme conditions when traffic flow stops completely, usually in the range of 115–160 vehicles per km per lane.

When traffic is operating below capacity, the effect of flow on speeds can be represented by a simple speed-flow mathematical relationship. However, when traffic is operating as interrupted flow, the relationship between flow and delay becomes more complex, queues are formed and dispersed, and a speed-flow relationship is only a coarse approximation.

Good grade-separated junctions allow long weaving sections so that vehicles can move across lanes without causing significant interference to the flow. However, this is expensive and takes up a large amount of land seldom available in urban areas. At-grade junctions require rules specifying priorities or traffic lights giving way to traffic streams in a sequence.

The maximum flow per unit of time that can clear a signal-controlled approach to a junction during the green phase is called *saturation flow*; although this can be measured, it is often estimated. The capacity of an approach controlled by traffic lights is the saturation flow (vehicles/second) times the length of the green phase. When demand is higher than this capacity, queues are formed and whenever this happens delays depend on when each vehicle joins the queue and how quickly demand becomes less than capacity. Some traffic light systems are able to respond to traffic demand by adjusting the length of the green phases in real time to reduce overall delay. This traffic responsiveness can be provided at isolated junctions or at a network of junctions (e.g. SCAT and SCOOT, see Heydecker 2004).

Most transport models assume *steady-state* conditions, that is regular uniform flow, speeds, and delays during modelled periods. In reality, traffic is much more variable with significant changes in flows during an hour and subsequent variations in delays and queue formation. These can only be handled well by micro-simulation models.

3.2.2 Network Details

The transport network may be represented at different levels of aggregation in a model. At one extreme, there are models with no specific links at all; they are based on continuous representations of transport supply (Smeed 1968). These models may provide, for example, a continuous equation of the average traffic capacity per unit of area instead of discrete elements or links. At a slightly higher level of disaggregation, one can consider individual roads but include speed-flow properties taken over a much larger area; see, for example, Wardrop (1968).

Normal practice, however, is to model the network as a *directed graph*, that is, a system of nodes and links joining them (see Larson and Odoni 1981 or Willumsen 2007), where most nodes are taken to represent junctions and the links stand for homogeneous stretches of road between junctions. Links are characterised by several attributes, such as length, speed, number of lanes, and so on, and are normally unidirectional; for example, a single two-way link will be converted into two one-way links in the internal computer representation of the network. A subset of nodes is associated with zone centroids, and a subset of links to centroid connectors. Normally, the principal source of network data should be one of the many digital maps available for most cities. One should not assume, however, that they are error-free. They will need checking, updating, pruning (to focus on the network of interest), and complementing with observations on items like on-street parking, pedestrian friction, bus lanes, and other features that may affect their performance. It is not unusual to find one or two links coded in the wrong direction.

A simple network configuration of this type is presented in Figure 3.1. Note, however, that at least zones 2, 3, and 5 are (wrongly) connected to junctions, violating an earlier recommendation.

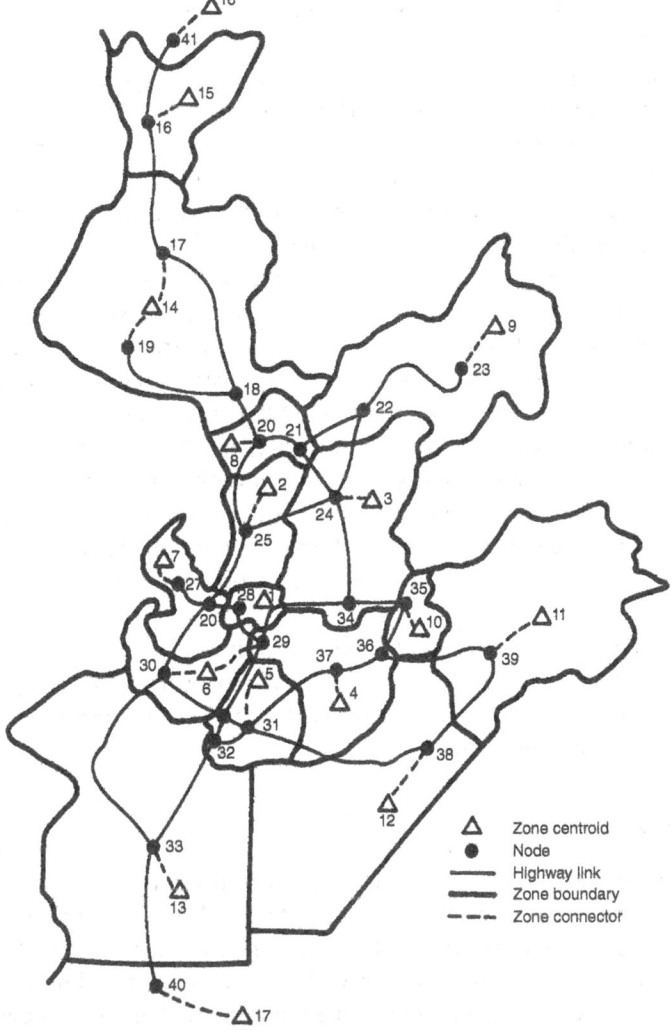

Figure 3.1 A road network coded as nodes and links

A problem with this scheme is that 'at-node' connectivity is offered to each link joining it at no cost. In practice, some turning movements at junctions may be more difficult to perform than others; indeed, some turning movements may not be allowed at all. To represent these features of real-road networks better, it is possible to penalise and/or ban some turning movements. This can be done manually, by expanding the junction providing separate (sometimes called dummy) links for each turning movement, and associating a different cost to each. Alternatively, some commercial computer programs are capable of performing this expansion in a semi-automatic way, following simple instructions from the user about difficult or banned movements.

The level of disaggregation can be increased further when detailed traffic simulation models are used. In these cases, additional links are used at complex junctions to account for the performance of reserved lanes, give-way lines, and so on.

Sometimes networks are subsets of larger systems; they may be cordoned off from them, defining access or cordon points where the network of interest is connected to the rest of the world. These points are sometimes called *gateways* and dummy links may be used to connect them to external zones.

A key decision in setting up a network is how many levels to include in the road hierarchy. If more roads are included, the representation of reality should be better; however, there is again a problem of economy versus realism, which forces the modeller to select some links for exclusion. Moreover, it does not make much sense to include a large number of roads in the network and then make coarse assumptions about turning movements and delays at junctions. It is not sensible either to use a detailed network with a coarse zoning system as then *spatial aggregation* errors (i.e. in terms of centroid connections to the network) will reduce the realism of the modelling process. This is particularly important in the case of public transport networks, as we will see in Chapter 10. What matters is to make route choices and flows as realistic as possible within the limitations of the study, bearing in mind the level of accuracy expected from future year networks.

Jansen and Bovy (1982) investigated the influence of network definition and detail over road assignment accuracy. Their conclusion was that the largest errors were obtained at the lower levels in the hierarchy of roads. Therefore, one should include in the network at least one level below the links of interest: for example, in a study of A (trunk) roads, one should also include B (secondary) roads.

In the case of public transport networks, an additional level of detail is required. The modeller must specify the network structure corresponding to the services offered. These will be coded as a sequence of nodes visited by the service (bus, rail), normally with each node representing a suitable stop or station. Junctions without bus stops can, therefore, be excluded from the public transport network. Two types of extra links are often added to public transport networks. These are walk links, representing the parts of a journey using public transport made on foot, and links to model the additional costs associated with transferring from one service (or mode) to another.

3.3 Link Properties and Functions

3.3.1 Link Properties

The level of detail provided about the attributes of links depends on the general resolution of the network and on the type of model used. At the very minimum, the data for each link should include its length, its travel speed (either free-flow speed or an observed value for a given flow level), and the capacity of the link, usually in passenger car units (pcu) per hour.

In addition to this, a cost-flow relationship is associated with each link as discussed below. In some cases, more elaborate models are used to relate delay to traffic flow, but these require additional information about links, for example:

- type of road (e.g. expressway, trunk road, local street);
- road width or number of lanes, or both;
- an indication of the presence or otherwise of bus lanes, or prohibitions of use by certain vehicles (e.g. lorries);
- tolls or other user charges;
- banned turns, or turns to be undertaken only when suitable gaps in the opposing traffic become available, and so on;
- type of junction and junction details including signal timings, and
- storage capacity for queues and their presence at the start of a modelling period.

Some research results have identified other attributes of routes as important to drivers, for example signposting and fuel consumption (see Outram and Thompson 1978; Wootton et al. 1981). Work in The Netherlands has shown that (weighted) time and distance explain only about 70% of the routes actually chosen. The category of the road (motorway, A road, B road), the predictability of the time taken, scenic quality, traffic signals, and capacity, are attributes that may help to explain the choice of additional routes. As our understanding of how these attributes influence route choice improves, we will be able to develop more accurate assignment models. The counterpart of this improvement will be the need to include other features of roads, like their scenic quality, number of junctions of each type, and so on.

In the case of public transport networks, the choice of route may be combined with elements of sub-mode choice (i.e. a Bus–metro combination), and further elements of importance may need to be incorporated. For example, recent work in this area has shown the importance of the *angular cost* in deciding the best route in a public transport network (Raveau et al. 2011).

3.3.2 Network Costs

Most current assignment techniques assume that drivers seek to minimise a linear combination of time and distance, the latter as a proxy for *vehicle operating costs* (VOC), sometimes referred to as *generalised cost* for route choice. This is known to be a simplifying assumption as there may be differences not only in the perception of time but also about its relative importance compared with other route features. The majority of network models in use today deal with travel time and distance plus tolls/road user charges if present.

When modelling travel time as a function of flow, one must distinguish two different cases. The first is when the assumption can be made that delays on a link depend only on the flow on the link itself; this is typical of long links away from junctions and, therefore, it has been used in most inter-urban assignment models so far. The second case is encountered in urban areas where the delay on a link depends in an important way on flows on other links, for example for non-priority traffic at a give-way or roundabout junction.

The introduction of general flow-delay formulations is not difficult until one faces the next issue, equilibration of demand and supply. There, the mathematical treatment of the first case (often called the *separable cost function* case) is simpler than the second; however, there are now techniques for balancing demand and supply in the case of link-delay models depending on flows on several links (i.e. when the effect of each link flow cannot be separated).

3.3.3 Definitions and Notation

Some further notation will be introduced as required but the basic elements used in this chapter are:

T_{ijr} is the number of trips between i and j via route r.

V_a is the flow volume on link a in vehicles per hour (vph), or passenger car units (pcu) per hour, where typically a bus is equivalent to between 2 and 3 pcu, and trucks between 3 and 4 pcu.

$C(V_a)$ is the cost-flow relationship for link a.

$c(V_a)$ is the actual cost for a particular level of flow V_a; in particular, the cost when $V_a = 0$ is referred to as *free-flow* cost.

c_{ijr} is the cost of travelling from i to j via route r.

$$\delta_{ijr}^a = \begin{cases} 1 & \text{if link } a \text{ is on path (or route) } r \text{ from } i \text{ to } j \\ 0 & \text{otherwise} \end{cases}$$

A superscript n will be used to indicate a particular iteration in iterative methods. A superscript $*$ will be used to indicate an optimum value, for example, c_{ij}^* is the minimum cost of travelling between i and j.

In many cases, it is important to recognise that there are different road users that may display different behaviour on the same link. Therefore, we further introduce an additional index (usually u) for *user class*. It is possible to have different user classes for each vehicle type (car, bus, truck) and for different types of drivers as a function of their journey purpose, willingness to pay (income) tolls and parking, and other characteristics relevant to the study, for example, the level of emissions of the vehicle.

3.3.4 Speed-Flow and Cost-Flow Curves

A familiar relationship in traffic engineering relates the speed on a link to its flow. This concept was originally developed for long links in motorways, tunnels, or trunk roads. A speed-flow relationship is usually presented as shown in Figure 3.2; as flow increases, speed tends to decrease after an initial period of little change; when flow approaches *capacity* the rate of reduction in speed increases. Maximum flow is obtained at capacity and when attempts are made to force traffic volumes beyond this value an unstable region with low flows and low speeds is reached.

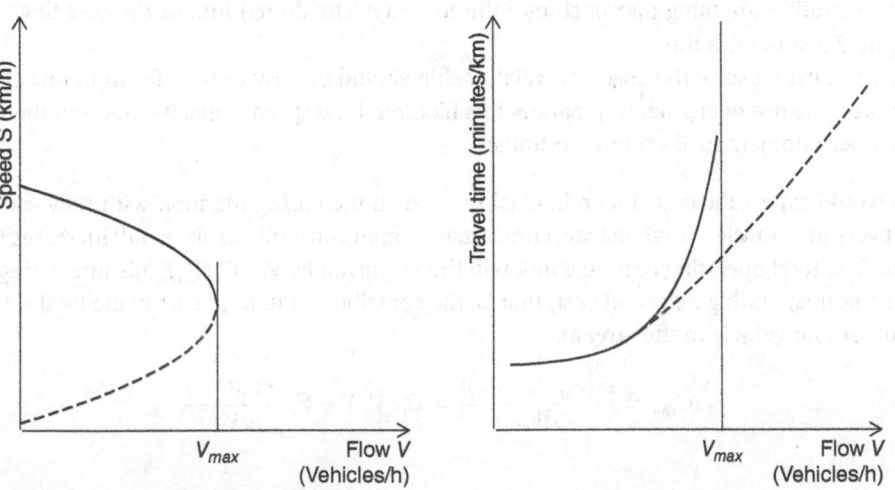

Figure 3.2 Typical speed-flow and cost-flow relationship for a long link

For practical reasons, in traffic assignment this type of relationship is handled in terms of travel time per unit distance versus flow, or more generally, as a travel time or cost-flow relationship, as also shown in Figure 3.2. Traffic assignment methods considering congestion effects need a set of suitable functions relating link attributes (capacity, free-flow speed) and the flow on the network with the resulting speeds or costs. This can be written in general terms as:

$$C_a = C_a(\{\mathbf{V}\}) \tag{3.1}$$

that is, the cost on link a is a function of all flows $\mathbf{V}$ in the network (i.e. not just the flow on the link itself). This general formulation is relevant in urban areas where there is a good deal of interaction between flows on different links and their corresponding delays, for example at priority junctions or roundabouts. However, this can be simplified if one considers long links, that is, links where most of the travel time takes place on the link rather than at the end junctions. In this case, the function is said to be *separable* and we can write:

$$C_a = C_a(V_a) \tag{3.2}$$

that is, the cost on the link depends just on its flow and the link characteristics. This assumption simplifies the estimation of these functions and the development and use of suitable trip assignment techniques. It must be recognised, however, that it becomes less realistic as one works with denser and more congested urban areas.

A number of general functional forms have been proposed to embody the general relationship in Eq. (3.2). The fact that our main concern in this section is traffic assignment permits us to concentrate on a smaller set of these functions, in particular those with good mathematical properties. The following are desirable properties from the point of view of traffic assignment:

1) Realism: The modelled travel times should be realistic enough.
2) The function should be nondecreasing and monotone; increasing flow should not reduce travel time. This is not only reasonable but also desirable, as we shall see below.
3) The function should be continuous and differentiable.
4) The function should allow for the existence of an overload region, that is, it should not generate infinite travel time, even when flow is equal or greater than capacity. This may happen as part of an iterative process when more traffic is assigned to a link than its capacity; a high positive value for travel time should be produced but infinity will generate overflow in computer programs, an undesirable occurrence. Moreover, short-term overload can certainly happen in practice without generating anything approaching infinite delay! The dotted line in the cost-flow curve in Figure 3.2 simulates this.
5) For practical reasons, the cost-flow relationship should be easy to transfer from one context to another; the use of engineering parameters like free-flow speed, capacity, and number of junctions per kilometre is therefore desirable.

One would expect the cost-flow relationship to be an increasing function with flow, except perhaps at very low flow levels when travel times may remain constant despite small increases in traffic volume. The total operating cost on a link will then be given by $V_a \cdot C_a(V_a)$; it is interesting to consider the corresponding marginal cost, that is, the contribution to total cost made by the marginal addition of one vehicle to the stream:

$$C_{ma} = \frac{\partial [V_a \cdot C_a(V_a)]}{\partial V_a} = C_a(V_a) + V_a \frac{\partial C_a(V_a)}{\partial V_a} \tag{3.3}$$

On the right-hand side, we have two terms, the first one corresponding to the average cost on the link and the second to the contribution to the delay of other traffic made by the marginal vehicle. This is an external effect and corresponds to the additional costs incurred by other users of the link when a new car is added to it. As the cost-flow curve is an increasing one, this contribution is always greater than zero. It is also clear that in economic terms the average and marginal costs will only be the same in the flat part of the cost-flow curve, if any.

A number of authors have suggested functional forms for cost-flow relationships. These usually rely on the assumption that one is trying to model steady-state conditions and some kind of average behaviour. Branston (1976) has produced a good review of the practical problems encountered when trying to calibrate these cost-flow functions:

- there are problems with the length of the observation period, in particular in congested areas and where an upstream junction acts as bottleneck; the exact location of flow and delay measuring areas plays a critical role in determining the quality of the results obtained;
- the assumption that delays depend only on flow on the link itself is unrealistic in most dense urban networks and this is particularly critical in trying to estimate cost-flow functions.

Branston (1976) also reviewed cost-flow curves proposed by other authors. Some of them have more historical than practical interest:

1) Smock (1962) for the Detroit Study:

$$t = t_0 \exp (V/Q_s) \tag{3.4}$$

where t is travel time per unit distance (min/km), t_0 is travel time per unit distance under free-flow conditions, and Q_s is the steady-state capacity of the link.

2) Overgaard (1967) generalised (3.4) as follows:

$$t = t_0 \alpha^{\beta(V/Q)} \tag{3.5}$$

where Q is the capacity of the link, and α and β are parameters for calibration.

3) The Bureau of Public Roads (1964) in the USA proposed what is probably the most commonly used function of this type to this day, the BPR curve:

$$t = t_0 \left[1 + \alpha (V/Q)^\beta\right] \tag{3.6}$$

4) The Department of Transport in the UK has produced a large number of cost-flow curves for a variety of link types in urban, sub-urban, and inter-urban roads. Some have a general form that considers first the speed-flow $s(V)$ curve:

$$s(V) = \begin{cases} S_0 & V < F_1 \\ S_0 - \dfrac{S_0 - S_1}{F_2 - F_1} (V - F_1) & F_1 \le V \le F_2 \\ S_1/[1 + (S_1/8d)(V/F_2 - 1)] & V > F_2 \end{cases} \tag{3.7}$$

where S_0 is the free-flow speed, S_1 is the speed at capacity flow F_2 (or Q), F_1 is the maximum flow at which free-flow conditions prevail, and d is the distance or length of the link.

Then the time-flow $T(V)$ relationship becomes:

$$T(V) = \begin{cases} d/S_0 & V < F_1 \\ d/S(V) = \dfrac{d}{S_0 + SS_{01} F_1 - SS_{01} V} & F_1 \le V \le F_2 \\ d/S_1 + (V/F_2 - 1)/8 & V > F_2 \end{cases} \tag{3.8}$$

with SS_{01} given by:

$$SS_{01} = \frac{S_0 - S_1}{F_1 - F_2} \qquad (3.9)$$

Typical values for these coefficients (Department of Transport 1985) are given in Table 3.2. In some cases, a cut-off point in speed reductions is assumed; for example, the speed may be assumed to remain at F_2 for $V > F_2$.

Table 3.2 Typical speed-flow curve coefficients in the UK

Type	S_0 (km/h)	S_1 (km/h)	F_1 (pcu/h/lane)	F_2 (pcu/h/lane)
Single 2 lane, rural	63	55	400	1400
Dual 2 lane, rural	79	70	1600	2400
Single 2 lane, urban, outer area	45	25	500	1000

5) **Akçelik function.** All the functions mentioned above tend to underestimate delays at junctions as they concentrate on the links characteristics. Moreover, they also tend to underestimate delays when demand is close or above the capacity of the link. They are less appropriate in urban conditions where junctions play a more important role in determining travel times than the speed mid-link. Akçelik (1991) has suggested a curve, based on earlier work by Davidson, which tackles these problems better. When considering conditions close or above saturation, the length of the modelling period matters considerably as it influences the length of the 'overflow' curve which in turn drives delay. Akçelik's function applies to volume capacity ratios above and below one:

$$t = t_0 + \left\{ 0.25T \left[(x-1) + \sqrt{(x-1)^2 + \frac{8J_A}{Q_j T} x} \right] \right\} \qquad (3.10)$$

where t_0 is travel time under free-flow conditions (hours/km); T is the flow modelling period (hours); Q_j is the capacity at the junction (vehicles/h); if the saturation flow is Q_s, then $Q_j = Q_s g/cy$; g is the length of the green period at the junction and cy is the cycle length in the same units; x is the degree of saturation $= V/Q_j$, and J_A is a delay parameter.

In principle, there is no upper limit on the value of x that could be input above, since this equation is designed to approximate delays due to queuing when demand exceeds capacity. The equation explicitly takes into account such delays and may be applied to any facility type. The assumptions are that there is no queue at the start of the analysis period, and there is no peaking of demand within the analysis period (T).

The delay parameter J_A is a function of the number of delay-causing elements in the section of road and the variability of the demand. Akçelik suggests lower values of J_A for freeways and coordinated signal systems. Higher values would apply to secondary roads and isolated intersections.

The value of J_A can be computed if the difference in the rate of travel (hours per km) between capacity and free-flow conditions on the facility is known. Substituting $x = 1$ in the above equation and solving for J_A yields:

$$J_A = \frac{2Q_j}{T} (t_c - t_0)^2 \qquad (3.11)$$

where t_c is the rate of travel at capacity (hours per km).

All the above speed or cost-flow curves produce information about travel time on a link. However, it is recognised that most users might wish to minimise a combination of link attributes including time and distance. Conventional practice recommends the use of a simplified version of the generalised cost concept, namely a linear weighted combination of time and distance:

$$C_a = \alpha(\text{travel time})_a + \beta(\text{link distance})_a \qquad (3.12)$$

This cost could be measured in generalised time or generalised money units. It is also possible to include an out-of-pocket expenditure element, for example, a toll to be applied on a given link.

The calibration of cost-flow relationships is time-consuming and requires a good deal of high-quality data (i.e. observations of travel times on links under different flow levels). For this reason, this is rarely attempted and many countries have developed their own standard functions for typical roads. See also the limitations of link-based cost-flow functions in urban areas as discussed in Section 10.7.

Suh et al. (1990) put forward an approach to estimate cost-flow curves based on traffic counts; they used a bi-level optimisation method that, in essence, seeks to establish the parameters for the cost-flow curves minimising a measure of difference between assigned and observed flows. The value of this approach is limited by the errors in the assignment process as discussed, again, in Section 10.7: for example, errors in the network, trip matrix, in the assumption of perfect information, and that all users perceive link costs in the same way. Thus the estimated cost-flow curves incorporate these errors and are, therefore, difficult to transfer to other areas or even schemes.

3.3.5 Public Transport Networks

Public transport networks are more complex than road networks. They require an identification of the route taken by each service as a unique sequence of links. It is also necessary to identify the locations of stops and also of those nodes where an interchange with other services is permissible. The frequency of the service, and in some cases the actual timetable and the fare, must also be specified and included in the network description. Access to stops may be on foot or by another mode and this can be represented by centroid connectors in the simpler models and by one or more auxiliary networks of access modes in more realistic ones. This is why the centroid connectors for public transport are always different from those used for the road network.

The public transport network could be entirely independent of the road network as in the case of most rail and metro services; alternatively, the speed of the service may be affected by road traffic as in the case of buses and on-street running of trams, even when priority measures to support them help to reduce the impact of road congestion.

In addition to the effect of road congestion, it is sometimes necessary to account for the issue of passenger congestion: crowding on buses and trains leading to discomfort and even having to miss a service because it was full and impossible to board.

Public-transport services can also become congested, and this can take two forms. In the first instance, the density of passengers in a unit may be such that it is not possible to find a seat, and may even become uncomfortable to stand so close to other passengers. As with traffic congestion, criteria that define the unacceptable crowding levels vary across regions. Li and Hensher (2013) state that four standees per m^2 is the benchmark for Europe and Australia, five for the US, and even eight per m^2 for China's bus sector. In several countries of Latin America, six passengers per m^2 is a usual maximum.

Exercises

3.1 Take the BPR function and using a spreadsheet plot and compare two versions of the curve for a road with free-flow speed of 120 km/h. Case 1 has $\alpha = 2$ and $\beta = 3$ and Case 2 with $\alpha = 2$ and $\beta = 6$; plot the curves for Flow to Capacity ratios of 0.1–1.3 and comment on their likely suitability for a long link in motorway.

3.2 Take the same two BPR curves in Exercise 3.1 but now the free-flow speed is only 60 km/h and comment on their suitability for an urban road.

3.3 Take now the Akçelik function (3.10) and plot it for a range of degrees of saturation x and for an urban road under the following conditions: free-flow speed 50 km/h, Q_s is 2000 vehicle/h, green length $g = 45$ s, cycle length $cy = 90$ s and $J_A = 0.7$. Plot it for two modelling periods, one and two hours, and compare the results commenting in their suitability for two medium size cities, such as Montevideo in Uruguay and Adelaide in Australia.

4

Trip Generation Modelling

As we saw in Chapter 1, the trip generation stage of the classical transport model aims at predicting the total number of trips generated by each zone (O_i) and the total attracted to each zone (D_j) in the study area. This stage determines most of the mobility activity in the study area; missing from this mobility will be the external trips (External–Internal, I–E, and E–E trips), which are normally treated separately. This level of mobility is often estimated, in aggregate models, at the level of trips; however, it could also be estimated at the level of (simplified) tours with some advantages in realism as we discuss later in this chapter.

The estimation of the future number of trips can be achieved in a number of ways, but most models are estimated on the basis of information about the trips made and the characteristics of the households in each zone. The subject has also been viewed as a *trip frequency* choice problem: how many shopping (or other purpose) trips will be carried out by a certain person type during a representative week? This is usually undertaken using discrete choice models, as discussed in Chapters 7–9, and it is then cast in terms like: what is the probability that this person type will undertake zero, one, two, or more trips with this purpose per week or day?

In this chapter, we concentrate on the first approach, that is, to estimate the total number of trips generated and attracted by each zone using data about household socio-economic attributes; we also give a glimpse about the second, which has advantages in some studies.

We will start by defining some basic concepts and will proceed to examine some of the factors affecting the generation and attraction of trips. Then we will review the main modelling approaches, starting with the simplest growth-factor technique. Before embarking on more sophisticated approaches, we present a reasonable review of linear regression modelling, which is the basic tool for these models.

We will then consider zonal and household-based linear regression trip generation models, giving some emphasis to the problem of non-linearities, which often arise in this case. We will also address, for the first time in this book, the problem of aggregation (e.g. obtaining zonal totals), which has a trivial solution here precisely because of the linear form of the model. Then we will move to cross-classification models, where we will examine not only the classical category analysis specification but also more modern approaches. We will also consider the estimation of the number and characteristics of *tours* generated by each person, household, or zone. We then examine the relationship between trip generation and accessibility including a short discussion on trip frequency models. The chapter ends with two short sections: the first discusses the problem of predicting future values for the explanatory variables in the models, and the second the problems of stability and updating of trip generation parameters.

Modelling Transport, Fifth Edition. Juan de Dios Ortúzar and Luis G. Willumsen.
© 2024 John Wiley & Sons Ltd. Published 2024 by John Wiley & Sons Ltd.
Companion website: www.wiley.com/go/ortuzar5e

4.1 Introduction

4.1.1 Some Basic Definitions

As always, definitions play an important role in our understanding of any phenomenon. Travel is no different. The question is, what is the basic event of interest for our models, activities involving a short period of stay in a location (sojourn), the displacement from one location to another (trip or journey), or a sequence of such trips and sojourns that starts and end at home and constitute an outing or a tour? What about tours that are not based at home? The modeller must bear all these options in mind while settling on an approach and the question is not academic. Focusing on activities enables the modeller to consider how different activities are shared and allocated among members of a household, something with a richer (albeit difficult to forecast) behavioural content. Focusing on trips is the simplest option, but may lead to inconsistencies when, for example, the return trip home is estimated as made by car when the individual cycled to work.

Trip or Journey. This is a one-way movement from a point of origin to a point of destination with a certain purpose. We are usually interested in all vehicular trips. Walking trips less than a certain study-defined threshold (say 300 metres or three blocks) have been often ignored as well as trips made by infants of less than five years of age. However, this has changed as greater attention is now normally paid to non-motorised trips and, as discussed in Chapter 2, due to the requirements of the recall activity framework recommended for mobility surveys.

Home-based (HB) Trip. This is one where the home of the trip maker is either the origin or destination of the journey. Note that for visitors from another city, for example, their hotel will act as a temporary home in most studies.

Non-home-based (NHB) Trip. This, conversely, is one where neither end of the trip is the home of the traveller.

Trip Production. This is defined as the home end of an HB trip or as the origin of an NHB trip (see Figure 4.1).

Trip Attraction. This is defined as the non-home end of an HB trip or the destination of an NHB trip (see Figure 4.1).

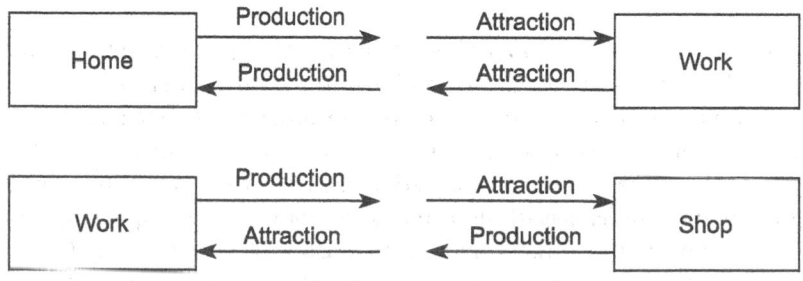

Figure 4.1 Trip productions and attractions

Trip Generation. This is often defined as the total number of trips generated by households in a zone, be they HB or NHB. This is what most models would produce and the task then remains of allocating NHB trips to other zones as trip productions.

Sojourn. A short period of stay in a particular location. It usually has a purpose associated with this stay: work, study, shopping, leisure, etc.

Activity. An endeavour or interest often associated with a purpose as above but not necessarily linked to a fixed location. One could choose to go shopping or to the cinema in different locations.

Tour or Trip Chain. A set of linked sojourns and trips. Concepts like activities, sojourns and tours correspond better to the idea of travel as derived demand (i.e. it depends strongly on the demand for other activities), but initially were used mainly by discrete choice modellers in practice (see Daly et al. 1983). Contemporary models, aggregate and disaggregate, tend to use tours rather than just trips.

4.1.2 Characterisation of Journeys

4.1.2.1 By Purpose

It has been found in practice that a better understanding of travel and trip generation models can be obtained if journeys by different purposes are identified and modelled separately. In the case of HB trips, a number of categories have been employed:

- travel to work: HBW;
- travel to school or college (sometimes distinguishing tertiary education): HBE;
- shopping trips: HBSh;
- social and recreational journeys: HBS;
- escort trips (to accompany or collect somebody else);
- other journeys: HBO.

The first two are usually called compulsory (or mandatory) trips and all the others are called discretionary (or optional) trips. The latter category encompasses all journeys made for less routine purposes, such as health and personal business (need to obtain a passport or a certificate). Note that social and cultural contexts may change the importance of different types of trips and, therefore, the most appropriate classification. NHB trips are sometimes separated into 'on business' and 'other' as the distinction is often needed for the appraisal of schemes and plans.

When dealing with tours it becomes necessary to distinguish a sub-set of simpler ones, for example, Home-Work-Home, Home-Work-Leisure-Home. In reality, people perform trips and tours with multiple purposes and activities at the same location, for example, Shopping and Social at a mall.

4.1.2.2 By Time of Day

Trips are sometimes classified into peak and off-peak period trips; the proportion of journeys by different purposes usually varies greatly with time of day. This type of classification, although important, gets more complicated when tours rather than trips are of interest, as a complete tour may comprise trips made at several times of the day.

Table 4.1 summarises data from the Greater Santiago 1977 O–D Survey (DICTUC 1978) as an example of good and bad traits; the morning (AM) peak period (the evening peak period is sometimes assumed to be its mirror image) occurred between 7:00 and 9:00 and the representative off-peak period was taken to be between 10:00 and 12:00.

Table 4.1 Example of trip classification

Purpose	AM peak		Off peak	
	No.	%	No.	%
Work	465 683	52.12	39 787	12.68
Education	313 275	35.06	15 567	4.96
Shopping	13 738	1.54	35 611	11.35
Social	7064	0.79	16 938	5.40
Health	14 354	1.60	8596	2.74
Bureaucracy	34 735	3.89	57 592	18.35
Accompanying	18 702	2.09	6716	2.14
Other	1736	0.19	2262	0.73
Return to home	24 392	2.72	130 689	41.65

Some comments are in order with respect to this table. Firstly, note that although the vast majority (87.18%) of trips in the AM peak are compulsory (i.e. either to work or education), this is not the case in the off-peak period. Secondly, a typical trait of a developing country emerges from the data: the large proportion of trips for bureaucratic reasons in both periods. Thirdly, a problem caused by faulty classification, or lack of forward thinking at the data-coding stage, is also clearly revealed: the *return to home* trips (which account for 41.65% of all off-peak trips) are obviously trips with another purpose; the fact that they were returning home is not as important as to why they left home in the first place. In fact, these data needed recoding in order to obtain adequate information for trip generation modelling (see Hall et al. 1987). This kind of problem used to occur before the concepts of trip productions and attractions replaced concepts such as origins and destinations, which did not explicitly address the generating capacity of home-based and non-home-based activities.

Tours can be classified by a combination of departure time and duration, together with an identifier of the main activities associated with them.

4.1.2.3 By Person Type

This is another important classification, as individual travel behaviour is heavily dependent on socioeconomic attributes. The following categories are usually employed:

- income level (e.g. three strata: low, middle, and high income);
- car ownership (typically three strata: 0, 1, and 2 or more cars);
- household size and structure (e.g. six strata in the classical British studies).

It is important to note that the total number of strata can increase very rapidly (e.g. 54 in the above example) and this may have strong implications in terms of data requirements, model calibration, and use. For this reason, trade-offs, adjustments, and aggregations are usually required (see the discussion in Daly and Ortúzar 1990).

4.1.3 Factors Affecting Trip Generation

In trip generation modelling we are typically interested not only in person trips but also in freight trips. For this reason, models for four main groups (i.e. personal and freight, trip productions, and attractions) have been usually required. In what follows we will briefly consider some factors which have been found important in practical studies. We will not discuss freight trip generation modelling, however, but postpone a discussion on the general topic of freight demand modelling until Chapter 15.

4.1.3.1 Personal Trip Productions

The following factors have been proposed for consideration in many practical studies:

- income;
- car ownership;
- family size;
- household structure;
- value of land;
- residential density;
- accessibility.

The first four (income, car ownership, household structure, and family size) have been considered in several household trip generation studies, while value of land and residential density are typical of zonal studies. The last one, accessibility, has rarely been used, although many studies have attempted to include it because it offers a way to make trip generation elastic (responsive) to changes in the transport system; we will come back to this issue in Section 4.5.

4.1.3.2 Personal Trip Attractions

For the journeys to work and education, the most obvious factors are employment and enrolment at educational establishments. For other purposes, the most widely used factor has been roofed space available for industrial, commercial, and other services. Some studies have attempted to incorporate an accessibility measure. However, it is important to note that in this case not much progress has been reported.

4.1.3.3 Freight Trip Productions and Attractions

These normally account for few vehicular trips; in fact, at most, they amount to 20% of all journeys in certain areas of industrialised nations, although they can still be significant in terms of their contribution to congestion. Important variables include:

- number of employees;
- sales (number of items or revenue);
- roofed area of firm;
- total area of firm.

Accessibility and type of firm are seldom considered as explanatory variables in urban studies; the latter is curious because it would appear logical that different products should have different transport requirements.

In urban areas, many freight movements are construction related, and, therefore, their origins and destinations change over time as buildings are raised in different parts of the city. One of the most difficult cases is that of delivery vans and couriers. They have variable tours depending on demand and how they will evolve in the future is difficult to predict. Demand has increased in recent years due to the popularity of internet shopping.

The growing importance of estimating the number of vehicular trips associated with pick up or delivery of supplies, and service activities such as any kind of repair technicians to both households and commercial establishments, is now well established (Holguín-Veras et al. 2017). Although these trips are a small portion of the total transport demand in metropolitan areas, they are important to the economy, the environment, and the quality of life. Commercial trips perform a crucial service to the economy by delivering the supplies needed by commercial establishments and households. In fact, deliveries to households represent the fastest-growing segment of transport demand in the world; in many places, the number of deliveries to households is larger than the number of deliveries to commercial establishments, although different vehicle types are used. Moreover, service activities and repairs typically take longer than deliveries creating a tough competition for parking spaces, sometimes resulting in double-parking (Holguín-Veras et al. 2021). Thus, efforts to forecast demand for suitable parking facilities for commercial vehicles can make a positive contribution to reducing congestion and emissions.

Recent research on freight and service trip generation indicates that traditional explanatory variables such as roofed area or the total area of the establishment are poor predictors of delivery and service trip generation (Holguín-Veras et al. 2017). Instead, it has been shown that it is better to model first the number of shipments sent and deliveries received (in the case of freight activity), as well as the number of service visits (in the case of services). The next step is to estimate the corresponding number of tours dividing these numbers by the average number of shipments and deliveries made on a single tour, and the average number of services that can be performed in a single tour. Among other benefits, decoupling the demand and the vehicle trips enable analysts to consider the effects of novel delivery technologies.

The correct model specification is critical, as some industry sectors receive and produce a constant number of deliveries and shipments, regardless of the number of employees (Holguín-Veras et al. 2011). In these cases, using a generation rate per employee will overestimate the freight trips for large establishments, and underestimate those for small establishments. Although freight and service activities are influenced by spatial factors (Sánchez-Díaz et al. 2016), these factors are difficult to account for in practical applications.

4.1.4 Growth-Factor Modelling

Since the early 1950s, several techniques have been proposed to model trip generation. Most methods attempt to predict the number of trips produced (or attracted) by a household or zone as a function of (generally linear) relations to be defined from available data. Prior to any comparison of results across areas or time, it is important to be clear about the following aspects mentioned above:

- what trips will be considered (e.g. only vehicle trips and walking trips longer than three blocks);
- what is the minimum age to be included in the analysis (i.e. five years or more).

In what follows we will briefly present a technique that may be applied to predict the future number of journeys by any of the categories mentioned above. Its basic equation is:

$$T_i = F_i t_i \qquad (4.1)$$

where T_i and t_i are respectively future and current trips in zone i, and F_i is a growth factor.

The only problem with the method is the estimation of F_i, the rest is trivial. Normally the factor F_i is related to variables, such as population (P), income (I), and car ownership (C), in a function such as:

$$F_i = \frac{f\left(P_i^d, I_i^d, C_i^d\right)}{f\left(P_i^c, I_i^c, C_i^c\right)} \qquad (4.2)$$

where f can even be a direct multiplicative function with no parameters, and the superscripts d and c denote the design and current years, respectively.

Example 4.1 Consider a zone with 250 households with car and 250 households without car. Assume we know the average trip generation rates of each group:

car-owning households produce: 6.0 trips/day

non-car-owning households produce: 2.5 trips/day

then, we can easily deduce that the current number of trips per day is:

$$t_i = 250 \times 2.5 + 250 \times 6.0 = 2125 \text{ trips/day}$$

Let us also assume that in the future all households will have a car; therefore, assuming that income and population remain constant (a safe hypothesis in the absence of other information), we could estimate a simple multiplicative growth factor as:

$$F_i = C_i^d / C_i^c = 1/0.5 = 2$$

and applying Eq.(4.1) we could estimate the number of future trips as:

$$T_i = 2 \times 2125 = 4250 \text{ trips/day}$$

However, the method is obviously very crude. If we use our information about average trip rates and make the assumption that these will remain constant (which is actually the main assumption behind one of the most popular forecasting methods, as we will see below), we could estimate the future number of trips as:

$$T_i = 500 \times 6 = 3000$$

which means that the growth-factor method would overestimate the total number of trips by approximately 42%. This is very serious because trip generation is the first stage of the modelling process; errors here are carried through the entire process and may invalidate work in subsequent stages.

In general, growth-factor methods are mostly used in practice to predict the future number of *external* trips to an area; this is because they are not too many in the first place (so errors cannot be too large) and also because there are no simple ways to predict them. In some cases, they are also used, at least as a sense check, for interurban toll road studies. Growth-factor methods may be also used to project vehicular trips, in contrast with person trips. This has occurred in many toll-road studies as it avoids the need to estimate occupancy rates, always a difficult task because of its high variability in space and time.

In the following sections, we will discuss other (superior) methods that can also be used to model freight trip productions and attractions. However, we will just make explicit reference to the case of personal trip productions as this is the area where there is more practical experience, and also where the most interesting findings have been reported.

4.2 Regression Analysis

The next subsection provides a brief introduction to linear regression modelling. The reader familiar with this subject can proceed directly to subsection 4.2.2.

4.2.1 The Linear Regression Model

4.2.1.1 Introduction

Consider an experiment consisting in observing the values that a certain variable $\mathbf{Y} = \{Y_i\}$ takes for different values of another variable $\mathbf{X}$. If the experiment is not deterministic, we would observe different values of Y_i for the same value of X_i.

Let us call f_i $(Y|X)$ the probability distribution of Y_i for a given value X_i; thus, in general, we could have a different function f_i for each value of **X** as shown in Figure 4.2.

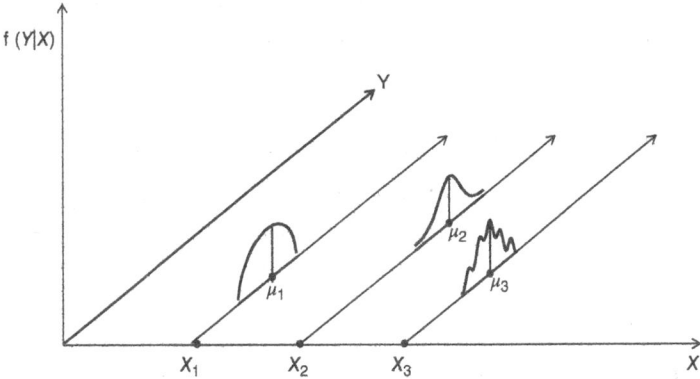

Figure 4.2 General distributions of Y given X

However, such a completely general case is intractable; to make it more manageable certain hypotheses about population regularities are required. Let us assume that:

1) The probability distributions f_i $(Y|X)$ have the same variance σ^2 for all values of **X**.
2) The means $\mu_i = E\ (Y_i)$ form a straight line known as the *true regression line* and given by:

$$E(Y_i) = a + bX_i \tag{4.3}$$

where the population parameters a and b, defining the line, must be estimated from sample data.
3) The random variables **Y** are statistically independent; this means, for example, that a large value of Y_1 does not tend to make Y_2 large.

The above *weak set of hypotheses* (see for example Wonnacott and Wonnacott 1990) may be written more concisely as:

The random variables Y_i are statistically independent with mean $a + b\ X_i$ and variance σ^2.

With these, Figure 4.2 changes to the distribution shown in Figure 4.3.

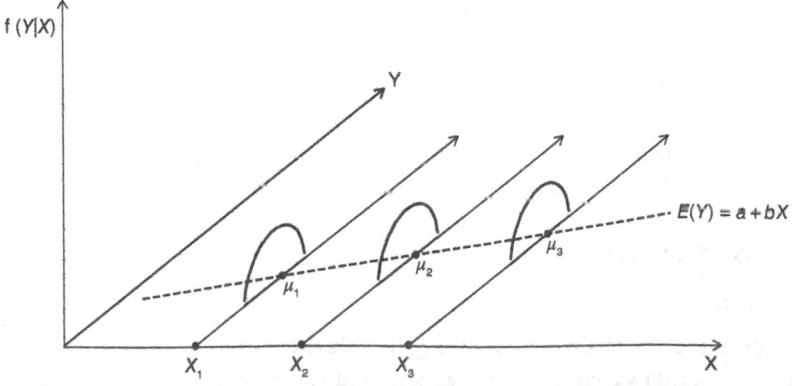

Figure 4.3 Distribution of Y assumed in linear regression

It is convenient to describe the deviation of Y_i from its expected value as the error or disturbance term e_i, so that the model may also be written as:

$$Y_i = a + bX_i + e_i \qquad (4.4)$$

Note that we are not making any assumptions yet about the shape of the distribution of **Y** (and **e**, which is identical except that their means differ) provided it has a finite variance. These will be needed later, however, in order to derive some formal tests for the model. The error term is, as usual, composed of measurement and specification errors (recall the discussion in Chapter 2).

4.2.1.2 Estimation of a and b

Figure 4.4 can be labelled the *fundamental graph* of linear regression. It shows the true (dotted) regression line $E(Y) = a + bX$, which is of course unknown to the analyst, who must estimate it from sample data about **Y** and **X**. It also shows the estimated regression line $\hat{Y} = \hat{a} + \hat{b}X$; as is obvious, this line will not coincide with the previous one unless the analyst is extremely lucky (though he will never know it). In general, the best he can hope is that the parameter estimates will be close to the target.

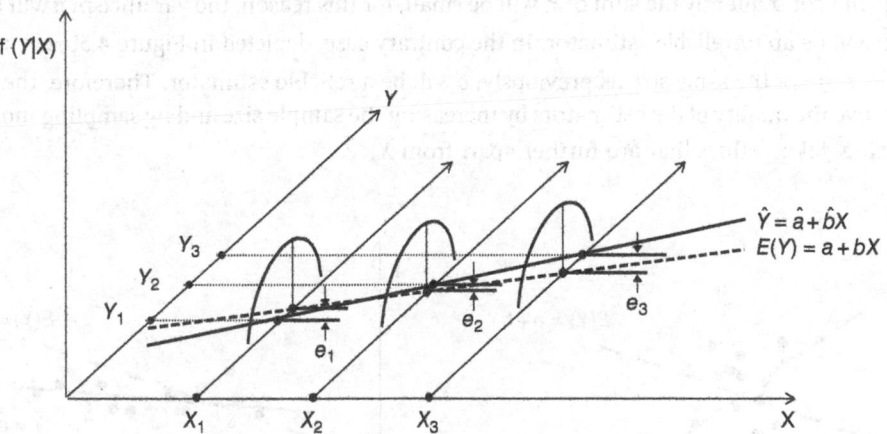

Figure 4.4 True and estimated regression lines

It is important to distinguish between the errors e_i, which are not known and pertain to the true regression line, and the differences ε_i, between observed (Y_i) and fitted values ($\hat{Y}_i$). Least squares estimation, which is the most attractive line-fitting method in this case, results from the minimization of ε_i^2.

If we make the following change of variables $x_i = X_i - \overline{X}$, where $\overline{X}$ is the mean of **X**, it is easy to show that the previous regression lines keep their slopes (b and $\hat{b}$, respectively) but obviously change their intercepts (a and $\hat{a}$, respectively) in the new axes (Y, x). The change is convenient because the new variable **x** has the following property: $\sum_i x_i = 0$.

Under this transformation, the least square estimators are given by:

$$\hat{a} = \overline{Y} \qquad (4.5)$$

which ensures that the fitted line goes through the *centre of gravity* $(\overline{X}, \overline{Y})$ of the sample of n observations, and

$$\hat{b} = \frac{\sum_i x_i Y_i}{\sum_i x_i^2} \tag{4.6}$$

It worth noting that if X_i is equal to some constant for all i, expression (4.6) would be undefined since x_i would be equal to zero for all i, and, therefore, its denominator would be equal to zero.

These estimators have the following interesting properties:

$$E(\hat{a}) = a \quad \text{Var}(\hat{a}) = \sigma^2/n$$
$$E(\hat{b}) = b \quad \text{Var}(\hat{b}) = \sigma^2 \Big/ \sum_i x_i^2$$

In passing, the formula for the variance of $\hat{b}$ has interesting implications in terms of experimental design. First, it can be noted that the variances of both estimators decrease with the sample size n. Also, if the values $\mathbf{X}$ are too close together, as in Figure 4.5a, their deviations from the mean $\overline{X}$ will be small and consequently the sum of x_i will be small; for this reason, the variance of $\hat{b}$ will be large and so $\hat{b}$ will be an unreliable estimator. In the contrary case, depicted in Figure 4.5b, even though the errors e are of the same size as previously, $\hat{b}$ will be a reliable estimator. Therefore, the analyst can improve the quality of the estimators by increasing the sample size and by sampling more cases for which $\mathbf{X}$ takes values that are further apart from $\overline{X}$.

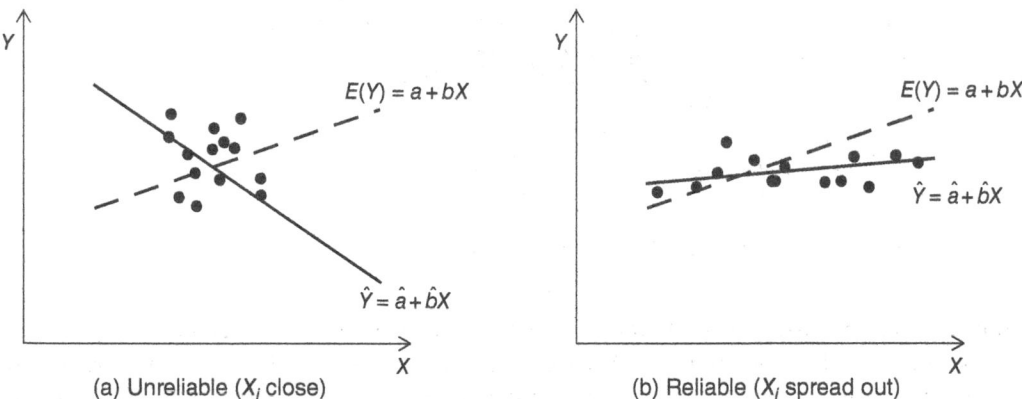

(a) Unreliable (X_i close)   (b) Reliable (X_i spread out)

Figure 4.5 Goodness-of-fit and experimental design

If a fourth hypothesis is considered, that the expected value of e conditional on $\mathbf{X}$ is zero, the least squares estimators (4.5) and (4.6) acquire some desirable statistical properties. In this case (i.e. when the mean of e weighed by the probability of occurrence of $\mathbf{X}$ is zero), the least square estimators are said to be *unbiased* (i.e. their expected values are equal to the true values *a* and *b*), and *consistent* (i.e. they can be as near as desired to the true values as the sample size goes to infinity).

This important assumption may be easily violated if a relevant variable is omitted from the model and is correlated with the observed **X**. For example, consider that the number of trips generated depends on household's income and number of cars, variables that are positively correlated since it is more likely to have a car as income grows. If the number of cars is omitted from the model, the least square estimator of the effect of income will account for both the effect of income and of the number of cars and will therefore be larger than the true coefficient of income. Methods to test and to correct for the violation of this assumption, the problem of *endogeneity*, exist in the literature and we refer to them below.

If in addition to hypothesis 4, hypotheses 1 and 3 hold, the least squares estimators (4.5) and (4.6) are not only consistent and unbiased, but are also the *Best* (i.e. most efficient, and with the smallest variance) among all possible *Linear* and *Unbiased Estimators* (BLUE). This result is known as the Gauss–Markov theorem; methods to test and correct for violations of hypotheses 1 and 3 can be found in the literature. In Section 4.2.2, we present a practical case where the failure of hypothesis 3 is corrected. For further examples and applications, the interested reader is referred to Greene (2008).

4.2.1.3 On Endogeneity in Linear Regression Models

The term *endogeneity* is used in the literature when there is correlation between one or more observed explanatory variables and the error term of an econometric model. As a result, the parameter estimates of these variables may be inconsistent. Endogeneity may be caused by omitted variables, measurement or specification errors, simultaneous determination, and/or self-selection (Guevara 2015).

The correction of endogeneity has been studied in depth in the case of linear models, where the problem has been considered from several viewpoints (Stock and Yogo 2005; Ebbes et al. 2011). Two main methods have been proposed to treat this problem:

1) **Control Function (CF) method** (Heckman 1978; Rivers and Vuong 1988). It consists in identifying an auxiliary variable (or control function), such that when it is added to the regression it makes the error of the model uncorrelated with the observed variables. This auxiliary variable or CF is constructed by means of an *instrumental variable* (IV) or *instrument*. To be valid, the IV must fulfil two properties: (i) be correlated with the endogenous variable, and (ii) be uncorrelated with the regression's error. We will come back to this method in Chapter 8.

2) **Multiple indicator solution method** (MIS, Wooldridge 2010). It uses indicators (instead of IV) to correct the model and achieve consistent estimators. Guevara (2015) shows some advantages of this method, such as its easy applicability in practice. In several situations, the indicators may be easier to obtain than the IV in the CF method, because they are of a different nature (i.e. they come from questions in a survey designed for knowing the attitudes and/or perceptions that respondents have regarding their decision making). In the literature, indicators have been collected in several ways, such as Likert (1932) scales or verbal scales (Glerum et al. 2014), but, theoretically, to apply the MIS method we need them to be continuous. We will come back to the use of indicators to model with latent variables in Chapter 8.

Interested readers may consult, again, the book by Greene (2008) for further details.

4.2.1.4   Hypothesis Tests for b̂

To carry out these tests we need to know the distribution of $\hat{b}$ and for this, we need to consider the *strong hypothesis* that the variables **Y** are distributed Normal. Note that in this case, the least squares estimators will not just be BLUE, but BUE (i.e. Best Unbiased Estimators) among all possible linear and non-linear estimators. This assumption may be strong when the sample is small, but as the sample size increases it begins to hold no matter which is the true distribution thanks to the *Law of Large Numbers*.

Now, as $\hat{b}$ is just a linear combination of the Y_i, it follows that it is also distributed $N(b, \sigma^2/\sum_i x_i^2)$. This means we can standardise it in the usual way, obtaining

$$z = \frac{\hat{b} - b}{\sigma / \sqrt{\left(\sum_i x_i^2\right)}} \tag{4.7}$$

which is distributed $N(0,1)$. However, we do not know σ^2, the variance of **Y** with respect to the true regression. A natural estimator is to use the *residual variance* s^2 around the fitted line:

$$s^2 = \frac{\sum_i (Y_i - \hat{Y}_i)^2}{n - 2}$$

We divide by $(n-2)$ to obtain an unbiased estimator because two degrees of freedom have been used to calculate $\hat{a}$ and $\hat{b}$ which define $\hat{Y}_i$ (see Wonnacott and Wonnacott 1990).

However, if we substitute s^2 by σ^2 in (4.7) the standardised $\hat{b}$ becomes distributed Student (or t) with $(n-2)$ degrees of freedom:

$$t = \frac{\hat{b} - b}{s / \sqrt{\left(\sum_i x_i^2\right)}} \tag{4.8}$$

The denominator of (4.8) is usually called *standard error* of $\hat{b}$ and is denoted by s_b, hence $t = (\hat{b} - b)/s_b$.

The *t*-test. A typical null hypothesis is H_0: $b = 0$; in this case (4.8) reduces to:

$$t = \hat{b}/s_b \tag{4.9}$$

and this value needs to be compared with the critical value of the Student t-statistics for a given significance level α and the appropriate number of degrees of freedom. One problem is that the alternative hypothesis H_1 may imply unilateral ($b > 0$) or bilateral (b not equal 0) tests; this can only be determined by examining the phenomenon under study.

Example 4.2   Assume we are interested in studying the effect of income (I) in the number of trips by non-car-owning households (T), and that we postulate the following relation:

$$T = a + bI$$

As in theory, we can conclude that any influence must be positive (i.e. higher income always means more trips) in this case we should test H_0 against the unilateral alternative hypothesis H_1: $b > 0$. If H_0 is true, the t-value from (4.9) is compared with the value $t_{\alpha;d}$, where d denotes the appropriate number of degrees of freedom, and the null hypothesis is rejected if $t > t_{\alpha;d}$ (see Figure 4.6).

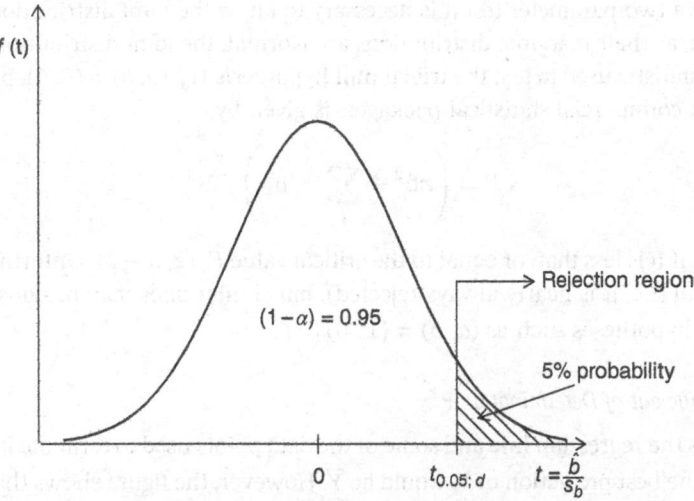

Figure 4.6 Rejection region for $\alpha = 5\%$

On the other hand, if we were considering incorporating a variable the effect of which in either direction was not evident (for example, number of female workers, as these may or may not produce more trips than their male counterparts), the null hypothesis should be the bilateral H_1: $b \neq 0$, and H_0 would be rejected if 0 is not included in the appropriate confidence interval for $\hat{b}$.

The F-test for the Complete Model. Figure 4.7a shows the set of values $(\hat{a}, \hat{b})$ for which null hypotheses such as the one discussed above are accepted individually. If we were interested in testing the hypothesis that both estimators are equal to 0, for example, we could have a region such as that depicted in Figure 4.7b; that is, accepting that each parameter is 0 individually does not necessarily mean accepting that both should be 0 together.

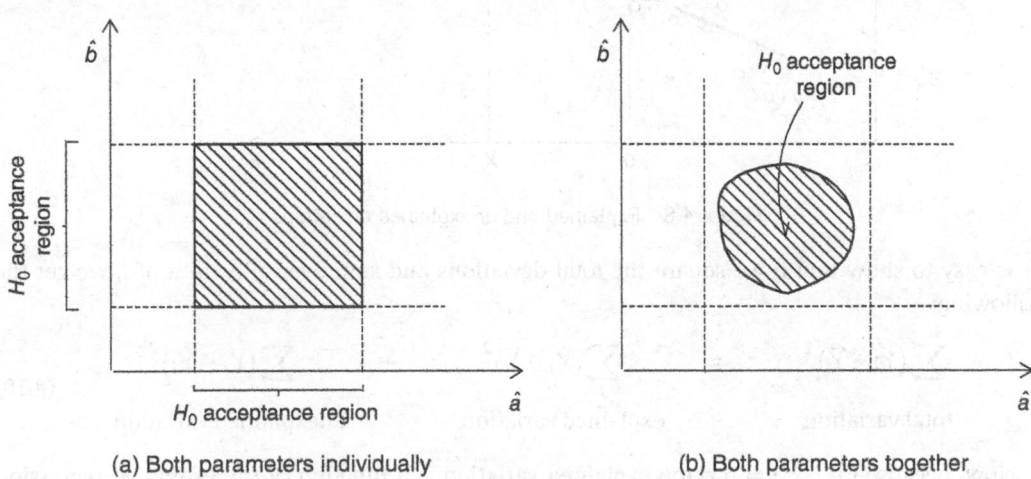

(a) Both parameters individually   (b) Both parameters together

Figure 4.7 Acceptance regions for null hypothesis

Now, to make a two-parameter test it is necessary to know the joint distribution of both estimators. In this case, as their marginal distributions are Normal, the joint distribution is also bivariate Normal. The F-statistic used to test the trivial null hypothesis H_0: $(a, b) = (0, 0)$, provided as one of the standards in commercial statistical packages, is given by:

$$F = \left(n\hat{a}^2 + \sum_i x_i^2 \hat{b}^2 \right) \Big/ 2s^2$$

H_0 is accepted if F is less than or equal to the critical value $F_\alpha(2, n-2)$. Unfortunately, the test is not very powerful (i.e. it is nearly always rejected), but similar ones may be constructed for more interesting null hypotheses such as $(a, b) = (\overline{Y}, 0)$.

4.2.1.5 The Coefficient of Determination R^2

Figure 4.8 shows the regression line and some of the data points used to estimate it. If no values of x were available, the best prediction of Y_i would be $\overline{Y}$. However, the figure shows that for a particular value x_i the error of this method could be high: $(Y_i - \overline{Y})$. When x_i is known, on the other hand, the best prediction for Y_i is $\hat{Y}_i$ and this reduces the error to just $(Y_i - \hat{Y}_i)$, that is, a large part of the original error has been explained. From Figure 4.8 we have:

$$(Y_i - \overline{Y}) \quad = \quad (\hat{Y}_i - \overline{Y}) \quad + \quad (Y_i - \hat{Y}_i), \qquad \forall i$$

total deviation $\qquad$ explained deviation $\qquad$ unexplained deviation

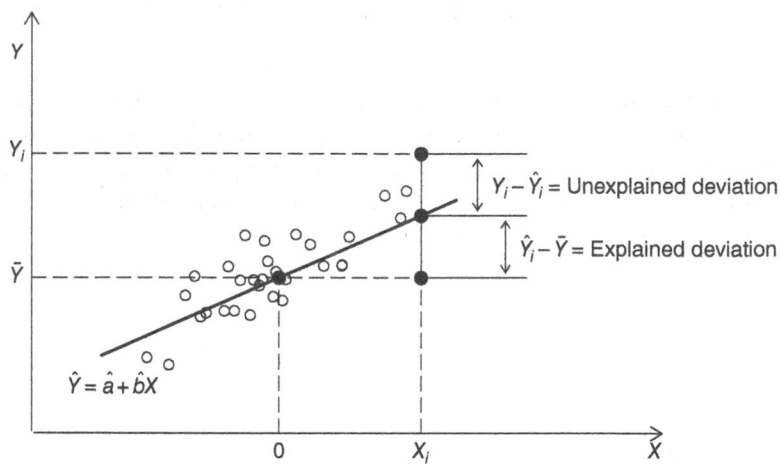

Figure 4.8 Explained and unexplained deviations

It is easy to show that if we square the total deviations and sum over all values of i, we get the following:

$$\sum_i (Y_i - \overline{Y})^2 \quad = \quad \sum_i (\hat{Y}_i - \overline{Y})^2 \quad + \quad \sum_i (Y_i - \hat{Y}_i)^2 \tag{4.10}$$

total variation $\qquad$ explained variation $\qquad$ unexplained variation

Now, because $(\hat{Y}_i - \overline{Y}) = \hat{b} x_i$ the explained variation is a function of the estimated regression coefficient $\hat{b}$. The process of decomposing the total variation into its parts is known as *analysis of variance* of the regression, or ANOVA (note that variance is just variation divided by degrees of freedom).

The coefficient of determination is defined as the ratio of explained to total variation:

$$R^2 = \frac{\sum(\hat{Y}_i - \overline{Y})^2}{\sum(Y_i - \overline{Y})^2} \tag{4.11}$$

It has limiting values of 1 (perfect explanation) and 0 (no explanation at all); intermediate values may be interpreted as the percentage of the total variation explained by the regression. The index is trivially related to the sample correlation R, which measures the degree of association between X and Y (see Wonnacott and Wonnacott 1990).

4.2.1.6 Multiple Regression

This is an extension of the above for the case of more explanatory variables and, obviously, more *regressors* ($\hat{b}$ parameters). The solution equations are similar, although more complex, but some extra problems arise which are usually important, such as the following:

1) **Multicollinearity**. This occurs when there is a linear relation between the explanatory variables. Equivalently with what occurred when one explanatory variable was a linear function of the intercept in (4.6), in this case, the equations for the regressors $\hat{\mathbf{b}}$ are not independent and cannot be solved uniquely.

2) **How many regressors to include**. To make a decision in this case, several factors have to be taken into consideration:

   - are there strong theoretical reasons to include a given variable, or is it important for policy testing with the model?
   - is the estimated sign of the coefficient consistent with theory or intuition and is the variable significant (i.e. is H_0 rejected in the t-test?)?

   If in doubt, one way forward is to take out the variable in question and re-estimate the regression to examine the effect of its removal on the rest of the coefficients; if this is not too important the variable can be left out for *parsimony* (the model is simpler and the rest of the parameters can be estimated more accurately). Commercial software packages provide an 'automatic' procedure for tackling this issue (the *stepwise* approach); however, this is not recommended as it may induce problems, as we will comment below. We will come back to this general problem in Section 8.4 (Table 8.3) when discussing discrete choice model specification issues.

3) **Coefficient of determination**. This has the same form as (4.11). However, in this case, the inclusion of another regressor always increases R^2; to eliminate this problem the *corrected* R^2 is defined as:

$$\overline{R}^2 = [R^2 - k/(n-1)][(n-1)/(n-k-1)] \tag{4.12}$$

where n stands for sample size as before and k is the number of regressors $\hat{\mathbf{b}}$.

In trip generation modelling, the multiple regression method has been used both with aggregate (zonal) and disaggregate (household and personal) data. The first approach has been practically abandoned in the case of trip productions, but it is still the premier method for modelling trip attractions. In this sense, it is worth noting that expressions (4.11) and (4.12) will have values between 0 and 1 if and only if the least square model considers an intercept, that is if model (4.4) is not forced to consider $a = 0$. Also, (4.12) is a good tool to compare models as long as the variables **Y** used for the cases under analysis are the same. For example, if the analyst wants to compare a model for the number of trips as a function of zone attributes and another one using the logarithm of the number of trips, the measures are not comparable since the denominator in (4.11) is not the same for both models.

4) **Hypothesis testing**. If the analyst is interested in testing a hypothesis regarding a specific estimator, the t-test described in (4.8) may be used. However, if the hypothesis involves a linear restriction between many estimators, an F-test should be used instead. In this case, we need to estimate first a *restricted* model, where the restrictions to be tested hold and calculate the *Sum of Squared Residuals* of the restricted model (SSR_R) which is equal to $\Sigma e_i^2 = \Sigma (Y_i - \hat{Y}_i)^2$ and is often an output of regression software. Second, we need to estimate an *unrestricted* model (i.e. where the restrictions are not imposed) and calculate the SSR_U. Then, the F statistic is computed as follows, where k is the number of variables in the unrestricted model, and r is the number of restrictions imposed:

$$\hat{F} = \frac{\{SSR_R - SSR_U\}}{SSR_U} \frac{(n-k)}{r} \sim F_{r,n-k}$$

This statistic follows an F distribution with r and $n-k$ degrees of freedom. The intuition of the test is as follows: if the restrictions are true, SSR_R should be similar to SSR_U and the statistic should be near to zero. On the contrary, if the statistic is larger than $F_{r,n-k}$ the null hypothesis can be rejected for some desired confidence level.

4.2.2 Zonal-Based Multiple Regression

In this case, an attempt is made to find a linear relationship between the number of trips produced or attracted by the zones and average socioeconomic characteristics of the households in each one. The following are some important considerations:

1) **Zonal models can only explain the variation in trip-making behaviour between zones**. For this reason, they can only be successful if the inter-zonal variations adequately reflect the real reasons behind trip variability. For this to happen it would be necessary that zones not only had a homogeneous socioeconomic composition but represented as wide as possible a range of conditions. A major problem is that the main variations in person trip data tend to occur at the intra-zonal level.

2) **Role of the intercept**. One would expect the estimated regression line to pass through the origin; however, large intercept values (i.e. in comparison with the product of the average value of any variable and its coefficient) have often been obtained. If this happens the equation may be rejected; if on the contrary, the intercept is not significantly different from zero, it might be informative to re-estimate the line, forcing it to pass through the origin.

3) **Null zones**. It is possible that certain zones do not offer information about certain dependent variables (e.g. there can be no HB trips generated in non-residential zones). Null zones must be excluded from the analysis; although their inclusion should not greatly affect the coefficient estimates (because the equations should pass through the origin), an arbitrary increment in the number of zones that do not provide useful data will tend to produce statistics which overestimate the accuracy of the estimated regression.

4) **Zonal totals versus zonal means**. When formulating the model, the analyst appears to have a choice between using aggregate or *total* variables, such as trips per zone and cars per zone, or *rates*, such as trips per household per zone and cars per household per zone. In the first case, the regression model would be:

$$Y_i = \theta_0 + \theta_1 X_{1i} + \theta_2 X_{2i} + \dots + \theta_k X_{ki} + E_i$$

whereas the model using rates would be:

$$y_i = \theta_0 + \theta_1 x_{1i} + \theta_2 x_{2i} + ... + \theta_k x_{ki} + e_i$$

with $y_i = Y_i/H_i$; $x_i = X_i/H_i$; $e_i = E_i/H_i$ and H_i the number of households in zone i.

Both equations are almost identical, in the sense that they seek to explain the variability of trip-making behaviour between zones, and in both cases, the parameters have the same meaning. Their unique and fundamental difference relates to the error-term distribution in each case; it is obvious that the constant variance condition of the model cannot hold in both cases unless H_i was itself constant for all zones i.

Now, as the aggregate variables directly reflect the size of the zone, their use should imply that the magnitude of the error actually depends on zone size; this *heteroskedasticity* (variability of the variance) has indeed been found in practice. Using multipliers, such as $1/H_i$, allows heteroskedasticity to be reduced because the model is made independent of zone size. In this same vein, it has also been found that the aggregate variables tend to have higher intercorrelation (i.e. multi-collinearity) than the rates. In this sense, it is important to note that models using aggregate variables often yield higher values of R^2, but this is just a spurious effect because zone size obviously helps to explain the total number of trips (see Douglas and Lewis 1970). What is certainly unsound is the mixture of rates and aggregate variables in a single model.

To end this theme, it is important to remark that even when rates are used, zonal-based regression is conditioned by the nature and size of zones (i.e. the spatial aggregation problem). This is clearly exemplified by the fact that inter-zonal variability diminishes with zone size as shown in Table 4.2, constructed with data from Perth (Douglas and Lewis 1970).

Table 4.2 Inter-zonal variation of personal productions

Zoning system	Mean value of trips/ household/zone	Inter-zonal variance
75 small zones	8.13	5.85
23 large zones	7.96	1.11

4.2.3 Household-Based Regression

Intra-zonal variation may be reduced by decreasing zone size, especially if zones are homogeneous. However, smaller zones imply a greater number of them and this has two consequences:

- more expensive models in terms of data collection, calibration, and operation;
- larger sampling errors, which are assumed non-existent by the multiple linear regression model.

For these reasons, it seems logical to postulate models which are independent of zone boundaries. Since the 1970s it has been argued that the most appropriate analysis unit, in this case, is the household (and not the individual), because a series of important interpersonal interactions inside a household cannot be incorporated even implicitly in an individual model (e.g. car availability, that is, which member has use of the car).

In a household-based application, each home is taken as an input data vector in order to bring into the model all the range of observed variability about the characteristics of the household and its travel behaviour. Care has to be taken with automatic stepwise computer packages when calibrating the model, because they may leave out variables that are slightly worse predictors than others, left in the model, but which may prove much easier to forecast.

In actual fact, stepwise methods are not recommended; it is preferable to proceed the other way around, that is, to test a model with all the variables available and take out those which are not essential (on theoretical or policy grounds) and have low significance or an incorrect sign.

Example 4.3 Consider the variables trips per household (Y), number of workers (X_1), and number of cars (X_2). Table 4.3 presents the results of successive steps of a stepwise model estimation; the last row also shows (in parenthesis) values for the t-ratio Eq. (4.9). Assuming large sample size, the appropriate number of degrees of freedom ($n - 2$) is also a large number so the t-values may be compared with the critical value 1.645 for a 95% significance level on a one-tailed test (we know the null hypothesis is unilateral in this case as Y should increase with both X_1 and X_2).

Table 4.3 Example of stepwise regression

Step	Equation	R^2
1	$Y = 2.36 X_1$	0.203
2	$Y = 1.80 X_1 + 1.31 X_2$	0.325
3	$Y = 0.91 + 1.44X_1 + 1.07X_2$	0.384
	(3.7) (8.2) (4.2)	

The third model is a reasonable equation in spite of its low R^2. The intercept 0.91 is not large (i.e. compare it with 1.44 times the number of workers, for example) and the regression coefficients are significantly different from zero (H_0 is rejected in all cases). The model could probably benefit from the inclusion of further variables if they were available.

An indication of how good these models are may be obtained from comparing observed and modelled trips for some groupings of the data (see Table 4.4). This is better than comparing totals because in such cases different errors may compensate and the bias would not be detected. As can be seen, the majority of cells show a reasonable approximation (i.e. errors of less than 30%). If large bias were spotted it would be necessary to adjust the model parameters; however, this is not easy as there are no clear-cut rules to do it, and depends heavily on context.

Table 4.4 Comparison of trips per household (observed/estimated)

| No. of cars | Number of workers in household | | | |
	0	1	2	3 or more
0	0.9/0.9	2.1/2.4	3.4/3.8	5.3/5.6
1	3.2/2.0	3.5/3.4	3.7/4.9	8.5/6.7
2 or more	—	4.1/4.6	4.7/6.0	8.5/7.8

4.2.4 The Problem of Non-Linearity

As we have seen, the linear regression model assumes that each independent variable exerts a linear influence on the dependent variable. It is not easy to detect non-linearity because apparently linear relations may turn out to be non-linear when the presence of other variables is allowed for in the model. Multivariate graphs are useful in this sense; the example of Figure 4.9 presents data for households stratified by car ownership and number of workers. It can be seen that travel behaviour is clearly non-linear with respect to family size.

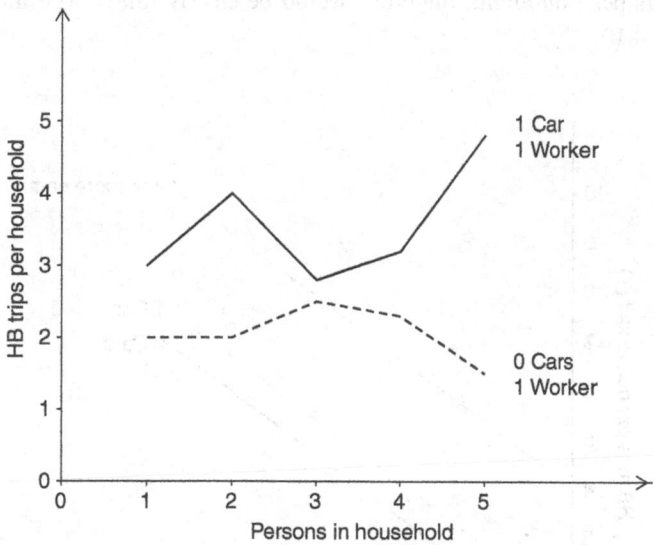

Figure 4.9 An example of non-linearity

It is important to mention that there is a class of variables, those of a qualitative nature, which usually show non-linear behaviour (e.g. type of dwelling, occupation of the head of the household, age, sex). In general, there are two methods to incorporate non-linear variables into the model:

1) Transform the variables in order to linearise their effect (e.g. take logarithms, raise to a power). However, selecting the most adequate transformation is not an easy or arbitrary exercise, so care is needed; also, if we are thorough, it can take a lot of time and effort;
2) Use *dummy* variables. In this case, the independent variable under consideration is divided into several discrete intervals and each of them is treated separately in the model. In this form, it is not necessary to assume that the variable has a linear effect, because each of its portions is considered separately in terms of its effect on travel behaviour. For example, if car ownership was treated in this way, appropriate intervals could be 0, 1, and 2 or more cars per household. As each sampled household can only belong to one of the intervals, the corresponding dummy variable takes a value of 1 in that class and 0 in the others. It is easy to see that only $(n - 1)$ dummy variables are needed to represent n intervals.

Example 4.4 Consider the model of Example 4.3 and assume that variable X_2 (number of cars) was replaced by the following dummies:

Z_1, which takes the value 1 for households with one car and 0 in other cases;
Z_2, which takes the value 1 for households with two or more cars and 0 in other cases.

It is easy to see that non-car-owning households correspond to the case where both Z_1 and Z_2 are 0. The corresponding model to the third step in Table 4.3 would now be:

$$Y = 0.84 + 1.41X_1 + 0.75Z_1 + 3.14Z_2 \quad R^2 = 0.387$$
$$(3.6) \quad (8.1) \quad (3.2) \quad (3.5)$$

Even without the better R^2 value, this model would be preferable to the previous one just because the non-linear effect of X_2 (or Z_1 and Z_2) is clearly evident and cannot be ignored. Note that if the coefficients of the dummy variables were for example, 1 and 2, and if the sample never contained more than two cars per household, the effect would be clearly linear. The model is graphically depicted in Figure 4.10.

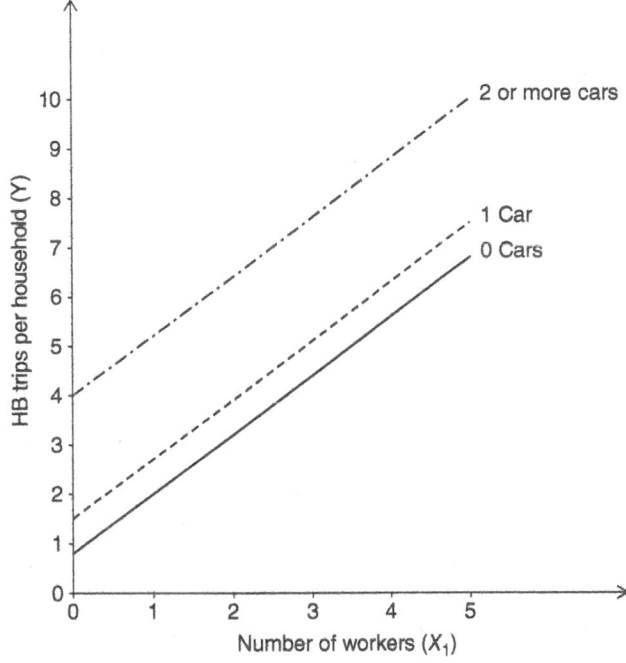

Figure 4.10 Regression model with dummy variables

Looking at Figure 4.10, the following question arises: would it not be preferable to estimate separate regressions for the data on each group, as in that case, we would not require each line to have the same slope (i.e. the coefficient of X_1)? The answer is in general *no* unless we had a reasonable amount of data for each class. The fact is that the model with dummies uses all the data, while each separate regression would use only part of it, and this is in general disadvantageous. It is also interesting to mention that the use of dummy variables tends to reduce problems of multicollinearity in the data (see Douglas and Lewis 1971).

4.2.5 Obtaining Zonal Totals

In the case of zonal-based regression models, this is not a problem as the model is estimated precisely at this level. In the case of household-based models, though, an aggregation stage is required. Nevertheless, precisely because the model is linear the aggregation problem is trivially solved by replacing the average zonal values of each independent variable in the model equation and then multiplying it by the number of households in each zone. However, it must be noted that the aggregation stage can be a very complex matter in non-linear models, as we will see in Chapter 9.

Thus, for the third model of Table 4.3 we would have:

$$T_i = H_i(0.91 + 1.44\overline{X}_{1i} + 1.07\overline{X}_{2i})$$

where T_i is the total number of HB trips in zone i, H_i is the total number of households in it and $\overline{X}_{ji}$ is the average value of variable X_j for the zone.

On the other hand, when dummy variables are used, it is also necessary to know the number of households in each class for each zone; for instance, in the model of Example 4.4 we would require:

$$T_i = H_i(0.84 + 1.41\overline{X}_{1i}) + 0.75H_{1i} + 3.14H_{2i}$$

where H_{ji} is the number of households of class j in zone i.

This last expression allows us to appreciate another advantage of using dummy variables over separate regressions. To aggregate the models in that latter case, it would be necessary to estimate the average number of workers per household (X_1) for each car-ownership group in each zone, and this may be complicated for long-term forecasts.

4.2.6 Matching Generations and Attractions

It might be obvious to most readers that the models above do not guarantee, by default, that the total number of trips originating (the *origins* O_i) at all zones will be equal to the total number of trips attracted (the *destinations* D_j) to them, that is the following expression does not necessarily hold:

$$\sum_i O_i = \sum_j D_j \tag{4.13}$$

The problem is that this equation is implicitly required by the next sub-model (i.e. trip distribution) in the classic structure; it is not possible to have a trip distribution matrix where the total number of trips (T) obtained by summing all rows is different to that obtained when summing all columns (see Chapter 5).

The solution to this difficulty is a pragmatic one that takes advantage of the fact that, normally, the trip generation models are far 'better' (in every sense of the word) than their trip attraction counterparts. The first usually are fairly sophisticated household-based models with typically good explanatory variables. The trip attraction models, on the other hand, are at best estimated using zonal data. For this reason, normal practice considers that the total number of trips arising from summing all origins O_i is in fact the correct figure for T; therefore, all destinations D_j are multiplied by a factor f given by:

$$f = T \Big/ \sum_j D_j \tag{4.14}$$

which obviously ensure that their sum also adds to T.

4.3 Cross-Classification or Category Analysis

4.3.1 The Classical Model

4.3.1.1 Introduction

Although linear regression was the early recommended approach for trip generation, from the late 1960s an alternative method for modelling trip generation appeared and quickly became established as the preferred one in the United Kingdom. The method was known as *category analysis* in the UK (Wootton and Pick 1967) and *cross-classification* in the USA; there it went through a similar development process as the linear regression model, with earliest procedures being at the zonal level and subsequent models based on household information.

The method is based on estimating the response (e.g. the number of trip productions per household for a given purpose) as a function of household attributes. Its basic assumption is that trip generation rates are relatively stable over time for certain household stratifications. The method finds these rates empirically and for this, it typically needs large amounts of data; in fact, a critical element is the number of households in each class. Although the method was originally designed to use census data in the UK, a serious problem of the approach remains the need to forecast the number of households in each stratum in the future.

4.3.1.2 Variable Definition and Model Specification

Let $t^p(h)$ be the average number of trips with purpose p (and at a certain time period) made by members of households of type h. Types are defined by the stratification chosen; for example, a cross-classification based on m household sizes and n car ownership classes will yield mn types h. The standard method for computing these cell rates is to allocate households in the calibration dataset to the individual cell groupings and total, cell by cell, the observed trips $T^p(h)$ by purpose group. The rate $t^p(h)$ is then the total number of trips in cell h, by purpose, divided by the number of households $H(h)$ in it. In mathematical form it is simply as follows:

$$t^p(h) = T^p(h)/H(h) \tag{4.15}$$

The 'art' of the method lies in choosing the categories such that the standard deviations of the frequency distributions depicted in Figure 4.11 are minimised.

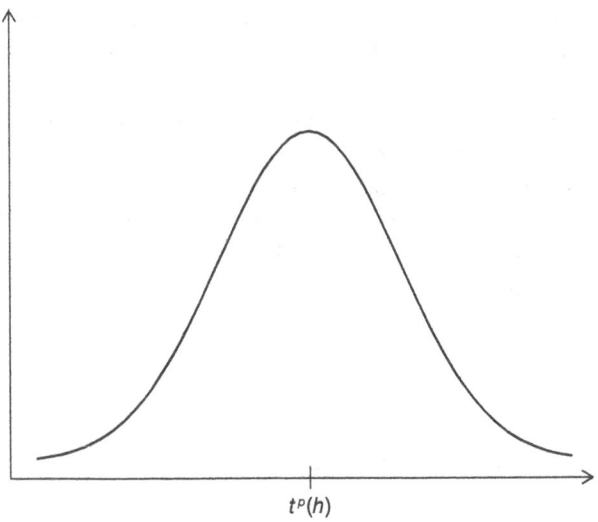

$t^p(h)$

Figure 4.11 Trip-rate distribution for household type

The method has, in principle, the following advantages:

1) Cross-classification groupings are independent of the zone system of the study area.
2) No prior assumptions about the shape of the relationship are required (i.e. they do not even have to be monotonic, let alone linear).

3) Relationships can differ in form from class to class (e.g. the effect of changes in household size for one or two car-owning households may be different).

And in common with traditional cross-classification methods, it has also several disadvantages:

1) The model does not permit extrapolation beyond its calibration strata, although the lowest or highest class of a variable may be open-ended (e.g. households with two or more cars and five or more residents).
2) There are no statistical goodness-of-fit measures for the model, so only aggregate closeness to the calibration data can be ascertained.
3) Unduly large samples are required; otherwise, cell values will vary in reliability because of differences in the numbers of households being available for calibration at each one. For example, in the Monmouthshire Land Use/Transportation Study (see Douglas and Lewis 1971) the estimators for 108 categories (six income levels, three car ownership levels and six household structure levels) shown in Table 4.5 were found, using a sample of 4000 households.

Table 4.5 Household frequency distribution

	No. of categories				
	21	69	9	7	2
No. of households surveyed	0	1–49	50–99	100–199	200+

Accepted wisdom suggests that at least 50 observations per cell are required to estimate the mean reliably; thus, this criterion would be satisfied in only 18 of the 108 cells for a sample of 4000 households. There may be some scope for using stratified sampling to guarantee more evenly distributed sample sizes in each category. This involves, however, additional survey costs.

4) There is no effective way to choose among variables for classification, or to choose best groupings of a given variable; the minimisation of standard deviations hinted at in Figure 4.11 would require an extensive 'trial and error' procedure, which may be considered infeasible in practical studies.

4.3.1.3 Model Application at Aggregate Level

Let us denote by q the person type (i.e. with and without a car), by $a_i(h)$ the number of households of type h in zone i, and by $H^q(h)$ the set of households of type h containing persons of type q. With this, we can write the trip productions with purpose p by person type q in zone i, O_i^{qp}, as follows:

$$O_i^{qp} = \sum_{h \in H^q(h)} a_i(h) t^p(h) \tag{4.16}$$

To verify how the model works, it is possible to compare these modelled values with observed values from the calibration sample. Inevitable errors are due to the use of averages for the rates $t^p(h)$; one would expect a better stratification (in the sense of minimising the standard deviation in Figure 4.11) to produce smaller errors.

There are various ways of defining household categories. The first application in the UK (Wootton and Pick 1967) employed 108 categories as follows: six income levels, three car ownership levels (0, 1, and 2 or more cars per household), and six household structure groupings, as in Table 4.6.

Table 4.6 Example of household structure grouping

Group	No. employed	Other adults
1	0	1
2	0	2 or more
3	1	1 or less
4	1	2 or more
5	2 or more	1 or less
6	2 or more	2 or more

The problem is clearly how to predict the number of households in each category in the future. The most common method (see Wilson 1974) consists in, firstly, defining and fitting to the calibration data, probability distributions for income (I), car ownership (C) and household structure (S); secondly, using these to build a joint probability function of belonging to household type $h = (I, C, S)$. Thus, if the joint distribution function is denoted by $\phi(h) = \phi(I, C, S)$, the number of households in zone i belonging to class h, $a_i(h)$, is simply given by:

$$a_i(h) = H_i \cdot \phi(h) \tag{4.17}$$

where H_i is the total number of households in the zone. This household estimation model may be partially tested by running it with the base-year data used in calibration. The total trips estimated with Eq. (4.16), but with simulated values for $a_i(h)$, can then be checked against the actual observations.

One further disadvantage of the method can be added at this stage:

5) If it is required to increase the number of stratifying variables, it might be necessary to increase the sample enormously. For example, if another variable was added to the *original* application discussed above and this was divided into three levels, the number of categories would increase from 108 to 324 (and recall the discussion on Table 4.5).

4.3.2 Improvements to the Basic Model

4.3.2.1 *Equivalence Between Category Analysis and Linear Regression*

Some of the limitations of the basic model above may be overcome by noting that Category Analysis estimators can be obtained using a linear regression model with dummy variables representing each category (Goodman 1973). This result can be easily shown recalling that if a dummy variable is defined for each category this is equivalent to running separate least squares models with no more variables than an intercept. Then, from Eq. (4.15), it follows directly that the least square estimator is identical to the average of **Y** for each category.

Given this equivalence, the disadvantage of not having statistical goodness-of-fit measures for the model disappears. One can use, for example, the $\overline{R}^2$ measure (4.12) to compare different potential category structures; however, a small shift is needed. To make $\overline{R}^2$ comparable among different models and be constrained between 0 and 1, we need to consider an intercept, but in that case, the model would not be identifiable because the category dummies will add up to one (i.e. equal to the intercept). As noted by Guevara and Thomas (2007), this can be solved by setting one of the categories as the base and using dummies for all the others. Then, the model intercept corresponds to the estimated trip rate of the base category and the estimators associated with each other dummy variable will correspond to the difference between the trip rate of the respective category and that one used as a base. The $\overline{R}^2$ calculated in this way and also the F-test may be used to compare different groupings of alternative variables for stratification.

Equally, the analyst may use the t-test (4.8) as a statistical measure of the reliability of the estimates in each category. The analyst may consider valid, for example, only stratifications for which the estimators are different from zero at the 95% confidence level.

Another practical limitation of the basic model is that in some cases the number of observations in the sample is too small or even inexistent, precluding the estimations of trip rates for some categories. However, if the analyst is interested in having an estimate for such categories there is the following alternative: if it is assumed that the impact of an additional variable level in the number of trips is independent of other variables, one can formulate a linear model that depends on each of the variables' levels. For example, if we consider that the number of trips depends on Income and Car ownership and that those variables are divided into two levels (Low and High income; 0 and 1+ cars), the following linear model could be formulated:

$$\text{Trips}_i = \theta_{IL} I_{\text{Low}\_i} + \theta_{IH} I_{\text{High}\_i} + \theta_{M0} M_{0\_i} + \theta_{M1} M_{1\_i} + e_i$$

where I_{Low} is equal to 1 if the household belongs to the low-income category and zero otherwise. Other variables are defined equivalently and e_i corresponds to an error term.

However, it can be noted that this model is not estimable since there is a problem of multi-collinearity, as $I_{\text{Low}} + I_{\text{High}} = M_0 + M_1 = 1$. To achieve estimation the model needs to be normalised, that is, to use some of the categories as a base or reference. This can be achieved, for example, by considering model (4.18), where we also included an intercept to make the $\overline{R}^2$ comparable among different models:

$$\text{Trips}_i = \alpha + \alpha_{IH} I_{\text{High}\_i} + \alpha_{M1} M_{1\_i} + e_i \tag{4.18}$$

It follows that even if we do not have, for example, observations for households of low income and high motorization, their trip rates may be calculated as $\hat{\alpha} + \hat{\alpha}_{M1}$ if we accept the hypothesis of a linear effect of income and car ownership in the number of trips.

On the other hand, even if there are enough observations to estimate the trip rates for all categories, the analyst may be interested in estimating a model such as (4.18); as it involves the estimation of fewer parameters with the same dataset, it would result in estimators with smaller variance. This hypothesis can be tested, for example, by adding a non-linear interaction calculated as the product of I_{Low} times M_1 or, equivalently, a dummy variable for the combined effect of belonging to a high-income motorised household. This alternative model can be shown to be equivalent to considering one dummy variable per category.

$$\text{Trips}_i = \alpha + \alpha_{IH} I_{\text{High}\_i} + \alpha_{M1} M_{1\_i} + \alpha_{MI} I_{\text{High}\_i} \cdot M_{1\_i} + e_i'$$

The validity of this linear assumption can be tested by checking the joint significance of the interaction variables through an F-test, as described in Section 4.2.1.5. In this case, $k = 4$ and $r = 1$, because the unrestricted model involves the estimation of four coefficients, and only one of them is constrained to zero in the restricted model (4.18). The SSR_R corresponds to model (4.18) and the SSR_U to the model including $I_{\text{Low}} \cdot M_1$. If the statistic is smaller than the critical value $F_{r,n-k}$, the null hypothesis is accepted meaning that the linear model (4.18) is acceptable. If the statistic is larger than $F_{r,n-k}$ the model with interactions should be considered. It is worth noting that since in this case, the interaction term involves the inclusion of only one additional variable, the *t*-statistic (4.8) could also be used but the *F*-test is the tool applied in general.

Another interesting test, beyond the linearity assumption, has to do with the possibility of using an additional variable for classification say, household size. In such case, the following model may be estimated, where S_L would take the value one if the household size is large and zero otherwise, say:

$$\text{Trips}_i = \alpha + \alpha_{IH} I_{\text{High\_}i} + \alpha_{M_1} M_{1\_i} + \alpha_{S_L} S_{L\_i} + e_i''$$

To find out whether the inclusion of household size is a good idea, an *F*-test can be used in general and a *t*-test in the particular case of needing just one additional variable to do it; the significance level to be used in this latter case deserves some attention, though. The usual procedure is to consider it as small as possible, generally 5%, to reduce Type I errors (i.e. rejecting the null hypothesis when it is true). However, as excluding household size may cause endogeneity, because the variable may be correlated with income or car ownership, the cost of excluding it when it should be there (Type II error) may be higher than the cost of including it if it should not be considered (Type I error). Since there is a trade-off between Type I and Type II errors, it may be advisable to consider a significance level of 10% or even 20%. in this case.

Another possibility is that the analyst may want to explore the validity of the threshold used to define Low- and High-income strata, say. This can be easily achieved by running a regression considering the alternative thresholds and comparing the $\bar{R}^2$ of both models. In general, the model with the larger $\bar{R}^2$ should be preferred, as it will explain a larger portion of the variance. However, if the model with the larger $\bar{R}^2$ results in unreasonable signs or size of coefficients, or if it affects their significance, the alternative should be chosen instead.

Guevara and Thomas (2007) point out that even after all the potential improvements to Category Analysis described above, the $\bar{R}^2$ of this type of model tends to be very low (i.e. less than 0.2 or 0.3). This is not surprising since the model is indeed extremely simplistic. The consideration of more realistic relationships between explanatory variables and the number of trips may be attained by means of linear regression. The models described in the next subsection represent an improvement in that direction.

4.3.2.2 Regression Analysis for Household Strata

A mixture of cross-classification and regression modelling of trip generation may be the most appropriate approach on certain occasions. For example, in an area where the distribution of income is unequal, it may be important to measure the differential impact of policies on different income groups; therefore, it may be necessary to model travel demand for each income group separately throughout the entire modelling process. Assume now that in the same area car ownership is increasing fast and, as usual, it is not clear how correlated these two variables are; a useful way out may be to postulate regression models based on variables describing the size and make-up of different households, for a stratification according to the two previous variables.

Example 4.5 Table 4.7 presents the 13 income and car-ownership categories (C_i) defined in ESTRAUS (1989) for the Greater Santiago 1977 O–D data. As can be seen, the bulk of the data corresponds to households with no cars and low income. Also, note that categories 7 and 10 have rather few data points; this is, unfortunately, a general problem of this approach. Even smaller samples for very low income and high car ownership led to the aggregation of some categories at this range.

Table 4.7 Stratification of the 1977 Santiago sample

Household income (US$/month)	Household car ownership			
	0	1	2+	Total
<125	6564 (C_1)	215 (C_2)		6779
125–250	4464 (C_3)	627 (C_4)		5091
250–500	1532 (C_5)	716 (C_6)	87 (C_7)	2334
500–750	305 (C_8)	436 (C_9)	118 (C_{10})	859
>750	169 (C_{11})	380 (C_{12})	301 (C_{13})	790
Total	12 974	2373	506	15 853

The independent variables available for analysis (i.e. after leaving out the stratifying variables) included variables of the *stage in the family cycle* variety, which we will discuss in Section 4.4. However, after extensive specification searches, it was found that the most significant variables were: number of workers (divided into four classes depending on earnings and type of job), number of students, and number of residents.

Linear regression models estimated with these variables for each of the 13 categories were judged satisfactory on the basis of correct signs, small intercepts, reasonable significance levels, and R^2 values (e.g. between 0.401 for category 4, and 0.682 for category 7; see Hall et al. 1987).

Finally, one assumption that may be lifted in this case and that may improve even more the adjustment and the quality of the models, is to accept that some coefficients may be the same across categories. This will involve the joint estimation of the 13 models and necessarily produce and increase in efficiency, that is, a reduction in the variance of the estimators.

4.4 Other Trip Generation Formulations

The traditional model formulations discussed above do not recognise explicitly the non-negativity and integer nature of trips. In such a case count-data models (Cameron and Trivedi 1998) offer an attractive alternative, and there are many examples of using these in the literature (Jang 2005; Badoe 2007). In what follows, we will present briefly two such models and also discuss how their results compare to using the classical linear regression approach, for modelling the generation of different types of trips (i.e. HBW, vs HBO and NHB).

4.4.1 Alternative Model Formulations

4.4.1.1 The Negative Binomial (NB) Approach

The negative binomial (NB) model for household trip generation can be derived by assuming that the distribution of the number of trips made by a household, within a certain time such as a day, is Poisson distributed. The mean of this Poisson distribution is assumed to be random with an error following a Gamma distribution. On the basis of these assumptions, the probability mass function can be written as (Cameron and Trivedi 1998):

$$P(Y_i) = \frac{\Gamma(\alpha^{-1} + Y_i)}{\Gamma(\alpha^{-1})\Gamma(Y_i + 1)} \left(\frac{1}{\mu_i \alpha + 1}\right)^{\alpha^{-1}} \left(\frac{\mu_i \alpha}{\mu_i \alpha + 1}\right)^{Y_i}$$

where the expected value $E(Y_i) = \mu_i$ is the exponential of a sum of parameters β multiplying explanatory variables X, and it's variance Var $(Y_i) = \mu_i + \alpha\mu_i^2$, and α is often referred to as an over dispersion parameter. Note that if α tends to zero, the mean and variance become equal (i.e. unobserved heterogeneity vanishes) and the NB model collapses into a Poisson model (see Lawless 1987 for details).

4.4.1.2 The Ordinal Probit Model

In this approach, we postulate a latent (i.e. unobserved) propensity of a household to make trips, which is assumed related to a vector of explanatory variables as follows (McKelvey and Zavoina 1975):

$$U_i = \beta_0 + \beta_1 X_{1i} + ... + \beta_k X_{ki} + \epsilon_i$$

where ϵ_i distributes standard Normal. This unobserved propensity is related to the observed number of trips (Y_i) through a set of threshold parameters, such that:

$$Y_i = n, \; if \; \lambda_{n-1} < U_i \leq \lambda_n \; for \; n = 0, ..., N$$

and to estimate the model, the thresholds λ_{-1} and λ_N are set equal to minus and plus infinity, respectively, and the threshold λ_0 is assumed to be zero. Given this, the probability that a household is observed to make exactly n trips is given by:

$$P(Y_i = n) = \Phi(\lambda_n - U_i) - \Phi(\lambda_{n-1} - U_i)$$

where Φ is the standard Normal cumulative distribution function, with tabulated values.

4.4.1.3 Comparing the Performance of Count Data and Linear Regression Models

Not many such comparisons have been reported in the literature. Here we will refer to an interesting paper by Lim and Srinivasan (2011), who reviewed earlier model comparisons, finding that as they were mainly based on a single type of trip the conclusions were specific to the distributional pattern of the selected trip purpose. For this reason, these authors used three different trip purposes (HBW, HBO, and NHB) with significantly different distribution patterns, from a large dataset corresponding to the US national Household Travel Surveys (NHTS) of 2001 (for model estimation) and 2009 (for model validation).

The NHTS collects detailed travel information for one day, from a nationally representative sample of households, plus several socioeconomic and location characteristics of the households involved. To estimate the models, they used a sample of almost 40 000 households (and over 400 thousand trips) who travelled on a weekday in 2001, where all members of the family responded to the survey. To validate the models, they used another sample of over 86 000 households (and more than 720 thousand trips) than answered the survey in 2009.

The estimated models used six socioeconomic factors (household size, number of workers, vehicle share, presence of children, household income, and housing tenure), two location descriptors (residential area and census region), and two temporal descriptors (day of the week and month of the year). All estimated coefficients had the correct sign and their statistical significance was stable across the estimated model structures.

It is not simple to compare models with different model structures. In this case, the authors used the Akaike information criterion (AIC), where a lower value indicates a superior model in terms of fit (Greene 2008). They found that for each of the three trip purposes examined, the Ordinal Probit specification performed the best and the linear regression model the worst, with the NB model in between.

Lim and Srinivasan (2011) went a bit further by making aggregate and even household-based comparisons of the number of trips predicted by the various models and the trips observed in their large validation sample. In this exercise, they found that, again, the Ordinal Probit models were able to replicate the trip generation patterns better than the linear regression and NB models for the three different trip purposes analysed.

4.5 Trip Generation and Accessibility

As mentioned in Chapter 1, the classical specification of the urban transport planning (four/five-stage) model incorporates an iterative process between trip distribution and assignment which leaves trip generation unaltered. This is true even in the case of more contemporary forms which attempt to solve the complex supply–demand equilibration problem appropriately, as we will discuss in Chapter 12. A major disadvantage of this approach is that changes to the network are not allowed to have any effects on trip productions and attractions. For example, this would mean that the extension of an underground line to a location that had no service previously would not generate more trips between that zone and the rest. Although this may be the case for compulsory trips, it may not be correct in the case of discretionary trips (e.g. consider the case of shopping trips and a new line connecting a low-income zone with the city's central market, which features more competitive prices than the zone's local shops).

To solve this problem, modellers have attempted to incorporate a measure of accessibility (i.e. ease or difficulty of making trips to/from each zone) into trip generation equations; the aim is to replace $O_i^n = f\left(H_i^n\right)$ by $O_i^n = f\left(H_i^n, A_i^n\right)$, where H_i^n are household characteristics and A_i^n is a measure of accessibility by person type.

Typical accessibility measures take the general form:

$$A_i^n = \sum_j f\left(E_j^n, C_{ij}\right)$$

where E_i^n is a measure of attraction of zone j and C_{ij} the generalised cost of travel between zones i and j. A typical analytical expression used to this end has been:

$$A_i^n = \sum_j E_j^n \exp\left(-\beta C_{ij}\right)$$

where β is a calibration parameter from the trip distribution model, as discussed in Chapter 5.

Unfortunately, this procedure has seldom produced the expected results in the case of aggregate urban modelling applications because the estimated parameters of the accessibility variable have either been non-significant or of the wrong sign. This issue has remained highly topical for many years and it is clearly related to two interesting and yet unresolved problems: model dynamics and modelling with longitudinal instead of cross-sectional data (Chapter 1). Ortúzar et al. (2000b) give an interesting discussion about the problem and offer an example of what can be gained by using stated preference data in this context.

New emphasis was given to elastic trip generation models by work done in the UK on induced traffic in the assessment of trunk road schemes (Department of Transport 1997). This work led to the study of trip generation methods which were sensitive to changes in accessibility, as it was recognised that the classical methods were inadequate in this sense. Daly (1997) proposed a three-component framework for trip generation:

1) Individuals in their household context formulate an activity pattern for the period to be modelled, say a day. Out-of-home activities are, of course, the only activities that generate trips.

2) The out-of-home activities are organised into 'sojourns', which are defined as stays at a specific location, each of which has a primary purpose (and possibly secondary purposes at the same location too).

3) Formulate a travel plan to link the sojourns, particularly deciding which must be visited by separate home-based tours (two or more trips) and which can be linked with other sojourns by non-home-based trips.

From here it appears reasonable to try and model the number of sojourns generated by a household or person, and to split those sojourns between home-based tours and non-home-based trips. One practical modelling point that follows immediately from this framework is that the dependent variables (i.e. number of tours and/or trips, or alternatively number of sojourns that can be reached) will be integers: 0, 1, 2, 3, etc. Moreover, the decision between whether to travel or not (i.e. between 0 and 1) can be expected to be taken on a different basis from the decision on whether to make more than one trip (the former decision is whether to take part in an activity *at all*, and the latter is how to organise the time and location *given* that some participation will take place). Another point is that travel by all modes needs to be included to ensure that all out-of-home activities are considered; thus, the exclusion of short trips or trips by non-motorised modes will detract from the quality of the model. This is therefore consistent with the contemporary approach to O–D data collection discussed in Chapter 2.

The variables to be included in the model are the same as in the classical methods discussed above, but it is hoped that accessibility can be incorporated. However, there may be a negative cross-influence between home-based and non-home-based accessibilities; for example, if home-based trips can be made easily (i.e. in a small town) then fewer non-home-based trips will be needed.

Predictions of number of sojourns must be made for each travel purpose (and note that the variables influencing each type may vary). Logically we should model compulsory purposes first and then the non-mandatory purposes, conditional upon the decisions made for the compulsory purposes. Similarly, the choice between meeting a travel need by a home-based tour or a non-home-based trip as a detour on a previously planned tour should be modelled explicitly (Algers et al. 1995), but independent models are conceivable in the interest of simplicity. Finally, although trip frequency models may be set up to describe the behaviour of complete households (i.e. considering all the interactions that may be relevant to the number of trips made), the development of person-based models should be much simpler in practice, given the data that is usually available.

4.6 The Frequency Choice Logit Model

Daly (1997) discussed several models, concluding that the most adequate was one with a Logit form (see Chapter 7) and which would predict the total number of trips by first calculating the probability that each individual would choose to travel or not. The total travel volume can be then obtained by multiplying the number of individuals of each type by their probabilities of making a trip. The extension needed to deal with individuals that make more than one trip is presented subsequently.

If V is the utility of making a trip (assuming that the utility of not travelling is zero, with no loss of generality), the probability of making a trip is given by:

$$P_1 = \frac{1}{1 + \exp(-V)}$$

where V is usually specified as being a linear function of unknown parameters θ:

$$V = \sum_k \theta_k X_k$$

where X are measured data items such as income, car ownership and household size, and accessibility, which needs to be input in a form consistent with utility-maximising theory (i.e. the theory behind the Logit model). For this, the preferred form is to use a result by Williams (1977) made popular by Ben-Akiva and Lerman (1979), which states that the correct form of accessibility is the 'logsum' of the destination (or mode) choice model; furthermore, and as we discuss in Section 7.4, because in this case we have a Nested Logit structure, the parameter multiplying the logsum accessibility variable must lie between 0 and 1. If this condition is not met, the model predictions may violate common sense (Williams and Senior 1977).

The Logit model represents the choice of each individual whether or not to make a trip, and this means it is particularly suited to dealing with disaggregate data. But aggregate data can also be used; then the probabilities P represent proportions rather than probabilities. However, to obtain the best chance of finding a significant relationship between accessibility and number of trips use disaggregate data as it preserves the maximum amount of variance. To model higher trip frequencies, Daly (1997) proposed the use of a hierarchical structure representing an indefinite number of choices (Figure 4.12).

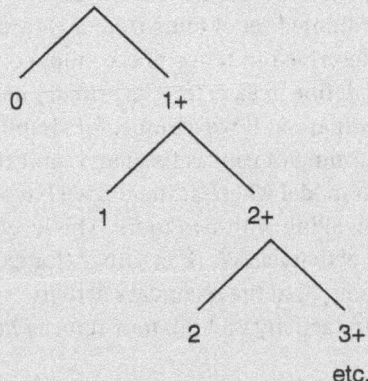

Figure 4.12 'Stop-go' trip generation model

At each hierarchical level, the choice is whether to make further journeys or to stop at the present number (hence the name 'stop-go' model). Because of the possibly strong difference in behaviour between the 0/1+ choice and the remaining choices, it has been found preferable to model the first choice using a separate model. However, because there are often little data on travellers making multiple journeys, it is also necessary to model the remaining choices with a single 'stop-go' model (i.e. which predicts the same probability of stopping at every level of the hierarchy).

It has been found that applying this model system is straightforward. If the probability of making any journeys is p (from the 0/1+ model) and the probability of choosing the 'go' option at each subsequent stage is q (from the stop-go model), then the expected number of journeys is simply:

$$t = p/(1-q)$$

The method has been applied in several studies in Europe (Daly 1997), obtaining coefficients for the accessibility variable ranging from 0.07 to 0.33 for various trip purposes. A more aggregate version of this model, using linear regression on trips observed at an intercept survey of most roads to the North of Chile, also gave good results yielding accessibility measures (of the log-sum type) with significant coefficients of the proper sign and magnitude (Iglesias et al. 2008).

4.7 Tour Generation

People make a wide variety of tours in the real world. Some of them are simple: Home to Work and Return Home, others are more complex including lunch breaks (at a location more than 300 metres away if walking), visits to the gym and shopping, cinema and meals out. Some of them will be recurrent, perhaps twice-weekly visits to the gym, others less frequent or more irregular. It is not realistic to try to model all of these and, therefore, usually a selection of the most common ones is incorporated at this stage.

In practice, tour generation models are often implemented as disaggregate models based on individuals or households. Disaggregate models are described in detail in Chapters 7–9 and, therefore, here we only discuss some of the basic principles involved in tour generation; these can be seen as an extension of the Trip Frequency models just discussed.

The first step is to identify a limited set of tours from a Household Mobility Survey or another useful source. The tours are described in terms of the stops (or activities) linked and their timing. A practical approach is to define in each tour its *primary activity*, more often than not Work, and its associated primary destination. If we assume, for simplicity, that tours have at most two secondary activities, that the timing of trips is fixed and omit the segmentation by person type, we can build a tour generation model where at the top we have the probability of not travelling, travelling with one tour or travelling with two tours. The level below will have the probability that the tour has the primary activity as Work or Other (for example shopping). Below each of these, there will be the probability that the secondary activity is S_0, S_1, or S_2; S_0 is the probability that there is no other secondary activity and the tour returns home. This structure is illustrated in Figure 4.13.

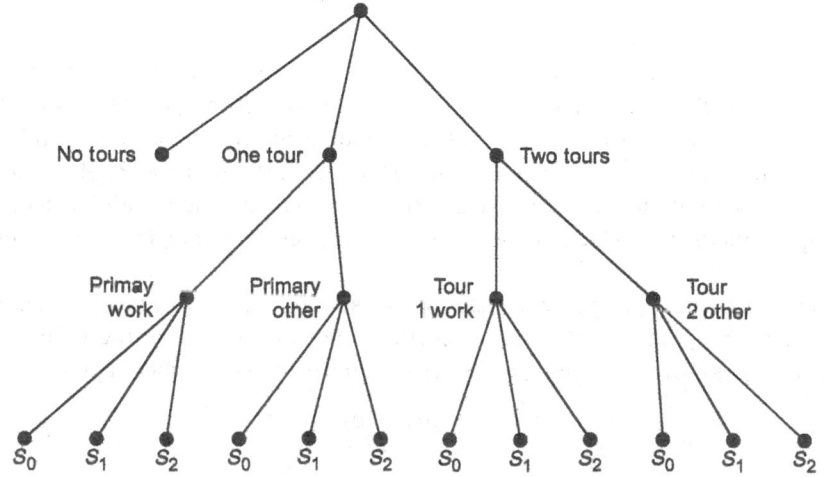

Figure 4.13 Structure of a simple tour generation model

This model can be extended to accept more secondary activities, have variable timing for the activities and trips, and even change modes increasing its complexity and realism, see for example Section 4.4 of Cascetta (2009).

4.8 Forecasting Variables in Trip Generation Analysis

The choice of variables used to predict (household) trip generation rates has long been an area of concern for transportation planners; these variables typically include household numbers, household size (and/or structure), number of vehicles owned, and income. However, interest arose in the early 1980s in research aimed at enriching trip generation models with theories and methods from the behavioural sciences. The major hypothesis behind this work was that the social circumstances in which individuals live should have a considerable bearing on the opportunities and constraints they face in making activity choices; the latter in turn may lead to differing travel behaviour. For example, it is clear that whether a person lives alone or not should affect the opportunities to coordinate and trade-off activities with others to satisfy their travel necessities. Thus, a married couple with young pre-school children will generally find themselves less mobile than a similar couple without children or with older children who require less intensive care. Elderly and retired persons living with younger adults are likely to be more active outside the home than elderly people living alone.

At the household level the situation is quite similar: households of unrelated individuals, for example, tend to follow a pattern of activities that are less influenced by the presence of other household members (and which normally leads to more frequent trips) than is the case of households of related individuals (obviously with similar size, and other characteristics). This is due to the reduced coordination among different members and also to the fact that their activity patterns typically involve fewer home-centred activities.

One way of introducing these notions into trip generation modelling is to develop a set of household types that effectively captures these distinctions and then add this measure to the equations predicting household behaviour. One possible approach considers the age structure of the household and its lifestyle. The approach is consistent with the idea that travel is a *derived demand* and that travel behaviour is part of a larger allocation of time and money to activities in separate locations. For example, the concept of *lifestyle* can be made operational as the allocation of varying amounts of time to different (activity) purposes both within and outside the home, where travel is just part of this time allocation (see Allaman et al. 1982). It appears that the time allocation of individuals varies systematically across various segments of the population, such as age, sex, marital status, and even race; this may be because different household structures place different demands on individuals.

One set of hypotheses that can be tested empirically is whether the major break points (or stages) in the life (or family) cycle are consistent with major changes of time allocation. For example, the break points may be:

- the appearance of pre-school children;
- the time when the youngest child reaches school age;
- the time when a youth leaves home and either lives alone, with other young adults or marries;
- the time when all the children of a couple have left home but the couple has not yet retired, and
- the time when all members of a household have reached retirement age.

It is usually illuminating to compare households at one stage of this life cycle with households of the immediately preceding stage.

The concepts of lifestyle and stage of family cycle are important from two points of view: first, that of identifying stable groupings (based on age or sex) with different activity schedules and consequently demands for travel; second, that of allowing the tracing of systematic changes that may be based on demographic variations (e.g. changes in age structure, marital or employment status). Numerous demographic trends of significance in terms of travel behaviour have been receiving increasing attention since the early 1980s (see Spielberg et al. 1981). One of the most significant for predicting travel behaviour is the changing ratio of households to population, particularly in industrialised nations. Although the rate of population growth has been falling steadily for over 40 years, the rate of household formation has increased in many cases. This is due, among other reasons, to increases in the number of single-parent households and the number of persons who are setting up individual households. Therefore, travel forecasting methodologies that implicitly assume stable ratios of households to population (as was often the case) should be severely affected by this structural shift in the demographic composition of society.

Another trend that has been well discussed is the overall ageing of the population, again, particularly in industrialised nations. This is important because age tends to be associated with a decline in mobility and a change in lifestyle. It is interesting to note though, that differences in trip generation by age may reflect in part the so-called *cohort effects*. This means that older people may travel less, simply because they always did so, rather than because of their age. However, this effect may be largest for people over 65 and declining trip generation rates for other age groups probably reflect a true decrease in the propensity of travel.

A more recent trend, accelerated during the 2020 pandemic, is an increased use of the Internet to facilitate remote working and the procurement of goods and services. Hybrid approaches, where the employee works some days from home and others at work are likely to continue to be attractive. This reduces the average number of HBW trips but may generate some additional local trips that did not exist before (Elldér 2020).

Finally, another trend worth noting is the increase in the proportion of women joining the labour force. Its significance for transportation planning and forecasting stems from two effects. The first is simply the direct employment effect, where time allocation and consequently travel behaviour are profoundly influenced by the requirements of actually being employed. The second one is more subtle and concerns changes in household roles and their impacts on lifestyle, particularly for couples with children.

To end this section, it is interesting to mention that the ideas discussed above led to a proposal for incorporating a household structure variable in trip generation modelling, which was tested with real data (Allaman et al. 1982). The household structure categories proposed were based on the age, gender, marital status, and last name of each household member. These variables allowed the determination of the presence or absence of dependents in the household, the number and type of adults present, and the relationship among household members. However, although models using this variable were pronounced a considerable improvement over traditional practice by Allaman et al. (1982), further tests with a different data set performed by McDonald and Stopher (1983) led to its rejection. This was not only on the basis of statistical evidence but also on policy sensitivity (i.e. it is difficult to use household structure as a policy variable) and ease of forecasting grounds (i.e. forecasting at zone level, particularly to obtain a distribution of households by household structure category, appears to be very problematic). McDonald and Stopher (1983) argue that in these two senses a variable of the housing type variety should be preferred and it is bound to be easier to use by a local government planning agency.

Whatever the case, the accuracy of any future year forecast will depend significantly on the accuracy of the population synthesis produced to estimate the type of households, activities, and individuals in the different zones of the study area.

4.9 Stability and Updating of Trip Generation Parameters

4.9.1 Temporal Stability

Transport models, in general, are developed to assist in the formulation and evaluation of transport plans and projects. Although on many occasions use has been made of descriptive statistics for examining travel trends, most developments have used cross-sectional data to express the amount of travel in terms of explanatory factors; these factors need to be both plausible and easy to forecast for the model to be policy sensitive in the design-year. A key (often implicit) assumption of this approach is that the model parameters will remain constant (or stable) between base and design years.

Several studies have examined this assumption in a trip generation context, finding in general that it cannot be rejected when trips by all modes are considered together (see Kannel and Heathington 1973; Smith and Cleveland 1976), even in the case of the rather crude zonal-based models, although these are not recommended anyway, for reasons similar to those discussed in Section 4.2.2 (see Downes and Gyenes 1976). However, later analyses reported different results. For example, Hall et al. (1987) compared observed trip rates and regression coefficients of models fitted to household data collected for Santiago in 1977 and 1986 and found them significantly different. Copley and Lowe (1981) reported that although trip rates by bus for certain types of household categories seemed reasonably stable over time, car trip rates appeared to be highly correlated with changes in real fuel prices. The latter has the following potential implications:

1) If there is non-zero elasticity of car trip rates to fuel prices, the usual assumption of constant trip rates in a period of rapidly increasing petrol prices could lead to serious over-provision of highway facilities. If, on the other hand, fuel prices were to fall in real terms, the constant trip rates assumption would lead to under-provision (which is precisely what was experienced in the UK and other industrialised countries towards the end of the 1980s).
2) Furthermore, the balance between future investments in public and private transport facilities may be judged incorrectly if based on the assumption of constant trip rates over time.

Clearly then, the correct estimation of the effect of energy prices on trip rates (and of any other similar *longitudinal* effects) is of fundamental importance for policy analysis. Unfortunately, it cannot be tackled with the cross-sectional data sets typically available for transportation studies.

Another factor affecting the stability of trip generation models over time is the evidence available on changes in travel behaviour. We do change our mind and the set of activities we would like to achieve is not fixed. Behavioural change programmes have an effect in reducing the number of trips (often by combining them into more efficient tours on a different day of the week) and in transferring some of them to more environmentally friendly modes. These techniques work because the individual does benefit from saving time spent travelling. There is some evidence that even without these interventions people do change their travel behaviour in response to easier home working and greater awareness of health and environmental issues. The opportunity for changes seems to be most clear when a major intervention into the transport and activity system takes place. This would partially explain, for example, the greater-than-expected shift to public transport when congestion charging was first introduced in London.

4.9.2 Geographic Stability

Temporal stability is often difficult to examine because data (of similar quality) are required for the same area at two different points in time. Thus, on many occasions, it may be easier to examine geographic stability (or transferability) as data on two different locations might become available (for example, if two institutions located in different areas decide to conduct a joint research project). Geographic transferability should be seen as an important attribute of any travel demand model for the following reasons:

1) It would suggest the existence of certain repeatable regularities in travel behaviour that can be picked up and reflected by the model;
2) It would indicate a higher probability that temporal stability also exists; this, as we saw, is essential for any forecasting model;
3) It may allow reducing substantially the need for costly full-scale transportation surveys in different metropolitan areas (see the discussion in Chapter 9).

Not all travel characteristics can be transferable between different areas or cities; for example, the average work trip duration is obviously context dependent, that is, it should be a function of area size, shape, and the distributions of workplaces and residential zones over space. However, transferability of trip rates should not be seen as unrealistic: trips reflect needs for individuals' participation in various activities outside the home and if trip rates are related to homogeneous groups of people, they can be expected to remain stable and geographically transferable within the same cultural context.

The transferability of trip generation models (typically trip rates on a household-category analysis framework) has been tested relatively rarely, producing normally unsatisfactory results (see Caldwell and Demetski 1980; Daor 1981); the few successful experiences have considered only part of the trips, for example, trips made by car (see Ashley 1978). On the other hand, Supernak (1979, 1981) reported the successful transferability of a personal-category trip generation model, both for Polish and American conditions. Finally, Rose and Koppelman (1984) examined the transferability of a discrete choice trip generation model, allowing for adjustment of modal constants using local data. One of their conclusions was that context similarity appeared to be an important determinant of model transferability; also, because their results showed considerable variability, they caution that great care must be taken to ensure that the transferred model is usable in the new context.

4.9.3 Bayesian Updating of Trip Generation Parameters

Assume we want to estimate a trip generation model but lack funds to collect appropriate survey data; a possible (but inadequate) solution is to use a model estimated for another (hopefully similar) area directly. However, it would be highly desirable to adjust it in order to reflect local conditions more accurately.

This can be done by means of Bayesian techniques for updating the original model parameters using information from a small sample in the application context. Bayesian updating considers a *prior* distribution (i.e. that of the original parameters to be updated), new information (i.e. to be obtained from the small sample), and a *posterior* distribution corresponding to the updated model parameters for the new context. Updating techniques are very important in a continuous planning framework; we will see this theme appearing in various parts of this book.

Consider, for example, the problem of updating trip rates by household categories; following Mahmassani and Sinha (1981) we will employ the notation in Table 4.8.

Table 4.8 Bayesian updating notation for trip generation

Variable	Prior information	New information
Mean trip rate	t_1	t_s
No. of observations	n_1	n_s
Trip rate variance	S_1^2	S_s^2

The mean trip rate of a category (or cell), is of course the average of a sample of household trip rates. According to the *Central Limit Theorem*, if the number of observations in a cell is at least 30, the sample distribution of the cell (mean) trip rates may be considered distributed Normal independently of the distribution of the household trip rates. Therefore, the prior distribution of the cell trip rates for the original model is $N(t_1, S_1^2/n_1)$, because t_1 and S_1^2/n_1 are unbiased estimators of its mean and variance. Similarly, the cells for the small sample (new information) may be considered distributed Normal with parameters t_s and S_s^2/n_s.

Bayes' theorem states that if the prior and sample distributions are Normal with known variances σ^2, then the posterior (updated) distribution of the mean trip rates is also Normal with the following parameters:

$$t_2 = \frac{1/\sigma_1^2}{1/\sigma_1^2 + 1/\sigma_s^2}t_1 + \frac{1/\sigma_s^2}{1/\sigma_1^2 + 1/\sigma_s^2}t_s \tag{4.19}$$

$$\sigma_2^2 = \frac{1}{1/\sigma_1^2 + 1/\sigma_s^2} \tag{4.20}$$

which, substituting by the known values S^2 and n, yield:

$$t_2 = \frac{n_1 S_s^2 t_1 + n_s S_1^2 t_s}{n_1 S_s^2 + n_s S_1^2} \tag{4.21}$$

$$\sigma_2^2 = \frac{S_1^2 S_s^2}{n_1 S_s^2 + n_s S_1^2} \tag{4.22}$$

It is important to emphasise that this distribution is not that of the individual trip rates of each household in the corresponding cell, but that of the mean of the trip rates of the cell. In fact, the distribution of the individual rates is not known; the only information we have is that they share the same (posterior) mean t_2.

Example 4.6 The mean trip rate, its variance, and the number of observations for two household categories, for a study undertaken 10 years ago are shown in Table 4.9:

Table 4.9 Prior data for Example 4.6

	Household categories	
Variable (prior data)	1	2
Trips per day	8	5
No. of observations	65	300
Trip rate variance	64	15
Mean trip variance	0.98	0.05

It is felt that these values might be slightly out of date for direct use today, but there are not enough funds to embark on a full-scale survey. A small stratified sample is finally taken, which yields the values shown in Table 4.10:

Table 4.10 Data for stratified sample in Example 4.6

	Household categories	
Variable (new data)	1	2
Trips per day	12	6
No. of observations	30	30
Trip rate variance	144	36
Mean trip variance	4.80	1.20

The reader can check that by applying Eqs. (4.21) and (4.22) it is possible to estimate the trip rate values and variances shown in Table 4.11:

Table 4.11 Posterior results for Example 4.6

	Household categories	
Posterior	1	2
Trip rate (trips/day)	8.68	5.04
Variance	0.82	0.05

Exercises

4.1 Consider a zone with the characteristics shown in Table 4.12:

Table 4.12 Zonal characteristics for Exercise 4.1

Household type	No.	Income ($/month)	Inhabitants	Trips/day
0 cars	180	4000	4	6
1 car	80	18 000	4	8
2 or more cars	40	50 000	6	11

Due to a decrease in import duties and a real income increase of 30% it is expected that 50% of households without car would acquire one in the next five years. Estimate how many trips would the zone generate in that case; check whether your method is truly the best available.

4.2 Consider the following trip attraction models estimated using a standard computing package (*t*-ratios are given in parentheses);

$$Y = 123.2 + 0.89X_1 \qquad\qquad R^2 = 0.900$$
$$\quad\;\; (5.2) \qquad (7.3)$$

$$Y = 40.1 + 0.14X_2 + 0.61X_3 + 0.25X_4 \quad R^2 = 0.925$$
$$\quad (6.4) \qquad (1.9) \qquad (2.4) \qquad (1.8)$$

$$Y = -1.7 + 2.57X_1 - 1.78X_4 \qquad R^2 = 0.996$$
$$\quad (-0.6) \qquad (9.9) \qquad\;\; (-9.3)$$

where Y are work trips attracted to the zone, X_1 is total employment in the zone, X_2 is industrial employment in the zone, X_3 is commercial employment in the zone and X_4 is service employment.

Choose the most appropriate model, explaining clearly why (i.e. considering all its pros and cons).

4.3 Consider the following two AM peak work trip generation models, estimated by household linear regression:

$$y = 0.50 + 2.0x_1 + 1.5x_2 \qquad\qquad R^2 = 0.589$$
$$\quad (2.5) \qquad (6.9) \qquad (5.6)$$

$$y = 0.01 + 2.3x_1 + 1.1Z_1 + 4.1Z_2 \quad R^2 = 0.601$$
$$\quad (0.9) \qquad (4.6) \qquad (1.9) \qquad (3.4)$$

where y are household trips to work in the morning peak, x_1 is the number of workers in the household, x_2 is the number of cars in the household, Z_1 is a dummy variable that takes the value of 1 if the household has one car and Z_2 is a dummy that takes the value of 1 if the household has two or more cars.

a) Choose one of the models explaining clearly the reasoning behind your decision.
b) Graphically depict both models using appropriate axis.

c) If a zone has 1000 households (with an average of two workers per household), of which 50% have no cars, 35% have only one car and the rest exactly two cars, estimate the total number of trips generated by the zone, O_i, with both models. Discuss your results.

4.4 Table 4.13 presents data collected in the last household O–D survey (made 10 years ago) for three particular zones:

Table 4.13 Last household O–D survey data for Exercise 4.4

Zone	Residents/HH	Workers/HH	Mean income	Population
I	2.0	1.0	50 000	20 000
II	3.0	2.0	70 000	60 000
III	2.5	2.0	100 000	100 000

Ten years ago, two household-based trip generation models were estimated using this data. The first was a linear regression model given by:

$$y = 0.2 + 0.5x_1 + 1.1Z_1 \qquad R^2 = 0.78$$

where y are household peak hour trips, x_1 is the number of workers in the household and Z_1 is a dummy variable that takes the value of 1 for high-income (>70 000) households and 0 in other cases.

The second was a category analysis model based on two income strata (low and high income) and two levels of family structure (1 or less and 2 or more workers per household). The estimated trip rates are given in Table 4.14:

Table 4.14 Estimated trip rates for Exercise 4.4

Family structure	Income	
	Low	High
1 or less	0.8	1.0
2 or more	1.2	2.3

If the total number of trips generated today during the peak hour by the three zones are given in Table 4.15:

Table 4.15 Total number of trips during the peak hour

Zone	Peak hour trips
I	8200
II	24 300
III	92 500

and it is estimated that the zone characteristics (income, number of households, and family structure) have remained stable, decide which model is best. Explain your answer.

5

Trip Distribution Modelling

We have seen how trip generation models can estimate the total number of trips or tours emanating from a zone (origins, productions) and those attracted to each zone (destinations, attractions). Productions and attractions provide an idea of the level of trip-making in a study area, but this is seldom enough for modelling and decision-making. What is needed is a better idea of the pattern of trips and or tours, from where to where trips take place, the modes of transport chosen and, as we shall see in Chapter 10, the routes taken. The pattern of travel is essential to identify the potential modes and routes available and their respective levels-of-service.

The pattern of travel can be represented, at this stage, in at least two different ways. The first one is as a 'trip matrix' or 'trip table' storing the trips made from each origin to each destination during a particular time period; it is also called an origin–destination (O–D) matrix and may be disaggregated by person type and purpose or the activity undertaken at each end of the trip. This representation is needed later for all assignment models.

The second way of presenting a trip pattern is to consider the factors that generate and attract trips, that is, on a production–attraction (P–A) basis, with *Home* generally being treated as the 'producing' end, and *Work*, *Shop*, etc. as the 'attracting' end. By necessity, a P–A matrix will cover a longer time span, (usually a day) than an O–D matrix. Take for example a journey to school and back; on an O–D basis this will generate one trip in the morning from Home to School and a return trip in the afternoon. On a P–A basis the Home end will generate two school trips and the School end will attract two school trips during the day. Note that the P–A treatment is closer, but not equivalent, to the idea of tours.

Trip patterns obtained through intercept surveys (i.e. roadside interviews or public transport questionnaires) will result in O–D matrices that are probably partial; not all O–D pairs would have been sampled. Even a combination of intercept and home interview surveys will fail to produce matrices where all cells have been sampled. Modelling is required to generate fuller matrices in either P–A or O–D format. Alternatively, data from digital sources may be used to provide fuller trip matrices.

Several methods have been put forward over the years to allocate or distribute trips (from a trip generation model) among destinations; some of the simplest are only suitable for short-term, tactical studies where no major changes in the accessibility provided by the network are envisaged. Others were designed to respond better to changes in network cost and are, therefore, suggested for longer-term strategic studies or tactical ones involving significant changes in relative transport prices; these are often P–A based.

Trip Distribution is often seen as an aggregate problem with an aggregate model for its solution. In fact, most of its treatment in this chapter shares that view. However, the same problem can also

Modelling Transport, Fifth Edition. Juan de Dios Ortúzar and Luis G. Willumsen.
© 2024 John Wiley & Sons Ltd. Published 2024 by John Wiley & Sons Ltd.
Companion website: www.wiley.com/go/ortuzar5e

be considered as a discrete choice of destination (i.e. at a disaggregate level), and treated with models at the level of the individual. This is discussed in greater detail in subsequent chapters.

This chapter starts by detailing additional definitions and notation used; these include the idea of *generalised costs* of travel. The next section introduces methods that respond only to relative growth rates at origins and destinations; these are suitable only for short-term trend extrapolation. Section 5.3 discusses a family of synthetic models, the best known being the Gravity model. Approaches to model generation, in particular the entropy-maximising formalism, are presented in Section 5.4. An important aspect of the use of synthetic models is their calibration, that is, the task of finding values for their parameters so that the base-year travel pattern is well represented by the model; this is examined in Section 5.5. Section 5.6 presents a variation on the Gravity model calibration theme that enables more general forms for the model. Other synthetic models have also been proposed and the most important of them, the Intervening Opportunities Model, is explored in Section 5.7. Finally, the chapter concludes with some practical issues associated with distribution modelling.

5.1 Definitions and Notation

It is customary to represent the trip patterns in a study area by means of a trip matrix. This is essentially a two-dimensional array of cells where rows and columns represent each of the z zones in the study area (including external zones), as shown in Table 5.1.

The cells of each row i contain the trips originating in that zone which have as destinations the zones in the corresponding columns. The main diagonal corresponds to intra-zonal trips. T_{ij} is the number of trips between origin i and destination j; the total array is $\{T_{ij}\}$ or $\mathbf{T}$; O_i is the total number of trips originating in zone i, and D_j is the total number of trips attracted to zone j.

Table 5.1 A general form of a two-dimensional trip matrix

Origins	Destinations					$\sum_j T_{ij}$
	1	**2**	**3**	**...j**	**...z**	
1	T_{11}	T_{12}	T_{13}	$... T_{1j}$	$... T_{1z}$	O_1
2	T_{21}	T_{22}	T_{23}	$... T_{2j}$	$... T_{2z}$	O_2
3	T_{31}	T_{32}	T_{33}	$... T_{3j}$	$... T_{3z}$	O_3
$\vdots$						
i	T_{i1}	T_{i2}	T_{i3}	$... T_{ij}$	$... T_{iz}$	O_i
$\vdots$						
z	T_{z1}	T_{z2}	T_{z3}	$... T_{zj}$	$... T_{zz}$	O_z
$\sum_i T_{ij}$	D_1	D_2	D_3	$... D_j$	$... D_z$	$\sum_{ij} T_{ij} = T$

In a P–A format, P_i would be the number of trips produced or generated in a zone i and Q_j those attracted to zone j.

We shall use lower case letters, t_{ij}, o_i, and d_j to indicate observations from a sample or from an earlier study; capital letters will represent our target, or the values we are trying to model for the corresponding modelling period. The matrices can be further disaggregated, for example, by person type (n) and/or by mode (k). For example, T_{ij}^{kn} are trips from i to j by mode k and person type n; O_i^{kn} is the total number of trips originating at zone i by mode k and person type n, and so on.

Summation over sub or superscripts will be indicated implicitly by omission, e.g.

$$T_{ij}^n = \sum_k T_{ij}^{kn}$$

$$T = \sum_{ij} T_{ij} \quad \text{and} \quad t = \sum_{ij} t_{ij}$$

In some cases, it may be of interest to distinguish the proportion of trips using a particular mode and the cost of travelling between two points:

p_{ij}^k is the proportion of trips from i to j by mode k;

c_{ij}^k is the cost of travelling between i and j by mode k.

The sum of the trips in a row should equal the total number of trips emanating from that zone; the sum of the trips in a column should correspond to the number of trips attracted to that zone. These conditions can be written as:

$$\sum_j T_{ij} = O_i \qquad (5.1a)$$

$$\sum_i T_{ij} = D_j \qquad (5.1b)$$

If reliable information is available to estimate both O_i and D_j, then the model should aim to satisfy both conditions; in this case, the model would be *doubly constrained*. In some cases, however, there will be good information only about one of these constraints, for example, to estimate all the O_i's, and, therefore, the model will be *singly constrained*. Thus, a model can be origin or production constrained if the O_i's, are available, or destination or attraction constrained if the D_j's are at hand.

The cost element may be expressed in distance, time, or money units. It is often convenient to use a measure combining all the main attributes related to the disutility of a journey and this is normally referred to as the *generalised cost* of travel. This is typically a linear function of the attributes of the journey weighted by coefficients that attempt to represent their relative importance as perceived by the traveller (this generalises the simple concept we used in Chapter 3). One possible representation of this for mode k is (omitting superscript k for simplicity):

$$C_{ij} = a_1 t_{ij}^v + a_2 t_{ij}^w + a_3 t_{ij}^t + a_4 t_{ij}^n + a_5 F_{ij} + a_6 \phi_j + \delta \qquad (5.2)$$

where t_{ij}^v is the in-vehicle travel time between i and j; t_{ij}^w is the walking time to and from stops (stations) or from parking area/lot; t_{ij}^t is the waiting time at stops (or time spent searching for a parking space); t_{ij}^n is the interchange time, if any; F_{ij} is a monetary charge, for example, the fare charged to travel between i and j or the cost of using the car for that journey, including any tolls or congestion charges (note that car operating costs are often not well perceived and that electronic means of payment tend to blur somehow the link between use and payment); ϕ_j is a terminal (typically parking) cost associated with the journey from i to j; δ is a *modal penalty*, a parameter representing all other attributes not included in the generalised measure so far (e.g. safety, comfort, and convenience); $a_{1...6}$ are weights attached to each element of cost; they have dimensions appropriate for conversion of all attributes to common units (e.g. money or time).

If the generalised cost is measured in money units ($a_5 = 1$) then a_1 is sometimes interpreted as the *value of time* (or more precisely the *value of in-vehicle time savings*) as its units are money/time.

In that case, a_2 and a_3 would be the values of walking and waiting time respectively, and in many practical studies, they have been taken or found to be of the order of two or three times the expected value of a_1.

The generalised cost of travel, as expressed here, represents an interesting compromise between subjective and objective disutility of movement. It is meant to represent the disutility of travel as perceived by the trip maker; in that sense, the value of time should be a perceived value rather than an objective, resource-based, value. However, the coefficients $a_1 \ldots a_6$ are often provided externally to the modelling process, sometimes specified by government. This presumes stability and transferability of values for which there is only limited evidence.

As generalised costs may be measured in money or time units it is relatively easy to convert one into the other. For example, if the generalised cost is measured in time units, a_1 would be 1.0, $a_{2\ldots3}$ would probably be between 2.0 and 3.0, and $a_{5\ldots6}$ would represent something like the 'duration of money'.

There are theoretical and practical advantages in measuring generalised cost in time units. Consider, for example, the effect of income levels increasing with time; this would increase the *value of time* and, therefore, increase generalised costs and make the same destination appear more expensive. If, on the other hand, we measure generalised costs in time units, increased income levels would reduce the cost of reaching the same destination, and this seems intuitively more acceptable. There are formal reasons in evaluation to prefer expressing generalised cost in time units; we refer the interested reader to the excellent book by Jara-Díaz (2007). Moreover, it is easier to compare parameters across cities and countries when expressed in time units, thus reducing the importance of issues of exchange rates and purchase parity.

A distribution model tries to estimate the number of trips in each of the matrix cells based on any information available. Different distribution models have been proposed for different sets of problems and conditions. We shall explore, first, models which are mainly useful in updating a trip matrix, or in forecasting a future trip matrix, where information is only available in terms of future trip rates or growth factors. We shall then study more general models, in particular the Gravity model family. We shall finally explore the possibility of developing modal-split models from similar principles.

5.2 Growth-Factor Methods

Consider a situation where we have a trip matrix **t**, obtained from a previous study, or estimated from recent survey data. We want to estimate the matrix corresponding to the design year, say ten years into the future. We may have information about the expected growth rate in this 10-year period for the whole study area; alternatively, we may have information on the likely growth in the number of trips originating and/or attracted to each zone. Depending on this information we may be able to use different growth-factor methods in our estimation of future trip patterns.

5.2.1 Uniform Growth Factor

If the only information available is about a general growth rate τ for the whole of the study area, then we can only assume that it will apply to each cell in the matrix:

$$T_{ij} = \tau \cdot t_{ij} \text{ for each pair } i \text{ and } j \qquad (5.3)$$

Of course, $\tau = T/t$, that is, the ratio of expanded over previous total number of trips.

Example 5.1 Consider the simple four-by-four base-year trip matrix of Table 5.2. If the growth in traffic in the study area is expected to be of 20% in the next three years, it is a simple matter to multiply all cell values by 1.2 to obtain a new matrix as in Table 5.3.

Table 5.2 Base-year trip matrix

	1	2	3	4	$\sum_j$
1	5	50	100	200	355
2	50	5	100	300	455
3	50	100	5	100	255
4	100	200	250	20	570
$\sum_i$	205	355	455	620	1635

Table 5.3 Future estimated trip matrix with $\tau = 1.2$

	1	2	3	4	$\sum_j$
1	6	60	120	240	426
2	60	6	120	360	546
3	60	120	6	120	306
4	120	240	300	24	684
$\sum_i$	246	426	546	744	1962

The assumption of uniform growth is generally unrealistic except perhaps for very short time spans of, say, one or two years. In most other cases one would expect differential growth for different parts of the study area.

5.2.2 Singly Constrained Growth-Factor Methods

Consider the situation where information is available on the expected growth in trips originating in each zone, for example, shopping trips. In this case, it would be possible to apply this origin-specific growth factor (τ_i) to the corresponding rows in the trip matrix. The same approach can be followed if the information is available for trips attracted to each zone; in this case, the destination-specific growth factors (τ_j) would be applied to the corresponding columns. This can be written as:

$$T_{ij} = \tau_i \cdot t_{ij} \text{ for origin-specific factors} \tag{5.4}$$

$$T_{ij} = \tau_j \cdot t_{ij} \text{ for destination-specific factors} \tag{5.5}$$

Example 5.2 Consider Table 5.4, a revised version of Table 5.2 with growth predicted for origins:

Table 5.4 Origin-constrained growth trip table

	1	2	3	4	$\sum_j$	Target O_i
1	5	50	100	200	355	400
2	50	5	100	300	455	460
3	50	100	5	100	255	400
4	100	200	250	20	570	702
$\sum_i$	205	355	455	620	1635	1962

This problem can be solved immediately by multiplying each row by the ratio of target O_i over the base year total (Σ_j), thus giving the results in Table 5.5.

Table 5.5 Expanded origin-constrained growth trip table

	1	2	3	4	$\sum_j$	Target O_i
1	5.6	56.3	112.7	225.4	400	400
2	50.5	5.1	101.1	303.3	460	460
3	78.4	156.9	7.8	156.9	400	400
4	123.2	246.3	307.9	24.6	702	702
$\sum_i$	257.7	464.6	529.5	701.2	1962	1962

5.2.3 Doubly Constrained Growth Factors

A more interesting problem is generated when information is available on the future number of trips originating and terminating in each zone. This implies different growth rates for trips in and out of each zone and consequently having two sets of growth factors for each zone, say τ_i and Γ_j. The application of an 'average' growth factor, say $F_{ij} = 0.5\,(\tau_i + \Gamma_j)$ is a poor compromise as none of the two targets or trip-end constraints would be satisfied. Historically a number of iterative methods have been proposed to obtain an estimated trip matrix that satisfies both sets of trip-end constraints, or the two sets of growth factors, which is the same thing.

All these methods involve calculating a set of intermediate correction coefficients which are then applied to cell entries in each row or column as appropriate. After applying these corrections to say, each row, the totals for each column are calculated and compared with the target values. If the differences are significant, new correction coefficients are calculated and applied as necessary.

In the transport field, these methods are known by the names of their authors, that is, as *Fratar* in the US and *Furness* elsewhere. For example, Furness (1965) introduced 'balancing factors' A_i and B_j as follows:

$$T_{ij} = t_{ij} \cdot \tau_i \cdot \Gamma_j \cdot A_i \cdot B_j \tag{5.6}$$

or incorporating the growth rates into new variables a_i and b_j:

$$T_{ij} = t_{ij} \cdot a_i \cdot b_j \tag{5.7}$$

with $a_i = \tau_i A_i$ and $b_j = \Gamma_j B_j$.

The factors a_i and b_j (or A_i and B_j) must be calculated so that the constraints (5.1) are satisfied. This is achieved in an iterative process which in outline is as follows:

1) Set all $b_j = 1.0$ and solve for a_i; in this context, 'solve for a_i' means find the correction factors a_i that satisfy the trip generation constraints.
2) With the latest a_i solve for b_j (e.g. satisfy the trip attraction constraints).
3) Keeping the b_j's fixed, solve for a_i and repeat steps (2) and (3) until the changes are sufficiently small.

This method produces solutions within 3–5% of the target values in a few iterations when certain conditions are met. A tighter degree of convergence may be important from the perspective of model system consistency as we will see in Chapter 11. This method is often called a 'bi-proportional algorithm' because of the nature of the corrections involved. The problem is not restricted to transport; techniques to solve it have also been 'invented', among others, by Kruithof (1937) for telephone traffic and Bacharach (1970) for updating input–output matrices in economics. The best treatment of its mathematical properties seems to be due to Bregman (see Lamond and Stewart 1981).

It will be shown below that this method is a special case of entropy-maximising models of the Gravity type if the effect of distance or separation between zones is excluded. But in any case, the bi-proportional algorithm tries to produce the minimum corrections to the base-year matrix **t** necessary to satisfy the future year trip-end constraints.

A key condition for the convergence of this method is that the growth rates produce target values T_i and T_j such that their sums are both equal to T, the total number of trips in the design year.

$$\sum_i \tau_i \sum_j t_{ij} = \sum_j \Gamma_j \sum_i t_{ij} = T \tag{5.8}$$

Enforcing this condition may require correcting trip-end estimates produced by the trip generation models.

Example 5.3 Table 5.6 represents a doubly constrained growth factor problem:

Table 5.6 Doubly constrained matrix expansion problem

	1	2	3	4	$\sum_j$	Target O_i
1	5	50	100	200	355	400
2	50	5	100	300	455	460
3	50	100	5	100	255	400
4	100	200	250	20	570	702
$\sum_i$	205	355	455	620	1635	
Target D_j	260	400	500	802		1962

The solution to this problem, after three iterations on rows and columns (three sets of corrections for all rows and three for all columns), is shown in Table 5.7.

Table 5.7 Solution to the doubly constrained matrix expansion problem

	1	2	3	4	$\sum_j$	Target O_i
1	5.25	44.12	98.24	254.25	401.85	400
2	45.30	3.81	84.78	329.11	462.99	460
3	77.04	129.50	7.21	186.58	400.34	400
4	132.41	222.57	309.77	32.07	696.82	702
$\sum_i$	260.00	400.00	500.00	802.00	1962	
Target D_j	260	400	500	802		1962

Note that this estimated matrix is within 1% of meeting the target trip ends, more than enough accuracy for this problem.

5.2.4 Advantages and Limitations of Growth-Factor Methods

Growth-factor methods are simple to understand and directly use observed trip matrices and forecasts of trip-end growth. They preserve the observations as much as is consistent with the information available on growth rates. This advantage is also their limitation, as they are probably only reasonable for short-term planning horizons or when no changes in transport costs are expected.

Growth-factor methods require the same database as synthetic methods, namely an observed (sampled) trip matrix; this is an expensive data item. The methods are heavily dependent on the accuracy of the base-year trip matrix. As we have seen, this is never high for individual cell entries, and therefore, the resulting matrices are no more reliable than the sampled or observed ones. Applying successive correction factors may well amplify any base-year errors. Moreover, if parts of the base-year matrix are unobserved, they will remain so in the forecasts. Therefore, we cannot use these methods to fill unobserved cells of partially observed trip matrices.

But the most important limitation is that these methods do not consider changes in transport costs due to improvements (or new congestion) in the network. Therefore, they are of limited use in the analysis of policy options involving new modes, new links, pricing policies, and new zones.

5.3 Synthetic or Gravity Models

5.3.1 The Gravity Distribution Model

Distribution models of a different kind have been developed to assist in forecasting future trip patterns when important changes in the network take place. They start from assumptions about group trip-making behaviour and the way this is influenced by external factors such as total trip ends and distance travelled. The best known of these models is the *Gravity model*, originally generated from an analogy with Newton's gravitational law. This model estimates trips for each cell in the matrix without directly using the observed trip pattern; therefore, it is sometimes called synthetic as opposed to the previous growth-factor models.

Probably the first rigorous use of a Gravity model was by Casey (1955), who suggested the approach to synthesise shopping trips and catchment areas between towns in a region. In its simplest formulation, the model had the following functional form:

$$T_{ij} = \frac{\alpha P_i P_j}{d_{ij}^2} \qquad (5.9)$$

where P_i and P_j are the populations of the towns of origin and destination, d_{ij} is the distance between i and j, and α is a proportionality factor (with units: trips $\cdot$ distance^2/population^2).

This was soon considered to be too simplistic an analogy with the gravitational law, and early improvements included the use of total trip ends (O_i and D_j) instead of total populations, and a parameter n for calibration as the power for d_{ij}. This new parameter was not restricted to being an integer and different studies estimated values between 0.6 and 3.5.

The model was further generalised by assuming that the effect of distance or 'separation' could be modelled better by a decreasing function, to be specified, of the distance or travel cost between the zones. This can be written as:

$$T_{ij} = \alpha O_i D_j f(c_{ij}) \qquad (5.10)$$

where $f(c_{ij})$ is a generalised function of the travel costs with one or more parameters for calibration. This function often receives the name of '*deterrence function*' because it represents the disincentive to travel as distance (time) or cost increases. Popular versions for this function are:

$$f(c_{ij}) = \exp(-\beta c_{ij}) \qquad \text{exponential function} \qquad (5.11)$$

$$f(c_{ij}) = c_{ij}^{-n} \qquad \text{power function} \qquad (5.12)$$

$$f(c_{ij}) = c_{ij}^n \exp(-\beta c_{ij}) \qquad \text{combined function} \qquad (5.13)$$

The general form of these functions for different values of their parameters is shown in Figure 5.1.

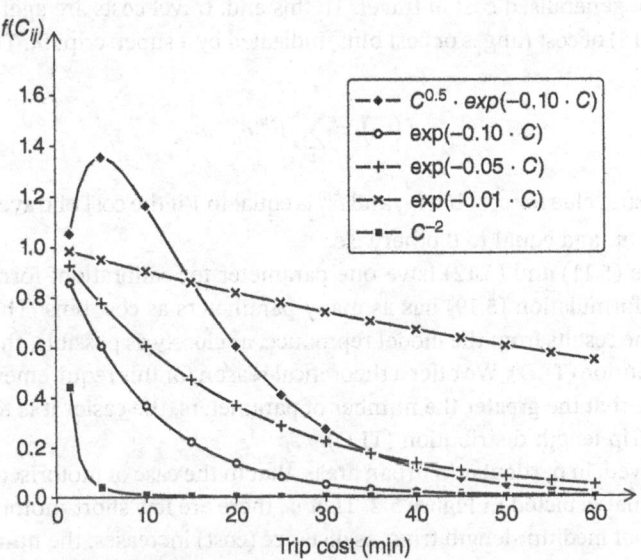

Figure 5.1 Different deterrence functions

5.3.2 Singly and Doubly Constrained Models

The need to ensure that the restrictions (5.1) are met requires replacing the single proportionality factor α by two sets of balancing factors A_i and B_j as in the bi-proportional method, yielding:

$$T_{ij} = A_i O_i B_j D_j f(c_{ij}) \tag{5.14}$$

In a similar vein, one can again subsume O_i and D_j into these factors and rewrite the model as:

$$T_{ij} = a_i b_j f(c_{ij}) \tag{5.15}$$

The expression in (5.14) or (5.15) is the classical version of the doubly constrained Gravity model. Singly constrained versions, either origin or destination constrained, can be produced by making one set of balancing factors A_i or B_j equal to one. For an origin-constrained model, $B_j = 1.0$ for all j, and

$$A_i = 1 \bigg/ \sum_j D_j f(c_{ij}) \tag{5.16}$$

In the case of the doubly constrained model, the values of the balancing factors are:

$$A_i = 1 \bigg/ \sum_j B_j D_j f(c_{ij}) \tag{5.17}$$

$$B_j = 1 \bigg/ \sum_i A_i O_i f(c_{ij}) \tag{5.18}$$

The balancing factors are, therefore, interdependent; this means that the calculation of one set requires the values of the other set. This suggests an iterative process analogous to the bi-proportional approach that works well in practice: given a set of values for the deterrence function $f(c_{ij})$, start with all $B_j = 1$, solve for A_i and then use these values to re-estimate the B_j's; repeat until convergence is achieved.

A more general version of the deterrence function accepts empirical values for it, and these depend only on the generalised cost of travel. To this end, travel costs are aggregated into a small number (say 10 or 15) of cost ranges or cost bins, indicated by a superscript m. The deterrence function then becomes:

$$f(c_{ij}) = \sum_m F^m \delta_{ij}^m \tag{5.19}$$

where F^m is the mean value for cost bin m, and δ_{ij}^m is equal to 1 if the cost of travelling between i and j falls in the range m, and equal to 0 otherwise.

The formulations (5.11) and (5.12) have one parameter for calibration; formulation (5.13) has two, β and n, and formulation (5.19) has as many parameters as cost bins. These parameters are estimated so that the results from the model reproduce, as closely as possible, the observations' trip length (cost) distribution (TLD). We offer a theoretical reason for this requirement below. However, it is enough to note that the greater the number of parameters, the easier it is to obtain a closer fit with the sampled trip length distribution (TLD).

It has been observed, in particular in urban areas, that in the case of motorised trips, the TLD has a shape similar to that depicted in Figure 5.2. That is, there are few short-motorised trips, followed by a larger number of medium-length trips; as distance (cost) increases, the number of trips decays again with a few very long trips. The negative exponential and power functions reproduce reasonably well the second part of the curve but not the first. That is one of the reasons behind the 'combined formulation', which is more likely to fit better both parts of the TLD. The greater flexibility of the cost-bin formulation permits an even better fit. However, the approach requires the assumption that the same TLD will be maintained in the future; this is similar but more stringent to requiring β to be the same for the base and the forecasting years.

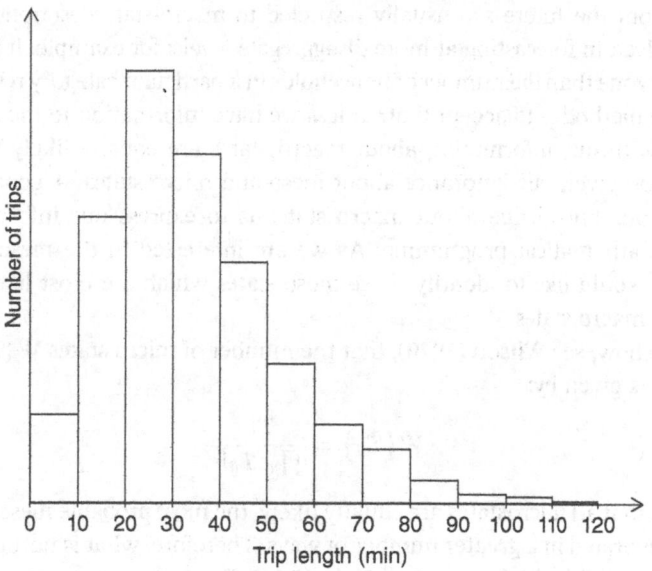

Figure 5.2   Typical trip length distribution in urban areas

Interestingly, the bulk of the representational and policy relevance advantages of the Gravity model lies in the deterrence function; the rest is very much like the Furness method.

5.4 The Entropy-Maximising Approach

5.4.1 Entropy and Model Generation

We shall now introduce the entropy-maximisation approach, which can be used to generate a wide range of models, including the Gravity, shopping, and location models. The approach has many followers and detractors, but it is generally acknowledged as one of the important contributions to improved modelling in transport. There are several ways of presenting the approach; we have chosen an intuitive rather than strictly mathematical formulation. For a stricter presentation and references to related and alternative approaches, see the seminal book by Wilson (1974).

Consider a system made up of a large number of distinct elements. A full description of such a system requires the complete specification of its *micro* states, as each is distinct and separable. This would involve, for example, identifying each individual traveller, its origin, destination, mode, time of journey, and so on. However, for many practical purposes it may be sufficient to work on the basis of a more aggregate or *meso* state specification; following our example, a meso state may just specify the *number* of trips between each origin and each destination. In general, there will be numerous and different micro states which produce the same meso state: Anne Smith and Pedro Pérez, living in the same zone, may exchange destinations generating different micro states but keeping the same meso state.

There is always an even higher level of aggregation, a *macro* state, for example, the total number of trips on particular links, or the total trips generated and attracted to each zone. To obtain reliable measures of trip-making activity it is often easier to make observations at this higher level of aggregation. In fact, most of our current information about a system is precisely at this level. In a similar way, estimates about the future are usually restricted to macro-state descriptions because of the uncertainties involved in forecasting at more disaggregate levels: for example, it is easier to forecast the population per zone than the number of households in a particular category residing in each zone.

The basis of the method is to accept that, unless we have information to the contrary, all micro states consistent with our information about macro states are equally likely to occur. This is a sensible assumption given our ignorance about meso and micro states. A good way of enforcing consistency with our knowledge about macro states is to express our information as equality constraints in a mathematical programme. As we are interested in the meso-state descriptions of the system, we would like to identify those meso states which are most likely, given our constraints about the macro states.

It is possible to show, see Wilson (1970), that the number of micro states $W\{T_{ij}\}$ associated with the meso state T_{ij} is given by:

$$W\{T_{ij}\} = \frac{T!}{\prod_{ij} T_{ij}!} \tag{5.20}$$

As it is assumed that all microstates are equally likely, the most probable meso state would be the one that can be generated in a greater number of ways. Therefore, what is needed is a technique to identify the values of $\{T_{ij}\}$ which maximise W in (5.20). For convenience, we seek to maximise a monotonic function of W, namely $\log W$, as both problems have the same maximum. Therefore:

$$\log W = \log \frac{T!}{\prod_{ij} T_{ij}!} = \log T! - \sum_{ij} \log T_{ij}! \tag{5.21}$$

Stirling's (short) approximation for $\log X! = X \log X - X$, can be used to make it easier to optimise:

$$\log W = \log T! - \sum_{ij} \left(T_{ij} \log T_{ij} - T_{ij} \right) \tag{5.22}$$

Usually the term $\log T!$ is a constant, therefore it can be omitted from the optimisation problem. The rest of the equation is often referred to as the *entropy function*:

$$\log W' = - \sum_{ij} \left(T_{ij} \log T_{ij} - T_{ij} \right) \tag{5.23}$$

Maximising $\log W'$, subject to constraints corresponding to our knowledge about the macro states, enables us to generate models to estimate the most likely meso states, in our case the most likely matrix **T**. The key to this model generation method is, therefore, the identification of suitable micro, meso, and macro state descriptions, together with the macro level constraints that must be met by the solution to the optimisation problem.

In some cases, there may be additional information in the form of prior or old values for the meso states, for example, an outdated trip matrix **t**. The problem may be recast with this information and the revised objective function becomes:

$$\log W'' = - \sum_{ij} \left(T_{ij} \log T_{ij}/t_{ij} - T_{ij} + t_{ij} \right) \tag{5.24}$$

This is an interesting function in which each element in the summation takes the value zero if $T_{ij} = t_{ij}$ and otherwise is a positive value that increases with the difference between **T** and **t**.

Therefore $-\log W'''$ is a good measure of the difference between **T** and **t**; it can further be shown that

$$-\log W'' \approx 0.5 \sum_{ij} \frac{(T_{ij} - t_{ij})^2}{t_{ij}} \tag{5.25}$$

where the right-hand side is another good measure of the difference between prior and estimated meso states. Models can be generated minimising $-\log W''$ subject to constraints reflecting our knowledge about macro states. The resulting model is the one with the meso states closest to the prior meso states, in the sense of Eq. (5.24) or approximately (5.25), and which satisfies the macro state constraints.

5.4.2 Generation of the Gravity Model

Consider the definition of micro, meso, and macro states from the discussion above. The problem becomes the maximisation of $\log W'$ subject to the following two sets of constraints corresponding to the meso states:

$$O_i - \sum_j T_{ij} = 0 \tag{5.26}$$

$$D_j - \sum_i T_{ij} = 0 \tag{5.27}$$

These two sets of constraints reflect our knowledge about trip productions and attractions in the zones of the study area. We are only interested in matrix entries that can be interpreted as trips; therefore, we need to introduce the additional constraint that:

$$T_{ij} \geq 0$$

The constrained maximisation problem can be handled forming the Lagrangian:

$$L = \log W' + \sum_i \alpha_i' \left\{ O_i - \sum_j T_{ij} \right\} + \sum_j \alpha_j'' \left\{ D_j - \sum_i T_{ij} \right\} \tag{5.28}$$

Taking the first partial derivatives with respect to T_{ij} and equating them to zero we obtain:

$$\frac{\partial L}{\partial T_{ij}} = -\log T_{ij} - \alpha_i' - \alpha_j'' = 0 \tag{5.29}$$

therefore

$$T_{ij} = \exp(-\alpha_i' - \alpha_j'') = \exp(-\alpha_i') \exp(-\alpha_j'')$$

The values of the Lagrange multipliers are easy to find; making a simple change of variables:

$$A_i O_i = \exp(-\alpha_i') \quad \text{and} \quad B_j D_j = \exp(-\alpha_j'')$$

we obtain

$$T_{ij} = A_i O_i B_j D_j \tag{5.30}$$

On the other hand, the use of $-\log W''$ as an objective function generates the model:

$$T_{ij} = A_i O_i B_j O_j t_{ij} \tag{5.31}$$

which is, of course, the basic Furness model. The version resulting in Eq. (5.30) corresponds to the case when there is no prior information (e.g. all $t_{ij} = 1$). These two models are close to but not yet the Gravity model. What is missing is the deterrence function term. Its introduction requires an additional constraint:

$$\sum_{ij} T_{ij} \cdot c_{ij} = C$$

where C is the (unknown) total expenditure in travel in the system (in generalised cost units if they are in use). Restating this constraint as

$$C - \sum_{ij} T_{ij} \cdot c_{ij} = 0 \tag{5.32}$$

one can maximise log W' subject to (5.26), (5.27), and (5.32), and using the same constrained optimisation technique it is possible to obtain the Lagrangian:

$$L = \log W' + \sum_i \alpha_i' \left\{ O_i - \sum_j T_{ij} \right\} + \sum_j \alpha_j'' \left\{ D_j - \sum_i T_{ij} \right\} + \beta \left\{ C - \sum_{ij} T_{ij} \cdot c_{ij} \right\} \tag{5.33}$$

Again, taking its first partial derivatives with respect to T_{ij} and equating them to zero gives

$$\frac{\partial L}{\partial T_{ij}} = -\log T_{ij} - \alpha_i' - \alpha_j'' - \beta c_{ij} = 0 \tag{5.34}$$

therefore

$$T_{ij} = \exp(-\alpha_i' - \alpha_j'' - \beta c_{ij}) = \exp(-\alpha_i')\exp(-\alpha_j'')\exp(-\beta c_{ij}) \tag{5.35}$$

Making the same change of variables as before one obtains:

$$T_{ij} = A_i O_i B_j D_j \exp(-\beta c_{ij}) \tag{5.36}$$

which is the classic Gravity model. The values for the balancing factors can be derived from the constraints as:

$$A_i = \left(\sum_j B_j D_j \exp(-\beta c_{ij}) \right)^{-1} \quad \text{and} \quad B_j = \left(\sum_i A_i O_i \exp(-\beta c_{ij}) \right)^{-1}$$

If one of (5.26) or (5.27) is omitted from the constraints a singly constrained Gravity model is obtained.

The Lagrange multipliers α_i' and α_j'' are the dual variables of the trip generation and attraction constraints and relate to the variations in entropy for a unit variation in trip generation and attraction. The value of β is related to the satisfaction of condition (5.32). In general, C can only be estimated and therefore β is left as a parameter for calibration in order to adjust the model to each specific area. Values of β cannot, therefore, be easily borrowed from one place to another. A useful first estimate for the value of β is one over the average travel cost; in effect, β is precisely measured in inverse of travel cost units.

The use of a different cost constraint, such as (5.37) instead of (5.32),

$$C' - \sum_{ij} T_{ij} \cdot \log c_{ij} = 0 \tag{5.37}$$

results in a model of the form

$$T_{ij} = A_i O_i B_j D_j \exp(-\beta' \log c_{ij}) = A_i O_i B_j D_j c_{ij}^{-\beta'} \tag{5.38}$$

that is, the Gravity model with an inverse power deterrence function!

The reader can verify that the simultaneous use of constraints (5.32) and (5.37) leads to a Gravity model with a combined deterrence function. A further interesting approach is to disaggregate constraint (5.32) into several trip cost groups or bins indicated, as before, by a superscript m:

$$C^m - \sum_{ij} T_{ij} \cdot c_{ij} \cdot \delta_{ij}^m = 0 \qquad \text{for each } m \qquad (5.39)$$

The maximisation of (5.23) subject to (5.26), (5.27), and (5.39) leads to:

$$T_{ij} = A_i O_i B_j D_j \sum_m F^m \delta_{ij}^m = a_i b_j \sum_m F^m \delta_{ij}^m \qquad (5.40)$$

which is, of course, the Gravity model with a cost-bin deterrence function. This model has some attractive properties, which will be discussed in Section 5.6.

5.4.3 Properties of the Gravity Model

As can be seen, entropy maximisation is a flexible approach for model generation. A whole family of distribution models can be generated by casting the problem in a mathematical programming framework: the maximisation of an entropy function subject to linear constraints representing our level of knowledge about the system. The use of this formalism has many advantages:

1) It provides a more rigorous way of specifying the mathematical properties of the resulting model. For example, it can be shown that the objective function is always convex; it can be shown also that, provided the constraints used, say (5.26) and (5.27) have a feasible solution space, the optimisation problem has a unique solution even if the values of all the parameters A_i and B_j are not unique (because one is redundant).
2) The use of a mathematical programming framework also facilitates the application of a standard tool-kit of solution methods and the analysis of the efficiency of alternative algorithms.
3) The theoretical framework used to generate the model also assists in providing an improved interpretation of the solutions generated by it. We have seen that the Gravity model can be generated from analogies with the physical world or from entropy-maximising considerations; the latter is closely related to information theory, to error measures, and maximum likelihood in statistics, and the three provide alternative ways of generating the same mathematical form of the model. However, although the functional form is the same, each theoretical framework provides a different interpretation of the problem and the solution found. Each may be more appropriate in specific circumstances. We shall come back to this *equifinality issue* in Chapter 8.
4) The fact that the Gravity model can be generated in a number of different ways does not make it 'correct'. The appropriateness of the model depends on the acceptability of the assumptions required for its generation and their interpretation. No model is ever appropriate or correct in itself, it can only be more or less suitable to handle a decision question given our understanding of the problem, of the options or schemes to be tested, the information available or collectable at a justifiable cost, and the time and resources securable for analysis; see the discussion on calibration and validation below.

It is interesting to contrast the classical Gravity model as in Eq. (5.36) with Furness's method as derived above in Eq. (5.31). We can see that one possible interpretation of the deterrence function is to provide a synthetic set of prior entries for each cell in the trip matrix (i.e. use of $\exp\left[-\beta c_{ij}\right]$ instead of t_{ij}). Both the deterrence function and the prior matrix t_{ij} take the role of providing 'structure' to the resulting trip matrix. This can be seen more clearly if one multiplies and divides the right-hand side of Eq. (5.31) by T and subsumes this constant in the balancing factors:

$$T_{ij} = T a_i b_j t_{ij} / T = a_i' b_j' p_{ij} \qquad (5.41)$$

where $p_{ij} = t_{ij}/T$, thus giving a better-defined meaning to 'structure' as the proportion of the total trips allocated to each origin–destination pair.

Example 5.4 It is useful to illustrate the Gravity model with an example related to the problem of expanding a trip matrix. Consider the cost matrix of Table 5.8 together with the total trip ends as in Table 5.6, and attempt to estimate the parameters a_i and b_j of a Gravity model of the type:

$$T_{ij} = a_i b_j \exp\left(-\beta c_{ij}\right)$$

Table 5.8 A cost matrix and trip-end totals for a Gravity model estimation

	Cost matrix (min)				
	1	2	3	4	Target O_i
1	3	11	18	22	400
2	12	3	12	19	460
3	15.5	13	5	7	400
4	24	18	8	5	702
Target D_j	260	400	500	802	1962

given the information that the best value of β is 0.10. The first step would be to build a matrix of the values $\exp(-\beta c_{ij})$, as in Table 5.9.

Table 5.9 The matrix $\exp(-\beta c_{ij})$ and sums to prepare for a Gravity model run

	$\exp(-\beta c_{ij})$				
	1	2	3	4	$\sum_j$
1	0.74	0.33	0.17	0.11	1.35
2	0.30	0.74	0.30	0.15	1.49
3	0.21	0.27	0.61	0.50	1.59
4	0.09	0.17	0.45	0.61	1.31
$\sum_i$	1.34	1.51	1.52	1.36	5.74

Base	1	2	3	4	$\sum_j$	Target	Ratio
1	253.12	113.73	56.48	37.86	461.19	400	0.87
2	102.91	253.12	102.91	51.10	510.04	460	0.90
3	72.52	93.12	207.23	169.67	542.54	400	0.74
4	31.00	56.48	153.52	207.23	448.23	702	1.57
$\sum_i$	459.54	516.45	520.15	465.87	1962.00		
Target	260	400	500	802			
Ratio	0.57	0.77	0.96	1.72			

With these values, we can calculate the resulting total 'trips' (5.74) and then expand each cell in the matrix by the ratio 1962/5.74 = 341.67. This produces a matrix of base trips that now has to be adjusted to match trip-end totals. This process is the same as Furness iterations. The values for a_i and b_j are the product of the corresponding correction factors; these will then be multiplied by the basic expansion factor 341.67. The resulting Gravity model matrix is given in Table 5.10.

Table 5.10 The resulting gravity model matrix with trip length distribution

	1	2	3	4	$\sum_j$	Target	Ratio	a_i
1	155.73	99.00	64.46	74.17	393.36	400	1.02	1.17
2	57.54	200.22	106.73	90.98	455.56	460	1.01	1.07
3	25.87	47.01	137.16	192.77	402.81	400	0.99	0.68
4	20.86	53.77	191.65	444.08	710.37	702	0.99	1.28
$\sum_i$	260.00	400.00	500.00	802.00	1962.00			
Target	260	400	500	802				
Ratio	1.00	1.00	1.00	1.00				
b_j	179.17	253.50	332.37	570.53				

Ranges (min)							
Cost	1.0–4.0	4.1–8.0	8.1–12.0	12.1–16.0	16.1–20.0	20.1–24	Sum
Trips	355.9	965.7	263.3	72.9	209.2	95.0	1962

The reader may wish to verify that the balancing factors a_i and b_j are only unique to a multiplicative constant. It is also possible to calculate, as usual, the standard balancing factors A_i and B_j dividing each corresponding a_i and b_j by the target values O_i and D_j.

5.4.4 Production–Attraction Format

Note that the Gravity model can also be used within a production–attraction format. In fact, there are some very good reasons to prefer the P–A format in demand modelling. The P–A approach is closer to dealing with simple tours (from and to home) rather than trips. In choosing their destination, travellers would consider the cost of getting there *and* returning home and not just the outward journey. The Gravity model is then treated in the same way as for trips although travel costs and the interpretation of the results are, of course, different. In this case, one should use an average (or the total) of the costs of travelling between the two zones. These costs should correspond to the correct time periods: the generalized cost of the outward and inward journeys. These times will depend on the trip's purpose. In an aggregate model, these times can only be an average as some travellers will have the two legs of the tour earlier and others later. The correct average measure, an inclusive value or *logsum*, will be discussed later.

The resulting P–A matrix will have to be converted into a directional O–D matrix to perform the assignment procedure. To achieve this, it is essential to have the distribution of the times for outbound and inbound trips, the best source of which will come from a good set of home interviews; during intercept surveys, the answers to the question about 'return' trips are less reliable. If we are only interested in the 24-hour case, the two demand matrices are practically the same as it is assumed that each production–attraction trip is made once in each direction during the day. This is, of course, an approximation but probably a reasonable one.

However, when a shorter-period O–D matrix is required, some trips will be made in the production-to-attraction direction while others only in the opposite one. Two different approaches can be used to overcome this problem. The first is very simplistic and requires producing a matrix for just a single purpose, typically 'to work', and then assuming that these trips follow just one direction of travel, thus producing, for example, the morning journey to work from production to attraction. Survey data must be used to correct shift work, flexible working hours, and trips for other purposes being made during the morning peak; however, the pattern of the morning peak is still dominated by this journey-to-work purpose. The second and better approach is to use survey data directly to determine the proportions of the matrices for each purpose that are deemed appropriate for the part of the day under consideration. For example, a typical morning peak matrix may consist of 70% production-to-attraction movements and only 15% of attraction-to-production movements.

There is a case for handling the mode choice model also in a P–A format. The same argument used for the Gravity model applies here. The choice of mode of travel is surely dependent on *all* trips of the tour; at least the P–A format captures the attributes of two of these trips. This argument is even stronger for more advanced 'time-of-day' choice models as we will discuss later.

5.4.5 Segmentation

The Gravity model can be applied with different levels of segmentation. The most obvious one is by journey purpose as different 'generators and attractors' will apply for journeys to Work, to School, Shopping, and Other.

It may also be desirable to segment by person type, at least 'car owners' and 'non car owners', as they are likely to have different influences in trip patterns and would probably perceive costs in different ways. Most non car owners will perceive public transport costs as the measure of separation. Car owners, on the other hand, will be influenced by a combination of car and public transport costs, their two basic options. In this case, an appropriate average of these should be incorporated in the model, again as a *logsum* as we will discuss later.

Although this segmentation, car and non-car owners, is possible at the production end it is not quite appropriate for the attraction end, especially in forecasting mode. Therefore, we will have these two segments competing for a set of job (and education) places at the attraction end. This requires a simple extension of the Gravity model equivalent of using an asymmetric matrix of $2N \times N$.

5.5 Calibration of Gravity Models

5.5.1 Calibration and Validation

Before using a Gravity distribution model, it is necessary to calibrate it; this just makes sure that its parameters are such that the model comes as close as possible to reproduce the base-year trip pattern. Calibration is, however, a very different process from model validation.

In the case of *calibration*, one is conditioned by the functional form and the number of parameters of the chosen model. For example, the classical Gravity model has the parameters A_i, B_j, and β (that is $Z + Z + 1$ parameters, Z being the number of zones). The sets of parameters A_i and B_j are calibrated during the estimation of the model, as part of the direct effort to satisfy constraints (5.1). Recall that at least one of the A_i or B_j is redundant as there is an additional condition $\sum_i O_i = \sum_j D_j = T$, and therefore one of the (5.1) constraints is linearly dependent on the rest. The parameter β, on the other hand, must be calibrated independently, as we do not have complete information about the total expenditure C in the study area. If we had this information, we could have used it directly without having to estimate β by other means. If the combined deterrence function (5.13) is used, we would have an additional parameter and therefore some additional flexibility in calibrating the Gravity model.

The *validation* task is different. In this case, one wants to make sure that the model is appropriate for the decisions likely to be tested with it. It may be that the Gravity model is not a sufficiently good representation of reality for the purpose of examining a particular set of decisions. It follows from this that the validation task depends on the nature of the policies and projects to be assessed.

A general strategy for validating a model would then be to check whether it can reproduce a known state of the system with sufficient accuracy. As the future is definitively not known, this task is sometimes attempted by trying to estimate some well-documented state in the past, say a matrix from an earlier study. However, it is seldom the case that such a past state is sufficiently well documented. Therefore, less demanding validation tests incorporating data not used during estimation are often employed, for example: to check whether the number of trips across important screen lines or along main roads is well reproduced.

5.5.2 Calibration Techniques

As we have seen, the parameters A_i and B_j are estimated as part of the bi-proportional balancing factor operations. The parameter β, on the other hand, is calibrated to make sure that the trip length distribution (TLD) is reproduced as closely as possible. This is a tall order for a single parameter. We shall see later how to improve on this but meantime, what is needed is a practical technique to estimate the best value for β, say β^*.

A naive approach to this task is simply to 'guess' or to 'borrow' a value for β, run the Gravity model and then extract the modelled trip length distribution (MTLD). This should be compared with the observed trip length distribution (OTLD). If they are not sufficiently close, a new guess for β can be used and the process repeated until a satisfactory fit between MTLD and OTLD is achieved; this would then be taken as the value β^*. Note that a set of home or roadside interviews will produce OTLDs with much greater accuracy than that of individual cell entries in the trip matrix because the sampling rate for trip lengths is in effect much higher in this case.

However, this naive approach is not very practical. Running a doubly constrained Gravity model is time-consuming and the approach provides no guidance on how to choose a better value for β if the current one is not satisfactory. Conventional curve-fitting techniques are unlikely to work well because the Gravity model is not just non-linear but also complex analytically; the A_i's and B_j's are also functions of β through the two sets of Eqs. (5.17) and (5.18).

A number of calibration techniques have been proposed and implemented in different software packages. The most important ones were compared by Williams (1976), who found that a technique due to Hyman (1969) was particularly robust and efficient. We shall describe briefly here Hyman's method.

At any stage in the calibration process, a trip matrix $\mathbf{T}(\beta)$, function of the current estimate of β, is available. This matrix also defines the total number of trips $\sum_{ij} T_{ij}(\beta) = T(\beta)$. Hyman's method is based on the following requirement for β:

$$c(\beta) = \sum_{ij} \left[T_{ij}(\beta) c_{ij} \right] / T(\beta) = c^* = \sum_{ij} \left(N_{ij} C_{ij} \right) / \sum_{ij} N_{ij} \tag{5.42}$$

where c^* is the mean cost from the OTLD and N_{ij} is the observed (and expanded) number of trips for each origin–destination pair. The method can be described as follows:

1) Start the first iteration making $m = 0$ and an initial estimate of $\beta_0 = 1/c^*$.

2) Using β_0 calculate a trip matrix using the standard Gravity model. Obtain the mean modelled trip cost c_0 and estimate a better value for β as follows:

$$\beta_m = \beta_0 c_0 / c^*$$

3) Make $m = m + 1$. Using the latest value for β (i.e. β_{m-1}) calculate a trip matrix using a standard Gravity model and obtain the new mean modelled trip cost c_{m-1} and compare it with c^*. If they are sufficiently close, stop and accept β_{m-1} as the best estimate for this parameter; otherwise, go to step 4.

4) Obtain a better estimate of β as:

$$\beta_{m+1} = \frac{(c^* - c_{m-1})\beta_m - (c^* - c_m)\beta_{m-1}}{c_m - c_{m-1}}$$

5) Repeat steps 3 and 4 as necessary (i.e. until the last mean modelled cost c_{m-1} is sufficiently close to the observed value c^*).

The recalculations in step 3 are made to approximate closer to the equality in (5.42). A few improvements can be introduced to this method, in particular from the computational point of view. But Hyman's approach has been shown to be robust and to offer, in general, advantages over alternative algorithms.

5.6 The Tri-Proportional Approach

5.6.1 Bi-Proportional Fitting

We have seen in Section 5.4.2 how the bi-proportional method can be derived from a mathematical programming framework. This non-linear mathematical program can be solved by a number of algorithms, including Newton's method. The simplest approach involves successive corrections by rows and then columns to satisfy the constraints; the algorithm stops when the corrections are small enough (i.e. when the constraints are met within reasonable tolerances).

The conditions necessary for the existence of a unique solution are that the set of constraints (5.26) and (5.27) define a feasible solution space in non-negative T'_{ij}s. This requires $\sum_i O_i = \sum_j D_j$ but this is not a sufficient condition. The model has a multiplicative form and therefore it preserves the zeros present in the prior matrix $\{t_{ij}\}$. The existence of many zero entries in the prior matrix may prevent the satisfaction of one or more constraints. In summary, the product $a_i\, b_j\, c_k$ is unique but not each individual factor; there are two-degrees of indeterminacy (say α and β) that can have arbitrary values without affecting the value of the product:

$$a_i \alpha b_j \beta c_k / \alpha\beta = a_i b_j c_k$$

Example 5.5 Consider the case where a previously empty zone k is expected to see development in the future, thus originating and attracting trips. The cell entries for t_{ik} and t_{kj} would have been zero whilst the future O_k and D_k are non-zero. Therefore, in this case, there are no possible multiplicative correction factors capable of generating a matrix satisfying the constraints for zone k. It may be possible, however, to replace these empty cell values by 'guesses' (i.e. suitable values borrowed from similar zones).

Nevertheless, the presence of zeros in the prior matrix may cause subtler but no less difficult problems. For example, if we try to solve the problem in Example 5.1 but with the prior matrix in Table 5.11, we will find that this problem has no feasible solution in non-negative T_{ij}; there are only 11 unknowns and 7 independent constraints but the position of the zeros is such that there is no feasible solution and the bi-proportional algorithm oscillates without converging.

Table 5.11 A revised version of the doubly constrained growth factor problem in Table 5.6

	1	2	3	4	$\sum_j$	Target O_i
1	5	50	100	200	355	400
2	0	50	0	0	50	460
3	50	100	5	100	255	400
4	100	200	250	20	570	702
$\sum_i$	155	400	355	320	1230	
Target D_j	260	400	500	802	1962	

Readers familiar with linear algebra will be able to describe this problem in terms of the rank of the original and an augmented matrix containing the last column in Table 5.7. Furthermore, the reader may verify that after 10 iterations with this problem, the corrected matrix stands as in Table 5.12:

Table 5.12 The matrix from problem in Table 5.11 after 10 Furness iterations

	1	2	3	4	$\sum_j$	Target O_i
1	3.4	0.7	61.0	355.3	420	400
2	0	388.2	0	0	388	460
3	65.5	2.8	5.9	345.7	420	400
4	191.2	8.3	433.1	101.0	734	702
$\sum_i$	260	400	500	802	1962	
Target D_j	260	400	500	802		1962

Several comments can be made at this stage:

1) The matrix after 10 iterations looks quite different from the prior one, thus casting some doubt about the realism, either of the old matrix, its zeros, or the new trip-end totals.
2) The main problem seems to be in the second row, where there is a big difference (about 20%) between target and modelled total. There is no way this row can add up to 460 as the only non-zero cell entry has a maximum of 400 trips. The constraints do not generate a feasible solution space.
3) The problem is ill-conditioned (e.g. a small change in a cell entry can make the problem a feasible one and produce a fairly different trip matrix). For example, the zero in cell $t_{2,4}$ could have arisen because of the sample used; replacing this zero by a 1 produces the matrix in Table 5.13 after the same 10 iterations. This is a much-improved match with a fairly different matrix. In fact, it matches the targets with better than 1% accuracy. There is now a feasible solution space.

Real matrices are often sparse and this type of difficulty cannot be discarded as an academic problem. Failure to converge in a few iterations may well indicate that the presence and location of zeros in the prior matrix prevents the existence of a feasible solution with the new trip ends.

Table 5.13 Matrix from Table 5.11 plus a single trip in (2, 4) after 10 iterations

	1	2	3	4	$\sum_j$	Target O_i
1	4.1	4.5	76.2	315.4	400	400
2	0	339.2	0	119.1	458	460
3	77.3	17.0	7.2	298.5	400	400
4	178.6	39.3	416.6	68.9	703	702
$\sum_i$	260	400	500	802	1962	
Target D_j	260	400	500	802		1962

5.6.2 A Tri-Proportional Problem

The Gravity model with a flexible deterrence function that takes discrete values constrained by a functional form for each cost bin was written in Eq. (5.40) as:

$$T_{ij} = a_i b_j \sum_m F^m \delta_{ij}^m$$

The main advantages of this model are its flexibility and ease of calibration. In effect, we can define any number of cost bins and the deterrence function can take any positive value for them; we could even represent situations where, for example, there are few short trips, many intermediate trips, few long trips, and again a larger number of long-distance commuting trips.

The calibration of this model requires finding suitable values for the deterrence factor F^m for each cost bin so that the number of trips undertaken for that distance is as close as possible to the observed number. This task is, in fact, very similar to the problem of grossing up a matrix to match trip-end totals. In this case, we can start with a unity value for the deterrence factors and then correct these and the parameters a_i and b_j until the trip ends and the TLD constraints are met. It seems natural to extend the bi-proportional algorithm to handle this third dimension (cost bins) and utilise a tri-proportional method to calibrate the model.

The principles behind the technique were proposed by Evans and Kirby (1974). Murchland (1977) has shown that the application of successive corrections on a two-, three- or multi-dimensional space conforms to just one of a group of possible algorithms to solve this type of problems; furthermore, the method is simple to program and does not make excessive demands on computer memory.

Example 5.6 The tri-proportional algorithm can be illustrated with the problem stated in Table 5.8 and with the trip length distribution (cost-bin) targets of Table 5.14.

Table 5.14 TLD target values for a tri-proportional gravity model calibration

	Ranges					
	1.0–4.0	4.1–8.0	8.1–12.0	12.1–16.0	16.1–20.0	20.1–24+
TLD	365	962	160	150	230	95

The model can then be solved using balancing operations to match trip targets by origin, destination, and cost bin. After five complete iterations, the matrix and modelled trips by cost bin T_k shown in Table 5.15 are obtained.

Table 5.15 The matrix from Table 5.8 with costs in Table 5.14 after five iterations

	1	2	3	4	$\sum_j$	a_i
1	161.6	102.5	60.8	72.5	397.4	1.27
2	56.5	199.4	101.2	101.0	458.0	1.13
3	18.9	48.7	116.7	217.1	401.4	0.60
4	23.0	49.5	221.3	411.5	705.3	1.14
$\sum_i$	260	400	500	802	1962	
b_j	0.57	0.70	0.87	1.63		

Ranges						
	1.0–4.0	4.1–8.0	8.1–12.0	12.1–16.0	16.1–20.0	20.1–24+
TLD	365	962	160	150	230	95
T_k	360.9	966.5	159.0	149.8	230.3	95.5
F_k	224.55	220.13	87.54	102.05	54.66	34.90

Of course, in this case, the balancing factors are again not unique, at least up to two arbitrary multiplicative constants. Another way of expressing this is to say the balancing factors have two *degrees of indeterminacy*, the two multiplicative constants. It is easy to see that if we multiply each a_i by a factor Γ and each b_j by another factor τ, and then divide each F^k by the product $\Gamma \tau$, the modelled matrix will remain unchanged.

5.6.3 Partial Matrix Techniques

The tri-proportional calibration method has been used with a full trip length distribution (i.e. one that has an entry from observations in each cell). It would certainly be advantageous if one could calibrate a suitable Gravity model without requiring a complete or full trip matrix. This is particularly important as we know that the cost of collecting data to obtain a complete trip matrix is rather high; furthermore, the accuracy of some of the cell entries is not very high and in calibration we actually use aggregations of the data, namely the TLD and the total trip ends O_i and D_j. Having explored the preferred methods for calibration, it should be clear that the possibility of calibrating a Gravity model with an incomplete or partial matrix does actually exist. For example, we can calibrate a model with exponential cost function just with the total trip ends and a good estimate of the average trip cost, c^*.

The calibration of a Gravity model with general deterrence function using the tri-proportional method is even more attractive in this case, as we could use just roadside interviews on cordons and screen-lines to obtain good TLDs and trip ends for some but not all the zones in the study area. There would be no need to use trip generation models except for forecasting purposes.

Example 5.7 The basic idea above can be described with the aid of a 3 × 3 matrix. Consider first a bi-proportional case where the full matrix-updating problem is to adjust a base-year matrix as follows:

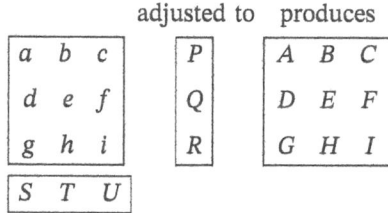

In the case of a partial matrix, for example, a survey where entries a and h cannot be observed, we would adjust only to trip ends excluding the corresponding total:

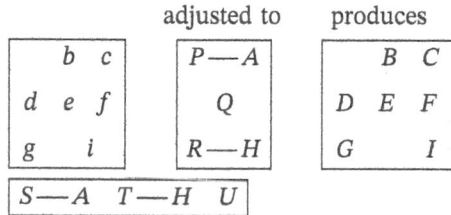

To fill in the missing cells we could use a Gravity model; in the case of this example, one without deterrence function:

$$T_{ij} = a_i b_j$$

The estimated values of a_i and b_j (using data from the observed cells) would then be used to fill in these cells.

An extension to the tri-proportional case is almost trivial. Kirby (1979) has shown that there are two basic conditions required for a valid application of this approach:

1) The Gravity model must fit both the available data and the data that are not available (i.e. the model must be a good model for the two regions of the matrix: the observed and the unobserved).
2) The two regions of the matrix should not be separable, that is, it should not be possible to split the matrix into two or more independent matrices, typically:

	Internal	External
Internal	× ×	* *
External	* *	× ×

The problem is that each separate area has the two (or three in the tri-proportional case) degrees of indeterminacy and therefore the balancing factors cannot produce unique products, and hence trip estimates. This problem is also referred to as the *non-identifiability* of unique products for unobserved cell entries. As the figure above shows, this is likely to occur when roadside interviews take place only on a cordon to a study area. The provision of interviews on a screen-line will probably eliminate the problem as it would generate observations for the 'internal-internal' matrix

5.7 Other Synthetic Models

5.7.1 Generalisations of the Gravity Model

The classic Gravity model is the most commonly used aggregate trip distribution model. It has several theoretical advantages, and there is suitable software to calibrate and use it. It can be extended further to incorporate more than one person type and even be used to model specific types of freight movements. However, the classic Gravity model only exhausts some of the theoretical possibilities. We explore here three other approaches which, although less used, offer natural alternatives to it. The first one is simply a generalisation of the Gravity model itself; the second one is the *Intervening-opportunities* model, and the third one the family of *Direct demand* models that will be discussed in Chapter 6.

Several authors have suggested extending the classic Gravity model to consider not just the deterrent effect of distance but also the fact that the farther away people are willing to travel the greater the number of opportunities to satisfy their needs.

Fang and Tsao (1995) suggested an entropy distribution model with quadratic costs:

$$T_{ij} = A_i B_j O_i D_j e^{-\beta C_{ij} - \lambda T_{ij} \cdot C_{ij}} \tag{5.43}$$

$$A_i = \frac{1}{\sum\limits_j B_j D_j e^{-\beta C_{ij} - \lambda T_{ij} \cdot C_{ij}}}, \quad B_j = \frac{1}{\sum\limits_i A_i O_i e^{-\beta C_{ij} - \lambda T_{ij} \cdot C_{ij}}} \tag{5.44}$$

They call it a *Self-deterrent gravity* model. The inclusion of a 'congestion term' $\lambda T_{ij} C_{ij}$ in the exponent is the main extension to the classic model. The parameters β and λ are expected to have the same sign; if they have different signs this would indicate that certain trips have economies of scale: they become more attractive with the greater number of people undertaking them. If $\lambda = 0$ we have the classic Gravity model.

De Grange et al. (2010) generalised this approach and proposed to:

$$\frac{\min}{\{T_{ij}\}} \quad Z = \sum_{ij} T_{ij} \cdot C_{ij} + \frac{1}{\beta} \sum_{ij} T_{ij} (\ln T_{ij} - 1) - \frac{\rho}{\beta} \sum_{ij} T_{ij} \cdot \ln S_{ij} + \frac{\lambda}{2\beta} \sum_{ij} C_{ij} \cdot T_{ij}^2$$

$$s.t.$$

$$\sum_j T_{ij} = O_i \qquad (\mu_i) \tag{5.45}$$

$$\sum_i T_{ij} = D_j \qquad (\gamma_j)$$

where:

$$S_{ij} = \sum_{\substack{k=1 \\ k \neq i, k \neq j}}^{n} D_k e^{-C_{jk}} \tag{5.46}$$

and the term S_{ij} represents the accessibility to destinations as perceived from the origin i. Applying the optimality conditions to (5.45) results in

$$T_{ij} = A_i B_j O_i D_j \left(S_{ij}\right)^\rho e^{-\beta C_{ij} - \lambda T_{ij} \cdot C_{ij}} \tag{5.47}$$

$$A_i = \frac{1}{\displaystyle\sum_j B_j D_j \left(S_{ij}\right)^\rho e^{-\beta C_{ij} - \lambda T_{ij} \cdot C_{ij}}} \tag{5.48}$$

$$B_j = \frac{1}{\displaystyle\sum_i A_i O_i \left(S_{ij}\right)^\rho e^{-\beta C_{ij} - \lambda T_{ij} \cdot C_{ij}}} \tag{5.49}$$

and the best-fit parameters can be found using maximum likelihood.

Here if calibration results in $\rho = 0$, we find Fang and Tsao's model. This general model was estimated by De Grange et al. (2010) for Santiago using different levels of aggregation.

5.7.2 Intervening Opportunities Model

The basic idea behind this model is that trip making is not explicitly related to distance but to the relative accessibility of opportunities for satisfying the objective of the trip. The original proponent of this approach was Stouffer (1940), who also applied his ideas to migration and the location of services and residences. But it was Schneider (1959) who developed the theory in the way it is presented here.

Consider first a zone of origin i and rank all possible destinations in order of increasing distance from i. Then look at one origin–destination pair (i, j), where j is the mth destination in order of distance from i. There are $m - 1$ alternative destinations actually closer (more accessible) to i. A trip maker would certainly consider those destinations as possible locations to satisfy the need giving rise to the journey: these are the *intervening opportunities* influencing a destination choice. Let α be the probability of a trip maker being satisfied with a single opportunity; the probability of her being attracted by a zone with D opportunities is then αD.

Consider now the probability q_i^m of not being satisfied by any of the opportunities offered by the mth destinations away from i. This is equal to the probability of not being satisfied by the first, nor the second, and so on up to the mth:

$$q_i^m = q_i^{m-1} \left(1 - \alpha D_i^m\right) \tag{5.50}$$

therefore, omitting the subscript i for simplicity we get

$$\frac{q^m - q^{m-1}}{q^m} = -\alpha D^m \tag{5.51}$$

Now, if we make x_m the cumulative attractions of the intervening opportunities at the mth destination:

$$x_m = \sum_m D^m$$

we can rewrite (5.50) as

$$\frac{q^m - q^{m-1}}{q^{m-1}} = -\alpha[x_{m-1} - x_m] \tag{5.52}$$

The limit of this expression for infinitesimally small increments is, of course,

$$\frac{dq(x)}{q(x)} = -\alpha \, dx \qquad (5.53)$$

Integrating (5.53) we obtain:

$$\log q(x) = -\alpha x + \text{constant}$$

or

$$q(x) = A_i \exp(-\alpha x) \qquad (5.54)$$

where A_i is a parameter for calibration. This relationship expresses the chance of a trip purpose not being satisfied by any of the m destinations ($m = 1,...,M$) from i as a negative exponential function of the accumulated or intervening opportunities at that distance from the origin. The trips T_{ij}^m from i to a destination j (which happens to be the mth away from i) is then proportional to the probability of not being satisfied by any of the $m - 1$ closer opportunities minus the probability of not being satisfied by any of the opportunities up to the mth destination:

$$T_{ij}^m = O_i[q_i(x_{m-1}) - q_i(x_m)]$$
$$T_{ij}^m = O_i A_i[\exp(-\alpha x_{m-1}) - \exp(-\alpha x_m)] \qquad (5.55)$$

It is easy to show that the constant A_i must be equal to

$$A_i = 1/[1 - \exp(-\propto x_M)] \qquad (5.56)$$

to ensure that the trip-end constraints are satisfied. The complete model then becomes:

$$T_{ij}^m = O_i \frac{[\exp(-\propto x_{m-1}) - \exp(-\propto x_m)]}{[1 - \exp(-\propto x_M)]} \qquad (5.57)$$

Wilson (1970) has shown that this expression can also be derived from entropy-maximisation considerations.

The Intervening-opportunities model is interesting because it starts from different first principles in its derivation: it uses distance as an ordinal variable instead of a continuous cardinal one as in the Gravity model. It explicitly considers the opportunities available to satisfy a trip purpose at increased distance from the origin. However, the model is not often used in practice, probably for the following reasons:

- the theoretical basis is less well-known and possibly more difficult to understand by practitioners;
- the idea of matrices with destinations ranked by distance from the origin (the nth cell for origin i is not destination n but the nth destination away from i) is more difficult to handle in practice;
- the theoretical and practical advantages of this function over the Gravity model are not overwhelming;
- lack of suitable software.

In Chapter 14 we will discuss a more general version of this model that combines gravity and intervening-opportunities features. This is due to Wills (1986) and lets the data decide which combination of the two models fits reality better. However, the computational complexity of this new model is considerable.

5.7.3 Disaggregate Approaches

The whole discussion about distribution models has been cast in terms of zonal-based productions–attractions and origins–destinations. We may have increased disaggregation by considering journey purposes and simple person types (with and without a car). Couched in these terms we obtain the number of trips undertaken between each O–D pair. It can be argued, as in Chapters 7–9, that this is too coarse to capture the rich characteristics of travel behaviour; to achieve this capture we need to model to the level of individuals, or at least, representative individuals.

In this context, we do not deal with the number of trips to a particular destination but rather with the probability that a (representative) individual would choose a particular destination to satisfy some basic need. These 'disaggregate models' are probabilistic although they may share apparently similar functional forms.

For example, a disaggregate model that would consider the choice of destination as discussed in Section 8.3.3, is likely to have a Multinomial Logit model structure (as we will discuss in Chapter 7) with a form similar to a singly constrained Gravity model.

5.8 Practical Considerations

We have discussed several frequently used models to associate origins and destinations and estimate the number of trips between them. While doing so, we have omitted several practical considerations that must necessarily affect the accuracy attainable from the use of such models. These stem from the inherent limitations of our modelling framework and our inability to include detailed descriptions of reality in the models.

It must be recognised that the actual task of modelling how people choose their destination (in practice a place of work, study, shopping, and so on) using a few variables is an extremely difficult one. People choose a place of residence or work based on several factors that are too difficult to forecast. For residence, for example, it may depend on the quality of schools for their children, where friends live, the quality of any parks nearby, etc. Jobs are similarly specific; they must be appropriate to the skills, qualifications, and experience and from an interesting company. Distance clearly will have a role but it will be one of many. Choosing where to shop or visit the cinema must be more flexible, although the type of shops in each location will have an influence. Moreover, these models based on cross-section data implicitly assume instant response to changes in conditions. As soon as a new road is opened people would instantly change homes, jobs, and schools; this is clearly unrealistic.

Indeed, the practical experience of assigning a synthetically generated trip matrix, whatever its model form, onto a network, produces flows that are significantly different from observations. It is important, therefore, to develop ways to deal with this mismatch and improve upon the results of trip distribution or destination choice models.

We shall discuss these practical aspects under the general headings below.

5.8.1 Sparse Matrices

Observed trip matrices are almost always sparse (i.e. they have a large number of empty cells, and it is easy to see why). A study area with 500 zones (250 000 cells) may have some 2.5 million expected total trips during a peak hour. This yields an average of 10 trips per cell; however, some O–D pairs are more likely to contain trips than others, in particular from residential to high employment areas, thus leaving numerous cells with a very low number of expected trips. Consider now the method used to observe this trip matrix, perhaps roadside interviews. If the sampling rate is 20% (1 in 5) then the chances of making no observations on a particular O–D pair are very high.

This sampled trip matrix will then be expanded, probably using information about the exact sampling ratios in each interview station. The problem generated when expanding empty cells has already been alluded to in Section 5.2.3. It may be possible to fill in gaps in the matrix through the use of a partial matrix approach; alternatively, it may be desirable to 'seed' empty cells with a low number and use an alternative matrix expansion method such as that discussed in Chapter 14. It is important to realise, however, that 'observed' trip matrices normally contain a large number of errors and that these will be amplified by the expansion process.

5.8.2 Treatment of External Zones

It may be quite reasonable to postulate the suitability of a synthetic trip distribution model in a study area, in particular for internal-to-internal trips. However, a significant proportion of the trips may have at least one end outside the area. The suitability of a model which depends on trip distance or cost, a variable essentially undefined for external trips, is thus debatable. Moreover, external trips will be mainly by non-residents and therefore not included in the trip generation model.

Common practice in such cases is to take these trips outside the synthetic modelling process: roadside interviews are undertaken on cordon points at the entrance/exit to the study area. The resulting matrix of external–external (E–E) and external–internal (E–I) trips is then updated and forecasted using growth factor methods, in particular, Furness. However, a number of trip-ends from the trip generation/attraction models correspond to the E–I trips and these must be subtracted from the trip-end totals for inclusion as constraints to the synthetic models.

5.8.3 Special Generators

Some trips cannot be expected to be well represented by trip generation and attraction models: trips to and from airports, train and coach stations, hotel clusters, and ports in the case of freight. They are certainly attractors for employment but their trip generation needs to be handled separately with models developed *ad hoc* for that purpose. The trips emanating from these special generators can then be distributed using conventional singly constrained Gravity models; these are likely to need k factors (see Section 5.8.6) to represent, for example, the special relationships between international business travellers and certain areas of interest to them.

5.8.4 Intra-Zonal Trips

A similar problem occurs with intra-zonal trips. Given the limitations of any zoning system, the cost values given to centroid connectors are a crude but necessary approximation to those experienced in reality. The idea of an intra-zonal trip cost is then poorly represented by these centroid connector costs. Some commercial software packages allow the user to add/subtract terminal costs to facilitate better modelling of these trips; the idea is that by manipulating these intra-zonal costs one would make the Gravity model fit better. However, this is not very good; it is actually preferable to remove intra-zonal trips from the synthetic modelling process and to forecast those using even simpler approaches. This typically assumes that intra-zonal trips are a fixed proportion of the trip ends calculated by the trip generation models.

Moreover, intra-zonal trips are not normally loaded onto the network as they 'move' from a centroid to itself. This makes it less essential to model them in detail. However, in reality, some of these trips use the modelled network. Nevertheless, this problem is probably of significance only for rather coarse zoning systems.

5.8.5 Journey Purposes

Different models are normally used for different trip purposes and/or person types. Typically, the journey to work will be modelled using a doubly constrained Gravity model while almost all other purposes will be modelled using singly constrained models. This is because it is often difficult to estimate trip attractions accurately for shopping, recreational and social trips, and therefore, proxies for trip attractiveness are used: retail floor space, recreational areas, and population.

Some trip purposes may be more sensitive to cost and therefore deserve the use of different values for the deterrence function.

5.8.6 *K* Factors

The Gravity model can provide a reasonable representation of trip patterns provided they can be explained mainly by the size of the generation and attraction power of zones and the deterrence to travel generated by distance (generalised cost). However, we recognise that most individual decisions on residential location and/or choice of employment may incorporate many other factors; therefore, the Gravity model could only model destination choice at an aggregate level if the importance of these other factors were much reduced on aggregation. Notwithstanding, there are always aggregate effects that do not conform to a simple Gravity model. In some circumstances, there may be pairs of zones that have a special association in terms of trip making; for example, a major manufacturer may be located in one zone and most of its employees in another, perhaps as a result of a housing estate developed by the company. In this case, it is likely that more trips will take place between these two points than predicted by any model failing to consider this association. This has led to the introduction of an additional set of parameters K_{ij} to the Gravity model as follows:

$$T_{ij} = K_{ij}A_iO_iB_{ij} \exp\left(-\beta c_{ij}\right) \tag{5.58}$$

Practical studies have used these *K* factors in an attempt to improve the calibration of the model. Of course, there cannot be factors for each O–D pair as allowed in the equation above. In practice, they are chosen to affect only groups of zones, keeping them as a relatively low number. They could be used to represent how strong the psychological barrier of a river or railway track represents, or the stronger association between zones that grew simultaneously. This will improve the model fit.

Whether these *K* factors will remain constant in the future is a difficult matter; associations between zones are likely to change over time, and the impact of a river may be diminished as more bridges are built. Duffus et al. (1987) investigated this issue using data from several comprehensive surveys of the same city in Canada. They concluded that while *K*-factors are very meaningful in theory, they were inconsistent from one prediction period to the next, casting doubts about their stability over time.

Note that *K* factors are also related to the use of incremental models, that is when the Gravity model, for example, is used to estimate not the absolute number of trips but only the incremental change on trips over an observed trip matrix; this application is discussed in Chapter 14.

The best advice that can be given in respect of *K* factors is to use them with care, in limited quantity, or to prefer incremental approaches.

5.8.7 Adjusting Trip Matrices

Many studies adjust the trip matrices resulting from synthetic models to better represent observations. This is often undertaken using traffic counts measured individually or at screen lines; techniques for doing this are discussed in Section 14.4. The result is a new trip matrix that is more closely aligned with the observed traffic and public transport usage. It is then possible to estimate, for each O–D cell, the difference between the synthetic model results and the corrected trip matrix, a set of Δ_{ij}. These can be positive or negative and can be preserved in future years' forecasts.

$$T_{ij}^{\text{future}} = \text{Model } T_{ij}^{\text{future}} + \Delta_{ij} \tag{5.59}$$

It is possible, and it happens, that (5.59) generates negative trips in the future, so care must be taken when applying it. This is, again, an issue related to incremental modelling as discussed in Chapter 14.

5.8.8 Errors in Modelling

Many of the above practical issues reduce the accuracy of the modelling process. This is a reflection of the contrast between the limitations of the state-of-the-art in transport modelling and the complexities and inherent uncertainties of present and future human behaviour. These practical issues are not restricted to distribution models; they are present, in one form or another, in other parts of the modelling process.

Because many cells in a trip matrix will have small values, say between 0 and 5 in the sample and perhaps 20–30 in the expanded or synthesised matrix, their corresponding errors will be relatively large. A few studies have tackled the task of calibrating synthetic models and then compare the resulting trip matrices with observed ones. An investigation by Sikdar and Hutchinson (1981) used data from 28 study areas in Canada to calibrate and test doubly constrained Gravity models. They found that the performance of these models was poor, equivalent to a randomly introduced errors in the observations of about 75–100%; these results reinforce the call for caution in using the results of such models. This should not be entirely surprising; to model a trip matrix with the use of a few parameters (twice the number of zones for an exponential deterrence function) is a very tall order. This is certainly one of the reasons why few studies nowadays make use of the Gravity model in its conventional form. In many cases, however, it is desirable to consider how changes in transport costs would influence trip patterns, in particular for more optional purposes like shopping and recreation. In these cases, the idea of using 'pivot point' or 'incremental' versions of the Gravity model becomes more attractive, as discussed in Chapter 14.

The treatment of errors in modelling has received attention for some time. There seem to be two methods deserving consideration in this field: statistical and simulation approaches. Statistical methods are very powerful but they are not always easy to develop or implement. They follow the lines suggested in Chapter 2 when discussing the role of data errors in the overall accuracy of the modelling process. Errors in the data are then traced through to errors in the outputs of the models. The UK Department of Transport provides advice in the *Traffic Appraisal Manual* (Department of Transport 1985) on the sensitivity of distribution models to errors in the input data. To some extent, the simplest problem is to follow data errors, in particular those due to sampling, through the process of building matrices.

A more demanding problem is to follow these errors when a synthetic distribution model is used. One of the advances of the early 1980s was the development of approximate analytical techniques to estimate the output errors due to sampling variability. For example, the work of Gunn et al. (1980) established approximate expressions for the confidence interval for cell estimates for the tri-proportional formulation of the Gravity model. For example, the 95% confidence interval for the number of trips in a cell (i, j) is given by the range $\{C_{ij}/T_{ij}$ to $T_{ij} C_{ij}\}$, where C_{ij} is a *confidence factor* given by:

$$C_{ij} = \exp\left(2\left[1\bigg/\sum_{ij} n_{ijk} + 1\bigg/\sum_{jk} n_{ijk} + 1\bigg/\sum_{ki} n_{ijk}\right]^{0.5}\right) \tag{5.60}$$

and n_{ijk} are the number of trips sampled in the observed cells; therefore, the summations are over observed cells only. This expression covers only errors due to sampling; data collection and processing errors are likely to increase the range. Moreover, there are other sources of error in the model estimates, which are more difficult to quantify; these are mis-specification errors, due to the fact that the model is only a simplified and imperfect representation of reality. Mis-specification errors will, again, increase the range for any confidence interval estimates.

Simulation techniques may be helpful in cases where analytical expressions for confidence intervals of model output do not exist or are difficult to develop. One can calibrate a model assuming that the data available contain no errors; one would then introduce controlled but realistic variability in the data and recalibrate the model. We can repeat this process several times to obtain a range of parameters, each calibrated with a slightly different set of 'survey' data. This process is, of course, quite expensive in terms of time and computer resources, and it is therefore attempted chiefly for research purposes. However, this type of research can provide valuable insights into the stability of model parameters to data error.

A more straightforward use of Monte Carlo simulation is in testing the sensitivity of model output to input data in a forecasting mode. One knows that future planning data are bound to contain errors; simulation, in this case, involves the introduction of reasonable 'noise' into these data and then running the model with each of these future data sets. The results provide an idea of the sensitivity of model output to errors in these planning variables. As no recalibration is involved (we assume the model is calibrated with no errors in the base year), the demand on time and resources, although significant, is less than in the previous case.

5.8.9 The Stability of Trip Matrices

The stability of trip matrices over time is an issue seldom discussed in transport demand modelling. We know from experience that reality is not entirely repeatable from day to day. We can observe significant day-to-day variations at the level of traffic flows on any link in a network. One would typically expect some 10% variation on flow levels on similar days and on the same day of the week over similar weeks (i.e. excluding seasonal variations). These variations are easily observed, as permanent and semi-permanent automatic traffic counters are easy to install and maintain and are mostly reliable. These variations in traffic flows may result from at least two sources: variations in the trip matrices that originate them, and day-to-day changes in route choice. The question arises, therefore, about the extent of day-to-day variations at the level of trip matrix cell values. This information is much more difficult to come by as very rarely repeated data is collected about trip matrices, in the same location, on different days.

Traffic counts are the result of an aggregation of trips into trip matrices and therefore this aggregation process will tend to compensate some of the random variations at the trip matrix level. Leonard and Tough (1979) report on the collection of detailed O–D (trip table and traffic count) data on four consecutive days in the centre of Reading, UK. The data was collected to help developing a detailed simulation model. Observers recorded car number plates, thus tracking the route vehicles took through the centre of Reading together with their points of entry/exit and parking. Therefore, there were no interview or reporting errors but only a 10% sample was collected over four days (Monday to Thursday) for some 80 links and 40 zones. This data was independently analysed by Willumsen (1982) to look at day-to-day variations at link flow and O–D matrix level. He used the percentage mean absolute error (%MAE) for both traffic levels and trip matrices:

$$\%MAE = 100\% \cdot \left(\sum_a |V^a - V^b| \Big/ \sum_a V^a \right) \tag{5.61}$$

and

$$\%MAE = 100\% \cdot \left(\sum_{ij} |T_{ij}^a - T_{ij}^b| \Big/ \sum_{ij} T_{ij}^a \right) \tag{5.62}$$

where the indices a and b relate to *observed* flows V and O–D trips T_{ij} on different days. Willumsen (1982) found that typical variations were as shown in Table 5.16:

Table 5.16 Typical trip and flow variations on different days

% MAE	Tuesday		Wednesday		Thursday	
	Link	Matrix	Link	Matrix	Link	Matrix
Monday	11	76	11	72	12	75
Tuesday			13	68	14	85
Wednesday					12	70

The day-to-day variations at flow level are consistent with expectations, whereas those at the trip matrix level are much larger. This is partly because, at trip matrix level, we are dealing with small values and sparse matrices, but even then, the evidence suggests that variations at this level can be quite significant.

This is a rather 'inconvenient truth', an indication that the Activity Regularity Assumption is not valid; it weakens the case for collecting travel data on different days and putting it all together in an 'average (usually working) day'. The representativeness of this average day is seldom questioned and we have few tools to consider it seriously. This limitation applies to all our approaches: aggregate, disaggregate, and activity-based models.

These results suggest that efforts to obtain an accurate trip matrix may not be warranted as it will only be a snapshot. The objective for a destination choice model in this context should not be to replicate an observed or underlying trip matrix but to estimate one that captures the main features of the underlying trip matrices that, when loaded onto the network, produce link flows consistent with observations.

The results also suggest that one should be more careful when testing how the value of a scheme or plan changes with variations in the estimated trip matrix used during assessment. Sensitivity analysis seems a particularly appropriate way to investigate the effects of varying the trip matrix.

5.8.10 Sense Checks

It is always important to perform 'sense checks' about the results of any transport model before embarking into its use in travel forecasting. These sense checks are done against experience and judgment and also contrasted with known observations.

A first sense check should be to compare the trip cost (length) distribution in the base year with that in the future conditions. They should not be too different, and if they are, there should be a reasonable explanation for this. This explanation should be based on an understanding of the underlying forces behind movement in the study area, not that ... 'the model equations say so'.

Another sense check is to plot where are the most significant differences in trip making, perhaps aggregating zones into some 20 regions to see the effects more clearly. This should also be contrasted with any expected changes in land use, new developments, and economic regeneration. Good use of graphics and GIS pay handsomely here.

Note that extensive use of K factors tends to retain the pre-existing patterns and therefore judgment should be applied in interpreting results.

Finally, one should consider whether all these expected changes will happen in the relevant forecasting year or some of them will be lagged in time, for example, changes in residence and jobs/educational places. It may be possible to allow a few years' lag in these responses without abandoning the synthctic models.

Exercises

5.1 A small study area has been divided into four zones and a limited survey has resulted in the trip matrix presented in Table 5.17:

Table 5.17 Trip matrix for Example 5.1

	1	2	3	4
1	—	60	275	571
2	50	—	410	443
3	123	61	—	47
4	205	265	75	—

Estimates for future total trip ends for each zone are as in Table 5.18:

Table 5.18 Future trip ends by zone for Example 5.1

Zones	Estimated future origins	Estimated future destinations
1	1200	670
2	1050	730
3	380	950
4	770	995

Use an appropriate growth-factor method to estimate future inter-zonal movements. *Hint*: check conditions for convergence of the chosen method first.

5.2 A study area has been divided into three large zones, A and B on one side of a river and C on the other side. It is thought that travel demand between these zones will depend on whether or not the O–D pair is at the same side of the river. A small sample home interview survey has been undertaken with the results shown in Table 5.19:

Table 5.19 Trip matrix for Example 5.2

	Destination		
Origin	A	B	C
A	12	10	8
B		5	3
C	4		7

where blank entries indicate unobserved cells.

Assume a model of the type $T_{ij} = R_i S_j F_k$ where the parameter F_k can be used to represent the fact that the O–D pair is on the same side of the river or not. Calibrate such a model using a tri-proportional algorithm and fill the empty cells in the matrix above.

5.3 A transport study is being undertaken incorporating four cities A, B, C, and D. The travel costs between these cities in generalised time units are given in Table 5.20; note that intra-urban movements are excluded from this study:

Table 5.20 Cost matrix for Example 5.3

	Destination			
Origin	A	B	C	D
A	—	1.23	1.85	2.67
B	1.23	—	2.48	1.21
C	1.85	2.48	—	1.44
D	2.67	1.21	1.44	—

Roadside interviews have been undertaken at several sites and the number of drivers interviewed is shown in Table 5.21 together with their respective origins and destinations. Blank entries indicate unobserved cells.

Table 5.21 Number of drivers interviewed

	Destinations			
Origin	A	B	C	D
A	—	6		2
B		—	1	4
C	8		—	8
D	6	18	6	—

Assume now that a gravity model of the type $Tij = R_i S_j F_k$ will be used for this study area with only two cost bins. The first cost bin will cover trips costing between 0 and 1.9 and the second will consider trips costing more than 1.9. Calibrate such a model using a tri-proportional method on this partial matrix. Provide estimates of the parameters R_i, S_j, and F_k and of the missing entries in the matrix, excluding intra-urban trips. Are these estimates unique?

6

Modal Split and Direct Demand Models

6.1 Introduction

In this chapter, we discuss mode choice as an aggregate problem. It is interesting to see how far we can get using similar approaches to those pursued in deriving and using trip distribution models. Then, we will examine the need for consistency between the parameters and structure of distribution and mode choice models, a topic often disregarded by practitioners at their peril.

The choice of transport mode is probably one of the most important classic model stages in transport planning. This is because of the key role played by public transport in policy-making. Almost without exception, travelling by public transport uses road space more efficiently and produces fewer accidents and emissions than using a private car. Furthermore, underground and other rail-based modes do not require additional road space (although they may require a reserve of some kind) and, therefore, do not contribute to road congestion. Moreover, if some drivers could be persuaded to use public transport instead of cars, the rest of the car users would benefit from improved levels of service. It is unlikely that all car owners wishing to use their cars could be accommodated in urban areas without sacrificing large parts of the urban fabric to roads and parking space.

The issue of mode choice is, therefore, probably the single most important element in transport planning and policy-making. It affects the general efficiency with which we can travel in urban areas, the amount of urban space devoted to transport functions, and whether a range of choices is actually available to all travellers. The issue is equally important in inter-urban transport, as, again, rail modes can provide a more efficient mode of transport (in terms of resources consumed, including space), but there is also a trend to increase travel by road.

Therefore, it is important, to develop and use models that are sensitive to those attributes of travel that influence individual choices of mode. We will see how far this necessity can be satisfied using aggregate approaches, where alternative policies need to be expressed as modifications to useful, if rather inflexible, functions like the generalised cost of travel.

6.2 Factors Influencing the Choice of Mode

The factors influencing mode choice may be classified into three groups:

1) Characteristics of the trip maker. The following features are generally believed to be important:
   - car availability and/or ownership;
   - possession of a driving licence;

Modelling Transport, Fifth Edition. Juan de Dios Ortúzar and Luis G. Willumsen.
© 2024 John Wiley & Sons Ltd. Published 2024 by John Wiley & Sons Ltd.
Companion website: www.wiley.com/go/ortuzar5e

- household structure (young couple, couple with children, retired, singles, etc.);
- income;
- decisions made elsewhere, for example the need to use a car at work, take children to school, etc.;
- residential density;
- availability of parking for residents.

2) Characteristics of the journey. Mode choice is strongly influenced by:
- The purpose of the trip; for example, the journey to work is normally easier to undertake by public transport than other journeys because of its regularity and the adjustment possible in the long run;
- time of the day, when the journey is undertaken; late trips are more difficult to accommodate by public transport;
- whether the trip is undertaken alone or with others.

3) Characteristics of the transport facility. These can be divided into two categories. Firstly, quantitative factors such as:
- components of travel time: in-vehicle, waiting, and walking times by each mode;
- components of monetary costs (fares, tolls, fuel, and other operating costs);
- availability and cost of parking at destinations;
- reliability of travel time and regularity of service.

Secondly, qualitative factors which are less easy (or impossible) to measure in practice, such as:

- comfort and convenience;
- safety, protection, and security;
- the demands of the driving task;
- opportunities to undertake other activities during travel (use the phone, read, etc.).

Note that we have described these in terms of journeys or trips. A richer concept is that of tours with trips as their components. It is clear that the choice of mode is made more on a tour basis (that is, considering the requirements of all trips) than on the basis of a single trip. If one chooses the car for the first leg of a tour, this is likely to remain the choice for the other legs. A good mode-choice model would be based at least on simple tours (from home and back) and should include the most important of these factors. It is easy to visualise how the concept of generalised cost can be used to represent several of the quantitative factors included under point 3 above.

Mode choice models can be *aggregate* if they are based on zonal (and inter-zonal) information. We can also have *disaggregate* models if they are based on household and/or individual data (see Chapter 7).

A simplistic but useful way to think about mode choice is as follows. Given that somebody knows where she is going (Destination) this person has many alternative 'routes' to get there; some involve just driving a car, whereas others may require to walk to a subway station, take the train, alight at some other station, and, say, walk to the final destination (plus many other combinations of modes and routes). This person can then choose the option with the lowest generalised cost, among all of these, and in doing so, the physical route and the combination of modes would be found. If all people think the same, we would have an *all-or-nothing* route and mode choice model. Alas, life is not so simple for a number of reasons:

- some people do not have a car available, so their choice set will be more limited; this would be the minimum segmentation required;
- as we will see, congestion, both on roads and in public transport, makes the choice of more than one route a necessity;

- generalised costs cannot hope to capture all the relevant elements that determine mode choice; this is particularly relevant in the case of the choice between cars and public transport, which focus on parameters additional to those relevant for route choice;
- different people would perceive costs in different ways and would seek to minimise a different 'version' of generalised costs (time versus money minimisers, for example); we must allow, therefore, for a degree of dispersion in choices to consider other factors, not fully visible to the analyst, in mode preferences;
- the modelled costs in a zonal-based model are only centroid-to-centroid averages of the actual costs (time and money) actually perceived by individuals; for example, some may live closer to a rail station and therefore be more inclined to use public transport.

The combined effect of these influences is dispersion in the choices of mode made at each origin–destination (O–D) pair. The nature of this dispersion is influenced by the three groups of factors and conditions mentioned above.

6.3 Trip-End Modal-Split Models

In the past, in particular in the USA, personal characteristics were thought to be the most important determinants of mode choice, and therefore attempts were made to apply modal-split models immediately after trip generation. In this way, the different characteristics of the individuals could be preserved and used to estimate modal split: for example, the different groups after a category analysis model. Note that, as at that level there was no indication of where those trips might go, the characteristics of the journey and modes were omitted from these models.

This was consistent with the general view that as income grew most people would acquire cars and would want to use them. The objective of transport planning was to forecast this growth in demand for car trips so that investment could be planned to satisfy it. This was characterised as the *Predict and Provide* approach to transport planning, which is today considered a blind and dangerous alley. The modal-split models of that time related the choice of mode only to features like income, residential density, and car ownership. In some cases, the availability of reasonable public transport was included in the form of an accessibility index.

In practice, there is always a group of people that appears to be captives of a mode of transport, usually car; they organise the location of home and workplace indifferent to the provision of public transport services. The opposite may also be found, that is, people who search first for locations with good public transport or seek employment such that they can walk or cycle from home; higher urban densities facilitate this choice.

In the short run, these models may be fairly accurate, in particular if public transport is available in a similar way throughout the study area and there is little congestion. However, this type of model is, to a large extent, defeatist in the sense of being insensitive to policy decisions; it appears that there is nothing that the decision maker can do to influence the choice of mode. Improving public transport, restricting parking, or charging for the use of roads would not have any effect on modal split according to these trip-end models.

6.4 Trip Interchange Modal-Split Models

Modal-split modelling in Europe was dominated, almost from the beginning, by post-distribution models; that is, models applied after the Gravity or another distribution model. This had the advantage of facilitating the inclusion of the characteristics of the journey and those of the alternative modes available to undertake it. However, it made it more difficult to include the characteristics of the trip maker as they may have already been aggregated into the trip matrix (or matrices).

The first models of this type included only one or two characteristics of the journey, typically (in-vehicle) travel time. It was observed that an S-shaped curve seemed to represent this kind of behaviour better, as in Figure 6.1, showing the proportion of trips by mode 1 (T_{ij}^1/T_{ij}) against the cost or time difference.

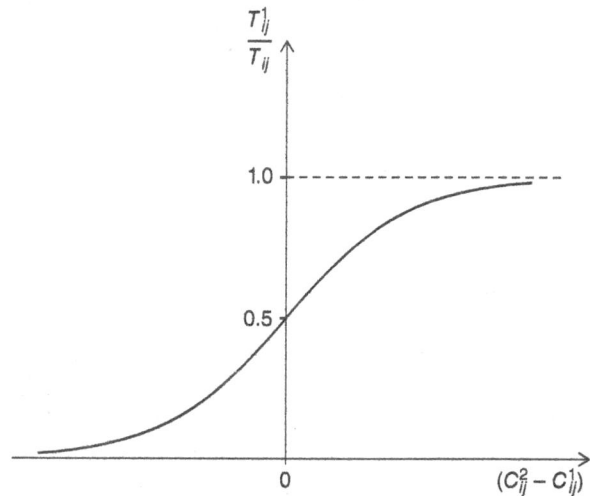

Figure 6.1 Empirical modal-split curve

These were empirical curves, obtained directly from the data and following a similar approach to the curves used to estimate what proportion of travellers would be diverted to use a (longer but faster) bypass route; hence their name as *diversion curves*. For example, the London Transportation Study (Phase III) used diversion curves for trips to the central area and non-central trips (the former trips were more likely to be made by public transport) and for different trip purposes.

Another approach was to use, by analogy, a version of Kirchhoff's formulation in electricity. The proportion of trip makers between origin i and destination j that choose mode k as a function of the respective generalised costs by mode k, C_{ij}^k is given by:

$$P_{ij}^k = \frac{\left(C_{ij}^k\right)^{-n}}{\sum_m \left(C_{ij}^m\right)^{-n}} \tag{6.1}$$

where n is a parameter to be calibrated or transferred from another location or time (values for n between 4 and 9 were suggested for both mode and route choice models of this nature). Note that as the value of n increases, the costs become more important in determining the choice of mode. With a judicious choice of n this formulation produces a curve not too dissimilar from the Logit equation (see Section 7.3). Kirchhoff's model can be derived from entropy maximisation principles, assuming that the generalised costs are perceived in a logarithmic fashion as in Eqs. (5.37) and (5.38). The interested reader can also verify that this formulation is consistent with the Box-Cox transformation on the utility function of a Logit model when $\tau = 0$ (see Section 8.3.1.1).

Model (6.1) is sometimes considered attractive because the choice of mode (or route) depends on the ratio of costs (to a power) and not on their difference. As we will see, one of the issues with most

Logit models is that a 5-minute difference in a 30-minute journey has the same effect as a 5-minute difference in a 6-hour trip.

A standard limitation of these models is that we can only apply them to trip matrices of travellers with an available choice, which often means the matrix of people with a car available. Notwithstanding, we can also apply them to choosing between different public transport modes as a sub-mode choice.

The above models have limited theoretical basis, and therefore their interpretation and forecasting abilities must be in doubt. Furthermore, as these models are aggregated, they are unlikely to model in full the constraints and characteristics of the modes available to individual households.

6.5 Synthetic Models

6.5.1 Distribution and Modal-Split Models

The entropy-maximising approach can be used to generate models of distribution and mode choice simultaneously. To do this, we need to cast the entropy-maximising problem in terms of, for example, two modes as follows:

$$\text{Maximise } \log W\left\{T_{ij}^k\right\} = -\sum_{ijk}\left(T_{ij}^k \cdot \log T_{ij}^k - T_{ij}^k\right) \tag{6.2}$$

subject to

$$\sum_{jk} T_{ij}^k - O_i = 0 \tag{6.3}$$

$$\sum_{ik} T_{ij}^k - D_j = 0 \tag{6.4}$$

$$\sum_{ijk} T_{ij}^k \cdot C_{ij}^k - C = 0 \tag{6.5}$$

It is easy to see that this problem leads to the solution:

$$T_{ij}^k = A_i O_i B_j D_j \exp\left(-\beta C_{ij}^k\right) \tag{6.6}$$

$$P_{ij}^1 = \frac{T_{ij}^1}{T_{ij}} = \frac{\exp\left(-\beta C_{ij}^1\right)}{\exp\left(-\beta C_{ij}^1\right) + \exp\left(-\beta C_{ij}^2\right)} \tag{6.7}$$

where P_{ij}^1 is the proportion of trips travelling from i to j via mode 1. The functional form in (6.7) is known as Logit, and it is discussed in greater detail in Chapter 7. However, it is useful to reflect here on some of its properties:

- it generates an S-shaped curve, similar to some of the empirical diversion curves in Figure 6.1;
- if $C_1 = C_2$, then $P_1 = P_2 = 0.5$;
- if C_2 is much greater than C_1, then P_1 tends to 1.0;
- the model can easily be extended to multiple modes

$$P_{ij}^1 = \frac{\exp\left(-\beta C_{ij}^1\right)}{\sum_k \exp\left(-\beta C_{ij}^k\right)} \tag{6.8}$$

It is obvious that in this formulation β plays a double role. It acts as the parameter controlling dispersion in mode choice and also in the choice between destinations at different distances from the origin. This is probably asking too much of a single parameter, even if underpinned by a known theoretical basis. Therefore, a more practical joint distribution/modal-split model was proposed and used in many studies. This has the form (Wilson 1974):

$$T_{ij}^{kn} = A_i^n O_i^n B_j D_j \exp\left(-\beta_n K_{ij}^n\right) \frac{\exp\left(-\lambda_n C_{ij}^k\right)}{\sum\limits_{k'} \exp\left(-\lambda_n C_{ij}^{k'}\right)} \tag{6.9}$$

where K_{ij}^n is the *composite cost* of travelling between i and j as perceived by person type n. In principle, this composite cost may be specified in different ways; for example, in earlier studies, it was taken to be the minimum of the two costs, and in others, it was taken to be weighted average of these:

$$K = \sum_k P^k \cdot C^k \quad (i,j \text{ and } n \text{ omitted for simplicity})$$

However, it is interesting to note that some of these formulations are, in fact, inappropriate, as shown in Example 6.1.

The mode choice or mode split component of (6.9) in a binary choice situation takes the form:

$$P_{ij}^1 = \frac{\exp\left(-\lambda C_{ij}^1\right)}{\exp\left(-\lambda C_{ij}^1\right) + \exp\left(-\lambda C_{ij}^2\right)} = \frac{1}{1 + \exp\left(-\lambda\left(C_{ij}^2 - C_{ij}^1\right)\right)} \tag{6.10}$$

The right-hand side of the equation shows clearly that choice depends only on the differences in generalised costs. This property suggests, in some cases, segmenting the demand by trip length so that, say, a 5-minute saving is more important in a short trip than in a long one, as discussed above. There are, however, other ways of compensating for this.

The proportion of trips using one mode, say mode 1, is shown in Figure 6.2 as a function of the scaling parameter λ and the differences in costs between the two modes. The figure also includes the plot of the Kirchhoff model with a power value of -8. Note that the greater the value of λ, the closer the Logit model is to an 'all or nothing' allocation of trips to the cheapest mode.

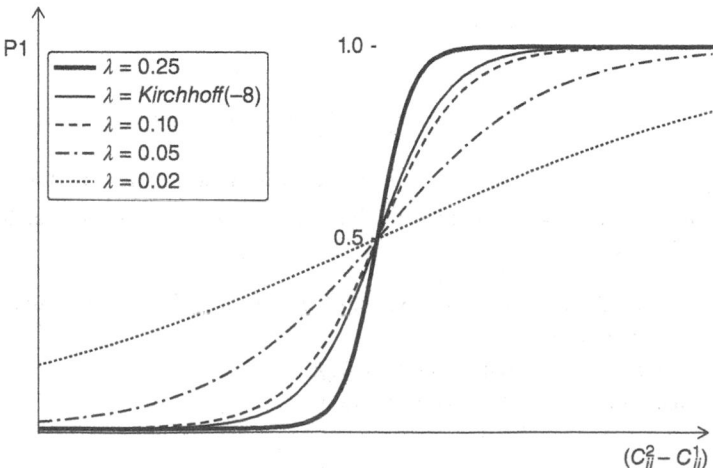

Figure 6.2 Proportion selecting mode 1 for Logit model with different λ and Kirchhoff with power of -8

Note also that with either model, there is a certain proportion of travellers who seem to make a poor choice, that is, they select the most expensive mode. As these are aggregate models, part of the explanation may be that some people live closer to a commuter rail station and others further away. The average zone-to- zone cost difference may favour cars, but for residents close to the station, the opposite is true. The larger the zones, the more significant will be this effect; this suggests that larger zones would warrant a smaller scaling parameter, particularly if the model was calibrated with individual data. Another part of the explanation could be that some travellers give different weights to each of the variables in the generalised cost expression or that they consider other factors (perhaps the need to do some shopping after work or to store golf clubs in the car to play in the evening).

Example 6.1 Consider the weighted average form of the composite costs K_{ij}^n and examine what happens when a new, more expensive mode ($C_2 > C_1$) is added to an existing unimodal system. In the initial state we would have:

$$K = \sum_k P^k \cdot C^k = C^1$$

and in the final state, that is, after the introduction of mode 2:

$$K^* = P^1 \cdot C^1 + P^2 \cdot C^2$$

However, by definition $P^1 + P^2 = 1$ and therefore:

$$K^* = \left(1 - P^2\right)C^1 + P^2 \cdot C^2 = C^1 + P^2\left(C^2 - C^1\right)$$
$$K^* = K + P^2\left(C^2 - C^1\right)$$

Now, as both P^2 and $(C^2 - C^1)$ are greater than zero, we can conclude that $K^* > K$, which is nonsensical as the introduction of a new option, even if it is more expensive, should not increase the composite costs; at worst they should remain the same. The use of the wrong composite costs will lead to mis-specified models.

6.5.2 Distribution and Modal-Split Structures

Williams (1977) has shown that the only correct specification, consistent with the prevailing theory of rational choice behaviour (see Section 7.2), is to define:

$$K_{ij}^n = \frac{-1}{\lambda_n} \log \sum_k \exp\left(-\lambda_n C_{ij}^k\right) \tag{6.11}$$

where the following restriction must be satisfied if destination choice is specified above mode choice as in (6.9):

$$\beta_n \leq \lambda_n \tag{6.12}$$

We will come back to this restriction in Chapter 7. Intuitively, it means that the importance of costs is more critical in the choice of mode than in the choice of destination. If this is not the case, the model structure in (6.9), simultaneous or sequential, would be inappropriate. The composite cost measure (6.11) has the following properties:

- $K \leq \mathrm{Min}_k \{C^k\}$
- $\displaystyle\lim_{\lambda \to \infty} K = \mathrm{Min}_k \{C_k\}$, that is 'all-or-nothing' mode choice
- $\dfrac{\partial K}{\partial C^k} = P^k$

The first of these properties means that when a new alternative is added, even if it is very unattractive in principle, the composite costs will either reduce (somebody must like it) or at most remain the same. The second property highlights the importance of λ_n as a weight attached to generalised costs in the choice of mode. For a very large λ_n the model will predict an all-or-nothing choice of the least generalised cost alternative, as suggested in Figure 6.2.

The model (6.9–6.12) is frequently found in aggregate applications, in particular in urban areas. One of the problems in practice, however, is that modellers sometimes fail to check whether the restriction (6.12) is satisfied. As the destination and mode choice models may have been calibrated independently, it is quite possible that the restriction is not satisfied. If this was the case, the combined models (Gravity and then mode choice) produce pathological results. This structure, often described as G/D/MS/A (Generation, Distribution, Mode Split, and Assignment), would be wrong if $\beta > \lambda$; in that case the structure G/MS/D/A would be probably the correct one.

It has been found in practice that the structure itself may be different for different journey purposes. Typically, the correct structure would be G/D/MS/A for journey to work/education and G/MS/D/A for other purposes. This would reflect a condition where it is easier to change destinations for, say, a shopping trip than to change modes.

Note that in the case of the G/MS/D/A structure one starts by calculating the composite cost by mode n of reaching all destinations from each origin i:

$$K_i^n = \frac{-1}{\beta} \log\left(\sum_j \exp\left(-\beta C_{ij}^n\right)\right) \tag{6.13}$$

Then, these composite costs are used to obtain mode splits by origin i.

$$P_i^1 = \frac{1}{1 + \exp\left(-\lambda\left(K_i^2 - K_i^1\right)\right)} \tag{6.14}$$

Separate Gravity models are developed using the costs of each mode. Although this is a generation-based mode choice model, it fully takes into account the costs of reaching each destination by each mode.

For many applications, these aggregate models remain valid and in use. However, for a more refined handling of personal characteristics and preferences, we now have disaggregate models that respond better to the key elements in mode choice and make a more efficient use of data collection efforts; these are discussed in Chapters 7–9.

6.5.3 Multimodal-Split Models

Figure 6.3 depicts possible model structures for choices involving more than two modes. The N-way structure, which became very popular in disaggregate modelling work, as we will see in Chapter 7, is the simplest; however, because it assumes that all alternatives have equal 'weight', it can lead to problems when some of the options are more similar than others (i.e. they are correlated), as demonstrated by the famous blue bus-red bus example (Mayberry 1973).

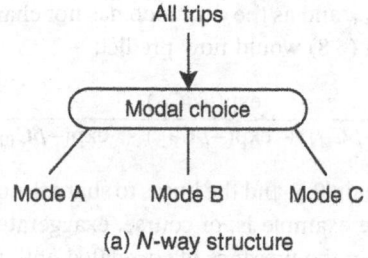

(a) N-way structure

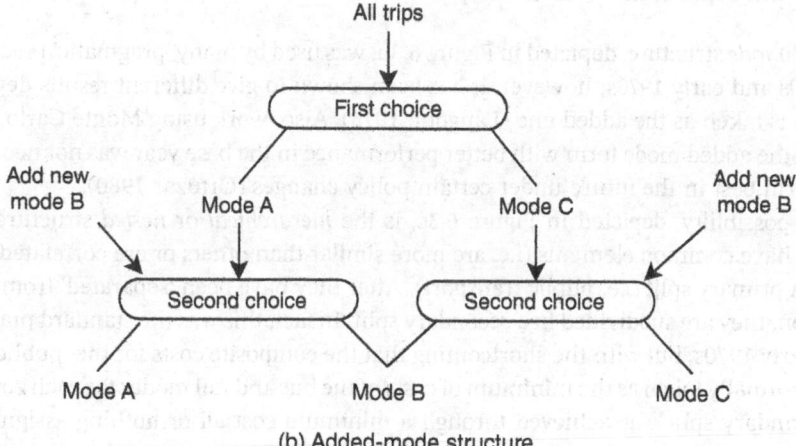

(b) Added-mode structure

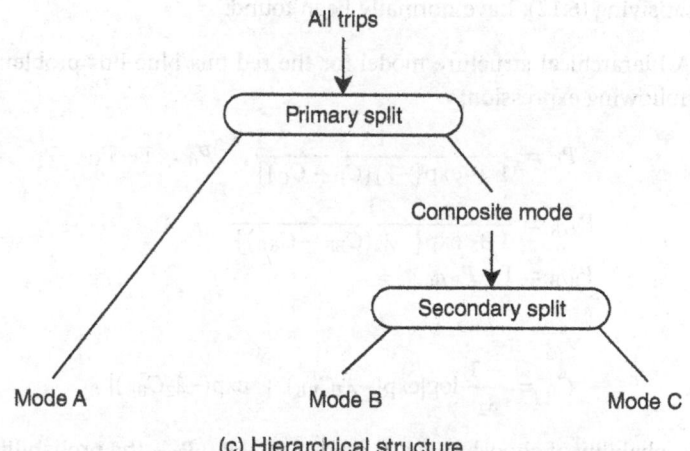

(c) Hierarchical structure

Figure 6.3 Multimodal model structures

Example 6.2 Consider a city where 50% of travellers choose cars (C) and 50% choose buses (B). In terms of model (6.8), which is a N-way structure, this means that $C_C = C_B$. Let us now assume that the manager of the bus company, in a stroke of marketing genius, decides to paint half the buses red (RB) and half of them blue (BB), but manages to maintain the same level of service as before. This means that $C_{RB} = C_{BB}$, and as the car mode has not changed, this value is still equal to C_C. Note that in this case, model (6.8) would now predict:

$$P_C = \frac{\exp(-\beta C_C)}{\exp(-\beta C_C) + \exp(-\beta C_{RB}) + \exp(-\beta C_{BB})} = 0.33$$

when one would expect P_C to remain 0.5, and the buses to share the other half of the market equally between red and blue buses. The example is, of course, exaggerated but serves well to show the problems of the N-way structure in the presence of correlated options (in this case completely correlated). We will come back to this in Chapter 7.

The *added-mode* structure, depicted in Figure 6.3b, was used by many 'pragmatic' practitioners in the late 1960s and early 1970s; however, it has been shown to give different results depending on which mode is taken as the added one (Langdon 1976). Also, work using Monte Carlo simulation showed that the added mode form with better performance in the base year was not necessarily the one to perform best in the future under certain policy changes (Ortúzar 1980).

The third possibility, depicted in Figure 6.3c, is the *hierarchical* or *nested* structure. Here the options that have common elements (i.e. are more similar than others or are correlated) are taken together in a primary split (i.e. public transport). After they have been 'separated' from the uncorrelated option, they are subdivided in a secondary split. In fact, this was the standard practice in the 1960s and early 1970s, but with the shortcoming that the composite costs for the 'public-transport' mode were normally taken as the minimum of costs of the bus and rail modes for each zone pair and that the secondary split was achieved through a minimum-cost all-or-nothing assignment. This 'pragmatic' procedure essentially implies an infinite value for the dispersion parameter of the sub modal-split function, whereas values of the same order as the dispersion parameter in the primary split, but satisfying (6.12), have normally been found.

Example 6.3 A hierarchical structure model for the red bus/blue bus problem of Example 6.2 would have the following expression:

$$P_C = \frac{1}{1 + \exp\{-\lambda_1(C_B - C_C)\}}; \quad P_B = 1 - P_C$$

$$P_{R/B} = \frac{1}{1 + \exp\{-\lambda_2(C_{BB} - C_{RB})\}}$$

$$P_{B/B} = 1 - P_{R/B}$$

with

$$C_B = \frac{-1}{\lambda_2} \log[\exp(-\lambda_2 C_{RB}) + \exp(-\lambda_2 C_{BB})]$$

where P_C is the probability of choosing car, as before, $(1 - P_C) P_{R/B}$ the probability of selecting red bus and $(1 - P_C) P_{B/B}$ the probability of selecting blue bus; λ_1 and λ_2 are the primary and secondary split parameters. It is easy to see that, if $C_B = C_C$, this model correctly assigns a probability of 0.5 to the car option and 0.25 to each of the bus modes.

However, the value of the composite bus cost C_B is not the same as the cost of the red or blue buses (C_{RB} and C_{BB}). The former depends on the value of λ_2 and for the red bus/blue bus problem it would be:

$$C_B = C_{BB} - \frac{1}{\lambda_2} \log 2$$

Therefore, the composite cost of bus will always be cheaper than the cost of the blue or red buses. The dispersion parameter λ_2 allows users to choose options that do not minimise the observed part of the generalised cost function because of other variables not included in the model.

Consider now an O–D pair where the costs of travelling by bus (red or blue) and by car are all the same and equal to 50 generalised minutes. Assume also that λ_2 is 0.9. In this case, the value of the composite cost C_B is not 50 but 49.23, and the proportion choosing car will depend on the value of λ_1 as shown in Table 6.1.

Table 6.1 Proportion choosing car as a function of λ_1

λ_1	P_C
0.001	0.500
0.005	0.499
0.010	0.498
0.050	0.490
0.100	0.481
0.500	0.405
0.600	0.386
0.700	0.368
0.800	0.351
0.900	0.333

It can be seen that if $\lambda_1 = \lambda_2$ then $P_C = 1/3$, the same result as in a trinomial Logit model; this is expected because the nested structure collapses to the simple Logit model in this case (Section 7.4). However, for small values, say $\lambda_1 = 0.1$, the hierarchical or nested structure predicts proportions, choosing car (48%) closer to the expected 50%. Is this 2% loss due to travellers with a strong colour preference or those influenced by any change (new paint) in the supply of a service?

6.5.4 Calibration of Binary Logit Models

Consider a model of choice between car and public transport with generalised costs of travel, C_{ij}^k, given by an expression such as (5.2). As discussed in Chapter 5, the weights a attached to each element of cost are considered given and calibration only involves finding the 'best-fit' values for the dispersion parameter λ and modal penalty δ (assumed associated with the second mode).

Let us assume that we have C_{ij}^1 and C_{ij}^2 as the 'known' part of the generalised cost for each mode and O–D pair. If we also have information about the proportions choosing each mode for each (i, j) pair, P_{ijk}^*, we can estimate the values of λ and δ using linear regression as follows. The modelled proportions P for each (i, j) pair, dropping the (i, j) indices for convenience, are:

$$P_1 = \frac{1}{1 + \exp\{-\lambda(C_2 + \delta - C_1)\}}$$

$$P_2 = 1 - P_1 = \frac{\exp\{-\lambda(C_2 + \delta - C_1)\}}{1 + \exp\{-\lambda(C_2 + \delta - C_1)\}}$$

(6.15)

Therefore, taking the ratio of both proportions yields:

$$P_1/(1 - P_1) = 1/\exp\{-\lambda(C_2 + \delta - C_1)\} = \exp\{\lambda(C_2 + \delta - C_1)\}$$

and taking logarithms of both sides and rearranging, we get:

$$\log[P_1/(1 - P_1)] = \lambda(C_2 - C_1) + \lambda\delta$$

(6.16)

where we have observed data for P and C, and therefore the only unknowns are λ and δ (this is known as the Berkson-Theil transformation). These values could be estimated by linear regression with the left-hand side of (6.16) acting as the dependent variable and $(C_2 - C_1)$ as the independent one; then λ is the slope of the line and $\lambda\delta$ is the intercept. Note that if we assume the weights $\boldsymbol{a}$ in the generalised cost function to be unknown, we can still calibrate the model using (6.16) and multiple linear regression. In this case, the calibrated weights would include the dispersion coefficient λ. Other and often better calibration methods are discussed in the next section.

Example 6.4 Data about aggregate mode choice between five zone pairs is presented in the first four columns of Table 6.2; the last column of the table gives the values needed for the left-hand side of Eq. (6.16).

Table 6.2 Aggregate binary split data

Zone pair	P_1 (%)	P_2 (%)	C_1	C_2	$\log [P_1/(1 - P_1)]$
1	51.0	49.0	21.0	18.0	0.04
2	57.0	43.0	15.8	13.1	0.29
3	80.0	20.0	15.9	14.7	1.39
4	71.0	29.0	18.2	16.4	0.90
5	63.0	37.0	11.0	8.5	0.53

This information can be plotted following (6.16) as in Figure 6.4, where it can be deduced that $\lambda \approx 0.72$ and $\delta \approx 3.15$.

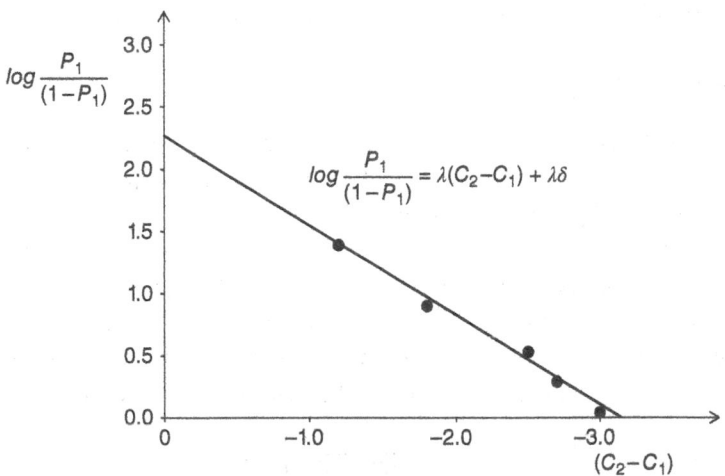

Figure 6.4 Best-fit line for the data in Table 6.1

6.5.5 Calibration of Hierarchical Modal-Split Models

This was usually performed in a heuristic or recursive fashion, starting with the sub modal split and proceeding upwards to the primary split. A general discussion on the merits of this approach in comparison with theoretically better simultaneous estimation is postponed until Chapter 7.

Within this general approach, there are several possible calibration procedures. It has been shown (see Domencich and McFadden 1975) that maximum likelihood estimates are preferable to least squares estimates, both on theoretical and practical grounds. This is particularly true when working with large data sets. However, when dealing with aggregate data sources, it is usually convenient to group the information into suitable classes for analysis (i.e. cost-difference bins). More importantly, the normally available 'factored-up' data are, by definition, raw sampled data that have been manipulated and multiplied by some empirically derived factors. This can cause discrepancies when several data sources with different factors are employed, but the important point at this stage is that the real data set is usually very small. Hartley and Ortúzar (1980) compared various calibration procedures, and found that maximum likelihood produced not only the most accurate calibration results but also the most efficient ones in terms of computer time.

Example 6.5 Consider a trinomial problem involving choice between, for example, cars, buses, and rail. Let us also assume that the last two modes are suspect of being correlated due to their 'public-transport' nature. In this case, the heuristic calibration would proceed as follows:

- find λ_2 for the sub modal split (bus vs. rail) as explained in Example 6.4 and use its value to calculate the public-transport composite costs needed for the primary split using an expression such as that in Example 6.3;
- for zone pairs where there is a choice of mode (e.g. trips by both modes are possible), classify trips into cost- difference bins of a certain minimum size; those trips with no choice of mode are excluded from the calibration;
- as there can be cost bins without any trips, bins should be aggregated into bigger bins until each bin contains some trips; then, a weighted representative cost is calculated for each bin;
- if N is the total number of bins, n_k the observed number of trips in cost-difference bin k, r_k the observed number of trips by the first mode in the interval, and

$$P_k = 1/[1 + \exp(-Y_k)]$$

the probability of choosing the first mode in interval k, with $Y_k = ax_k + b$, x_k the representative cost of bin k, and a and b parameters to be estimated (i.e. $\lambda = a$ and the modal penalty $\delta = b/a$), then the logarithm of the likelihood function (see Chapter 8 for more details) can be written as:

$$L = \text{Constant} + \sum_k [(n_k - r_k)\log (1 - P_k) + r_k \log P_k] \qquad (6.17)$$

- to maximise L, we need the first and second derivatives of (6.17) with respect to the parameters, which in this simple case have straightforward analytical expressions:

$$\frac{\partial L}{\partial a} = \sum_k (r_k - n_k P_k)x_k$$

$$\frac{\partial L}{\partial b} = \sum_k (r_k - n_k P_k)$$

$$\frac{\partial^2 L}{\partial a^2} = -\sum_k n_k P_k(1 - P_k)x_k^2$$

$$\frac{\partial^2 L}{\partial b^2} = -\sum_k n_k P_k(1 - P_k)$$

$$\frac{\partial^2 L}{\partial a \partial b} = -\sum_k n_k P_k(1 - P_k)x_k$$

- knowing the values of the derivatives, any search algorithm will find the maximum without difficulty;
- note that the maximisation routines require starting values for the parameters, together with an indication of how far they are from the optimum; the efficiency of calibration typically depends, strongly, upon the accuracy of these estimates, and one procedure for generating close first estimates is to find the 'equiprobability' cost (see Bates et al. 1978), where the probability of choosing either mode is 0.5.

Before closing this chapter, we will consider an alternative approach that offers to consolidate into a single model the features of two or three of the classic sub-models.

6.6 Direct Demand Models

6.6.1 Introduction

The conventional sequential methodology of the classic transport model requires the estimation of relatively well-defined sub-models. An alternative approach is to develop, directly, a model subsuming trip generation, distribution, and mode choice. This is, of course, very attractive in principle, as it would avoid some of the pitfalls of the sequential approach. For example, Gravity models may suffer from the problem of having to cope with errors in trip-end totals and those generated by poorly estimated intra-zonal trips. A Direct Demand model, calibrated simultaneously for the three sub-models, would not suffer from this drawback.

Direct Demand models can be of two types: purely direct, which use a single estimated equation to relate travel demand to mode, journey, and person attributes; and quasi-direct approaches, which employ a form of separation between mode split and total (O–D) travel demand. Direct Demand models are closely related to general econometric models of demand and are inspired by research in that area.

6.6.2 Direct Demand Models

The earliest forms of Direct Demand models were of the multiplicative kind. The SARC model (Kraft 1968), for example, estimates demand as a multiplicative function of activity and socioeconomic variables for each zone pair and level-of-service attributes of the modes serving them:

$$T_{ijk} = \phi\left(P_i P_j\right)^{\theta_{k1}} \left(I_i I_j\right)^{\theta_{k2}} \prod_m \left[\left(t_{ij}^m\right)^{\alpha_{km}^1} \left(c_{ij}^m\right)^{\alpha_{km}^2}\right] \tag{6.18}$$

where P is population, I income, t and c travel time and cost of travel between i and j by mode k, and ϕ, θ, and α parameters of the model. It is easy to see that ϕ is just a scale parameter that depends on the purpose of the trips examined. θ_{k1} and θ_{k2} are elasticities of demand with respect to population and income, respectively; we would expect them to be of positive sign. α_{km}^1 and α_{km}^2 are demand elasticities with respect to time and cost of travelling; the direct elasticities (i.e. when $k = m$) should be negative, and the cross-elasticities of positive sign.

This model should allow to handle generation, distribution, and modal split simultaneously, including attributes of competing modes and a wide range of level of service and activity variables. Unfortunately, it requires a large number of parameters to cash in on these advantages.

Example 6.6 Consider the following demand function for a bus service between cities 1 and 2:

$$T_{12} = 10\,000 t_{12}^{\alpha} c_{12}^{\beta} q_{12}^{\mu}$$

where time t is measured in hours, the fare c in dollars, and the service frequency q in trips/day. The estimated parameter values are $\alpha = -2$, $\beta = -1$, and $\mu = 0.8$ (note that all the signs are correct according to intuition). The bus operator wants to increase the fares by 20%; what changes should he make to the level of service in order to keep the same volume of trips if all other things remain equal?

Let us define $L_{12} = L = t^{-2} c^{-1} q^{0.8}$; we know that if L remains constant the total volume T_{12} will not vary (*ceteris paribus*). We also know that the elasticities $E(L, x)$ of the level of service (and hence demand) with respect to each attribute x (time, cost, and frequency) are respectively -2, -1, and 0.8.

Now, if only c varies, we have that $L = k/c$, where k is a constant; therefore, a 20% increase in c means a new level of service $L' = k/1.2c$ or $L'/L = 0.833$. That is, a decrease of 16.67% in L. In order to offset this, the operator must introduce changes to the travel time, frequency of service, or both. Now, from the definition of elasticity, we have that:

$$\Delta L^{(c)} \approx E(L,c) L \Delta c/c \approx -L \Delta c/c$$
$$\Delta L^{(t)} \approx E(L,t) L \Delta t/t \approx -2 L \Delta t/t$$
$$\Delta L^{(q)} \approx E(L,q) L \Delta q/q \approx 0.8 L \Delta q/q$$

Therefore, if we want $\Delta L^{(c)}$ to be equal to $- \Delta L^{(q)}$, we require:

$$-L \Delta c/c \approx -0.8 L \Delta q/q$$

that is:

$$\Delta q/q \approx 1.25 \Delta c/c \approx 1.25 \times 0.20 = 0.25 \text{ or } 25\%$$

If we are prepared to vary both frequency and travel time, we would require:

$$\Delta L^{(c)} = -\left(\Delta L^{(q)} + \Delta L^{(t)}\right)$$

that is

$$2\Delta t/t = 0.8\Delta q/q - 0.20$$

which is a straight line of feasible solutions as shown in Figure 6.5.

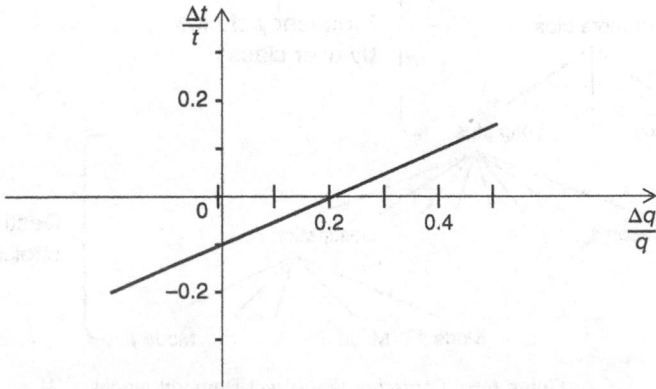

Figure 6.5 Feasible solutions for Example 6.5

Direct Demand models may be an attractive proposition in areas where the zones are large, for example in inter-urban studies. Timberlake (1988) has discussed the use of Direct Demand models in developing countries, finding them better than conventional approaches. For example, in a corridor in Sudan, the Direct Demand model gave a better fit than a Gravity model because of the unique traffic characteristics exhibited by Khartoum and Port Sudan in comparison with the rest of the country.

6.6.3 An Improvement on Direct Demand Modelling

Recent versions of the Direct Demand model bring it closer to the demand component of the classic transport model, albeit still in an interurban context, and use the choice paradigm explained in Chapter 7 to a full extent. Data, coming typically from an intercept O–D survey (supplemented by any household data available) may be used to estimate a combined frequency-mode-destination choice model where the structure is of Nested Logit form. In particular, a disaggregate version of the combined distribution-modal split model of Section 6.5.1 is coupled, through a composite 'accessibility' variable, to the choice of frequency (or trip generation). This has allowed for the successful incorporation of a measure of accessibility (i.e. related to the ease or difficulty of travelling from a given zone to the rest of the study area) at the trip generation stage, solving the problem of inelastic demand discussed in Section 1.5.1 (see RAND 2004).

Example 6.7 A Direct Demand model for the North of Chile macro-zone (i.e. 117 zones corresponding to local authorities in a territory of some 1800 km length housing 67% of the country's population) was estimated using intercept survey data (Iglesias et al. 2008). The model structure is shown in Figure 6.6; the composite accessibility measure in the trip frequency choice component was the *logsum* (see Section 7.4) of the destination-mode choice component, and estimation yielded correct coefficients (i.e. greater than zero and less than 1.0) for this variable in all user classes. A distinction was made between home-based and non-home-based trips as well as between trips of three different class lengths: short trips (less than 150 km), medium (between 150 and 500 km), and long trips (greater than 500 km); the probability of belonging to a length of trip class was modelled as a trinomial Logit with utilities depending on the zone characteristics and its accessibility.

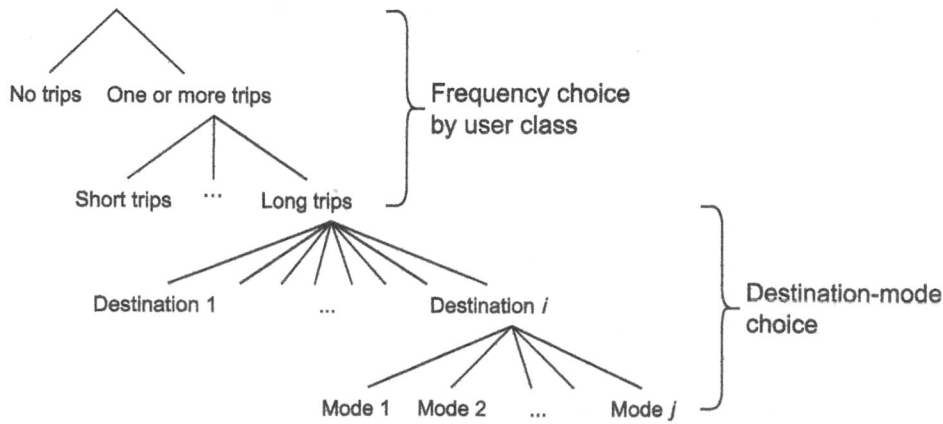

Figure 6.6 Contemporary Direct Demand model

The destination-mode choice component (where eight modes were considered: four types of buses, car, shared taxi, train, and airplane) was a Nested Logit model with the following general utility form:

$$V^*_{M_{zj}} = V^{g,a,l}_{M_{zj}} + \log(S_{1j} + \gamma_2 S_{2j} + \dots + \gamma_{117} S_{117j})$$

where $V^*_{M_{zj}}$ stands for the utility of travelling by mode M ($M = 1, \dots, 8$) from origin z to destination j ($j = 1, \dots, 117$). The first term on the right includes individual characteristics (size of group g, possession of car in the household a, and possession of driving license l) and the level-of-service variables for each destination-mode combination (measured at the zone level). The second term introduces *size variables* S related to the destination attractiveness as a weighted sum inside a logarithm (Daly 1982a).

Based on (6.19), the representative accessibility for a given zone (Z) and trip length (L) by user class (g, a, l) was defined as shown in (6.20) following Williams (1977). The zones included in the summation over j are exclusively those corresponding to the length of trip considered (i.e. the set J_L) for each given origin; the structural parameter φ has to be greater than zero and less than or equal to one (the reason for this is discussed in Section 7.4):

$$Acc^{g,a,l}_{Z,L} = \frac{1}{\varphi} \cdot \log\left(\sum_{\forall j \in J_L; j \neq Z} e^{\varphi \log\left(\sum_{M=1,8} \exp\left(V^*_{M_{zj}}\right) \right)} \right) \tag{6.19}$$

As the application used aggregate data, the accessibility measure for each zone and length of trip was calculated as the weighted sum of representative accessibilities, as follows:

$$Acc_{Z,L} = \sum_{g=1}^{9} \sum_{a=1}^{2} \sum_{l=1}^{2} \left[(P_{G=g} \cdot P_{A=a} \cdot P_{Lic=l}) \cdot Acc^{g,a,l}_{Z,L} \right] \tag{6.20}$$

where P_G is the probability of having a given group size (1–9 people), which was calculated on the basis of the observed distribution of group sizes by time of year (normal and summer) and type of trip (to work and other); P_A is the probability of having a car in the household; and P_{Lic} the probability of having a driving license. The model was applied successfully, and details may be consulted in Iglesias et al. (2008).

6.7 Sense Checks

The estimation of a model with satisfactory goodness of fit is not enough to trust it. It is important to make sure that the model makes sense, that the parameters are not too wildly different from others known from experience, and that when used in earnest, it will not produce strange or exaggerated (or insensitive) responses.

One must pay particular attention when the model is operating close to the tails. Consider Figure 6.7, depicting a well-estimated Logit model for the choice between Options 1 and 2, with costs measured in generalised minutes. If Option 1 is more expensive than Option 2, more people will choose the latter.

It can be seen that there is a bias in favour of Option 1 of about 10 generalised minutes; Option 1 can be 10 minutes more expensive and still half of the travellers will select it. This may be because Option 1 is more convenient; for example, it is a car. A lot depends on what the model is representing and how the constant is estimated.

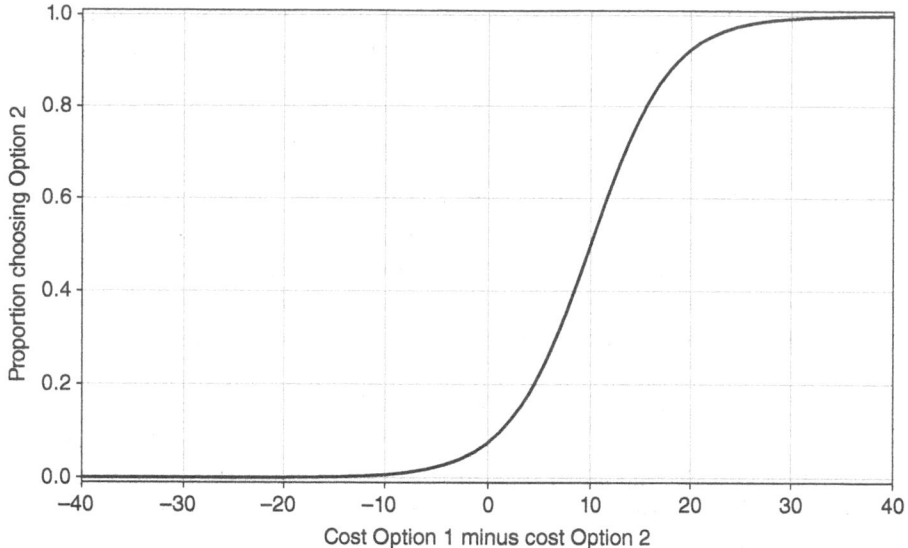

Figure 6.7 A Logit choice model between two alternatives

Example 6.8 Consider a trip from A to B that is currently undertaken 95% by car (Option 1). A bus lane is proposed that would reduce bus travel times between A and B by 10 minutes. This would bring the market share of cars down to 82%. Is this reasonable? This is, in principle, a judgment call and is likely to depend on whether the original bus travel time was 30 or 90 minutes.

On the other hand, assume that the model represents the choice between similar tolled (Option 2) and untolled (Option 1) roads, where the cost of the toll has been converted into 10 generalised minutes (i.e. the average value of time is $6.00/h and the toll is $1.00). Now the toll road between A and B has to be 10 minutes faster to attract half of the traffic. However, the toll road may be no faster than the untolled one, and still, according to the model, about 8% of the traffic would prefer it. In other words, 8% of the traffic would be willing to pay $1.00 for nothing.

One may truncate the curve at that point to prevent drivers from paying $1.00 to lose time, but the Logit curve will still look much less reasonable than in the previous car/bus case. Of course, other explanations may be possible and make the curve more realistic. For example, the toll road may be considered safer, or it may provide more reliable travel times (and these factors are not considered in the model).

This is an area where international benchmarking can help, despite the difficulties of finding comparable contexts. Sensitivity analysis, discussed later in this text, can help to gain confidence in the robustness of the results of any application of mode choice modelling. Small changes in one mode should not have a disproportionate impact on mode choice. Sensitivity analysis can provide elasticity estimates, and these can be compared with elasticities found elsewhere (see the compilation by Litman (2009) at the Victoria Transport Policy Institute). For example, it has been found that fare elasticities for transit ridership are in the range of −0.3 to −0.5 in the short-run and −0.6 to −0.7 in the long-run (based on British and French data). However, these values depend on the quality of the alternatives, the actual mode share, and, of course, on congestion levels and even cultural values. Another useful source of elasticities is Balcombe et al. (2004).

The distinction between short- and long-term elasticities is valid as certain behavioural responses take longer to materialize, as commented with respect to job and residential changes. For example, the shift from car to bus after a bus lane is introduced may take longer than any shift on a toll road when prices are doubled. It is unclear which elasticity is implied in the classic models. They are estimated with cross-sectional data containing a mixture of short- and longer-term changes, as reality is never in equilibrium. This is, again, a case where noise in the data detracts from the quality of the models, in particular the ability to distinguish between short- and long-term travel responses.

Exercises

6.1 A mode choice survey has been undertaken on a corridor connecting four residential areas A, B, C, and D with three employment areas U, V, and W. The corridor is served by a good rail link and a reasonable road network. The three employment zones are in a heavily congested area, and therefore journeys by rail are often faster than by car. The information collected during the survey is summarised in Table 6.3:

Table 6.3 Survey data for Exercise 6.1

O–D pair	By car				By rail			Proportion by car
	X_1	X_2	X_3	X_4	X_1	X_2	X_3	
A–U	23	3	120	40	19	10	72	0.82
B–U	20	3	96	40	17	8	64	0.80
C–U	18	3	80	40	14	10	28	0.88
D–U	15	3	68	40	14	12	20	0.95
A–V	26	4	152	60	23	10	104	0.72
B–V	19	4	96	60	18	9	72	0.90
C–V	14	4	60	60	11	9	36	0.76
D–V	12	4	56	60	12	11	28	0.93
A–W	30	5	160	80	25	10	120	0.51
B–W	20	5	100	80	16	8	92	0.56
C–W	15	5	64	80	12	9	36	0.58
D–W	10	5	52	80	8	9	24	0.64

where the costs per trip per passenger were defined as follows:

X_1 = in-vehicle travel time in minutes (line haul plus feeder mode, if any);
X_2 = excess time (walking plus waiting) in minutes;
X_3 = out-of-pocket travel costs (petrol or fares), in cents;
X_4 = parking costs associated with a one-way trip, in cents.

a) Calibrate a Logit modal-split model assuming that the value of travel time is 8 $/min and that the value of excess time is twice as much.
b) Estimate the impact on modal split on each O–D pair of an increase in petrol prices which doubles the perceived cost of running a car (X_3).
c) Estimate the shift in modal split that could be obtained if no fares were charged on the rail system.

6.2 An inter-urban mode choice study is being undertaken for people with a choice between car and rail. The figures in Table 6.4 were obtained as a result of a survey on five origin-destination pairs A–E:

Table 6.4 Survey data for Exercise 6.2

	Elements of cost by each mode				
	Car		Rail		
O–D	X_1	X_2	X_1	X_2	Proportion choosing car
A	3.05	9.90	2.50	9.70	0.80
B	4.05	13.10	2.02	14.00	0.51
C	3.25	9.30	2.25	8.60	0.57
D	3.50	11.20	2.75	10.30	0.71
E	2.45	6.10	2.04	4.70	0.63

where X_1 is the travel time (in hours) and X_2 the out-of-pocket cost (in $). Assume that the 'value of time' coefficient is 2.00 $/h and calculate the generalised cost of travelling by each mode.

a) Calibrate a binary Logit modal-split model with these data, including the mode-specific penalty.
b) An improved rail service is to be introduced, which will reduce travel times by 20% on every journey; by how much could the rail mode increase its fares without losing customers at each O–D pair?
c) How would you model the introduction of an express coach service between these cities?

6.3 Consider the following trip distribution/modal-split model:

$$V_{ij}^n = A_i O_i B_j D_j \exp\left(-\beta M_{ij}^n\right)$$

where

$$M_{ij}^n = -(1/\tau^n) \log \sum\nolimits_k \exp\left(-\tau^n C_{ij}^k\right)$$

and $n = 1$ stands for persons with access to car, $n = 2$, persons without access to car, $k = 1$ stands for car and $k = 2$ for public transport.

If the total number of trips between zones i and j is $V_{ij} = 1000$, compute how many will use car and how many will use public transport according to the model. The estimated parameter values were found to be: $\tau^1 = 0.10$, $\tau^2 = 0.05$, and $\beta = 0.04$; also, for trips between i and j the modal costs were calculated as: $C_{ij}^1 = 30$ and $C_{ij}^2 = 40$.

6.4 Consider the following modal-split model:

$$P_k = \exp\left(-\tau C_{ij}^k\right) \Big/ \sum_m \exp\left(-\tau C_{ij}^m\right)$$

with generalised costs given by the following expression:

$$C_{ij}^k = \sum_p \theta_{kp} x_{kp}$$

where θ are parameters weighing the model explanatory variables (time, cost, etc).

a) Write an expression for the elasticity of P_k with respect to x_{kp}.
b) Consider now a binary choice situation where the generalised costs have the following concrete expressions:

$$C_{car} = 0.2tt_{car} + 0.1c_{car} + 0.3et_{car}$$
$$C_{bus} = 0.2tt_{bus} + 0.1c_{bus} + 0.3et_{bus} + 0.3$$

where tt is in-vehicle travel time (min), c is travel cost ($), and et is access time (walking and waiting, min). Assume we know average data for the modes (Table 6.5):

Table 6.5 Average data for modes in Exercise 6.4

Mode	Variable		
	tt	c	et
Car	20	50	0
Bus	30	20	5

Calculate the proportion of people choosing car if $\tau = 0.4$.

6.5 The railway between the towns of A and B spans 800 km through mountainous terrain. The total one-way travel time, t_r, is 20 h and currently the fare, c_r, is 600 $/ton. As the service is used at low capacity, t_r is a constant, independent of the traffic volume V_r.

There is a lorry service competing with the railway in an approximately parallel route; its average speed is 50 km/h and it charges a fare of 950 $/ton. There is a project to build a highway in order to replace the present road; it is expected that most of its traffic will continue to be heavy trucks.

The level-of-service function of the new highway has been estimated as:

$$t_t = 7 + 0.08V_t \text{ (h)}$$

where V_t is the total flow of trucks per hour.

On the other hand the railway has estimated its demand function as follows:

$$(V_r/V_t) = 0.83(t_r/t_t)^{-0.8}(c_r/c_t)^{-1.6}$$

and it is expected that the total volume transported between the two towns, $V_r + V_t$, will remain constant and equal to 200 truck loads/h in the medium term.

a) Estimate the current modal split (i.e. volumes transported by rail and lorry).
b) Estimate modal split if the highway is built.
c) What would be the modal split if:
  • the railway decreases its fare to 450 $/ton?
  • the lorries were charged a toll of 4 $/ton in order to finance the highway?
  • both changes applied simultaneously?

7

Discrete Choice Models

In this chapter, we provide a comprehensive introduction to *discrete choice* (i.e. when individuals have to select an option from a finite set of alternatives) modelling methods. We start with some general considerations and move on to explain the theoretical framework, random utility theory, in which these models are cast. This serves us to introduce some basic terminology and to present the individual-modeller 'duality', which is so useful to understanding what the theory postulates. Next, we introduce the two most popular discrete choice models: Multinomial Logit (MNL) and Nested Logit (NL), which taken as a family provided practitioners with a powerful modelling tool set for over four decades. We also discuss other choice models, in particular Mixed Logit (ML) which is now recognised as one of the standards in the field, and consider the benefits and special problems involved when modelling with panel data and when *latent variables* are incorporated. These are two important subjects that have also become standard practice. Finally, we briefly look at other choice paradigms that offer an alternative perspective to the classical utility-maximising approach.

The problems of model specification and estimation, both with revealed- and stated-preference data, are considered in sufficient detail for practical analysis in Chapter 8; we provide information about certain issues, such as validation samples, which are seldom found in texts on this subject. The problem of aggregation, from various perspectives, and the important question of model updating and transference (particularly for those interested in a continuous planning approach to transport), are tackled in Chapter 9.

7.1 General Considerations

Aggregate demand transport models, such as those we have discussed in the previous chapters, are either based on observed relations for groups of travellers or on average relations at the zone level. On the other hand, disaggregate demand models are based on observed choices made by individual travellers or households. It was expected that the use of this framework would enable more realistic models to be developed.

Despite the pioneering work of researchers such as Warner (1962) or Oi and Shuldiner (1962), who drew attention to serious deficiencies in the conventional methodologies, aggregate models continued to be used, almost unscathed, in the majority of transport projects until the early

Modelling Transport, Fifth Edition. Juan de Dios Ortúzar and Luis G. Willumsen.
© 2024 John Wiley & Sons Ltd. Published 2024 by John Wiley & Sons Ltd.
Companion website: www.wiley.com/go/ortuzar5e

1980s. In fact, only then did discrete choice models start to be seriously considered as a modelling option (see Williams 1981). In general, discrete choice models postulate that:

the probability of individuals choosing a given option is a function of their socioeconomic characteristics and the relative attractiveness of the option.

To represent the attractiveness of the alternatives, the concept of *utility* (which is a convenient theoretical construct defined as what the individual seeks to maximise) is used. Alternatives, *per se*, do not produce utility; this is derived from their characteristics (Lancaster 1966) and those of the individual; for example, the *observable utility* is usually defined as a linear combination of variables, such as:

$$V_{\text{car}} = 0.25 - 1.2 \cdot \text{IVT} - 2.5 \cdot \text{ACC} - 0.3 \cdot \text{C/I} + 1.1 \cdot \text{NCAR} \tag{7.1}$$

where each variable represents an attribute of the alternative or of the traveller. The relative influence of each attribute, in terms of contributing to the overall satisfaction produced by the alternative is given by its coefficient. For example, a unit change on *access time* (ACC) in (7.1) has approximately twice the impact of a unit change on *in-vehicle travel time* (IVT) and more than seven times the impact of a unit change on the variable *cost/income* (C/I), which is also expressed in min. But the variables can also represent characteristics of the individual; for example, we would expect that an individual belonging to a household with a large *number of cars* (NCAR), would be more likely to choose the car option than another belonging to a family with just one vehicle. The *alternative-specific constant*, 0.25 in Eq. (7.1), is normally interpreted as representing the net influence of all unobserved or not explicitly included characteristics of the individual and the alternative in its utility function. For example, it could include elements such as comfort and convenience, which are not easy to measure or observe.

To predict if an alternative will be chosen, according to the model, the value of its utility must be contrasted with those of other alternatives available and transformed into a probability value between 0 and 1. For this, a variety of mathematical transformations exist that are typically characterised by having an S-shaped plot, such as:

Logit $\qquad P_1 = \dfrac{\exp(V_1)}{\exp(V_1) + \exp(V_2)}$

Probit $\qquad P_1 = \displaystyle\int_{-\infty}^{\infty} \int_{-\infty}^{V_1 - V_2 + x_1} \dfrac{\exp\left\{ -\dfrac{1}{2(1-\rho^2)} \left[\left(\dfrac{x_1}{\sigma_1}\right)^2 - \dfrac{2\rho x_1 x_2}{\sigma_1 \sigma_2} + \left(\dfrac{x^2}{\sigma^2}\right)^2 \right] \right\}}{2\pi \sigma_1 \sigma_2 \sqrt{(1-\rho^2)}} \, dx_2 \, dx_1$

where the completely general covariance matrix of the Normal distribution associated with this latter model has the form:

$$\Sigma = \begin{pmatrix} \sigma_1^2 & \rho\sigma_1\sigma_2 \\ \rho\sigma_1\sigma_2 & \sigma_2^2 \end{pmatrix}$$

that is, it allows for correlation (i.e. $\rho \neq 0$) and heteroskedasticity (i.e. different variances; see McCulloch 1985 for a little divertimento) among alternatives.

In general, discrete choice models cannot be calibrated using standard curve-fitting techniques, such as least squares, because their dependent variable P_i is an un-observed probability (between 0 and 1) and the observations are the individual choices (which are either 0 or 1). The only exceptions to this are models for homogeneous groups of individuals, or when the behaviour of every individual is recorded on several occasions, because observed frequencies of choice are also variables between 0 and 1.

Some useful properties of these models were summarised by Spear (1977):

1) Disaggregate demand models (DM) are based on theories of individual behaviour and do not constitute physical analogies of any kind. Therefore, as an attempt is made to explain individual behaviour, an important potential advantage over aggregate models is that it is more likely that DM models are stable (or transferable) in time and space.

2) DM models are estimated using individual data, and this has the following implications:

   • DM models may be more efficient than aggregate models in terms of information usage because fewer data points are required as each individual choice is used as an observation. In aggregate modelling one observation is the average of (sometimes) hundreds of individual observations;

   • as individual data are used, all the inherent variability in the information can be utilised;

   • DM models may be applied, in principle, at any aggregation level; however, although this appears obvious, the aggregation processes are not trivial, as we will discuss in Chapter 9;

   • DM models are less likely to suffer from biases due to correlation between aggregate units; a serious problem when using aggregate information is that individual behaviour may be hidden by unidentified characteristics associated with the zones; this is known as *ecological correlation*; the example in Figure 7.1 shows that if a trip generation model was estimated using zonal data, we would obtain that the number of trips decreases with income; however, the opposite would be shown to hold if the data were considered at a household level; this phenomenon, which is of course exaggerated in the figure, might occur for example if the land-use characteristics of zone B are conducive to more trips on foot.

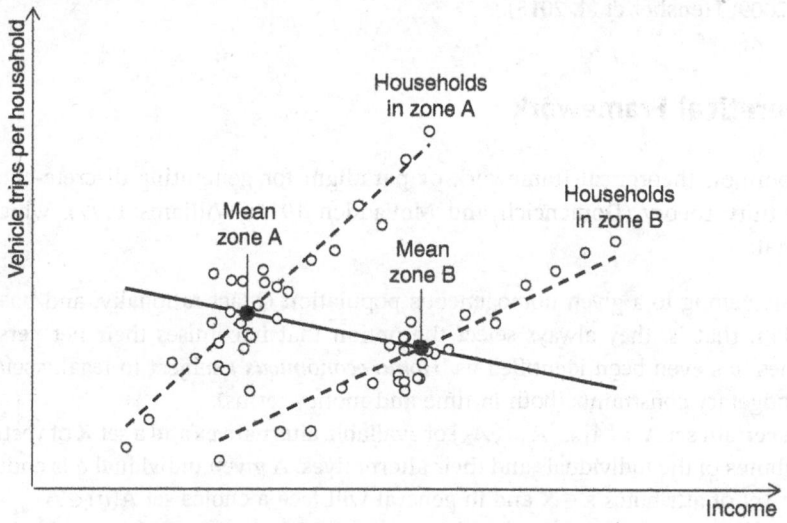

Figure 7.1 Example of ecological fallacy

3) Disaggregate models are probabilistic; furthermore, as they yield the probability of choosing each alternative but do not indicate which one is selected, use must be made of basic probability concepts such as:

- the expected number of people using a certain travel option equals the sum over each individual of the probabilities of choosing that alternative:

$$N_i = \sum_n P_{in}$$

- an *independent* set of decisions may be modelled separately, considering each one as a conditional choice; then the resulting probabilities can be multiplied to yield joint probabilities for the set, such as in:

$$P(f, d, m, r) = P(f)\, P(d/f)\, P(m/d,f)\, P(r/m,d,f)$$

with f = frequency; d = destination; m = mode; r = route.

4) The explanatory variables included in the model can have explicitly estimated coefficients. In principle, the utility function allows any number and specification of the explanatory variables, as opposed to the generalised cost function in aggregate models, which is generally limited and has several fixed parameters. This has implications such as the following:

- DM models allow a more flexible representation of the policy variables considered relevant for the study;
- the coefficients of the explanatory variables have a direct marginal utility interpretation (i.e. they reflect the relative importance of each attribute).

In the sections that follow and in the next two chapters, we will examine in some detail several interesting aspects of discrete choice models, such as their theoretical base, structure, specification, functional form, estimation, and aggregation. Notwithstanding, interested readers are advised that there are at least three good books dealing exclusively with this subject (Ben-Akiva and Lerman 1985; Train 2009; Hensher et al. 2015).

7.2 Theoretical Framework

The most common theoretical framework or paradigm for generating discrete-choice models is random utility theory (Domencich and McFadden 1975; Williams 1977), which basically postulates that:

1) Individuals belong to a given homogeneous population Q, act rationally, and possess perfect information, that is, they always select the option that maximises their net personal utility (the species has even been identified as '*Homo economicus*') subject to legal, social, physical, and/or budgetary constraints (both in time and money terms).
2) There is a certain set $\mathbf{A} = \{A_1,..., A_j,..., A_N\}$ of available alternatives and a set $\mathbf{X}$ of vectors of measured attributes of the individuals and their alternatives. A given individual q is endowed with a particular set of attributes $\mathbf{x} \in \mathbf{X}$ and in general will face a choice set $\mathbf{A}(q) \in \mathbf{A}$.

   In what follows, we will assume that the individual's choice set is predetermined; this implies that the effect of the constraints has already been taken care of and does not affect the process of selection among the available alternatives. Choice-set determination will be considered, together with other important practical issues, in Chapter 8.

3) Each alternative $A_j \in \mathbf{A}$ has associated a net utility U_{jq} for each individual q. The modeller, who is an observer of the system, does not possess complete information about all the elements considered by the individual making a choice; therefore, the modeller assumes that U_{jq} can be represented by two components:

- a measurable, systematic, or representative part V_{jq}, which is a function of the measured attributes $\mathbf{x}$; and
- a random part ε_{jq}, which reflects the idiosyncrasies and particular tastes of each individual, together with any measurement or observational errors made by the modeller.

  Thus, the modeller postulates that:

$$U_{jq} = V_{jq} + \varepsilon_{jq} \tag{7.2}$$

which allows two apparent 'irrationalities' to be explained: (i) that two individuals with the same attributes and facing the same choice set may select different alternatives, and (ii) that some individuals may not always select what appears to be the best alternative from the point of view of the attributes considered by the modeller.

For the decomposition (7.2) to be correct, we need certain homogeneity in the population under study. In principle, we require that ... 'all individuals share the same set of alternatives and face the same constraints' (see Williams and Ortúzar 1982a), and to achieve this, we may need to segment the market.

Although we have termed $\mathbf{V}$ *representative*, it carries the subscript q because it is a function of the attributes $\mathbf{x}$, and this may vary from individual to individual. On the other hand, without loss of generality, it can be assumed that the residuals ε are random variables with mean 0 and a certain probability distribution to be specified. A popular and simple expression for $\mathbf{V}$ is:

$$V_{jq} = \sum_k \theta_{kj} x_{jkq} \tag{7.3}$$

where the parameters $\boldsymbol{\theta}$ are assumed to be constant for all individuals in the homogeneous set (i.e. a fixed-coefficients model) but may vary across alternatives. Other possible forms, together with a discussion on how each variable should enter the utility function, will be presented in Chapter 8.

It is important to emphasise the existence of two points of view in the formulation of the above problem: firstly, that of the individual who calmly weighs all the elements of interest (with no randomness) and selects the most convenient alternative; and secondly, that of the modeller who by observing only some of the above elements, needs the residuals ε to explain what otherwise would amount to non-rational behaviour.

4) The individual q selects the maximum-utility alternative; that is, the individual chooses A_j if and only if:

$$U_{jq} \geq U_{iq}, \forall A_i \in \mathbf{A}(q) \tag{7.4}$$

that is

$$V_{jq} - V_{iq} \geq \varepsilon_{iq} - \varepsilon_{jq} \tag{7.5}$$

As the analyst ignores the value of $(\varepsilon_{iq} - \varepsilon_{jq})$ it is not possible to determine with certitude if (7.5) holds. Thus, the probability of choosing A_j is given by:

$$P_{jq} = \text{Prob}\{\varepsilon_{iq} \leq \varepsilon_{jq} + (V_{jq} - V_{iq}), \forall A_i \in \mathbf{A}(q)\} \tag{7.6}$$

and as the joint distribution of the residuals ε is not known, it is not possible at this stage to derive an analytical expression for the model. What we do know, however, is that the residuals are random variables with a certain distribution which we can denote by $f(\boldsymbol{\varepsilon}) = f(\varepsilon_1,..., \varepsilon_N)$. Let us note in passing that the distribution of $\mathbf{U}$, $f(\mathbf{U})$, is the same but with different mean (i.e. $\mathbf{V}$ rather than 0).

Therefore, we can write (7.6) more concisely as:

$$P_{jq} = \int_{R_N} f(\boldsymbol{\varepsilon})\, d\boldsymbol{\varepsilon} \tag{7.7}$$

where

$$R_N = \begin{cases} \varepsilon_{iq} \leq \varepsilon_{jq} + (V_{jq} - V_{iq}), & \forall A_i \in \mathbf{A}(q) \\ V_{jq} + \varepsilon_{jq} \geq 0 \end{cases}$$

and different model forms may be generated depending on the distribution of the residuals ε.

An important class of random utility models is that generated by utility functions with independent and identically distributed (IID) residuals. In this case, $f(\varepsilon)$ can be decomposed into:

$$f(\varepsilon_1, ..., \varepsilon_N) = \prod_n g(\varepsilon_n)$$

where $g(\varepsilon_n)$ is the utility distribution associated with alternative A_n, and the general expression (7.7) reduces to:

$$P_j = \int_{-\infty}^{\infty} g(\varepsilon_j)\,d(\varepsilon_j) \prod_{i \neq j} \int_{-\infty}^{V_j - V_i + \varepsilon_j} g(\varepsilon_i)\,d\varepsilon_i \tag{7.8a}$$

where we have extended the range of both integrals to $-\infty$ (a slight inconsistency) in order to solve them.

A two-dimensional geometric interpretation of this model, together with extensions to the more general case of correlation and unequal variances, was presented and discussed by Ortúzar and Williams (1982). Equation (7.8a) can also be expressed as:

$$P_j = \int_{-\infty}^{\infty} g(\varepsilon_j)\,d\varepsilon_j \prod_{i \neq j} G(\varepsilon_j + V_j - V_i) \tag{7.8b}$$

with

$$G(x) = \int_{-\infty}^{X} g(x)\, dx$$

and it is interesting to mention that a large amount of effort was spent just trying to find out appropriate forms for g that would allow to solve (7.8b) in closed form.

Note that the IID residuals requisite means that the alternatives should, in fact, be independent. Mixed-mode alternatives, for example, car-rail combinations, will usually violate this condition.

7.3 The Multinomial Logit (MNL) Model

This is the simplest and was, for decades, the most popular practical discrete choice model (Domencich and McFadden 1975). It can be generated assuming that the random residuals in (7.7) distribute IID Extreme Value Type I (also known as Gumbel or Weibull distribution).

The cumulative distribution function of the Extreme Value Type I (EV1) is given by the following expression:

$$F(\varepsilon) = \exp[-\exp(-\lambda(\varepsilon - \eta))]$$

and deriving it we get the following density function:

$$f(\varepsilon) = \lambda \exp[-\lambda(\varepsilon - \eta)] \exp[-\exp(-\lambda(\varepsilon - \eta))]$$

where η is the mode of the function and λ is a scale factor; these two parameters allow to represent the EV1 function completely, so it is generally said that $\varepsilon \sim (\eta, \lambda)$.

The mean of the distribution is at $\eta + \gamma/\lambda$ where γ is Euler's constant ($\gamma \approx 0.577$), and the variance is given by $\pi^2/6\lambda^2$. The shape of the distribution is shown in Figure 7.2 for conditions that allow the mean to be zero.

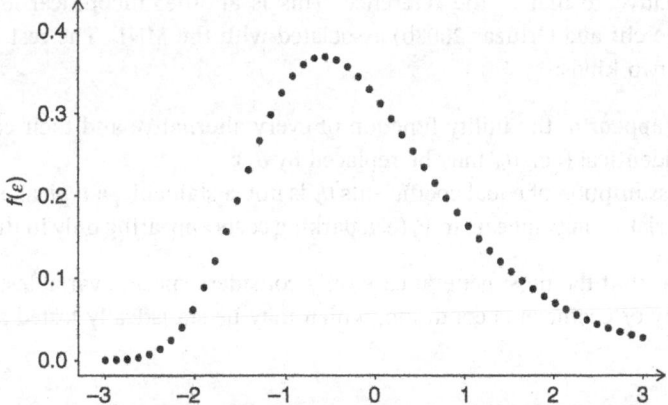

Figure 7.2 EV1 density function for $\lambda = 1$ and $\eta = -0.577$

An important characteristic of the EV1 distribution is that it is closed to maximisation (i.e. as the Normal distribution is closed to summation or subtraction), and this property came in very handy for deriving the MNL model. Also, the difference of two IID EV1 variables follows the logistic distribution, a property that we will refer to below when discussing the NL model.

So, if the residuals ε distribute IID EV1 in (7.7), the choice probabilities are given by:

$$P_{iq} = \frac{\exp(\beta V_{iq})}{\sum_{A_j \varepsilon A(q)} \exp(\beta V_{jq})} \tag{7.9}$$

where the utility functions usually have the linear form in the parameters (7.3), and the scale factor β (which is usually normalised to one in practice as it cannot be estimated separately from the θ) is related to the common standard deviation of the EV1 variate by:

$$\beta = \pi/\sigma\sqrt{6} \tag{7.10}$$

In Chapter 9, we will use (7.10) to discuss the problem of potential bias in forecasts when use is made of data at different levels of aggregation. The fact that β cannot be estimated separately from the parameters θ in V_{iq} is known as *theoretical identification*. All discrete choice models have identification problems, which require setting certain parameters to a given value in order to estimate the model uniquely (see Walker 2002). We will come back to this important issue several times in this chapter.

7.3.1 Specification Searches

To decide which variables $x_k \in \mathbf{x}$ enter the utility function and whether they are of generic type or specific to a particular alternative, a search process is normally employed, starting with a theoretically appealing specification (Ortúzar 1982). Then variations are tested at each step to check whether they add explanatory power to the model; we will examine methods for doing this in Chapter 8.

If we define one of the values of $\mathbf{x}$ equal to one for all individuals q that have available a given alternative A_j, the coefficient θ_k corresponding to that variable is interpreted as an *alternative specific constant* (ASC). Although we may specify an ASC for every alternative, it is not possible to estimate their N parameters individually due to the way the model works (as shown in Example 7.1). For this reason, one alternative is taken as reference (fixing to 0 the value of its parameter without loss of generality), and the remaining $(N-1)$ values obtained in the estimation process are interpreted as relative to that of the reference. This is another theoretical identification issue (Walker 2002; Cherchi and Ortúzar 2008b) associated with the MNL. The rest of the variables $\mathbf{x}$ may be of one of two kinds:

- generic, if they appear in the utility function of every alternative and their coefficients can be assumed to be identical (i.e., θ_{jk} may be replaced by θ_k);
- specific, if the assumption of equal coefficients θ_k is not sustainable, a typical example occurring when the kth variable only appears in V_j (e.g. parking costs appearing only in the car alternative).

It must be noted that the most general case only considers specific variables; the generic ones impose an equality of coefficients condition, which may be statistically tested as we will discuss in Chapter 8.

Example 7.1 Consider the following binary Logit model:

$$P_1 = \exp(V_1)/[\exp(V_1) + \exp(V_2)] = 1/[1 + \exp(V_2 - V_1)]$$

where the observable utilities are postulated as linear functions of two generic variables x_1 and x_2, and two constants (with coefficients θ_3 and θ_4) as follows:

$$V_1 = \theta_1 x_{11} + \theta_2 x_{12} + \theta_3$$
$$V_2 = \theta_1 x_{21} + \theta_2 x_{22} + \theta_4$$

As can be seen from the model expression, the relevant factor is the difference between both utilities:

$$V_2 - V_1 = \theta_1(x_{21} - x_{11}) + \theta_2(x_{22} - x_{12}) + (\theta_4 - \theta_3)$$

and this allows us to deduce the following:

1) It is not possible to estimate both θ_3 and θ_4, only their difference; for this reason, there is no loss of generality if one is taken as 0, and the other is estimated relative to it (this, of course, applies to any number of alternatives).
2) If either x_{1j} or x_{2j} have the same value for both options (as in the case of variables representing individual attributes, such as income, age, sex, or number of cars in the household), a generic

coefficient cannot be estimated as it would always multiply a zero value. This also applies to level-of-service variables that happen to share a common value for two or more alternatives (for example, public-transport fares in an integrated system). In either case, they can only appear in some (but not all) alternatives or need to be entered as specific variables (i.e. with different coefficients for each alternative).

The problem posed by individual attributes is further compounded by the fact that it is not always easy or clear to decide in which alternative utility(ies) the variable should appear. Consider the case of a variable such as SEX (i.e. 0 for males, 1 for females) in a mode choice study; if we believe, for example, that males have first call on access to the car for commuting purposes, we will not enter the variable in the utilities of both car driver and car passenger, say. However, we may have no insights on whether to enter it or not in the utilities of other modes such as, for example, bus or metro. The problem is that entering the variable in different ways usually yields different estimation results, and choosing the optimum may become a hard combinatorial problem, even for a small number of alternatives and attributes. If we lack insight and there are no theoretical grounds for preferring one form over another, the only way out may be trial and error.

7.3.2 Universal Choice Set Specification

When individuals have different choice sets, it is useful to rewrite the model based on the universal choice set formulation introducing availability variables into the utility function (see Bierlaire et al. 2010). Let A_{iq} be 1 if individual q has alternative A_i available and 0 otherwise. For example, if walking is considered unavailable for distances longer than 3 km, we would have:

$$A_{iq} = \begin{cases} 1 & \text{if } d_q < 3 \\ 0 & \text{if } d_q \geq 3 \end{cases}$$

where d_q is the distance to be travelled by individual q. It is possible to write any choice model based on this idea of a universal choice set:

$$P_{iq}\{\mathbf{A}(q)\} = \text{Prob}\{U_{iq} \geq U_{jq}, \quad \forall A_j \in \mathbf{A}(q)\}$$
$$= \text{Prob}\{U_{iq} + \log A_{iq} \geq U_{jq} + \log A_{jq}, \quad \forall A_j \in \mathbf{A}\}$$

Thus, when one of the A_{iq} is equal to 1, the additional term does not play any role. But if $A_{iq} = 0$, the inequality is never verified, and the probability of choosing the alternative is 0, which makes sense as it is not available. Finally, when $A_{jq} = 0$, for $j \neq i$ the right-hand side of the above equation is trivially lower than anything else.

Using this formulation, the MNL expression becomes:

$$P_{iq}\{\mathbf{A}(q)\} = \frac{\exp(V_{iq} + \log A_{iq})}{\sum\limits_{A_j \in \mathbf{A}} \exp(V_{jq} + \log A_{jq})} = \frac{A_{iq} \exp(V_{iq})}{\sum\limits_{A_j \in \mathbf{A}} A_{jq} \exp(V_{jq})}$$

and this helps to generalise some properties that were originally applicable only to cases where all individuals had the same choice set, as we will see below.

7.3.3 Some Properties of the MNL

The model satisfies the axiom of *independence of irrelevant alternatives* (IIA) which can be stated as:

> *Where any two alternatives have a non-zero probability of being chosen, the ratio of one probability over the other is unaffected by the presence or absence of any additional alternative in the choice set* (Luce and Suppes 1965).

As can be seen, in the MNL case the ratio

$$\frac{P_j}{P_i} = \exp\{\beta(V_j - V_i)\}$$

is indeed a constant independent of the rest of the alternatives. Initially, this was considered an advantage of the model, as it would allow to treat quite neatly the *new alternative* problem (i.e. being able to forecast the share of an alternative not present at the calibration stage, if its attributes are known); however, this property is now perceived as a potentially serious disadvantage that makes the model fail in the presence of correlated alternatives (recall the red bus–blue bus problem of Chapter 6). We will come back to this in Section 7.4.

If there are too many alternatives, such as in the case of destination choice, it can be shown (McFadden 1978) that unbiased parameters are obtained if the model is estimated with a random sample of the available choice set for each individual (for example, seven destination options per individual). Models without this property may require, even if their estimation process is not complex, a large amount of computing time for more than, say, 50 options. Unfortunately, such a figure is not uncommon in a destination-choice context; if one thinks of zoning systems of normal size, even if the combinatorial problem of forming destination/mode choice options is bypassed.

If the model is estimated with information from a sub-area or with data from a biased sample, it can be shown (Cosslett 1981) that if a complete set of mode-specific constants is specified, an unbiased model may be obtained just by correcting the constants according to the following expression:

$$K_i' = K_i - \log(q_i/Q_i) \tag{7.11}$$

where q_i is the market share of alternative A_i in the sample and Q_i its market share in the population. All constants must be corrected, including the reference one that is made equal to 0 during estimation.

It is possible to derive simple equations for the direct and cross-elasticities of the model. For example, the direct point elasticity, that is, the percentage change in the probability of choosing A_i with respect to a marginal change in a given attribute X_{ikq}, is simply given by:

$$E_{P_{iq}, X_{ikq}} = \theta_{ik} X_{ikq} \left(1 - P_{iq}\right) \tag{7.12}$$

while the cross-point elasticity is also simply given by:

$$E_{P_{iq}, X_{jkq}} = -\theta_{jk} X_{jkq} \cdot P_{jq} \tag{7.13}$$

that is, the percentage change in the probability of choosing A_i with respect to a marginal change in the value of the kth attribute of alternative A_j, for individual q. Note that as this value is independent of alternative A_i, the cross-elasticities of any option A_i with respect to the attributes X_{jkq} of alternative A_j are equal. This seemingly peculiar result is also due to the IIA property, or more precisely, to the need for IID utility functions in the model generation.

7.4 The Nested Logit Model (NL)

7.4.1 Correlation and Model Structure

In the last section, we discussed the MNL model, which has a very simple covariance matrix. For example, in the trinomial case it is of the form:

$$\sum = \sigma^2 \begin{pmatrix} 1 & 0 & 0 \\ 0 & 1 & 0 \\ 0 & 0 & 1 \end{pmatrix}$$

This simplicity may give rise to problems in any of the following cases:

- When alternatives are not independent (i.e. there are groups of alternatives that are more similar than others, such as public-transport modes vs. the private car);
- When the variances of the error terms ε are not equal, that is, when there is heteroskedasticity (e.g. *between observations*, if some individuals possess a GPS device and are thus able to measure their times more precisely than others; or *between alternatives*, when certain alternatives have more precise attributes, say waiting times of metro and bus, see Munizaga et al. 2000).
- When there are taste variations among individuals (i.e. if the perception of costs varies with income but we have not measured this variable), in which case we may require random coefficient models rather than fixed coefficient models as the MNL.
- When there are multiple responses per individual, as in the case of panel data or stated preference observations; these introduce problems associated with dependency between observations, violating one of the assumptions underpinning the MNL model (see Chapter 8).

In these four senses more flexible models such as the Multinomial Probit (MNP) model, which can be derived from a multivariate Normal distribution (rather than IID EV1), or the Mixed Logit (ML) model (we will discuss both models in Sections 7.5 and 7.6), are completely general because they are endowed with an arbitrary covariance matrix. However, as we will see below, the MNP is not easy to solve except for cases with up to three alternatives (see Daganzo 1979), and the ML requires a more involved estimation method, but it is now recognized as one of the standards in the field.

Notwithstanding, there are certain situations where even if these more powerful models were available, their full generality could be an unnecessary luxury because specific forms for the utility functions suggest themselves. A good example are cases of bi-dimensional choices, such as the combination of destination (D) and mode (M) choices, where alternatives are correlated but taste variations or heteroskedasticity need not be a problem. In these cases, the alternatives at each dimension can be denoted as $(D_1,..., D_D)$ and $(M_1,..., M_M)$ with their combination yielding the choice set **A**, whose general element $D_d M_m$ may be a specific destination-mode option to carry out a certain activity.

In this type of context, it is interesting to consider functions of the following types (Williams and Ortúzar 1982a):

$$U(d, m) = U_d + U_{dm}$$

where, for example, U_d could correspond to that portion of utility specifically associated with the destination and U_{dm} to the disutility associated with the cost of travelling. If we write the expression above following our previous notation, we get:

$$U(d, m) = V(d, m) + \varepsilon(d, m)$$

where

$$V(d, m) = V_d + V_{dm}$$

and

$$\varepsilon(d, m) = \varepsilon_d + \varepsilon_{dm}$$

It can be shown that if the residuals ε are separately IID, under certain conditions the Hierarchical or Nested Logit (NL) model (Williams 1977; Daly and Zachary 1978) is formed:

$$P(d, m) = \frac{\exp\{\beta(V_d + V_d^*)\} \exp(\lambda V_{dm})}{\sum_{d'} \exp\{\beta(V_{d'} + V_{d'}^*)\} \sum_{m'} \exp(\lambda V_{dm'})}$$

with

$$V_d^* = (1/\lambda) \log \sum_{m'} \exp(\lambda V_{dm'})$$

This is precisely the model form used in the destination-mode choice component of contemporary Direct Demand models, as discussed in Section 6.6.3. Furthermore, it can be easily shown that if $\beta = \lambda$ (which occurs when $\varepsilon_d = 0$) the NL collapses, as special case, to the single parameter MNL. To understand why this is so, let us first write in full the utility expressions for the first destination in a simple binary mode case:

$$U(1, 1) = V_1 + V_{11} + \varepsilon_1 + \varepsilon_{11}$$
$$U(1, 2) = V_1 + V_{12} + \varepsilon_1 + \varepsilon_{12}$$

As can be seen, the source of correlation is the residual ε_1, which can be found in both $U(1, 1)$ and $U(1, 2)$; therefore, when ε_d becomes 0, there is no correlation left and the model is indistinguishable from the MNL.

Finally, it can also be shown that for the model to be internally consistent, we require that the following condition holds (Williams 1977):

$$\beta \leq \lambda$$

Models that fail to satisfy this requirement have been shown to produce elasticities of the wrong size and/or sign (Williams and Senior 1977).

7.4.2 Fundamentals of Nested Logit Modelling

In his historical review of the NL model, Ortúzar (2001) mentions several authors whose work predates the model's actual theoretical formulation. Wilson (1970, 1974), Manheim (1973), and Ben-Akiva (1974) all used intuitive versions that – although based on concepts such as marginal probabilities and utility maximisation – did not have a rigorous construction of the functional forms and a clear interpretation of all the model parameters. Domencich and McFadden (1975) generated structured models of Nested Logit form but had an incorrect definition of *composite utilities*.

Williams (1977) was the first to make an exhaustive analysis of the NL properties, especially composite utilities (or *inclusive values*), showing that all previous versions had important inconsistencies with micro-economic concepts. He also reformulated the NL and introduced structural conditions associated with its inclusive value parameters, which are necessary for the NL's compatibility with utility maximising theory. With these, he formally derived the NL model as a *descriptive* behavioural model completely coherent with basic micro-economic concepts. Other authors, whose seminal work completed the fundamental theoretical development of the NL, are Daly and Zachary (1978), who worked simultaneously and totally independently from Williams, and McFadden (1978, 1981), who generalised the work of both Williams, and Daly and Zachary. Unfortunately, the latter has given rise to some confusion in terms of estimation and interpretation of results, which we discuss below. In what follows, we will draw heavily on the work of Carrasco and Ortúzar (2002).

7.4.2.1 The Model of Williams and of Daly–Zachary

As mentioned above, Williams (1977) initially worked with a two-level model in the context of two-dimensional situations, such as destination-mode choice, defining utility functions of the following form:

$$U(i,j) = U_j + U_{i/j} \tag{7.14}$$

where i denotes alternatives at a lower-level nest and j the alternative at the upper level that represents that lower-level nest. In terms of the representative utility and stochastic terms, (7.14) becomes:

$$U(i,j) = V(i,j) + \varepsilon(i,j)$$

where $V(i,j) = V_j + V_{i/j}$ and $\varepsilon(i,j) = \varepsilon_j + \varepsilon_{i/j}$.

Williams' definition of the stochastic errors may be synthesised as follows:

- the errors ε_j and $\varepsilon_{i/j}$ are independent for all (i,j);
- the errors $\varepsilon_{i/j}$ are identically and independently distributed (IID) EV1 with scale parameter λ;
- the errors ε_j distribute with variance σ_j^2 and such that the sum of U_j and the maximum of $U_{i/j}$ distributes EV1 with scale parameter β. It is interesting to mention that such a distribution may not exist; also, this derivation is sufficient, but many other formulations could lead to the same model.

These assumptions have as a consequence the following relationship for the error variances:

$$Var(\varepsilon(i,j)) = Var(\varepsilon_j) + Var(\varepsilon_{i/j}) \tag{7.15}$$

which in our case, using (7.10), may be expressed as

$$\frac{\pi^2}{6\beta^2} = \sigma_j^2 + \frac{\pi^2}{6\lambda^2}$$

leading to
$$\frac{\beta}{\lambda} = \left(1 + \frac{6\sigma_j^2 \lambda^2}{\pi^2}\right)^{-\frac{1}{2}} \tag{7.16}$$

The above implies the *structural condition* that we had anticipated:

$$\beta \leq \lambda \tag{7.17}$$

Now, if we define the *structural parameter* $\phi = \beta/\lambda$, condition (7.17) becomes:

$$\phi \leq 1 \tag{7.18}$$

and when $\beta = \lambda$ (i.e. $\phi = 1$), the NL collapses to the MNL (7.9), as the reader can easily check; but if $\beta > \lambda$ (i.e. $\phi > 1$), the hierarchical structure postulated is incompatible with the utility theoretic basis of this formulation.

The above construction may be generalised in two directions:

- allowing for a different scale parameter λ_j, associated with each nest j, as proposed by Daly and Zachary (1978);
- allowing for an increase in the number of levels in series and parallel (Williams 1977; Daly and Zachary 1978; Sobel 1980).

A very popular NL specification in practice, is one with just two levels of nesting and different scale parameters λ_j in each nest (Figure 7.3), the functional form of which is given by:

$$P(i,j) = \frac{\exp(\lambda_j V_{i/j})}{\sum_{i' \in j} \exp(\lambda_j V_{i'/j})} \cdot \frac{\exp \beta \left\{ \frac{1}{\lambda_j} \log \left(\sum_{i \in j} \exp(\lambda_j V_{i/j}) \right) \right\}}{\sum_{j'=1}^{m} \exp \beta \left\{ \frac{1}{\lambda_{j'}} \log \left(\sum_{i \in j'} \exp(\lambda_{j'} V_{i/j'}) \right) \right\}} \tag{7.19}$$

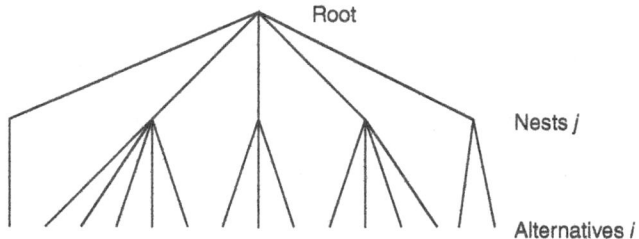

Figure 7.3 A general Nested Logit structure with two levels

In this case, the structural conditions of the model become:

$$\beta \leq \lambda_j \quad \text{for all } j \qquad \Leftrightarrow \qquad \phi_j = \frac{\beta}{\lambda_j} \leq 1 \quad \text{for all } j \tag{7.20}$$

The above model (7.19) and conditions (7.20) allow a range of complex choice processes to be modelled, such as location-mode and multi-mode contexts, which allow for different degrees of substitution (responses to policies) within and between the nests.

7.4.2.2 The Formulation of McFadden: The GEV Family

McFadden (1981) generated the NL model as one particular case of the Generalised Extreme Value (GEV) discrete choice model family (which is considered further in Section 7.7). The members of this family come from a non-negative function $G(Y_1, Y_2..., Y_M)$, with $Y_1, Y_2..., Y_M \geq 0$, which is homogeneous of degree $\mu > 0$; the function approaches to infinite as any Y_i does and has m cross-partial derivatives which are non-negative for odd m and non-positive for even m. As an aside, note that McFadden originally considered $\mu = 1$, but this was later generalized by Ben-Akiva and Lerman (1985).

If we consider the utility function $U_i = V_i + \varepsilon_i$ for M elemental alternatives, the choice probability may be written as:

$$P_i = \int_{\varepsilon = -\infty}^{\varepsilon = \infty} F_i(V_i - V_1 + \varepsilon, ..., V_i - V_M + \varepsilon) \ d\varepsilon$$

where F is the cumulative distribution function of the errors ($\varepsilon_1, ..., \varepsilon_M$) and $F_i = \partial F / \partial \varepsilon_i$. Thus, defining the extreme value multivariate distribution:

$$F(\varepsilon_1, ..., \varepsilon_M) = \exp\{-G(e^{-\varepsilon_1}, ..., e^{-\varepsilon_M})\}$$

P_i, the probability of choosing alternative A_i, is given by:

$$P_i = \frac{e^{V_i} G_i(e^{V_1}, e^{V_2}, ..., e^{V_M})}{\mu G(e^{V_1}, e^{V_2}, ..., e^{V_M})} \tag{7.21}$$

where G_i is the first derivative of G with respect to $Y_i = \exp(V_i)$. Using the above, McFadden showed that the NL probability function is obtained from the following G function:

$$G(e^{V_1}, e^{V_2}, ..., e^{V_M}) = \left(\sum_{j=1}^{J} \left(\sum_{i \in j} e^{V_{(i,j)}} \right)^{1/\mu_j} \right)^{\mu_j} \tag{7.22}$$

$$\text{leading to} \quad P(i,j) = \frac{\exp\left(\dfrac{V_{(i,j)}}{\mu_j}\right)}{\displaystyle\sum_{i \in j} \exp\left(\dfrac{V_{(i,j)}}{\mu_j}\right)} \cdot \frac{\exp \mu_j \ln\left(\displaystyle\sum_{i \in j} \exp\left(\dfrac{V_{(i,j)}}{\mu_j}\right)\right)}{\displaystyle\sum_{j=1}^{J} \exp \mu_{j'} \ln\left(\displaystyle\sum_{i \in j'} \exp\left(\dfrac{V_{(i,j')}}{\mu_{j'}}\right)\right)} \tag{7.23}$$

This probability density function is well-defined (i.e. positive) on the real numbers if the parameter μ_j of the G function (7.22) satisfies the following restriction (McFadden 1981):

$$\mu_j \leq 1 \ \forall j$$

Note that this is equivalent to Williams' structural condition (7.20). Furthermore, functional form (7.19) of Williams is equivalent to McFadden's functional form (7.23) if the following relations are established:

$$\beta = 1$$
$$\frac{1}{\lambda_j} = \phi_j = \mu_j \ \forall j$$

But although the conditions are numerically equivalent, they have different meanings. In Williams' theory, the restriction stems from the definition of the error as the sum of two *separable* terms, one of which, EV1, distributed with lower variance than that of the total error, allowing the NL function to satisfy the basic integrability conditions required to be consistent with utility maximisation.

On the other hand, McFadden's condition is directly related to the restriction that the GEV-based probability density function has to be compatible with random utility theory; thus, in his context, the definition of the error as the sum of two independent components and the condition this imposes on their variances *are not necessary*. This aspect was mentioned – in an indirect way – by Daganzo and Kusnic (1993), who stated that although the conditional probability may be derived with a Logit form, it is not necessary that the conditional error distribution should be EV1.

7.4.3 The NL in Practice

As a modelling tool, the NL may be usefully presented in the following fashion (Ortúzar 1980; Sobel 1980):

1) Its structure is characterised by grouping all subsets of correlated (or more similar) alternatives in hierarchies or nests. Each nest, in turn, is represented by a *composite alternative* that competes with the others available to the individual (the example in Figure 7.3 considers two levels of nesting and four nests).

2) The introduction of information from lower nests into the next higher nests is done by means of the utilities of the composite alternatives; these are, by definition, equal to the expected maximum utility (EMU) of the options belonging to the nest and have the following expression:

$$EMU_j = \log \sum_k \exp\left(V_k/\phi_j\right)$$

Therefore, the composite utility of nest j is:

$$V_j = \phi_j \cdot EMU_j$$

where ϕ_j are *structural* parameters to be estimated.

3) The probability that individual q selects option A_i in nest j may be computed as the product of the marginal probability of choosing the composite alternative N_J (in the higher nest) and the conditional probability of choosing option A_i in the lower nest, given that q selected the composite alternative.

4) If there is only one nest, the internal diagnosis condition (7.17) is expressed in this new notation as:

$$0 < \phi \leq 1 \tag{7.24}$$

and let us briefly see why it needs to hold. If $\phi < 0$, an increase in the utility of an alternative in the nest, which should increase the value of EMU, would actually diminish the probability of selecting the nest; if $\phi = 0$, such an increase would not affect the nest's probability of being selected, as EMU would not affect the choice between car and public transport.

On the other hand, if $\phi > 1$, an increase in the utility of an alternative in the nest would tend to increase not only its selection probability but also those of the remaining alternatives in the nest (but note that the real reason is that β cannot be greater than λ as shown in expression 7.16). Finally, if $\phi = 1$, which is the equivalent to $\beta = \lambda$, the NL model becomes mathematically equivalent to the MNL; in such cases (i.e. when $\phi \approx 1$) it is more efficient to recalibrate the model as an MNL, as the latter has fewer parameters.

But NL models are not limited to just two hierarchical levels; in cases with more nesting levels, such as in Figure 7.4, we need at each branch of the structure conditions such as:

$$0 < \phi_1 \leq \phi_2 \leq \ldots \leq \phi_s \leq 1 \tag{7.25}$$

where ϕ_1 corresponds to the most inclusive parameter and ϕ_s to that of the highest-level Note that there are no relationship to be expected between the structural parameters pertaining to different branches.

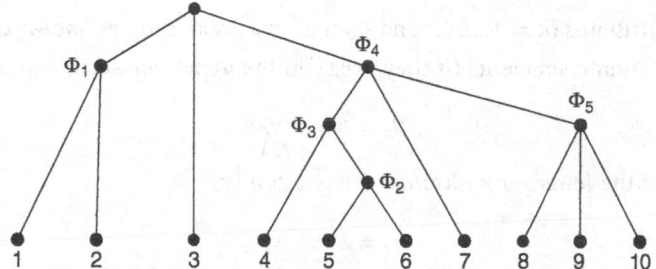

Figure 7.4 Nested Logit model with several nests

7.4.3.1 Limitations of the NL

1) In common with the MNL, the NL is not a random coefficients model, so it cannot cope with taste variations among individuals without explicit market segmentation. Neither can it treat heteroskedastic options, as the error variances of each alternative are assumed to be the same.
2) It can only handle as many interdependencies among options as nests have been specified in the structure; furthermore, alternatives in one nest cannot be correlated with alternatives in another nest (this cross-correlation effect, which might be important to test in a mixed-mode modal choice context, for example, can be handled by more general forms, as we will see below).
3) The search for the best NL structure may imply the tentative examination of many nesting patterns, as the number of possible structures increases geometrically with the number of options (Sobel 1980). Although *a priori* notions help greatly in this sense (i.e. only theoretically expected nesting patterns should be tried), the modelling exercise might take much longer than with the simple MNL.

7.4.4 Controversies About Some Properties of the NL Model

This section discusses some properties of the NL model that were the subject of some controversy in the literature in order to illuminate aspects that allow a correct use of the model in practice.

7.4.4.1 Specifications Which Address the Non-Identifiability Problem

As we have mentioned, all discrete choice models are subject to identifiability problems. The NL model has an additional degree of freedom over the MNL; to estimate it, we need (7.19) to 'fix' one of the scale factors.

Consider, without loss of generality, the two-level model where β is the parameter at the upper level and λ_j ($j = 1, ..., J$) are the corresponding J parameters of the nests. In this case, J structural coefficients may be defined as in (7.20), and the theoretical identifiability problem means that although the ϕ_j parameters can all be estimated, *one* of the $J + 1$ scale factors (i.e. the J parameters λ_j plus β) associated with the variance cannot be determined (Carrasco and Ortúzar 2002).

7.4.4.1.1 Upper and Lower Normalisations

According to the above definitions, two normalisations can be distinguished: the *upper* one, where β is chosen as the non-identifiable parameter, and the *lower* one, where *one* of the λ_j parameters (for example that for $j = r$, $1 \leq r \leq J$) is selected. Consider a typical (linear) representative utility function:

$$\hat{V}_{i/j} = \sum_{k=1}^{K} \hat{\theta}_k x_{(i,j)}^k \tag{7.26}$$

where $x_{(i,j)}^k$ are attributes ($k = 1,..., K$) and $\hat{\theta}_k$ their corresponding *estimated* coefficients. Estimated and population coefficients (if they exist) on the *upper normalisation* are related by:

$$\hat{\theta}_k = \beta \theta_k \quad \forall k \tag{7.27}$$

and in the case of the *lower normalisation* the relation is:

$$\hat{\theta}_k = \lambda_r \theta_k \quad \forall k \tag{7.28}$$

Equations (7.27) and (7.28) allow us to see more clearly that normalisation does not strictly mean to 'define' the parameter as unity (indirectly assuming the value of the variance), but that the 'normalised' parameter multiplies the coefficients of the utility function, 'mixing' with them rather than having a value defined *a priori*.

Now, the specification of the model using the upper normalisation is given by:

$$P(i,j) = \frac{\exp\left(\dfrac{1}{\phi_j}\hat{V}_{i/j}\right)}{\displaystyle\sum_{i' \in j}\exp\left(\dfrac{1}{\phi_j}\hat{V}_{i'/j}\right)} \cdot \frac{\exp\phi_j\left(\log\left(\displaystyle\sum_{i \in j}\exp\left(\dfrac{1}{\phi_j}\hat{V}_{i/j}\right)\right)\right)}{\displaystyle\sum_{j'=1}^{J}\exp\phi_{j'}\left(\log\left(\displaystyle\sum_{i \in j'}\exp\left(\dfrac{1}{\phi_{j'}}\hat{V}_{i/j'}\right)\right)\right)} \tag{7.29}$$

and using the lower normalisation, it would be:

$$P(i,j) = \frac{\exp\left(\dfrac{\phi_r}{\phi_j}\hat{V}_{i/j}\right)}{\displaystyle\sum_{i' \in j}\exp\left(\dfrac{\phi_r}{\phi_j}\hat{V}_{i'/j}\right)} \cdot \frac{\exp\phi_j\left(\log\left(\displaystyle\sum_{i \in j}\exp\left(\dfrac{\phi_r}{\phi_j}\hat{V}_{i/j}\right)\right)\right)}{\displaystyle\sum_{j'=1}^{m}\exp\phi_{j'}\left(\log\left(\displaystyle\sum_{i \in j'}\exp\left(\dfrac{\phi_r}{\phi_{j'}}\hat{V}_{i/j'}\right)\right)\right)} \tag{7.30}$$

Equations (7.29) and (7.30) show that, from a practical point of view, the resulting specifications are equivalent to defining the corresponding normalising parameter as one in the NL general functional form (7.19). However, there are two problems here, as we discuss in more detail below. First, the option of normalising at the lower level raises the problem of which lower-level nest to use. Second, confusion is added when the scales (the parameters λ_j, associated with the EV1 distribution) are unequal, as can be seen by comparing Eqs. (7.29) and (7.30). Since an important aspect of modelling is communicating the results to decision makers, this is no non-trivial issue.

7.4.4.1.2 Theoretical Considerations
Eqs. (7.31) and (7.32) below describe the relation between both normalisations (Carrasco and Ortúzar 2002):

$$\hat{\phi}_j^{up} = \hat{\phi}_j^{low} \forall j \tag{7.31}$$

$$\hat{\theta}^{up} = \hat{\phi}_r \hat{\theta}^{low} \tag{7.32}$$

where *up* and *low* denote the corresponding upper and lower normalisations. The equations show that *both* specifications are equivalent and therefore compatible with utility maximising principles. Nevertheless, it is interesting to note that depending on the chosen normalisation there will be differences between the estimated values of the coefficients. However, this dissimilarity is not relevant in cases where the main interest is the ratio of coefficients, such as the *value of time* (Gaudry et al. 1989), as the scale factors cancel out and therefore the same result, independent of the normalisation, is obtained. Also, the model elasticities are indistinguishable if the normalisations are executed properly (Daly 2001).

However, the dissimilarity may be important if we wish to compare a given NL coefficient, such as the marginal utility of income (i.e. the coefficient of the cost variable with a minus sign in the typical wage rate specification, see Section 8.3.2), with its MNL counterpart. In this case it is *only possible to compare directly the MNL estimated coefficients $\hat{\theta}$ with the upper normalisation coefficients*; this is because the former are the product of the population coefficients θ and the scale parameter associated with the errors, as we already saw, and Eqs. (7.27) and (7.28) show that only $\hat{\theta}^{up}$ involves the parameter β associated with the *total* variance of the EV1 distributed errors in the NL case. Those of the lower normalisation are the product of θ and the parameter λ_r, which is only related to the variance of the normalised nest.

A final aspect to consider is the possibility of comparing the NL functional forms of Williams-Daly and Zachary and McFadden in this context; the equations above clearly show that only the upper normalisation allows a direct comparison between the coefficients of both specifications. In conclusion, although both normalisations are consistent with the theory, there are interesting reasons to prefer the upper normalisation:

- the possibility of establishing a direct comparison between NL and MNL coefficients;
- the simplicity of having the *only* parameter related to total variance as reference;
- the simpler functional form of the probability in this case.

7.4.4.2 On the Limits of the Structural Parameters

This section considers the controversy arising from a paper by Börsch-Supan (1990), who suggested that under special circumstances the structural parameter ϕ could be larger than one, violating the structural condition (7.24). The compatibility of the NL with the basic theoretical conditions is an important issue that has been extensively studied in the literature. Williams and Ortúzar (1982b) presented the necessity of these conditions as a rigorous and unambiguous argument to reject a model, where goodness-of-fit can be a necessary condition but not enough to validate a model. In fact, an important property of any discrete choice model is precisely the successful marriage between an explicit theory of behaviour and a micro- representation, allowing the constructive use of statistical goodness-of-fit measures for model specification and testing. Thus, an inconsistent model would be a theoretical setback of at least 40 years.

Then, it is important to study if the results of Börsch-Supan (BS) are consistent with theory, especially focusing on its general theoretical consequences and the interpretation in empirical cases. These aspects were not explored by the various authors who cited the BS conditions (for example, Herriges and Kling 1995, 1996; Koppelman and Wen 1998), but were dealt with conclusively by Carrasco and Ortúzar (2002).

7.4.4.2.1 BS Proposed Extension and Further Corrections The consistency conditions for utility maximisation analysed by BS, derive from the work of McFadden (1981). One of these conditions is:

$$\frac{(-1)^{I-1}\partial^{I-1}P_i(\boldsymbol{V})}{\partial V_1...\partial V_{i-1}\partial V_{i+1}...\partial V_I} \geq 0 \quad \forall \ \boldsymbol{V} \in R^I \tag{7.33}$$

where R is the set of real numbers, $\boldsymbol{V} = (V_1,...,V_I)$ is the vector of representative utilities of I alternatives, and P_i is the probability of choosing alternative A_i. In passing, note that BS did not consider that the sign alternates; this was corrected by Herriges and Kling (1995).

Equation (7.33) ensures that the probability density function cannot be negative and is equivalent to $0 < \phi \leq 1$ if the condition is valid for all the representative utilities $\boldsymbol{V} \in R^I$. BS argued that the need for condition (7.33) to hold for all R^I is overly restrictive because economic theory (and practical experience) would suggest that only a subset of data points is used for modelling ('relevant subset'). This subset should include the data points used to estimate the model and examine potential policy changes. As a consequence of this approach, it would become feasible that a NL with structural parameters larger than one could be consistent with utility maximising theory. However, it is nearly impossible to find a data set that allows a NL with structural parameters larger than one to be consistent with utility maximisation, as this only happens if the relevant subset is not empty.

Herriges and Kling (1995) not only corrected the omission of signs by BS in (7.33) but presented the necessary conditions for consistency with utility maximisation in two-level NL models, as follows:

$$P_j \geq \tau_j \tag{7.34}$$

$$2(\tau_j - P_j)^2 + \tau_j P_j \geq \tau_j \tag{7.35}$$

$$6(P_j - \tau_j)^3 + \tau_j[2(P_j - 1) - \tau_j](1 - P_j) \geq 0 \tag{7.36}$$

$$\text{with } \tau_j \equiv \frac{(\phi_j - 1)}{\phi_j}.$$

These conditions result from the differentiation of the NL functional form (7.23) using restriction (7.33) for the first, second, and third partial derivatives. Inequality (7.34) must be satisfied for nests with two or more options, (7.35) for nests with three or more alternatives, and (7.36) for nests with four options or more. Conditions (7.34)–(7.36) would replace (7.33) when testing consistency with utility maximisation under the framework of McFadden (1981).

7.4.4.2.2 Interpretation and Applicability of the BS Extension First note that this framework does not impede in any way the use of $\phi = 1$ as a method to test if a NL collapses to the MNL (which is curiously ignored sometimes). Further, having a structural parameter greater than one implies a greater degree of substitution between nests than within them. However, it may be possible that another tree (which correctly considers a greater degree of substitution within nests) could be postulated. It also implies negative values for the covariance and correlation between nested alternatives. However, both these interpretations seem to be more statistical than behavioural, as Train et al. (1987) argued.

Without doubt, the most important consequence of allowing the structural parameters to be larger than one is that it denies their use as a test for establishing a hierarchical relationship between the different nesting levels. This has consequences not only in behavioural interpretation, but also in terms of the search for the best tree structure (i.e. the information provided by the structural parameter values is quite useful to define upper and lower levels when this is not obvious). Thus, it is important to remark that the BS framework *is not possible* to use if the NL model postulates separability on choice levels (for example, destination-mode choice), where the variance condition (7.20) of Williams is a *fundamental* property to understand behaviour.

Herriges and Kling (1996) have made the only empirical investigation of the BS extension reported in the literature. To test model consistency with utility maximising, they explored three different procedures, each progressively more restrictive. The first two failed, and the third imposed restrictions (7.34) and (7.35) *ex ante*, estimating the coefficient vector from a Bayesian perspective. In practical terms, this is equivalent to estimating the NL without prior information, yielding a coefficient vector distributed Normal with mean and covariance matrix taken from the estimation. The procedure generated a large number of Normal distributed values, but only the draws satisfying the consistency conditions were retained. So, although by construction all parameters were consistent with theory, some were calculated with a very low percentage of the generated values.

Important objections about the real applicability of the above procedures (and in general about the applicability of the BS extension) were formulated by Carrasco and Ortúzar (2002). Furthermore, the same kind of choice context was later treated successfully using the more flexible Mixed Logit model (Train 1998).

7.4.4.3 Two Further Issues

In this section, we will consider two final, relatively minor controversies about the NL that have also been discussed in recent literature.

7.4.4.3.1 *Alternative Definition of Model Parameters* Hensher and Greene (2002) and Hunt (2000) proposed an alternative definition of the NL coefficients, which may lead to some confusion about certain properties of the model. They defined a scale parameter for each nest (γ_j) and another at the alternative level (μ_j), associated with each nest, as in Figure 7.5.

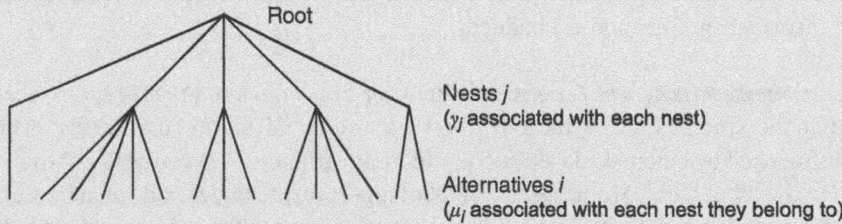

Figure 7.5 NL alternative parameter definition

The most important difference between this specification and the Nested Logit of Williams (1977) is the incorporation of parameters at the level of the elemental alternatives instead of the unique parameter associated with the root. Thus, this alternative specification has $2J$ scale parameters instead of the traditional $J + 1$ (with J being the number of nests).

In this alternative vision the choice probability is given by:

$$
P_{ij} = \frac{\exp\left(\mu_j V_{i/j}\right)}{\sum\limits_{i' \in j} \exp\left(\mu_j V_{i'/j}\right)} \cdot \frac{\exp \gamma_j \left\{ \dfrac{1}{\mu_j} \log \left(\sum\limits_{i \in j} \exp\left(\mu_j V_{i/j}\right) \right) \right\}}{\sum\limits_{j'=1}^{m} \exp \gamma_{j'} \left\{ \dfrac{1}{\mu_{j'}} \log \left(\sum\limits_{i \in j'} \exp\left(\mu_{j'} V_{i/j'}\right) \right) \right\}}
\tag{7.37}
$$

If we redefine the variables μ as λ and γ as β above – without considering their theoretical meaning – it is possible to get a very similar specification to that of Williams (7.19), except for the fact that here we get different β_j parameters for each nest

This specification would indirectly allow modelling heteroskedasticity between alternatives, generalising the NL of Williams-Daly and Zachary and McFadden (see Example 4 in Section 7.6). However, the specification is inconvenient in practice because it leads to some confusion when solving the identifiability problem. In this case, it would be necessary to normalise J parameters; that is, either define *all* parameters β_j (upper normalisation) or *all* parameters λ_j (lower normalisation) as non-estimable. This is different from the normalisations discussed before, where to solve the identifiability problem it was necessary to define *only one* parameter as non-identifiable. Another interesting point is that the alternative model's upper normalisation is the same as that of Williams (i.e. it is correct); but the lower normalisation leads to an incorrect specification.

In addition, a different definition of the scale parameters also implies some issues with 'partial degenerated' structures (i.e. trees with some nests containing only one alternative). Under the Williams-Daly and Zachary and McFadden theoretical frameworks, if an option A_k is 'degenerated' its corresponding structural parameter ϕ_k is equal to one because λ_k is equal to β (i.e. the nest that contains the alternative 'collapses' to the upper level). This result is *independent from the normalisation*. However, this basic theoretical interpretation is not possible if there is a parameter β_j specific to each nest. If the upper normalisation is used, Hunt (2000) argues that λ_k becomes 'non identifiable', but what actually occurs is that it has collapsed to β, which is non-identifiable by definition (Carrasco and Ortúzar 2002). This confusion becomes even worse if the lower normalisation is used because the parameter ϕ_k results non-estimable, being necessary to *define* its value as unity to get consistency with the theory (Hunt 2000); however, in this case, again, an incorrect specification is obtained.

7.4.4.3.2 *Heteroskedasticity and Correlation* Hensher and Louviere (1998) report results that suggest that the specification of the NL tree structure could be, in some cases, even more strongly influenced by heteroskedasticity (i.e. different variances between options) than by correlation. On the other hand, Munizaga et al. (2000) report surprisingly good behaviour of the NL in the presence of heteroskedasticity between alternatives (but not in the presence of heteroskedasticity between observations), showing a low predictive capacity only for radical policy changes in their Monte Carlo simulation study.

Furthermore, Hensher and Louviere (1998) and Hensher (1999), proposed a new method of specifying a NL tree structure (i.e. a way to define which alternative should belong to each nest) based on the scale differences between the options. They used the Heteroskedastic Extreme

Value model (Bhat 1995) as a 'search engine' in order to define nestings of alternatives with similar variance. It is interesting to note that the tree specification process in the NL does not have a rigorous procedure (with standard steps), and traditionally it has been based on the idea of grouping alternatives that theoretically (or intuitively) appear to be correlated (see Ortúzar 1982).

Hensher (1999) argues that a statistical rationale for nesting could be related to differential patterns of variance between subsets of alternatives. However, this argument is theoretically suspect because the differential patterns between subsets of options in the NL are based on the different value of *correlation* rather than variances. In fact, in a two-level model as (7.19) the scale parameter defining nesting is λ_j (which is *only* related to correlation) and not β (which is associated with both correlation and variance). Therefore, Hensher's tree specification method should be rejected as it is based on a property that the NL does not possess (i.e. heteroskedasticity).

Finally, if there are grounds to believe *a priori* that heteroskedasticity could be an important issue in modelling in a given context, there are more general models that can handle this effect in theory (and with even better results than the NL for simulated data, see Munizaga et al. 2000), such as Mixed Logit (Train 2009) or Multinomial Probit (Daganzo 1979), which are nowadays less problematic to estimate and a little less problematic to interpret than in the past.

7.5 The Multinomial Probit Model

As we mentioned in Section 7.4.1, in the MNP model, the stochastic residuals ε of (7.2) are distributed multivariate Normal with mean zero and an arbitrary covariance matrix, that is, in this case the variances may be different and the error terms may be correlated in any fashion. The problem is, of course, that this generality does not allow us to write the model in a simple closed form as in the MNL (except for the binary case); therefore, to solve it numerically, we need approximations or, more effectively, simulation.

7.5.1 The Binary Probit Model

In this case we can write the utility expressions (7.2) as:

$$U_1(\boldsymbol{\theta}, \mathbf{Z}) = V_1(\boldsymbol{\theta}, \mathbf{Z}) + \varepsilon_1(\boldsymbol{\theta}, \mathbf{Z})$$
$$U_2(\boldsymbol{\theta}, \mathbf{Z}) = V_2(\boldsymbol{\theta}, \mathbf{Z}) + \varepsilon_2(\boldsymbol{\theta}, \mathbf{Z})$$

where $\varepsilon\ (\boldsymbol{\theta}, \mathbf{Z})$ is distributed bivariate $N(0, \Sigma)$ with

$$\sum = \begin{pmatrix} \sigma_1^2 & \rho\sigma_1\sigma_2 \\ \rho\sigma_1\sigma_2 & \sigma_2^2 \end{pmatrix}$$

where ρ is the correlation coefficient between U_1 and U_2. From (7.6), the probability of choosing option 1 is given by:

$$P_1(\boldsymbol{\theta}, \mathbf{Z}) = \text{Prob}\{\varepsilon_2 - \varepsilon_1 \leq V_1 - V_2\}$$

but as the Normal distribution is closed to addition and subtraction (as the EV1 is closed to maximisation) we have that $\varepsilon_2 - \varepsilon_1$ distributes univariate $N(0, \sigma_\varepsilon)$, where:

$$\sigma_\varepsilon^2 = \sigma_1^2 + \sigma_2^2 - 2\rho\sigma_1\sigma_2$$

Dividing $(\varepsilon_2 - \varepsilon_1)$ by σ_ε we obtain a standard $N(0, 1)$ variable; therefore we can write the binary Probit choice probability concisely as:

$$P_1(\boldsymbol{\theta}, \mathbf{Z}) = \Phi[(V_1 - V_2)/\sigma_\varepsilon] \tag{7.38}$$

where $\Phi[x]$ is the cumulative standard Normal distribution which has tabulated values. Although this is indeed a simple model, it is completely general for binary choice. Note, however, that Eq. (7.38) is not directly estimable as the parameters $\boldsymbol{\theta}$ in the representative utilities $\mathbf{V}$ cannot be estimated separately from the standard deviation σ_ε. In fact, just as occurred in the MNL and NL models, there is an identifiability problem, and one would need to normalise before obtaining an estimate of the model parameters. Bunch (1991) looks at this problem for the general MNP model, and Walker (2002) provides a good discussion about the issue of identifiability in general.

7.5.2 Multinomial Probit and Taste Variations

As we noted in Sections 7.3 and 7.4, a potentially important problem of fixed-coefficient models, such as the MNL and NL, is their inability to treat the problem of random taste variations among individuals without explicit market segmentation. In what follows, we will first show with an example what is meant by this, and then we will proceed to show how the MNP handles the problem.

Example 7.2 Consider a mode choice model with two explanatory variables, cost (c) and time (t) and the following postulated utility function:

$$U = \alpha t + \beta c + \varepsilon$$

Let us suppose, however, that the perception of costs in the population varies with income (I), that is, poorer individuals are more sensitive to cost changes, such that the true utility function is:

$$U = \alpha t + \phi c/I + \varepsilon$$

It can easily be seen, comparing both expressions, that the model will be correct only if β can be considered a random variable with the same distribution as ϕ/I in the population; in this case, the model contains random taste variations.

The problem of random taste variations is normally very serious, as has been clearly illustrated by Horowitz (1981), and may be considered a special case of one well-known specification error, the omission of a relevant explanatory variable, which we discussed in Chapter 2.

Let us consider again the utility function (7.3), which is linear in the parameters as discussed in Section 7.2. Its most general case considers the parameter set $\boldsymbol{\theta}$ to be a random vector distributed across the population; in this case the residuals may be modelled as alternative specific parameters, hence the variables $\boldsymbol{\varepsilon}$ in (7.2) may be omitted without loss of generality, and the equation can be written more concisely as:

$$U_j = \sum_k \theta_k x_{jk} \tag{7.39}$$

which is a very general linear specification as it allows for taste variations across the population. If the vector θ is distributed multivariate Normal, the choice model resulting from (7.39) is of MNP form (see Daganzo 1979). Various procedures for estimating this model were discussed by Sheffi et al. (1982), Langdon (1984) and others, and we will look at some of these in Chapter 8.

7.5.3 Comparing Independent Probit and Logit Models

When estimating a MNP model (and it is easy to see it in the binary case) the parameters obtained are:

$$\beta_i^P = \frac{\theta_i}{\sigma_\varepsilon} \quad \text{with} \quad \sigma_\varepsilon^2 = \sigma_1^2 + \sigma_2^2 - 2\rho\sigma_1\sigma_2$$

On the other hand, we know from Eq. (7.10) that when estimating a MNL the parameters obtained are:

$$\beta_i^L = \lambda\theta_i \quad \text{with} \quad \lambda = \frac{\pi}{\sigma\sqrt{6}}$$

Therefore, we have that:

$$\beta_i^L = \frac{\theta_i\pi}{\sigma\sqrt{6}}$$

Now, to compare both models we need to estimate a MNP model with a covariance matrix similar to that of the MNL (i.e. an Independent and Identical Probit). In this case $\sigma_\varepsilon^2 = \sigma^2 + \sigma^2$, which implies that $\sigma_\varepsilon = \sigma\sqrt{2}$ and thus $\beta_i^P = \theta_i/\sigma\sqrt{2}$. Therefore, in order to compare both sets of parameters, we should multiply the β_i^P by a factor that makes them equal to $\beta_i^L = \theta_i/\sigma\sqrt{6}$. This is achieved using the factor:

$$K = \frac{\sigma\pi\sqrt{2}}{\sigma\sqrt{6}} = \frac{\pi}{\sqrt{3}} \tag{7.40}$$

Therefore, if one wants to compare the estimated coefficients of a MNL and an IID Probit model, those belonging to the second structure must be scaled by the factor $\pi/\sqrt{3}$. We have successfully used this method to test the correctness of an experimental code to estimate MNP models (Munizaga et al. 2000).

7.6 The Mixed Logit Model

Although its current form originated from the parallel work of two research groups in the 1990s (Ben-Akiva and Bolduc 1996; McFadden and Train 2000), the original formulation of the model, as Hedonic or Random Parameters Logit, was made much earlier (Cardell and Reddy 1977; Cardell and Dunbar 1980).

7.6.1 Model Formulation

The Mixed Logit (ML) model can be derived under several behavioural specifications, each providing a particular interpretation. Train (2009) correctly states that the model is *defined* on the basis of the functional form for its choice probabilities. As such, the ML label is applicable to any model the probabilities of which can be expressed as an integral of standard Logit probabilities over a distribution of the parameters, such as:

$$P_{iq} = \int L_{iq}(\theta) f(\theta) \, d\theta \tag{7.41}$$

where $L_{iq}(\theta)$ is typically an MNL probability evaluated at a set of parameters θ and their density function, $f(\theta)$, is known as 'mixing distribution'. If $f(\theta)$ is degenerate at fixed parameters $\mathbf{b}$ (i.e. it equals one for $\theta = \mathbf{b}$ and zero for $\theta \neq \mathbf{b}$), the choice probability (7.41) becomes the simple MNL.

If, on the other hand, the mixing distribution is discrete (i.e. if θ takes M values labelled $b_1,..., b_M$ with probabilities s_m that $\theta_m = b_m$), the ML becomes the *Latent Class* model with applications in many fields, including psychology and marketing; for one of the earliest examples in transportation, see Bhat (1997). This is useful when there are distinct segments in the population, each with their own choice behaviour (Train 2009). In Section 8.6.3.2, we will consider a class of ML models where the mixing distribution lies somewhere between the typical continuous form, below, and the latent class model.

Now, in most ML applications $f(\theta)$ has been taken as continuous with mean $\mathbf{b}$ and covariance matrix Σ, and modellers have attempted just to estimate these 'population parameters' without taking advantage of one of the most powerful features of the model, that is, estimating the θ that enter in the Logit component (kernel) for each individual; this can be done directly or conditional on the population parameters, $\mathbf{b}$ and Σ, as we will discuss in Chapter 8.

7.6.2 Model Specifications

7.6.2.1 Basic Formulations

The ML model has two basic forms. The first is the *error components* (EC) version, whose utility function is characterised by an error term with two elements. One (ε_{qjt}) allows the MNL probability to be obtained (and as such has the usual IID EV1 distribution), while the other has a distribution that can be freely chosen by the modeller, depending on the phenomenon he needs to reproduce. In this case the utility of option j ($j = 1,..., J$) for individual q in choice situation t ($t = 1,..., T$) is given by:

$$U_{jqt} = \theta_{jt}\mathbf{X}_{jqt} + \mathbf{\Omega}_{jqt}\mathbf{Y}_{jqt} + \varepsilon_{jqt} \tag{7.42}$$

where θ are fixed parameters and $\mathbf{X}$ are observable attributes, $\mathbf{\Omega}_{jqt}$ is a vector of random elements with a distribution specified by the modeller, with zero mean and unknown covariance matrix, and $\mathbf{Y}_{jqt}$ is a vector of attributes unknown (in value and nature) to the modeller. Thus, without loss of generality, they can be taken as equal to one for all alternatives or for groups of them. Given this, the covariance matrix of the model utilities is:

$$Cov(\mathbf{U}_{jqt}) = Cov(\mathbf{\Omega}_{jqt}) + (\pi^2/6\lambda^2) \cdot \mathbf{I}_J$$

where $\mathbf{I}_J$ is a $J{\times}J$ identity matrix. An adequate choice of $\mathbf{Y}_{jqt}$ allows different error structures such as correlation, cross-correlation, heteroskedasticity, dynamics, and even auto-regressive errors to be treated (Hensher and Greene 2003; Train 2009). Indeed, it has been proven that the ML can approximate any discrete choice model derived from a random utility maximisation model as closely as one pleases (Dalal and Klein 1988; McFadden and Train 2000); this, in fact, led to the final demise of the MNP model as a serious contender in this area. On the other hand, to obtain the simple MNL model $\mathbf{Y}_{jqt}$ has to be zero, such that there is no correlation among alternatives.

Example 7.3 To generate a heteroskedastic version of the MNL, one simply needs to specify the following utility function:

$$U_{iq} = \theta X_{iq} + \sigma_i \, \Omega_{iq} + \varepsilon_{iq} \quad \text{with} \quad \Omega_{iq} \sim \text{IID} \, N(0, 1)$$

and as the errors Ω and ε are independent, it is easy to see that the covariance matrix of the utilities **U** has the following form (for simplicity we are considering a trinomial case):

$$\Sigma = \begin{bmatrix} \sigma_1^2 + \pi^2/6\lambda^2 & 0 & 0 \\ 0 & \sigma_2^2 + \pi^2/6\lambda^2 & 0 \\ 0 & 0 & \sigma_3^2 + \pi^2/6\lambda^2 \end{bmatrix}$$

where λ is the scale factor associated with the EV1 errors.

To generate a heteroskedastic version of the NL model, one would need a similarly simple specification; assume a five-alternative case where the first two are correlated and the last two are also correlated (the third is independent of all others):

$$U_{1q} = \mathbf{X}_{1q}\theta + \sigma_1 \eta_{1q} + \varepsilon_{1q} \qquad U_{2q} = \mathbf{X}_{2q}\theta + \sigma_1 \eta_{1q} + \varepsilon_{2q}$$
$$U_{3q} = \mathbf{X}_{3q}\theta + \sigma_2 \eta_{2q} + \varepsilon_{3q}$$
$$U_{4q} = \mathbf{X}_{4q}\theta + \sigma_3 \eta_{3q} + \varepsilon_{4q} \qquad U_{5q} = \mathbf{X}_{5q}\theta + \sigma_3 \eta_{3q} + \varepsilon_{5q}$$

In this case, it is again easy to see that the covariance matrix of the model utilities is given by:

$$\Sigma = \begin{bmatrix} \sigma_1^2 + \pi^2/6\lambda^2 & \sigma_1^2 & 0 & 0 & 0 \\ \sigma_1^2 & \sigma_1^2 + \pi^2/6\lambda^2 & 0 & 0 & 0 \\ 0 & 0 & \sigma_2^2 + \pi^2/6\lambda^2 & 0 & 0 \\ 0 & 0 & 0 & \sigma_3^2 + \pi^2/6\lambda^2 & \sigma_3^2 \\ 0 & 0 & 0 & \sigma_3^2 & \sigma_3^2 + \pi^2/6\lambda^2 \end{bmatrix}$$

and correlation is due to the presence of the common unobservable elements in the utilities of the correlated alternatives; note that replicating the true NL, which is homoskedastic, is more involved (see Munizaga and Alvarez-Daziano 2005).

The second, more classical, version of the ML model considers a *random coefficients* (RC) structure, in which the marginal utility parameters are different for each sampled individual q, but do not vary across choice situations; this last assumption may be relaxed if choice situations are significantly separated along time, as taste parameters could then be presumed to alter (Hess and Rose 2009). So, in this case we have:

$$U_{jqt} = \theta_q \mathbf{X}_{jqt} + \varepsilon_{jqt} \tag{7.43}$$

and the parameters vary over individuals with density $f(\theta)$. This specification yields the choice probabilities (7.41) naturally. Note that the presence of the vector **X** in the covariance matrix does not allow the modeller to control for it but helps to ease an important problem of the model, its identification, which we discuss in Section 7.6.3.

The EC and RC specifications are formally equivalent as the coefficients θ_q can be decomposed into their means (**b**) and deviations, denoted s_q, such that:

$$U_{jqt} = \mathbf{b}\mathbf{X}_{jqt} + s_q\mathbf{X}_{jqt} + \varepsilon_{jqt} \tag{7.44}$$

which has error components defined by $\mathbf{Y}_{jqt} = \mathbf{X}_{jqt}$; conversely, we can also start from the EC specification and get the RC specification. However, though formally equivalent, the manner in which the modeller looks at the phenomenon under study affects the model specification. For example, if the main interest is to represent appropriate substitution patterns through an EC specification, the emphasis will be placed on specifying variables that can induce correlation in a parsimonious fashion, not necessarily considering taste variations or too many explanatory variables. In fact, Train (2009) wisely states that:

there is a natural limit on how much one can learn about things that are not seen

but this is sometimes overlooked by even the most skilful econometricians, who focus on the error terms at the expense of correct specification of the observed utility component and ensuring the data are of appropriate quality.

7.6.2.2 More Advanced Formulations

An important issue concerning the apparent dual representation of the model (i.e. EC or RC), as noted by many analysts, is that the two versions of the model may give rise to *confounding* effects. As most discrete choice models, the ML is based on the linear-in-parameters-with-additive-disturbances (LPAD) structure, where individuals are assumed to compensate (trade-off) the effects of good and bad attributes even when there are many situations where compensatory rules do not hold (see Cantillo and Ortúzar 2005). For example, an omitted structure (i.e. any interaction between two variables) will be captured by the error terms, but it may be confused with random heterogeneity for a given attribute in the RC version.

On the other hand, as the model works on the basis of differences between alternatives, it does not matter whether an attribute is included in one alternative or in all others except that one, as long as the relative difference between them does not change. This property, in conjunction with the compensatory rule, may lead to another confounding effect: between correlation and heterogeneity in tastes and responses, which can appear in estimated models and produce misleading forecasts, as discussed by Cherchi and Ortúzar (2008a). To understand the correct underlying structure and to test whether heterogeneity is really present, they recommend estimating alternative specifications and comparing results, looking carefully at the absolute value of the random parameters and the relative values of the alternative specific constants (ASC) and correlation coefficients. They also found that a significant specific random parameter may not actually reveal variation in tastes, but correlation among competing alternatives, cautioning that this is especially important if the model is intended as a forecasting tool. These findings complement the observation by Hess et al. (2005a) that the assumptions made with regard to error structure can have significant impacts on willingness-to-pay (WTP) indicators.

The RC and EC specifications can also be combined easily, allowing for the joint modelling of random taste heterogeneity and inter-alternative correlation. This, however, as mentioned above, comes at the cost of important issues in identification, and also a heightened cost of estimation and application when using error components for the representation of correlation. While integration over mixture distributions is necessary in the representation of continuous random taste heterogeneity, this is not strictly the case for inter alternative correlation. Indeed, just as, conditional on a given value of the taste coefficients, a typical RC specification allowing for random taste heterogeneity reduces to a MNL model, a model allowing for inter-alternative correlation in addition to random taste heterogeneity can in this case be seen to reduce to a given GEV model (assuming that an appropriate GEV model actually exists). As such, the correlation structure can be represented with the help of a GEV model, while the random taste heterogeneity is accommodated through integration over the assumed distribution of the taste coefficients. The use of the choice probability of a more complicated GEV model instead of the MNL as the integrand in (7.41), leads to a more general type of a GEV mixture model, of which the typical RC specification is simply the most basic form.

In a more general GEV mixture, we would simply replace the MNL choice probability inside the integral by, say, a NL choice probability. Such model can be estimated, and it is useful for cases where we need to allow for correlation between, say, train and bus in a car-train-bus mode choice context, while additionally allowing for random variations across respondents in the time and cost sensitivities; in such cases, a NL model could deal with the former, while a RC Mixed Logit could deal with the latter. A general GEV mixture ML can deal with both at the same time, without the need for additional error components.

Applications of this approach include Bhat and Guo (2004) and Hess et al. (2005a). In such a GEV mixture model, the number of random terms, and hence the number of dimensions of integration (and thus simulation), is limited to the number of random taste coefficients, whereas, in the EC specification, one additional random term is in principle needed for representing each separate nest.

Finally, it is interesting to mention that problems of a similar nature have been encountered when modelling jointly state dependence (i.e. the state at a given moment depends on the previous state(s) of the system) and preference heterogeneity. Smith (2005) concluded that one should be cautious in interpreting random parameters if researchers are unable to model state dependence. Nevertheless, he also stated that if a more elaborate parameterization of preference heterogeneity is used, excluding state dependence may magnify the apparent preference heterogeneity in the model but not necessarily generate it where none exists. To some extent, this could be viewed as the converse of the problem explored by Heckman (1981), where the emphasis was on the emergence of spurious state dependence if heterogeneity was not modelled properly.

7.6.3 Identification Problems

A seminal reference for the 'identification problem' is the work of Walker (2002), who noted in passing that even the most famous econometricians have been guilty of overlooking this issue in some applications. Nowadays, analysts are more cautious and test for this problem in their usual practice, but new evidence has appeared showing that it is multifaceted with no easy recipes available to avoid it.

The nature of the problem is that there are infinite sets of restrictions that can be imposed to identify a given set of parameters to be estimated. For example, in the case of the MNL model, the problem only relates to the impossibility of estimating the scale parameter β (which has to be normalised), and that one of the alternative specific constants (ASC) needs to be taken as zero (i.e. that of the *reference* alternative). Note that even in this simple case there are 'good practice' rules to follow, for example, choose as reference the alternative more universally available (Ortúzar 1982).

For more complex models, such as the ML, apart from the above considerations that apply to the vector θ, we also need to examine the identification of the unrestricted parameters of the error distribution. This could be done by studying the *Fisher information matrix* (i.e. the matrix of expected values of the second derivatives of the log-likelihood function), but this requires estimating the model, something which is not always possible. In fact, there are two types of identification problems: the *theoretical identification*, which is inherent to the model specification regardless of the data at hand, and the *empirical identification*, which depends on the information used to estimate the model. Although much has been written about the first, the second later surfaced as a serious problem deserving more attention.

7.6.3.1 Theoretical Identification

This problem is usually associated with the presence of too many parameters; that is, the model cannot be estimated simply because of its implicit structure. By looking at the covariance matrix of utility differences, Walker (2002) generalised the work of Bunch (1991) for the MNP and provided an outstanding analysis of the three conditions (order, rank, and equality) that must hold for the ML model to be identifiable.

In particular, the *order condition* is a necessary condition and establishes the maximum number of parameters that can be estimated based on the number of alternatives (J) in the choice set. In the EC version of the model, this condition states that there are at most $J(J-1)/2-1$ parameters estimable in the disturbance; this is equal to the number of unique elements in the covariance matrix of utility differences (as it is symmetric), minus one to set the scale.

The *rank condition* refers to the rank of the covariance matrix of utility differences. This is a sufficient condition and establishes the actual number of parameters that can be estimated as the number of independent equations available to do it. If this condition holds, the previous one necessarily holds, but it is trivial to apply and useful to highlight any obvious identification problems. In many cases, these two conditions can be applied by simple visual inspection of the covariance matrix of utility differences; Walker (2002) also notes that when restrictions are needed for the covariance matrix terms it is desirable that these point to the MNL being a special case of the ML (i.e. if only two variances can be estimated, the restriction on the third is that it should equal zero; furthermore, the choice of which variance should be zero is not arbitrary – she recommends choosing that which obtained the lowest value in an estimation run without considering the identifiability problem).

Finally, the *equality condition* is used to verify that the chosen normalisation, based on the identification restrictions imposed by the rank and order conditions, is valid in the sense that the resulting unique solution does in fact maximise the log-likelihood. Walker et al. (2007) note that this condition is particular to the ML model due to the special structure of its disturbance (the sum of an IID EV1 component and another with a different distribution).

It is important to note that the theoretical identification problem is only crucial for the EC version of the ML model, and does not exist when the RC version is specified for continuous attributes of the competing alternatives. In the RC model the random parameters are associated with some known (by the modeller) attributes, and thus there is always some information that allows theoretically identifying extra parameters. But whether the full covariance structure can be estimated or not, will depend on the quality of the information as discussed below.

7.6.3.2 Empirical Identification

This problem, instead, occurs when the model is estimable in principle but the data cannot support it. In theory, the parameters can be empirically identified if the number of observations and draws in the simulated maximum likelihood procedure required to estimate the model (which we will discuss in Chapter 8) are sufficiently large to provide enough information.

However, in practice, researchers face datasets with a limited number of observations and must apply a finite number of draws. Therefore, it becomes an empirically important question to check whether a given dataset can support the model at hand.

Ben-Akiva and Bolduc (1996) and Walker (2002), noted that an identification problem can arise when a low number of draws is used, and they and others, such as Hensher and Greene (2003), emphasized the necessity of verifying the stability of parameter estimates as the number of draws increased (thereby ensuring that the bias was sufficiently reduced). Later on, Munizaga and Alvarez-Daziano (2005) confirmed, using simulated data, that small sample sizes could lead to erroneous conclusions about the model's covariance structure (a warning in relation to the sample size required to recover parameters by simulation was given nearly 25 years ago by Williams and Ortúzar 1982a). Chiou and Walker (2007) demonstrated that a low number of draws in the simulation process could mask identification issues, leading to biased estimation results, even when a large number of draws (i.e. 1000) was used.

Finally, Cherchi and Ortúzar (2008b) used simulated data to analyse the extent to which the empirical identification problem depended on the variability of the data, the degree of heterogeneity of the taste parameters, the sample size, and the number of choice tasks for each individual. They found that identification problems appeared if a variable had low variability between alternatives; they also found that models may be quite unstable in the case of low variability, and – deceptively – very often cannot be estimated unless very few draws (i.e. as low as 30) are used, clearly a problem of identification and a procedure that results in a suspicious model. Contrariwise, if the difference in attributes has a high standard deviation (i.e. four times the mean), the identification problem disappears for any number of draws. Also, the identification problem does not depend on the degree of variability inherent in the random parameters but only on the richness of the associated data. Finally, they found that the capability of the ML to reproduce random heterogeneity increases when more than one choice is available for each individual (as in SP or panel data, except when there are identical repeated observations), and in that case the effect of sample size on empirical identification reduced considerably.

7.7 Other Choice Models and Paradigms

7.7.1 Other Choice Models

As we saw in Section 7.4, each alternative in a Nested Logit (NL) model belongs to only one nest. This is a restriction that may be inappropriate sometimes as, for example, mixed modes (such as park and ride) could be correlated both to cars and to rail.

To tackle this problem various types of GEV models have been formulated with what Train (2009) calls 'overlapping nests', such that a given alternative can belong to more than one nest. For example, Vovsha (1997), Bhat (1998), and Ben-Akiva and Bierlaire (1999) developed a Cross-Nested Logit (CNL) model, managing to implement an original idea of Williams (1977), the Cross-Correlated Logit model, that was solved numerically by Williams and Ortúzar (1982a) but was not used ever since.

Chu (1989) proposed the Paired Combination Logit (PCL), where each pair of alternatives constitutes a nest with its own correlation; thus, each alternative is a member of $J-1$ nests. Koppelman and Wen (2000) examined this relatively simple but flexible structure and found that it outperformed both NL and MNL in their application. All these models can be derived as members of the GEV family (McFadden 1981), as shown for the NL in Section 7.4.2.2.

As in general, all these models can be approximated by the ML, we will leave this topic here and refer readers to the excellent book by Train (2009) for more details.

7.7.2 Choice by Elimination and Satisfaction

In Chapter 8, we discuss the problem of specification and functional forms giving particular emphasis to the linear-in-the-parameters form, which has accompanied the vast majority of disaggregate demand (normally of MNL structure) applications. Owing to a growing body of criticism directed at linear-in-the-parameters forms, the early 1980s witnessed an interest in the specification and estimation of non-linear formulations of varying designs. Commentary on the functional characteristics of these forms was intertwined with statements about alternative models of the decision process considered to underpin choice models.

One typical view was that because linear-in-the-parameters forms were associated with a compensatory decision-making process (i.e. a change in one or more of the attributes may be compensated by changes in the others), models could not be appropriately specified for decision processes characterised by perception of discontinuities, which are more plausibly of a non-compensatory nature. For example, where good aspects of an alternative may not be allowed to compensate for bad aspects, which are ranked higher in importance in the selection procedure, simply because that alternative may be eliminated earlier in the search process, see the discussion in Golob and Richardson (1981).

Example 7.4 Let us consider a set of individuals, confronted by a choice implying a set of objectives **G** and a set of constraints **B**. A general multi-criterion problem can then be formally stated as:

$$\text{Max}_{(\text{options})}\left\{F_1\left(Z_1^1\right)...F_1\left(Z_N^1\right)\right\}$$

$$\vdots$$

$$\text{Max}_{(\text{options})}\left\{F_k\left(Z_1^k\right)...F_k\left(Z_N^k\right)\right\} \tag{7.45}$$

$$\vdots$$

$$\text{Max}_{(\text{options})}\left\{F_K\left(Z_1^K\right)...F_K\left(Z_N^K\right)\right\}$$

subject to the vector of constraints:

$$\mathbf{f(Z)} \leq \mathbf{B} \tag{7.46}$$

in which $F_k\left(Z_j^k\right)$ is the value of the criterion function associated with attribute Z_j^k of option A_j. For example, we might be interested in finding a mode, in a choice set of size N, which minimises travel time and cost, maximises comfort and safety, and so on. These attributes associated with any particular alternative might, in addition, be required to satisfy absolute constraints such as (7.46).

If a single alternative is simultaneously found to satisfy these optimality criteria (i.e. it optimises the K functions in expression 7.45) and the attributes of which are feasible in terms of (7.46), then an unambiguous optimal solution is obtained. In general, however, there will be conflicts between objectives (i.e. options that are superior in some respects and inferior in others).

A number of important questions can be posed before a choice model based on this multi-criterion problem may be constructed:

- what strategies might be adopted to solve the problem?
- are there differences in the strategies adopted by different individuals in a given population?
- how can these strategies be formally represented?
- how should the aggregation over the population be conducted to produce a model that can be estimated with individual data.

The last point is especially important because choice models are derived by aggregating over the actions of individuals within the population, and while any or all of them may indulge in a non-compensatory decision process, it may or may not be appropriate to characterise the sum total of these decisions and the resultant choice model in these terms (see the discussion in Williams and Ortúzar 1982a).

We will just refer here to the first of these issues, namely, how an individual confronted by a hypothetical decision context may resolve the multi-criterion problem. There is of course a wide literature dispersed over several fields, which involves the application of decision theory to problems of this kind. We will mention three methods, starting with the best-known, simplest, and most widely used approach: the trade-off strategy which forms the basis for compensatory decision models.

7.7.2.1 *Compensatory Rule*

Here, the preferred option is selected by optimising a single objective function $O = O(F_1, F_2, ..., F_K)$. If the F_k functions are simply the attributes $\mathbf{Z}^k$, or linear transformations of them, O may be written as:

$$O = O\left(\sum_k \theta_k Z_1^k, ..., \sum_k \theta_k Z_j^k, ..., \sum_k \theta_k Z_N^k \right) \qquad (7.47)$$

and the conventional linear trade-off problem is addressed. The parameters θ are determined from either stated or revealed preferences of the individual decision maker. One of the characteristics of this trade-off approach is its symmetric treatment of the objective functions.

7.7.2.2 *Non-Compensatory Rules*

An alternative general approach is to treat the objective functions (7.45) asymmetrically by either ranking them or converting some or all to constraints by introducing norms or thresholds. That is, we might require that any acceptable alternative has, for example, an associated travel cost not exceeding a particular amount; formally the restriction is imposed that:

$$Z_1^k, ..., Z_j^k, ..., Z_N^k \le Z^k \qquad (7.48)$$

in which Z^k is a maximum (or minimum when the inequality sign is reversed) satisfactory value for the attribute. The creation of norms or thresholds restricts the range of feasible alternatives that individuals are considered to impose on their decision-process.

Choice by Elimination. In this case, it is assumed that individuals possess both a ranking of attributes (e.g. cost is more important than waiting time, which in turn is more important than walking time, etc.) and minimum acceptable values or thresholds (7.48) for each. For example, the decision process may solve the multi-criterion problem in the following fashion: first the highest ranked attribute is considered and all alternatives not satisfying the threshold restriction are eliminated (even though they may excel in lesser ranked attributes); the process is repeated until only one option is left, or a group which satisfies all the threshold constraints among which one is selected in a compensatory manner (see Tversky 1972; Cantillo and Ortúzar 2005).

Satisficing Behaviour. There are, however, a great many ways in which the above search strategy may be organised; for example, it might be that a complex cyclic process is used by the individual whereby the thresholds become sequentially modified until a unique alternative is found. Equally, a *satisficing* mechanism might operate in which the individual might be prepared to curtail the search at any point according to a pre-specified rule, in which case some or all of the attributes or alternatives may not be considered. Indeed, when the notion of satisficing (Simon 1957; Eilon 1972) is applied to travel-related decisions involving location, the decision model is closely associated with the acquisition of information in the search process (González-Valdés and Ortúzar 2017).

As Young and Richardson (1980) remarked, a search may be characterised by an elimination process based on attributes or based on alternatives. In the former, attributes are selected in turn and options are processed, and maintained or rejected depending on the values of these attributes; in the latter, alternatives are considered in turn and their bundle of attributes examined. At any stage of the process, options that do not satisfy norms or other constraints are eliminated. A more detailed consideration of decision strategies is given by Foerster (1979) and Williams and Ortúzar (1982a). Denstadli et al. (2012) discuss different decision strategies and go on to characterise the decision process of individuals confronted with different types of choice tasks by recording their verbalised thoughts while completing them.

7.7.3 Habit and Hysteresis

At the end of the 1970s, there was considerable interest in the relevance and role of *habit* in travel choice behaviour, particularly in cases of relocation (i.e. migration) or other phenomena granting a fresh look at the individual's choice set. Empirical evidence (Blase 1979) suggested that the effect of habit could be of practical significance and the problem should be treated seriously. The interest on this issue has not abated as most commuter trips have a tendency to be repeated over time, thus acquiring a potentially important *inertia* component (Pendyala et al. 2001; Lanzendorf 2003; Cantillo et al. 2007; Gardner 2009; Cherchi et al. 2013).

The existence of habit, or what might be considered inertia accompanying the decision process of an individual, is possibly the most insidious of behavioural aspects that represent divergences from the traditional assumptions underpinning choice models, for it appears directly in the response context. To examine the effects and implications of habit it is appropriate to return to the assumptions behind the conventional cross-sectional approach.

Figure 7.6a reproduces the S-shaped curve relevant to binary choice. For a given difference in utility $(V_2 - V_1)$ there exists a certain unique probability of choice; under conditions of change $(V_2' - V_1')$, the probability will correspond to that observed for that utility difference in the base year, that is, the response is determined from the cross-sectional dispersion. An implication of this assumption is that response to a particular policy or change will be exactly reversed if the stimulus is removed; the stimulus–response relation is symmetric with respect to the sign and size of the stimulus.

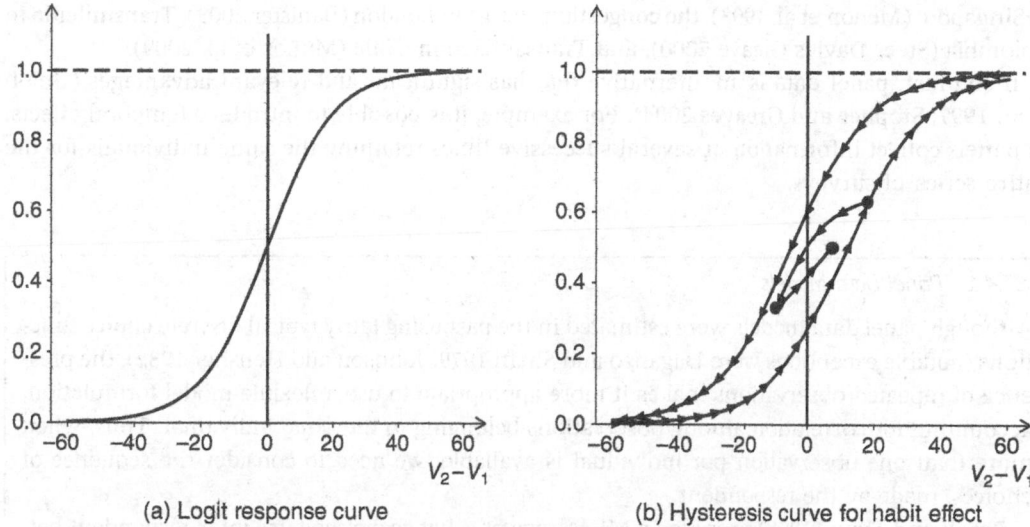

(a) Logit response curve   (b) Hysteresis curve for habit effect

Figure 7.6 Influence of habit in cross-sectional models

However, if habit exists, it will affect those members of the population who are currently associated with an option and are experiencing a stimulus to the relative advantage of another alternative. This introduces a basic asymmetry into response behaviour and gives rise to the phenomenon of *hysteresis* (Goodwin 1977), as pictured in Figure 7.6b. In this case, the present state of the population identified in terms of the market share of each alternative depends not only on the utility values V_2 and V_1, but also on how these variables attained their current value.

Formally, the state of the system P may be expressed as a path integral in the space of utility components $\mathbf{V}$; the value of the integral is path independent when habit is absent but path dependent when it is present (see the discussion in Williams and Ortúzar 1982a). These ideas have been taken into an operational model by Cantillo et al. (2007), which is a precursor to models for panel data that we will examine in the following section.

Another approach to dealing with habit is to try and measure its existence using a set of indicators purposely designed for the task (see for example, Verplanken and Orbell 2003), that can later be used to estimate habit as a latent variable in a hybrid choice model setting, as Gutierrez et al. (2020) did for bicycle choice.

7.7.4 Modelling with Panel Data

The long-term planning of transport systems, especially when decisions about substantial changes are involved, requires special demand models. However, most demand models to date have used readily available cross-sectional data, which do not allow for an appropriate consideration of temporal effects as information is considered only for a single point in time. This limitation may be especially restrictive when personal routines are habitual (such as in the cases discussed in the previous section) or when a substantial intervention is planned for a system.

Unfortunately, in transport this is more the norm than the exception, as, on one hand, mode choice in a stable context (particularly for trips to work or study) is a process that is more of a habit than a plan (Wood et al. 2002). On the other hand, changes introduced to transport systems are becoming more common every day. Some famous examples are the electronic road pricing system in Singapore (Menon et al. 1993), the congestion charge in London (Banister 2003), Transmilenio in Colombia (Steer Davies Gleave 2000), and Transantiago in Chile (Muñoz et al. 2009).

In contrast, panel data is an alternative that has significant and relevant advantages (Golob et al. 1997; Stopher and Greaves 2004). For example, it is possible to introduce temporal effects, as panels collect information at several successive times retaining the same individuals for the entire series of surveys.

7.7.4.1 Panel Data Models

Although panel data models were estimated in the past using fairly typical discrete choice functions (notable exceptions were Daganzo and Sheffi 1979; Johnson and Hensher 1982), the presence of repeated observations makes it more appropriate to use a flexible model formulation, accounting for correlation among observations belonging to the same individual. Thus, when more than one observation per individual is available, we need to consider the sequence of choices, made by the respondent.

Revelt and Train (1998) proposed a ML framework that accommodates inter-respondent heterogeneity but assumes intra-respondent homogeneity in tastes (i.e. it includes the effect of repeated choices by assuming that tastes vary across respondents, but stay constant across observations for the same respondent); this ML *panel probability*, is given by the following product of ML probabilities:

$$P_{jq} = \int_{\theta_q} \prod_{t=1}^{T} \left(\frac{e^{V_{jqt}(\theta_q)}}{\sum_{A_i \in \mathbf{A}^t(q)} e^{V_{iqt}(\theta_q)}} \right) f(\theta_q | \mathbf{b}, \boldsymbol{\Sigma}) d\theta_q \tag{7.49}$$

where V_{iqt} is the observable component of the utility of option A_i for individual q at time t; $\mathbf{A}^t(q)$ is the choice set of individual q at time t; T is the number of periods (*waves*) in the panel, and $f(\cdot)$ is the mixing distribution, with means $\mathbf{b}$ and covariance matrix $\boldsymbol{\Sigma}$ (i.e. the population parameters) of the coefficients to be estimated in $\mathbf{V}$.

Hess and Rose (2009) relaxed the assumption of intra-respondent homogeneity of tastes, proposing a choice probability with the following form:

$$P_{jq} = \int_{\alpha_q} \prod_{t=1}^{T} \left(\int_{\gamma_{q,t}} \frac{e^{V_{jq}(\theta_q)}}{\sum_{A_i \in \mathbf{A}^t(q)} e^{V_{iq}^t(\theta_q)}} g\left(\gamma_{q,t} \mid \boldsymbol{\Sigma}_\gamma\right) d\gamma_{q,t} \right) h(\alpha_q | \boldsymbol{\Sigma}_\alpha) d\alpha_q \tag{7.50}$$

where θ are a function of α_q that varies over respondents with density $h(\alpha_q | \boldsymbol{\Sigma}_\alpha)$, and $\gamma_{q,t}$, which varies over all choices with density $g(\gamma_{q,t} | \boldsymbol{\Sigma}_\gamma)$. This model has integrals inside and outside the product over periods; the latter accounts for inter-respondent heterogeneity as in the previous model (Revelt and Train 1998), while the inside integral accounts for intra-respondent heterogeneity. However, this formulation is more demanding in terms of estimation time and most packages just allow using a simplified version of it (see Hess and Rose 2009). Fortunately, the need for assuming intra-respondent heterogeneity is not that pressing, as it is reasonable to expect that in the short to medium term respondent tastes will probably stay the same.

On the other hand, several empirical applications have shown that the inclusion of inter-respondent heterogeneity in the random parameters leads to significant model fit improvements and a greater ability to retrieve taste heterogeneity. In fact, this is also the most common approach to deal with stated preference data that includes multiple choices for each respondent as we discuss in Section 8.7.2.8 (this approach has been implemented in the majority of estimation packages). Although the influence of repeated observations (i.e. inter-respondent heterogeneity in tastes) can be considered directly via the estimation of random parameters, there might be extra correlation across repeated observations besides their effect. Thus, even though random parameters and error components might induce confounding, they might also account for slightly different effects. In fact, as long as both effects are significant, the pure error-panel component accounts for correlation in the preference for alternatives, while the random parameters account for correlation in tastes (Yáñez et al. 2011).

7.7.4.2 *Efficiency and Repeated Observations*

Efficiency, in general, can be measured by the Fisher information matrix $\mathbf{I}$ (see Example 2.15); this is inversely related to sample size, the attribute values associated with the estimated parameters and the probability associated with the chosen alternative (McFadden 1974). Rose and Bliemer (2008) analysed the effect of the number of alternatives, attributes, and attribute levels on the optimal sample size for SC experiments in MNL models, as part of their search for the design with highest asymptotic efficiency of the estimated parameters. They found that only the range of attribute levels could offer an explanation for some problems of convergence encountered in their experiments. Cherchi and Ortúzar (2008b) demonstrated that while efficiency clearly improves with sample size, data variability does not always increase it.

In contrast, repeated observations in a short-survey panel, for example, will increase the number of observations but might reduce data variability, because observations that are identical do not bring new information about attribute trade-offs. Thus, when using panel data, it is important to understand how efficiency is influenced by repeated observations and up to what point these are actually beneficial. This is also crucial to determine the length of a multi-day panel survey, which is something that has not been explored much up to date. Moreover, as in panel data each individual provides more than one observation, it is necessary to account for correlation among these and this has a different effect depending on how the repeated observations are treated. Cherchi et al. (2017) found that the effect of correlation is, to a large extent, given by the repeated observations.

In Chapter 8, we will see that when the parameters of a discrete choice model are estimated by maximum likelihood, the expected value of the variance of the *k*th estimated parameter (i.e. the *k*th element of the diagonal of the Fisher information matrix) is given by:

$$E\left[\frac{\partial^2 \ell(\theta)}{\partial \theta_k^2}\right] \cong \sum_{q=1}^{Q} \sum_{A_j \in \mathbf{A}(q)} \left[\frac{\partial^2 \left(g_{jq} \ln P_{jq}(\mathbf{x}_{jq}, \theta)\right)}{\partial \theta_k^2}\right]_{\theta = \hat{\theta}} \tag{7.51}$$

where $\ell(\theta) = \ln \prod_q P_{jq}^{g_{jq}}$ is the log-likelihood function with respect to the parameters θ evaluated at their estimated values, P_{jq} is the probability that individual q chooses alternative A_j among the alternatives belonging to her choice set $\mathbf{A}(q)$, $\mathbf{x_{jq}}$ are the level-of-service and socio-economic attributes, and g_{jq} equals one if A_j is the alternative actually chosen by individual q and zero otherwise.

Equation (7.51) shows that the efficiency of the estimated parameters depends on sample size, the values of the attributes associated with the estimated parameters, and the probability of the chosen alternative. As the Logit probability also depends, among other things, on the data variability and on the variance of the error term (through the scale factor), understanding the sensitivity of the efficiency of the estimated parameters is a complex task. Cherchi and Ortúzar (2008b) analysed how the efficiency of the estimated parameters varied for revealed preference (RP) and stated choice (SC) data.

Looking at expressions for a single element of the Fisher information matrix, as above, is convenient for a theoretical discussion of the efficiency issue, because they illustrate what elements influence the matrix $\mathbf{I}$. However, in practice it would be better to measure the statistical efficiency of the expected outcomes of models as in the experimental design literature (Rose and Bliemer 2009), by computing the negative inverse of the Fisher information matrix (i.e. the asymptotic covariance matrix, $\mathbf{S}^2$) and then computing the D-error (see Section 2.4.2.3); a smaller D-error yields estimates that are more efficient.

Let us consider, for simplicity, a binary Logit model (i.e. with 'fixed' parameters). The variance of the parameters estimated with panel data is given by:

$$\text{var}\left(\hat{\theta}\right) = -\frac{1}{\sum_q \sum_t \Delta x_{jqt}^2 \hat{P}_{jqt}\left(1 - \hat{P}_{jqt}\right)} \qquad (7.52)$$

where Δx_{jqt}^2 is the attribute difference between both alternatives in period t. However, in contrast to the case of, for example SC data, when using information from a short survey panel the attribute values will be identical for the same individual in the period (i.e. five days of the week). Thus, in such cases we will have that $\Delta x_{jqt} = \Delta x_{jq} \forall t$ and the variance of the parameters will simplify to:

$$\text{var}\left(\hat{\theta}\right) = -\frac{1}{\sum_q \sum_t \Delta x_{jq}^2 \hat{P}_{jqt}\left(1 - \hat{P}_{jqt}\right)}$$

These equations show that the variance depends clearly on the number of repeated observations as well as on the data variability and number of observations. However, the efficiency of the parameters increases with the variability of the attributes only for scale factors over 0.5. This, which might seem counterintuitive, is due to the effect that the scale factor has on the variability of the data, because efficiency reduces as data variability diminishes; and is also due to the second order function of the probability, that tends to zero as the probability of the chosen alternative approximates one.

It is important to note that a panel with identical repeated observations for each individual is a special case. In fact, in terms of the above discussion having equal observations repeated a certain number of times increases only marginally the variability of the attributes. In particular, if N is the number of observations and R is the number of times these are repeated for each individual, the variance of the attributes (Δx_{jq}) for N and RN observations is related by the following expression (Yáñez et al. 2011):

$$\frac{\text{var}\left(R\Delta x_{jq}\right)}{\text{var}\left(\Delta x_{jq}\right)} = \frac{(RN - R)}{(RN - 1)} \qquad (7.53)$$

Hence, identical repeated observations should not in theory influence the efficiency of the estimated parameters. This result may be confirmed by computing the D-error.

The extension of this result to the ML case (which we need to properly estimate models with panel data) is not difficult. In the ML model, the variance of the mean of the random parameters is more complex, but the structure is basically the same (Cherchi and Ortúzar 2008b). It is still inversely related to the square value of the attributes associated with each parameter (as in the case of the fixed parameters model), to the number of repeated observations, and is also a function of the probabilities (Bliemer and Rose 2010).

Using observed data, Yáñez et al. (2011) found that the inclusion of intra-respondent heterogeneity requires more observations, which means that the repeated observations can affect the definition of model structure. Therefore, a potential benefit of considering a longer multi-day survey in a short-survey panel context is the highest probability of capturing different kinds of heterogeneity among observations.

Complementary, results from simulated data have shown that having repeated observations in a data panel increases the efficiency of the estimated parameters only because this increases the sample dimensions. Therefore, based on the results from real and synthetic data, it is possible to say that there is a trade-off between the higher probability of capturing effects (different types of heteroskedasticity) in a longer multi-day-panel sample, and the risk of a decreased capability of reproducing true phenomena (as this worsens in the presence of repeated observations).

Finally, a suggestion for the definition of the length of a multi-day-panel survey would be to consider not only the number of individuals, but also the level of routine expected. This last factor seems to be especially important in a short-survey panel context, as these panels commonly feature a large proportion of identical observations, which are actually harmful, as they reduce the capability of reproducing the true phenomenon. Thus, even though having more observations per respondent requires smaller sample sizes to establish the statistical significance of the parameter estimates derived from choice data (Rose et al. 2009b), Yáñez et al. (2011) show that this is effectively true if and only if the level of routine is not strong.

7.7.4.3 Dealing with Temporal Effects

One of the temporal effects more often discussed in the literature is *habit*, leading to *inertia* (Goodwin 1977; Blase 1979; Williams and Ortúzar 1982a); Daganzo and Sheffi (1979) proposed a MNP formulation to treat this phenomenon, which was later implemented by Johnson and Hensher (1982) for a two-wave panel in Australia. Later on, the discrete choice modelling field saw significant advances in terms of incorporating inertia, examples of that are: a model including prior behaviour on a time-series context (Swait et al. 2004), a model including inertia on a two-wave panel formulation (Cantillo et al. 2007), and a *planning-and-action* model considering inertia as an effect of previous plans (Ben-Akiva 2009). All these studies refer to cases where there are no changes in the transport system (i.e. a stable choice environment).

The changing choice environment defined by the *Santiago Panel* (Yáñez et al. 2010b), with data before and after the introduction of Transantiago (Muñoz et al. 2009), acted like a *shock* to the system and required the introduction of another temporal effect beyond inertia. Assuming that the shock effect could reduce or even overcome the effect of inertia, Yáñez et al. (2009) formulated a model incorporating the effects of three forces involved in the choice process: (i) the relative values of the modal attributes, (ii) the inertia effect, and (iii) the shock resulting from an abrupt policy intervention.

In their model, inertia was a function of the previous valuation of the options and its effect was allowed to vary for each wave and among individuals due to systematic or purely random effects. Furthermore, the effect might be positive or negative; the former representing the 'typical' inertia effect in the absence of changes, the latter indicating the preference for changing that might occur after a significant variation in the system.

On the other hand, individuals may modify their valuation process after a shock, altering their utility function. The shock effect is a function of the difference between the utility of an option evaluated at the current wave w, and its utility evaluated at the preceding wave $(w-1)$; hence, the effect is expected to be negative when the alternative worsens (making its utility lower), and positive when it improves. The perception of the shock may also be different for each wave and may vary among individuals due to systematic or random effects. In particular, the shock effect should have the highest value immediately after the introduction of the new policy, and then its magnitude should attenuate.

According to these assumptions, let the utility associated with each option A_j at wave $w = 1$ (i.e. the base situation) be the sum of observable (V_{jq1}) and non-observable components (ζ_{jq1}). Then, the probability of choosing option $A_j \in \mathbf{A}^1(q)$ at wave $w = 1$ will be, as usual:

$$P_q\left(A_j^1\right) = \text{Prob}\left((V_{jq1} + \zeta_{jq1}) - (V_{iq1} + \zeta_{iq1}) \geq 0, \forall A_i^1 \in \mathbf{A}^1(q)\right) \tag{7.54}$$

where $\mathbf{A}^1(q)$ is the choice set of individual q in wave $w = 1$. In subsequent waves, the alternative chosen in the previous wave will be denoted by A_r; temporal effects will be also included to detect how the choices in a given wave (w) are influenced by the choices made in a previous one $(w-1)$.

If $\tilde{U}_{jqw}$ denotes the utility that individual q associates to a generic alternative A_j on wave w $(w = 2, 3, \text{etc.})$. This utility will include inertia and shock effects, such that:

$$\tilde{U}_{jqw} = U_{jqw} - I_{jrq}^w + S_{jq}^w \tag{7.55}$$

where I denotes inertia and S shock, and there are several ways to express them. In particular, Yáñez et al. (2009) proposed the following general expressions:

$$I_{jrq}^w = \left(\theta_{Ij}^w + \delta_{Iq} \cdot \sigma_{Ij}^w + \theta_{I\_SE} \cdot SE_I\right) \cdot \left(V_{rq(w-1)} - V_{jq(w-1)}\right) \tag{7.56}$$

$$S_{jq}^w = \left(\theta_{Sj}^w + \delta_{Sq} \cdot \sigma_{Sj}^w + \theta_{S\_SE} \cdot SE_S\right) \cdot \left(V_{jqw} - V_{jq(w-1)}\right) \tag{7.57}$$

where θ_{Ij}^w and θ_{Sj}^w are the population means, and σ_{Ij}^w and σ_{Sj}^w the standard deviations of the inertia and shock parameters respectively, for option A_j on wave w; SE_I and SE_S are socioeconomic variables, with parameters θ_{I_SE} and θ_{S_SE} respectively; these allow for systematic variations of the inertia and shock parameters, δ_{Iq}, δ_{Sq} are the standard factors to introduce panel correlation (note that these could be included either as random parameters or error components), and $\mathbf{V}$ are the observable components of the utility function without temporal effects.

Note that if I_{jrq}^w is greater than zero inertia exists; while, if I_{jrq}^w is negative, it implies that the individual has a high disposition to change. Also, note that (7.56) assumes a zero-inertia effect on wave w for the alternative chosen on wave $(w-1)$. It means: $\tilde{U}_{rqw} = U_{rqw} + S_{rq}^w$.

In the presence of inertia and shock, the probability to change from A_r (i.e. the alternative chosen in the previous wave) to A_j (i.e. the 'candidate alternative') for individual q on wave w, is given by:

$$P_{jqw} = \text{Prob}\left(\tilde{U}_{jqw} - \tilde{U}_{rqw} \geq 0 \quad \text{and} \quad \tilde{U}_{jqw} - \tilde{U}_{iqw} \geq 0,\right.$$
$$\left.\forall A_i^w \in \mathbf{A}^w(q), \text{except } r = j\right) \tag{7.58}$$

while, the probability to remain with A_r is given by:

$$P_{rqw} = \text{Prob}\big(\tilde{U}_{rqw} - \tilde{U}_{jqw} \geq 0\big)$$

In this formulation, and as usual in current practice, alternative attributes and socioeconomic characteristics are associated with parameters that could be either fixed or random; on the other hand, the non-observable component ζ_{jqw} is a random error term that can be formulated as $\zeta_{jqw} = v_q + \varepsilon_{jqw}$, where v_q is a random effect specific to the individual and ε_{jqw} is, once more, the typical random error distributed IID EV1.

With all the above, the probability of choosing alternative A_j on wave w, $(\forall w > 1)$ can be written as:

$$
\begin{aligned}
P_{jqw} = \exp\Big(& V_{jqw} - \big(\theta_{Ij}^w + \delta_{Iq} \cdot \sigma_{Ij}^w + \theta_{I\_SE} \cdot SE_I\big) \cdot \big(V_{rq(w-1)} - V_{jq(w-1)}\big) \\
& + \big(\theta_{Sj}^w + \delta_{Sq} \cdot \sigma_{Sj}^w + \theta_{S\_SE} \cdot SE_s\big) \cdot \big(V_{jqw} - V_{jq(w-1)}\big)\Big) \\
& \cdot \Big[\sum_i \big(\exp\big(V_{iqw} - \big(\theta_{Ii}^w + \delta_{Iq} \cdot \sigma_{Ii}^w + \theta_{I\_SE} \cdot SE_I\big) \cdot \big(V_{rq(w-1)} - V_{iq(w-1)}\big) \\
& + \big(\theta_{Si}^w + \delta_{Sq} \cdot \sigma_{Si}^w + \theta_{S\_SE} \cdot SE_s\big) \cdot \big(V_{iqw} - V_{iq(w-1)}\big)\big)\big)\Big]^{-1}
\end{aligned}
\tag{7.59}
$$

where if $j = r$, then $(V_{rq(w-1)} - V_{jq(w-1)}) = 0$ and, as previously discussed, inertia is zero while the shock effect would still be active. Actually, the shock effect S_{jq}^w is null if either the shock parameter is itself null $(\theta_{Sj}^w = 0)$ or if the utility of alternative A_j does not change between consecutive waves, that is, $V_{jqw} = V_{jq(w-1)}$.

Note that Eq. (7.59) is a general formulation that can accommodate panel correlation either in the representative utility V_{jqw} (using random parameters), error term (as an error component), or in the inertia and shock effects (again using random parameters). But for empirical estimation it is not possible to consider all these panel correlation forms at the same time. In fact, since the inertia and shock parameters multiply the expressions $\Delta V_I = (V_{rq(w-1)} - V_{jq(w-1)})$ and $\Delta V_S = (V_{jqw} - V_{jq(w-1)})$ respectively, randomness cannot be added in the representative utility and temporal effects at the same time; we will come back to these issues in Chapter 8.

As individual responses present panel correlation given a sequence of choices A_j^w, one for each wave, the probability that a person follows this sequence is given by:

$$P_q\big(A_j^1 \wedge A_j^2 \wedge ... A_j^W\big) = \prod_{w=1}^{W} P_{jqw} \tag{7.60}$$

and as inertia, shock and panel correlation are actually unknown, the probability of this sequence of choices is of ML form; we will look at ways to estimate this model in Chapter 8.

7.7.5 Hybrid Choice Models Incorporating Latent Variables

The inclusion of subjective elements in discrete choice models re-emerged recently as an analysis and discussion topic, after losing some of the importance that made it a subject in the early 1980s (see for example Ortúzar and Hutt 1984; McFadden 1986). Thus, Hybrid Choice models have been proposed that consider not only tangible attributes of the alternatives (classic explanatory variables) as in traditional choice models, but also more intangible elements associated with users' perceptions and attitudes (including happiness), expressed through latent variables (Morikawa and Sasaki 1998; Ashok et al. 2002; Abou-Zeid and Ben-Akiva 2009; Bahamonde-Birke et al. 2017a).

To estimate models with both kinds of variables, two methods were developed: the sequential approach, in which the latent variables were constructed before entering the discrete choice model as a further regular variable (Ashok et al. 2002; Vredin-Johansson et al. 2006); and the simultaneous approach, in which both processes are done at once (Bolduc et al. 2008; Raveau et al. 2010). The second approach results in more efficient estimators of the involved parameters (Ben-Akiva et al. 2002), and now dominates because currently available software allows to exploit the full capabilities of the base discrete choice model. We will come back to these issues in Chapter 8.

7.7.5.1 Modelling with Latent Variables

Latent variables are factors that, although they influence individual behaviour and perceptions, cannot be quantified in practice (e.g. safety, comfort, and reliability). This is because of either their intangibility, as these variables do not have a measurement scale, or their intrinsic subjectivity (i.e. different people may perceive them differently). Identification of latent variables requires supplementing a standard survey with questions that capture users' perceptions about some aspects of the alternatives (and the choice context). The answers to these questions generate 'perception indicators' that serve to identify the latent variables. Otherwise, these latent variables could not be measured.

To make use of latent variables a multiple indicator multiple cause (MIMIC) model (Bollen 1989) is estimated, where the latent variables (η_{ilq}) are explained by characteristics s_{iqr} from the users and from the alternatives through *structural equations* such as (7.61); at the same time, the latent variables explain the perception indicators (y_{ipq}) through *measurement equations* as (7.62):

$$\eta_{ilq} = \sum_r \alpha_{ilr} \cdot s_{iqr} + \nu_{ilq} \tag{7.61}$$

$$y_{ipq} = \sum_l \gamma_{ilp} \cdot \eta_{ilq} + \zeta_{ipq} \tag{7.62}$$

where the index i refers to an alternative, q to an individual, l to a latent variable, r to an explanatory variable and p to an indicator; α_{ilr} and γ_{ilp} are parameters to be estimated, while ν_{ilq} and ζ_{ipq} are error terms with mean zero and standard deviation to be estimated. As the η_{ilq} terms are unknown, both equations must be considered jointly in the parameter estimation process.

7.7.5.2 Hybrid Discrete Choice Model

When latent variables η_{ilq} are considered, the systematic or representative utility V_{iq} in Eq. (7.2) incorporates them together with the objective attributes x_{ikq} (i.e. travel time or fare, as well as socioeconomic characteristics of the individual), leading to a utility function such as:

$$V_{iq} = \sum_k \theta_{ik} \cdot x_{ikq} + \sum_l \beta_{il} \cdot \eta_{ilq} \tag{7.63}$$

where θ_{ik} and β_{il} are parameters to be estimated. However, as the η_{ilq} variables are unknown, the model must be estimated jointly with the MIMIC model's structural (7.61) and measurement (7.62) equations. Finally, to characterise individual decisions binary variables g_{iq}, that take values according to (7.64), have to be defined:

$$
g_{iq} = \begin{cases} 1 & \text{if} \quad U_{iq} \geq U_{jq}, \quad \forall j \in \mathbf{A}(q) \\ 0 & \text{in other case} \end{cases} \tag{7.64}
$$

where, as usual, $\mathbf{A}(q)$ is the set of available alternatives for individual q.

Note that as the latent variables are on the right-hand side (i.e. as explanatory or independent variables) both in the utility function (7.63) and in the measurement Eq. (7.62) of the MIMIC model, there will not be endogeneity for simultaneous determination even if the errors (of either equation) are correlated (see Guevara and Ben-Akiva 2006).

In Chapter 8, we will discuss the method currently used to estimate these hybrid models in practice, and comment on some interesting findings.

7.7.6 Attribute Non-Attendance and Other Heuristics

The process used to make a decision is just as relevant as the outcome. Traditional discrete choice studies rely on very specific behavioural assumptions when specifying and estimating a choice model. Even though different heuristics have been discussed in behavioural studies since the 1950s (Simon 1957), as we saw in Section 7.7.2, only in the past few decades alternative decision process strategies have gained importance in discrete choice studies (McFadden 2001). These decision processes, often referred to as 'choice heuristics', challenge the traditional decision-making assumptions.

One process strategy that has proven to have a significant influence on the way we understand and interpret preferences is referred to as Attribute-Non-Attendance (ANA). In what follows we give advice based on comments by Dr. Camila Balbontín, a leading expert in this theme.

The ANA process strategy states that, to make a decision, individuals only evaluate a subset of the attributes, that is, they do not consider certain attributes (Hensher 2006b). ANA can be incorporated when specifying the utility function of a choice model in two different ways: (i) using stated ANA or (ii) inferring ANA.

Stated ANA. In this case, individuals are directly asked which attributes they did not consider, typically in a stated preference (SP) context, but this could also be incorporated in revealed preference (RP) studies. As in SP studies, respondents normally answer several choice tasks, they could be asked, after each one, which attributes they did not attend to. Alternatively, they could be asked only once at the end of the survey (i.e. assuming individuals did not attend to the same attributes across all choice tasks).

ANA can certainly depend on the attribute levels presented in each choice task and, hence, it is preferred to ask about it at the end of each choice task (Puckett and Hensher 2009), but this depends on the attribute levels' variation and the type of study.

Figure 7.7 presents an illustrative example of a SP choice task and how could stated *ANA* be incorporated in the survey (Balbontín and Hensher 2020).

	Car	Public transport
Cost (US$)	$10.00	$2.00
Time (min)	30.00	60.00
Which would be your preferred mode of transport for your daily commute?	O	O

(a) A stated preference choice task

Thinking about the previous choice task, did you ignore any of the attributes when making a decision as to your preferred mode of transport?

	Select all attributes that you ignored
Cost (US$)	☐
Time (min)	☐
I did not ignore any attribute	☐

(b) Questioning ANA in the previous choice task

Figure 7.7 Illustrative example of SP choice task and stated ANA

Each attribute k is associated with a dummy variable φ_{kq} which takes the value of 1 if individual q attended to attribute k when making a decision, and 0 if she ignored it. As mentioned above, this dummy variable could also be identified by individuals during the data collection process of an RP survey. Having done the above, the deterministic part of the utility function would be defined as:

$$V_{jq} = \sum_k \theta_{jk} x_{jkq} \varphi_{kq} \tag{7.65}$$

where x_{jkq} is the kth attribute of alternative j for individual q, and θ_{jk} it's associated parameter j.

The ANA dummy variable, φ_{kq}, can vary across choice tasks or be the same for the individual, depending on if the ANA question was included after each choice task or at the end of the experiment. Both options will result in the same utility specification presented in (7.65).

Inferred ANA. A second option is to infer ANA analytically through an appropriately specified model. This will be the case when a stated ANA response is not available (i.e. an RP dataset that did not initially consider ANA). Most studies that have included inferred *ANA* have used a latent class specification (see Section 8.6.3.2), where each class represents a group of individuals that attended to a different subset of attributes (Hensher 2008).

For example:

Class 1: $\{\psi_{1q}, \psi_{2q}, ..., \psi_{kq}\} = \{1, 1, 1, ..., 1\}$ → All attributes are attended to
Class 2: $\{\varphi_{1q}, \varphi_{2q}, ..., \varphi_{kq}\} = \{0, 0, 1, ..., 0\}$ → All attributes except the third are ignored
Class 3: $\{\varphi_{1q}, \varphi_{2q}, ..., \varphi_{kq}\} = \{1, 0, 0, ..., 1\}$ → Only the first and last attributes are attended to and the utility specification would be equivalent to that in (7.65) for each class.

The probability of belonging to each class can be, estimated and, as usual, it may depend on socio-demographic characteristics of the individuals. The results of such a model would link sociodemo-graphic characteristics with the ANA process strategies, showing, for example, if individuals of a certain age group are more or less likely to attend to all attributes.

Other ways that have been used to include ANA relate to the number of alternatives in each choice task (Collins and Hensher 2015), or to the levels of the attribute (Campbell et al. 2014). In these cases, the ANA dummy variable, φ_{kq}, is defined depending on the choice task or attribute characteristics. For example, φ_{kq} equals to 1 if the level of the attribute is higher than a certain threshold, and 0 otherwise.

ANA results and interpretation. When including ANA, individuals stating that they ignored a given attribute are not included in the estimation of its parameter. Therefore, it is expected that its level of significance increases (i.e. its standard error to decrease), as only the individuals that actu-ally considered it are involved in the estimation. Note, however, that the estimate then represents only those individuals that attended to the attribute.

The same should be noted for willingness-to-pay (WTP) estimates in ANA model results. If, for example, the value of travel time savings (VTTS) was estimated for everyone in the sample, includ-ing those that ignored the time attribute, say, then those estimates may be compared with other studies that did not include ANA. However, if the VTTS were estimated only for those individuals that attended to ANA (i.e. including the ANA dummy variable φ_{kq} as a weighting factor in the VTTS), then the VTTS results should be interpreted carefully as they would not be representing the whole sample.

Other heuristics. Several heuristics have been proposed in the literature that challenge the tra-ditional discrete choice modelling studies' assumptions. They can be divided into three cate-gories: (i) context-free heuristics; (ii) local choice context dependent; and (iii) choice set interdependent. The first category includes all heuristics that assume that the utility function of each alternative is only a function of the characteristics of that alternative. Attribute non-attendance is the most commonly used heuristic in this category, but others include the non-compensatory heuristics we saw in Section 7.7.2.2: Satisficing, which assumes that individuals choose the first alternative which is good enough according to certain criteria (Simon 1957; González-Valdés and Ortúzar 2017), and Elimination by Aspects (EBA), which assumes that indi-viduals eliminate alternatives that fail to meet certain threshold requirements for each attribute, starting by the most important one and then proceeding following a predetermined ranking (Tversky 1972).

The second category considers that the characteristics of some alternatives may also affect the utility of competing alternatives, generating a dependence between the local choice context. One of the most popular heuristics in this category is Random Regret Minimisation (RRM), which assumes that individuals minimise their regret instead of maximising their utility (Chorus et al. 2008, 2014). A similar heuristic to RRM is Relative Advantage Maximisation (RAM), which assumes that individuals maximise the advantage of the chosen alternative (Leong and Hensher 2014).

The third category considers that past decisions (e.g. in previous choice tasks) might also influ-ence current decision-making. One of the most commonly used heuristics in this category is Value Learning, which assumes that preferences are not stable and might change when an individual has to make similar decisions repeatedly (DeShazo 2002; Balbontín et al. 2019).

Exercises

7.1 There is interest to study the behaviour of a group of travellers in relation to two transport options A and B, with travel times t_a and t_b respectively. It has been postulated that each traveller experiments the following net utilities from each option:

$$U_a = \alpha t_a + \beta I$$
$$U_b = \alpha t_b$$

where α and β are known parameters and I is the traveller's personal income.

Although there is no reliable data about the income of each traveller, it is known that the variable I has the distribution shown in Figure 7.8 in the population:

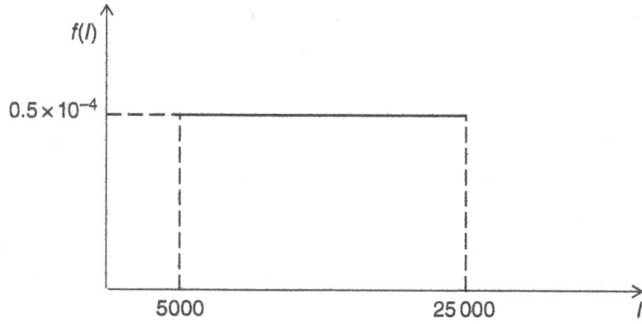

Figure 7.8 Distribution of income for Exercise 7.1

If $\alpha = -0.5$ and $\beta = 2\ 10^{-4}$, find out the probability function of choosing option A for a given traveller, as a function of the value of $(t_b - t_a)$; sketch the function in appropriate axis.

7.2 Consider a binary Logit model for car and bus, where the following representative utility functions have been estimated with a sample of 750 individuals belonging to a particular sector of an urban area:

$$V_c = 3.5 - 0.25t_c - 0.42e_c - 0.1c_c$$
$$V_b = -0.25t_b - 0.42e_b - 0.1c_b$$

where t is in-vehicle travel time (min), e is access time (min) and c is travel cost ($). Assume the average data shown in Table 7.1 is known:

Table 7.1 Average data for Exercise 7.2

Mode	Variable		
	t	e	c
Car	25	5	140
Bus	40	8	50

If you are informed that the number of individuals choosing each option in the sector and in the complete area is respectively, as in Table 7.2:

Table 7.2 Number of choices for Exercise 7.2

	Number of individuals choosing option *i*	
Option	Sample	Population
Car	283	17 100
Bus	467	68 900

(a) Indicate what correction would be necessary to apply to the model and write its final formulation.
(b) Calculate the percent variation in the probability of choosing car if the bus fares go up by 25%.
(c) Find out what would happen if, on the contrary, the car costs increase by 100%.

7.3 Compute the probabilities of choosing car, bus, shared taxi and underground, according to the Nested Logit model depicted in Figure 7.9:

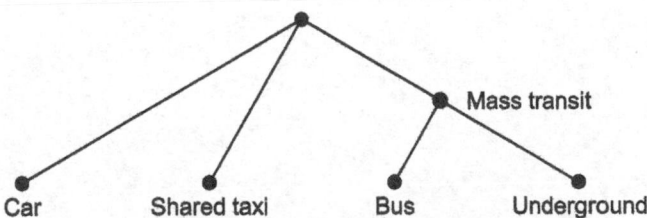

Figure 7.9 Nested Logit model for Exercise 7.3

with the following utility functions:
(a) High nest

$$V_c = -0.03t_c - 0.02c_c + 1.25$$

$$V_{st} = -0.03t_{st} - 0.02c_{st} - 0.20$$

$$V_{mt} = 0.60 \text{EMU}$$

(b) Mass transit nest

$$V_b = -0.04t_b - 0.03c_b + 0.5$$
$$V_u = -0.04t_u - 0.03c_u$$

and for the average variable values presented in the Table 7.3:

Table 7.3 Average variable values for Exercise 7.3

Mode	Time (*t*)	Cost/income (*c*)
Car	4.5	23.0
Shared taxi	5.5	15.0
Bus	7.5	5.5
Underground	5.5	3.6

7.4 The binary Probit model has the following expression:

$$P_1 = \Phi\left\{ (V_1 - V_2)/\sqrt{\sigma_1^2 + \sigma_2^2 - 2\rho\sigma_1\sigma_2} \right\}$$

Using this result write down the probability of choosing option one in the following binary model:

$$U_i = \theta X_i + \varepsilon_i$$

where the ε are distributed IID standard Normal, for the following cases:

(a) If the value of θ is fixed and equal to 3

(b) If θ is distributed Normal $N(3, 1)$ and is independent of the ε.

8
Specification and Estimation of Discrete Choice Models

8.1 Introduction

The previous chapter provided an overview of discrete choice modelling and an introduction to different model forms and theoretical frameworks for individual decisions. This chapter is devoted to a discussion of two key issues: how to fully specify a discrete or disaggregate model (DM) and how to estimate such a model once properly specified.

The search for a suitable model specification involves selecting the structure of the model (MNL, NL, ML, etc.), the explanatory variables to consider, the form in which they enter the utility functions (linear, non-linear), and the identification of the individual's choice set (alternatives perceived as available). In broad terms the objectives of a specification search include realism, economy, theoretical consistency, and policy sensitivity. In other words, we search for a realistic model, which does not require too many data and computer resources, does not produce counter-intuitive results and is appropriate to the decision context in which it is to be used. Early aggregate models such as those discussed in Chapters 5 and 6 were often critically portrayed as policy-insensitive, either because key variables were not included or because important model components were specified as insensitive to certain policies (e.g. consider the problem of inelastic trip generation). Most of the features of model specification are susceptible to analysis and experimentation (see Leamer 1978), but they are also strongly dependent on study context and data availability.

In this chapter, we start by considering how to identify the set of options available to individuals: choice-set determination. This is a key problem as we usually estimate DM by means of the (generally) observed individual choices between alternatives. These should be the alternatives actually considered, consciously or unconsciously, by the individual. The omission of seemingly unimportant options on the grounds of costs may bias results. For example, in the vast majority of aggregate studies only binary choice between car and public transport has been considered, with the consequence that the multimodal problem cannot be treated seriously; in fact, in many cases the consideration of alternative public-transport alternatives was relegated to the assignment stage employing multipath allocation of trips to sub-modal network links. In the same vein, the inclusion of alternatives that are actually ignored by certain groups (say walking more than 500 metres for high-income individuals), could also bias model estimation.

Then, in Section 8.3 we consider other elements of model specification and in particular functional form and model structure. The criteria of economy, realism, theoretical consistency, and decision-making context play an essential role in complementing the experience and intuition of the modeller during the specification searches. An additional, and often over-riding element, is the availability of specialised software. In fact, one of the reasons behind the immense popularity of the linear-in-the-parameters Multinomial Logit (MNL) model was that it could be easily

Modelling Transport, Fifth Edition. Juan de Dios Ortúzar and Luis G. Willumsen.
© 2024 John Wiley & Sons Ltd. Published 2024 by John Wiley & Sons Ltd.
Companion website: www.wiley.com/go/ortuzar5e

estimated with normally available software; this was not the case, for many years, for more general structures or functional forms which presented much greater difficulties (Daganzo 1979; Liem and Gaudry 1987; Train 2009).

The increasing availability of good software to select and estimate these models has certainly alleviated this problem. However, one issue to which we will return is that although we may be able to successfully estimate the parameters of widely different models with a given dataset, these (and their implied elasticities) will tend to be different, and we often lack the means to discriminate between different models, at least with cross-sectional data. Another important issue is the interpretation of results. More complex/richer models are even more dependent on data quality than their simpler counterparts, and the insights they offer on individual behaviour often require experienced analysts to interpret them correctly.

The final specification will then depend heavily on the modeller's experience and theoretical understanding, and context-specific factors such as: time and resources available for the modelling activity, degree of correlation among alternatives, heterogeneity of preferences and required degree of accuracy of the forecasts. It must be borne in mind that using an inadequate model, such as the MNL when the hypotheses needed to generate it do not hold, may lead to serious errors (Williams and Ortúzar 1982a).

Section 8.4 concentrates on the statistical estimation of discrete choice models using data from random and choice-based samples, and including methods to validate models and compare different model structures; we also consider here the estimation of Hybrid Choice (HC) models with latent variables, how to correct for endogeneity, and to model using stochastic variables (for example to solve the problem of errors in measurements). Section 8.5 discusses two methods available to estimate the Multinomial Probit model, and Section 8.6 discusses, in depth, the estimation of the Mixed Logit model, including the estimation of individual (rather than just population) parameters, and its application to modelling with panel data. The chapter concludes with considerations relevant to model estimation and forecasting with stated preference (SP) data and the estimation of joint models using revealed preference (RP) and SP data.

8.2 Choice-Set Determination

One of the first problems an analyst has to solve, given a typical RP cross-sectional data set, is deciding which alternatives are available to each individual in the sample. It has been noted that this is a difficult issue to resolve, because it reflects the dilemma the modeller has to tackle in arriving at a suitable trade-off between the model's relevance and its complexity; usually, however, data availability acts as a yardstick.

8.2.1 Choice-Set Size

It is hard to decide on an individual's choice set unless one asks the respondent directly; therefore, the problem is closely connected with the dilemma of whether to use reported or measured data, as discussed in Chapter 2. Although in mode choice modelling the number of alternatives is usually small, rendering the problem less severe, in other cases such as destination choice, the identification of alternatives in the choice set is a more complex matter. This is not simply because the total number of alternatives is usually very high, but because we face the added problem of how to measure/represent the attractiveness of each alternative. Ways of managing a large choice set include:

1) considering only subsets of the alternatives that are effectively chosen in the sample (i.e. in a sampling framework such as the one used by Ben-Akiva 1977);
2) using the *brute force* method, which assumes that everybody has all alternatives available and, hence, lets the model decide that the choice probabilities of unrealistic alternatives are low or zero.

Both approaches have disadvantages. For example, in case 1, it is possible to miss realistic alternatives that are not chosen owing to the specific sample or sampling technique; in case 2, the inclusion of too many alternatives may affect the discriminatory capacities of the model, in the sense that a model capable of dealing with unrealistic options may not be able to describe adequately the choices among the realistic ones (see Ruijgrok 1979). Other methods to deal with the choice-set-size problem are:

- the aggregation across alternatives, such as in a destination choice model based on zonal data;
- assuming continuity across alternatives, such as in the work of Ben-Akiva and Watanatada (1980).

8.2.2 Choice-Set Formation

Another problem in this realm is that the decision-maker being modelled may well choose from a relatively limited set; in this sense, if the analyst models choices that are actually ignored by the individual, some alternatives will be given a positive probability even if they have no chance of being selected in practice. Moreover, consider the case of modelling the behaviour of a group of individuals who vary a great deal in terms of their knowledge of potential destinations (owing perhaps to varying lengths of residence in the area); because of this, model coefficients, which attempt to describe the relationship between predicted utilities and observed choices may be influenced as much by variation in choice sets among individuals (which are not fully accounted for in the model) as by variations in actual preferences (which are accounted for). Because changes in the nature of the destinations may affect choice set and preferences to different degrees, this confusion may play havoc with the use of the model in forecasting or with the possibility of transferring it over time and space.

Ways to handle this problem include:

1) The use of heuristic or deterministic choice-set generation rules that permit the exclusion of certain alternatives (i.e. bus is not available if the nearest stop is more than some distance away) and which may be validated using data from the sample.
2) The collection of choice-set information directly from the sample, simply by asking respondents about their perception of available options (it has been found preferable to ask which options, out of a previously researched list, are not available and why).
3) The use of random choice sets, whereby choice probabilities are considered to be the result of a two-stage process: firstly, a choice-set generating process in which the probability distribution function over all possible choice sets is defined; and secondly, conditional on a specific choice set, a probability of choice for each alternative is defined (see the discussions by Lerman 1984 and Richardson 1982).

Non-compensatory protocols, such as satisfaction, lexicographic behaviour and elimination by aspects, may often be more appropriate than compensatory behaviour, as we saw in Chapter 7. In fact, many choice processes may be seen as a mixture of compensatory and non-compensatory protocols, and this is especially the case when the number of physically available options is large. In this context, Morikawa (1996) developed a hybrid model that applies compensatory and non-compensatory decision rules with a relatively large number of alternatives in a model where the decision process was divided into a choice-set formation stage and a choice stage. Choice-set formation was modelled by a random constraints model that had a non-compensatory nature among constraints, and the choice stage was described by a MNL model. This approach gave good results when applied to destination choice of vacation trips with up to 18 alternatives.

8.3 Specification and Functional Form

The search for the best model specification is also related to functional form. Although it may be argued that the linear function (7.3) is probably adequate in many contexts, there are others, such as destination choice, where non-linear functions are deemed more appropriate (Foerster 1981; Daly 1982a). The problems in this case are: firstly, that in general there is no guarantee that the parameter-estimation routine will converge to unique values and, secondly, that suitable software may not be readily available. Another specification issue related to functional form is how the explanatory variables should enter the utility function, even if this is linear in the parameters.

Three approaches have been proposed in the literature to handle the functional form question:

1) The use of SP data from real or laboratory experiments to determine the most appropriate form of the utility function (Lerman and Louviere 1978); we will briefly come back to this in Section 8.7.
2) The use of statistical transformations, such as the Box–Cox method, letting the data 'decide' to a certain extent an appropriate form (Gaudry and Wills 1978).
3) The constructive use of econometric theory to derive functional form (Train and McFadden 1978; Jara-Díaz and Farah 1987; Jara Díaz 2007); this is perhaps the most attractive proposition, as the final functional form can be tied up to evaluation measures of user benefit.

As we will see later, it is important to note that, in general, non-linear forms imply different trade-offs to those normally associated with concepts such as the value of time (Bruzelius 1979; Gaudry et al. 1989); also, it is easy to imagine that model elasticities and explanatory power may vary dramatically with functional form.

8.3.1 Functional Form and Transformations

Linear-in-the-parameters expressions such as (7.3) usually contain a mixture of quantitative and qualitative variables (where the latter are normally specified as dummies, i.e. gender, age, income level), and the problems are how to enter both kinds and where to enter the latter, as we have already discussed. In other words, it might be more appropriate to write (7.3) as:

$$V_{jq} = \sum_k \theta_{kj} f_{kj}(x_{kjq}) \tag{8.1}$$

which is still linear in the parameters, but makes it explicit that the functional form of the **x** variables is somewhat arbitrary. Usual practice consists in entering the variables in raw form (i.e. time rather than 1/time or its logarithm) but this could have some consequence if the model response is sensitive to functional form.

If we do not have theoretical reasons to back up a given form, it appears interesting to let the data indicate which could be an appropriate one. A class of transformations widely used in econometrics has been successfully adapted for use in transport modelling (see Gaudry and Wills 1978; Liem and Gaudry 1987). We will review two examples, the second one being a generalisation of the first.

8.3.1.1 Basic Box–Cox Transformation

The transformation $x^{(\tau)}$ of a positive variable x, given by:

$$x^{(\tau)} = \begin{cases} (x^\tau - 1)/\tau, & \text{if } \tau \neq 0 \\ \log x, & \text{if } \tau = 0 \end{cases} \tag{8.2}$$

is continuous for all possible τ values. With this we can rewrite Eq. (8.1) as:

$$V_{jq} = \sum_k \theta_{kj} \, x_{kjq}^{(\tau_k)} \tag{8.3}$$

and it is easy to see that if $\tau_1 = \tau_2 = \ldots = \tau_k = 1$, (8.3) reduces to the typical linear form (7.3); furthermore, if all $\tau_k = 0$, we obtain the widely used log-linear form. Therefore, both traditional forms are only special cases of (8.3).

8.3.1.2 Box–Tukey Transformation

The basic transformation (8.2) is only defined for $x > 0$; a more general form, for variables that may take negative or zero values, is given by:

$$(x + \mu)^{(\tau)} = \begin{cases} [(x + \mu)^\tau - 1]/\tau, & \text{if } \tau \neq 0 \\ \log (x + \mu), & \text{if } \tau = 0 \end{cases} \tag{8.4}$$

where μ is just a translational constant chosen to ensure that $(x + \mu) > 0$ for all observations.

The values of τ must satisfy certain conditions if the model is to be consistent with microeconomic theory. In particular, it is instructive to derive what restrictions exist in the case of attributes such as travel time (which produce disutility) or the number of cars in the household (which should increase the probability of choosing car), to ensure decreasing marginal utilities as the theory demands. This small challenge is left for the interested reader.

It can be shown that if an MNL is specified with functional form (8.4) and restricting all τ to be equal, its elasticities are given by:

$$E_{P_J, x_{kl}} = (\delta_{ji} - P_j) x_{ki} \theta_k (x_{ki} + \mu)^{\tau - 1} \tag{8.5}$$

with δ_{ji} equal to 1 if $j = i$ and 0 otherwise. Although it is obvious from (8.5) that the elasticities depend on the values of τ and μ, it is not clear how large the effect might be as the values of θ also vary.

In Chapter 18, we will discuss the consequences of using Box–Cox models in the derivation of subjective values of time (Gaudry et al. 1989).

8.3.2 Theoretical Considerations and Functional Form

Although we have made it clear that in any particular study, data limitations and resource restrictions often play a vital role, it is important to consider the influence of theory in the construction of a demand function. In what follows, we will show how the constructive use of economic theory helps to solve the important problem of how to incorporate a key variable, such as income, in a utility function. Throughout, we will assume a linear-in-the-parameters form and will not be concerned with model structure, but the analysis may be generalised at a later stage.

The conventional approach to understanding the roles of income, time and cost of travel within the discrete choice framework, is based on the work of Train and McFadden (1978); they established the microeconomic foundations of the theory by considering the case of individuals who choose between leisure (L) and goods consumed (G); the trade-off appears once the link between G and income (I) is formulated: they assumed that I was related with the number of hours worked (W). Thus, increasing W allowed G to increase, diminishing L. More formally the problem was stated as follows:

$$\text{Max } U(G, L)$$

subject to:

$$\left. \begin{array}{c} G + c_i = wW \\ W + L + t_i = T \end{array} \right\} \forall A_i \in \mathbf{A} \tag{8.6}$$

where U is the individual utility function, w is the real *wage rate* (the amount the individual gets paid per hour), c_i and t_i are the money and time spent per trip respectively, $\mathbf{A}$ is the choice set and T is a reference period; the unknowns are G, L, and W.

If U in problem (8.6) is given a fairly general form, such as Cobb–Douglas, finding its maximum with respect to $A_i \in \mathbf{A}$ is equivalent to finding the maximum of $(-c_i/w - t_i)$ among other possibilities. This is the origin of the widely used cost/wage rate variable in discrete-mode choice models, for which cost/income has been used as a proxy in many applications. The possibility of adapting working hours to attain a desired level of income plays a key role in the above derivation; thus, as W is endogenously determined and w is given exogenously, income becomes endogenous. This formulation assumes that the cost of travelling is negligible in relation to income (i.e. that there is no 'income effect').

However, for many individuals (particularly in emerging countries) both income and working hours are fixed and there may be income effects. In such cases it can be shown that the maximum of U depends on the value of $(-c_i/g - t_i)$ among other possibilities (Jara-Díaz and Farah 1987), where g is an *expenditure rate* defined in general by:

$$g = I/(T - W) \tag{8.7}$$

The presence of such an income variable, reflecting purchasing power in the utility specification, indicates that the marginal utility of income varies with income (i.e. the model allows for an income effect). Besides, it is interesting to mention that empirical tests have shown that this new specification consistently outperforms the conventional wage-rate specification, even for individuals with no income effect (Jara-Díaz and Ortúzar 1989). More complex theoretical derivations of functional form, even for general joint models of activities (time use) and mode choice can be derived in similar fashion (see Munizaga et al. 2006; Jara-Díaz 2007).

8.3.3 Intrinsic Non-Linearities: Destination Choice

Let us treat the singly constrained version of the Gravity model (5.14)–(5.18) we examined in Chapter 5 in a disaggregate manner by considering each individual trip maker in zone i as making one of the O_i trips originating in that zone. In this case, the probability that a person will make the choice of travelling to zone j is simply:

$$P_j = T_{ij}/O_i = D_j f_{ij} / \sum_d D_d f_{id} \tag{8.8}$$

Now if we define:

$$V_d = \log(D_d f_{id}) = \log D_d + \log f_{id} \tag{8.9}$$

the model is seen to be exactly equivalent to the MNL model (7.9). Thus, the conventional origin-constrained Gravity model may be represented by the disaggregate MNL without any loss of generality (Daly 1982a); note that (8.9) imposes no restrictions on the specification of the separation function f_{ij}. On the other hand, as we saw in Chapter 5, probably the most common function used in practice is the negative exponential of c_{ij}, the generalised cost of travelling between zones i and j; it is interesting to mention that when this form is substituted in (8.9) we obtain:

$$V_d = \log D_d - \beta c_{id} \tag{8.10}$$

which is in fact linear in the parameter β. The problem of non-linearity arises due to the presence of D_d, which may contain variables of the *size* variety that describe not the quality but the number of elementary choices within k and are typical of cases, such as choice of destination, where aggregation of alternatives is required (Daly 1982a). An example of this type of form was presented in Example 6.6.

8.4 Statistical Estimation

This section considers methods for estimating DM together with the goodness-of-fit statistics to be used in this task. Model estimation methods need to be adapted to the sampling framework used to generate the observations. This is necessary to improve estimation efficiency and avoid bias.

8.4.1 Estimation of Models from Random Samples

To estimate the coefficients θ_k in (7.3) the maximum likelihood method is normally used. The main hypotheses of this method are:

- the sample is random, all draws x_i ($i = 1, ..., n$) taken from the population are independent of each other and the whole sample corresponds to the same population;
- the distribution function is known in the population, with the exception of the parameters θ.

If every value x_i is assumed to have a density function $f(x_i, \theta)$, as they are independent the joint density function for all x can be written as:

$$g(x_1, x_2, ..., x_n, \theta) = \prod_n f(x_i, \theta)$$

Note that the usual interpretation of this density function is that x are as unknown variables and the parameters θ are fixed. Inverting the process, the previous equation can be interpreted as a likelihood function $L(\theta)$; if we maximise it with respect to θ, the result $\hat{\theta}$ is called maximum likelihood estimate, because it corresponds to the set of parameter values that have the greatest probability of having generated the observed sample. Note that in Linear Regression models (see Section 4.2.1) it can be shown that the least squares coefficients are in fact maximum likelihood estimates. We will come back to these in Section 8.4.1 and others.

In calculating the maximum, it is easier to work with the logarithm of $L(\theta)$, which is called *log-likelihood* function $l(\theta)$; as the logarithm of a function increases with x the maximisation procedure yields the same results. In this latter case then, we would maximise the function:

$$l(\theta) = \ln g(x_1, x_2, ..., x_n, \theta) = \prod_n \ln f(x_n, \theta)$$

Example 8.1   We wish to estimate the mean μ of a $N(\mu, \sigma^2)$ distribution from a random sample of size n. If each $x_i \sim N(\mu, \sigma^2)$, we have that:

$$f(x_i, \mu) = \frac{1}{\sqrt{2\pi}\sigma} e^{-\frac{1}{2}\left(\frac{x-\mu}{\sigma}\right)^2}$$

then

$$g(x_1, x_2, ..., x_n, \hat{\mu}) = \left(\frac{1}{\sqrt{2\pi}\sigma}\right)^n e^{-\frac{1}{2}\sum_{i=1}^{n}\left(\frac{x-\hat{\mu}}{\sigma}\right)^2}$$

Taking logarithms, we get:

$$L(\hat{\mu}) = \ln\left(\frac{1}{\sqrt{2\pi}\sigma}\right)^n - \frac{1}{2}\sum_{i=1}^{n}\left(\frac{x-\hat{\mu}}{\sigma}\right)^2$$

deriving, and equalising to zero, we get:

$$\frac{\partial L}{\partial \mu} = 0 = \sum x_i - n\hat{\mu}$$

so finally: $\hat{\mu} = \dfrac{\sum x_i}{n} = \bar{x}$ a well-known result in statistics.

The maximum likelihood method is based on the idea that although a sample could originate from several populations, a particular sample has a higher probability of having been drawn from a certain population than from others. Therefore, the maximum likelihood estimates are the set of parameters that will generate the observed sample most often.

Let us assume a sample of Q individuals for which we observe their choice (0 or 1) and the values of x_{jkq} for each available alternative, such that for example:

> individual 1 selects alternative 2
> individual 2 selects alternative 3
> individual 3 selects alternative 2
> individual 4 selects alternative 1, etc.

As the observations are independent, the likelihood function is given by the product of the model probabilities that each individual chooses the option she actually selected:

$$L(\theta) = P_{21}P_{32}P_{23}P_{14}...$$

Defining the following dummy variable:

$$g_{jq} = \begin{cases} 1 & \text{if } A_j \text{ was chosen by } q \\ 0 & \text{otherwise} \end{cases} \tag{8.11}$$

the above expression may be written more generally as:

$$L(\theta) = \prod_{q=1}^{Q} \prod_{A_j \in A(q)} (P_{jq})^{g_{jq}} \tag{8.12}$$

To maximise this function we proceed as usual, differentiating $L(\theta)$ partially with respect to the parameters θ and equating the derivative to 0. As we saw above, we normally maximise $l(\theta)$, the natural logarithm of $L(\theta)$, which is more manageable and yields the same optima θ^*.

Therefore, the function we seek to maximise is (Ortúzar 1982):

$$l(\theta) = \log L(\theta) = \sum_{q=1}^{Q} \sum_{A_j \in A(q)} g_{jq} \log P_{jq} \qquad (8.13)$$

When $l(\theta)$ is maximised, a set of estimated parameters θ^* is obtained which is asymptotically distributed $N(\theta, S^2)$ where:

$$S^2 = -\left(E\left(\frac{\partial^2 l(\theta)}{\partial \theta^2}\right)\right)^{-1} \qquad (8.14)$$

Also $LR = -2 \cdot l(\theta)$ is asymptotically distributed χ^2 with Q degrees of freedom (see Ben-Akiva and Lerman 1985). All this indicates that even though θ^* may be biased in small samples, the bias is small for large enough samples (normally, samples of 500–1000 observations are more than adequate).

Now, although we have an explicit expression for the covariance matrix S^2, determining the parameters θ^* involves an iterative process. In the case of linear-in-the-parameters MNL models, the function is well behaved, so the process converges quickly and always to a unique maximum; this explains why software to estimate this model is so easily available. Unfortunately, this is not the case for other discrete choice models, whose estimation processes are more involved; therefore, in what follows in this section, we will mainly refer to this simpler model.

Substituting the MNL expression (7.9) in (8.13), it can be shown that if the variable set includes an alternative specific constant (ASC) for alternative A_j we get:

$$\sum_q g_{jq} = \sum_q P_{jq}$$

and this allows us to deduce that the ASCs ensure that the model always reproduces the aggregate market shares of each alternative. Therefore, it is not appropriate to compare, as a goodness-of-fit indicator, the sum of the probabilities of choosing one option with the total number of observations that selected it, because this condition will be satisfied automatically by a MNL model with a full set of constants. As it is also inappropriate to compare the model probabilities with the g_{jq} values (which are either 0 or 1), a goodness-of-fit measure such as R^2 in ordinary least squares, which is based on estimated residuals, cannot be defined.

Example 8.2 Consider a simple binary-choice case with a sample of just three observations (as proposed by Lerman 1984); let us also assume that there is only one attribute x, such that:

$$P_{1q} = 1/\{1 + \exp[\theta(x_{2q} - x_{1q})]\}; \quad P_{2q} = 1 - P_{1q}$$

and also, that we observed the choices and values in Table 8.1:

Table 8.1 Observed choices and values for Example 8.2

Observation (q)	Choice	x_{1q}	x_{2q}
1	1	5	3
2	1	1	2
3	2	3	4

In this case for any given value of θ, the log-likelihood function for the sample is given by:

$$l(\theta) = \log(P_{11}) + \log(P_{12}) + \log(P_{23})$$

and replacing the values we obtain:

$$l(\theta) = 10\theta - \log(e^{5\theta} + e^{3\theta}) - \log(e^{\theta} + e^{2\theta}) - \log(e^{3\theta} + e^{4\theta})$$

Figure 8.1 shows the results of plotting $l(\theta)$ for different values of θ.

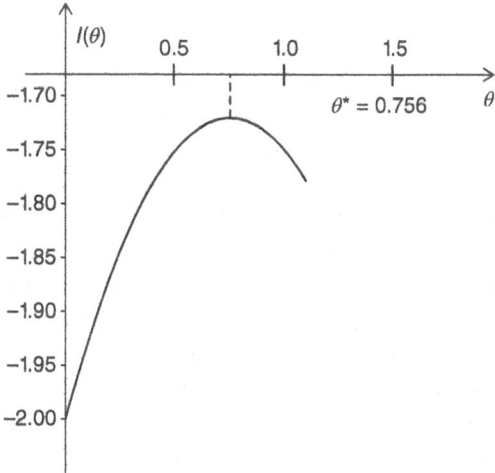

Figure 8.1 Variation of $l(\theta)$ with θ

The optimum, $\theta^* = 0.756$, allows us to predict the probabilities shown in Table 8.2:

Table 8.2 Predicted probabilities for Example 8.2

Observation (q)	P_{1q}	P_{2q}
1	0.82	0.18
2	0.32	0.68
3	0.32	0.68

Therefore, if we adopt the criterion that individuals are assigned to that option which has maximum utility, this would result in an incorrect prediction for the second observation.

We mentioned that the maximum likelihood parameters $\boldsymbol{\theta}^*$ are asymptotically distributed Normal with covariance matrix $\mathbf{S}^2$. In general, the well-understood properties of the maximum likelihood method for well-behaved likelihood functions allow, as in multiple regression, a number of statistical tests which are of major importance:

8.4.1.1 The t-test for Significance of any Component θ_k^* of $\boldsymbol{\theta}^*$

Equation (8.14) implies that θ_k^* has an estimated variance s_{kk}^2, where $\mathbf{S}^2 = \{s_{kk}^2\}$, which is calculated during estimation. Thus, if its mean $\theta_k = 0$, then:

$$t = \theta_k^*/s_{kk} \tag{8.15}$$

has a standard Normal distribution $N(0,1)$. For this reason, it is possible to test whether θ_k^* is significantly different from zero (it is not exactly a t-test as we are taking advantage of a large-sample approximation and t is tested with the Normal distribution). Sufficiently large values of t (typically bigger than 1.96 for 95% confidence levels in the case of two-sided test) lead to the rejection of the null hypothesis $\theta_k = 0$ and hence to accepting that the kth attribute has a significant effect.

The variable selection process followed during the specification searches of discrete choice models normally considers both formal statistical tests, such as the above one, and more informal (but even more important) tests, such as examining the sign of the estimated coefficients to judge whether they conform to *a priori* notions or theory. In this sense, it is worth noting that rejection of a variable with a proper sign crucially depends on its importance; for example, let us note that the set of available explanatory variables can be usefully divided into two classes:

- highly relevant or policy variables, which have either a solid theoretical backing and/or are crucial to model forecasting;
- other explanatory variables, which are either not crucial for policy evaluation (for example, gender), or for which there are no theoretical reasons to justify or reject their inclusion.

Table 8.3 depicts the cases that might occur when considering the possible interactions in the above framework, and the solutions recommended by current practice. Consider first the case of rejecting a variable of type Other with correct sign; this may depend on its significance level (i.e. it may only be significant at the 85% level), and usual practice is to leave it out if it is not significant at the 80% level.

Table 8.3 Variable selection cases

		Variable	
		Policy	**Other**
Correct sign	Significant	Include	Include
	Not significant	Include	May reject
Wrong sign	Significant	Big problem	Reject
	Not significant	Problem	Reject

Current practice also recommends including a relevant (i.e. Policy type) variable with the correct sign even if it fails any significance test. The reason is that the estimated coefficient is the best approximation available for its real value; the lack of significance may just be caused by an insufficient number of observations in the dataset.

Variables of the Other class with a wrong sign are always rejected; however, as variables of the Policy type must be included at almost any cost, current practice dictates in their case model re-estimation, fixing their value to an acceptable one obtained in a study elsewhere. This will be an easy task if the variable is also non-significant, but might be very difficult otherwise as the fixed value will tend to produce important changes in the rest of the model coefficients.

Let us now move to consider the role of socio-economic variables like gender, age, profession, and occupation in discrete choice models. The usual way of introducing these variables was as additive constants, to one or more of the utilities of the alternatives (but not to all, unless they had specific coefficients), based on the modeller's experience and common sense, as in:

$$V_{1q} = \alpha t_{1q} + \beta c_{1q} + \gamma f_{1q} + \dots + \sum_l \theta_l \cdot s_{lq}$$

$$V_{2q} = \alpha t_{2q} + \beta c_{2q} + \gamma f_{2q} + \dots$$

(8.16)

where, for example, t is time, c is cost, f is frequency, and the dummy variables s_{lq} represent socio-economic characteristics of the individuals q. In this case the socio-economic data served to improve the explanation of choice but did not provide any bonus in terms of using the model to estimate subjective values or willingness-to-pay, that is, the ratio of the parameters of time and cost (Gaudry et al. 1989). It is also normally found that very few of these Other type variables provide enough explanation to be kept in the models.

An alternative and much better procedure is to parameterise the coefficients of each attribute in the model using socio-economic variables; in this case (8.16) changes to:

$$V_{iq} = \left(\alpha_0 + \sum_l \alpha_l s_{lq} \right) t_{iq} + \left(\beta_0 + \sum_l \alpha_l s_{lq} \right) c_{iq} + \left(\gamma_0 + \sum_l \gamma_l s_{lq} \right) f_{iq} \quad (i = 1, 2) \quad (8.17)$$

Now, dummy variables s_{lq} refer to the socio-economic characteristic l (i.e. gender) of individual q. This is both a simple and interesting manner of incorporating socio-economic variables, while at the same time helping in computing value functions that vary for each individual. Fowkes and Wardman (1988) proposed this method as a way of segmenting individual tastes. Equation (8.17) states that given the characteristics of the individual, different coefficients will be obtained for a given attribute; note that the same socio-economic variable can appear in the expression corresponding to each coefficient. And note how this formulation does not imply that tastes are randomly distributed in the population; on the contrary, it assumes that the taste parameters (α, β, and χ) depend on the individual characteristics in a deterministic manner; it has been popularised as *systematic taste variations* (Rizzi and Ortúzar 2003). This parameterisation allows for the incorporation of taste heterogeneity in an economical way, using computer programs widely available, rather than having to rely on a more complex function such as the Mixed Logit model.

Example 8.3   Table 8.4 presents two models. The first uses the method explained in Eq. (8.16) and the second uses the new method of Eq. (8.17). The sample size was 1631 SP observations (Rizzi and Ortúzar 2003) about route choice in the presence of the following attributes: accident risk, toll charge and travel time.

The socio-economic variables considered were sex (one for males), age (three dummies, with value one if the person's age was in the range considered) and night/day (one if the person travelled by day) in the case of the accident risk variable; for the toll variable the only SE variable that resulted significant was high income (one if the respondent's income was high) These binary variables were entered in the utility function of the safest route in the first model, but were assumed to interact with the base coefficients of either risk or toll in the case of model 2.

Looking at the results, it is obvious that the more flexible parameterisation of model 2 is superior to the traditional way of incorporating SE variables. Note how the results suggest that women value safety more than men, as do people with progressively older age; on the other hand, if the trip takes place at night, the value of safety also should increase according to model 2. Finally, it is worth noting that the marginal utility of income (i.e. the toll coefficient with the opposite sign) correctly decreases for high-income individuals.

Table 8.4 Alternative ways of entering socio-economic variables

Variables (t-ratios)	Model 1	Model 2
Risk of death	-2.41×10^5	-2.18×10^5
	(−5.6)	(−3.4)
Sex	−0.4233	1.29×10^5
	(−3.3)	(2.9)
Age$_1$ (30–49)	0.4605	-1.76×10^5
	(3.5)	(−3.5)
Age$_2$ (50–65)	1.02	-3.75×10^5
	(5.8)	(−6.0)
Age$_3$ (>65)	1.48	-5.49×10^5
	(2.8)	(−3.0)
Day/night	0.2097	-8.45×10^4
	(2.5)	(−3.1)
Travel time (h)	−3.318	−3.738
	(−13.9)	(−14.0)
Toll charge (US$)	−0.702	−0.826
	(−9.9)	(−10.8)
High income	—	$4. \times 10^{-4}$
		(3.4)
ρ^2 (c)	0.1545	0.1703

8.4.1.2 The Likelihood Ratio Test

A number of important model properties may be expressed as linear restrictions on a more general linear in the parameters model. Some important examples are:

1) Are attributes generic? As mentioned in Section 7.3, there are two main types of explanatory variables, generic, and specific; the former variables have the same weight or meaning in all alternatives, whereas the latter have a different, specific, meaning in each of the choice alternatives and therefore can take on a zero value for certain elements of the choice set.
2) Sample homogeneity. It is possible to test whether or not the same model coefficients are appropriate for two subpopulations (say living north and south of a river). For this a general model using different coefficients for the two populations is formulated and equality of coefficients may be tested as a set of linear restrictions.

Example 8.4 Let us assume a model with three alternatives, car, bus and rail, and the following choice influencing variables: travel time (TT) and out-of-pocket cost (OPC). Then a general form of the model would be:

$$V_{car} = \theta_1 TT_{car} + \theta_2 OPC_{car}$$
$$V_{bus} = \theta_3 TT_{bus} + \theta_4 OPC_{bus}$$
$$V_{rail} = \theta_5 TT_{rail} + \theta_6 OPC_{rail}$$

However, it might be hypothesised that costs (but not times, say) should be generic. This can be expressed by writing this hypothesis as two linear equations in the parameters:

$$\theta_2 - \theta_4 = 0$$
$$\theta_2 - \theta_6 = 0$$

In general, it is possible to express the possibility of having generic attributes as linear restrictions on a more general model. For extensive use of this type of test, refer to Dehghani and Talvitie (1980). Some programs, for example Biogeme (Bierlaire 2016), present as a standard output a covariance/correlation analysis of pairs of estimated parameters θ_i and θ_j, sorted according to a t-test value constructed as follows:

$$t^* = \frac{\theta_i - \theta_j}{\sqrt{\sigma_i^2 + \sigma_j^2 + 2\rho\sigma_i^2\sigma_j^2}}$$

where σ are their standard errors and ρ the correlation coefficient between both estimates; if this test is accepted (i.e. when the value of t^* is less than a critical value, say 1.96 for the typical 95% level) the two parameters are not significantly different and are, thus, candidates for being treated as generic (if they refer to the same attribute in two alternatives)

Because of the properties of the maximum likelihood method, it is very easy to test any such hypotheses, expressed as linear restrictions, by means of the well-known likelihood ratio test (LR). To perform the test the estimation program is first run for the more general case to produce estimates θ^* and a log-likelihood at convergence $l^*(\theta)$. It is then run again to attain estimates θ_r^* of θ and the new log-likelihood at maximum $l^*(\theta_r)$ for the restricted case. Then, if the restricted model under consideration is a correct specification, the LR statistic,

$$-2\{l^*(\theta_r) - l^*(\theta)\}$$

is asymptotically distributed χ^2 with r degrees of freedom, where r is the number of linear restrictions; rejection of the null hypothesis implies that the restricted model is erroneous. It is important to note that to carry out this test we require one model to be a restricted or nested version of the other. Train (1977) offers examples of using this test to study questions of non-linearity, non-generic attributes, and heterogeneity. Horowitz (1982) has discussed the power and properties of the test in great detail and should be consulted for further reference.

8.4.1.3 The Overall Test of Fit

A special case of likelihood ratio test is to verify whether the estimated model is superior to a model where all the components of θ are equal to zero. This model is known as the equally likely (EL) model and satisfies:

$$P_{jq} = 1/N_q$$

with N_q the choice set size of individual q. The test is not helpful in general because we already know that a model with alternative-specific constants (ASC) will reproduce the data better than a purely random function. For this reason, a more rigorous test of this class is to verify whether all variables, except the ASC, are 0. This better *reference* or *null* model is the market shares (MS) model, where all the explanatory variables are 0 but the model has a full set of ASC; in this case we get:

$$P_{jq} = MS_j$$

where MS_j is the market share of option A_j.

Let us first look at the test for the EL model because it is simpler than that for the MS model. Consider a model with k parameters and with, as usual, a log-likelihood value at convergence of $l^*(\theta)$, and denote by $l^*(0)$ the log-likelihood value of the associated EL model; then under the null hypothesis $\theta = 0$ we have that the LR statistic:

$$-2\{l^*(0) - l^*(\theta)\}$$

is distributed χ^2 with k degrees of freedom; therefore, we can choose a significance level (say 95%) and check whether LR is less than or equal to the critical value of $\chi^2(k, 95\%)$, in which case the null hypothesis would be accepted.

However, we already hinted that the test is weak because as it is always rejected, it only means that the parameters θ explain the data better than a model with no significant explanatory power. Actually, the best feature of this test is its low cost as $l^*(0)$ does not require a special program run since it is usually computed as the initial log-likelihood value by most search algorithms.

To carry out the test with the market shares model, we must compute $l^*(C)$, it's log-likelihood value at convergence; if there are $(k - c)$ parameters that are not specific constants, the appropriate value of LR is compared with $\chi^2(k - c, 95\%)$ in this case. In general, an extra run of the estimation routine is required to calculate $l^*(C)$ except for models where all individuals face the same choice set, in which case it has the following closed form equation:

$$l^*(C) = \sum_j Q_j \log (Q_j/Q) \tag{8.18}$$

where Q_j is the number of individuals choosing option A_j.

Figure 8.2 shows the notional relation between the values of the log-likelihood function, for the set of parameters that maximise it, $l^*(\theta)$, for the two previous models, $l^*(0)$ and $l^*(C)$ respectively, and for a fully saturated (perfect) model with an obvious value $l^* = 0$.

$l^*(0)$ $l^*(C)$ $l^*(\theta)$ $l^* = 0$

Figure 8.2 Notional relation between log-likelihood values

8.4.1.4 The ρ^2 Index

Although it is not possible to build an index such as R^2 in this case, it is always interesting to have an index that varies between 0 (no fit) and 1 (perfect fit) in order to compare alternative models. An index that satisfies some of the above characteristics was initially defined as:

$$\rho^2 = 1 - \frac{l^*(\theta)}{l^*(0)} \tag{8.19}$$

However, although its meaning is clear in the limits (0 and 1) it does not have an intuitive interpretation for intermediate values as in the case of R^2; in fact, values around 0.4 are usually considered excellent fits.

Because a ρ^2 index may in principle be computed relative to any null hypothesis, it is important to choose an appropriate one. For example, it can be shown that the minimum values of ρ^2 in (8.19), in models with specific constants, vary with the proportion of individuals choosing each alternative. Taking a simple binary case, Table 8.5 shows the minimum values of ρ^2 for different proportions choosing option 1 (Tardiff 1976). It can be seen that ρ^2 is only appropriate when both options are chosen in the same proportion.

Table 8.5 Minimum ρ^2 for various relative frequencies

Sample proportion selecting the first alternative	Minimum value of ρ^2
0.50	0.00
0.60	0.03
0.70	0.12
0.80	0.28
0.90	0.53
0.95	0.71

These values mean, for example, that a model estimated with a 0.9/0.1 sample yielding a ρ^2 value of 0.55, would be undoubtedly much weaker than a model yielding a value of 0.25 from a sample with an equal split. Fortunately, Tardiff (1976) proposed a simple adjustment that allows us to solve this difficulty; it consists of calculating a similar index but with respect to the market shares model:

$$\bar{\rho}^2 = 1 - \frac{l^*(\theta)}{l^*(C)} \tag{8.20}$$

This *corrected* ρ^2 lies between 0 and 1, is comparable across different samples and is related to the χ^2 distribution.

Ben-Akiva and Lerman (1985) proposed another correction to the ρ^2 index; this is usually referred as *adjusted* ρ^2 and it is defined as:

$$\rho_{adj}^2 = 1 - \frac{l^*(\theta) - K}{l^*(0)}$$

which takes into account the number of parameters estimated. However, it is still based on the likelihood of the equally likely model so it maintains the main problems of the original ρ^2. A more useful version of this idea would be to apply the correction to the corrected ρ^2 index, where K should, of course, be replaced by $K - C$.

8.4.1.5 The Percentage Right or First Preference Recovery (FPR) Measure

This is an aggregate measure that simply computes the proportion of individuals effectively choosing the option with the highest modelled utility. FPR is easy to understand and can readily by compared with the chance recovery (CR) given by the equally likely model:

$$CR = \sum_q (1/N_q)/Q$$

Note that if all individuals have a choice set of equal size N, then $CR = 1/N$. FPR can also be compared with the market share recovery (MSR) predicted by the best null model (Hauser 1978):

$$MSR = \sum_{Aj} (MS_j)^2$$

Disadvantages of the index are exemplified by the fact that although an FPR of 55% may be good in general, it is certainly not so in a binary market; also, an FPR of 90% is normally good in the binary case, but not if one of the options has a market share of 95%. Another problem with the index, worth noting in the sense of not being an unambiguous indicator of model reliability, is that too high a value of FPR should lead to model rejection as well as a too low value; to understand this point it is necessary to define the expected value of FPR for a specific model as:

$$ER = \sum_q P_q \tag{8.21}$$

where P_q is the calculated (maximum) probability associated with the best option for individual q. Also, because FPR is an independent binomial random event for individual q, occurring with probability $1/N_q$ in the CR case and P_q in the ER case, their variances are given respectively by:

$$Var\,(CR) = (1/N_q)(1 - 1/N_q) \tag{8.22}$$

and

$$Var(FPR) = P_q(1 - P_q) \tag{8.23}$$

Thus, a computed value of FPR for a given model can be compared with CR and ER; if the three measures are relatively close (given their estimated variances) the model is *reasonable but uninformative*; if FPR and ER are similar and larger than CR, the model is *reasonable and informative*; finally, if FPR and ER are not similar, the model does not explain the variation in the data and should be rejected whether FPR is larger or smaller than ER (see Gunn and Bates 1982).

8.4.1.6 *Working with Validation Samples*

As we mentioned in Chapter 5, the performance of any model should be judged against data other than that being used to specify and estimate it and, ideally, taken at another point in time (perhaps after the introduction of a policy in order to assess the model's response properties). This is true for any model. We will define a subsample of the data, or preferably, another sample *not used* during estimation, as a *validation sample*.

We will first briefly describe a procedure to estimate the minimum size of such a validation sample (ideally to be subtracted from the total sample available for the study) conditional on allowing us to detect a difference between the performance of two or more models, when there is a true difference between them. The method, which is based on the FPR concept, was devised by Hugh Gunn and first applied by Ortúzar (1983).

Consider the layout shown in Table 8.6, where n_{ij} is the number of individuals assigned to cell (i, j).

Table 8.6 2 × 2 table for comparing models

		Model 2	
		Not FPR	FPR
Model 1	Not FPR	n_{11}	n_{12}
	FPR	n_{21}	n_{22}

For all individuals in a validation sample, choice probabilities and FPR are calculated for each of two models under investigation, and the cells of the table are filled appropriately (e.g. assigning to cell (1,1) if there is no FPR in both models, and so on). We are interested in the null hypothesis that the probabilities with which individuals fall into cells (1,2) and (2,1) are equal, for in that case the implication on simple FPR is that the two models are equivalent; on this null hypothesis the following statistic M is distributed χ^2 with one degree of freedom (see Foerster 1979):

$$M = \frac{(n_{12} - n_{21})^2}{n_{12} + n_{21}} \tag{8.24}$$

Thus, a test of the equivalence of the two models in terms of FPR is simply given by computing M and comparing the result with χ^2 (1, 95%); if M is less than the appropriate critical value of χ^2 (3.84 for the usual 95% confidence level) we cannot reject the null hypothesis and we conclude that the models are equivalent on these terms.

Given this procedure, we can select whichever level of confidence seems appropriate for the assertion that the two models under comparison differ in respect of the expected number of FPR. This gives us control over the fraction of times that we will incorrectly assert a difference between similar models. As usual, the aim of choosing a particular sample size is to ensure a corresponding control over the proportion of times we will make the other type of error, namely incorrectly concluding that there is no difference between different models.

Now, to calculate the probability of an error of the second type, we need to decide what the minimum difference is that we would like to detect; with this, we can calculate the sample size needed to reduce the chance of errors of the second kind to an acceptable level for models that differ by exactly this minimum amount, or more.

Example 8.5 Consider the case of two models such that, on average, model 2 produces 10 extra FPR per 100 individuals modelled as compared to model 1. Note that here it does not matter whether this arises as a result of model 1 having 20% FPR and model 2 having 30% FPR, or the first 80% and the second 90%; in other words, both models can be inadequate.

In this simple case n_{21} is zero and M simply becomes n_{12}. If we are ensuring 95% confidence that any difference we establish could not have arisen by chance from equivalent models, we will compare n_{12} with the χ^2 value for one degree of freedom (3.84); for any given sample size n, the probability that r individuals will be assigned to cell (1, 2) is simply the binomial probability $\binom{n}{r} p^r (1-p)^{(n-r)}$ where p denotes the probability of an individual chosen at random being assigned to cell (1, 2), that is, the minimum difference we wish to detect.

Given n and taking $p = 0.05$ as usual, we can calculate the probabilities of 0, 1, 2, and 3 individuals being assigned and sum these to give the total probability of accepting the null hypothesis (i.e. committing an error of the second kind). Table 8.7 gives the resulting probabilities for different sample sizes.

Table 8.7 Type II error probabilities and sample size

Sample size	Minimum difference 5% Prob (error II)
50	0.75
100	0.26
150	0.05
200	0.01
250	0.00

It is clear that the required validation sample size needs to be relatively large given that typical estimation data sets have only a few hundred observations. Also recall that Table 8.4 is for the simple case of one model being better than or equal to the other in each observation; the method of course may easily be extended to cases where both the (1, 2) and (2, 1) cells have non-zero probability.

An especially helpful feature of validation samples is that provided their size is adequate the issue of ranking non-nested models (see Section 8.4.3) is easily resolved, as likelihood ratio tests can be performed on the sample regardless of any difference in model structure parameters. This is because the condition of one model being a generalisation of the other is only required for tests with the same data used for estimation (Gunn and Bates 1982; Ortúzar 1983).

Example 8.6 Let us assume that we are interested in an option with low market share at present and that we have two model specifications (models A and B) for a six-alternative choice situation. The two models have similar FPR but one always predicts that option badly and the others a bit better, compared with the second model that gives reasonable predictions for all options. In this case, we can use a validation sample and estimate, for each individual in it, the choice probabilities for each option by the two models; the alternative actually chosen is, as usual, an observed piece of information. In order to investigate the consistency of the predictions with the data, we can compare them with proportions calculated from the sample.

Table 8.8 presents the values N_{ij} and O_{ij} (where i indicates a probability band and j an option). N_{ij} is the number of observations to which the model assigned a probability in band i to alternative A_j; O_{ij} is the observed number of choices of option A_j to which the model assigned a probability in that band.

Table 8.8 Modelled choices by probability band

Predicted probability band (*i*)	0–0.1		0.1–0.2		0.2–0.3		...		0.9–1.0	
Alternative (*j*)	N_{1j}	O_{1j}	N_{2j}	O_{2j}	N_{3j}	O_{3j}	...	...	N_{10j}	O_{10j}
Model A										
1	0	0	8	0	11	0	...	...	0	0
2	40	0	0	0	0	0	...	...	0	0
3	94	0	0	0	0	0	...	...	0	0
...	...	...	...	...	...	...	...	...	...	...
...	...	...	...	...	...	...	...	...	...	...
6	55	6	11	3	58	14	...	...	0	0
Total		6		6		24	...	...		0
Model B										
1	9	0	5	0	0	0			0	0
2	36	0	4	0	0	0			0	0
...	...	...	...	...	...	...	...	...	...	...
...	...	...	...	...	...	...	...	...	...	...
...	...	...	...	...	...	...	...	...	...	...
6	43	3	44	7	18	8			0	0
Total		6		11		13	...	...		15

Table 8.9 builds on the previous table and presents the values E_{ij} and O_{ij}, where E_{ij} is given by:

$$E_{ij} = N_{ij} \times \bar{p}_i$$

which corresponds to the expected value of the number of individuals choosing option A_j with probability in the band i, associated with a mean probability $\bar{p}_i$. For example, in the case highlighted in the table we have that $E_{36} = 58 \times 0.25 = 14.5$, as 0.25 is the mean value of probability band 3 (i.e. between 0.2 and 0.3).

Table 8.9 Expected proportions by probability band

Predicted probability band (i)	0–0.1		0.1–0.2		0.2–0.3		...		0.9–1.0	
Alternative (j)	E_{1j}	O_{1j}	E_{2j}	O_{2j}	E_{3j}	O_{3j}	...	...	E_{10j}	O_{10j}
Model A										
1	0	0	1.2	0	2.75	0	...	...	0	0
2	2	0	0	0	0	0	...	...	0	0
3	4.7	0	0	0	0	0	...	...	0	0
...	...	...	...	...	...	...	...	...	...	...
...	...	...	...	...	...	...	...	...	...	...
6	2.75	6	1.65	3	14.5	14	...	...	0	0
Total	9.45	6	4.05	6	29	24	...	...	0	0
Model B										
1	0.45	0	0.75	0	0	0			0	0
2	1.8	0	0.6	0	0	0			0	0
3	4.05	0	1.65	0	0.5	0	...	...	0	0
...	...	...	...	...	...	...	...	...	...	...
...	...	...	...	...	...	...	...	...	...	...
6	2.15	3	6.6	7	4.5	8	...	...	0	0
Total	9.5	6	11.7	11	7.5	13	...	...	15.2	15

To compare the values in Table 8.9 one may apply a χ^2 test defined as follows (Gunn and Bates 1982):

$$\chi^2_{\text{cell}} = \sum_{ij} \frac{(O_{ij} - E_{ij})^2}{E_{ij}} \text{ with } ij - 1 \text{ degrees of freedom}$$

It is possible in principle to apply the test to each cell in the matrix if $E_{ij} > 5$, as the test is not valid otherwise. For this reason, and given the limited size of validation samples, it may be necessary to aggregate cells but, unfortunately, there are no clear-cut methods to do it. The reader may check that different aggregation strategies lead to different results.

A less informative case, but one that it is usually possible to carry out, is to compare expected and observed totals for each column in Table 8.9, where $E_i = \sum_j E_{ij}$ and $O_i = \sum_j O_{ij}$ respectively, using the index:

$$\chi^2_{\text{FPR}} = \sum_{i=1}^{m} \frac{(O_i - E_i)^2}{E_i} \tag{8.25}$$

where m is the number of columns with $E_i > 5$. In this case the appropriate number of degrees of freedom is $m - 1$, and χ^2_{FPR} may be compared with the critical value $\chi^2_{m-1;\,0.95}$. If $\chi^2_{\text{FPR}} < \chi^2_{m-1;\,0.95}$ then the null hypothesis that the model is consistent with the data is accepted.

If, according to the previous test, two or more models are acceptable then it is possible to discriminate between them using the direct likelihood ratio test (Gunn and Bates 1982; Ortúzar 1983):

$$\frac{L_A}{L_B} = \frac{\prod_i \bar{p}_i^{O_i}(\text{model A})}{\prod_i \bar{p}_i^{O_i}(\text{model B})} \tag{8.26}$$

If we applied this test to the data of Table 8.9, we would get:

$$\frac{L_A}{L_B} = \frac{(0.05)^6 \times (0.15)^6 \times (0.25)^{24} \times \ldots \times (0.95)^0}{(0.05)^6 \times (0.15)^{11} \times (0.25)^{13} \times \ldots \times (0.95)^{15}} = 0.0455$$

and we would say that the data are approximately 22 times (that is 1/0.0455) more probable under model B that under model A. This means that we would prefer the second model although both yield predictions which are consistent with the data.

8.4.2 Estimation of Models from Choice-based Samples

As mentioned in Chapter 2, estimating a model from a choice-based sample may be of great interest because the data-collection costs are often considerably lower than those for typical random or stratified samples. The problem of finding a tractable estimation procedure possessing desirable statistical properties is not an easy one, and the state of practice has been provided by the excellent papers of Cosslett (1981) and Manski and McFadden (1981).

It has been found, in general, that maximum likelihood estimators specific for choice-based sampling are impractical, except in very restricted circumstances, due to computational intractability. However, if it can be assumed that the analyst knows the fraction of the decision-making population selecting each alternative, then a tractable method can be introduced. The approach modifies the familiar maximum likelihood estimator of random sampling by weighting the contribution of each observation to the log-likelihood by the ratio Q_i/S_i, where the numerator is the fraction of the population selecting option A_i and the denominator is the analogous fraction for the choice-based sample.

Manski and Lerman (1977) have shown that the un-weighted random-sample maximum likelihood estimator is generally inconsistent in the case of choice-based samples and in most choice models this inconsistency affects all parameter estimates. However, as we saw in Section 7.3.2, for simple MNL models with a full set of alternative-specific constants the inconsistency is fully confined to the estimates of these dummy variables. In this case, the estimates obtained without weighting are more efficient than the estimates obtained with the weighted sample. Therefore, it is good practice to estimate the parameters of the MNL model without weighting the sample, and to correct the constants afterwards. Bierlaire et al. (2008) show that this property does not apply to Generalized Extreme Value models (including Nested Logit and Cross-Nested Logit models). They propose a simple estimator for these models that does not require the weighting of the sample nor knowledge of the actual market shares.

8.4.3 Estimation of Hybrid Choice Models with Latent Variables

Originally, Hybrid Choice models were estimated sequentially, in two stages. First, the MIMIC model (Bollen 1989) discussed in Section 7.7.5.1 was solved to obtain parameter estimators for the equations relating the latent variables with the explanatory variables and the perception indicators. Then, using these parameters in Eq. (7.61), expected values for the latent variables of each individual and alternative were obtained. In turn, these expected latent variable values were added to the set of typical variables of the discrete choice model, as in Eq. (7.63), and their parameters estimated together with those of the traditional variables in a second stage.

This method has the disadvantage of not using all the available information jointly, but its application was clear and simple (Ashok et al. 2002; Vredin-Johansson et al. 2006; Raveau et al. 2010). Nevertheless, a potentially serious problem of the approach was that it could result in biased estimators for the parameters involved (Bollen 1989); similarly, it was noted that the method tended to underestimate the parameters' standard deviations, resulting in estimators with a statistical significance higher than their real contribution to the model.

In the current simultaneous estimation approach, we seek to maximise the likelihood of the probability of replicating the individual choices based on the representative utility proposed by the modeller; that is, Prob $(g_{iq}|V_{iq})$, where g_{iq} is equal to one if individual q chooses option A_i. Now, recall Eq. (7.63) for the hybrid discrete choice model:

$$V_{iq} = \sum_k \theta_{ik} \cdot x_{ikq} + \sum_l \beta_{il} \cdot \eta_{ilq}$$

where as usual, $\mathbf{x}$ are level-of-service attributes, $\mathbf{\eta}$ are the latent variables, to be estimated jointly with the structural (7.61) and measurement (7.62) equations, and $\mathbf{\theta}$ and $\mathbf{\beta}$ are parameters to be estimated.

From (7.63), the conditional probability can be expressed in terms of the variables and parameters of the discrete choice model. However, as the latent variables are not observed it is necessary to integrate over their whole variation range, conditioning them by their explanatory variables. Thus, the choice probability is given by (8.27), where $h(\cdot)$ is the probability density function of the latent variables:

$$\text{Prob}\left(g_{iq}|x_{ikq}, s_{iqr}, \theta_{ik}, \beta_{il}, \alpha_{ilr}\right) = \int_{\eta_{ilq}} \text{Prob}\left(g_{iq}|x_{ikq}, \eta_{ilq}, \theta_{ik}, \beta_{il}\right) \cdot h\left(\eta_{ilq}|s_{iqr}, \alpha_{ilr}\right) \cdot d\eta_{ilq} \quad (8.27)$$

and $\mathbf{s}$ and $\boldsymbol{\alpha}$ are the socio-economic variables (and their parameters) explaining the latent variables in structural Eq. (7.61). However, to estimate the model it is necessary also to introduce the information provided by the perception indicators $\mathbf{y}$ in the measurement Eq. (7.62), since otherwise the model would not be identifiable. The indicators are not explanatory variables of the model; instead, they are endogenous to the latent variables. This implies that the choice probability used during estimation is given by (8.28), where $f(\cdot)$ is the probability density function of the indicators.

$$\begin{aligned}
&\text{Prob}\left(g_{iq}, y_{ipq} \mid x_{ikq}, s_{iqr}, \theta_{ik}, \beta_{il}, \alpha_{ilr}, \gamma_{ipq}\right) \\
&= \int_{\eta_{ilq}} \text{Prob}\left(g_{iq} \mid x_{ikq}, \eta_{ilq}, \theta_{ik}, \beta_{il}\right) \cdot f\left(y_{lpq} \mid \eta_{ilq}, \gamma_{ipq}\right) \cdot h\left(\eta_{ilq} \mid s_{iqr}, \alpha_{ilr}\right) \cdot d\eta_{ilq}
\end{aligned} \quad (8.28)$$

Once the functional form of the discrete choice model is defined, the simulated maximum likelihood method can be used for estimation (Bolduc and Alvarez-Daziano 2009; Bolduc and Giroux 2005); we will examine the method in depth in Section 8.5.2, but as we will see there are difficult practical problems in this case due to the particular form of the estimation problem (see Hess and Rose 2009).

Example 8.7 A recent urban mode choice study considered ten transport modes, both pure and combined, for journey to work trips: car-driver, car-passenger, shared taxi, bus, underground, combinations of the previous four with underground and shared taxi/bus. For each available mode, information was precisely measured about walking, waiting and in-vehicle time, trip cost, and number of transfers made. Regarding users' information, socio-economic variables such as age, sex, education level and income, were obtained. Respondents were also asked to evaluate different characteristics of the modes, generating perception indicators to allow the inclusion of latent variables in the model.

Three latent variables were considered: *accessibility/comfort*, *reliability*, and *safety*; the effects of each were captured through seven perception indicators based on the evaluation of several aspects of the pure modes: (i) safety regarding accidents, (ii) security regarding theft, (iii) ease of access, (iv) comfort during the trip, (v) availability of suitable information, (vi) possibility of calculating the travel time prior to the trip, and (vii) possibility of calculating the waiting time prior to the trip.

Four explanatory variables were finally included in the MIMIC model: education level, age, sex, and monthly income. The MIMIC model's structural relations were studied using factor analysis to guarantee their correct specification. Figure 8.3 illustrates the results of this process (Raveau et al. 2010).

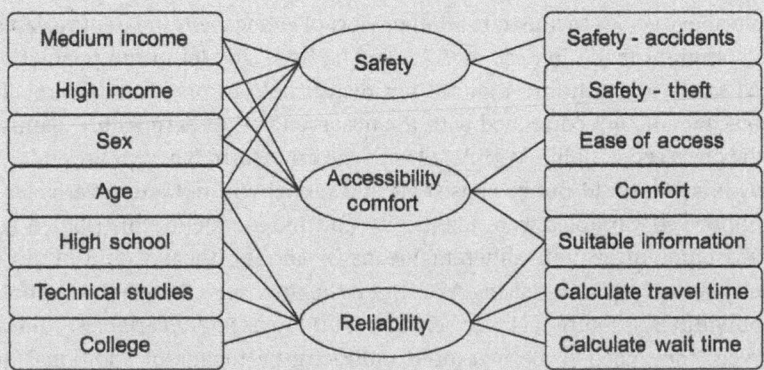

Figure 8.3 Latent variables model relationships

The representative utility function included the number of transfers during the trip as well as the different time variables obtained from the survey; in the case of travel time, systematic taste variations according to the respondent's sex were found (the variable Sex, took the value one for males). Travel cost was standardized by the individual's wage rate. This was the best specification obtained among several formulations studied. The model also included a complete set of alternative specific constants (ASC).

Table 8.10 presents the results together with those of an equivalent MNL model without latent variables (the ASC are not reported, interested readers are referred to Raveau et al. 2010).

The signs of all estimated parameters are consistent with microeconomic theory. In the Hybrid Choice (HC) model all variables are statistically significant at least at the 90% confidence level, but not all variables are statistically significant in the MNL model without latent variables; the waiting time variable is especially problematic.

The HC model implies that men are slightly more sensitive to travel time than women, but the model without latent variables shows precisely the opposite effect and the large difference (in magnitude) of the marginal utility of travel time for women is certainly suspect. The ASC obtained for the model without latent variables were more significant than those obtained

Table 8.10 Hybrid Choice Model estimation results

Parameter	MNL	Hybrid Choice Model
Cost/wage rate	−0.027 (−8.13)	−0.032 (−7.32)
Travel time	−0.033 (−4.82)	−0.006 (−4.67)
Sex interaction with time	0.030 (2.98)	−0.001 (−3.01)
Waiting time	−0.009 (−0.53)	−0.015 (−1.69)
Walking time	−0.016 (−1.80)	−0.022 (−2.89)
No. of transfers	−1.110 (−8.20)	−1.102 (−8.21)
Accessibility-comfort	—	0.622 (3.79)
Reliability	—	0.441 (2.70)
Safety	—	0.613 (1.87)

for the HC model. This is an expected result since the model without latent variables has fewer explanatory variables, and so the constants must explain (as far as possible) the missing information according to the individual choice patterns.

The example above serves to illustrate another view of *endogeneity* in latent variable modelling (recalled the comment made in Section 7.7.5.2). The 'true' model in the population has attributes, such as safety and comfort, that are not measurable in practice, are probably relevant in decision making, and are correlated with the observed variables (possibly mainly with cost). This makes the observed variables correlated with the error term (i.e. they are endogenous), and hence their estimates should not be consistent in a model without latent variables.

If these unobserved variables were identically and independently distributed (IID) among modes and attributes (even with different means by mode), there would be no problem as the endogeneity could be resolved using ASC. But if the unobserved variables are not IID among modes and individuals, a solution is precisely to treat them as latent variables. And, as they cannot be observed, they need to be measured indirectly by means of additional questions or 'indicators'.

A note on modelling with latent variables. Prior to discussing the way in which latent variables (LV) may be considered in a HC model, it is necessary to understand the difference between *attitudes*, which are individual-specific, and *perceptions*, which are alternative-specific latent attributes. The former may be considered a tendency to act in a particular way based on the individual's experience and temperament (Pickens 2005). Therefore, indicators representing individual-specific latent constructs should be considered constant for all alternatives (i.e. they depend only on the individuals), and one set of attitudinal indicators should be enough to describe an individual. Contrariwise, as perceptions depend on both the person and the stimuli, perceptual indicators should be a function of both the individual and the alternatives. Therefore, to analyse the role of perceptions and perceptual indicators, it should be necessary to gather a set of indicators for every alternative in the individual's choice set.

This last issue can lead to a significant increase in the amount of information to be collected, as normally the alternatives consist of attribute packages that are subject to variations. Therefore, certain simplifying assumptions are required. First, it may be assumed that certain attributes will not affect the way in which a given alternative is perceived and, consequently, that dimension may be excluded from the design (e.g. price may not affect accessibility indicators). Second, it may be assumed that the model is valid across individuals (avoiding the need that all state their perceptions for every combination of attributes, as long as some are faced with the remaining combinations).

On another hand, the treatment of both kinds of variables (attitudes and perceptions) in a DM should not be equal, as some attitudes, as socio-economic variables (which affect the way in which the attributes of the alternatives are perceived), should be incorporated through systematic taste variations (i.e. as in Eq. 8.17) and not linearly into the utility function. Bahamonde-Birke et al. (2017a) distinguish three kinds of LV:

(a) Non-alternative related individual-specific latent attributes (e.g. attitudes). Most researchers working with HC models consider this kind of variable (Bolduc and Alvarez-Daziano 2009; Abou-Zeid et al. 2010; Jensen et al. 2014, among many others). Note that even when using variables that may be understood as perceptions, such as comfort and security, the modeller is, in fact, dealing with a non-alternative related attitude, that stands for the importance assigned by the individual to a given aspect and not to the perception of the alternative. Thus, inferences such as ... 'Alternative A is perceived as more comfortable' would not be accurate, but rather ... 'Individuals caring for comfort favour alternative A', which is not equivalent.

Chorus and Kroesen (2014) have rightly argued that this kind of models do not allow deriving policy implications, as the attitudinal variables are intrinsic characteristics of the individuals (like sex or age) and, therefore, not sensitive to changes in the alternatives. In the same line, they are concerned with the causal relation between attitudes and choices, as attitudes and stated attitudinal indicators may be indeed affected by the individual's choices (e.g. the stated attitude towards the comfort provided by a given alternative may depend on whether the alternative is selected). However, this criticism may be substantially reduced if no direct link can be established between indicators and choices. In that case, both would be an expression of deeper underlying attitudes, such as in the case of environmental attitudes, political views, values, and so on. As these variables resemble socio-economic characteristics, to identify the DM they must be considered together with alternative-specific attributes in the utility function. However, they are typically considered in conjunction with alternative-specific constants (Vredin-Johansson et al. 2006; Bolduc et al. 2008). This restriction may neglect important aspects of the decision process, as individuals with different attitudes towards life may exhibit a different valuation of the attributes of the alternatives, and therefore systematic taste variations should be allowed for.

Furthermore, and similar to socio-economic variables, it is not clear if attitudes should have a linear impact on utility. Therefore, a categorization of the LV should be considered. For instance, it is plausible that a low or intermediate appraisal of security (or safety) may have no effect whatsoever over the decision, but a high concern could lead to a significantly different valuation. If this were the case, treating the variable linearly would not properly reflect individual behaviour. Categorizing the LV offers also significant advantages in terms of flexibility, as it allows estimating different utility functions for every category, resembling a latent class model, but expanding it in order to account for the behavioural information. This categorization may be performed using a latent variable-latent class structure (Hess et al. 2013; Bahamonde-Birke and Ortúzar 2020) or attempting a direct categorization (i.e. a dummy variable taking a positive value if the LV surpasses a certain threshold).

(b) Alternative related individual-specific latent attributes (e.g. attitudes). These variables are similar to the previous ones, with the exception that attitudes are unequivocally related to a given alternative. Thus, they must be considered in conjunction with the alternative-specific constants. As in the previous case, systematic taste variations (within the same alternative) and a possible categorization should also be analysed.

(c) Alternative-specific latent attributes (e.g. perceptions). These variables exhibit a different valuation depending on the alternative considered and, as such, they resemble observed attributes such as price or travel time; hence, both kinds of variables (observed and alternative-specific latent attributes) should be treated in the same fashion. In this case, both alternative-specific and generic estimators may be considered. These variables allow evaluating how changes in alternatives may affect the perceptions, and thus the choices. As in this case the perceptions and indicators are driven by exogenous attributes of the alternatives, causality issues as described by Chorus and Kroesen (2014) may be overcome.

Regarding the model's identifiability, necessary and sufficient conditions have not yet been developed. Therefore, most studies using HC models (notable exceptions are Atasoy et al. 2010; Kamargianni et al. 2014) achieve identification simply by not letting some explanatory variables impact the utility of a given alternative, both directly and through a LV (Bhat 2015). This is a sufficient but restrictive condition and, especially when dealing with perceptual LV, the modeller may be forced to use the same attributes in the structural equations and in the utility functions to represent behaviour properly (e.g. air conditioning may have an effect over perceived comfort, but still have a direct impact on the decision due to other considerations). Under these circumstances identification must be analysed on a case-by-case basis (see Vij and Walker 2014, for a good discussion about the identifiability of HC models).

8.4.4 Comparison of Non-Nested Models

The likelihood ratio (LR) test outlined in Section 8.4.1.2 requires testing a model against a parametric generalisation of itself, that is, it requires the model to be *nested*. Models with utility functions having significantly different functional forms, or models based on different behavioural paradigms, cannot be compared by this test.

It is easy to conceive of situations in which it would be useful to test a given model against another that is not a parametric generalisation of itself. The following example, provided by Horowitz (1982), is very illustrative.

Example 8.8 Consider one model with a representative utility function specified as:

$$V = \theta_1 x_1 + \theta_2 x_2$$

and another with a representative utility function given by:

$$W = \theta_3 x_3 x_4$$

and assume we want to test both models to determine, which explains the data best. Clearly, there is no value of θ_3 that causes V and W to coincide for all values of θ_1 and θ_2 and the attributes $\mathbf{x}$. If both models belong to the same general family, however, it is possible to construct a hybrid function; for example, in our case we could form a model with a measured utility Z containing both V and W as special cases:

$$Z = \theta_1 x_1 + \theta_2 x_2 + \theta_3 x_3 x_4$$

and using a LR tests both models could be compared against the hybrid; the first one would correspond to the hypothesis $\theta_3 = 0$ and the second to the hypotheses $\theta_1 = \theta_2 = 0$.

Horowitz (1982) discusses several other tests at length, including cases where the competing models do not belong to the same general family. But recall that in the presence of a validation sample, the issue may be particularly easily resolved, as discussed by Gunn and Bates (1982).

8.4.5 Correcting for Endogeneity in Discrete Choice Models

As we saw in Chapter 4, endogeneity arises when one or more of the explanatory variables of an econometric model are correlated with the model's error term. This may be caused by measurement/specification errors, omitted attributes, and self-selection, among other reasons (Guevara 2015). The main problem of models affected by endogeneity is that their parameters may be inconsistent, leading to faulty policy analysis, forecasts, conclusions and/or behavioural assessments (Guevara and Ben-Akiva 2006).

The correction of endogeneity has been studied in depth in the case of linear models, as we saw in Section 4.2.1.3. In the case of DM, which are highly non-linear, some research gaps have been closed in recent years (Guevara and Polanco 2016); further, Lurkin et al. (2017) and Guerrero et al. (2020) have successfully applied practical corrections for endogeneity in the transport field.

Several methods have been proposed to correct for endogeneity in DM, such as Berry–Levinsohn–Pakes – BLP (Berry et al. 1995), Latent Variables – LV (Ben-Akiva et al. 2002), Maximum Likelihood (Park and Gupta 2009), Control Function – CF (Petrin and Train 2010), Proxies (Guevara 2015), and Multiple Indicator Solution (Guevara and Polanco 2016). To decide which method is preferable, the analyst must consider the information available, the assumptions he is willing to make, and the associated computational costs. In what follows we will concentrate on the last approach as it has been found advantageous over the other methods in the case of DM (Guevara 2015).

Multiple Indicator Solution (MIS). This approach is based on the use of indicators, such as those coming from surveys designed for knowing respondents' attitudes and/or perceptions. Indicators are typically collected using Likert (1932) or verbal scales (Glerum et al. 2014), and can be *attitudinal* if they represent the characteristic of the individuals toward life (Walker and Ben-Akiva 2002; Bahamonde-Birke et al. 2017a) or *perceptual* if they are exclusively related to the way certain alternatives are perceived (Bolduc and Alvarez-Daziano 2009).

An important advantage of the MIS method is that it allows to gather proper instrumental variables (IV) directly from the respondent, avoiding the need to collect or build them from existing information, which is always hard and sometimes even controversial. The difficulty resides in that the IV must satisfy two conflicting criteria:

- to be exogenous (independent of the error term of the model), and
- to be relevant (strongly correlated with the endogenous variable).

In practice, finding proper IV may be impossible and if they are endogenous or *weak* (i.e. not strongly correlated with the endogenous variable), their use can yield results as bad as when using an endogenous model (see, for example, Guevara 2018; Guerrero et al. 2020).

However, the relative easiness in the collection of IV for the MIS approach comes with a caveat. In theory, to apply the method the indicators are mathematically required to be continuous (Wooldridge 2010); but, since they are typically obtained through Likert scales, they tend to be discrete. To apply the MIS method at least two indicators are needed for each endogenous variable considered (Wooldridge 2010; Guevara and Polanco 2016), as we show in the following example.

Example 8.9 Consider a departure time choice decision represented by a DM with utility function (U_{in}) for alternative i and individual n, given by:

$$U_{in} = \text{ASC}_i + \beta_t t_{in} + \beta_c c_{in} + \beta_x x_{in} + e_{in}, \tag{8.29}$$

where ASC_i is an alternative specific constant, β_t, β_c, and β_x are parameters to be estimated, t_{in} and c_{in} represent, for example, time and cost, x_{in} represents a qualitative attribute (e.g. schedule delay, additional travel time, safety, comfort, or reliability), and e_{in} is an exogenous error term.

Let us assume that t_{in} and c_{in} are known (measurable) attributes whereas the variable x_{in} is unknown to the modeller and correlated with t_{in}. Therefore, the modeller's specification of the DM would be as follows:

$$U_{in} = \text{ASC}_i + \beta_t t_{in} + \beta_c c_{in} + \varepsilon_{in} \tag{8.30}$$

where the new error term ε_{in} contains both $\beta_x x_{in}$ and e_{in}. Thus, as the error term ε_{in} is correlated with t_{in} in (8.30) through x_{in}, the DM will suffer from endogeneity as a result of omitting x_{in}; therefore, by definition, t_{in} is endogenous.

Now, let us suppose that the variable x_{in} and the error terms ε_{in} can explain two indicators (I_{1in} and I_{2in}), as shown:

$$I_{1in} = \alpha_1 + \alpha_{1x} x_{in} + \epsilon_{1in} \tag{8.31}$$

$$I_{2in} = \alpha_2 + \alpha_{2x} x_{in} + \epsilon_{2in} \tag{8.32}$$

For consistency, to apply the MIS method we need also to assume that the pairs of variables (x_{in}, ϵ_{1in}),(t_{in}, ϵ_{1in}), (x_{in}, ϵ_{2in}), (t_{in}, ϵ_{2in}), and $(\epsilon_{1in}, \epsilon_{2in})$ are mutually independent, and that the coefficients α_{1x} and α_{2x} are not null; α_1 and α_2 are intercepts to be estimated.

Although x_{in} in (8.30) is unknown to the analyst, we need to represent its effect in the model and also to correct for the endogeneity resulting from its omission. In this case, the correction is guaranteed because I_{1in} and I_{2in} can be expressed as a linear combination of x_{in} and other elements (α_1, α_2, ϵ_{1in}, and ϵ_{2in}). For this, we can rewrite x_{in} as a function of I_{1in} using (8.31), that is, $x_{in} = \left(\dfrac{1}{\alpha_{1x}}\right)(I_{1in} - \alpha_1 - \epsilon_{1in})$ and by defining $\theta_x = \dfrac{\beta_x}{\alpha_{1x}}$, a new expression for the utility function is given by (8.33):

$$
\begin{aligned}
U_{in} &= \text{ASC}_i + \beta_t t_{in} + \beta_c c_{in} + \beta_x \left[\left(\frac{1}{\alpha_{1x}}\right)(I_{1in} - \alpha_1 - \epsilon_{1in}) \right] + e_{in} \\
&= \text{ASC}_i + \beta_t t_{in} + \beta_c c_{in} + \theta_x I_{1in} \underbrace{- \theta_x(\alpha_1 + \epsilon_{1in}) + e_{in}}_{\omega_{in}}
\end{aligned}
\tag{8.33}
$$

The model in (8.33) attempts to correct for the endogeneity caused by omitting x_{in} in (8.30); however, this modified model suffers from a different source of endogeneity because the term I_{1in} is correlated with ω_{in} through ϵ_{1in}. The variable t_{in} is not endogenous in (8.33) because x_{in} is no longer part of the error term; so, the only endogenous variable is I_{1in}, and to correct for this endogeneity we need another instrumental variable. This may come from the second indicator I_{2in}, which by construction is correlated with I_{1in} only through x_{in}, but it is independent of the error term ω_{in} in (8.33).

In linear models, this final correction for the MIS method is performed through Two Stage Least Squares (Wooldridge 2010), but in DM this is done using the Control Function (CF) method

(Guevara and Polanco 2016). In practice, for the endogenous DM model (8.30), the MIS method may be applied in two-stages as follows:

1) Apply an ordinary least squares (OLS) regression to I_{1in} on I_{2in}, t_{in}, and c_{in}, and obtain the residuals $\hat{\delta}_{in}$. Both t_{in} and c_{in} must be included in this auxiliary regression.
2) Estimate the DM considering $\hat{\delta}_{in}$ and I_{1in} within the utility function.

Therefore, the DM corrected for endogeneity using the MIS method (Guevara and Polanco 2016) in two-stages would be as shown in (8.34) and (8.35):

$$I_{1in} = \alpha_t t_{in} + \alpha_c c_{in} + \alpha_{I_{2in}} I_{2in} + \delta_{in} \overset{OLS}{\rightarrow} \hat{\delta}_{in} = I_{1in} - \hat{I}_{1in} \tag{8.34}$$

$$U_{in} = \widehat{ASC}_i + \hat{\beta}_t t_{in} + \beta_c c_{in} + \hat{\beta}_{I_{1in}} I_{1in} + \hat{\beta}_{\hat{\delta}} \hat{\delta}_{in} + \tilde{\varepsilon}_{in} \tag{8.35}$$

Note that in (8.35), the term $\hat{\beta}_{\hat{\delta}} \hat{\delta}_{in} + \tilde{\varepsilon}_{in}$ is an orthogonal decomposition of ω_{in} in (8.33), where $\hat{\beta}_{\hat{\delta}} \hat{\delta}_{in}$ captures the endogenous effect of ω_{in} that was present in (8.33). In this way, no term in (8.35) is correlated with the error term $\tilde{\varepsilon}_{in}$, and therefore, the endogeneity problem is accounted for.

Empirical studies have suggested that discrete indicators could be as good as continuous ones for correcting endogeneity with the MIS approach (Guevara and Polanco 2016; Fernández-Antolín et al. 2016). However, as this is only an approximation, Guerrero et al. (2022) compared the results of using discrete and continuous indicators, using real and simulated data. They were able to identify some of the conditions under which discrete indicators could work adequately, providing support for or refuting the previous empirical evidence. In particular, using real data, they concluded that:

1) The correction with continuous indicators worked better, in general, than the correction with discrete indicators; therefore, they recommend that in experiments where endogeneity is expected, data collection should consider gathering indicators on a more continuous scale; if discrete grades are inevitable, respondents should be required to use wider ranges to achieve ratings throughout the entire scale.
2) In particular, by checking parameter ratios (to bypass scaling problems), they found that the subjective value of time (SVT) for a model corrected with continuous indicators was closer to that of a benchmark model than those computed from the model corrected with discrete indicators; notwithstanding, the SVT achieved with an endogenous (i.e. uncorrected) model was much worse, having a very significant bias with respect to the benchmark model.

8.4.6 Accounting for Stochastic Variables in Choice Models

The estimation of DM requires data such as socio-economic characteristics of individuals and attributes of the alternatives within their choice sets. These explanatory variables are usually assumed to be inherently deterministic, that is, that they would yield the same values if measured repeatedly. The problem is that some are intrinsically stochastic (e.g. travel times under congested conditions). Furthermore, even deterministic variables can be measured inaccurately producing measurement errors, which induce a particular kind of randomness from the modeller's point of view.

For instance, in strategic model applications (as we discussed in Chapter 5) it is common practice to use zone-based network models to obtain level-of-service attributes, such as travel times, instead of measuring the data at an individual level due to the high collection costs involved. Also, trips with different levels-of-service are usually aggregated, both temporally and spatially (i.e. the set of trips between two specific zones at a peak hour) and a single 'average' level-of-service value is assigned to them, which may obviously differ from the true values experienced by the users (Train 1978). Measurement errors also occur when values are directly provided by the individual in a revealed preference (RP) survey (e.g. waiting time to board a bus, income, or preferred departure time). In the latter case, the difference between the reported value and the real one can be significant due to cognitive issues or even policy bias (Daly and Ortúzar 1990).

When there is a discrepancy between the 'true' value and the value measured by the modeller, an estimation bias arises. To see this, consider the simple MNL model (7.9) derived in Section 7.3, with a typical utility function $\mathbf{U} = \boldsymbol{\theta}\boldsymbol{x} + \boldsymbol{\varepsilon}$, where $\boldsymbol{\theta}$ are parameters to be estimated, $\boldsymbol{x}$ are measured attributes and $\boldsymbol{\varepsilon}$ are IID EV1 error terms with mean zero and standard deviation σ_ε. Recall, also, the inverse relation (7.10) that the scale parameter β has with the standard deviation σ_ε of the residuals ε. It is easy to see that it is not possible to estimate β separately from the parameters $\boldsymbol{\theta}$; in fact, the estimation process will yield estimates:

$$\hat{\boldsymbol{\theta}} = \beta\boldsymbol{\theta} \tag{8.36}$$

which correspond to the marginal utilities $\boldsymbol{\theta}$ deflated by σ_ε.

Assume now that there is a difference between the attribute values as perceived by the modeller (x^*) and the true values (x), such that: $x = x^* + \eta$, where η distributes with mean zero (i.e. no systematic bias exists) and standard deviation σ_η.

In this case the utility function becomes: $\mathbf{U} = \boldsymbol{\theta}(x^*+\eta) + \boldsymbol{\varepsilon}$, that is: $\mathbf{U} = \boldsymbol{\theta}x^* + (\boldsymbol{\varepsilon} + \boldsymbol{\theta}\eta) = \boldsymbol{\theta}x^* + \boldsymbol{\delta}$. The outcome of this is that in the original model, an estimated parameter θ'_k would be:

$$\theta'_k = \frac{\pi}{\sqrt{6}\sigma_\varepsilon}\theta_k$$

whilst in the second model the estimated parameter θ''_k would be (assuming that $\delta = \varepsilon + \theta\eta$ also follows an IID EV1 distribution, for illustrative purposes):

$$\theta''_k = \frac{\pi}{\sqrt{6}\sigma_\delta}\theta_k$$

and the standard deviation of the distribution function of the new error component δ would be:

$$\sigma_\delta = \sqrt{\sigma_\varepsilon + \theta^2 \cdot \sigma_\eta^2} \tag{8.37}$$

Hence $\theta'' < \theta'$ and this estimation bias may affect the model forecasts as we will see in Section 9.2.

There is also experimental evidence about bias estimation and miscalculation of marginal rates of substitution when measurement errors occur. For instance, Train (1978) explored the use of more accurate data in the estimation of mode choice models, concluding that it could be advisable to carry out an additional effort to correct for the measurement bias of some attributes, such as transit transfer time, when analysing transport policies. Ortúzar and Ivelic (1987) showed that use of very precise real data, measured at the individual level, resulted in better fit and clearly different subjective values of time when estimating mode choice models in comparison with models estimated with aggregate data. Bhatta and Larsen (2011) have shown, using synthetic data, how measurement bias may induce biased parameter estimates on a MNL model, apart from miscalculation of marginal rates of substitution. Therefore, the use of more accurate (but more expensive) data results in better parameter estimates and this clearly establishes a trade-off between data quality and data collection costs as we saw in Section 2.2.1.6.

The effort to specify stochastic variables when estimating econometric models has mainly arisen from the need to solve the 'errors in variables' (EIV) problem. Although there is a vast literature in the case of regression models, research concerning EIV within DM is scarce. Steinmetz and Brownstone (2005) proposed a model that considers EIV using multiple imputations, which can be used when there is accurate information for a subsample of observations. Yamamoto and Komori (2010) estimated a latent class model for handling errors when measuring access distances to public transport. Walker et al. (2010) proposed a HC model including a latent variable to account for travel time measurement errors as these were obtained from a network zone-based model. In their specification, the true travel time was treated as a latent variable, using the measured travel time as an indicator. Although their theoretical specification is consistent, it is not easy to justify the formulation in practice, especially when high dispersion exists in travel time values, which is the most frequent case.

Other examples of HC models used to deal with the EIV problem are those of Bolduc and Alvarez-Daziano (2009) and Brey and Walker (2011), where latent variables were used to account for measurement errors in variables such as income in a vehicle choice experiment, or preferred departure time in an airline itinerary choice context. In both cases, the structural equations of the latent variable component were a function of individual characteristics, allowing their use in forecasting scenarios. Nevertheless, the specification of a structural equation associated with individual characteristics in the case of an exogenous variable such as travel time is less natural.

8.4.6.1 *Econometric Analysis*

Let us consider the typical additive and linear-in-parameters specification for the utility function in a discrete choice context based on random utility theory:

$$U_{in} = \sum_{k=1}^{K} \theta_{ink} x_{ink} + \varepsilon_{in} \tag{8.38}$$

where, as usual, U_{in} represents the utility of alternative A_i for individual n, x_{ink} refers to the value of the kth explanatory attribute of alternative A_i for individual n, θ_{ink} is an unknown parameter to be estimated and ε_{in} is an error term that distributes IID Extreme Value type I (EV1). This formulation corresponds to the classical MNL model (7.9).

The stochastic nature of the kth explanatory variable can be expressed as following:

$$x_{ink} = \bar{x}_{ink} + \eta_{ink} \tag{8.39}$$

where $\bar{x}_{ink}$ is the mean measured value of the variable (i.e. the value that the modeller would typically use) and η_{ink} is the discrepancy between this mean measured value and the true value, perceived by the individual. Equation (8.39) can be seen as a measurement equation in the context of a HC model that includes latent variables in the discrete choice component (Bolduc et al. 2008). The discrepancy η_{ink} is a stochastic component that follows a certain distribution. Replacing (8.39) in (8.38) we get:

$$U_{in} = \sum_{k=1}^{K} \theta_{ink} (\bar{x}_{ink} + \eta_{ink}) + \varepsilon_{in} \tag{8.40}$$

which is equivalent to a Mixed Logit (ML) formulation, because the random error component is a mix of an EV1 distribution and some other distribution contained in η_{ink} (McFadden and Train 2000). It suggests that the errors in variables problem can be approximated through a particular specification of the ML model, and depends on the definition of η_{ink}. Note, therefore, that using a MNL model when stochastic variables are present is not appropriate because the IID EV1 error term of the model cannot represent the full error structure in the data. In what follows we discuss two particular ML formulations for dealing with the EIV problem (Díaz et al. 2015).

8.4.6.2 Stochastic Variables Model

The stochastic variations η_{ink} can be specified as follows:

$$\eta_{ink} = \sigma_{ik} \cdot u_{ink} \tag{8.41}$$

where u_{ink} are independently distributed variables with zero mean and unitary standard deviation; σ_{ik} is a fixed real number representing the standard deviation of the probability function related with the stochastic variation; this value is alternative-specific but constant among individuals. Replacing (8.41) in (8.40), and assuming generic tastes in the population, the Stochastic Variables (SV) model can be written as:

$$U_{in} = \sum_{k=1}^{K} \theta_{ink}(\bar{x}_{ink} + \sigma_{ink} \cdot u_{ink}) + \varepsilon_{in} \tag{8.42}$$

and the probability P_n that individual n chooses alternative A_i for given values u_{ink}^d can be computed as a class of *logit* model:

$$P_n\left(A_i/\boldsymbol{\theta}, \boldsymbol{\sigma}, \mathbf{u}^d\right) = \frac{\exp\left\{\beta\left(\sum_k \theta_{ik}\left(\bar{x}_{nik} + \sigma_{ik} \cdot u_{nik}^d\right)\right)\right\}}{\sum_{A_j \in C_n}\left[\exp\left\{\beta\left(\sum_k \theta_{jk}\left(\bar{x}_{njk} + \sigma_{jk} \cdot u_{njk}^d\right)\right)\right\}\right]} \tag{8.43}$$

where $\boldsymbol{\theta}$ and $\boldsymbol{\sigma}$ are vectors with elements θ_{ik} and σ_{ik}, β is the scale factor (which is usually normalized to 1), and C_n the individual's choice set. The final choice probability can be obtained by integrating (8.43) over the range of $\mathbf{u}$ values. This is a typical ML formulation and, therefore, can be computed using simulated maximum log-likelihood techniques as we will see in Section 8.6.

It can be shown that in this case, due to identification issues, it is only possible to estimate $I-1$ covariance matrix parameters if there are I alternatives. As a consequence, the modeller has to choose a specification where all the correspondent covariance matrix parameters are normalised; the normalisation is not simple as it must guarantee that these parameters are fixed at a sufficiently large value in relation to $\boldsymbol{\sigma}$. In the most general case, these values are unknown, and a trial-and-error procedure would need to be undertaken.

Equation (8.42) can be re-arranged as an error components ML (ECML) structure, frequently used to account for heteroskedasticity among alternatives (Díaz et al. 2015). Under some circumstances and specially when treating the EIV problem, the SV and ECML structures are mathematically equivalent (i.e. in practice, this equivalence may be interpreted as meaning that the stochastic variables cause heteroskedasticity among alternatives). Furthermore, the effect of the stochastic variables can also be confounded with other sources of heteroskedasticity or heterogeneity in the estimation process.

As before, in the ECML structure only $I-1$ error component variances can be estimated and one of those needs to be normalised. As we will see in Section 8.6, the usual normalisation method involves estimating a non-identifiable model (i.e. with I variances), and later estimate a new model but fixing to zero the lowest variance in the preliminary estimation (Walker 2002). As this normalisation is easier in the ECML model than in the SV model, it would appear that estimating the former structure would be preferable.

8.4.6.3 Random Coefficients Model

The stochastic variations in the SV model are assumed to be independent of the corresponding attribute values. Furthermore, the standard deviations σ_{ik} are equal for all individuals implying homoskedasticity across observations. However, under some circumstances it is reasonable to expect that the higher an attribute value the larger should be its level of randomness. In such cases, a proportional direct relationship may be specified:

$$\eta_{ink} = \lambda_{ink} \cdot \bar{x}_{ink} \tag{8.44}$$

where λ_{ink} follows a probability distribution function with zero mean and unknown standard deviation γ_{ik}:

$$\lambda_{ink} = \gamma_{ink} \cdot u_{ink}$$

In this case then, heteroskedasticity across both respondents and alternatives would be found in the model. Replacing (8.44) in (8.40), and assuming taste homogeneity, we can write:

$$U_{in} = \sum_{k=1}^{K} \alpha_{ink} \cdot \bar{x}_{ink} + \varepsilon_{in} \tag{8.45}$$

where:

$$\alpha_{ink} = \theta_{ik}(1 + \gamma_{ik} \cdot u_{ink})$$

Even though the taste variations across respondents are fixed, Eq. (8.45) represents a random coefficients (RC) model. If stochastic variables are included in the formulation, and if the randomness is directly proportional to the variable size, it is possible that the RC model estimates could be confounded with (apparent) random taste heterogeneity in the population (Díaz et al. 2015). Confounded effects in ML estimates would be also possible if the stochastic variables in the model are interpreted as taste heterogeneity (Cherchi and Ortúzar 2008b; Swait and Bernardino 2000), even if the stochastic variations are independent from the size of the variables, as shown before.

8.5 Estimating the Multinomial Probit Model

Flexible choice models, such as Multinomial Probit (MNP) and Mixed Logit (ML) do not have a closed form, so their choice probabilities are characterised by a multiple integral that is not easy to solve efficiently. The MNP model is generated by assuming that the residuals ε in (7.2) distribute Normal with zero mean and an arbitrary covariance matrix.

The Normal distribution. The probability density function of a *standard* Normal random variable Z is defined as:

$$f(Z) = \frac{1}{\sqrt{2\pi}} \exp\left(-\frac{1}{2}Z^2\right)$$

where the constant $\frac{1}{\sqrt{2\pi}}$ ensures that the integral of $f(Z) = 1$.

It is easy to show that the mean μ_z of $f(Z)$ is equal to zero and that its variance σ_z^2 is equal to one; for this reason the standard Normal is also known as N(0, 1). In general a variable X distributes N(μ, σ^2) if its density function is given by:

$$f(Z) = \frac{1}{\sqrt{2\pi}\sigma} \exp\left(-\frac{1}{2}\left(\frac{x-\mu}{\sigma}\right)^2\right) \tag{8.46}$$

and from here it is easy to see that X can be 'standardised' by applying the transformation $Z = \dfrac{X-\mu}{\sigma}$. The advantage of standardising is that one can use tabulated values of the function. Note that, conversely, if one wants to generate Normal values with mean b and variance s^2 these are given as $X = b + s \cdot Z$; this will come handy when we need to generate draws of Normal (and other) distributions for various simulation procedures.

Some useful properties of this well-known bell-shaped function are that:

$$\text{between} \begin{cases} \mu \pm \sigma \\ \mu \pm 2\sigma \\ \mu \pm 3\sigma \end{cases} \text{we find} \begin{cases} 68.20\% \text{ of the distribution} \\ 95.44\% \text{ of the distribution} \\ 99.73\% \text{ of the distribution.} \end{cases}$$

and precisely 95% of the distribution lies in the range $\mu \pm 1.96\sigma$. Another interesting property is that if we have n variables $\mathbf{x}$ that distribute with any distribution with finite variance, according to the Central Limit Theorem (see Wonnacott and Wonnacott 1990) it can be shown that:

$$\frac{\bar{x}-\mu}{\sigma/\sqrt{n}} \sim N(0,1) \qquad \text{if } n \geq 30 \tag{8.47}$$

Also, and as we had already mentioned, the Normal distribution is *closed* to algebraic summation, that is, the sum and the difference (and indeed any linear combination) of Normal variables is also Normal distributed.

Quadratic form. The portion $\left(\dfrac{x-\mu}{\sigma}\right)^2$ in (8.46) is known as quadratic form (*QF*) and distributes Chi-squared with one degree of freedom (χ_1^2). For the bivariate Normal distribution, we have:

$$\mathbf{x} = (x_1, x_2)^{\mathrm{T}} \sim N\left[\bar{\mathbf{x}} = (\bar{x}_1 \ \bar{x}_2)^{\mathrm{T}}, \Sigma_x\right]$$

where Σ_x is the covariance matrix. In this case, the quadratic form is given by:

$$QF_2 = (\mathbf{x}-\bar{\mathbf{x}})^{\mathrm{T}}\Sigma_x^{-1}(\mathbf{x}-\bar{\mathbf{x}})$$

and this distributes χ_2^2 (i.e. with two degrees of freedom). In this case the density function is given by:

$$f(x_1, x_2) = \frac{|\Sigma_x|^{1/2}}{2\pi^{2/2}} \exp\left(-\frac{1}{2}QF_2\right)$$

where $|\Sigma_x|$ is the determinant of the covariance matrix (note that if it was a p-variate Normal the denominator above would be $2\pi^{p/2}$). The covariance matrix in this case is:

$$\Sigma_x = \begin{pmatrix} \sigma_1^2 & \rho\sigma_1\sigma_2 \\ \rho\sigma_1\sigma_2 & \sigma_2^2 \end{pmatrix}$$

and the quadratic form can, for the last time, be written in extended (rather than matrix) form:

$$QF_2 = \frac{1}{1-\rho^2} \left[\left(\frac{x_1 - \bar{x}_1}{\sigma_1} \right)^2 - \frac{2\rho \, (x_1 - \bar{x}_1)(x_2 - \bar{x}_2)}{\sigma_1 \sigma_2} + \left(\frac{x_2 - \bar{x}_2}{\sigma_2} \right)^2 \right]$$

Note that the following theorem holds for QF in multivariate Normal distributions. If $\mathbf{X} = (x_1, ..., x_n)$ distributes multivariate Normal with mean $\bar{\mathbf{x}}$ and non-singular covariance matrix $\boldsymbol{\Sigma}$ (i.e. $|\boldsymbol{\Sigma}| \neq 0$), then the random scalar variable QF_n, defined by the quadratic form $QF_n = (\mathbf{x}\text{-}\bar{\mathbf{x}}) \, \boldsymbol{\Sigma}^{-1}(\mathbf{x}\text{-}\bar{\mathbf{x}})^{\mathrm{T}}$ distributes χ^2 with n degrees of freedom.

Choleski decomposition for the multivariate Normal. As described above, a univariate Normal variable with mean b and variance s^2 is obtained as $x = b + s \cdot z$ where z is a standard Normal. An analogous procedure can be used to take draws from a multivariate Normal distribution.

Let $\mathbf{x}$ be a vector with n elements distributed $N(\mathbf{b}, \boldsymbol{\Sigma})$. A Choleski transformation (or factorisation) of the matrix $\boldsymbol{\Sigma}$ is defined as a lower triangular matrix $\mathbf{L}$ such that $\mathbf{L} \cdot \mathbf{L}^{\mathrm{T}} = \boldsymbol{\Sigma}$ (see Daganzo 1979). It can also be called generalised square root of $\boldsymbol{\Sigma}$ or generalised standard deviation of $\mathbf{x}$; in fact, when $n = 1$ the Choleski factor is precisely s (see Train 2009). Nowadays, most statistical packages have routines to calculate a Choleski factorisation for any positive definite symmetric matrix.

8.5.1 Numerical Integration

The choice probability for a general random utility model may be expressed as follows:

$$P_i(\boldsymbol{\theta}, \mathbf{x}) = \int_{u_1 = -\infty}^{u_i} \int_{u_2 = -\infty}^{u_i} \cdots \int_{-\infty}^{\infty} \cdots \int_{u_J = -\infty}^{u_i} f(\mathbf{u}) \, du_J ... du_1 \qquad (8.48)$$

where $f(\mathbf{u})$ is the joint distribution function of the alternative utilities. For example, in the case of the MNP model we have:

$$f(\mathbf{u}) = \mathrm{MVN}(\mathbf{V}, \boldsymbol{\Sigma}) = \left[(2\pi)^J |\boldsymbol{\Sigma}| \right]^{-1/2} \exp\left\{ -\frac{1}{2}(\mathbf{u} - \mathbf{V}) \, \boldsymbol{\Sigma}^{-1}(\mathbf{u} - \mathbf{V})^{\mathrm{T}} \right\} \qquad (8.49)$$

To integrate numerically, the region of integration must first be divided into a series of elements of differential size. Then the area under the curve is approximated, for each element, as the equivalent mean rectangle (given the element and its height); finally, the value of the integral is the sum of these areas. Although the difficulty of the problem increases geometrically with the dimensionality of the integral, in the majority of cases this dimensionality may be reduced for the MNP because:

(a) If a change of variables is made, expressing all elements of the integral as the difference between the utility of the alternative under consideration and the others, this yields a vector $\hat{\mathbf{u}}$ of just $J - 1$ components (that are also distributed Normal) given by:

$$\hat{u}_k = u_k - u_i$$
$$...$$
$$\hat{u}_{J-1} = u_J - u_i$$

(assume we are evaluating P_i)

Then the probability of choosing A_i will be:

$$P_i(\hat{\mathbf{u}}, \hat{\boldsymbol{\Sigma}}) = \mathrm{Prob}\{\hat{u}_k < 0, \forall A_k \in \mathbf{A}\}$$

and the integral reduces to:

$$\int_{\hat{u}_1 = -\infty}^{0} \cdots \int_{\hat{u}_{J-1} = -\infty}^{0} \left(2\pi^{J-1} | \hat{\Sigma} |\right)^{-1/2} \exp\left\{ -\frac{1}{2} (\hat{u} \cdot \hat{V}) \hat{\Sigma}^{-1} (\hat{u} \cdot \hat{V})^{\mathrm{T}} \right\}$$

with $\hat{V}$ and $\hat{\Sigma}$ the vector of means and the covariance matrix of the new variables.

(b) Make a Choleski decomposition, which in practical terms also reduces the integral dimensionality by one, because it allows us to separate the integrals and the first, corresponding to A_i, is equal to one.

Numerical integration is the most accurate method to solve these problems, but it is only feasible at a reasonable cost for problems with a maximum of four alternatives. It may also have problems of (computer) approximation if one or more choice probabilities are close to zero. For these reasons it has generally been used only as a standard of comparison for the other methods.

8.5.2 Simulated Maximum Likelihood

8.5.2.1 The Basic Approach

Lerman and Manski (1981) originally proposed the evaluation of the MNP choice probability $P_i (V, \Sigma)$ by generating a number of draws U, from MVN (V, Σ), counting a *success* when U_i was the highest value. For a sufficiently large number of draws, the proportion of successes approximates the choice probability (see Figure 8.4). Thus, the method was theoretically simple but unfortunately had several problems in practice:

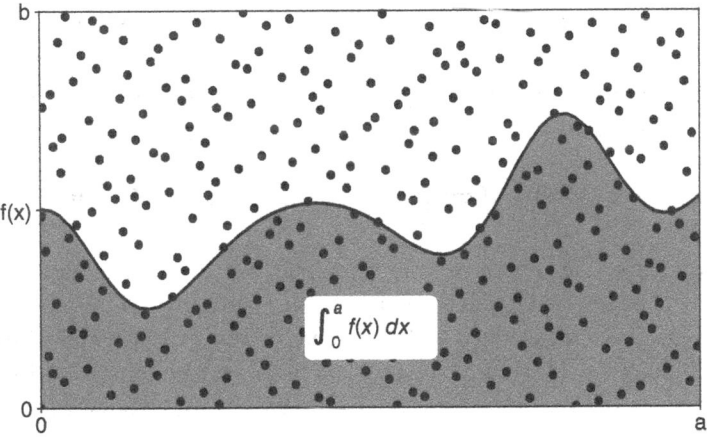

Figure 8.4 Solving an integral through Monte Carlo simulation

(a) If the number of successes is equal to zero (an event that could occur in certain circumstances), the log-likelihood tends to infinity, and the method collapses. To solve this problem, Lerman and Manski (1981) suggested replacing the ratio of the number of successes over the total number of draws (i.e. the estimate of the choice probability) by the quantity $(N_i + 1) / (N + J)$ where N_i is the number of successes, N the sample size (number of draws) and J the number of options. However, this introduces bias (as the correct estimator of P_i is obviously N_i/N). The bias is small in large problems but it could be considerable in more practical problems.

(b) The relative error associated with this simulation method is inversely proportional to the square root of the number of successes. This implies that many draws have to be made, and this was computationally too demanding for real-life problems in the 1970s.

However, at the beginning of the 1990s this approach found renewed favour through a series of advances in the simulation of multivariate processes in discrete choice models (Börsch-Supan and Hajivassiliou 1993). There is also an alternative approach (McFadden 1989; Pakes and Pollard 1989) that avoids evaluating the multiple integral by replacing the choice probability in the moments' equation with an unbiased simulator. This method of 'simulated moments' may be considered a precursor of the Mixed Logit or 'error components' model, the estimation of which we will review in Section 8.6.

8.5.2.2 Advanced Techniques

Using advanced integration techniques based on Monte Carlo simulation developed by several authors, Börsch-Supan and Hajivassiliou (1993) proposed the GHK simulator. This has the essential property of producing unbiased simulated probabilities that lie strictly between zero and one, and that are also continuous and differentiable functions of the model parameters. Furthermore, the computational effort increases only linearly with the dimensionality of the integral and is independent of the true probabilities. The simulator is based on recursively decreasing the problem dimension, and for this it has to generate repetitions of a truncated uni-dimensional Normal distribution.

For the MNP model, the method started with the model reduced in one dimension after subtracting the utility of the chosen option from the remaining utilities for each observation (i.e. $U_1 - U_c$ where c is the chosen option); as we already mentioned, this transformed utility is also Normal distributed. In mathematical terms the transformation simply consists of pre-multiplying the vector of utilities by a matrix $\mathbf{A}$, which is equal to minus the identity matrix and incorporating a column of ones in the position corresponding to the chosen option. Then the resulting vector can be freed of the row corresponding to the chosen alternative because it only contains zeros.

In this way, the transformed systematic utility is given by $\mathbf{V}^* = \mathbf{AV}$, and the error term distributes Normal with zero mean and covariance matrix given by $\mathbf{M} = \mathbf{A\Sigma A'}$. In turn, $\mathbf{M}$ can be decomposed by applying the Choleski decomposition to produce a lower triangular matrix $\mathbf{L}$ and a superior matrix $\mathbf{L'}$, such that $\mathbf{LL'} = \mathbf{M}$. The GHK simulator was developed to simulate the probability that a Normal random variable lies within limits a and b:

$$\mathbf{u} \sim N\left(\mathbf{\theta} \cdot \mathbf{x}, \sum\right) \quad \text{subject to} \quad a \leq \mathbf{AU} \leq b$$

Instead of simulating for these variables, the process is performed for:

$$\mathbf{e} \sim N(0, \mathbf{I}) \quad \text{subject to} \quad a^* \equiv a - \mathbf{A\theta x} \leq \mathbf{Le} \leq b^* \equiv b - \mathbf{A\theta x}$$

and, thanks to the triangular structure of $\mathbf{L}$, the restrictions are recursive:

$$e_1 \sim N(0, \mathbf{I}) \quad \text{subject to} \quad a_1^* \leq l_{11} e_1 \leq b_1^* \quad \Leftrightarrow \quad a_1^*/l_{11} \leq e_1 \leq b_1^*/l_{11}$$

$$e_2 \sim N(0, \mathbf{I}) \quad \text{subject to} \quad a_2^* \leq l_{21} e_1 + l_{22} e_2 \leq b_2^* \quad \Leftrightarrow \quad (a_2^* - l_{21} e_1)/l_{22} \leq e_2 \leq (b_2^* - l_{21} e_1)/l_{22}$$

etc.

In this form, the e_i values can be generated sequentially with a univariate truncated simulator. Finally, the random vector of interest, $\mathbf{u}^*$, can be defined as:

$$\mathbf{u}^* = \mathbf{\theta x} + \mathbf{A}^{-1}\mathbf{Le}$$

This vector has a covariance matrix given by $\mathbf{A}^{-1}\mathbf{LL'A}^{-1'} = \mathbf{A}^{-1}\mathbf{A\Sigma A'A}^{-1'} = \mathbf{\Sigma}$ and is subject, by construction, to the condition $a \leq \mathbf{Au}^* \leq b$. Börsch-Supan and Hajivassiliou (1993) show that although the generation of draws of $\mathbf{u}^*$ is biased, the contribution of each observation to the likelihood function (i.e. the probability that $\mathbf{Au}$ is between a and b) is simulated correctly by the probability that $a^* \leq \mathbf{Le} \leq b^*$.

To speed the process and reduce the variance of the choice probabilities that are eventually calculated, it is possible to use *antithetic draws* (see the discussion by Train 2009). If we consider that P_{iq} can be approximated as the average of the probabilities $\left(P_{iq}^0 \text{ and } P_{iq}^1\right)$ corresponding to two sets of repetitions of random variables, then it can be seen that:

$$\text{Var}\left(P_{iq}\right) = \text{Var}\left(\frac{P_{iq}^0 + P_{iq}^1}{2}\right) = \frac{1}{4}\text{Var}\left(P_{iq}^0\right) + \frac{1}{4}\text{Var}\left(P_{iq}^1\right) + \frac{1}{2}\text{Cov}\left(P_{iq}^0, P_{iq}^1\right)$$

Thus, if both sets are independent then the covariance is zero, but if they are negatively correlated then the covariance will be less than zero. This suggests the ideal situation of generating a series of random numbers to calculate probabilities and then, as an antithetic, to use the same series but with the opposite sign to generate the new set of probabilities. This does not only achieve savings in random number generation, but it also computes choice probabilities with a smaller variance.

In the case of the MNP, where we are interested in evaluating the probability that the utility of the chosen option is higher than those of the remaining alternatives in the choice set of each individual, the lower limit a^* is equal to zero and the upper limit b^* is infinity. The likelihood function is, as usual, the product of the probabilities of choosing the chosen option for each individual. Programs to estimate the MNP model using the GHK simulator are available in GAUSS (aptech.com/resources/manuals/), and have been validated using simulated data (e.g. Munizaga et al. 2000).

The optimisation problem to solve in this case is not necessarily convex, so convergence to a unique optimum is not guaranteed. For example, among the routines offered in GAUSS, the Newton–Raphson method was the more robust in convergence terms (although somewhat slow) and the fastest method was the Berndt–Hall–Hall–Hausman algorithm (Berndt et al. 1974), although it did not always converge.

A practical issue of interest is that it is highly convenient to start by considering a very simple model, where only the parameters of the representative utility function are estimated (even starting with initial values taken from an MNL), and then re-estimate the model liberating the covariance matrix parameters one by one. This is not a sequential estimation, because at the last iteration the complete model is estimated, but a useful strategy to obtain the best initial point for what is in general a very complex optimisation problem (among other things, the log-likelihood surface is relatively flat and full of local optima).

8.6 Estimating the Mixed Logit Model

In Section 7.6 we presented the Mixed Logit (ML) model and made reference to its two specifications, as an error components (EC) model (7.45) and as a random coefficients (RC) model (7.46). However, we also saw that both were formally equivalent (7.47) and the manner in which the modeller looks at the phenomenon under study allows to decide which form is more appropriate for a given case. Interestingly, there are also two general methods for estimating the model, the *classical* approach (using simulated maximum likelihood) and the *Bayesian* approach. Also, recall that there are two sets of parameters that in principle can be estimated, the *population* parameters (i.e. the vector of means and the covariance matrix associated with the mixing distribution), and the *individual* parameters, which have a distribution over the population conditional on the former.

First, we present the classical approach, incorporating developments in the field of estimation via simulated maximum likelihood methods (Bhat 2001; Train 2009), including the framework by which population distribution parameters combined with information from individual choices can lead to consistent estimates of individual marginal utilities (Revelt and Train 2000). Secondly, we present the hierarchical Bayes estimation procedure, which saw remarkable development in the 2000s (Allenby and Rossi 1999; Huber and Train 2001; Andrews et al. 2002; Sillano and Ortúzar 2005; Godoy and Ortúzar 2008).

8.6.1 Classical Estimation

By classical estimation we refer to the maximum likelihood procedure commonly used to estimate flexible discrete choice models (Train 2009).

8.6.1.1 Estimation of Population Parameters

Consider the most general case, of having available a sequence of T choices per individual (i.e. as in stated choice or panel data), denoted by $\mathbf{c}_q = (c_{1q}, ..., c_{Tq})$, where $c_{tq} = i$ if $U_{iqt} > U_{jqt} \; \forall A_j \neq A_i$. In a typical ML model, the conditional probability of observing an individual q stating a sequence $\mathbf{c}_q$ of choices, given *fixed* values for the model parameters $\overline{\mathbf{\theta}}_q$, is given by a product of Logit functions:

$$\Lambda(\mathbf{c}_q \mid \mathbf{\theta}_q) = \prod_{t=1}^{T} \frac{\exp(\lambda \cdot \overline{\mathbf{\theta}}_q \cdot \mathbf{x}_{iqt})}{\sum_{j=1}^{J} \exp(\lambda \cdot \overline{\mathbf{\theta}}_q \cdot \mathbf{x}_{jqt})} \tag{8.50}$$

where λ is the MNL's scale factor that has to be normalised as usual.

Now since $\mathbf{\theta}_q$ is unknown, the unconditional probability of choice is given by the integration of (8.50) weighted by the density distribution of $\mathbf{\theta}_q$ over the population:

$$\mathbf{P}_q(\mathbf{c}_q) = \int \Lambda(\mathbf{c}_q \mid \mathbf{\theta}_q) f(\mathbf{\theta}_q \mid \mathbf{b}, \mathbf{\Sigma}) \, d\mathbf{\theta}_q \tag{8.51}$$

where $f(\cdot)$ is the multivariate distribution of $\mathbf{\theta}_q$ over the sampled population. If covariance terms are not specified, $\mathbf{\Sigma}$ is a diagonal matrix. Note that the majority of applications use diagonal matrices as results seem not to be affected strongly by this assumption (Sillano and Ortúzar 2005).

The log-likelihood function in $\mathbf{b}$ and $\mathbf{\Sigma}$ is:

$$l(\mathbf{b}, \mathbf{\Sigma}) = \sum_{q=1}^{Q} \log \mathbf{P}_q(\mathbf{c}_q)$$

but as the probabilities $\mathbf{P}_q$ do not have a closed form they are approximated through simulation ($\mathbf{SP}_q$), where draws are taken from the mixing distribution $f(\cdot)$ weighted by the logit probability, and then averaged up (McFadden and Train 2000):

$$\mathbf{SP}_q\left(\mathbf{c}_{qt}\,|\,f(\cdot|\mathbf{b},\boldsymbol{\Sigma})\right) = \frac{1}{R}\sum_r \left(\prod_t \frac{\exp\left(\boldsymbol{\theta}_q^r \cdot x_{iqt}\right)}{\sum\limits_{A_j \in \mathbf{A}(q)} \exp\left(\boldsymbol{\theta}_q^r \cdot x_{jqt}\right)}\right)$$

The issue of how many draws R use, and how should they be generated to improve the efficiency of the simulation is discussed below (Bhat 2003; Hess et al. 2006). The simulated log-likelihood function is given by:

$$sl(\mathbf{b},\boldsymbol{\Sigma}) = \sum_{q=1}^{Q} \log \mathbf{SP}_q(\mathbf{c}_q)$$

Under regularity conditions this estimator is consistent and asymptotically Normal; furthermore, when the number of repetitions grows more rapidly than the square root of the number of observations, the estimator is asymptotically equivalent to the maximum likelihood estimator (Hajivassiliou and Ruud 1994). Other useful properties of the estimator are being twice differentiable (which helps in the numerical search of the optimum) and being strictly positive, so the log-likelihood function is always defined. Note that the same procedure would be followed if the ML had another Logit kernel, say a NL function or any more general and appropriate GEV model, instead of the MNL.

Different forms of 'smart' drawing techniques (i.e. Halton or other low discrepancy sequences, antithetic draws, quasi-random sampling, etc.) can be used to reduce the simulation variance and to improve the efficiency of the estimation (Hajivassiliou and Ruud 1994, Bhat 2003; Hensher and Greene 2003); we will refer briefly to this issue in Section 8.6.4. Train (1998) presents a good example of the use of this model and offers an experimental estimation code, written in GAUSS, which can be downloaded from his web page. Two other pieces of free software available for estimating the ML are Biogeme (Bierlaire 2016) and Apollo (Hess and Palma 2019), which offer many capabilities and allows the estimation of several other discrete choice models. Finally, new releases of the packages ALOGIT (https://www.alogit.com/) and LIMDEP (https://www.limdep.com/features/documentation.php) also include modules to estimate ML models, and these are generally faster than the more experimental codes available, definitely making the model now a practical proposition.

8.6.1.2 Estimating Individual Parameters

Numerical procedures are used to find the maximum likelihood estimators for $\mathbf{b}$ and $\boldsymbol{\Sigma}$ above. These parameters define a frequency distribution for the $\boldsymbol{\theta}_q$ over the population. To obtain actual point estimates for each $\boldsymbol{\theta}_q$ a second procedure, described originally by Revelt and Train (2000), is required as follows.

The conditional density $h\left(\boldsymbol{\theta}_q|\mathbf{c}_q,\mathbf{b},\boldsymbol{\Sigma}\right)$ of any $\boldsymbol{\theta}_q$ given a sequence of T_q choices $\mathbf{c}_q$ and the population parameters $\mathbf{b}$ and $\boldsymbol{\Sigma}$, may be expressed by Bayes' rule as:

$$h\left(\boldsymbol{\theta}_q\,|\,\mathbf{c}_q,\mathbf{b},\boldsymbol{\Sigma}\right) = \frac{\mathbf{P}_q\left(\mathbf{c}_q\,|\,\boldsymbol{\theta}_q\right)f\left(\boldsymbol{\theta}_q\,|\,\mathbf{b},\boldsymbol{\Sigma}\right)}{P_n\left(\mathbf{c}_q\,|\,\mathbf{b},\boldsymbol{\Sigma}\right)} \tag{8.52}$$

The conditional expectations of θ_q result from integrating over its domain. This integral can be approximated by simulation, averaging weighted draws $\theta_q{}^r$ from the population density function $f(\theta_q|\mathbf{b}, \boldsymbol{\Sigma})$. The simulated expectations $\mathbf{SE}$ of the individual parameters are then given by:

$$\mathbf{SE}(\theta_q|\mathbf{c}_q, \mathbf{b}, \boldsymbol{\Sigma}) = \frac{\sum\limits_{r=1}^{R} \theta_n^r \mathbf{P}_q(\mathbf{c}_q|\theta_n^r)}{\sum\limits_{r=1}^{R} \mathbf{P}_q(\mathbf{c}_q|\theta_n^r)}$$

Revelt and Train (2000) also proposed, but did not apply, an alternative simulation method to condition individual level choices. Consider the expression for $h(\theta_q|\mathbf{c}_q, \mathbf{b}, \boldsymbol{\Sigma})$ in (8.52). The denominator is a constant value since it does not involve θ_q, so a proportionality relation can be established as:

$$h(\theta_q|\mathbf{c}_q, \mathbf{b}, \boldsymbol{\Sigma}) \propto \mathbf{P}_q(\mathbf{c}_q|\theta_q)f(\theta_q|\mathbf{b}, \boldsymbol{\Sigma})$$

Draws from the posterior $h(\theta_q|\mathbf{c}_q, \mathbf{b}, \boldsymbol{\Sigma})$ can then be obtained using the Metropolis-Hastings algorithm (Chib and Greenberg 1995), with successive iterations improving the fit of the θ_q to the observed individual choices. During this process the prior $f(\theta_q|\mathbf{b}, \boldsymbol{\Sigma})$, that is, the parameter distribution obtained by maximum likelihood, remains fixed; it provides information about the population distribution of θ_q. After a sufficient number of 'burn-out' iterations to ensure that a steady state has been reached (typically a few thousand, Kass et al. 1998; Godoy and Ortúzar 2008), only one every m of the sampled values generated is stored to avoid potential correlation among them; m is a result of the analysis of convergence (Raftery and Lewis, 1992).

From these values a sampling distribution for $h(\theta_q|\mathbf{c}_q, \mathbf{b}, \boldsymbol{\Sigma})$ can be built, and inferences about the mean and standard deviation values can be obtained (Godoy and Ortúzar 2008). Sillano and Ortúzar (2005) favoured this latter procedure for implementation purposes and used WinBUGS (Spiegelhalter et al. 2003), a software package that can also be freely downloaded from the web.

Thus, the outcome of the estimation process is two sets of parameters: $\mathbf{b}$ and $\boldsymbol{\Sigma}$, the population parameters obtained by simulated maximum likelihood and θ_q, the individual parameters for $q = 1,..., Q$, estimated via conditioning the observed individual choices on the estimated population parameters. It is surprising that the large majority of applications of ML models stop short of reaching the full capability of the model, by not going to this second stage.

8.6.2 Bayesian Estimation

Use of the Bayesian statistic paradigm for estimating the ML model gained much interest at the beginning of the century (Train 2001; Huber and Train 2001; Sillano and Ortúzar 2005) but has surprisingly lost appeal in recent years, together with the possibility of estimating individual rather than just population parameters. In fact, the ability to estimate individual part-worths appeared initially as its main appeal, but the estimation procedure has subsequently shown further advantages (Godoy and Ortúzar 2008). The Bayesian approach considers the parameters as stochastic variables so applying Bayes' rule of conditional probability, a posterior distribution for θ_q conditional on observed data and prior beliefs about these parameters can be estimated; let us denote this distribution by $\pi(\mathbf{b}, \boldsymbol{\Sigma}|\mathbf{c}_q)$.

Now, let $\psi(\mathbf{b}, \boldsymbol{\Sigma})$ represent the analyst's prior knowledge about the distribution of $\mathbf{b}$ and $\boldsymbol{\Sigma}$; typically, a Normal distribution is used for the means $\mathbf{b}$ and an Inverted Wishart distribution for the variances in $\boldsymbol{\Sigma}$ (Allenby 1997). Then, consider a likelihood function for the observed sequence of choices conditional on fixed values of **b** and **Σ**. By Bayes' rule, the posterior distribution for $\boldsymbol{\theta}_q$, **b**, and **Σ** must be proportional to:

$$\prod_{q=1}^{Q} \Lambda(\mathbf{c}_q | \boldsymbol{\theta}_q) f(\boldsymbol{\theta}_q | \mathbf{b}, \boldsymbol{\Sigma}) \, \psi(\mathbf{b}, \boldsymbol{\Sigma}) \tag{8.53}$$

Although it is possible to draw directly from $\pi(\mathbf{b}, \boldsymbol{\Sigma} | \mathbf{c}_q)$ with the Metropolis–Hastings (MH) algorithm, this would be computationally very slow. Indeed, it would be necessary to calculate (8.53) at every iteration of the MH algorithm, but the choice probability inside is an integral without a closed-form resolution and must be approximated through simulation; thus, an iteration of the MH algorithm would require simulation for each individual q. That could be time consuming and affect the properties of the resulting estimator.

Drawing from $\pi(\mathbf{b}, \boldsymbol{\Sigma} | \mathbf{c}_q)$ becomes fast and simple if each $\boldsymbol{\theta}_q$ is considered to be a parameter along with **b** and **Σ**, and Gibbs sampling is used for the three sets of parameters for each individual. The posterior distribution in this case is:

$$\pi(\mathbf{b}, \boldsymbol{\Sigma} | \mathbf{c}_q) \; \propto \prod_{q=1}^{Q} \Lambda(\mathbf{c}_q | \boldsymbol{\theta}_q) f(\boldsymbol{\theta}_q | \mathbf{b}, \boldsymbol{\Sigma}) \, \psi(\mathbf{b}, \boldsymbol{\Sigma})$$

The sequential procedure simulates each set of parameters from the following conditional posterior distributions:

- the conditional posterior for **b** is $\pi(\mathbf{b} | \boldsymbol{\Sigma}, \boldsymbol{\theta}_q \; \forall q)$ and this distributes $N(\overline{\boldsymbol{\theta}}, \boldsymbol{\Sigma} | Q)$ where $\overline{\boldsymbol{\theta}} = \sum_q \boldsymbol{\theta}_q / Q$;
- the conditional posterior for **Σ** is $\pi(\boldsymbol{\Sigma} | \mathbf{b}, \boldsymbol{\theta}_q \; \forall q)$ which distributes inverted Wishart $\text{IW}\left(K + Q, \dfrac{K \cdot J + Q \cdot \overline{S}}{K + Q}\right)$ with $\overline{S} = \sum_q (\boldsymbol{\theta}_q - \mathbf{b})(\boldsymbol{\theta}_q - \mathbf{b})^T / Q$;
- the conditional posterior for $\boldsymbol{\theta}_q$ is given by $\pi(\mathbf{b}, \boldsymbol{\Sigma} | \mathbf{c}_q) \; \propto \prod_{q=1}^{Q} \Lambda(\mathbf{c}_q | \boldsymbol{\theta}_q) f(\boldsymbol{\theta}_q | b, \boldsymbol{\Sigma})$.

Then, the rth iteration of the Gibbs sampler consists on the following steps:

1) Draw $\mathbf{b}^r$ from $N\left(\overline{\boldsymbol{\theta}}^{r-1}, \boldsymbol{\Sigma} | Q\right)$;

2) Draw $\boldsymbol{\Sigma}^r$ from $\text{IW}\left(K + Q, \dfrac{K \cdot J + Q \cdot \overline{S}^{r-1}}{K + Q}\right)$;

3) For each individual q draw $\boldsymbol{\theta}_q^r$ using one iteration of the MH algorithm, starting from that at the previous iteration and using the Normal density $f(\boldsymbol{\theta}_q | \mathbf{b}, \boldsymbol{\Sigma})$.

These three steps are repeated many times. The resulting values converge to draws from the joint posterior of **b**, **Σ**, and $\boldsymbol{\theta}_q \, \forall q$. Once the converged draws from the posterior are obtained, the mean and standard deviation of the draws can be calculated to obtain estimates and standard errors of the parameters.

Train (2001) discusses how the posterior means from the Bayesian estimation can be analysed from a classical perspective. This is thanks to the Bernstein–von Mises theorem, which states that, asymptotically, the posterior distribution of a Bayesian estimator converges to a Normal distribution which is the same as the asymptotic distribution of the maximum likelihood estimator (e.g. the standard deviation of the posterior distribution of the Bayesian estimator can be taken as the classical standard error of a maximum likelihood estimator). This means that classical statistical analysis (for example the construction of *t*-statistics to analyse the significance of an estimated parameter) can be performed on Bayesian estimators without compromising the interpretation of the results.

Bayesian estimation has certain advantages over the classical approach:

1) No numerical maximization routines are necessary; rather, draws from the posterior distribution are taken until convergence is achieved.
2) As the number of attributes considered in the utility expression grows, the number of elements in the covariance matrix Σ rises exponentially increasing computation time in the classical approach. However, the Bayesian method can handle a full covariance matrix almost as easily as a restricted one, with computation time rising just as the number of parameters.
3) Identification issues are related to the lack of orthogonality in the effects of the random variables and not with the number of independent equations representing these. This means that an identification problem may arise when the effect of a certain variable in the structural utility formulation is confused with the effect of another variable, but not because of insufficient sample points.

The Bayesian estimation procedure is available as experimental code on Ken Train's website, and can also be implemented in WinBUGS. This package incorporates Gibbs sampling protocols and the Metropolis–Hastings sampling algorithm but lacks a convergence analysis that has to be performed separately (Godoy and Ortúzar 2008 provide useful advice on how to do this properly).

Example 8.10 Stated preference data on residential location choice considered the following attributes: travel time to work (by the parents), travel time to study (by the children), rent or mortgage of the flat, and a variable related to atmospheric pollution in the zone (days of alert, see Ortúzar and Rodríguez 2002). 75 families were asked to express their location preferences for a flat of otherwise exactly the same characteristics, finally yielding 648 usable responses (i.e. some observations were discarded in the data cleaning process). MNL and ML models were estimated with this data, and Table 8.11 shows the results for the classical estimation of population parameters in the ML (Sillano and Ortúzar 2005).

In model ML1, the nine choices from each household were considered, correctly, as repeated choice observations, and it was assumed that the parameters distributed IID Normal; the Inertia variable (a dummy that took the value one if the household ranked their current location first) received a non-significant standard deviation, and for that reason the model was re-estimated with Inertia as a fixed parameter for all individuals.

A number of issues are important from this table. First, although the MNL model would be judged adequate by any seasoned analyst, there is a notable increase in log-likelihood (more than a hundred points) for the addition of only four parameters in ML1; a large part of this increase is due to the proper consideration of repeated observations in ML1 (further evidence for this fact has already been discussed above).

Second, note the substantial increase in parameter values from the MNL model to ML1; this (expected) result is due to the 'lurking' scale factor λ corresponding to the IID EV1 error in both

Table 8.11 Model results for location choice analysis

Attributes		Parameters (*t*-test)	
		MNL	ML1
Travel time to work	Mean	−0.0042 (−10.6)	−0.0099 (−7.9)
	Std. dev.		0.0057 (4.5)
Travel time to study	Mean	−0.0025 (−7.8)	−0.0058 (−8.2)
	Std. dev.		0.0027 (2.7)
Days of alert (environment)	Mean	−0.2737 (−11.0)	−0.4786 (−6.8)
	Std. dev.		0.4057 (4.7)
Rent/mortgage	Mean	−0.0264 (−12.5)	−0.0574 (−7.0)
	Std. dev.		0.0475 (6.2)
Inertia	Mean	0.8969 (5.9)	1.0532 (5.5)
Log-likelihood		−849.6	−747.0

models, and it is illustrative to discuss it. The MNL parameters are not allowed to vary among individuals (when it is clear from ML1 that this should be the case); so, as the MNL EV1 error is picking up this, its variance is high and, correspondingly, the MNL scale factor is small. Conversely, the ML EV1 error only has to pick up other, remaining, sources of error so its variance is low, and its scale factor large.

Third, the estimated standard deviations are not only significant but relatively large in magnitude (in comparison to the mean parameter estimates). So, the portion of the population for which the model would assign an incorrect parameter sign can be estimated as the cumulative mass function of the frequency distribution of the parameter evaluated at zero (i.e. for supposedly negative parameters, the area under the frequency curve between zero and positive infinity). In this case, ML1 would account for 4% of the population having positive *Time-to-work* parameters, 1% of the population having positive *Time-to-study* parameters, 12% of the population having positive *Days-of-alert* parameters, and 11% of the population having positive *Rent* parameters. This problem may be overcome using, for example, a log-normal distribution (effectively constraining the parameters to be negative), but this has a series of undesirable properties as we will discuss in more depth below.

Sillano and Ortúzar (2005) went on to estimate individual parameters. Figure 8.5, shows the results for *Time-to-work* and *Days-of-alert*. As can be seen, the above expected proportions were overestimated (e.g. none in the first case and only three out of 75 households in the second); furthermore, the individual parameters for the offending households were not significantly different from zero; hence they could be considered as null values for those households, and the sign assumptions could be maintained.

A final issue relating to Example 8.10 is that while the distribution in Figure 8.5a looks acceptably like a Normal distribution (given the small sample size), that in Figure 8.5b certainly does not. This means that a certain amount of error must be expected when analysing a discrete set of values using a continuous distribution.

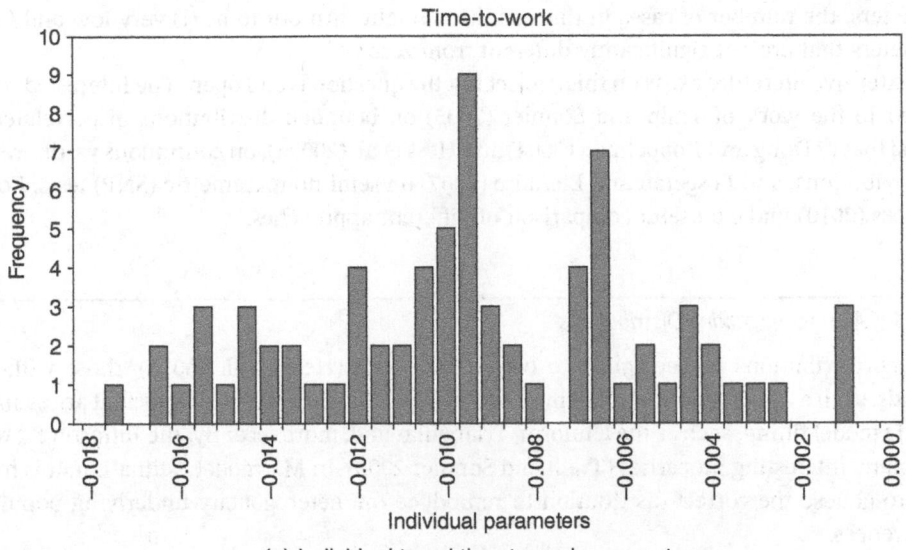

(a) Individual travel time to work parameters

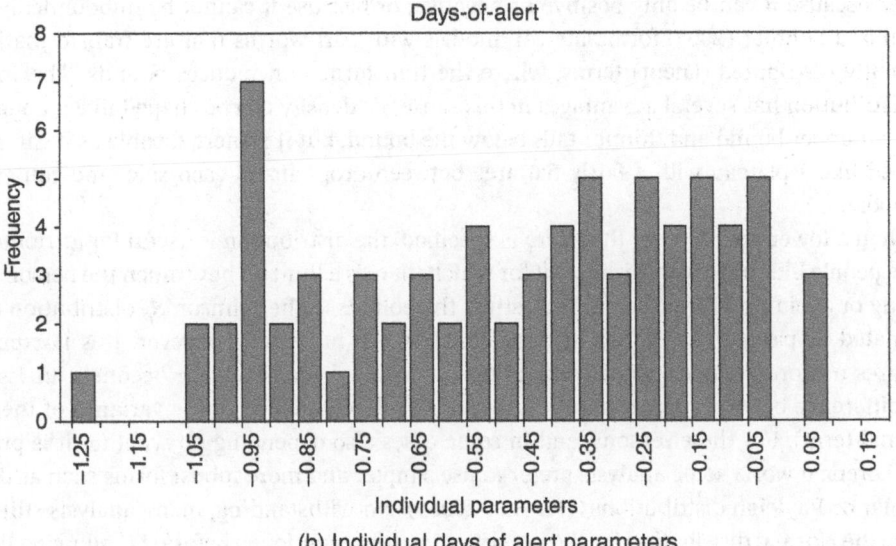

(b) Individual days of alert parameters

Figure 8.5 Distribution of individual parameters in the population for ML1

8.6.3 Choice of a Mixing Distribution

The most popular distribution in ML applications has been the Normal, but some analysts have claimed that the fact it is unbounded imposes unacceptable conditions on the signs of the estimated parameters. This a debatable issue, as precisely because it is unbounded, it may help to reveal 'outliers' or observations that are plainly wrong (in some sense, for example, badly coded or not consistent with the compensatory choice paradigm, see Sælensminde 2001). Furthermore, as we just saw in Example 8.10, even if a proportion of individuals would appear to receive a wrong sign given the estimated population parameters, when we move to the stage of estimating individual parameters, the number of cases in this condition might turn out to be (i) very low and (ii) with parameters that are not significantly different from zero.

An extensive literature exists on this subject but the question is still open. The interested reader is referred to the work of Train and Sonnier (2005) on bounded distributions of correlated part-worths; that of Dong and Koppelman (2004) and Hess et al. (2007a), on continuous vs. discrete mixing distributions, and Fosgerau and Bierlaire (2007) on semi-nonparametric (SNP) tests. Fosgerau and Hess (2010) make a useful comparison of different approaches.

8.6.3.1 *Alternative Mixing Distributions*

Mixing distributions can be split into two main groups (Hess et al. 2005b): those with fixed bounds, such as the Lognormal, Gamma and Rayleigh, and those with bounds that are estimated during model fitting, such as the Uniform, Triangular and, more recently, the Johnson S_β, which has many interesting properties (Train and Sonnier 2005). In ML model estimation, it is important to choose the correct distribution to reproduce the heterogeneity underlying population preferences.

In real cases, almost all attributes have associated a parameter which is logically bounded, either because it can be only positive or negative, or because it cannot be unboundedly large. Train and Sonnier (2005) formulate ML models with part-worths that are transformations of normally distributed (latent) terms, where the transformation induces bounds. The Johnson S_β distribution has several advantages in this sense; its density can be shaped like a Lognormal with an upper bound and thinner tails below the bound, but it is more flexible as it can also be shaped like a plateau with a fairly flat area between drop-offs on each side, and can even be bi-modal.

When a lower bound other than zero is specified, the distribution is useful for attributes that some people like and others dislike but for which there is a limit on how much the person values having or avoiding it. Even more interesting, the bounds of the Johnson S_β distribution can be estimated as parameters, rather than specified by the modeller. However, this last property requires a more complex model estimation process and identification becomes an issue (as the difference between upper and lower bounds is closely related to the variance of the latent Normal term). For these reasons, and in some cases also depending on whether it is practical or theoretical work, some analysts prefer to use simpler and more robust forms such as the Triangular or Rayleigh distributions (Hensher 2006a). Notwithstanding, many analysts still prefer to use the Normal distribution for the important reasons mentioned before (i.e. allowing them to detect potential outliers) and because it tends to faster convergence of the estimation procedure than other distributions.

Finally, a number of applications have also looked at incorporating deterministic heterogeneity components into the distribution of the random terms, either in the mean or the standard deviation, hence allowing the modeller to relate the variation of random coefficients to individual-specific observed attributes, that is, akin to what we called 'systematic taste variations' in Eq. (8.17). As an example, in a standard framework we would possibly use $\theta \sim N(\mathbf{b}, \Sigma)$, but here we would additionally specify $\mathbf{b} \sim f(\mathbf{s})$ and $\Sigma \sim g(\mathbf{s})$, making the parameters of the distribution a function of socio-demographic variables $\mathbf{s}$. This can be useful either in a random coefficients as well as in an error components context; an example of such an approach is given by Greene et al. (2006).

8.6.3.2 Discrete Mixtures and Latent Class Modelling

Dong and Koppelman (2004) represented heterogeneity with a discrete distribution with a finite number of supports; in this case $f(\theta)$ is replaced by a mass-point distribution with weight at mass point m given by π_m. Then, by replacing the integration in (8.51) with a sum over a finite number of mass points M, the ML model can be expressed as:

$$\mathbf{P}_{iq} = \sum_{m=1}^{M} \frac{\exp\left(\theta_i^m \mathbf{x}_{iq}\right)}{\sum_{A_j \in A(q)} \exp\left(\theta_j^m \mathbf{x}_{jq}\right)} \cdot \pi_m \tag{8.54}$$

so, it is a weighted average of logit probabilities computed at each possible value of θ (the weights are the probabilities of θ to be at each value θ^m), and can be estimated by maximum likelihood *without* simulation. Using simulated data, they found that model (8.54) was inferior to the conventional ML whether the true distribution was continuous or discrete; furthermore, they found that the model estimates could be misleading if the true distribution was, in fact, continuous; however, their model allowed the identification of heterogeneity which was not discovered by the continuous version of the ML.

Hess et al. (2007a) generalised this approach by letting the MNL probability be any more general GEV function; they divided the set θ into two parts, one, $\bar{\theta}$, containing deterministic parameters and another, $\hat{\theta}$, with K random parameters with a discrete distribution; within this set, $\hat{\theta}_k$ has m_k mass points, $\hat{\theta}_k^n$, $n = 1, ..., m_k$, each of them associated with a probability π_k^n, where the following conditions must be imposed:

$$0 \leq \pi_k^n \leq 1, \qquad k = 1, ..., K; \quad n = 1, ..., m_k$$

and

$$\sum_{n=1}^{m_k} \pi_k^n = 1, \quad k = 1, ..., K$$

They discussed several extensions that offer more modelling flexibility, but noted that some may lead to parameter over-specification, impairing estimation. They also noted that the non-concavity of the log-likelihood function in this case does not allow the identification of a global maximum, even for discrete mixtures of the simple MNL model; thus, they advised the performance of several estimations from various starting points and recommended, as good practice, the use of staring values different from 0 or 1 for the π_k^n parameters.

If the class allocations are linked to socio-demographic variables, we obtain a Latent Class (LC) model (see, for example, Hess et al. 2009). In a LC model the heterogeneity in tastes across respondents is accommodated by making use of separate classes with different values for the vector of taste coefficients θ. Specifically, in a LC model with M classes, we would have M instances of the vector θ, say θ^1 to θ^M, with a possibility of some of the elements in θ staying constant across some of the classes.

A LC model uses a probabilistic class allocation model, where respondent q belongs to class m with probability $\pi_{q,m}$ and where $0 \leq \pi_{q,m} \leq 1$, for all m and $\Sigma_m \pi_{q,m} = 1$. LC models are generally specified with an underlying MNL model, but can easily be adapted for more general underlying structures such as Nested Logit (NL) or Cross-Nested Logit (CNL).

Let $P_{iq}(\theta_m)$ give the probability of respondent q choosing alternative A_i conditional on her falling into class m. The unconditional (on m) choice probability for alternative A_i and respondent q is then given by:

$$P_q(A_i \mid \theta_1, ..., \theta_M) = \sum_{m=1}^{M} \pi_{q,m} P_{iq}(\theta_m) \tag{8.55}$$

that is, the weighted sum of choice probabilities across the M classes, with the class allocation probabilities being used as weights. Unlike with the ML model, no simulation is required in the estimation of LC models.

This specification can easily be extended to a situation with multiple choices per respondent, where, when making the same assumption of intra-respondent homogeneity as in the Revelt and Train (1998) work for continuous ML, we obtain:

$$P_q(A_i \mid \theta_1, ..., \theta_M) = \sum_{m=1}^{M} \pi_{q,m} \left(\prod_{t=1}^{T_q} P_{iqt}(\theta_m) \right)$$

In the most basic version of an LC model, the class allocation probabilities are constant across respondents such that $\pi_{q,m} = \pi_m$ for all q. The resulting model then corresponds to a discrete mixture analogue to the ML model, as discussed above.

The real flexibility however arises when the class allocation probabilities are not constant across respondents but a class allocation model is used to link these probabilities to characteristics of the respondents. Typically, these characteristics would take the form of socio-demographic variables such as income, age, and employment status. With $\mathbf{s}_q$ giving the concerned vector of characteristics for respondent q, and the class allocation model taking on a MNL form, the probability of respondent q falling into class m would be given by:

$$\pi_{q,m} = \frac{\exp\left(\delta_m + g(\beta_m, \mathbf{s}_q)\right)}{\sum_{l=1}^{M} \exp\left(\delta_l + g(\beta_l, \mathbf{s}_q)\right)}$$

where δ_m is a class-specific constant, β_m a vector of parameters to be estimated and $g(\cdot)$ gives the functional form of the *utility* function for the class allocation model.

Here, a major difference arises between class allocation models and choice models. In a choice model, the attributes vary across alternatives, while the estimated coefficients (with a few exceptions) stay constant across alternatives. In a class allocation model, the attributes normally stay constant across classes while the parameters vary across classes. This allows the model to allocate respondents to different classes depending on their socio-demographic characteristics. For example, a situation where high- and low-income respondents are allocated differently to two classes, could be represented with a positive income coefficient for the first class and a negative income coefficient for the second class. Finally, we can mention that it is possible to combine latent class and ML structures, leading to latent class models with some continuous elements, as for example done by Walker and Li (2007).

It is interesting to note the work of Fosgerau and Bierlaire (2007), who proposed the use of semi-nonparametric (SNP) techniques to test if a given mixing distribution is appropriate. The SNP models offer the advantage, over conventional ML, that the structure does not need to be specified *a priori*. In particular, they introduce parametric assumptions like the specification of some relationship to be a linear combination of independent variables while perhaps the errors remain nonparametric. SNP models are not based on local approximations; instead, they use series to approximate functions such as densities. The number of SNP terms must be chosen in advance; increasing this number makes the model more general but increases the demand on the data. Fosgerau and Bierlaire (2007) found that two or three terms give a large degree of flexibility, which may be sufficient for most purposes, while one SNP term is not always sufficient to reject a false null hypothesis.

8.6.3.3 *Empirical Identifiability of Latent Class Models*

Latent class (LC) models have been widely applied either exclusively with exogenous variables (Swait and Adamowicz 2001; Rossetti et al. 2018) or in conjunction with latent variables in a MIMIC model (Hess and Stathopoulos 2013), and either using only utility maximisation heuristics or adopting a different choice heuristic for each latent class (Hess et al. 2012; González-Valdés and Raveau 2018).

A key issue concerning the use of LC models is their identifiability, which is related to the possibility of drawing inferences from observed samples about an underlying theoretical structure that is observationally unique. Rothenberg (1971) examined the identifiability of parametric models, concluding that this required the information matrix to be non-singular. Walker and Ben-Akiva (2002) investigated theoretical and empirical identifiability. Here, we focus on the latter, where the model can be identified in theory, but due to the data and model structure, the Hessian matrix is singular or nearly so (Chiou and Walker 2007; Cherchi and Ortúzar 2008b), leading to poor estimates of model parameters and impeding empirical identification.

In LC models, identifiability informs about distinguishing different behaviour types and estimating the parameters that govern them, with the behaviour of each individual in the population being described as a linear combination of the theoretical constructs. The identifiability of LC models has been studied to varying extents. Huang and Bandeen-Roche (2004) explored theoretical identifiability in LC models, specifying conditions of the components of a latent class – latent variable choice model required to achieve it. Requirements for the empirical identifiability of models that have no latent variables were addressed recently by González-Valdés et al. (2022). The application that motivated their study involved the potential existence of multiple-choice heuristics (such as those discussed in Section 7.7.2.2) in a sample of consumers. Previous studies had considered, successfully, LC models under a single heuristic with multiple parameter sets (Greene and Hensher 2003). In the case of multiple heuristics, LC models resorted to latent variables (Hess and Stathopoulos 2013) and normalisations (Leong and Hensher 2012) for identifiability. González-Valdés et al. (2022) were the first to establish analytical conditions for identifiability, and showed how they apply in practice to the challenge of identifying multiple heuristics.

8.6.4 Binary Choice Case

Suppose that individuals align their behaviour to one of two latent classes, denoted as a and b, with probabilities π_a and $(1 - \pi_a)$ respectively. Let $P_{cni}(\theta)$ be the probability that according to class $c \in \{a, b\}$ with parameters θ, individual n chooses alternative i. Then, $P_{ni}(\theta, \pi_a)$, the probability of individual n choosing alternative i under the LC model, is given by:

$$P_{ni}(\theta, \pi_a) = \pi_a P_{ani}(\theta) + (1 - \pi_a)P_{bni}(\theta) \tag{8.56}$$

The log-likelihood function of this model is given by (8.57), where $P_{cn*}(\theta)$ represents the probability that individuals n would have chosen their selected alternative aligning their behaviour to latent class c:

$$l(\theta, \pi_a) = \sum_q \log_e(\pi_a P_{an*}(\theta) + (1 - \pi_a) P_{bn*}(\theta)). \qquad (8.57)$$

The maximum value of this function could arise either at a boundary or at an interior value of π_a. In the case of a boundary solution, the optimal model consists of a single latent class: a when $\pi_a = 1$, or b when $\pi_a = 0$. By contrast, in the case of an interior solution (i.e. $\pi_a \in (0, 1)$), both classes coexist in a mixture model. Thus, when an interior solution arises, it reflects theoretical identifiability.

The solution (interior or boundary) depends upon the losses and gains in likelihood associated with including an additional class in the model and, therefore, reducing the proportion of the complementary one. Class a may perform better than class b for some observations, with the reverse occurring for other observations. Including a second class, b, would improve the likelihood of the latter observations. However, in cases where the first class a performs better, there would be a loss of likelihood due to the reduction of its proportion in the model. The balance between these two changes in performance determines the type of solution obtained (i.e. whether the solution is a boundary or an interior one).

A boundary solution will be obtained when it is optimal for the model to consider a single class of individuals, corresponding to the case where the improvement in likelihood from the inclusion of a second class does not compensate for the associated losses. In the case of an interior solution, when identifiability of the class membership component is possible, likelihood is maximised when the likelihood function is stationary with respect to variations in the class membership probability π_a. This can be detected as an interior point at which the derivative of the log-likelihood function equals zero. Among the variables to examine, an interesting one is precisely π_a, because it indicates the proportions of the two classes and, therefore, connects them in the model.

González-Valdés et al. (2022) first show that when the class membership function π_a is constant across the population (i.e. the probability of class membership is the same for every individual), two latent classes coexist optimally if the vector θ of estimated parameters satisfies the balance specified by (8.58):

$$\sum_n \frac{P_{an*}(\theta)}{P_{n*}(\theta)} = \sum_n \frac{P_{bn*}(\theta)}{P_{n*}(\theta)} \qquad (8.58)$$

where $P_{n*}(\theta, \pi_a) = \pi_a P_{an*}(\theta) + (1 - \pi_a) P_{bn*}(\theta)$ denotes the modelled probability that individual n effectively chooses the chosen alternative (consistent with (8.57)).

They further proved that the two latent classes will coexist optimally in a discrete choice model in this case, if the balance quantity in (8.58) is equal to the sample size N. These findings can be helpful for practitioners to explain the lack of theoretical identifiability in their models. If only one class is identified, this is not sufficient to establish that the other behaviour is absent from the data but only shows that the single class can interpret the behaviour exhibited by the other class adequately. This arises when the gain in likelihood of including a second class does not compensate the loss in likelihood for the observations that are aligned more closely with the first class.

If the class membership function π_a is not constant but is instead some function $\pi_a(\theta)$, there is also a balance – albeit more complicated – for the parameter vector θ. In particular, if the set of parameters β, of the class membership function, is disjoint from the set θ affecting the choices, then, the balance required of sensitivity of class membership is given by (8.59):

$$\sum_n \frac{\frac{\partial \pi_a(\beta)}{\partial \beta} P_{an*}(\theta)}{P_{n*}(\theta, \beta)} = \sum_n \frac{\frac{\partial \pi_b(\beta)}{\partial \beta} P_{bn*}(\theta)}{P_{n*}(\theta, \beta)} \tag{8.59}$$

The analysis so far has identified when it is optimal for the model to include more than one latent class. Nevertheless, the coexistence of latent classes (i.e. theoretical identifiability) does not guarantee that the estimated parameters will have reasonable standard deviations (i.e. empirical identifiability).

Conditions for empirical identifiability. For a parametric model to be theoretically identifiable, the Fisher information matrix F given in (8.60) must be non-singular (Rothenberg 1971).

$$F = -\mathbb{E}\left(\frac{\partial^2 l(\theta)}{\partial \theta_x \partial \theta_y}\right) \tag{8.60}$$

Moreover, for greater precision in the parameter estimates, their covariance matrix Σ should have values on the principal diagonal with small square roots compared to the corresponding point estimates of the parameters. The covariance matrix is related to the Fisher information matrix by:

$$\Sigma \approx F^{-1}.$$

The elements on the principal diagonal of F^{-1} provide the Cramér–Rao lower bound on the variance of the parameters θ in the corresponding elements of Σ. Thus, to obtain higher precision in the estimation, the determinant of the information matrix F should be large, hence requiring large values of $-\mathbb{E}\left(\frac{\partial^2 l(\theta)}{\partial \theta^2}\right)$ on its principal diagonal.

Analysing the information matrix at the point determined by π_a when the class membership is constant, González-Valdés et al. (2022) show that the second derivative of the log-likelihood function with respect to π_a is directly proportional to the difference $(P_{an*} - P_{bn*})^2$. Thus, for F to have a large determinant, and for the standard errors of the estimators to be small, the magnitude of this difference must be large and this happens when the classes exhibit disparate behaviour.

Extension to several latent classes. Extending the previous notation, let π_c be the probability that individual behaviour aligns to class $c \in C$, so that $\sum_{c \in C} \pi_c = 1$ and $\pi_c \geq 0 \; \forall c \in C$. In this case, the joint log-likelihood function $l(\pi, \theta)$ of the model is given by (8.61):

$$l(\pi, \theta) = \sum_n \log_e \left(\sum_{c \in C} \pi_c P_{cn*}(\theta)\right) \tag{8.61}$$

This log-likelihood is maximised, subject to the constraint $\sum_{c \in C} \pi_c = 1$ (with Lagrange multiplier λ) and positivity constraints on the probabilities $\pi_c \geq 0 \; \forall c \in C$ (with Lagrange multipliers η_c), when the Lagrangian (8.62) is stationary with respect to variations in $\pi_c \forall c \in C$:

$$L = -l(\pi, \theta) - \lambda\left(1 - \sum_{c \in C} \pi_c\right) - \sum_{c \in C} \eta_c \pi_c \tag{8.62}$$

Differentiating the Lagrangian $\mathcal{L}$ with respect to π_c and equating this to zero for stationarity, allowed González-Valdés et al. (2022) to show that a necessary condition for several latent classes $c \in C$ to coexist optimally in a model requires that each of them achieves the same aggregated ratio $\sum_n \dfrac{P_{cn*}}{P_{n*}} = N$.

Again, this theorem presents the balance conditions for the optimal combination of latent classes, but does not guarantee their empirical identifiability. For the class membership probabilities π to be identifiable, the information matrix F should be non-singular and because it is real and symmetric, the Hessian matrix of the Lagrangian should be positive and definite. This requires that all principal submatrices of the Hessian, that correspond to the second derivatives with respect to the proportions, should have positive determinants. Working on these lines, González-Valdés et al. (2022) derived a final theorem: "If several latent classes coexist in an identifiable model, empirical identifiability improves as the covariance of the latent classes decreases", that is, empirical identifiability increases as the behavioural difference of the latent classes increases as quantified by a decreasing covariance among them.

To conclude this section, let us summarise its important findings:

1) There must be a balance between the latent classes for theoretical identifiability.
2) The behaviour of the classes must differ sufficiently so that they can be identified empirically with acceptable accuracy in estimates of their parameters. The balance required for joint estimation requires that one latent class is not sufficiently good at explaining the behaviour of members of other classes, as quantified in the balance equation.
3) On empirical identifiability, latent classes must differ sufficiently in their typical behaviour, and the data used in estimation must include sufficient cases that expose this difference. If either of these conditions is not satisfied, simultaneous identification of the latent classes in a single model will not be possible.

8.6.5 Random and Quasi Random Numbers

The multidimensional integral (8.50) has to be solved via simulated maximum likelihood, and this can be very time-consuming in real large-scale model estimations. As a consequence, several methods have been devised to help in this task, including the use of cheaper (in time) quasi-Monte Carlo approaches, based on the generation of 'low discrepancy' or 'quasi-random' sequences (see Niederreiter 1992), as they allow more accurate integration approximations than classical Monte Carlo sampling (Train 2009).

Figure 8.6a shows the uneven coverage of the space of integration by the typical pseudo-random numbers generated automatically by computers (300 points in two dimensions). Figure 8.6b shows the much better coverage of Halton numbers, one of the early sequences used by researchers, in this case.

In fact, it has been reported that only 125 Halton numbers can provide the equivalent coverage of 1000 pseudo-random numbers (see Train 2009). Now, although Halton sequences ruled for a while, it was soon shown that their coverage of the integration domain rapidly deteriorated for higher integration dimensions (Silva 2002); for example, Figure 8.7a shows the Halton sequence pattern for an example with several dimensions, and this should be compared with the Sobol sequence pattern for the same number of dimensions (Figure 8.7b).

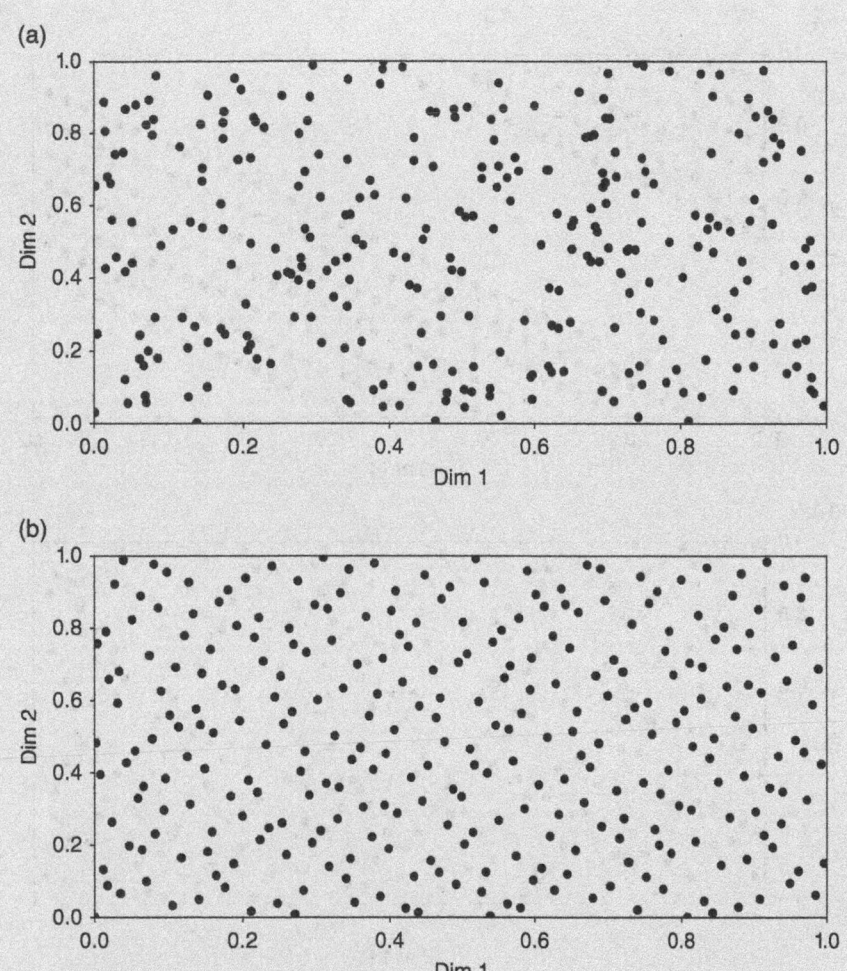

Figure 8.6 Pseudorandom and Halton coverage in two dimensions

This fostered a search for new sequences, including work on scrambled and shuffled Halton sequences (Bhat 2003), Sobol sequences (shown to be superior to the former by Silva 2002), and, shortly afterwards, Modified Latin Hypercube sampling (Hess et al. 2006).

A related but different approach was taken by Bastin et al. (2006), who capitalised on the desirable aspects of pure Monte Carlo techniques while significantly improving their efficiency. They proposed a new algorithm for stochastic programming based on 'trust-region' techniques (a well-known method in nonlinear non-concave optimization that proved reliable and efficient for both constrained and unconstrained problems). They also allowed for an adaptive variation of the number of draws used in successive iterations, yielding an algorithm with comparable execution time to existing methods for similar results.

Numerical experimentation suggests that the choice of optimisation framework is of crucial importance; also, the strategy of using a variable number of draws in the estimation of choice

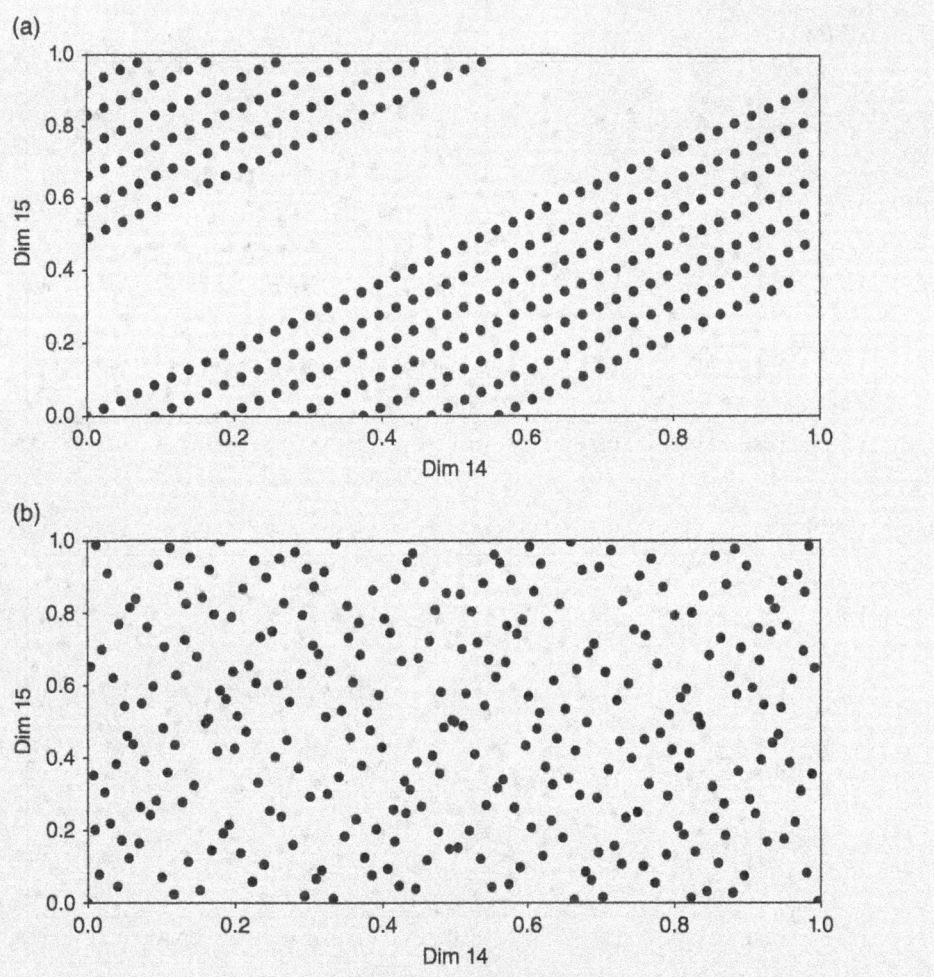

Figure 8.7 Halton and Sobol coverage in many dimensions

probabilities gives significant gains in optimisation time (compared with the usual approach of using fixed draws) and additional information on the closeness between the Monte Carlo approximation and the true function whilst not suffering from non-uniform coverage in high integration dimensions. However, the field is still young in this sense and many research directions remain wide open. The interested reader should look at what is in offer in the practical packages available to estimate these complex models (i.e. ALOGIT, Apollo, Biogeme, LIMDEP), as they are continuing to incorporate advances in the field.

8.6.6 Estimation of Panel Data Models

As we saw in Chapter 7, panel data offer major advantages over cross-sectional data; in particular, having repeated observations from the same individual generally allows for more accurate measurement of changes in individual mobility. Furthermore, as we commented in

Section 7.7.4.1, the inclusion of intra-respondent heterogeneity, which is only possible if there are multiple observations per individual, leads to significant improvements in model fit.

Given the potential of panel data structures, the challenge is to make the most of this potential by capturing as many effects as possible. A general formulation for panel data model estimation is that proposed by Hess and Rose (2009) in Eq. (7.50); this considers not only inter-respondent heterogeneity as in the more classical specification (7.49), but also intra-respondent heteroge-neity of tastes.

Surprisingly, even this general panel formulation (that considers two dimensions of hetero-geneity) accommodates heterogeneity only via the estimation of random parameters. Thus, random parameters θ_q account also for correlation in tastes. However, as we commented in Section 7.7.4.1, there might be extra correlation across multiple observations besides the effect of the random parameters. For example, the inclusion of pure panel component errors may also account for correlation in the preferences for alternatives, as proven by Yáñez et al. (2011). They analysed the impact of panel sample size and repeated observations on both the model's capa-bility to reproduce the true phenomenon and the probability of capturing different kinds of heterogeneity among observations. They found that their best model accommodated inter-respondent heterogeneity through random parameters and intra-respondent heterogeneity through pure-error components.

For practical purposes, another important issue regarding panel correlation has to do with model estimation using available software. The usual way to incorporate panel correlation under the pure error-components approach consists of adding an error component to (J-1) of the available alternatives; otherwise, for identifiability reasons, the model cannot be estimated (Walker 2002). However, this methodology may lead to biased results as it requires choosing, arbitrarily, a single reference alternative for the error component (i.e. one not having a pure panel error component) in all cases. The reason is that this is equivalent to assuming that this reference alternative has the same alternative specific constant (ASC) for all observations, whilst the remaining ones have different ASC values among observations. Moreover, even using the best recommended normalization (i.e. selecting as *error-component reference alternative* that option obtaining the minimum variance in a model run without considering identifiability, see Section 7.6.3.1), this approach leads to a heteroskesdastic Nested Logit model, as it correlates the (J-1) alternatives including error components.

One way to avoid this problem is to modify this traditional estimation method by randomly selecting the error-component reference alternative for each individual (or each observation in the case of allowing for intra-respondent heterogeneity).

For this we need first to choose randomly and exogenously (i.e. before model estimation) an error-component reference alternative for each individual (or for each observation in the case of needing to accommodate intra-respondent heterogeneity). Then, we need to create J binary vari-ables that take the value one only for the error-component reference alternative in either case of respondent heterogeneity. Finally, the pure error-component term is included in the utility function of each alternative multiplied by the corresponding binary variable.

Example 8.11 Consider the following utility function:

$$U_{iqd} = \alpha_i + \sum_j X_{iqk}^d \cdot \theta_{iqk} + \zeta_{iqd}$$

where the error component has the form $\zeta_{iqd} = \upsilon_{qd} + \varepsilon_{iqd}$. Here ε_{iqd} is a random term distributed IID EV1, as usual, and υ_{qd} is a random effect that may be specific to the individual (i.e. just υ_q), in which case we assume panel correlation as inter-respondent heterogeneity. But we could also make it variable among observations (v_{qd}), in which case we would assume intra-respondent heterogeneity.

As the θ vector has means θ_{ik} and standard deviations σ_{iqk} the utility function can be rewritten as:

$$U_{iqd} = (\alpha_i + v_{qd}) + \sum_k (\theta_{ik} + \sigma_{iqk})X_{iqk}^d + \varepsilon_{iqd}$$

where X_{iqk}^d is the kth level-of-service attribute of option A_i for individual q on day d.

This equation shows that both random coefficients and error components are separable. Indeed, the random coefficients allow tastes to vary across respondents in the sample, but stay constant across observations for the same respondent (Revelt and Train 1998). On the other hand, the 'pure' error components (which also capture heterogeneity) affect the values of the alternative specific constants (ASC). Thus, the error component v_{qd} has the power to increase/decrease the relative weight of the ASC in relation to the explanatory variables in the utility function.

Now, confounding effects are implicit in the ML structure as we saw in Section 7.6.2 and should not strictly depend on whether they do or do not account for random tastes. On the contrary, Cherchi and Ortúzar (2008b) have shown that decomposing randomness in as many components as possible, helps to reveal the confounding effects.

8.7 Modelling with Stated-Preference Data

In Chapter 2, we discussed the experimental design and the data collection process of stated choice (SC) data in some detail; we made scant reference to traditional conjoint analysis (rank and rating data) and left contingent valuation for Chapter 18. In Section 8.3, we noted that stated preference (SP) experiments could be instrumental in helping to decide the most appropriate functional form to model a given choice situation. In this section, we will first briefly review how this can be done and then we will proceed to discuss what changes are introduced to discrete choice modelling estimation by the use of SP data.

8.7.1 Identifying Functional Form

The travel-demand model estimation literature is heavily oriented towards the problem of estimating a set of model parameters given a functional specification; only occasionally are alternative model structures tested. The favoured functional forms are those which can be deduced from (economic) first principles and also satisfy the condition of being easily estimable; for this reason, the vast majority of studies has considered linear (in the parameters) utility functions. A notable exception to this rule is the increasing use of transformations to search for functional form but, as we saw in Section 8.3, in these cases the computational problem associated with estimating the model increases.

In contrast, the literature in the area of psychological measurement procedures that use laboratory or interview data, has been deeply concerned with questions of functional form for a long time (see Louviere 1988a). In these studies, subjects are asked to make judgements about hypothetical alternatives; for example, in a mode choice context they may be asked to select the preferred alternative from a hypothetical set, or to rank the options, or to associate a level of utility to each of them.

Because an individual can be asked to make a fairly large number of judgements in a single interview, the experiment designer can explore, for example, the effects on response of changes to one variable while keeping all the others constant. This allows a much more detailed assessment of functional form, since the analyst can almost trace the shape of response with respect to each variable. An interesting finding of such studies is that for any particular decision, functional forms tend to be fairly stable across the population even though the values of their parameters can vary widely (see Meyer et al. 1978).

Let us assume that travel behaviour is influenced by a set of independent factors, which may be quantitative or qualitative in nature. Following Lerman and Louviere (1978), let us denote the set of G quantitative factors for option A_i by $\mathbf{D}_i = \{D_{ig}\}$ and the set of H qualitative factors by $\mathbf{E}_i = \{E_{jh}\}$. The total number of factors is $K = G + H$, and the entire attribute vector $\mathbf{X}_i = \{X_{ki}\}$ is simply $\mathbf{D}_i$ and $\mathbf{E}_i$.

Let us also assume that each factor has associated with it a certain value (which may be obtained by some or other measurement process) and that the utility of this quantity as perceived by the individual is $u_{ki} = f_{ki}(X_{ki})$, where f is a perception function.

Consider now an experimental context where we observe the response to a combination of $(D_{1i}, ..., D_{Gi}; E_{1i}, ..., E_{Hi})$ on a psychological measurement scale. If we assume that this response measure is connected to the utility U_i of option A_i by some algebraic combination rule, we can write:

$$U_i = p_i(u_{1i}, ..., u_{Ki}) \tag{8.63}$$

Finally, if we postulate that the vector of responses $\mathbf{U} = \{U_i\}$ is connected to non-experimental (i.e. observed) behaviour B by another algebraic function, we can write:

$$B = w(\mathbf{U})$$

and by substituting, we get:

$$B = w\{p[f(\mathbf{D}, \mathbf{E})]\}$$

As this is too general a formulation for modelling purposes, in practical applications one must make explicit assumptions about the functions f, p, and w, and deduce their consequences.

Now, for the purposes of developing an appropriate functional form, the critical component of this approach is the specification of Eq. (8.63). Alternative forms, such as multiplicative or linear cases, may be tested and selected by means of analysis of variance; however, to successfully apply the approach, two conditions must be satisfied: first, the pattern of statistical significance of the utility responses to various combinations of the independent variables must be of a specific nature to permit diagnosis or testing of model form; second, corresponding graphical evidence must support the diagnosis or test.

Example 8.12 Consider a residential location model where individuals are assumed to trade off the total cost of travel (including travel times) with house price, independently of one another (i.e. it is assumed that they combine the effects of the two variables linearly). This hypothesis may be tested directly by an analysis of variance. Suppressing the option index i for simplicity, we can write:

$$U_{mn} = U_m^1 + U_n^2 + \varepsilon_{mn}$$

where U_l^k are utility values assigned to the lth level of the kth attribute in a factorial design, U_{mn} stands for the overall utility assigned by individuals to combinations of levels of both attributes, and ε_{mn} is a random term with zero mean.

A test for independence of the two effects corresponds to a test of the significance of the interaction effect $U_m^1 U_n^2$. As Lerman and Louviere (1978) point out, in an analysis of variance this is a global test for any and all interactions between both variables; thus, if the interaction effect is not significant, the hypothesis of linear form cannot be rejected. If the interaction is significant, on the other hand, it implies that a simple linear combination is not appropriate.

This test should be accompanied by a graphical plot of the interaction. If the linear hypothesis (no interaction) is correct, the data should plot as a series of parallel lines when plotted against either utility value. It can be shown that this is true regardless of the form assumed for the marginal relationships (8.48); it can also be shown that this is true for any multi-linear utility model and for any forms less restrictive than simple addition or multiplication.

8.7.2 Stated Preference Data and Discrete Choice Modelling

There are two particular features of SP data that lend the approach to different analysis methods, *vis à vis* other sources of disaggregated data: first, the fact that each respondent may contribute more than one observation; and second, the different forms in which preferences can be expressed. In Chapter 2, we mentioned in passing that traditional conjoint analysis considered two types of responses: ratings and rankings, but that the field has been clearly dominated by SC data. In the case of ratings, the subject is asked to rate each option using a number between 1 and 5 or 10. The result of this exercise may be interpreted as the strength of the individual preference for each alternative. Therefore, normal algebraic operations can be carried out on them, for example extracting a ratio or subtracting one from another. However, this is now believed to be a weak element in SP work, as there is no evidence to support the assertion that individual preferences can be elicited and translated into cardinal scales of this kind.

Simpler, and more reliable, tasks are to ask individuals to rank alternatives in order of preference or, much simpler, to make several choices between hypothetical alternatives. In the case of *ranking* experiments, the individual is asked to rank a set of N alternatives in order of preference. If r_i denotes the alternative ranked in the ith position, the response implies that:

$$U(r_1) \geq U(r_2) \geq \ldots \geq U(r_N) \tag{8.64}$$

In the case of SC exercises individuals are only asked to choose their preferred alternative from those (two or more) presented in the choice set; therefore, in this case the response corresponds with the usual discrete choice RP approach, except for the fact that both alternatives and choices are hypothetical. Note, however, that this type of exercise can be extended and enriched by allowing respondents to express their degree of confidence in the SC. To this end, the respondent is offered a semantic scale, the most typical having five points (*(i): Definitively prefer first option; (ii) Probably prefer first option; (iii) Indifferent; (iv) Probably prefer second option; and (v) Definitively prefer second option*). This exercise is sometimes also called *rating* in the transport literature although it is actually a generalization of a choice experiment (see, for example, Ortúzar and Garrido 1994a,b; Cherchi and de Ortúzar 2008b). This generalisation offers advantages and disadvantages: on the one hand it permits a richer range of modelling techniques to be applied to the data; on the other hand, it may weaken the specificity of the choice and that of the response, increasing the difference between experiment and behaviour.

Taking advantage of the special features of SP data, there are four broad groups of techniques for analysis:

- naive or graphical methods;
- least square fitting, including linear regression;
- non-metric scaling; and
- Logit and Probit analysis.

These methods can be used to provide different levels of analysis of SP experiments. In general, all seek to establish the weights attached to each attribute in an (indirect) utility function estimated for each alternative. These weights are sometimes referred to as 'preference weights', 'part utilities', 'part-worths', or simply 'coefficients' associated with each attribute. Once these have been estimated they can be used for various purposes:

1) To determine the relative importance of the attributes included in the experiment.
2) To estimate the rate at which one attribute is traded off with another (a typical example is the estimation of 'values-of-time' when both time and cost attributes have been included in the experiment); it is also possible to estimate the value of more qualitative attributes like reliability, security levels, and so on; we will come back to this in Section 18.4.
3) To specify utility functions for forecasting models, including questions of model structure.

The nature of SP data and the objective of the analysis will be determining factors in the choice of model estimation techniques.

8.7.2.1 Naive Methods

The naive or graphical methods use a simple approach based on the fact that in many designs each level of each attribute appears the same number of times. Therefore, some indication of the relative utility of that attribute-level pair can be obtained by computing the mean average rank, rating or choice score for each option in which it was included and comparing that with similar mean averages for other levels and attributes. In effect, just plotting these means on a graph often gives useful indications about the relative importance of the various attributes included in the experiment. This model does not make use of any statistical theory and therefore fails to give us an indication of the statistical significance of the results.

Example 8.13   Consider a SP exercise comparing three alternative modes of transport, a traditional diesel bus (DB), a modern mini bus (MB), and an electric light rail vehicle (LRT). The attributes included in the SP experiments are in-vehicle travel time, the headway, the fare, and, of course, the vehicle type. Table 8.12 shows the different levels to be tested for each attribute:

Table 8.12 Attribute levels for Example 8.13

	Level 1	Level 2	Level 3
Travel time (min)	25	15	35
Fare (£)	1.30	1.00	1.50
Headway (min)	5	10	20
Vehicle type	DB	LRT	MB

A fractional factorial design was used, and respondents were asked to rate, or score, the alternatives (10 as the highest or best service), with the results shown in Table 8.13:

Table 8.13 Attribute values and scores for Example 8.13

Travel time	Fare	Headway	Vehicle type	Score
25	1.30	5	DB	8
25	1.00	10	MB	9
25	1.50	20	LRT	4
15	1.30	10	LRT	10
15	1.00	20	DB	7
15	1.50	5	MB	8
35	1.30	20	MB	4
35	1.00	5	LRT	4
35	1.50	10	DB	1

It is now possible to calculate a 'naive' value for each attribute, by calculating the average score for that level and attribute and comparing it with the difference in values. For instance, in the case of travel time Table 8.14 can be constructed:

Table 8.14 Travel time naïve values for Example 8.13

Travel time level	Value (min)	Difference in values	Average rating	Difference in rating	Rating per min
1	25	—	21/3	—	—
2	15	−10 (2 − 1)	25/3	4/3 (2 − 1)	−0.133
3	35	20 (3 − 2)	9/3	−16/3 (3 − 2)	−0.267

and in the case of fares, Table 8.15:

Table 8.15 Fare naïve values for Example 8.13

Fare level	Value (£)	Difference in value	Average rating	Differences in rating	Rating per £
1	1.3	—	22/3	—	—
2	1.00	−0.3 (2 − 1)	20/3	−2/3	2.22
3	1.50	0.5 (3 − 2)	13/3	−7/3	−4.67

From this we can estimate the subjective value of time (SVT) as follows: SVT is equal to $(-0.267)/(-4.67) = 0.057$, that is the ratio of ratings per min over ratings per £. The reader can calculate the values of headway and vehicle type in the same way.

Two reflections follow from this very simple example: the values of time or other attributes depend on the 'difference' being considered; for instance, moving from 15 to 25 minutes does not produce the same SVT as moving from 25 to 35 minutes. The second comment is that we may estimate the values of these coefficients using the scores, as in this case, produced by a single respondent; that is, as each interview generates several observations in many cases, we can estimate individual rather than sample-based models.

However, the naive method is seldom used in practice, except as a quick way of estimating indicators like the value of time to provide an initial, 'in the field' validation of an experiment. Notwithstanding, this example has served to illustrate some of the ideas behind SP data analysis.

8.7.2.2   Discrete Choice Modelling with Rating Data

The objective of the rating data analyst is to find a quantitative relation between the set of attributes and the response expressed in the semantic scale. For this he needs first to associate a numerical value R_m to each sentence m ($m = 1, ..., M$) of the scale and postulate a linear model such as:

$$\theta_0 + \theta_1 X_1 + \theta_2 X_2 + ... + \theta_K X_k = r_j \tag{8.65}$$

where θ_0 is a constant, X_k is typically the difference between the kth attributes of two competing options in the situation considered; θ_k is the coefficient of X_k and r_j represents a transformation of the response of individual j (i.e. it defines a unique correspondence between the semantic scale and the numerical scale R_m).

Thus, when the questionnaire is completed, the analyst obtains the chosen values of the dependent variable R_m and knowing the attribute values X_k he can perform a multiple regression analysis to estimate the values of θ_k.

Ordinary least squares or weighted and generalised least squares have been used to this end. One of the advantages of using these techniques is the ability to obtain goodness-of-fit indicators and measures of the significance of the model parameters. The main problem with this approach is that there are innumerable numerical scales that could be associated with the response scale. It may occur, therefore, that the results of the analysis (estimated coefficients, their ratios, and model goodness of fit) will depend on the definition of R_m; this hints at the importance of choosing the scale correctly. This issue will be discussed in greater detail when considering the analysis of extended choice data below.

8.7.2.3   Discrete Choice Modelling with Rank Data

Rank data is arguably simpler and more reliable than rating data. Individuals are expected to be able to say that they prefer A to C and C to B with greater confidence and consistency than they can have in assigning scores to each alternative. There are several ways of exploiting rank data.

Monotonic Analysis of Variance or MONANOVA (Kruskal 1965) has been used for many years as a method for non-metric scaling. MONANOVA is a decomposition technique specifically developed to analyse rank order data. The method estimates part utilities iteratively thus estimating 'utility values' corresponding to each alternative. The first of these part utility estimates is generated using the naive method just discussed. These utilities permit the modelling of a ranking of alternatives; a 'stress' measure is used to indicate how much the modelled ranking differs from the ranking actually elicited from each individual. MONANOVA then seeks to improve the estimates of the 'part utilities' in order to reduce the stress (or badness-of-fit) indicator. MONANOVA, as in the naive method, is also capable of generating one model for each individual. Despite its uses, the approach lacks a robust statistical grounding and fails to provide global goodness-of-fit and measures of significance indicators; it also restricts the type of utility function that can be specified and it is less well suited to the development of forecasting models.

A more interesting form of analysing rank data is to convert them into implicit choices. In the case above, the rank ACB would be converted into the choices A is better than C, C is better than B, and A is also better than B. The data, thus transformed, can now be analysed using Logit or Probit discrete choice modelling software. For the MNL model this can be done using the following theorem (Luce and Suppes 1965).

$$\text{Prob}(r_1, r_2, r_3, \ldots) = \text{Prob}(r_1/\mathbf{C})\text{Prob}(r_2, r_3, \ldots)$$

where Prob $(r_1, r_2, r_3, \ldots)$ is the probability of observing that the ranking indicates that r_1 is preferred to r_2 and so on, and $\text{Prob}(r_1/\mathbf{C})$ is the probability of r_1 being chosen from the choice set $\mathbf{C} = \{r_1, r_2, r_3, \ldots\}$.

If the theorem is applied recursively, an expression for the probability of the ranking in terms of $N-1$ probabilities of choice is obtained:

$$\text{Prob}(r_1, r_2, r_3, \ldots) = \text{Prob}(r_1/\mathbf{C})\,\text{Prob}(r_2/\mathbf{C} - \{r_1\})\ldots$$

where, for instance, $\mathbf{C} - \{r_1\}$ indicates the choice set excluding alternative r_1. Using this theory, Chapman and Staelin (1982) proposed that the content of a ranking of choices (8.51) can be 'exploded' into $N-1$ statistically independent choices as:

$$(U_1 \geq U_n, n = 1, 2, \ldots, N)(U_2 \geq U_n, n = 2, 3, \ldots, N)\ldots(U_{N-1} \geq U_N) \tag{8.66}$$

and these data can be estimated simply by a MNL routine. However, care must be taken with the following potential problems.

1) As the ranking considers hypothetical alternatives it is likely that the information will contain some noise. This may be particularly serious in the case of less attractive alternatives which are often treated with less care by respondents and bunched together at the bottom of the ranking. This type of behaviour is not consistent with the independence of irrelevant alternatives axiom of the Logit model, so its occurrence must be statistically tested.
2) The rankings must be constructed in decreasing order of preference (i.e. from the best to the worst alternative) by each respondent; failure to do this might generate noisy data, which can invalidate the modelling results.

As ranking a set of N options is a difficult task, that is, it requires $\frac{1}{2}\left(N^2 + N\right) - 1$ comparisons, respondents are typically asked to divide the set (normally 9–12 options). First into three subsets (i.e. the better, medium, and worst alternatives), then to rank the alternatives in each subset, and finally to exchange, say, the last of the first set with the first of the second, if appropriate. This algorithm has been found to ease considerably respondent burden in practice (Galilea and Ortúzar 2005; Ortúzar and Rodríguez 2002).

Problems with this approach have been reported by Ben-Akiva et al. (1992). They found that the response data from different depths of the ranking (i.e. not exploding the full rank) were not equally reliable in the sense of producing statistically significantly different utility estimates. However, this may depend on how carefully designed and conducted the SP experiment is, as Ortúzar and Palma (1992) found that models for the full depth of the ranking consistently produced better results.

To treat this problem in a less *ad hoc* manner, Bradley and Daly (1994) proposed separating the data into $N-1$ different groups (n), each corresponding to a level of depth in the ranking (i.e. the first contains the individual preferences when all alternatives are available, and so on). Once the groups are identified, a joint estimation is performed considering different scale factors for each one (i.e. consistent with different variances for the error terms of their utilities). For this, one group has to be defined as reference and the scale factors (μ_n) associated with the rest of the groups represent the ratio between the variance of the error term corresponding to the reference group and that associated with the group under consideration (see the discussion in Section 8.7.2.7). Thus, if the error variance associated with group n is the same as that corresponding to the reference group, the scale factor of group n will be equal to one.

Bradley and Daly (1994) arbitrarily defined group one as reference and reached the following important conclusions:

- the magnitude of the scale factors diminished with ranking depth (i.e. the error variance was higher in the case of the less preferred options);
- a likelihood ratio test confirmed that the model with scale factors was superior to the simple Logit model;
- the t-ratios of the explanatory variables fell to about one-third of their values in the simple Logit model (see the discussion in Section 8.7.2.6), and
- the subjective values of the various attributes in their experimental design changed by as much as 50% when scale factors were considered.

Ortúzar and Rodríguez (2002) tested this approach, finding that results changed significantly in their case depending on which level of ranking depth was selected as reference; however, in all cases the model with scale factors was statistically superior to the simple MNL model. They considered a group-based ranking experiment designed to study the willingness-to-pay for reductions in atmospheric pollution in a residential location context. The attributes were travel time to work, travel time to study, number of days of environmental alert in the area, and value of the house rent.

Two important findings were that if the fourth depth level (rather than the first, say) was chosen as reference, not only the t-ratios changed (the attribute values and log-likelihood at convergence remained constant), but also the number of significantly different scale factors. In fact, they finally reached the conclusion that the preferred modelling technique was one with only two scale factors: if the first three options are taken as reference, there was one large scale factor for the second set of four options, and a smaller one (i.e. closer to one) for the last three options. This is consistent with the way in which the options were ranked by the individuals and suggests that households were clearer about the extreme alternatives rather than the 'middle-of-the-road' options.

8.7.2.4 Modelling with Stated Choice Data

In this case, we are able to use the whole range of analysis tools available for RP discrete choice modelling; for example, this includes Nested Logit because we are not restricted to only two options nor do we require the IIA property to hold (as in rank orderings) in order to exploit the data fully and also Mixed Logit, which is now the preferred option. We will come back to this issue in more depth below.

An interesting difference between RP and SC data is that the latter, by design, lacks some sources of error. In particular, there is no measurement error since all attribute values are *presented* to respondents (although there may be some perception problems). However, we have already discussed other features of SC surveys that weaken the behavioural value of the data: lack of realism in the decision context and artificiality of the alternatives.

Apart from specification error, which clearly does still apply, there is another potentially serious source of error related to the response itself. Although practical results are generally encouraging, in terms of suggesting that most respondents do understand what it is expected of them, there is no guarantee that they are able to complete an SC experiment with complete accuracy. In fact, a good review by Bates (1988) discussed the following types of potential error applying to all types of SP data:

- respondent fatigue, which obviously increases with the complexity of the experimental design (see the discussion in Chapter 2);
- policy response bias, which might occur if the respondent is interested in affecting the outcome of the analysis;
- self-selectivity bias, when respondents either inadvertently or on purpose, cast their existing behaviour in a better light.

The outcome of all this is that we may have measurement error in the dependent variable, that is, instead of getting a true estimate of the utility U, we are obtaining some pseudo utility $\ddot{U}$, which can be linked to our general formulation (7.2) by:

$$U_i = V_i + \epsilon_i = \ddot{U}_i + \tau_i \qquad (8.67)$$

Assuming homoskedastic τ_i (although it is quite possible that their variance varies across experiments either due to fatigue or learning), the estimation of the parameters of **V** would present no problems as (8.67) can be rewritten as:

$$\ddot{U}_i = V_i + (\varepsilon_i - \tau_i) \qquad (8.68)$$

and the normal estimation methodology may be employed. The problem comes in forecasting, because in that case we are interested in making estimates of **U**, and what we would get from applying this model are estimates of $\ddot{\mathbf{U}}$ provided the same distribution of errors apply in the design year. In other words (Bates 1988):

> ...we are making estimates of relative preferences as expressed in a Stated Preference experiment rather than of what would occur in the market.

The only way to get around this problem is to apportion the error between ε_i and τ_i, using both SC and RP data to estimate the models, and this is somewhat similar to the problem of using aggregate data in model estimation, which we discuss in Chapter 9. Bates (1988) also notes that an understanding of the magnitude of τ_i is crucial to the use of SC in forecasting. Only if it is insignificant in relation to ε_i, could the estimated model be used directly to give forecasts. This calls for special care in the design of SC experiments to reduce respondent fatigue, enhance realism, prevent policy-response bias, and minimise self-selectivity bias. However, the problem remains normally serious and so current practice recommends mixed estimation with RP data whenever possible (see Bradley and Daly 1997).

8.7.2.5 Model Estimation with Generalised Choice Data

In the case of generalised or extended choice surveys, respondents are allowed to express degrees of confidence in their choices. If conventional Logit modelling is used, two models can be estimated, one including only the 'definitely choose' responses and another including also the 'probably choose' responses and the results compared for goodness-of-fit and parameter significance. But note that in either case we would lose the responses marked 'indifferent' and if the choice tasks have been designed to make respondents really think, there might be many in this class and such data loss would be unfortunate.

Alternatively, one can research more closely what is the best transformation of the semantic scale into a numerical one, in the sense of producing the best possible models. Several practitioners have used the following symmetric scale: $R_1 = 2.197$, $R_2 = 0.847$, $R_3 = 0.000$, $R_4 = -0.847$, $R_5 = -2.197$, which corresponds to the Berkson–Theil transformation of the following choice probabilities: 0.1, 0.3, 0.5, 0.7, and 0.9 (see, for example, the review in Bates and Roberts 1986) and became almost standard practice among transport practitioners in the 1990s. However, this is not necessarily the most 'appropriate' scale for any given study and it is important to investigate if scale selection may have a significant effect on the results of the analysis.

Example 8.14 A group of staff and students participated in a generalised SC experiment comparing two options in the following context: a morning trip from home to the university (about 10 km away), involving choice between bus and light rail (an option which did not exist at the time). For simplicity, the experimental design considered only four attributes:

- travel cost (varying at three levels);
- travel time (varying at two levels);
- walking distance (varying at three levels); and
- waiting time, estimated as half of the public transport headway (varying at two levels).

Thus, we had a $3^2 2^2$ factorial design and since we were looking for main effects only, we just required nine options in the simple orthogonal case. Table 8.16 shows the attribute differences (instead of their absolute values) between the two options; the design (in terms of the options offered) was based on combinations of such differences. This implicitly assumed that the resulting model was generic (e.g. same coefficient for in-vehicle time for each mode) helping to reduce the size of the design.

Table 8.16 Attribute differences between options for Example 8.14

Bus attribute minus LRT attribute	Attribute level difference		
	Low	Medium	High
Travel cost (Ch$)	−10	60	80
In-vehicle time (min)	15	25	na
Walking distance (blocks)	−7	−3	0
Headway (min)	−3	2	na

Consider now the four probability scales presented in Table 8.17:

Table 8.17 Tested probability scales in Example 8.14

	Scale 1	Scale 2	Scale 3	Scale 4
R_1	0.100	0.010	0.300	0.200
R_2	0.300	0.400	0.450	0.400
R_3	0.500	0.500	0.500	0.500
R_4	0.700	0.600	0.850	0.880
R_5	0.900	0.990	0.950	0.970

Table 8.18 presents SVT (i.e. coefficient ratios of the parameters of time and cost) derived from multiple regression models estimated after applying the Berkson–Theil transformation to the four probability scales:

Table 8.18 SVT for the four probability scales in Example 8.14

Value of time	Scale 1	Scale 2	Scale 3	Scale 4
In-vehicle travel	4.01	1.73	3.98	4.11
Waiting	20.68	18.67	23.89	23.24
Walking	23.68	21.63	24.91	24.74
R^2	0.48	0.44	0.46	0.45

As can be seen, scale selection does indeed influence the modelling results. The SVT values do not only differ but belong to models with different goodness-of-fit to the data. Furthermore, the differences do not seem to depend on whether the scale is symmetrical or not; that is, although one could expect a symmetric scale (like scales 1 and 2) to produce more reasonable results, the fitted models and estimated SVT values reject this notion (Ortúzar and Garrido 1994a).

One way of avoiding the problem described above would be to consider an approach not requiring the analyst to specify the numerical scale *a priori* in order to estimate the model. McKelvey and Zavoina (1975) developed an approach with this feature, the Ordinal Probit model (see Section 4.4.1.2), which can be easily used but requires specialised software.

Another possibility would be to estimate the response scale during the model fitting process, by effectively considering each value of the scale as an additional variable. In this case, a coordinate search method may be used, starting with the typical symmetric scale 1 in Example 8.14. The procedure consists simply of changing in turn each point of the scale (say R_i) by a small amount and estimating a linear regression model with the new values. The search continues until R^2 is maximised and the value of R_i is fixed. The procedure is repeated for each point of the scale (save for R_3 which is always kept as 0.5) in an iterative routine until a best fit is found in each case (that with the highest R^2). This process is repeated again to check for differences. Ortúzar and Garrido (1994a) found that the search never involved more than two iterations before convergence (for four different samples), but they could not prove, mathematically, that a global optimal solution is guaranteed. Indeed, the method was used later by Bianchi et al. (1998) who found that the method did not converge for their pricing study data.

Example 8.15 Table 8.19 shows the original symmetric scale and the scales found after performing the above 'optimal scale linear regression approach' on two samples for the rating experiment of Example 8.14.

The results suggest the possibility of testing whether the original number of points in the semantic scale is appropriate. If only one value was used for the first two points of the scale in the 'optimal scale' models (which appear strikingly close), it would be interesting to see what consequences this apparent loss of information brings about. On the plus side a four-point scale would have one less parameter to be estimated.

Table 8.19 Probability scales for Example 8.15

	Initial	Students	Staff
R_1	0.1	0.284	0.228
R_2	0.3	0.286	0.278
R_3	0.5	0.500	0.500
R_4	0.7	0.714	0.722
R_5	0.9	0.900	0.842

Table 8.20 shows the optimal values of the new scale obtained when R_1 and R_2 are replaced by a single point R_1'. In these scales, as in the previous ones, the probability value of R_3 was fixed to 0.5 as it corresponds to the point of indifference between both modes.

Table 8.20 Reduced scales for Example 8.15

	Students	Staff
R_1'	0.277	0.121
R_3	0.500	0.500
R_4	0.716	0.776
R_5	0.899	0.922

As can be seen, the scale values in both samples are further apart than in the previous table, which suggests that no other point fusion would be necessary. Also, all values appear reasonable in relative terms (i.e. they correspond to increasing probability values from R_1' to R_5).

8.7.2.6 Modelling with Indifference Alternatives

When dealing with SP data the choice set is established *a priori* and it is important to carefully define it preserving the realism of the choice situations. Thus, in many cases it might be necessary to consider an 'opt-out' (non-purchase) alternative (Carson et al. 1994; Olsen and Swait 1998), but whether this alternative should be included directly into the choice set or indirectly via dual response procedures (Dhar and Simonson 2003), remains a debatable point (see Schlereth and Skiera 2017 for a good discussion).

A similar but less analysed problem is the inclusion of indifference alternatives in the choice set. If the modeller does not allow respondents to state their indifference among two or more alternatives that have been purposely designed to be close, they may be forced to select one of them in a rather stochastic manner, adding white noise to the experiment. Additionally, doing so would provide less information about the individuals' preferences, leading to loss of efficiency. Indeed, Cantillo et al. (2010) used a synthetic dataset to show that assigning the preferences associated with an indifference alternative randomly significantly diminished the model's capability to recover the input parameters. They also considered real databanks, observing than offering the possibility of stating indifference may indeed affect the outcome of the experiment (i.e. the estimated parameter values).

Along these lines, empirical evidence (Fenichel et al. 2009) shows that in experiments including non-purchase options, indifference situations may artificially increase the probability of selecting an opt-out alternative, as a kind of cognitive bias. Nevertheless, including indifference alternatives might generate other kinds of complications, especially if individuals are burdened by the complexity of the experiment.

Both situations (opt-out and indifference alternatives) exhibit, however, substantial differences. Whilst the former suggests the existence of a reservation utility higher than that provided by the alternatives in the choice set, the latter indicates that individuals ascribe the same utility to two or more alternatives in it. Therefore, in the first case an extra alternative accounting for this reservation prize should be considered. Nevertheless, considering an extra alternative to reflect indifference choices does not appear appropriate, as it does not reflect the causes leading to the statement of indifference; in fact, by treating indifference as a new (opt-out) alternative, the modeller implicitly assumes that the utility ascribed to this new option would be greater than that of the competing alternatives, which is clearly not the case.

Despite the fact, that indifference situations should only arise if the expected utility of two or more alternatives is the same, the theory behind the indifference phenomena suggests the existence of perception thresholds, below which the individuals are not able to perceive differences between two stimuli (Coombs et al. 1970; Cantillo and Ortúzar 2006).

Krishnan (1977) developed an operational discrete choice model accounting for the existence of indifference thresholds. This approach (Minimum Perceivable Differences model, MPD) allows considering that observations falling into the indifference interval will be assigned stochastically to one alternative, in the context of a binary choice situation. Cantillo et al. (2010) expanded the MPD-approach to allow for individuals stating their indifference in stated-choice (SC) experiments. This way, the indifference alternative would be selected if the difference between the utility of both alternatives was smaller than a threshold, to be estimated.

The main limitation of the MPD-approach is that it only allows considering binary choice situations. Thus, it can neither consider situations where two alternatives exhibit a similar utility (which is superior to all other alternatives in the choice set), nor cases when three or more choices report an apparently identical utility (which may be of particular interest when considering alternatives to first-choice SP experiments, such as rankings). The same limitation arises, when considering approaches such as an Ordered Logit framework (with indifference being an intermediate choice between two binary alternatives).

Bahamonde-Birke et al. (2017b) proposed an approach to deal with indifference choices in multinomial choice situations. Their framework allows not only addressing first-choice SP experiments but also rankings, where indifference choices may be expected to appear even more often.

Stating indifference in multinomial choices. If the modeller allows for individual q to state his indifference between n alternatives belonging to an indifference-set B, which is a subset of the complete choice set A (consisting of m alternatives, with $m \geq n$), the choice probability of the alternatives belonging to B, may be written as follows:

$$P_i = P_j = P_k = \ldots \qquad\qquad \forall i,j,k,\ldots \in B \qquad (8.69)$$

As under random utility theory all alternatives belonging to B must maximize the individual's expected utility U, Eq. (8.69) can be rewritten in these terms:

$$U_i \approx U_j \approx U_k \approx \ldots \qquad\qquad \forall i,j,k,\ldots \in \boldsymbol{B} \qquad (8.70)$$

where the inequality accounts for the existence of utility differences (smaller than a given threshold) that are not perceived by the respondents.

If we introduce error terms ϕ guaranteeing that the utility differences within a given indifference-set are equal to zero, then Eq. (8.70) can be expressed in terms of equalities:

$$U_i + \phi_i = U_j + \phi_j = U_k + \phi_k = \dots \qquad \forall i,j,k,\dots \in B \qquad (8.71)$$

Then, expressing the expected utility in terms of a representative utility V and the error terms ε as usual (i.e. accounting for all unknown elements of the decision except the indifference thresholds) leads to the following expression:

$$V_i + \varepsilon_i + \phi_i = V_j + \varepsilon_j + \phi_j = V_k + \varepsilon_k + \phi_k = \dots \qquad \forall i,j,k,\dots \in B$$

Finally, the probability associated with an alternative in the indifference set would be given by:

$$\Pr_i = P\left(V_i - V_j > \varepsilon_j + \phi_j - \varepsilon_i - \phi_i\right) \qquad \forall j \neq i \in A \;\wedge\; \phi_j = 0, \; \forall j \notin B$$

where $\Pr_i \neq P_i$, as it is associated with a larger error component (due to the parameters ϕ).

When modelling with indifference alternatives, it cannot be assumed that all alternatives in the indifference-set are selected at the same time, but rather that all are selected with a frequency $1/n$. Hence, when considering indifference options, the likelihood of observing a given choice can be expressed as:

$$L = \prod_{j \in A} \Pr_j{}^{y_j} \qquad (8.72)$$

where y_j takes the value of $1/n$ if alternative $j \in B$ and zero otherwise. If we consider a group of individuals M selecting indifference-sets and another group of individuals L selecting unique preferred alternatives (where M and L are mutually exclusive), the general likelihood function for the population would take the following form:

$$L = \prod_{q \in M} \prod_{j \in A} P_{jq}{}^{y_j} \cdot \prod_{q \in L} \prod_{j \in A} \Pr_{jq}{}^{y_j}$$

It is important to notice that using a value of y_j equal to one for all alternatives in an indifference-set would result in overweighting all observations by respondents selecting indifference options. Also, note that the likelihood function would be maximised when all alternatives in an indifference-set are equally likely.

It is easy to extend this framework for ranking preference elicitation procedures, that is, when the indifference set is not necessarily selected as the first choice (e.g. when considering a full-ranking). In this case, the modeller would just rely on simulated choices based on exploiting the ranking (Chapman and Staelin 1982) and the framework would only apply to simulated choices, where the indifference-set is assumed to be selected (i.e. allowing to deal with ties).

Identification of the error term ϕ. Both the identification and the distribution of the error terms ϕ are complicated topics. First, the errors only appear in the utility function if the alternatives are selected as part of an indifference-set (i.e. as part or Pr_i), which may significantly decrease the number of observations available for estimation. Furthermore, this additional error term will only be associated with the alternatives that are indeed part of a given indifference set; thus, the utility functions of the remaining alternatives (i.e. those not selected as a part of the indifference set) will not be affected by this term. For this reason, non-symmetrical assumptions regarding its distribution are unsuitable (the estimator would diverge, as it would only be associated with selected alternatives).

Moreover, establishing an adequate functional form for ϕ is not an easy task, as these are error components that merely guarantee that the difference among two or more different utilities is equal to zero; that is, an error is added to the alternatives with lesser utilities or subtracted from the alternatives with higher utilities, so that the utility of all alternatives in the indifference-set adds up to the same amount. Hence, ϕ would not be properly represented by the usual assumptions concerning error terms.

If the analyst is prepared to assume that the parameters ϕ follow a distribution equal to the difference between two Logistic distributions with different scale parameters, the sum $\varepsilon_l - \varepsilon_j + \phi_{lj}$ would also be represented by a Logistic distribution, with a smaller scale parameter (i.e. with a larger standard deviation). As a consequence, the utility functions of alternatives selected as a part of an indifference-set would be equivalent to those of the single-choice framework (P_l), but would have a smaller scale parameter, reflecting the increased uncertainty. This framework resembles the structure used to deal simultaneously with RP and SP data (see Section 8.7.3).

A limitation of this approach is that it does not allow quantifying the thresholds leading to the statement of indifference; notwithstanding, the model still offers an adequate depiction of the underlying utility functions associated with the different alternatives as well as an appropriate functional form for forecasting. Additionally, the approach allows considering different scale parameters depending on the number of alternatives in the indifference-sets, or on which alternatives are being considered as part of the indifference-sets, as well as on the socio-economic characteristics on the individuals.

Due to identification issues, the analyst is forced to normalise one of the scale parameters (either that related to the equations for single choices – when the respondent selects a unique alternative – or that associated with alternatives selected as part of an indifference-set). However, both normalisations lead to different estimates; if the modeller normalises the error associated with the single choices, the estimates would be scaled upwards in comparison with those of a model with a scale parameter fixed at one for the whole sample, as single choices are related to a lesser degree of uncertainty (i.e. the differences between representative utilities should be larger, in average, when indifference is not chosen; which is equivalent to say that indifference cases should arise more often if the utility differences are smaller). Thus, when fixing the scale parameter for single choices the estimates are no longer comparable with alternative specifications.

In general, we will not attempt to estimate separate models for individuals choosing single alternatives or indifference sets, but rather to estimate a parsimonious model that is consistent with both kind of choices. A possible way to do this is to fix the scale parameter associated with the weighted average of the sample (normally at one, but other values may be chosen). In this form, the analyst would estimate $k-1$ scale parameters and express the last one in terms of the others to be estimated as in:

$$\lambda \cdot N = \lambda_1 \cdot n_1 + \lambda_2 \cdot n_2 + \lambda_3 \cdot n_3 + \dots \tag{8.73}$$

where N is the total number of observations in the sample, λ the weighted average of the scale parameters, λ_j the scale parameters to be estimated, and n_j the number of observations associated with a given scale parameter.

A final characteristic of the approach is that when considering only two alternatives, different scale parameters cannot be estimated. If all alternatives were equally probable, all estimates should necessarily be equal to zero and, therefore, any scale parameter multiplying a group of indifference sets containing all alternatives would be unidentified (and therefore can be excluded without loss of generality). As in binary choice situations every indifference-set must necessarily include both alternatives, considering different scale parameters in this case would be redundant. As a corollary, considering different scale parameters would only impact the model when these can be associated with at least one indifference-set (selected by at least a part of the population), which does not contain all the alternatives in the choice set.

Example 8.16 Consider a simulated choice situation with two alternatives, but where pseudo-individuals are also offered an opt-out alternative, with the following utilities:

$$U_1 = \beta_1 \cdot X_1 + \varepsilon_1$$

$$U_2 = ASC_2 + \beta_2 \cdot X_2 + \varepsilon_2$$

$$U_{\text{opt-out}} = ASC_{opt-out} + \varepsilon_3$$

Pseudoindividuals are supposed to choose the alternative with the highest expected utility or to state indifference if the utility difference between alternatives is smaller than a given threshold. They could be indifferent among: (i) both alternatives, (ii) one existing alternative and opting-out, and (iii) both existing alternatives and opting-out (the ASC associated with the opt-out alternative was fixed at zero).

The variables X_1 and X_2 were taken as random draws from Normal distributions with mean 0 and -1, and standard deviations 1 and 1.5, respectively. The error terms were assumed to be independently EV1 distributed with mean 0 and scale parameter 1; both β parameters as well as ASC_2 were fixed to 1, whilst three different values were tested for the threshold: 0.15, 0.3, and 0.45, representing small, medium, and large indifference intervals, respectively. In all three cases 5000 pseudoindividuals were generated.

In the first case (low threshold), 400 observations indicated indifference (124 between both existing alternatives, 131 between alternative 1 and opting out, 135 between alternative 2 and opting-out, and 10 between both alternatives and opting-out). In the second case (intermediate indifference threshold), 745 pseudoindividuals stated their indifference (200 between both existing alternatives, 260 between alternative 1 and opting-out, 230 between alternative 2 and opting-out, and 55 between both alternatives and opting-out), whilst in the third case (large indifference threshold) 1093 indifference observations were recorded (297 between both existing alternatives, 345 between alternative 1 and opting out, 319 between alternative 2 and opting out, and 132 between both alternatives and opting out).

Four different models were estimated. Model 1 presents the results of ignoring the indifference choices. Model 2 considers the proposed framework but ignores the error term ϕ and does not consider a different scale parameter for the observations associated with indifference-sets. Model 3 does not consider the fact that all alternatives in a selected indifference-set are not observed with the same frequency (if a given alternative is selected as a part of a binary indifference-set, in reality, it would only be selected with a frequency of 0.5) as alternatives being selected directly; hence, Model 3 assigns a weight (y_j) of 1 to all observations. Finally, Model 4 represents the proposed framework. The scale parameters were fixed in accordance with (8.72) and (8.73). All models were estimated using PythonBiogeme (Bierlaire 2016), and the results are presented in Table 8.21 (the standard deviation of the estimated parameters is shown in parenthesis).

All specifications except Model 1 (which ignores the indifference observations), allow for an adequate estimation of the parameters used in the generation of the dataset. Model 2 performs

Table 8.21 Estimated models for Example 8.16

Threshold	Variable	Model 1	Model 2	Model 3	Model 4
0.15	X_1	1.09 (0.043)	1.03 (0.040)	1.04 (0.040)	1.04 (0.041)
	X_2	1.05 (0.033)	0.99 (0.031)	1.00 (0.031)	1.00 (0.031)
	ASC_2	1.11 (0.049)	1.05 (0.046)	1.05 (0.046)	1.05 (0.047)
	$ASC_{opt-out}$	0.02 (0.042)	0.03 (0.040)	0.03 (0.039)	0.03 (0.04)
	$\lambda_{Indifference-set}$	—	—	0.44 (0.051)	0.43 (0.072)
	Final log-likelihood	3848.23	−4294.82	−4699.12	4268.10
	$N°$ of observations	4600	5000	5410	5000
0.3	X_1	1.08 (0.044)	0.98 (0.039)	0.98 (0.038)	0.98 (0.039)
	X_2	1.08 (0.036)	0.96 (0.031)	0.97 (0.031)	0.97 (0.032)
	ASC_2	1.08 (0.051)	1.00 (0.046)	0.99 (0.045)	0.98 (0.046)
	$ASC_{opt-out}$	0.04 (0.043)	0.03 (0.039)	0.03 (0.038)	0.04 (0.038)
	$\lambda_{Indifference-set}$	—	—	0.44 (0.039)	0.45 (0.054)
	Final log-likelihood	3572.90	−4404.24	−5202.83	−4356.51
	$N°$ of observations	4255	5000	5800	5000
0.45	X_1	1.17 (0.047)	0.98(0.039)	0.94 (0.038)	0.93 (0.039)
	X_2	1.13 (0.038)	0.78 (0.027)	0.86 (0.029)	0.87 (0.030)
	ASC_2	1.19 (0.056)	1.00 (0.045)	0.96 (0.045)	0.94 (0.046)
	$ASC_{opt-out}$	0.02 (0.042)	−0.08 (0.040)	0.03 (0.037)	0.03 (0.036)
	$\lambda_{Indifference-set}$	—	—	0.15 (0.036)	0.14 (0.048)
	Final log-likelihood	−3163.75	−4515.94	−5701.26	−4360.30
	$N°$ of observations	3907	5000	6225	5000

surprisingly well, but still offers a significantly worse goodness-of-fit than Model 4. Nevertheless, the fact that Model 2 performs acceptably while ignoring the different variability, suggest that considering different scale parameters may be omitted without major complications if the analyst does not count with enough indifference-set observations (to allow for a correct estimation of the λ parameter). Model 3 also performs well; the reason is that when considering different scale parameters, the weighting only affects the observations multiplied by the same scale parameter. In this case, the weights associated with the single choices are not relevant, and only a small number of observations with a weight of 1/3 (individuals stating their indifference among all choices) and a large majority associated with a weight of ½ were observed, which clearly diminished the bias associated with ignoring the weighting. In fact, if the scale parameter was ignored (analogously to Model 2 but ignoring the weighting), Model 3 would no longer be capable of recovering the parameters.

Finally, regarding the magnitude of the thresholds, all specifications (with the exception of Model 1) perform adequately as long as the indifference threshold is small or intermediate. When the indifference threshold is large (causing that an important part of the pseudoindividuals choose indifference alternatives), all models exhibit some difficulties in terms of parameter recovery. This may be explained by the fact that a larger indifference threshold may be indeed associated with a larger uncertainty and that the assumptions regarding the distribution of the error terms ϕ are not exactly accurate. But even in this case, Model 4 exhibits the best performance, and is capable of recovering three out of four parameters, clearly outperforming Models 1 and 2.

8.7.2.7 Interactions in SC Modelling

Next, we consider a potential although seldom exploited advantage of the SC approach: the possibility of estimating models with non-linear utility functions. The reason for not doing this in practice has been typically one of convenience. SC experiments allowing the incorporation of interactions (and not just main effects) are more complex to design and analyse, and require data that is more difficult to collect.

In discrete choice modelling many potential forms of the utility function can be transformed (e.g. even as a last resort using series approximations) into additive linear forms of the type:

$$V = \theta_1 X_1 + \theta_2 X_2 + \theta_3 X_1^2 + \theta_4 X_1 X_2 + \theta_5 X_1 X_2^3 + \theta_6 X_1 X_2 X_3$$

where X_i are attributes and θ_i are coefficients to be estimated. This function contains linear terms ($\theta_1 X_1$ and $\theta_2 X_2$), non-linear terms ($\theta_3 X_1^2$), interactions with linear effects ($\theta_4 X_1 X_2$ and $\theta_6 X_1 X_2 X_3$) and general interactions ($\theta_5 X_1 X_2^3$). The main effects can be defined as the response to passing to the next level of the variable when the rest of the attributes remain constant (all other things being equal); it is normally postulated that these are the main determinants of changes in choice. In fact, according to Louviere (1988b):

- the main effects explain 80% or more of the data variance;
- two-term interactions rarely explain more than 2% or 3% of the variance;
- three-term interactions explain even smaller proportions of the data variance, normally of the order of 0.5–1% and rarely over 2% or 3%;
- higher-order effects explain a minuscule proportion of the data variance.

For these reasons, only main effects have been normally considered in practice. On the other hand, there seems to be a consensus that interactions between more than two variables as well as interactions incorporating non-linear effects should be insignificant. Therefore, only two-term interactions are in a kind of limbo and require more attention. Note that if interactions are actually insignificant, a model incorporating only main effects will allow us to obtain precise measurements of individual preferences. However, if the interactions are significant and are not included in the utility specification, their effects will be erroneously attributed to the simple variables. Notwithstanding, as we shall see below, it may happen that when certain interactions are included, their effects dominate that of certain individual variables to the extent that the latter may be left out of the regression (i.e. the variable may end up with a non-significant coefficient or with a counterintuitive sign).

The cost of allowing for interactions in the experimental design is that it becomes more complex (i.e. it requires respondents to evaluate a higher number of hypothetical situations). A good solution in such cases is to use block designs as we saw in Section 2.4.2.3. The assumption is that consistent models will be obtained when the total number of responses is considered. To ensure compatible answers, the size of each subsample should guarantee that its socio-economic characteristics are representative.

Example 8.17 A generalised SC experiment using a five-point semantic scale was designed to study choice between car-alone and car-pool for campus students (Ortúzar et al. 2000c). After extensive piloting, the following attributes were selected:

- **Daily travel time**: this was always higher for car-pool as the student providing the car on the day needed to collect the members of the group in the morning and take them back home in the afternoon;
- **Weekly travel cost**: associated with fuel consumption and estimated on the basis of information about travel distance and type of car (in some cases this value included a parking charge); this was always smaller for car-pool as drivers did not need to use their cars every day of the week in this case;
- **Waiting time**: associated with sharing the trip with a group in the case of car-pool; waiting occurred because the proposed car-pool system implied the complete group arriving at and leaving the campus at the same time, and not all exit hours may coincide; note that this time may be used in other activities, because both its duration and day of occurrence were known in advance, given the fixed university schedules.

The attribute levels were defined on the basis of differences between travelling by car and by car-pool. Two levels were used in the case of travel time (i.e. 10 and 20 minutes more that in the case of car-pool); four levels in the case of travel cost (i.e. three-quarters and half the cost of the car in the case of car-pool, and 25% and 40% more than that cost if a parking charge was included), and three levels for waiting time. In this last case the levels were determined based on the possibility that the group members would not coincide in their lectures. So, waiting times of zero, 30 minutes (i.e. one member needed to do a small errand) and 90 minutes (i.e. the extent of a complete lecture module) were considered.

With this, 16 hypothetical situations were needed to estimate main effects only and 24 if two-term interactions were considered in a simple orthogonal design. Given these numbers, block designs were used in both cases given the results of Caussade et al. (2005). In fact, we tested using 16 options directly but found that this confused or bored respondents, leading to too many inconsistencies, confirming the findings of Carson et al. (1994).

To model, we first looked at the expected signs of the interaction terms (given the special characteristics of the competing alternatives), concluding that their most appropriate definition was as follows:

- T^*C represents the interaction between the ratios of travel time and cost by both modes; it should have a positive coefficient:

$$T^*C = \frac{\text{Travel time}^{\text{car}}\text{Cost}^{\text{cp}}}{\text{Travel time}^{\text{cp}}\text{Cost}^{\text{car}}}$$

- W^*T represents the interaction between the car-pool waiting time and the ratio of travel time by both modes: it should receive a negative coefficient:

$$W^*T = \text{Waiting time}^{\text{cp}}\frac{\text{Travel time}^{\text{cp}}}{\text{Travel time}^{\text{car}}}$$

- W*C represents the interaction between the car-pool waiting time and the ratio of travel cost by both modes: also a negative coefficient:

$$W^*C = \text{Waiting time}^{cp} \frac{\text{Cost}^{cp}}{\text{Cost}^{car}}$$

Table 8.22 shows the results of two Ordinal Probit specifications, the 'best model' (estimated) and the 'preferred model'. The *Inertia* dummy took the value of one if the respondent was a current car-pool user and g is the expenditure rate, that is, the ratio between income and free time; see for example Jara-Díaz and Ortúzar (1989). As can be seen only the variables *Sex* (dummy which took the value of one for males) and *Waiting time* were not significant at the 95% level in the first model; however, if the latter was removed (because its effect was considered by the strong interaction terms) the model improved.

Table 8.22 Ordinal Probit model considering interactions

Attributes (*t*-ratios)	Best model	Preferred model
Car-specific constant	1.65418	1.68808
	(9.89)	(10.17)
Travel time (min)	−0.00311	−0.00343
	(−4.57)	(−5.30)
Waiting time (min)	−0.00363	—
	(−1.53)	
Cost/g (min)	−0.06729	−0.06930
	(−7.54)	(−7.83)
Sex	0.11021	0.11372
	(1.92)	(1.98)
Car-pool inertia	−0.40907	−0.41113
	(−6.57)	(−6.61)
T*C	0.70067	0.73763
	(9.67)	(10.73)
W*T	−0.00629	−0.01038
	(−2.11)	(−8.04)
W*C	−0.00124	−0.00157
	(−3.23)	(−4.92)
R^2	0.543	0.541
Sample size	1640	1640

To verify the relative importance of the interactions in the utility function, the product of the average value of each normalised variable and its coefficient was calculated. This revealed that the interactions were undoubtedly important, especially T*C. This procedure was confirmed by calculating the elasticity of the probability of choosing car for various changes in the attribute values (Ortúzar et al. 2000c).

8.7.2.8 The Problem of Repeated Observations

One of the most important attractions of the SC approach is the generation of multiple observations by each individual. However, almost every application in the last millennium considered the responses by a given individual not only independent of those given by the rest of the sample members, but also independent of each other. Although this problem received more attention at the end of the 1990s, it was only later that it was handled correctly using Mixed Logit (ML) models.

In the 1990s, it was generally assumed that SC observations were independent, leading to the concept of *pseudoindividuals*. Clearly, this hypothesis cannot be valid and for many years it was hoped (and believed) that the problem was bounded to obtaining upward biased values of the *t*-ratios associated with the estimated parameters. In this way the initial solution consisted of proposing correction factors for the resulting *t*-ratios.

By the end of the 1990s more interesting approaches were proposed and partially tested. For example, Cirillo et al. (2000) proposed the use of re-sampling techniques, such as bootstrap and jackknife (Shao and Tu 1995), finding that the jackknife-estimated parameters did not vary much with respect to those estimated assuming independence and that the *t*-ratios diminished, as expected (the bootstrap results were similar but had more noise, particularly for low-re-sampling strategies). Ortúzar et al. (2000c) also tested these methods (with all their variations in re-sampling) for four different samples, finding that the parameter values remained practically identical to those estimated with the traditional approach in all cases. However, the standard errors varied inconsistently (i.e. they correctly increased in three cases but decreased in the other). They extensively checked the samples for either outliers or peculiarities in the originally estimated values and found nothing special. Thus, they were forced to conclude that the applicability of these techniques to solving the problem of repeated observations required further scrutiny.

Ouwersloot and Rietveld (1996), and independently, Abdel-Aty et al. (1997), suggested decomposing the total error ε in a random utility model into two mutually exclusive parts: an individual-specific effect that distributes independently among individuals, and an observation-specific effect that distributes independently among individuals and observations (i.e. very much in line with the error components specification of the ML model). Inevitably, the standard approach led to multiple integrals, which were hard to evaluate. To avoid this problem, Ouwersloot and Rietveld (1996) used the 'minimum distance' method proposed by Chamberlain (1984), which considers dividing the sample into T randomly selected independent subsamples containing only one observation per person (T is the number of repeated observations per individual). The coefficients of the models estimated for each subsample were then used in a rather complex algorithm to obtain the final model parameters and their variances.

Contrary to expectations, Ouwersloot and Rietveld (1996) found that the parameters of their Probit model were different to those of the classic method (although less than 27%) but the *t*-ratios remained practically invariant. Ortúzar et al. (2000c) also tested this method, finding that most parameter values decreased, and in some cases considerably, but sometimes they also increased. With respect to the *t*-ratios they found that, in general, they decreased as expected, but not always and particularly in the case of the specific constants.

Yen et al. (1998) developed another method to treat this problem using a generalised dynamic version of the Ordinal Probit model, which allows one to incorporate a measure of the correlation between the responses of a given individual. As comparative issues were not their main concern, Yen et al. (1998) did not report whether there were differences between their estimations and those obtained with a standard application of Ordinal Probit.

Current practice accepts that estimation can be handled without problems by a ML model, such as (8.34). We will look at the way to do it in the richer case involving joint estimation with RP data in the next section.

8.7.3 Model Estimation with Mixed SC and RP Data

Consider the MNL model (7.9) and the inverse relation that its scale parameter β has with the single standard deviation σ of the Gumbel residuals ε. This relation explains why it is not correct to postulate the same error distribution for estimation and forecasting as mentioned above; the near and extreme right hand side expressions in (8.67) should yield different values for β. This produces 'scale' differences on the parameters and if such equality is improperly assumed, we might finish by estimating pseudo utilities instead of 'true' utilities. To avoid this problem, it is necessary to adjust the SC data to actual behaviour, exploiting the advantages of the RP data in this sense, and estimating the parameters θ jointly.

In econometrics the estimation of models with different data sources is called 'mixed estimation'. Often these data are divided into two sets: *primary* and *secondary* data. The primary data provide direct information about the main modelling parameters. The secondary data provide additional (indirect) information about the parameters. For example, in discrete choice modelling the primary data could be information coming from a survey at the disaggregate level, and the secondary one could be data coming from an aggregate survey. In our case, RP data constitute the primary set, since these data capture the actual behaviour of the individuals, and SC data constitute the secondary set.

8.7.3.1 Estimation without Considering Correlation among Repeated Observations

Although we know that this is not correct nowadays, it is still informative to learn how this important task was first undertaken. Ben-Akiva and Morikawa (1990) developed a framework that postulates that the difference between the errors in the RP and SC domains may be represented as a function of the variances of the errors ε and η associated with each data set, respectively. This can be written as follows:

$$\sigma_\varepsilon^2 = \mu^2 \sigma_\eta^2 \tag{8.74}$$

where μ is an unknown *scale coefficient*. This leads to the following utility functions for a certain alternative A_i:

$$U_i^{RP} = \theta x_i^{RP} + \alpha y_i^{RP} + \varepsilon_i$$
$$\mu U_i^{SC} = \mu \left(\theta x_i^{SC} + \phi z_i^{SC} + \eta_i \right) \tag{8.75}$$

where α, ϕ, and θ are sets of parameters to be estimated; x^{RP} and x^{SC} are attributes (of both alternatives and individuals) at the RP and SC levels, respectively. y^{RP} and z^{SC} are attributes that only belong to the RP or SC sets respectively (notice that vector x is common to both types of data).

The consideration of the utility functions (8.75) allows homogenising the type of error, as multiplying the SC utility by μ makes the associated stochastic error (η_i) to have the same variance as the corresponding RP error (from 8.74). Thus, assuming that both stochastic errors have IID EV1 distributions with zero mean but with a different variance, the choice probabilities at each domain would be given by (Morikawa et al. 1992):

$$P_i^{RP} = \frac{\exp\left(\boldsymbol{\theta}\mathbf{x}_i^{RP} + \alpha\mathbf{y}_i^{RP}\right)}{\sum_j \exp\left(\boldsymbol{\theta}\mathbf{x}_j^{RP} + \alpha\mathbf{y}_j^{RP}\right)}$$

$$P_i^{SC} = \frac{\exp \mu\left(\boldsymbol{\theta}\mathbf{x}_i^{SC} + \boldsymbol{\phi}\mathbf{z}_i^{SC}\right)}{\sum_j \exp \mu\left(\boldsymbol{\theta}\mathbf{x}_j^{SC} + \boldsymbol{\phi}\mathbf{z}_j^{SC}\right)} \tag{8.76}$$

From these expressions, it is possible to postulate a joint likelihood function, which should be maximised to yield the parameter estimates. The reader might have noted that Eq. (8.76) have incorporated two assumptions:

- the scale parameter of the RP model has been normalised;
- the scale parameter of the SC model should be identical to μ.

In fact, the real assumptions are different but when the joint model is estimated we arrive at the same result. Yáñez et al. (2011) provide a good discussion on this issue in their analysis of mixed RP–SC models in forecasting.

Choosing the attributes with the same parameter in both domains. Deciding which attributes should belong to set $\mathbf{x}$ is not straightforward. In principle, though, the only candidates are those attributes that being measured in practice (i.e. travel time, cost, waiting time) also appear in the SC tasks. If all 'common' attributes are taken as members of $\mathbf{x}$ we speak of the *full data enrichment approach*; if only some common attributes end up belonging to $\mathbf{x}$ we have the *partial data enrichment approach*. To decide this matter, Louviere et al. (2000) recommend the following procedure:

1) First, estimate (separately) the two models associated with Eq. (8.70), under the assumption that the errors distribute IID EV1 (obviously without including the unknown scale factor μ in the second case); this will yield two sets of parameters, $\boldsymbol{\theta}^{RP}$ and $\boldsymbol{\theta}^{SC}$, for all the attributes that are common to both domains.
2) As we know, these two sets of parameter estimates cannot be equal, in principle, as they contain the unknown scale parameters β associated with the MNL model in each domain; however, the idea is to find out if they are different over and above this scale problem, in which case they should not be joined under set $\mathbf{x}$.
3) From Eq. (8.69), and recalling the inverse relation between the scale parameter of the MNL and the variance of its IID EV1 error (7.10), the reader can readily deduce that if both parameters were equal (apart from scale), their relationship should be: $\boldsymbol{\theta}^{SC} = \mu\,\boldsymbol{\theta}^{RP}$.
4) Based on this relation, if we plot the estimated parameters in the two domains, we should expect them to fall inside the elliptical region shown in Figure 8.8; and values outside it (as those shown in stars) would not be available for set $\mathbf{x}$.
5) However, note that even if some parameters fall outside this range, the attributes involved could still be considered part of $\mathbf{x}$. This would be the case if one of the estimates is not significantly different from zero, as in that case, fixing its value to be equal to its counterpart in the other domain would bring no problems. In fact, as we will see below, this can be tested using a LR test (as we saw in Section 8.4.1.2).

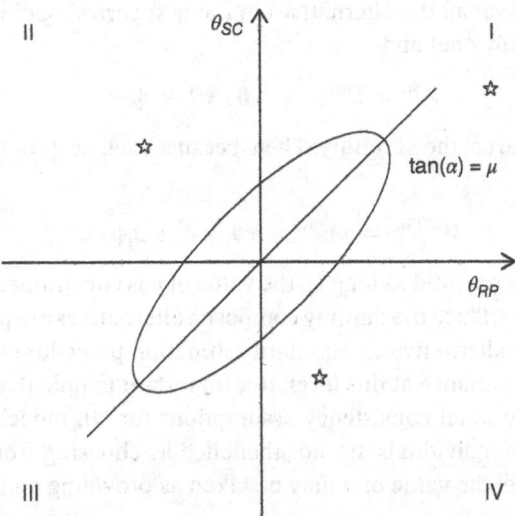

Figure 8.8 Plotting the parameters from both domains

The joint likelihood function incorporating simultaneously the two models in Eq. (8.76) is a highly non-linear function, because μ is multiplying not only the attributes but also the SC parameters. Two approaches were devised during the 1990s to solve this problem, the 'simultaneous estimation' method (Bradley and Daly 1997) and various forms of 'sequential estimation' method (Ben-Akiva and Morikawa 1990; Swait et al. 1994). We will only mention the former here as it was the most popular in practice until recently.

The simultaneous estimation method consists of constructing an artificial tree with more alternatives than there are in reality. Some of these are labelled RP alternatives, the others are SC alternatives. The utility functions are $\mathbf{U}^{RP}$ and $\mathbf{U}^{SC}$, as in (8.70).

As indicated in Figure 8.9, the RP alternatives are placed just below the root of the tree; however, the SC alternatives are each placed in a single-alternative nest; we will see now why this is so important. Observe that in this tree, for an RP observation the SC alternatives are unavailable and the choice is modelled as in a standard MNL or NL model. For an SC observation, the RP alternatives are also unavailable and the choice is modelled by a NL (tree) structure. For this reason, the method came to be known as the 'nested logit trick' (Louviere et al. 2000).

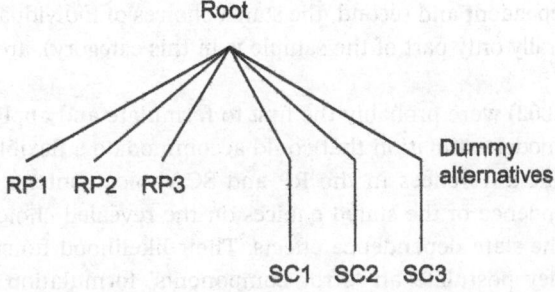

Figure 8.9 Artificial tree structure for joint RP and SC estimation

For the SC observations, the mean utility of each of the dummy composite alternatives is computed as usual (see Daly 1987):

$$V^{COMP} = \mu \log \sum e^{V^{SC}}$$

where the sum is taken over all the alternatives in the nest corresponding to the composite alternative (i.e. in this case only one) and

$$V^{SC} = U^{SC} - \eta = \boldsymbol{\theta} \cdot \boldsymbol{x}^{SC} + \boldsymbol{\phi} \cdot \boldsymbol{z}^{SC}$$

is simply the measured part of the SC utility. Then, because each nest contains only one alternative in this specification,

$$V^{COMP} = \mu V^{SC} = \mu \boldsymbol{\theta} \cdot \boldsymbol{x}^{SC} + \mu \boldsymbol{\phi} \cdot \boldsymbol{z}^{SC}$$

which is exactly the form required as long as the value of μ is constrained to be the same for each of the dummy alternatives. Since the dummy composite alternatives are placed just below the root of the tree, as are the RP alternatives, a standard estimation procedure will ensure that μ is estimated to obtain uniform variance at this level. It is important to note that this artificial construction does not require the usual consistency assumptions for NL models (i.e. that μ should not exceed one), because the individuals are not modelled as choosing from the whole choice set. However, as noted before, the value of μ may be taken as providing an indication of which data set is more accurate.

Partial or fuller data enrichment. To test whether a given common attribute to both datasets can form part of set $\mathbf{x}$, it is possible to use a likelihood ratio test. Let $l^* (\boldsymbol{\theta}^{RP}, \boldsymbol{\alpha})$ be the log-likelihood at convergence for the model with RP data only, $l^* (\boldsymbol{\theta}^{SC}, \boldsymbol{\phi})$ the same for the model with only SC data, and $l^* (\boldsymbol{\theta}, \boldsymbol{\alpha}, \boldsymbol{\phi}, \mu)$ the log-likelihood at convergence of the joint RP/SC model. If the k common parameters are equal, then:

$$LR = 2\{l^* (\boldsymbol{\theta}^{RP}, \alpha) + l^* (\boldsymbol{\theta}^{SC}, \phi) - l^* (\boldsymbol{\theta}, \alpha, \phi, \mu)\}$$

distributes χ^2 with k degrees of freedom. If LR is greater than the critical value of χ^2_k for the required confidence level, the test is rejected and one (or more) attribute(s) should be taken out of the set $\mathbf{x}$ and be specified with a different parameter in both domains for the joint estimation, and the test is repeated.

8.7.3.2 Joint Estimation Considering Correlation between Repeated Observations

The ML model offers much in terms of the appropriate mixing of revealed and stated preference data. The traditional approach above, employing an artificial NL structure (e.g. 'the nested logit trick'), suffers from at least two important deficiencies; first, the stated choices of the same individual are considered independent and second, the stated choices of individuals who also responded to the RP survey (generally only part of the sample is in this category), are unrelated to their RP choices.

Bhat and Castelar (2002) were probably the first to formulate and apply a unified ML framework for joint RP/SC model estimation that could accommodate a flexible competition pattern across alternatives, scale differences in the RP and SC choice contexts, heterogeneity across individuals, state dependence of the stated choices on the revealed choices, and heterogeneity across individuals in the state dependence effects. Their likelihood function has two levels of integration because they postulate an 'error components' formulation that generates inter-alternative correlation operating at the choice level, and also random coefficients that accommodate taste variations across individuals and operates at the individual level. Using real data, they found – among other things – that heterogeneity and state dependence effects were tempered when included simultaneously, indicating confounding of true and spurious state dependence. They also found that the better specified model significantly outperformed more restrictive structures.

Train and Wilson (2008) improved on the above by postulating a ML model that explicitly incorporated the fact that SC experiments are usually constructed on the basis of RP choices. Thus, they addressed an important issue that could be a source of inconsistency in estimation. For example, Bhat and Castelar (2002) included a state dependence variable in the form of a dummy for the choice in the RP setting that enters the SC model; however, they did not account for the fact that this variable is correlated with unobserved factors insofar as any unobserved factors from the RP setting carry over into the SC setting. This is equivalent to entering a lagged dependent variable in time series data and estimating by ordinary least squares (i.e. it is fine only if the unobserved factors are not correlated over time). Train and Wilson (2008) developed an appropriate method to use when the lagged dependent variable is included and unobserved factors are correlated over time. Thus, it is a discrete-choice-model analogue of the method of estimating regression with lagged dependent variables and serially correlated errors. Following this analogy, note that allowing for random coefficients and different variance of the error term does not change the fact that entering a lagged dependent variable (or variables created from it) is inconsistent when errors are serially correlated.

8.7.3.3   Forecasting with Joint RP–SC Models

A key issue in forecasting with joint RP–SC models is how to treat the alternative specific constants (ASC). Cherchi and Ortúzar (2006) provide an in-depth discussion of the problem for the following three cases:

- when the RP and SC alternatives are exactly the same;
- when the SC data include new alternatives (i.e. not present in the base year), and
- when the SC design implies substantial changes, such that alternatives sharing the same label (e.g. normal train and a substantially improved fast train service) could represent new alternatives.

They concluded that the first case is trivial as the ASC corresponding to the RP domain should be used without rescaling but, of course, adjusted to match the market shares of the base year. In the second case, if the analyst truly believes that the SC data reproduce correctly the market shares of the population in forecasting, then the ASC (both for the existing and for the new alternatives) should be adjusted to match the market shares in the SC data. Conversely, if the market shares to match are unknown then the analyst must rely on estimation results, that is, as long as the usual theoretical restrictions of the model are satisfied, it might be useful to draw further considerations on the ASC specification from the model that provides the best statistical fit. Finally, if the SC design implies substantial changes, such that alternatives sharing the same label could represent new options, and there is uncertainty as to the extent they are actually different, then best fit and analyst's judgment, seem to be the only guide.

On the other hand, regardless of the way the ASC are specified (i.e. generic or specific), depending on the results for each specific context the application of a mixed RP/SC model in forecasting implies some limitations on the scenarios to be tested. In particular:

- if ASC which are specific to RP and SC are estimated, forecasts can only be made for scenarios involving structural characteristics not inferior to those described in the SC design, and in that case rescaled SC-ASC should be used;

- if specific ASC for both domains are estimated and a scenario not involving structural changes is considered, the RP-ASC should be used;
- finally, if constrained generic ASC are estimated for both domains, scenarios involving structural changes (for those alternatives with constrained RP/SP ASC) should not be tested, unless we obtain ASC with a fairly close value from estimation.

Cherchi and Ortúzar (2011) considered the problem of forecasting with a joint RP–SC Mixed Logit model allowing for random taste heterogeneity. They note that although a basic assumption when pooling RP and SC data is that they share the same underlying behaviour, often the 'partial preference homogeneity' approach (i.e. the parameters are not constrained to be equal in both data sets) gives better results because some attributes can only be measured/estimated properly in one set, or because differences in the nature of attributes produce different, and highly significant, estimated parameters in both sets.

Note that the differences between RP and SC results might be implicit in the need for using SC data in the first place; indeed, they may represent exactly what we look for when using SC data. Consider the case when we want to forecast the effects of structural changes (i.e. departs from the current real market) and utilities are not linear in the attributes. The effect of the partial enrichment approach will be more evident when one attempts to consider the various components of individual heterogeneity, because to estimate complex behaviour we need datasets that are both fairly rich and fairly large; unfortunately, this is often *not* the case for RP data.

However, the partial preference homogeneity approach implies problems in forecasting, as the model used for prediction is not the same as the estimated one; thus, it is crucial to carefully check if the estimated model parameters fulfil the microeconomic conditions on the marginal utilities for any scenario to be tested (see the discussion in Cherchi and Ortúzar 2011).

Yáñez et al. (2010a) extended the analysis of partial preference homogeneity to the correlation structure among alternatives, that is, how to deal with the problem of finding different correlation structures revealed in the RP and SC data sets. They also discuss the problem of normalisation in the joint RP/SC model (i.e. defining an appropriate scale) and its effect in estimation and forecasting. They consider several cases from the simplest, when alternatives are independent in both the RP and SC datasets, to the most complicated one when the two datasets present different inter-alternative correlation structures. They show that from a theoretical point of view both the lower and upper normalizations of the NL model (see Section 7.4.4) are equivalent in this case, but their practical convenience is limited to the simple case of independent alternatives in both datasets; furthermore, they confirm the results of Carrasco and Ortúzar (2002) that the upper normalization is more intuitive. Moreover, although any inter-alternative correlation structure between the alternatives in the RP and SC domains can in principle be estimated, they recommend using a generic one for the existing alternatives in both datasets, if possible; otherwise, the model might not be consistent in prediction. Furthermore, assuming different structural parameters in the RP and SC data sets means that the unobserved components of the utilities (that make some alternatives to be perceived as more similar than others), are not the same in both cases. This can be justified when the systematic utilities are specified differently, but it should not occur when alternatives have the same specification.

Finally, in terms of model use in forecasting, Yáñez et al. (2010a) provide the following recommendations:

1) If the joint RP–SC model structure assumes the same structural parameters for both data environments, their estimated values can be used directly in forecasting.

2) If the joint RP–SC model structure assumes the same structural parameters for both environments except for alternatives present only in the SC case (i.e. usually new alternatives), the whole correlated (or uncorrelated) structure of the SC-alternatives needs to be moved into the RP domain to make forecasts. Notwithstanding, the structural parameter does not need to be scaled because it was already estimated scaled by the RP scale parameter, and it is associated with the EMU term.

3) If the joint RP–SC model structure allows for different structural parameters in both environments (the most general and most complicated case), the general advice is to use the structure estimated with each data set (RP or SC). However, to be consistent, the structural parameters should be associated with utilities measured in the same environment. This means that if we have more faith in the SC data, we should move both the SC structural parameters and the utilities associated with the alternatives in the nest to the RP environment.

Example 8.18 Consider the introduction of a new high-speed rail (HSR) interurban line to compete with car, bus and airplane. Furthermore, given the competitiveness among different services, the following groups of alternatives were identified: three bus alternatives (conventional bus, executive bus, sleeper bus), three plane alternatives (to represent different pairs of airports available at the two main cities affected by the new service), and two HSR alternatives (conventional and executive).

A RP–SC survey was conducted with the final aim of forecasting the demand for the new mode. After estimating separate models for each dataset, different inter-alternative correlation structures for the RP and SC data were found. In particular, the SC data presented a clear and strong correlation between the two HSR services and between the three bus alternatives; while the inter-alternative correlation among plane alternatives in both cases and between the RP bus alternatives was not significant.

Following the discussion about model consistency for prediction purposes above (Yáñez et al. 2010a), the correlation structure in both environments should be constrained to establish a unique and consistent structure. This offers three possibilities: (i) all alternatives are considered independent (model MNL below); (ii) all alternatives are considered independent except the two new ones in the SC environment (model MNL-NL Rail); and (iii) the bus alternatives are considered correlated with the same structural parameter in both data sets, and the two HSR alternatives are correlated in the SC case (model NL).

To evaluate the effect of establishing different inter-alternative correlation patterns on demand predictions, we can calculate the variation in aggregate market shares for various simple policies: Table 8.23 shows that all models predict a decrease in the market shares of the existing modes following a reduction in the HSR fares. However, there are important differences in the magnitude of the changes (the policy impacts are evidently greater for the MNL model). Indeed, for a 50% reduction in HSR fares, in the MNL case the estimated percent change in the aggregate HSR share (ΔP_j) is 50% larger than if both HSR alternatives are assumed to be correlated. However, the differences between the results for the two nested models are not large.

Based on this example, we could say that the models that simply follow the correlation structure detected for the RP data, without considering what the SC data might reveal in this sense, may overestimate the potential market shares of new alternatives. Thus, and as a conclusion, SC data may not only help to improve the specification of representative utility in estimation, but also to define the most appropriate correlation structure of a forecasting model.

Table 8.23 Forecast effects of including inter-alternative correlation

Model	Attribute	HSR fare			Airplane fare			Bus fare		
	% change	−50	−25	−10	−50	−25	−10	−50	−25	−10
MNL	Car	−0.460	−0.230	−0.090	−0.390	−0.180	−0.060	−0.077	−0.039	−0.016
	Airplane	−0.410	−0.210	−0.080	1.050	0.460	0.170	−0.048	−0.024	−0.009
	Bus	−0.490	−0.230	−0.080	−0.400	−0.160	−0.050	0.157	0.078	0.031
	HSR	0.630	0.310	0.120	−0.360	−0.170	−0.060	−0.057	−0.028	−0.011
MNL–NL rail	Car	−0.103	−0.051	−0.020	−0.139	−0.066	−0.026	−0.007	−0.003	−0.001
	Airplane	−0.081	−0.040	−0.016	0.200	0.093	0.036	−0.002	−0.001	−0.001
	Bus	−0.214	−0.108	−0.042	−0.215	−0.088	−0.031	0.048	0.024	0.010
	HSR	0.139	0.070	0.027	−0.138	−0.066	−0.026	−0.005	−0.003	−0.001
NL	Car	−0.100	−0.050	−0.020	−0.140	−0.060	−0.030	−0.007	−0.003	−0.001
	Airplane	−0.080	−0.040	−0.010	0.190	0.090	0.030	−0.003	−0.001	−0.001
	Bus	−0.200	−0.100	−0.040	−0.210	−0.090	−0.030	0.050	0.025	0.010
	HSR	0.130	0.070	0.030	−0.140	−0.070	−0.030	−0.006	−0.003	−0.001

Exercises

8.1 Consider the following mode choice model:

$$V_1 = \theta_1 t_1 + \theta_3 c_1 + \theta_4 Nc + \theta_7$$
$$V_2 = \theta_1 t_2 + \theta_2 e_2 + \theta_5 c_2 + \theta_8$$
$$V_3 = \theta_1 t_3 + \theta_2 e_3 + \theta_6 c_3$$

where t_k is in-vehicle travel time, e_k is access time, c_k is cost divided by income, and Nc is the number of cars in the household.

a) Indicate which variables are generic, which are specific and what is the real meaning of θ_7 and θ_8.

b) Discuss the implications of having obtained the following values during model estimation:

$$\theta_1 = -0.115 \qquad \theta_2 = -0.207 \qquad \theta_3 = -0.301$$
$$\theta_4 = 1.730 \qquad \theta_5 = 0.476 \qquad \theta_6 = -0.301$$
$$\theta_7 = -1.250 \qquad \theta_8 = 2.513$$

8.2 During specification searches you obtained the set of mode choice models for car (1), bus (2), and underground (3), shown in Table 8.24; the units of time and cost/income are minutes, sex is a dummy variable which takes the value of 1 for males and 0 for females; EMU is the expected maximum utility of the transit nest (bus-underground).

a) Indicate which model you prefer explaining very clearly why.
b) The sample you used for estimation comprised 1000 individuals having all alternatives available. If 250 choose car, 600 choose bus and the rest underground, compute $l^*(0)$, the log-likelihood value for the equally likely model, and $l^*(C)$, the log-likelihood for the constants only model.

Table 8.24 Mode choice models for Exercise 8.2

Variable (option entered)	Coefficient (*t*-ratio)			
	MNL-1	MNL-2	HL-1	HL-2
Car time (1)	−0.112	—	−0.114	—
	(−6.10)		(−6.00)	
Transit time (2,3)	0.006	—	−0.001	—
	(1.25)		(−0.94)	
Travel time (1–3)	—	−0.071	—	−0.083
		(−3.34)		(−3.60)
Cost/income (1–3)	−0.031	−0.040	−0.035	−0.033
	(−2.56)	(−3.52)	(−2.83)	(−3.10)
No. of cars (1)	1.671	1.823	1.764	1.965
	(4.21)	(4.80)	(4.12)	(5.14)
Sex (2,3)	−0.752	−0.776	−0.739	−0.701
	(−1.87)	(−1.98)	(−2.01)	(−1.83)
EMU	—	—	0.875	0.800
			(5.12)	(13.4)
ρ^2	0.412	0.284	0.376	0.315

8.3 You were asked to estimate a Multinomial Logit (MNL) model and an Independent Probit (IP) model with the same data set; imagine (as it is not possible to estimate σ in practice) that you obtained the values shown in Table 8.25:

Table 8.25 MNL and IP models for Exercise 8.3

Parameters	MNL	IP
θ_1	1.285	1.698
θ_2	−0.026	−0.034
θ_3	−0.123	−0.162
σ^2	Not applicable	2.870

Indicate whether these results appear to be consistent; if your answer is affirmative, explain which the cause of the differences is. If your answer is negative, explain why.

8.4 While conducting an SP survey you asked three individuals to rank the three options whose attributes are given in Table 8.26:

Table 8.26 Alternatives available for Exercise 8.4

Alternative	Travel time (min)	Fare ($)
1) High speed train	30	10
2) Express train	40	8
3) Luxury coach	60	5

After completing the survey, you obtained the results presented in Table 8.27:

Table 8.27 Ranking results for Exercise 8.4

Individual	Ranking
1	1, 2, 3
2	2, 3, 1
3	2, 1, 3

You are interested in estimating a MNL model with linear in the parameters utility function given by:

$$V_i = \theta_1 t_i + \theta_2 c_i$$

If you are told that $\theta_1 = -0.03$, find a maximum likelihood estimate for θ_2. Discuss your results.

9

Model Aggregation and Transferability

9.1 Introduction

The planning and appraisal of interventions to a transport system benefit from models to deliver forecasts. The forecasts themselves normally need to be aggregate, that is, to represent the behaviour of specific market segments or the entire population.

In many practical studies, the models have the classical aggregate five-stage form despite many (and sometimes justified) criticisms about their inflexibility, inaccuracy, and cost. One important reason for this persistence, apart from their familiarity (e.g. they have been considered accepted practice for many years), is that they offer a tool for the complete modelling process, from data collection through to the provision of forecasts of flows on links. This has not often been the case with, for example, disaggregate model approaches, perhaps because the data necessary to make aggregate forecasts with them are not readily available (see the discussion by Daly and Ortúzar 1990).

In an econometric interpretation of demand models, the aggregation over *unobservable* factors (either attributes or personal characteristics) results in a probabilistic decision model and the aggregation over the *distribution* of observables results in the conventional aggregate or macro relations (Williams and Ortúzar 1982b). Cast in these terms, the difficulty of the aggregation problem depends on how the components of the system are described within the frame of reference employed by the modeller; this framework will determine the degree of variability to be accounted for in a *causal* relation. To give an example, if the framework used by the analyst is that provided by the 'entropy-maximising' approach we saw in Chapter 5, the explanation of the statistical dispersion in a given data set will be very different from that provided by another modeller using a 'random utility' approach, even if they both finish with identical model functions; this *equifinality issue* is discussed by Williams (1981).

In the case of disaggregate random utility models, the aggregation problem is how to obtain aggregate measures, such as market shares of different modes and flows on links, from data at the level of the individual. This can be achieved in one of two ways, by having the process of aggregating individual data either before or after model estimation, as shown in Figure 9.1.

In the first case, we have variations of the classical aggregate approach, which can be easily criticised for being inefficient in the use of the data, not accounting for their full variability and for risking statistical distortion such as the 'ecological fallacy' discussed in Section 7.1. The second approach answers most of the above criticisms; the question that remains is how exactly to perform the aggregation operation over the micro-relations.

Modelling Transport, Fifth Edition. Juan de Dios Ortúzar and Luis G. Willumsen.
© 2024 John Wiley & Sons Ltd. Published 2024 by John Wiley & Sons Ltd.
Companion website: www.wiley.com/go/ortuzar5e

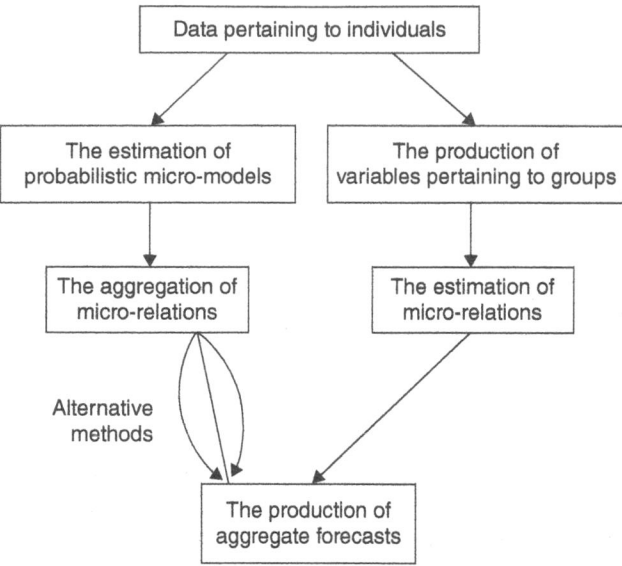

Figure 9.1 Alternative aggregation strategies

Daly and Ortúzar (1990) studied the problem of aggregation of exogenous data in some depth. They concluded that in the case of models representing the behaviour of more than one individual (as is the case with the classical aggregate model), some degree of aggregation of the exogenous data is inevitable and the issue becomes one of to what extent greater accuracy (i.e. smaller zones) is desirable. However, when the model represents the behaviour of a single individual, it is conceivable that exogenous data can be obtained and used separately for each traveller; therefore, the issue is whether it is preferable to use less accurate data on cost or other grounds. Their findings support the notion that the cost/accuracy trade-off is heavily dependent on context. For example, it is clear that for mode choice modelling and short-term forecasting, the use of highly disaggregate data is desirable; however, the plot thickens considerably for other choice contexts and long-term forecasting. The next two sections will consider aggregation bias and forecasting methods in greater detail.

9.2 Aggregation Bias and Forecasting

Recall the discussion in Section 8.4.6, and consider a well-specified MNL model (i.e. there are no taste variations or correlation problems). We are interested in examining the effect of the manner in which the attributes **x** (or at least some of them) are calculated, measured, or codified, on the estimated demand functions.

Let us assume that we replace one of the attributes, for example, x_1, by an aggregate estimate z_1, such that:

$$x_{1i} = z_{1i} + \tau_1 \tag{9.1}$$

where the τ_i are distributed with mean 0 and standard deviation σ_τ; then replacing (7.3) and (9.1) in (7.2), we get (note that we had dropped the individual index q for simplicity):

$$U_i = \theta_1 z_{1i} + \sum_k \theta_k x_{ki} + \delta_i \tag{9.2}$$

where the error term $\delta_i = \theta_1 \tau_1 + \varepsilon_i$ has variance $(\theta_1^2 \sigma_\tau^2 + \sigma^2)$. Thus, if we re-estimated the model in this case, the coefficient estimates would not be

$$\hat{\theta}_k = \frac{\pi \sqrt{6} \cdot \theta_k}{\sigma} \tag{9.3}$$

as before, but

$$\hat{\hat{\theta}}_k = \frac{\pi \sqrt{6} \cdot \theta_k}{\sqrt{(\theta_1^2 \sigma_\tau^2 + \sigma^2)}} \tag{9.4}$$

that is to say, $\hat{\hat{\theta}}_k \leq \hat{\theta}_k, \forall k$. This is normally known as *aggregation bias* and has led to the recommendation that using average zonal variables for estimating disaggregate demand models should be avoided whenever possible (see for example Horowitz 1981). The previous analysis may be extended to examine the consequences of this bias in forecasting, as in the following example taken from Gunn (1985a).

Example 9.1 Consider a choice situation modelled by an MNL model such as (7.9) and assume that attribute x_{1j} is doubled, *ceteris paribus*, for each option. It is clear that neither θ nor σ are affected by this; therefore, if the model was re-estimated with a new databank containing a consistent choice set, we would obtain exactly the same values $\hat{\theta}_k$ from the original context again, and so these would predict satisfactorily in the new context.

Consider now what would happen if after doubling x_{1j}, each of these values was replaced by its aggregate estimate z_{1j} (for example, the zonal average). In this case, we would obtain Eq. (9.2) again, but the variance of δ_i would now be $(\theta_1^2 4\sigma_\tau^2 + \sigma^2)$; in other words, if the model was re-estimated with the new data, it would yield coefficients with expected values given by

$$\hat{\hat{\theta}}_k' = \frac{\pi \sqrt{6} \cdot \theta_k}{\sqrt{(\theta_1^2 4\sigma_\tau^2 + \sigma^2)}} \tag{9.5}$$

that is $\hat{\hat{\theta}}_k > \hat{\hat{\theta}}_k'$ and $\hat{\hat{\theta}}$ would produce greater than normal predictions in these conditions. Alternatively, attribute reduction policies would imply under-predictions of the model calibrated with aggregate data (see Ortúzar and Ivelic 1987).

9.3 Confidence Intervals for Predictions

As we saw in Chapter 8, the maximum likelihood estimated parameters $\hat{\theta}$ of a discrete choice model are asymptotically distributed $N(\theta, S^2)$, where θ are the population parameters and S^2 is their covariance matrix given by (8.14):

$$S^2 = -\left(E\left(\frac{\partial^2 l(\theta)}{\partial \theta^2} \right) \right)^{-1}$$

that is, the negative inverse of the Fisher information matrix **I**. From this knowledge, it is straightforward to compute confidence intervals for the estimated parameters, on the basis of the well-known property that quadratic forms distribute χ^2 with degrees of freedom equal to the number of variables in the vector of interest (see Section 8.5).

Applying this to our estimated parameters, we obtain the following quadratic form:

$$QF\left(\hat{\boldsymbol{\theta}}, \boldsymbol{\theta}\right) = \left(\hat{\boldsymbol{\theta}} - \boldsymbol{\theta}\right) \cdot \mathbf{I} \cdot \left(\hat{\boldsymbol{\theta}} - \boldsymbol{\theta}\right)^{\mathrm{T}}$$

that distributes χ^2 with K degrees of freedom (K is the number of estimated parameters).

Therefore, a confidence region at the 95% level for the set of estimated parameters is given by the values of $\boldsymbol{\theta}$ that satisfy (9.6):

$$QF\left(\hat{\boldsymbol{\theta}}, \boldsymbol{\theta}\right) \leq \chi^2_{K,95\%} \tag{9.6}$$

However, converting the above region into a confidence region for the estimated probabilities is not easy, as the relation between parameters and probabilities is not linear.

It is interesting to mention that the immense majority of discrete choice model applications have failed to produce confidence intervals for the estimated probabilities, although two methods for doing it have been available for many years (Horowitz 1980):

1) Approximate the choice probabilities by a first-order Taylor series expansion; in practice, this is equivalent to assuming that the relation between probabilities and parameters is linear. This is a fairly usual approach in mathematical statistics because it is easy to implement and inexpensive in computational terms.
2) Solve a non-linear programming problem; although this is a bit more complex and expensive, it is subject to less errors than the previous method.

9.3.1 Linear Approximation

If $\hat{\mathbf{P}}$ is the estimated value and $\mathbf{P}$ is the true value of the choice probabilities, the Taylor series approximation is given by:

$$\hat{\mathbf{P}} = \mathbf{P} + \left(\hat{\boldsymbol{\theta}} - \boldsymbol{\theta}\right) \frac{\partial \mathbf{P}}{\partial \boldsymbol{\theta}} + \Delta \tag{9.7}$$

where the expected value of $\hat{\mathbf{P}}$ is equal to $\mathbf{P}$ and Δ is a residual term.

Now, as the parameters $\boldsymbol{\theta}$ distribute asymptotically Normal, $\hat{\mathbf{P}}$ also distributes asymptotically Normal as (9.7) is a linear transformation. Thus $\hat{\mathbf{P}} \sim N(\mathbf{P}, \mathbf{W})$, where:

$$\mathbf{W} = \left(\frac{\partial \mathbf{P}}{\partial \boldsymbol{\theta}}\right) \mathbf{S}^2 \left(\frac{\partial \mathbf{P}}{\partial \boldsymbol{\theta}}\right)^{\mathrm{T}}$$

and the numerical value of $\mathbf{S}^2$ can be estimated substituting $\hat{\boldsymbol{\theta}}$ by $\boldsymbol{\theta}$ in the derivatives. This approach is actually called the *Delta Method* in statistics (see Greene 2008) and has been used in practice for some time.

Given the above, if $Z_{\alpha/2}$ denotes the percentile $(1 - \alpha/2)$ of the standard Normal distribution, then a confidence region of $100(1 - \alpha)$ for $\mathbf{P}$ is given by:

$$\hat{\mathbf{P}} - Z_{\alpha/2}|\mathbf{W}|^{1/2} \leq \mathbf{P} \leq \hat{\mathbf{P}} + Z_{\alpha/2}|\mathbf{W}|^{1/2} \tag{9.8}$$

A problem with this quick and simple method is that it could lead to erroneous results, as shown in Example 9.2, but note that we are calculating errors based on the (incorrect) assumption that the model formulation is right. This is in the realm of the *Rumsfeld issue* (see Logan 2009); that is, we are calculating 'known unknowns', considering sampling error, but omitting the great sea of 'unknown unknowns', such as mistakes, misinterpretations, inappropriateness of approach and data, etc.

Daly et al. (2012b) have given a better interpretation of the method, based on the idea that at the maximum likelihood values of the parameters, the measure is no less exact than the original estimates; this notion can be applied to estimate the confidence intervals of various measures, such as willingness-to-pay, and it is also easy to use to estimate the errors of predicted market shares in MNL models. However, when the model is more flexible than the MNL and/or the if the sample being expanded is large, with complicated calculations for the weights attached to each observation in the aggregation procedure, the number of calculations involved can be prohibitive and a sampling approach may be necessary, as discussed by de Jong et al. (2007).

Example 9.2 Consider the following simple single-parameter Logit model:

$$P_1(x) = \frac{1}{1 + \exp{(\theta x)}}$$

where x is an independent variable and $P_1(x)$ is the probability of choosing the first option. Assume that the maximum likelihood estimate of θ was $\hat{\theta} = 3$ and that its sample variance was equal to 1. Then, if $x = 1$, the reader can check that Eq. (9.8) yields the following confidence interval at the 95% level:

$$-0.041 \leq \mathbf{P} \leq 0.136$$

This interval is clearly erroneous as it allows for negative values of $\mathbf{P}$ (in fact, in this extremely simple case, it is possible to calculate the exact interval, which is given by $0 \leq \mathbf{P} \leq 0.205$). Unfortunately, the problem of finding a good interval and – in fact – to know if Eq. (9.8) is a good approximation is fairly complex (the reader can check that if $x = 0.01$, we would get a consistent interval). We will come back to the Delta method in Chapter 18.

9.3.2 Non-Linear Programming

The simplest way to formulate this method is as follows. Let $P_i(\boldsymbol{\theta})$ be the probability of choosing alternative $A_i \in \mathbf{A}$, where J is the total number of alternatives in the choice set; the decision variables take fixed values $\mathbf{x}$, and the parameters are, as usual, $\boldsymbol{\theta}$. Consider that $b_i(\alpha)$ and $B_i(\alpha)$ are the results of the following non-linear problems:

$$b_i(\alpha) = \text{Min } P_i(\theta), i = 1, ..., J$$
$$B_i(\alpha) = \text{Max } P_i(\theta), i = 1, ..., J$$

$$\text{subject to } H(\hat{\boldsymbol{\theta}}, \boldsymbol{\theta}) \leq \chi^2_{K,(1-\alpha)}$$

where the maximisation and minimisations operations are done for different values of $\boldsymbol{\theta}$. In this case, the following inequalities define a rectangular confidence region for P_i at the $100(1-\alpha)$ level:

$$b_i(\alpha) \leq Pi \leq B_i(\alpha) \quad i = 1, ..., J \tag{9.9}$$

This method tends to produce larger confidence regions than the previous one, and it is also harder to implement; however, it has the advantage of yielding always reasonable (in the sense of not inconsistent) confidence intervals.

9.4 Aggregation Methods

While a disaggregate model allows us to estimate individual choice probabilities, we are normally more interested in the prediction of aggregate travel behaviour. If the choice model was linear, the aggregation process would be trivial, amounting only to replacing the average of the explanatory variables for the group in the disaggregate model equation; see for example the aggregation of household-based trip generation models in Chapter 4. However, if the model is non-linear, this method, called *naive aggregation*, will generally produce bias as shown in Figure 9.2.

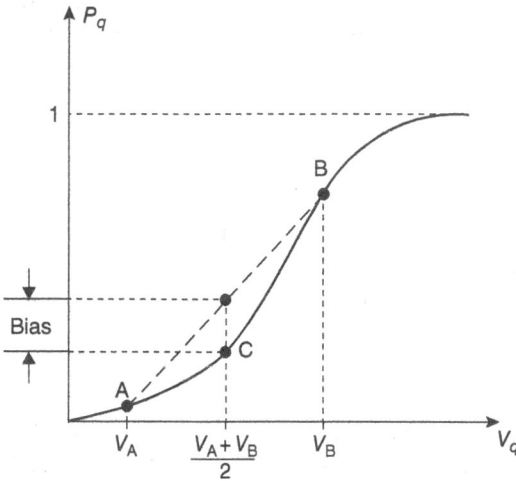

Figure 9.2 Bias of the naive aggregation method

The correct aggregate probability for a group of two individuals A and B is $(P_A + P_B)/2$; the naive method yields, instead, the probability $P[(V_A + V_B)/2]$. As can be seen, if the model was linear both values would coincide.

Discrete choice models, such as those we have discussed in Chapters 7 and 8, can be represented by a general expression of the form:

$$P_{jq} = f_j(\mathbf{x}_q)$$

where P_{jq} is the probability that individual q selects option A_j, $\mathbf{x}_q$ is the set of variables influencing her decision, and f_j is the choice function for A_j (for example, the MNL).

For a population of Q individuals, the aggregate proportion choosing A_j, according to the model, is the expected value (or 'enumeration') of the probabilities of each individual in the population:

$$P_{jQ} = \frac{1}{Q} \sum_q f_j(\mathbf{x}_q) \tag{9.10}$$

Unfortunately, this method would require an impossibly large data set. However, if we accept that the sample used to estimate the model is a good representative of the population, we can use a modified version of (9.10) and refer to *sample enumeration* as in (9.11):

$$MS_j = \sum_{q=1}^{Q_S} w_q f_j(\mathbf{x}_q) \tag{9.11}$$

where MS_j is the predicted market share of alternative A_j in the population, Q_S is the sample size, and w_q is the expansion factor corresponding to observation q in the sample.

This is a good practical method for moderate-size choice sets and is excellent for mode choice models in short-term predictions. However, the method is not so useful in the long term because it does not allow us to address overall contexts that are very different to that of the base year (i.e. it assumes that the distribution of the attributes will not differ from that of the sample in the future); it is also unable to produce aggregate zone-to-zone flows necessary for the estimation of demand at the link level.

Example 9.3 Consider the model of Example 9.2, with the same estimated parameter (i.e. $\hat{\theta} = 3$) but with a variance equal to 4.0; assume also that the sample was composed of five individuals with the values of x shown in Table 9.1:

Table 9.1 Sampled values of x for Example 9.3

Individual	x
1	0.89
2	0.75
3	−0.25
4	0.80
5	−0.40

In this case, and assuming that the expansion factors were all equal to 100, the reader can check that by applying (9.11), the market share of the first alternative would be approximately equal to 169 (i.e. out of the total 500). To estimate the precision of this estimate, we could apply the Delta method mentioned in the previous section. For the case of market shares estimated from a simple MNL model, Daly et al. (2012b) show that the appropriate expression is:

$$Var\left(MS_j\right) = \mathbf{MS}_j' \cdot \mathbf{S}^2 \cdot \mathbf{MS}_j'^{\mathrm{T}} \tag{9.12}$$

where MS_j' is the vector of first derivatives of the estimated market share with respect to the parameters; thus, its kth element is given by:

$$MS_{jk}' = \sum_q w_q \cdot \partial P_{jq}/\partial\theta_k$$

In the case of our example, the equations above simplify substantially as we have only one parameter. Thus, $\mathbf{S}^2$ is equal to the scalar 4.0, and considering the expression of our binary Logit model, we have that for alternative A_1 we would get:

$$MS_1' = 100^* \sum_q \partial P_{1q}/\partial\theta = 100^* \sum_q P_{1q}\left(1 - P_{1q}\right) x_q$$

The reader can easily check that $MS_1' = 5.40$ in this case, and so the estimated variance for the market shares of the first alternative equals 116.53. Thus, a 95% confidence interval for the market share would be approximately equal to:

$$169 - 1.96 \cdot \sqrt{116.53} \le MS_1 \le 169 + 1.96 \cdot \sqrt{116.53} \rightarrow 148 \le MS_1 \le 190$$

Note that if the same calculation was done for the second alternative, the confidence interval would be $310 \leq MS_2 \leq 352$, and so in this case, there would be a perfect match (which, in general, may not be the case).

To cope with the problem of having a sample that is only good for the relatively short term, the *artificial sample* enumeration approach may be used (see Daly and Gunn 1986; Daly 1998). An artificial sample is one in which personal characteristics of members of existing households (believed to be representative of the population of the study area) are matched with characteristics of a number of locations also believed to be typical of the area. Thus, the marginal distributions of both personal and location characteristics are, by construction, typical of the study area; the approximation made here is that the joint distribution of these characteristics can be represented by the product of the two marginal distributions.

Given suitable networks, zoning systems, and planning data, the marginal distributions of locations can be those of actual locations in the study area (details of their accessibility to available destinations are needed); if the locations are distributed over the whole of the study area, we can be reasonably confident of overall representativeness in large samples (see Gunn 1985b).

For personal characteristics, the following steps are needed to achieve realism:

1) Actual households' members are drawn at random from a large nationally representative data set (e.g. a Census).

2) For each zone of the study area, different expansion factors are found for each of these households such that the expanded sample corresponds as closely as possible to known or forecast aggregate totals (i.e. of variables such as numbers of workers, numbers of individuals by gender and age grouping, etc.).

3) The expansion factors, or more commonly the number of households in each group, are chosen such that the overall distribution of households in terms of a given stratification (say size, number of workers, and age of head of household) is not too different from the overall national average (note that when classifying data in this form, there are several impossible strata; for example, households of size 1 with more than 1 worker). Daly (1998) compares the two most used methods to do this, 'iterative proportional fitting' (what we called Furness method in Section 5.2.3), which has been extensively applied in practice (see the discussion by Beckman et al. 1995) and 'quadratic optimisation', as in (9.13), which was judged as an improvement over the previous one. Daly (1998) also provides details about the steps to follow in the construction and use of the 'prototypical sample' and several successful examples about its application:

$$S(N_i) = \sum_k W_k \left[\sum_i (X_{ik} \cdot N_i - Y_k)^2 \right] + \sum_i (N_i - R_i)^2 \tag{9.13}$$

Here, N_i is the required number of households in stratum i; W_k is a weight chosen to increase or decrease the importance of the fit to the kth variable (e.g. number of workers, number of males between 18 and 65, etc.); X_{ik} is the average value of variable k for household stratum i; Y_k is the average (observed) value of variable k for (each zone of) the study area; and R_i is the number of households in stratum i in the base-year sample. Various other constraints can be put on the process, as discussed by Gunn (1985b).

The artificial sample replicates the population of each zone of the study area; thus, aggregate forecasts can simply be obtained by applying the 'enumeration' method to these data. The interested reader can find more on the creation of synthetic samples, although not exactly for the same purposes as discussed here, in the work of Guo and Bhat (2007) and Ye et al. (2009). We also discuss the generation of synthetic samples in a little more detail in Section 16.5.

Another practical method is known as the *classification approach*, which consists in approximating (9.10) by a finite number of relatively homogeneous classes, as in:

$$P_{jQ} = \sum_c f_j(\mathbf{X}_c) \, Q_c/Q \qquad\qquad (9.14)$$

where $\mathbf{X}_c$ is the mean of the variable set vector for subgroup c and Q_c/Q is the proportion of individuals in the subgroup.

The accuracy of the method depends on the number of classes c and their selection criteria (in the limit it equals the naive method, when the number of subgroups $c = 1$, and the enumeration method, when $c = Q$). Interesting but not often practical methods to define the classes have also been reported (McFadden and Reid 1975), but, in general, the classification approach is recommended for cases where sample enumeration is not appropriate (Koppelman 1976).

An obvious method to define classes is to use as market segmenting variables those that present the greatest variance or those that limit in some way the available choice set of each individual. Thus, in the mode choice case, good variables are the number of cars per household and family income.

9.5 Model Updating or Transference

9.5.1 Introduction

During the 1980s, a substantial body of literature emerged with empirical evidence about the stability (or, in most cases, lack of it) of parameters of disaggregate travel demand models, across space, cultures, and time (see for example Gunn et al. 1985; Koppelman and Wilmott 1982; Koppelman et al. 1985a,b). The reasons were simple: firstly, evidence of stable values of estimated parameters could provide a direct indication of model validity; secondly, a model that is not stable over time is likely to produce inaccurate predictions; finally, and not less importantly, transferable models should allow for more cost-effective analyses of transport plans and policies.

Because it is unrealistic to expect an operational model in the social sciences to be perfectly specified, it is quite obvious that any estimated model is in principle context dependent. For this reason, it is not very useful to look for perfect model stability and to consider model transferability in terms of the equality of parameter values in different contexts.

A more appropriate view considers model transfer as a practical approach to the problem of estimating a model for a study area with little resources or a small available sample. In this sense, the model-transfer approach is based on the idea that estimated parameters from a previous study may provide useful information for estimating the same model in a new area, even when their true parameter values are not expected to remain the same. Now, as transferred models cannot be expected to be perfectly applicable in a new context, updating procedures are required to modify their parameters so that they represent behaviour in the application context more accurately. Depending on the information available in the new environment, different updating procedures may be applied (see Ben-Akiva and Bolduc 1987).

9.5.2 Methods to Evaluate Model Transferability

If we define transferability as the usefulness of a transferred model, information, or theory in a new context, we can attempt to measure it by comparing the model parameters and, more interestingly, its performance in the two contexts. For this, we will assume that we have estimated the parameters independently in the two contexts; we will also assume that we would like to measure the errors involved in using the first model in the second context. The following tests and measures were used in such analyses in many practical studies (Galbraith and Hensher 1982; Koppelman and Wilmott 1982; Ortúzar et al. 1986).

9.5.2.1 Test of Model Parameter for Equality

To evaluate the absolute difference between coefficients of a given model estimated in two different contexts, the t^*-statistics has been used; if (9.15) holds, the null hypothesis that this difference is zero cannot be rejected at the 95% level:

$$t^* = \frac{\theta_i - \theta_j}{\sqrt{(\theta_i/t_i)^2 + (\theta_j/t_j)^2}} < 1.96 \tag{9.15}$$

where θ_k denotes coefficients, t_k their t-ratios, i stands for the original context, and j for the new context; note that this is the same test we saw in Example 8.4, but here no correlation among the parameters is possible as they belong to two different contexts. Galbraith and Hensher (1982) recommended the application of this test only to parameters with low standard error (high t-ratio); otherwise, the t^*-statistic may reject the alternative hypothesis (i.e. that the parameters are different) even if they exhibit substantial differences. However, note that t^* suffers from the 'scale problem', as the variances of the error components in both contexts may not be the same; thus, one cannot be sure if differences are real or just a scaling problem. We will consider more appropriate methods below.

9.5.2.2 Disaggregate Transferability Measures

These are based on the ability of a transferred model to describe individual observed choices in the new context and rely on measures of log-likelihood as those that were depicted in Figure 8.2. In addition, we need to define $l_j^*(\theta_i)$ as the log of the likelihood that the observed data in the application context j were generated by the transferred model estimated in context i; note that we need to denote the measures previously used in Chapter 8 as $l_j^*(\theta_j)$, $l_j^*(C)$, and $l_j^*(0)$, respectively. Figure 9.3 shows the expected relation among these values.

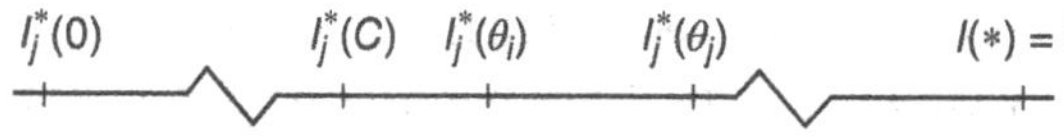

Figure 9.3 Expected relation between log-likelihood values

A natural measure of the transferability of a model estimated in context i for the application in context j is the difference in log-likelihood (i.e. likelihood ratio) between this model and one originally estimated in context j: $-\left\{ l_j^*(\theta_i) - l_j^*(\theta_j) \right\}$. This measure has been used to build two specific indices of transferability:

1) Transferability test statistics (TTS), defined by Atherton and Ben-Akiva (1976) as twice the difference in log-likelihood identified above:

$$\text{TTS} = -2\left\{ l_j^*(\theta_i) - l_j^*(\theta_j) \right\} \tag{9.16}$$

This statistic distributes χ^2 with degrees of freedom equal to the number of model parameters, under the assumption that the parameter vector of the transferred model is fixed. The test is not symmetric; therefore, it is both possible and reasonable to accept transferability in one direction, between a pair of contexts, but reject it in the other direction.

2) Transfer index (TI), which describes the degree to which the log-likelihood of the transferred model exceeds a null or reference model (such as the market shares model), relative to the improvement provided by a model developed in the new context. It was defined by Koppelman and Wilmott (1982) as:

$$
\mathrm{TI}_j(\theta_i) = \frac{l_j^*(\theta_i) - l_j^*(C)}{l_j^*(\theta_j) - l_j^*(C)}
\tag{9.17}
$$

TI has an upper bound of one (which is obtained when the transferred model is as accurate as the local one), but does not have a lower bound; negative values imply only that the transferred model is worse than the local reference model.

The two measures defined above are interrelated by their dependence on the difference in log likelihood between transferred and local models. However, they offer different perspectives on model transferability: TI provides a relative measure, and TTS is a statistical test measure (Koppelman and Wilmott 1982).

9.5.3 Updating with Disaggregate Data

The most general presentation of the MNL model (7.9) with linear utility functions **V** given by (7.3) considers not only the explicit inclusion of relation (7.10) – as we saw in Section 9.2 – but also the explicit inclusion of a set of location parameters w_i as in:

$$
P_{iq} = \frac{\exp\left[(w_i + \boldsymbol{\theta} \mathbf{X}_{iq})/\sigma\right]}{\sum_j \exp\left[(w_j + \boldsymbol{\theta} \mathbf{X}_{jq})/\sigma\right]}
\tag{9.18}
$$

where the location parameters represent the mode of the distribution of errors for each alternative, the scale parameter is, as usual, inversely related to σ, the standard deviation of the distribution of the error term (note that strictly speaking, we are missing the constant $\pi/\sqrt{6}$ in 9.18), and the parameters $\boldsymbol{\theta}$ are the attribute weightings employed by the individual in evaluating alternatives.

In his analysis of model mis-specification, Tardiff (1979) shows that the omission of explanatory variables should have the following effects:

- shift the mean of the error distribution, represented in the model by w_i, and increase its variance reflected by σ;
- bias the estimates of the parameters associated with the included variables.

When comparing models that are incompletely specified, in different contexts, it is expected that the differences in the mean values of the error distribution will be relatively large, the differences in the standard deviation of the errors will be smaller, and the differences in the parameter estimates the smallest. Thus, efforts to improve model transfer to a specific application environment should emphasise adjustment of constants first, parameter scale second, and relative values of the parameter last; this has been confirmed by several practical studies using both aggregate and disaggregate data (Dehghani and Talvitie 1983; Koppelman et al. 1985b; Gunn and Pol 1986).

The parameters in Eq. (9.18) are, of course, not uniquely identifiable and therefore cannot all be estimated; as we have seen, in the case of the alternative specific constants (ASC) one is arbitrarily (and with no loss of generality) set to 0. Also, it is not possible to estimate σ but only the ratios $\mathbf{w}/\sigma$ and $\mathbf{\theta}/\sigma$; defining these ratios by $\mathbf{\mu} = \mathbf{w}/\sigma$ and $\mathbf{\phi} = \mathbf{\theta}/\sigma$, we obtain the more familiar version of the MNL model as:

$$P_{iq} = \frac{\exp\left(\mu_i + \mathbf{\phi}\mathbf{X}_{iq}\right)}{\sum_{A_j \in A(q)} \exp\left(\mu_j + \mathbf{\phi}\mathbf{X}_{jq}\right)} \tag{9.19}$$

where one of the μ_i must be constrained to zero.

9.5.3.1 Updating the Constants

Parameter estimates for a choice model are obtained by maximising a log-likelihood expression such as (8.13), where embedded in the probability function P_{iq} are expressions for the representative utility of each option formulated as:

$$V_{iq} = \mu_i + \mathbf{\phi}\mathbf{X}_{iq} \tag{9.20}$$

Let us denote as $\mathbf{\phi}_T$ a set of parameters estimated in one context to be transferred to a new application context; in this case, the transferred portion of the utility function can be defined as (Koppelman et al. 1985b):

$$Z_{iq}^A = \mathbf{\phi}_T \mathbf{X}_{iq}^A \tag{9.21}$$

where $\mathbf{X}_{iq}^A$ is a vector of attributes of alternative A_i for individual q in the application context (A). The updating of the ASC is accomplished by modifying the utility function in Eq. (9.20) for the application context to:

$$V_{iq}^A = \mu_i^A + Z_{iq}^A \tag{9.22}$$

where V_{iq}^A is the representative utility of option A_i in the application context and μ_i^A its updated ASC. To estimate the updated value of the constants, it is necessary to maximise the log-likelihood function:

$$l(\mu^A) = \sum_q \sum_{Aj \in A(q)} g_{jq} \log P_{jq}\left(\mathbf{Z}_q^A, \mu^A\right) \tag{9.23}$$

where, as before, g_{jq} is defined by:

$$g_{jq} = \begin{cases} 1 & \text{if } A_j \text{ was chosen by } q \\ 0 & \text{otherwise} \end{cases}$$

9.5.3.2 Updating of Constants and Scale

The methodology just outlined can be trivially extended to adjust the scale of the transferred parameters as well as the constants. The coefficient of Z_{iq}^A in Eq. (9.22) was restricted to one in the preceding approach; to update the parameter scale, that restriction is relaxed yielding the following representative utility (Koppelman et al. 1985b):

$$V_{iq}^A = \mu_i^A + \lambda^A Z_{iq}^A \tag{9.24}$$

where λ^A is the scaling parameter for the application context relative to the estimation, or original, context. In this case, the log-likelihood function to be maximised is as (9.23) but includes the extra parameter λ^A. Note that this adjusts the scale of the explanatory variables but does not affect their relative importance. Practical applications of this method have been reported by Gunn et al. (1985), and a discussion of further refinements to this problem can be found in Ben-Akiva and Bolduc (1987).

9.5.4 Updating with Aggregate Data

Consider the same problem as before with the exception that no disaggregate data are available in the application context; however, assume we possess data on observed market shares P_{jq}^*, and also average values for the explanatory variables $\overline{X}_{jz}$, for certain groups $\mathbf{Z}$ (say residents of a given zone) in both contexts.

Consider a naive aggregation in the original context, where the measured utility of option A_j for a given group z is given by:

$$\overline{V}_{jz} = \mu_j + \phi\overline{\mathbf{X}}_{jz} \tag{9.25}$$

Updating both alternative constants and scale, in this case, requires first to compute non-constant utility for the application context as:

$$\overline{\mathbf{Z}}_{jz}^A = \phi_T\overline{\mathbf{X}}_{jz}^A \tag{9.26}$$

then postulate an expression for the representative utility of group z in the application context as:

$$\overline{V}_{jz}^A = \mu_j^A + \tau^A\overline{\mathbf{Z}}_{jz}^A \tag{9.27}$$

where μ^A and τ^A are chosen to maximise the following log-likelihood function (Koppelman et al. 1985a):

$$l(\mu^A, \tau^A) = \sum_z W_z \sum_j P_{jz}^* \cdot \log P_{jz}\left(\overline{\mathbf{Z}}_{jz}^A, \mu^A, \tau^A\right) \tag{9.28}$$

with W_z a weight, usually, the number of observations, which indicates the relative importance of the group in the dataset.

The aggregation issue in the presentation above is not trivial, as it is well-known that the 'naive method' may introduce severe bias. In this sense, it is interesting to mention that the methodology just discussed is wholly consistent with the aggregation approach implicit in most aggregate transport studies (recall Figure 9.1 and the discussion in Chapter 5). There, disaggregate model parameters have been traditionally used as fixed coefficients of generalised cost functions, and later *scale* and *bias* parameters have been fitted using aggregate data (Williams and Ortúzar 1982b).

It is also of interest to note that a more elaborate version of this approach has also been used in practice; for example, in the Greater Santiago Strategic Transport Study (ESTRAUS 1989), disaggregate mode choice parameters were firstly estimated with a mixture of data for 1983–1986 (Ortúzar and Ivelic 1988); these were used to build generalised cost functions, the scale and bias parameters of which were then calibrated using 1977 network and survey data (the only O–D and network data available at the time). Finally, the resulting aggregate distribution and modal-split models were validated using volume counts and other aggregate information for 1986.

An interesting alternative, if available, is the use of purposely designed synthetic samples in an enumeration approach, such as we discussed in Section 9.4 (Gunn et al. 1982). An important advantage of this method is that no major adjustments need to be made to the disaggregate models if the artificial sample provides unbiased information to the model system.

Exercises

9.1 A group of 800 heads of household, with different income levels and located in various parts of an urban area, are confronted with a choice between two transport services A and B, for travelling to the central business district. The first, which is more oriented to the population segment with higher income, has a cost of C_a, and the second has a cost of C_b.

It has been estimated that the utilities of each alternative are given by the following linear functions:

$$U_a = -0.30C_a + 3.23I$$
$$U_b = -0.30C_b$$

where I is family income (1000$/week).

Estimate the number of households that would choose service A using the data provided in Table 9.2:

Table 9.2 Information for Exercise 9.1

Family income (100$/week)	Number of households	C_a ($)	C_b ($)
Between 1 and 2	450	150	120
Between 2 and 3	250	175	145
Between 3 and 4	100	160	130

9.2 Consider the urban corridor depicted in Figure 9.4:

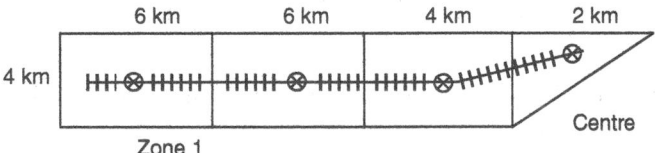

Figure 9.4 Simple corridor for Exercise 9.2

which has the following characteristics:

- underground and highway run parallel to each other,
- there are underground stations at each zone, and
- the households in the corridor have different income levels, different car ownership, and different access to the underground, as shown in Table 9.3.

We are interested in the trips between zone 1 and the centre of town. We are informed that a binary Logit model has been estimated yielding the following representative utilities:

$$V_c = -2.0 + 9 \times 10^{-5}I + 2.84CO - 0.03t_c - 0.68e_c/d - 50.0c_c/I$$
$$V_u = -0.03t_u - 0.68e_u/d - 50.0c_u/I$$

where t is in-vehicle travel time (min), e is access time (min), c is cost ($), d is distance (km), I is income ($/month), and CO is the number of cars divided by the number of licences in the household.

Table 9.3 Distribution of households with trips between zone 1 and the centre

CO	Access	Income			Total
		5000	10 000	15 000	
	U (DA)	0	0	350	350
1.0	U (CA)	0	50	150	200
	Total	0	50	500	550
	U (DA)	150	100	0	250
0.5	U (CA)	200	0	0	200
	Total	350	100	0	450
	U (DA)	150	100	350	600
Total	U (CA)	200	50	150	400
	Total	350	150	500	1000

Underground trips are divided according to access into U (DA), underground with direct access (i.e. on foot), and U (CA), underground with car access. The levels of service by individuals travelling between zone 1 and the centre are summarised in Table 9.4.

Table 9.4 Levels of service

	t_c	e_c	c_c	t_u	e_u	c_u	d
U (DA)	11.3	5	122.5	14	8	50	14.5
U(CA)	14.2	5	131.3	22	15	75	16.3

Find out, using an appropriate method, the aggregate probability (i.e. for the whole population) of choosing underground.

9.3 Consider a binary Logit model for car and bus with the following representative utility functions:

$$V_c = 1.35 - 0.03t_c - 0.15c_c$$
$$V_b = -0.03t_b - 0.15c_b$$

where t is total travel time (min) and c is travel cost divided by income (min). Assume the data in Table 9.5 is known about individuals from zone A travelling to work at zone C:

a) Find out the aggregate proportion choosing car by the naive aggregation method and by the sample enumeration method. Compute the naive aggregation error in this case.
b) Now find the aggregate proportion using the car by the classification method (using income as a stratification variable). Plot your results and those of the naive aggregation method; discuss your graph.
c) Compare all your results and discuss them critically.

Table 9.5 Individual data

Individual	Chosen option	Income level	t_c (min)	t_b (min)	c_c (min)	c_b (min)
1	Car	High	47.5	83.2	14.8	7.0
2	Car	High	30.2	45.0	10.4	5.0
3	Car	High	22.2	30.4	12.6	4.0
4	Bus	High	45.0	50.6	8.2	5.0
5	Bus	Low	15.3	20.5	50.0	17.0
6	Car	Low	34.8	50.2	55.0	35.0
7	Bus	Low	65.5	100.5	200.3	53.5
8	Bus	Low	12.0	14.0	44.6	17.0

9.4 You are interested in transferring the model of Exercise 9.3 to a new context, where you have taken a small sample of five individuals whose characteristics are presented in Table 9.6:

Table 9.6 Small sample data for Exercise 9.4

Individual	Chosen option	t_c (min)	t_b (min)	c_c (min)	c_b (min)
1	Car	37.5	70.2	16.8	10.0
2	Car	20.2	30.0	16.4	8.0
3	Car	12.0	15.4	18.6	7.0
4	Bus	35.0	35.6	14.2	8.0
5	Bus	5.3	6.5	56.0	20.0

Assuming there are no mode-specific constants, estimate the value of τ, the transfer scale parameter, using the data above. Discuss your result.

10

Static Assignment

10.1 Basic Concepts

10.1.1 Introduction

The last six chapters have dealt in detail with the key models currently in use to represent the demand for travel in a study area. This chapter will deal with the assignment of vehicles and people to road and public transport networks under the general assumption of steady-state flows. In these cases, we assume the existence of a trip matrix for a time period, say a morning peak hour, and that its trips are allocated (loaded) simultaneously onto the network so that all reach their destination within the peak. This assumption ignores some of the realities of traffic flow and demand, in particular that trips may peak at different times in different parts of the network and queues may remain at the end of a particular time period. This simplification, however, permits an efficient treatment of assignment and is commonly used in most models today. The next chapter will relax this assumption and discuss Dynamic Traffic Assignment.

The network system and, in the case of public transport, the characteristics of the services offered, such as frequency and capacity, represent the main elements of the supply side in transport. These are more or less fixed in the short run. Over a longer period, transport authorities and operators will change fares, frequencies, and vehicle types; road network managers will improve existing (and build new) roads, constrain parking, and introduce tolls and congestion charges. Although these are real representations of supply changes due to increased demand, we do not have good models to forecast this type of longer-term change in supply. Our network models fall short of that: they only show how transport costs will change with different levels of demand. The task of specifying a better longer-term supply system falls to the planner and analyst.

In conventional economic thinking, the actual exchanges of goods and services result from combining their demand with their supply. The equilibrium point defines the goods' prices and the quantities exchanged in the market. At equilibrium, the marginal cost of producing and selling the goods equals the marginal revenue. Economic theory admits that this equilibrium may never actually happen in practice as the system of prices and production levels is under permanent adjustment to cope with changes in purchasing power, tastes, technology, and production techniques. However, the equilibrium concept is still valuable in understanding the economy and forecasting its future states.

It is useful to consider the transport system within that context. The (short-term) supply side, or more precisely the cost model, is made up of a transport network $S(L, C)$ represented by links L (and their associated nodes) and their costs C. The costs are a function of a number of attributes associated with the links, such as distance, free-flow speed, capacity, and a *speed–flow relationship*; in the case of public transport, on-route attributes like fares, frequencies, and running times. The demand side indicates the number of trips by origin–destination (O–D) pair and mode expected for a given level of service (i.e., that assumed in their estimation). In this context, one of the main elements defining levels of service is travel time, but often monetary costs (fares, fuel) and features like comfort for the public may be relevant too. If the actual level-of-service (LOS) offered by the transport network turns out to be lower than estimated, then a reduction in demand and perhaps a shift to other destinations, modes, and/or times of day would be expected. The speed–flow (or generalised cost–flow) relationship is essential as it relates the use of the network to the LOS it can offer.

The public-transport network must be defined in similar terms to the private network. However, it should contain additional specifications of the services offered in terms of their routes, capacities, frequency, and ideally, though seldom in practice, their quality, reliability, and regularity.

In the case of a transport system, one can see 'equilibrium' taking place at several levels. The simplest one is equilibrium in the road network, where travellers from a fixed trip matrix seek routes to minimise their travel costs (times). This results in their trying alternative routes, exploring new ones and perhaps settling into a relatively stable pattern after much trial and error. This allocation of trips to routes yields a pattern of path and link flows which could be said to be in equilibrium when travellers can no longer find better routes to their destinations: they are already travelling on the best routes available. This is the *road network equilibrium*. A similar, but perhaps less dramatic, phenomenon takes place in public-transport networks where passengers may seek routes (i.e. combinations of services) to reduce their generalised journey costs as affected by overcrowding, waiting and walking times, and in-vehicle times.

There are, however, other (higher) levels of interaction. As car congestion increases, buses operating on the same roads will have their journey times increased as well. This may induce some public-transport users (and bus operators) to change their routes to avoid these delays. These choices interact with those of car drivers, as the new arrangements may provide additional capacity in some links and therefore new equilibrium points. These are *multimode network equilibrium* problems and are discussed in Chapter 12.

At an even higher level, the resulting flow pattern may affect choices of mode, destination, and time of day for travel. Each of these shifts in demand will induce, in turn, changes in the corresponding equilibrium points. In modelling terms, the new flow pattern produces LOS for routes and modes, which may or may not be consistent with those assumed in estimating the (presumed) fixed trip matrix. This requires re-estimating the matrix and therefore feeding back the new LOS into the estimation process to obtain a new one. The process may need to be repeated in a systematic way until the trip matrices (and therefore trip time, destination, and mode) are obtained with values for travel costs that are consistent with the flows estimated for each network. This higher level we shall call *system equilibrium* as opposed to *network equilibrium*.

The rest of the chapter is organised as follows. We consider first the problem of assigning a fixed trip matrix to a road network. To treat this problem, we consider typical characteristics of speed- or cost–flow curves. The assignment problem is split into a route choice model and the loading of the trip matrix onto the identified routes. Different conditions require different loading methods. Stochastic methods allow for variability in drivers' perception of route costs; these methods are discussed in Section 10.4. The most interesting deterministic assignment methods try to include consistently the effect of congestion on route choice. This chapter considers only pragmatic methods under the general title of *congested assignment* in Section 10.5; we leave a more rigorous treatment of equilibrium assignment for Chapter 12. Section 10.6 considers the problems and approaches required to model public-transport assignment.

10.1.2 Traffic and Queues

It is useful to consider traffic behaviour and how this is simplified, in different ways, when handling it in transport models. Human drivers have imperfect reaction times and limited knowledge about the traffic around and ahead of them. Therefore, it is necessary to keep a safe distance between vehicles. This separation is managed by 'rules of the road' that drivers must follow and by control devices such as traffic signals, stop/give way signs, and lane markings. On the other hand, traffic conditions are different in motorways/freeways with free-flowing conditions and grade-separated junctions than in urban networks with at-level junctions.

Consider first free-flowing conditions on a long stretch of road, sometimes called uninterrupted flow states. Here the main attributes of flow are speed (S, km/h), flow or volume (V, vehicles/h), and density (D, vehicles/km). Observations of these properties on a stretch of road will result in something like Figure 10.1.

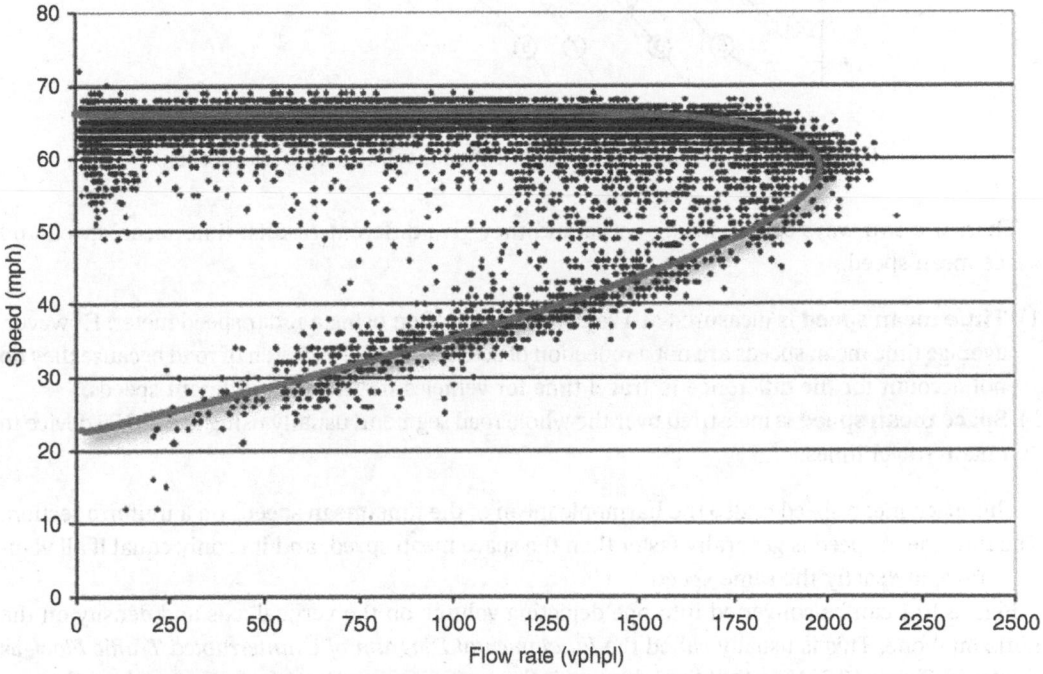

Figure 10.1 Volume and speed observations and a curve fitted to them

Any curve fitted to this 'cloud' of observations is bound to be an approximation. Note that the bottom part of the curve shows traffic behaviour where V and S are below optimal conditions; these are unstable conditions usually found upstream of an incident or simply a 'bottleneck' created by the merging of two or more traffic streams.

Traffic flow can be represented in a time-space diagram as shown in Figure 10.2, with time usually on the horizontal axis. In a free-flowing road, vehicles following each other along a lane will have parallel trajectories; these trajectories will cross when one vehicle overtakes another. In this diagram, each individual speed is the first derivative of the trajectory; the density D is given by the average separation or gap (g, metres) between vehicles and the volume by the inverse of the average headway (h, seconds).

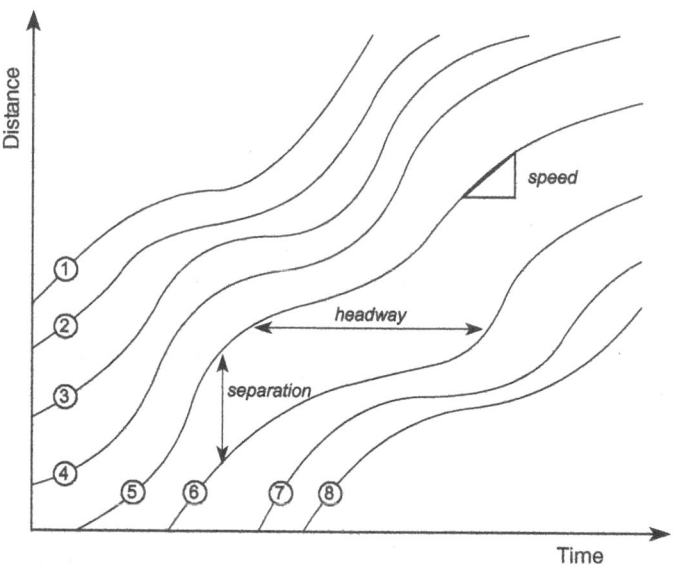

Figure 10.2 The trajectories of eight different vehicles on a road section

There are two ways of measuring speed and they give different results: time mean speed and space mean speed.

1) **Time mean speed** is measured at a reference point, often using a radar speed meter. However, average time mean speeds are not a reflection of actual speed on a stretch of road because they do not account for the difference in travel time for vehicles travelling at different speeds.
2) **Space mean speed** is measured over the whole road segment, usually using some GPS device to obtain travel times.

The space-mean speed is also the harmonic mean of the time-mean speeds on a uniform section. The time mean speed is generally faster than the space mean speed, and it is only equal if all vehicles travel at exactly the same speed.

Figure 10.1 can be converted into one depicting volume on the vertical axis and density on the horizontal one. This is usually called the *Fundamental Diagram of Uninterrupted Traffic Flow*, as shown in Figure 10.3. Note that 'jam density' is the maximum number of vehicles per lane that are found in a non-moving queue, some 115–156 vehicles/km (185–259 vehicles/mi).

The traffic behaviour in urban areas is more complex than that depicted in Figure 10.1. This is because delays at junctions are generally more important than travel times on their links. Therefore, it is sometimes desirable to consider the two types of delays separately. While delays when travelling on a given link depend mainly on traffic on the link itself (and therefore on a speed-flow relationship, as discussed in Chapter 3), the delay at an at-level junction depends also on traffic on other approaches; this is also true for merging movements from an access ramp at a motorway. The additional complexity of junction modelling is discussed in Chapter 11.

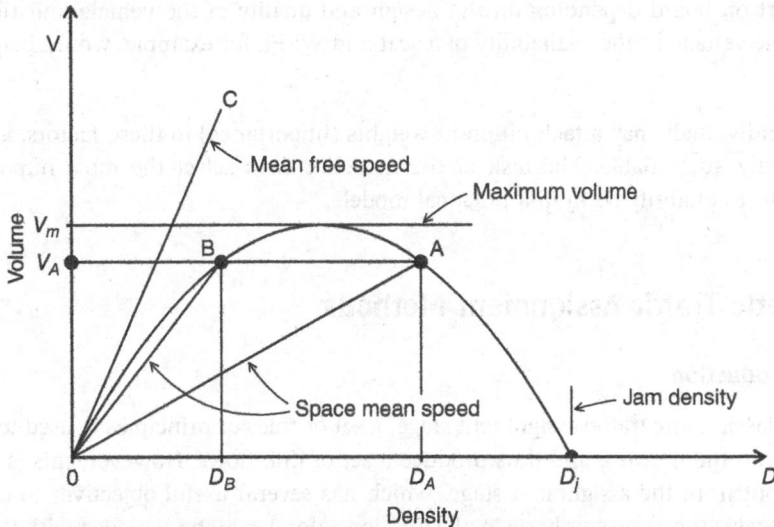

Figure 10.3 The fundamental diagram of traffic flow

10.1.3 Factors Influencing Route Choice

Before embarking on the mathematical models describing route choice and assignment, it is useful to consider the most likely factors considered by drivers and passengers in the selection of the best way to travel from origin to destination. In the case of light vehicle drivers, these are likely to be:

- travel time: it often matters whether this time is spent moving or stationary queueing; the reliability of travel time is increasingly important;
- distance or another proxy for vehicle operating costs (VOC); it is generally accepted that most drivers have only an imperfect perception of the costs incurred while driving; the main consumable perceived is fuel costs, and even these are only approximately related to distance travelled (tires, oil, and maintenance are rarely perceived); moreover, depreciation, insurance, and annual road taxes are generally perceived as fixed, sunk, costs related to car ownership but not to its use;
- money costs like tolls and any road user charges (RUC);
- the difficulty of the driving task: driving on an uncongested motorway may be perceived as easier than on congested urban roads with lateral friction from pedestrians and cyclists; traffic lights, roundabouts, priority junctions, mountain roads, and single or dual carriageways, pose different challenges to the driving task; beautiful scenery and environment may also add some pleasure;
- signage and ease of navigation; knowledge of the route.

In the case of public transport passengers, the factors may be:

- travel time spent in the vehicle (on-board time);
- access time to reach the bus stop/station;
- waiting time to the next service; partially dependent on the environment while waiting and the availability of real-time information;
- the reliability of the service and travel time;
- the fare payable; this may depend on the ticket type;

- the comfort on board depending on the design and quality of the vehicle and ride, and how crowded the vehicle is; the availability of a seat and Wi-Fi, for example, would help to improve comfort.

Different individuals may attach different weights (importance) to these factors, and not all of them are easily quantifiable. The task of the modeller is to select the most important factors and seek ways to quantify them in a practical model.

10.2 Static Traffic Assignment Methods

10.2.1 Introduction

During the classic static traffic assignment stage, a set of rules or principles is used to load a fixed trip matrix onto the network and thus produce a set of link flows. However, this is not the only relevant output from the assignment stage, which has several useful objectives to consider. Not all of them receive the same emphasis in all situations, nor can all be achieved with the same level of accuracy. The main objectives are:

1) Primary:
   - to obtain good *aggregate* network measures (e.g. total motorway flows, total revenue by bus service);
   - to estimate zone-to-zone travel costs (times) for a given level of demand;
   - to obtain 'reasonable' link flows and to identify heavily congested links.
2) Secondary:
   - to estimate the routes used between each O–D pair;
   - to analyse which O–D pairs use a particular link or route;
   - to obtain turning movements for the design of future junctions.

In general terms, we will be able to attain the primary objectives more accurately than the secondary ones. Even within objectives, we are likely to be more accurate with those earlier in the list. This is essentially because our models are more likely to estimate aggregate than disaggregate values correctly.

The basic inputs required for assignment models are:

- a trip matrix expressing estimated demand, normally a peak-hour matrix in congested urban areas, and perhaps other matrices for other peak and off-peak periods; a 24-hour matrix is sometimes used for assignment on uncongested networks, and its conversion into single hours is seldom satisfactory in congestion terms, as these matrices are symmetric and single-hour trips seldom are; as the matrices may be available in terms of person trips, they should be converted into vehicle trips as capacity and speed–flow relationships are described in these terms;
- a network, namely links and their properties, including speed–flow curves;
- principles or route selection rules thought to be relevant to the problem in question.

The *static traffic assignment* methods involve a set of rules on how to identify desirable routes (fastest, lowest generalised cost) to connect origins to destinations and then a systematic way of allocating O–D trips to these routes so that certain features of reality are achieved. In the next sections, we will discuss these methods from a practical viewpoint, identifying their strengths and weaknesses. In Chapter 12, we will adopt a more rigorous approach, setting up the assignment task as an optimisation problem and discussing solution algorithms in a more systematic way.

10.2.2 Modelling Route Choice

The basic premise of assignment is the assumption of a rational traveller, that is, one choosing the route that offers the lowest perceived (and anticipated) individual costs. The production of a 'generalised cost' expression incorporating all the route choice influences depicted in Section 10.1.3 is a difficult task. Furthermore, it is not practical to try to model all of them in a traffic assignment model, and therefore approximations are inevitable.

The most common approximation is to consider only two factors in route choice: time and monetary cost; further, monetary cost is often deemed proportional to travel distance plus any tolls. Most traffic assignment programs allow the user to allocate weights to travel time and distance to represent drivers' perceptions of these two factors. The weighted sum of these two values becomes a generalised cost used to estimate route choice. Evidence suggests that, at least for urban car traffic, time is the dominant factor in route choice. Outram and Thompson (1978) compared drivers' stated objectives with their actual performance in route choice. They found that the proportion of drivers who were successful in achieving their objectives was relatively low. They also found that the combination of time and distance gave the best explanation of route choice. However, even if we allow for the combination of time and distance in a generalised cost function, we can only explain something of the order of 60–80% of the routes actually observed in practice. As the marginal contribution of other factors in untangling route choice is small, the unexplained part must be attributed to factors like differences in perception, imperfect information on route costs, or simply errors.

The fact that different drivers often choose different routes when travelling between the same two points may be ascribed to three different types of reasons:

1) Differences in individual perceptions of what constitutes the 'best route'; some may wish to minimise time, others fuel consumption and many a combination of both, and this introduces a variety in route choices.
2) The level of knowledge of alternative routes varies, and this introduces apparent irrationality (from the point of view of the observer) in the choices.
3) Congestion effects, as these affect shorter routes first and make their generalised costs comparable to initially less attractive routes.

We normally handle the first issue through *multiple user classes*, the second through *stochastic effects*, and the third one via congested assignment and equilibrium.

Example 10.1 Consider an idealised town with a low-capacity through route (1000 vehicles/h, vph) and a high-capacity bypass, as in Figure 10.4. The bypass is a longer but faster route with a capacity of 3000 vph. Assume that during the morning peak 3500 drivers approach the town and that everyone would like to use the shortest route (i.e. via the town centre). It is clear that it would not be possible for all of them to do so, as the route would become too congested even before its ultimate capacity is reached. Many would then opt for the second route to avoid long queues and delays. Presumably drivers would experiment with the two routes until they found a more or less stable arrangement when neither could improve their time by switching to the other route. This is a typical case of 'Wardrop's equilibrium', which is discussed in greater detail below. Diversion across routes in this case is due to *capacity restraint*.

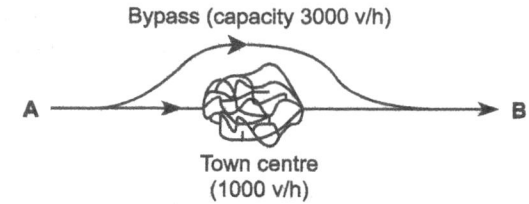

Figure 10.4   Town served by a bypass and a town centre route

However, surely not all 3500 drivers think alike; some may always prefer the bypass because of its uninterrupted flow conditions or its scenery, while others may value other features of the town-centre route. These differences in objectives can be modelled using multiple user classes. The differences in perceptions and knowledge would also lead to a spread of routes and such effect is customarily referred to as the *stochastic* element in route choice.

Particular types of models are more suited to representing one or more of the above influences. A possible classification of traffic assignment methods is given in Table 10.1. The details and characteristics of each method are discussed below.

Table 10.1   Classification scheme for static traffic assignment

		Stochastic effects included?	
		No	**Yes**
Single user class	No capacity restraint	All-or-nothing	Pure stochastic: Dial's, Burrell's
	With capacity restraint	Wardrop's equilibrium	Stochastic user equilibrium (SUE)
Multiple user classes	No capacity restraint	All-or-nothing with multiple user classes	Multiple user classes stochastic: Dial's, Burrell's
	With capacity restraint	Wardrop's equilibrium with multiple user classes	SUE with multiple user classes

Each assignment method has several steps that must be treated in turn. Their basic functions are:

- to identify a set of routes that might be considered attractive to drivers; these routes are stored in a particular data structure called a *tree* and therefore this task is often called the *tree-building stage*;
- to assign suitable proportions of the trip matrix to these routes or trees; this results in flows on the links in the network;
- to search for convergence; many techniques follow an iterative pattern of successive approximations to an ideal solution (e.g. Wardrop's equilibrium); convergence on this solution must be monitored to decide when to stop the iterative process.

10.2.3   Tree Building

Tree building is a key stage in any assignment method for two related reasons. First, it is performed many times in most algorithms, at least once per iteration. Second, a good tree-building algorithm can save a great deal of computer time and costs. By 'good algorithm' we mean an efficient one that is also well programmed in a suitable language. Van Vliet (1978) gives a good discussion of the most widely used algorithms for tree building, and this section is based on his paper.

There are two basic algorithms in general use for finding the shortest (cheapest) paths in road networks, one due to Moore (1957) and another due to Dijkstra (1959). We will discuss both using a more convenient node-oriented notation: the length (cost) of a link between A and B in the network is denoted by $d_{A,B}$. The path or route is defined by a series of connected nodes, A–C–D–H, etc., whilst the length of the path is the arithmetic sum of the corresponding link lengths in the path. Let d_A denote the minimum distance from the origin of the tree S to the node or centroid A; P_A is the *predecessor* or *backnode* of A so that the link (P_A, A) is part of the shortest path from S to A.

The procedure for building a minimum path tree from S to all other nodes may be described as follows:

Initialisation. Set all $d_A = \infty$ (a suitable large number depending on computer and compiler) except d_S, which is set equal to 0; set up a 'loose-end table' L to contain nodes already reached by the algorithm but not fully explored as predecessors for further nodes. They are the tip of the tree as branches grow to reach all nodes. Initialise all entries L_i in L to zero, and all P_A to a suitable default value.

Procedure. Starting with the origin S as the 'current' node $= A$:

1) Examine each link (A, B) from the current node A in turn and, if $d_A + d_{A,B} < d_B$ then set a new value for $d_B = d_A + d_{A,B}$, make $P_B = A$ and add B to L;
2) Remove A from L if the loose-end table is empty, stop; otherwise,
3) Select another node from the loose-end table and return to step 1 with it as the current node.

Three comments should be made at this stage. First, as in general, routes are not allowed to use centroids, in step 1, B would not be added to L if it was a centroid. Second, the essential difference between Moore's and Dijkstra's algorithms lies in the procedure for selecting a node from L. Moore selects the top entry, which is the oldest entry in the table; Dijkstra selects the node nearest to the origin (i.e. the node L_i such that d_{Li} is a minimum). This requires some additional calculations (including sorting of nodes), but ensures that each link is examined only once. It is well known that Dijkstra's algorithm is superior to Moore's, in particular for larger networks; it is, however, more difficult to program. Finally, trees are often stored in the computer in one of two forms: as a set of ordered backnodes in which A is the backnode of B if link (A, B) forms part of the tree; or as a set of 'backlinks' with a similar definition.

Van Vliet (1977) identified a lesser-known algorithm that performs very well even for large networks: D'Esopo's algorithm, as described and tested by Pape (1974). D'Esopo's uses a 'two-ended' loose-end table so that node B is entered at one or other end depending on its 'status'. If B had not been previously reached by the tree, then it is entered at the bottom of L; if it is currently in the table, no entry is made; but, if it has already been entered in L, examined, and removed from the table, then it is entered at the top. A simple array can be used to record the status, with three potential values ($+1$, 0, or -1) representing each case for each node. As shown by Van Vliet (1977), D'Esopo's algorithm can reduce CPU times by 50% relative to Moore's. Furthermore, its performance is very close and often better compared with that of the best implementations of Dijkstra's; finally, it has the added advantage of being much simpler to program.

Trees have two important additional uses in transport planning. They are often employed to extract cost information from a network. For example, the total travel time between two zones can be obtained by following the sequence of links in the tree connecting them and accumulating their travel times. This operation is often referred to as 'skimming' a tree. Trees built for, say, travel time can be skimmed for other attributes for example, generalised cost, distance, number of nodes, etc. We can also use trees to produce information on which O–D pairs are likely to use a particular link. This facility, often called a *selected link analysis*, permits the identification of who is likely to be affected by a network change. Moreover, it can also be used to 'cordon' a trip matrix for a smaller study area; in this case, the selected links are used to identify entry and exit points to the small study area, and the trees are used to combine the original zones into single external ones for the new sub-area.

10.3 All-or-Nothing Assignment

The simplest route choice and assignment method is 'all-or-nothing' assignment. This method assumes that there are no congestion effects, that all drivers consider the same attributes for route choice, and that they perceive and weigh them in the same way. The absence of congestion effects means that link costs are fixed, and the assumption that all drivers perceive the same costs means that every driver from i to j must choose the same route. Therefore, all drivers are assigned to one route between i and j and no driver is assigned to other, less attractive, routes. These assumptions are probably reasonable in sparse and uncongested networks where there are few alternative routes, and they are very different in cost.

The assignment algorithm itself is a procedure that loads the matrix **T** onto the shortest path trees and produces the flows $V_{A,B}$ on links (between nodes A and B). All load algorithms start with an initialisation stage, in this case making all $V_{A,B} = 0$, and then apply one of two basic variations: 'pair-by-pair' methods and 'once-through' approaches.

Pair-by-pair. This is probably the simplest but not necessarily the most efficient method. In this case, we start from an origin and take each destination in turn. First, we initialise all $V_{A,B} = 0$. Then for each pair (i, j):

1) Set B to the destination j;
2) If (A, B) is the backlink of B then increment $V_{A,B}$ by T_{ij}, that is, make $V_{A,B} = V_{A,B} + T_{ij}$;
3) Set B to A;
4) If $A = i$ terminate, that is, process the next (i, j) pair; otherwise, return to step 2.

Once-through. This is sometimes called a *cascade* method as it loads accumulated flow from nodes to links following the minimum cost trees from an origin, i. Let V_A be the cumulative flow at node A:

1) Set all $V_A = 0$ except for the destinations j for which $V_j = T_{ij}$;
2) Set B equal to the most distant node from i;
3) Increment V_A by V_B where A is the backnode of B, that is, make $V_A = V_A + V_B$;
4) Increment $V_{A,B}$ by V_B, that is, make $V_{A,B} = V_{A,B} + V_B$;
5) Set B equal to the next most distant node; if B = i the origin has been reached; so, begin processing the next origin; otherwise, proceed with step 3.

In this form, V_B represents the total number of trips from i passing through node B en route to destinations further away from i. By selecting nodes in reverse order of distance, each node is processed only once. This algorithm requires the trees to be stored in terms of backnodes ordered by distance from the origin.

Example 10.2 Consider the simple network in Figure 10.5 and its associated trip matrix: A–C = 400, A–D = 200, B–C = 300, and B–D = 100. Section (a) shows the travel costs (times) on each link; section (b) the corresponding trees based on these costs together with the contributions to the total flow after assignment; and the total flows on each link are shown in section (c).

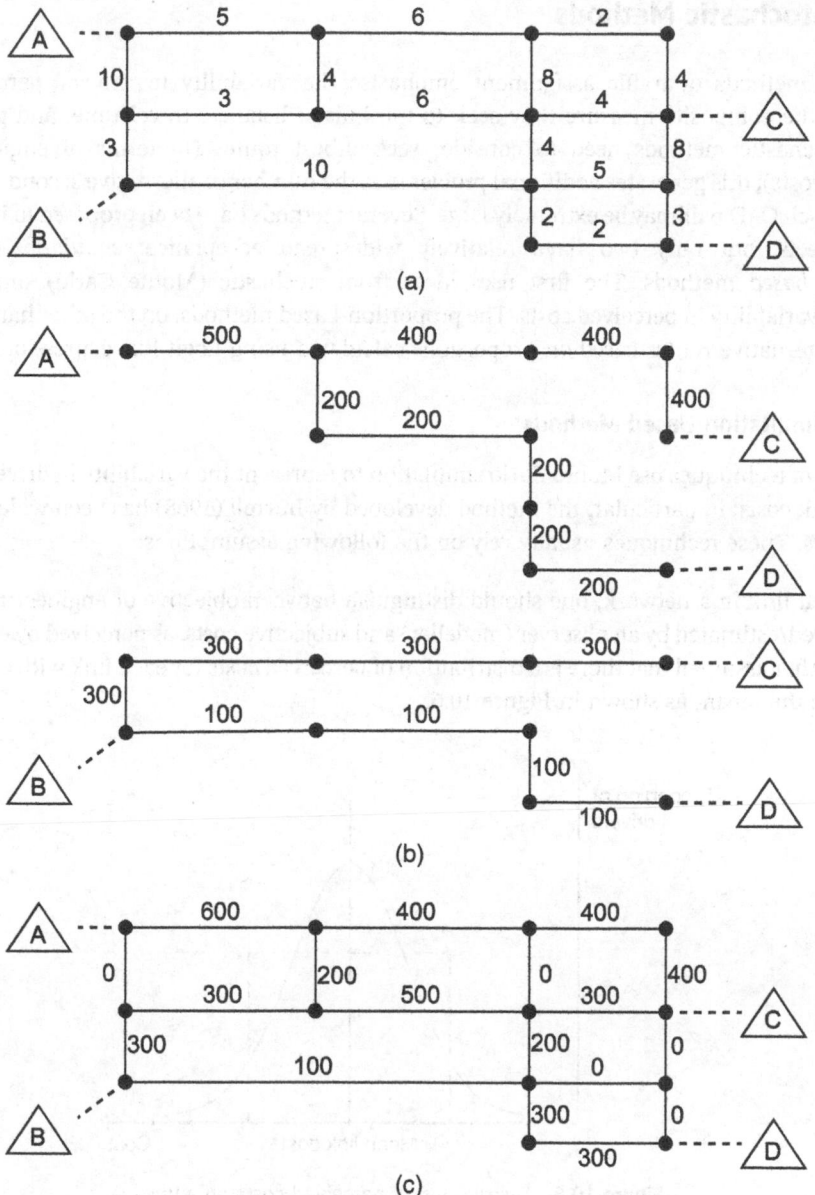

Figure 10.5 A simple network, its trees and flows from loading a trip matrix

All-or-nothing assignment is generally of limited interest to the planner; it may be used to represent some sort of 'desire line', that is, what drivers would like to do in the absence of congestion. However, its main usefulness is as a basic building block for other types of assignment techniques (e.g. *equilibrium* and *stochastic* methods).

10.4 Stochastic Methods

Stochastic methods of traffic assignment emphasise the variability in drivers' perceptions of costs and the composite measure they seek to minimise (distance, travel time, and generalised costs). Stochastic methods need to consider second-best routes (in terms of engineering or modelled costs); this generates additional problems as the number of alternative second-best routes between each O–D pair may be extremely large. Several methods have been proposed to incorporate these aspects, but only two have relatively widespread acceptance: *simulation-based* and *proportion-based* methods. The first uses ideas from stochastic (Monte Carlo) simulation to introduce variability in perceived costs. The proportion-based methods, on the other hand, allocate flows to alternative routes based on proportions calculated using Logit-like expressions.

10.4.1 Simulation-Based Methods

A number of techniques use Monte Carlo simulation to represent the variability in drivers' perceptions of link costs; in particular, the method developed by Burrell (1968) has been widely used for many years. These techniques usually rely on the following assumptions:

1) For each link in a network, one should distinguish between objective or engineering costs, as measured/estimated by an observer (modeller) and subjective costs, as perceived by each driver. It is further assumed that there is a distribution of perceived costs for each link with engineering costs as the mean, as shown in Figure 10.6.

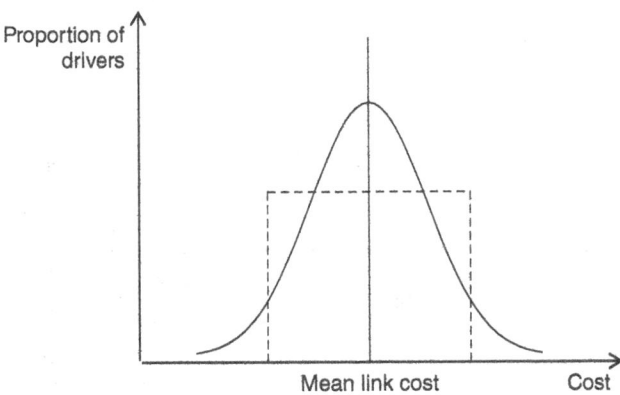

Figure 10.6 Distribution of perceived costs on a link

The various implementations of these ideas differ in their assumptions about the shape of these distributions. While Burrell's assumes a uniform distribution, other models hypothesise a Normal distribution. In either case, one also needs to assume or calibrate a standard deviation, or spread, for the distribution of perceived costs.

Further:

2) The distributions of perceived costs are assumed to be independent.
3) Drivers are assumed to choose the route that minimises their perceived route costs, which are obtained as the sum of the individual link costs.

A general description of these algorithms would be as follows. First, select a distribution (and spread parameter, σ) for the perceived costs on each link; then, split the population travelling along each O–D pair into N segments, each assumed to perceive the same costs; finally, apply the following procedure:

1) Make $n = 0$.
2) Make $n = n + 1$.
3) For each $i - j$ pair:
   - compute perceived costs for each link by sampling from the corresponding distributions of costs by means of random numbers;
   - build the minimum perceived cost path from i to j and assign T_{ij}/N trips to it, accumulating the resulting flows on the network.
4) If $n = N$ stop, otherwise go to step 2.
   In practice, many short-cuts are taken to reduce computation times; for example:
   - generate new sets of random costs per origin and not per O–D pair;
   - use N equal to just 3 or 5 and generate one set of random costs for each matrix and not for each O–D pair or origin;
   - use small values for N, even 1 in some circumstances.

This type of approach uses simulation to reduce the number of second-best routes to be considered. If a wider range of routes is thought necessary, one can increase the value of N and/or the spread parameter in the distribution of link costs. Burrell's approach has the advantage of generating cheap routes more often than more expensive ones: if a route is expensive, it is much less likely to be generated as the preferred one because of the stochastic variations in link costs. Although the Uniform distribution is efficient in computer time, it is not very realistic. A better function, but more expensive in terms of CPU time, is the Normal distribution with variance proportional to the mean engineering costs.

As in all Monte Carlo methods, the final results are dependent on the series of random numbers used in the simulation. Increasing the value of N reduces this problem. There are, however, more serious difficulties with this approach:

- the link perceived costs are not truly independent, as drivers usually have preferences, for example, for motorway links or to avoid priority junctions or minor roads; the assumption of independence in perceived costs may lead to unrealistic switching between parallel routes connected by minor roads;
- no explicit allowance is made for congestion effects.

In compensation, these methods often produce a reasonable spread of trips, are relatively simple to program, and do not require the choice or estimation of speed–flow relationships (which may turn out to be a problem in some cases).

10.4.2 Proportional Stochastic Methods

Virtually all these methods are based on a loading algorithm that splits trips arriving at a node between all possible exit nodes, as opposed to the all-or-nothing method, which assigns all trips to a single exit node. Very often the implementation of these methods reverses the problem so that the division of trips at a node is actually based upon where the trips are coming from rather than where they are going to. Consider node B in Figure 10.7; there are a number of possible entry points denoted by A_1, A_2, A_3, A_4, and A_5 for trips from I to J.

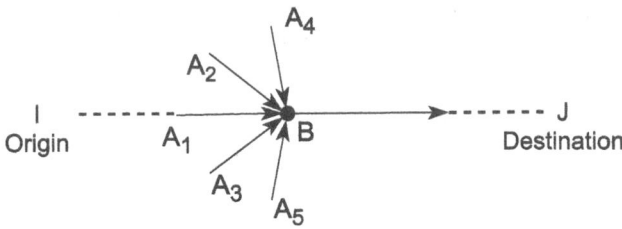

Figure 10.7 A node (B) and links feeding trips into it

The 'splitting factors' f_i are defined by:

$$f_i = 0 \qquad \text{if } d_{A_i} \geq d_B$$
$$0 < f_i \leq 1 \quad \text{if } d_{A_i} < d_{B_i}$$

where d_{Ai} represents the minimum cost of travel from the origin i to node A_i. The first condition requires that f_i should be zero if an entry node A_i is further from the origin than B, therefore ensuring that trips are allocated to routes which take them efficiently away from the origin. The trips T_B passing through B are divided according to the following equation:

$$F(A_i, B) = \frac{T_B f_i}{\sum_i f_i} \tag{10.1}$$

The assignment procedure is now equivalent to the cascade method for all-or-nothing assignment. Implementations of these ideas differ mainly in the way in which they define the splitting function f_i. The single-path method due to Dial (1971) requires that:

$$f_i = \exp(-\Omega \delta d_i) \tag{10.2}$$

where δd_i is the extra cost incurred in travelling from the origin to node B via node A_i rather than via the minimum cost route. In this way, if A_i is in the minimum-cost route, δd_i is equal to zero and $f_i = 1$. Nodes that lie on more expensive routes have $\delta d_i > 0$ and their f_i values are less than 1. In this way, shorter routes are favoured over more expensive ones. The parameter Ω can be used to control the spread of trips among routes.

Dial originally described a double-pass algorithm that effectively uses a Logit-type formulation to split trips from i to j among alternative routes r:

$$T_{ijr} = \frac{T_{ij} \exp(-\Omega C_{ijr})}{\sum_r \exp(-\Omega C_{ijr})} \tag{10.3}$$

The algorithm involves a forward and a backward pass:

1) Forward pass: take each node A in ascending order of d_A and define a weight for each exit link (A, B) such that:

$$w_{(A,B)} = W_A \exp\left(-\Omega \delta d_{(A,B)}\right) \text{ if } d_A < d_B \text{ or zero otherwise}$$

W_A is the accumulated weight at A defined as:

$$W_A = \sum_{A'} w_{(A',A)} \text{ and } W_I = 1 \quad [A' \text{ is a predecessor of A}]$$

2) Backward pass: identical to the single-pass algorithm with the exception that the weights $w_{(A, B)}$ are used to work out the split of trips rather than the splitting factors f_i.

Example 10.3 A practical problem with Dial's assignment is that it assumes that all routes are equally likely candidates, and for this reason, it is biased against trunk routes as opposed to secondary links. Consider the problem of a town served by a bypass and a town-centre route with three small variations, as illustrated in Figure 10.8. Assume also that there are 4000 trips from A to B and that all routes have approximately the same cost.

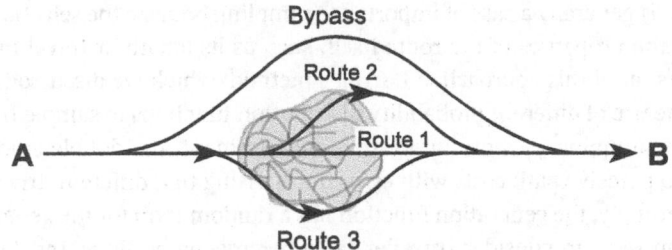

Figure 10.8 Town served by a bypass and three city-centre routes

In this case, Dial's algorithm would split the 4000 trips as follows: 1000 via the bypass and 1000 via each of the town-centre routes. However, most users would regard this problem as having only two alternatives: bypass or town centre. Recall the discussion about the 'independence of irrelevant alternatives' property of the MNL model in Chapter 7. Dial's algorithm runs into trouble when it considers every possible route, even if some permutations or combinations of links may differ just by a few percentage points of their total cost. In behavioural terms, the algorithm ignores the correlation between similar routes. In practice, it tends to allocate more traffic to dense sections of the network with short links, compared with sparser parts of the network with relatively longer links. In fact, coding strategies for networks can affect the allocation of flows.

10.4.3 Emerging Approaches

Research is still active seeking to integrate stochastic assignment methods closer with developments in discrete choice. The problem is generally split into three components: (i) how to identify a feasible, efficient, and distinct set of routes that would be considered by drivers when making their choices, (ii) how to estimate the parameters of route choice models, and (iii) how to integrate more efficiently the choice mechanism into an equilibrium assignment framework.

Prato (2009) reviewed approaches to accomplishing these three tasks. The methods discussed remain mostly in the research realm but take advantage of the advances in discrete choice modelling discussed in Chapters 7–9. One of the key problems is the difficulty associated with collecting good data, in particular for revealed preference (RP) choices. The provision of GPS units in an experimental setting may help to alleviate this constraint.

Let us assume that trip makers limit their choices among a certain number (K) of minimum-cost paths avoiding extremely costly alternatives. For both estimation and prediction purposes, we need to caution against the possibility of generating routes that are either too circuitous or very similar, as both types would be unattractive to drivers or not really perceived as different. A number of techniques have been developed to avoid these problems and generate acyclic and heterogeneous paths, but they still have to face two further problems. First, all drivers travelling between the same O–D pair will share the same generated choice set (and one would expect differences among them resulting from personal constraints and preferences). Second, the measures of route attractiveness are basically subjective and rely on the experience of the researcher that controls their inclusion.

An alternative choice set generation approach is based on doing repeated shortest path searches in the network using random extraction of link generalised costs and individual preferences from probability distributions. The various methods here (all of a heuristic nature) produce solutions that are stochastic and where O–D pairs are processed simultaneously (Prato 2009). Moreover, stochastic path generation is generally a case of importance sampling because the selection probability of a route depends on the properties of the route itself, such as its length or travel time.

The original version of this approach is Burrell's method, which we discussed earlier. Generalisations include the use of different probability distribution functions to sample from and different sequences for that sampling. A more interesting enhancement, the double stochastic approach, allows travellers to perceive path costs with error, recognising that different drivers have different preferences. Accordingly, the generation function has a random term for the generalized cost function and a random term to consider traveller taste heterogeneity. Bovy and Fiorenzo-Catalano (2007) proposed a trip utility function as the basis for a doubly stochastic generation function. Relevant routes are created through optimal path searches in the network by stochastically varying network attributes and attribute preferences. Variation in link impedances reflects differences in the knowledge and perception of link attributes among travellers. Variation in the parameter values reflects differences among travellers in their utility function.

Constrained enumeration methods rely on the behavioural assumption that drivers choose routes according to rules other than the minimum cost path. Prato and Bekhor (2006) proposed a 'branch and bound' algorithm where the branching rule requires the definition of thresholds. A directional threshold excludes from consideration links that take the driver significantly away from the destination and closer to the origin of the trip. A temporal threshold rejects paths that travellers would consider unrealistic because of excessive travel time. Other thresholds discard routes that include large detours or have overlapping paths that travellers would not consider separate alternatives.

The application of an assignment technique to these identified routes presents another problem. The sets of alternative routes generated with the described path generation techniques are usually quite large since all relevant routes are possibly included and some irrelevant routes are probably created. Intuitively, the number of alternatives in the choice set plays a role in the estimation of discrete choice models within the route choice context.

Accordingly, route choice models should exhibit robustness in utility parameter estimates with respect to the size of the choice set. For estimation purposes, this model requirement would allow the definition of choice sets with a reasonable number of attractive alternatives to secure reliable model estimates. Dense urban networks with many alternatives may show a high degree of similarity among alternative routes. For this reason, most of the literature focusses on the correlation between alternatives, which alters choice probabilities of overlapping routes. This problem has been discussed at length in Chapters 7 and 8; we only present here some of the relevant implications for assignment.

Cascetta et al. (1992) proposed a generalisation of the MNL model, in which a commonality factor measures the degree of similarity of each route with other routes in the choice set C. The expression of the probability P_k of choosing route k within the choice set C reflects the simple Logit structure of the model:

$$P_k = \frac{\exp(V_k + \beta_{CF}CF_k)}{\sum\limits_{l \in C} \exp(V_l + \beta_{CF}CF_l)} \qquad (10.4)$$

where V_k and V_l are the utility functions of routes k and l, respectively, CF_k and CF_l are the commonality factors, and β_{CF} is a parameter to be estimated.

Ben-Akiva and Bierlaire (1999) proposed the Path-Size Logit (PSL) model as an application of discrete choice theory for aggregate alternatives, which has been used in other transport contexts such as destination choice. In the PSL model, the expression of the probability of choosing route k within the alternative paths is:

$$P_k = \frac{\exp(V_k + \beta_{PS}PS_k)}{\sum\limits_{l \in C} \exp(V_l + \beta_{PS}PS_l)} \tag{10.5}$$

Despite its similarity with (10.4), the interpretation is different, and there are different expressions proposed for the path size. The CF factor reduces the attractiveness of a path because it shares elements of others, where the PS index identifies what proportion of a path is unique.

Generalised Extreme Value (GEV) models allow similarities within the stochastic part of the utility function and relate the network topology to the specific coefficients that characterize their tree structure. A small number of models using this approach have been suggested to handle probabilistic assignment, for example, a Cross Nested Logit (Prashker and Bekhor 2000) or a Generalised Nested Logit (Bekhor and Prashker 2001). However, they have significant computational demands and use complex nested structures, making them more difficult to implement.

One of the most attractive treatments of this problem uses the Mixed Logit (ML) model discussed in Chapter 7. Here, the unobserved factors can be decomposed into a part that contains correlation and heteroscedasticity and another part that is IID extreme value. The most straightforward derivation assumes that the probability for an individual n of choosing route k has the same form as the standard MNL, but it is conditional on the distribution of the coefficients β_n where $f(\beta)$ is the 'mixing distribution' of β over the population. The unconditional probability is computed by simulation, as we saw in Chapter 8:

$$P_{nk} = \frac{1}{D} \sum_{d=1}^{D} \frac{\exp(\beta'_d Xn_k)}{\sum\limits_{l \in C_d} \exp(\beta'_d X_{nl})} \tag{10.6}$$

where β_d indicates a draw d from the distribution of β and D is the number of draws.

An adaptation to route choice assumes that the covariance of path utilities is proportional to the length by which paths overlap (Bekhor et al. 2002). Extending from the derivation of the ML model with a factor analytic approach, the probability of choosing route k given a vector δ of standard Normal variables is given by:

$$P_k = \Lambda(k|\delta) = \frac{\exp(\mu(\beta X_k + F_k T\delta))}{\sum\limits_{l \in C_n} \exp(\mu(\beta X_l + F_l T\delta))} \tag{10.7}$$

where $\beta(1 \times B)$ is the column vector of parameters, X_k is the kth row of the matrix of explanatory variables $X(J \times B)$, F_k is the kth row of the factor loadings matrix $F_{(J \times M)}$ (J paths and M network elements), $T_{(M \times M)}$ is a diagonal matrix of covariance parameters σ_m and $\delta_{(M \times 1)}$ is a vector of standard Normal variables.

Bekhor et al. (2002) assume that the link-specific factors are IID Normal and that the variance is proportional to the link length; the F matrix corresponds to the link-path incidence matrix and the T matrix corresponds to the link-factor covariance matrix. Accordingly, the covariance parameter σ

shared by each link is estimated. Other variations on this theme have been proposed, but they are computationally demanding, and it is difficult to obtain significant parameter estimates.

In summary, currently emerging route choice models offer advantages and disadvantages. From a computational perspective, MNL generalisations, such as PSL (10.5), are not too challenging, but GEV models are more demanding because they require the estimation of structural coefficients within complex model structures. ML models introduce additional complexity because of the need to simulate choice probabilities and the absence of an equivalent mathematical formulation of the stochastic user equilibrium (SUE) problem. From a behavioural perspective GEV and ML models depend on theoretical formulations of the correlation structure among alternative routes. GEV models are competitive because of their superior theoretical foundation with respect to MNL-generalisations and relatively lighter computational demands compared to ML models.

10.5 Congested Assignment

10.5.1 Wardrop's Equilibrium

If one ignores stochastic effects and concentrates on capacity restraint as a generator of a spread of trips on a network, one should consider a different set of models. For a start, *capacity-restraint* models need functions relating flow to the cost (time) of travel on a link. These models usually attempt, with different degrees of success, to approximate the equilibrium conditions as formally enunciated by Wardrop (1952) as a 'criterion':

The journey times on all routes actually used are equal, and less than those which would be experienced by a single vehicle on any unused route.

This was later on expressed more formally as:

Under equilibrium conditions traffic arranges itself in congested networks in such a way that no individual trip maker can reduce his path costs by switching routes.

If all trip makers perceive costs in the same way and seek the same objective (single user class, no stochastic effects):

Under equilibrium conditions traffic arranges itself in congested networks such that all used routes between an O–D pair have equal and minimum costs while all unused routes have greater or equal costs.

This is usually referred to as Wardrop's *first principle*, or simply *Wardrop's equilibrium*. It is easy to see that if these conditions do not hold, at least some drivers would be able to reduce their costs by switching to other routes.

Example 10.4   Consider again the case of a bypass and a single town-centre route of Figure 10.4. Assume now that the absolute capacity restriction for each route is replaced with two corresponding time–flow relationships as illustrated in Figure 10.9.

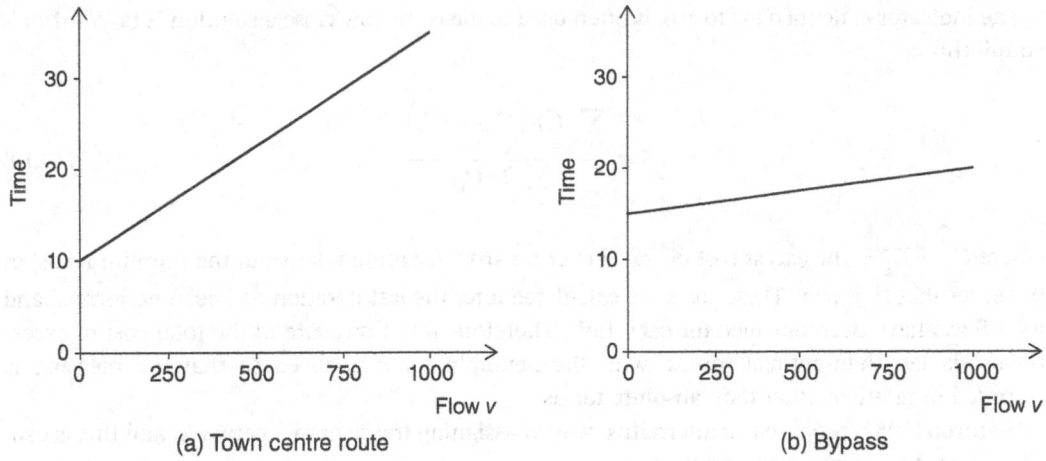

(a) Town centre route (b) Bypass

Figure 10.9 Time–flow relationships for Figure 10.4

The flows on the two routes will satisfy Wardrop's equilibrium when the corresponding 'costs' are identical. In this case it is relatively simple to write two equations for travel time vs. flow and equate them to find the equilibrium solution, for example:

$$t_b = 15 + 0.005V_b \tag{10.8a}$$
$$t_t = 10 + 0.02V_t \tag{10.8b}$$

where t_b and t_t are travel 'costs' (time in min) via the bypass and the town-centre routes respectively, and V_b and V_t are their corresponding flows.

By equating t_b to t_t it is possible to find, in this simple case, the direct solution to Wardrop's equilibrium as a function of the total flow $V_b + V_t = V$:

$$15 + 0.005V_b = 10 + 0.02 (V - V_b)$$

that is:

$$V_b = 0.8 V - 200 \tag{10.9}$$

Expression (10.9) has meaning only for non-negative flows, that is, for V greater than or equal to $200/0.8 = 250$. For $V < 250$, $C_t < C_b$, $V_b = 0$ and $V_t = V$ (i.e. all traffic chooses the town-centre route). For situations where $V > 250$ the two routes will be used; for example, the reader can verify that for $V = 2000$ the equilibrium flows are $V_b = 1400$ and $V_t = 600$ and the 'costs' by each route are 22 minutes.

The same idea would apply to flows on networks where the costs of travel by each of the routes used between two points are the same under Wardrop's equilibrium. The problem is, of course, that in anything but the simplest cases, it is not possible to solve the equilibrium flows algebraically; rather, an algorithmic solution method is required.

Several techniques have been proposed as reasonable approximations to Wardrop's equilibrium: some of them are simple heuristic approaches, and the most interesting ones follow a more rigorous mathematical programming framework. To compare these algorithms, the following properties are of interest:

- is the solution stable?
- does it converge to the correct solution (Wardrop's equilibrium)?
- is it efficient in terms of computational requirements?

The indicator δ, defined in (10.10), is often used to measure how close a solution is to Wardrop's equilibrium:

$$\delta = \frac{\sum\limits_{ijr} T_{ijr}\left(C_{ijr} - C_{ij}^*\right)}{\sum\limits_{ij} T_{ij}C_{ij}^*} \tag{10.10}$$

where $C_{ijr} - C_{ij}^*$ is the excess cost of travel over a particular route relative to the minimum cost of travel for that (i, j) pair. These costs are calculated after the last iteration has been performed and total flows have been obtained for each link. Therefore, δ is a measure of the total cost of excess travel via less-than-optimal routes, with the denominator introduced so that the measure is recorded in relative rather than absolute terms.

Wardrop (1952) proposed an alternative way of assigning traffic onto a network, and this is usually referred to as his *second principle*:

> *Under social equilibrium conditions traffic should be arranged in congested networks in such a way that the average (or total) travel cost is minimised.*

This is a 'design' principle, in contrast with his first principle, which endeavours to model the behaviour of individual drivers trying to minimise their own trip costs. The second principle is oriented towards transport planners and engineers trying to manage traffic to minimise travel costs and therefore achieve an optimum *social equilibrium*. In general, the flows resulting from the two principles are not the same, but one can only expect, in practice, traffic to be arranged following an approximation to Wardrop's first principle (i.e. *selfish* or *users' equilibrium*). It is interesting to note that the basic objective of congestion charging and road pricing is to get closer to Wardrop's second principle. Indeed, most methodologies for pricing congestion start by assessing what tolls should be charged on each link to achieve a social equilibrium.

10.5.2 Hard and Soft Speed-Change Methods

Some of the first heuristic methods still maintained the idea of assigning all trips per O–D pair to a single route (i.e. all-or-nothing assignment), but acknowledged the fact that speeds, and therefore travel times, responded to flow levels. The simplest of these methods involves just recalculating link travel times after an all-or-nothing assignment so that they are consistent with the current flow levels. A new all-or-nothing assignment is then performed with the new costs and trees. It is easy to see that this is a poor approach as the chosen routes will oscillate and the flow pattern will, in general, never converge. In the case of the town-centre bypass problem of Example 10.4 with, say, $V > 250$, the flows would oscillate between all trips via the town centre in one iteration and all via the bypass in the next one. This phenomenon will be repeated in larger networks, although in some cases it may be more difficult to identify.

In an attempt to dampen such route and flow oscillations, it has been proposed to use an average speed of two or more all-or-nothing assignments to perform the next iteration. This is often called a 'soft' speed change as opposed to the 'hard' speed change of the original method. However, this may only provide an apparent improvement, as the main weakness of these two approaches is that they still assign all traffic to a single route for each O–D pair, therefore contradicting Wardrop's principle. Taking again the case of Example 10.4, it can easily be seen that the soft speed-change method will still load all traffic alternatively via one route and then the other in the next iteration.

Both methods produce unstable solutions that are inherently non-convergent and the use of soft speed changes will only disguise this fact in larger networks.

10.5.3 Incremental Assignment

This is a more interesting and realistic approach. In this case, the modeller divides the total trip matrix T into a number of fractional matrices by applying a set of proportional factors p_n such that $\Sigma_n \, p_n = 1$. The fractional matrices are then loaded, incrementally, onto successive trees, each calculated using link costs from the last accumulated flows. Typical values for p_n are: 0.4, 0.3, 0.2, and 0.1. The algorithm can be written as follows:

1) Select an initial set of current link costs, usually free-flow travel times. Initialise all flows $V_a = 0$; select a set of fractions p_n of the trip matrix T such that $\Sigma_n \, p_n = 1$; make $n = 0$.
2) Build the set of minimum cost trees (one for each origin) using the current costs; make $n = n + 1$.
3) Load $T_n = p_n \, T$ all-or-nothing to these trees, obtaining a set of auxiliary flows F_a; accumulate flows on each link:

$$V_a^n = V_a^{n-1} + F_a$$

4) Calculate a new set of current link costs based on the flows V_a^n; if not all fractions of T have been assigned proceed to step 2; otherwise stop.

This algorithm does not necessarily converge to Wardrop's equilibrium solution, even if the number of fractions p is large and the size of the increments ($p_n \, T$) is small. Incremental loading techniques suffer from the limitation that once a flow has been assigned to a link, it is not removed and loaded onto another one; therefore, if one of the initial iterations assigns too much flow to a link for Wardrop's equilibrium conditions to be met (for example, because the link is short but has very low capacity), then the algorithm will not converge to the correct solution.

However, incremental loading has two advantages:

- it is very easy to program;
- its results may be interpreted as the build-up of congestion for the peak period.

Example 10.5 Consider again the problem of the two routes, town centre and bypass, of Example 10.4. We split the demand of 2000 trips into four increments of 0.4, 0.3, 0.2, and 0.1 of this demand, that is, 800, 600, 400, and 200 trips. At each increment we calculate the new travel costs using Eq. (10.4). Table 10.2 summarises the results of this algorithm:

Table 10.2 Flows and costs results for Example 10.5

N	Increment	Flow town	Cost town	Flow bypass	Cost bypass
0	0	0	10	0	15
1	800	800	26	0	15
2	600	800	26	600	18
3	400	800	26	1000	20
4	200	800	26	1200	21

It can be seen that the algorithm does not converge, in this case, to the correct equilibrium solution. This is because once the wrong flow (800) has been loaded onto the town-centre route, this method cannot reduce it; therefore, the flow and cost via the town centre remain overestimated. As a matter of interest, the value of the δ indicator for the solution above is:

$$\delta = [800\,(26-21) + 1200\,(21-21)]/(2000 \cdot 21) = 0.095$$

The reader can verify that using smaller increments would produce closer solutions to true equilibrium. Note that if one starts with an increment of 0.3 times the total demand, the solution is true equilibrium; however, this is just a chance occurrence in this case.

10.5.4 Method of Successive Averages

Iterative algorithms were developed, at least partially, to overcome the problem of allocating too much traffic to low-capacity links. In an iterative assignment algorithm, the 'current' flow on a link is calculated as a linear combination of the current flow on the previous iteration and an 'auxiliary flow' resulting from an all-or-nothing assignment in the present iteration. The algorithm can be described by the following steps:

1) Select a suitable initial set of current link costs, usually free-flow travel times. Initialise all flows $V_a = 0$; make $n = 0$.
2) Build the set of minimum cost trees with the current costs; make $n = n + 1$.
3) Load the whole of the matrix **T** all-or-nothing to these trees obtaining a set of auxiliary flows F_a.
4) Calculate the current flows as:

$$V_a^n = (1-\phi)V_a^{n-1} + \phi F_a$$
$$\text{with } 0 \le \phi \le 1 \tag{10.11}$$

5) Calculate a new set of current link costs based on the flows V_a^n. If the flows (or current link costs) have not changed significantly in two consecutive iterations, stop; otherwise proceed to step 2. Alternatively, the indicator δ in (10.10) could be used to decide whether to stop or not. Another, less good but quite common, criterion for stopping is simply to fix the maximum number of iterations; in this case δ should be calculated to know how close the solution is to Wardrop's equilibrium.

Iterative assignment algorithms differ in the methods used to give a value to ϕ. A simple rule is to make it constant, for example, $\phi = 0.5$. A much better approach due to Smock (1962), is to make $\phi = 1/n$. The reader may verify that equal weight is given to each auxiliary flow F_a in this case; for this reason, the algorithm is also known as the Method of Successive Averages (MSA). It has been shown (see, for example, Sheffi 1985) that making $\phi = 1/n$ produces a solution convergent to Wardrop's equilibrium, albeit not a very efficient one. As we shall see in Chapter 11, the Frank–Wolfe algorithm (Frank and Wolfe 1956) estimates optimal values for ϕ in order to guarantee and speed up convergence.

Example 10.6 Consider the same bypass versus town-centre problem of Example 10.5 and use $\phi = 1/n$. Table 10.3 summarises the steps in the MSA algorithm.

It can be seen that it takes several iterations to approximate to the right solution. Of course, the value of δ after iteration 10 is zero in this case. However, the reader will note that the algorithm was close to the correct equilibrium solutions in iterations 3, 6, and 9 but only reached it in iteration 10. This is due to the rigid nature of the rule used to calculate ϕ. For more realistic networks, the number of iterations needed to reach satisfactory convergence may be very high.

Table 10.3 Steps in MSA algorithm

Iteration		ϕ	Flow town	Cost town	Flow bypass	Cost bypass
1	F		2000		0	
	V^n	1	2000	50	0	15
2	F		0		2000	
	V^n	1/2	1000	30	1000	20
3	F		0		2000	
	V^n	1/3	667	23.3	1333	21.7
4	F		0		2000	
	V^n	1/4	500	20	1500	22.5
5	F		2000		0	
	V^n	1/5	800	26	1200	21
6	F		0		2000	
	V^n	1/6	667	23.3	1333	21.7
7	F		0		2000	
	V^n	1/7	572	21.4	1428	22.1
8	F		2000		0	
	V^n	1/8	750	25	1250	21.25
9	F		0		2000	
	V^n	1/9	667	23.3	1333	21.7
10	F		0		2000	
	V^n	0.1	600	22	1400	22

Another lesson from this simple example is that fixing the maximum number of iterations is not a good approach from the point of view of evaluation. Link and total costs can vary considerably in successive iterations, and this may affect the feasibility of a scheme.

10.5.5 Braess's Paradox

The basic ideas about Wardrop's first and second principles are often illustrated using Braess's Paradox; although strictly speaking not a paradox, it is nearly as famous as the 'blue bus/red bus' conundrum we saw in Chapter 6. The paradox was first proposed by Dietrich Braess in 1968, but it is mostly known through its translation into English in Braess et al. (2005). It demonstrates that under certain conditions adding capacity to a road network when drivers seek to minimise their own costs can actually make everybody worse off.

Consider the simple network depicted in Figure 10.10. The linear relationship associated with each link represents the travel time-flow formulation in minutes. Solid arrows indicate existing links, and the dotted arrow indicates a planned link. Assume first that there are 1000 cars wishing to travel between A and B and none from F. The logical route choice under these conditions is for 500 cars to use the ACB route and the other 500 the ADB route. Both costs are the same 38 minutes (2 + 10 + 25 + 1). Consider now what happens when a new, high-capacity link, is built between C and D. Under these conditions all drivers would choose to start on the AC path, as under the most loaded conditions it would cost 2 + 20 + 1 + 1 = 24 minutes to reach D when it takes at least

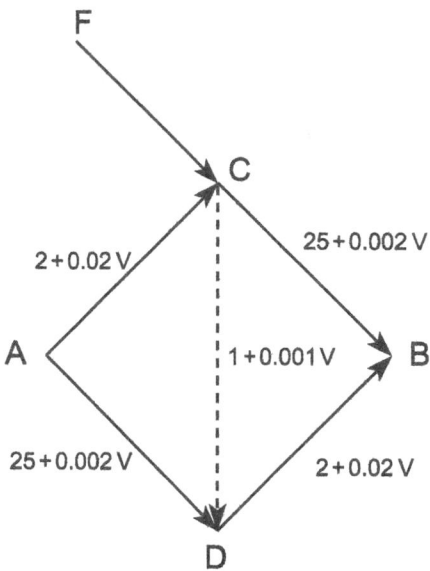

Figure 10.10 A simple network to illustrate Braess's Paradox

25 minutes if the AD route is used. At C, and for the same reasons, every rational driver would take the CD route as it would take at most $1 + 1 + 2 + 20 = 24$ minutes to reach B, one minute less than the most optimistic conditions for the CB route.

However, note that the total cost from A to B for each driver would then be $2 + 20 + 1 + 1 + 2 + 20 = 46$ minutes. In effect, eight minutes longer than before the link CD was built. If all drivers could agree not to use the CD link, they would all be better off. However, if starting from the original position (500 on each route) one driver chooses to use link CD he would be better off as from C it would only take $1 + 0.001 + 2 + 10.02 = 13.021$ to reach B, much less than the $25 + 1$ that the CB route offers.

So, the ACDB path represents a selfish equilibrium condition (Wardrop's first principle), but everybody is worse off than before the new link was built.

If, however, there are 1000 vehicles travelling from F to D, the travel time on that link would be at least two minutes. Now the choices open to our original drivers are a bit less clear. Starting from A and assuming the worst conditions (all drivers choose the ACD route) the cost of reaching D is practically the same via the new link or the old AD route (25 minutes). This would suggest that the original equilibrium could be restored and everybody benefit from 38 minutes journeys. However, if one driver at C chooses the new link, he would benefit again with $2 + 0.001 + 2 + 10.02 = 24.021$ minutes to reach B instead of the expected $25 + 1$ for the direct route.

This suggests that either drivers will find quite different conditions on their routes each day due to uncoordinated experimentation, with some days resulting in very poor choices, or that the new stable conditions will revert to all drivers using the centre route and spending now 47 minutes each day to reach B, a new equilibrium condition. Introducing a toll on link CDs could make things better. What is the minimum toll that would produce the best selfish and social (Wardrop's second principle) equilibrium conditions? Of course, to avoid charging drivers from F unnecessarily, the toll should be imposed only on vehicles taking the ACD turning at the top.

If the conditions that lead to Braess's paradox happen in practice as well as in textbooks, it would be interesting to identify the perverse links (perhaps built because they were feasible rather than desirable) and either toll them or close them to vehicular traffic. Steinberg and Zangwill (1983) developed necessary and sufficient conditions for Braess' paradox to occur when a new route or link is added. They concluded that these conditions were not unusual and that they were likely to occur in practice. Youn et al. (2008) studied the cities of New York, Boston, and London, established routes where these conditions were likely to be present, and pointed out roads that could be closed to traffic to reduce travel times.

10.6 Public-Transport Assignment

10.6.1 Introduction

In this section, the problems associated with route choice and assignment for passengers using public-transport networks will be discussed. These problems are, in many ways, more difficult than those encountered by private-transport assignment; computer requirements tend to be heavier, and even the best methods require some simplifying assumptions. Recent years have seen significant improvements in transit assignment techniques, leading to better public-transport service provision and operational efficiency.

We shall discuss first the issues that make public-transport assignment different from private vehicle route choice; then, we will outline some of the approaches that have been implemented to tackle them in practice. Each of the main software packages available offers particular ways of implementing these ideas. They are mostly based on fixed-schedule public transport services. The advent of demand-responsive public transport services poses additional and difficult challenges, which are discussed only in Chapter 20.

The list of characteristics influencing public transport route choice is probably longer than that for private cars, and some vary in importance by gender. They include, in addition to travel time and cost:

- number of transfers and the time spent transferring;
- waiting time is generally associated with the headway of each service;
- access time is often assumed to be walking time to the stop/station and from the last stop to the final destination; micro-mobility reduces this perception of access time;
- probability of finding a seat;
- level of crowding in the vehicle;
- reliability of service and travel time;
- perceived security.

It is difficult to incorporate all these attributes in a model, even when using a discrete choice modelling framework as discussed in Section 10.6.5. Therefore, most public transport assignment methods simplify the utility function to just two or three attributes.

10.6.2 Issues in Public-Transport Assignment

10.6.2.1 Supply

The network of public-transport services is different from that of private cars. It includes, as links, sections of the bus or rail service running between two stops or stations. The concept of link capacity is associated with the capacity of each unit (bus, train) and its corresponding frequency. The travel time has an in-vehicle component as well as components for waiting at stops and walking to and from them. Many of the public-transport sections will use road links, for example, most buses and some light rail-transit (LRT) services running on streets. There will be other public-transport sections or services that will use completely different links (e.g. busways, segregated rail tracks, and so on). The nature of these links generally produces a more complex network, an example of which is given in Figure 10.11.

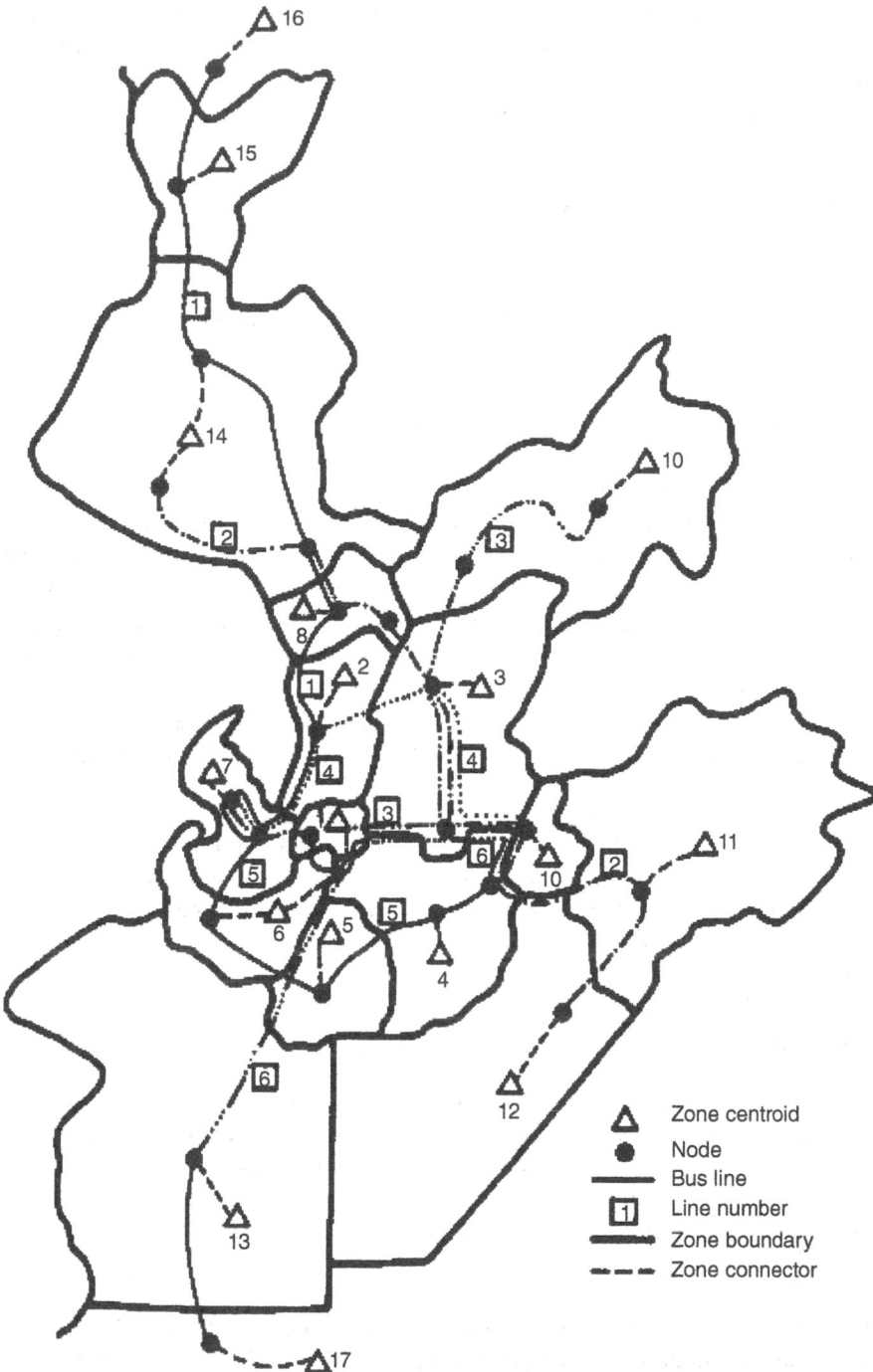

Figure 10.11 An example of a public-transport network

Zone centroid
Node
Bus line
Line number
Zone boundary
Zone connector

10.6.2.2 Passengers

In public-transport route choice, we need to deal with the movement of passengers and not of vehicles. Passengers can walk to a stop, interchange between two services, and even drive part of the way to board a public-transport service later. This calls for the need to provide and specify walk and transfer links between different services, different public-transport modes (bus, rail), and between public-and private-transport facilities (e.g. Park & Ride).

10.6.2.3 Monetary Costs

In private car networks, it is usually assumed that the monetary cost is directly associated with fuel consumption, which in turn is directly proportional to travel distance. These are both approximations, but they are usually accepted as drivers do not perceive these costs in such a direct way as a passenger buying a ticket when starting a bus journey. Modern payment systems based on smart cards or mobile phones allow more complex fare structures, and these have been introduced in many public-transport operations: fares variable with distance, flat fares (independent of distance travelled), zonal fares (for one or more specific geographic zones), combination and transfer tickets (valid for two or more services), time limit fares (e.g. valid for any number of boardings in an hour), daily, weekly, and other season tickets for a fixed service or covering one or more zones and modes. This wide range of fares places difficult requirements on route choice and assignment models, as monetary costs do not depend directly on distance but in general on the location of the origin and destination and on the route chosen.

10.6.2.4 The Definition of Generalised Costs

In the case of public-transport assignment the generalised cost of travelling may be defined as follows:

$$C_{ij} = a_1 t_{ij}^v + a_2 t_{ij}^w + a_3 t_{ij}^t + a_4 t_{ij}^n + a_1 \delta^n + a_5 F_{ij} \qquad (10.12)$$

where: t_{ij}^v is in-vehicle travel time between i and j; t_{ij}^w is walking time to and from stops (stations); t_{ij}^t is waiting time at stops; t_{ij}^n is interchange time; δ^n is an intrinsic 'penalty' or resistance to interchange, measured in time units (see Navarrete and Ortúzar 2013); F_{ij} is the fare charged to travel between i and j, and a_1 to a_5 are coefficients associated with the elements of generalised cost above.

Usually either a_1 or a_5 are equal to 1.0 in order to measure generalised costs in time or monetary units, respectively. Again, it is usual to find that a_2, a_3, and a_4 are taken to be two to three times the value of a_1 as passengers dislike a minute spent walking or waiting more than if spent travelling in-vehicle.

In modelling terms, the software should be able to handle these variables and produce good estimates of each of the component times (in-vehicle, walking, waiting, transfer) if they are not provided externally. In-vehicle travel time depends on the speed attainable and the number and duration of stops; walking time, which depends on proximity to the best stop or station, is in some cases approximated by an average value for a whole zone; interchange time depends on station/stop configuration and separation; waiting time depends essentially on the frequency of the service and its reliability. A general formulation for waiting time is:

$$t^w = \frac{(h^2 + \sigma^2)}{2h} \qquad (10.13)$$

where h is the expected headway of the service and σ its standard deviation (the less regular a service, the greater the expected waiting time). This formulation assumes that passengers arrive at the stop at random and that no passenger fails to board the next bus because of lack of space in it. This 'bus congestion' problem is difficult to solve, but algorithms incapable of handling it will tend to produce unrealistic loadings in terms of actual service capacity (De Cea and Fernández 1989). If the service is perfectly regular (i.e. $\sigma = 0$), then the expected waiting time is half of the headway. It is known, however, that if the frequency of the service is low, passengers will try to arrive just a few minutes before the next departure, thus setting an upper limit to the expected waiting time of perhaps 5–10 minutes; how close to the timetabled departure are passengers aiming to come will depend, of course, on the reliability of the service.

10.6.2.5 The Common Lines Problem

This is probably one of the most difficult and typical problems of public-transport assignment. The problem arises when, for at least some O–D pairs, there are sections in a path that have more than one parallel service offered, and passengers can choose the one suiting them better. This choice is often not trivial for passengers ('I wish I had known that an express service was going to come three minutes after the slow one I have taken!'), nor simple from a modelling point of view. We are used to the idea that a driver chooses a single path from a set of all possible paths. In the case of public-transport passengers, they may choose a 'set of paths' and let the vehicle that arrives first determine which of the paths they will actually use. The choice is therefore more complex and calls for a more detailed treatment.

A full review of the most suitable algorithms for public-transport assignment is outside the scope of this book. Instead, we shall discuss the main approaches to modelling route choice first and then assignment; not surprisingly, these different approaches result from the treatment they give to some of the issues above, in particular the parallel or common lines problem and the choice of all-or-nothing, stochastic, or capacity restraint-assignment methods.

10.6.2.6 Frequency or Schedule-Based Route Choice

When the frequency of a public transport service is reasonably high, say every five minutes for an urban context and 10 or 15 minutes for the inter-urban case, travellers will, in general, not use or memorise a timetable (if it exists), but just turn up at the stop for a short waiting time. In these cases, it may be quite appropriate to use the frequency of the services as sufficient descriptor to estimate waiting times. However, this approach would be less appropriate for larger headways where trip makers are more likely to plan their access to arrive just a few minutes before the bus/train is due, according to their schedule (timetable). This can be considered by capping the waiting time at a maximum of, say, eight minutes, depending on context. However, this also fails to take full account of two situations. The first one is the provision in practice of irregular frequencies, for example, a timetabled service at 5, 15, 20, 30, 45, and 55 minutes past the hour. The second is the opportunity to provide well-coordinated services even under low-frequency schedules; for example, timing a half-hourly bus service to a rail station to arrive there five minutes before the train departs for a main destination.

10.6.3 Modelling Public-Transport Route Choice

It is worthwhile defining some terms, such as route, line, and section in a bit more detail before embarking on a discussion of the route choice problem in the presence of common lines.

A *public-transport* (or *transit*) *line*, or simply a *line*, is a fleet of vehicles that run between two points (terminals) on a network. They generally have the same characteristics of size, capacity, speed, etc. Vehicles stop at each node in their path to allow passengers to alight and board. Therefore, each transit line is defined by the vehicle characteristics, the sequence of nodes it serves and its frequency. A *line section* is any portion of a public-transport line between two, not necessarily consecutive, nodes. A *public-transport route* is any path a user can follow on the transit network in order to travel between two nodes. The portion of a route between two consecutive transfer nodes is called a *route section*, and each route section has associated a set of *common (attractive) lines*.

Consider now the simple case of an origin A and destination J connected by three transit services: lines 1, 2, and 3 as in Figure 10.12; they follow different routes and offer travel times of 20, 17, and 18 minutes to reach the desired destination.

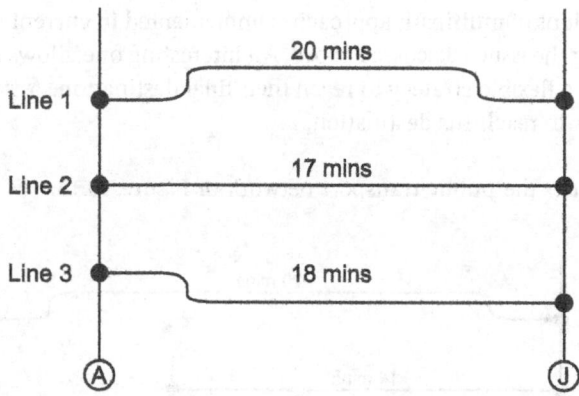

Figure 10.12 A basic section of public transport services showing travel times

The frequency of each line is six services/hour; this means an expected waiting time of five minutes assuming perfectly regular services and random arrival of travellers. A traveller will then face three alternative segments in his journey (either from origin to destination or as a stage of a trip with one or more transfers): Line 1 with an expected travel time of 25 minutes (20 plus 5), Lines 2 and 3 with 22 and 23 minutes, respectively.

A naive *all-or-nothing* route choice will assign all travellers to Line 2 to minimise travel time. On the other hand, a more realistic approach would be to allocate the probability of boarding proportional to its frequency given that travellers are faced with actually 18 useful services per hour. Now the average waiting time is three minutes and 20 seconds (18/60) and the average travel time is 18 minutes and 20 seconds. The total expected travel time is 21 minutes 40 seconds. This approach produces a smaller expected travel time than the naive one if the travel times are similar (as they are when the lines follow the same sequence of nodes) but a large difference in travel times will result in a larger expected value: check with travel times of 17, 20, and 30 minutes.

Note that one can also build the network recognising that waiting (and walking) times are valued as about twice in-vehicle times (IVT) producing slightly different results above in terms of generalised times. In a longer route over a transit network, we would also add transfer penalties and additional waiting times for some routes; moreover, in many cases, additional fares may be charged with each transfer, and this can also be added to the computation of generalised cost per link. Boarding penalties are often used to represent these effects.

Transit assignment methods can then be divided into:

- naive all-or-nothing approaches that would only be acceptable for sparse and long-distance travel networks;
- multi-path approaches, for example, the allocation of trips to paths proportional to the perceived service frequencies as outlined above;
- equilibrium assignment methods with or without a stochastic element in them; these focus on congestion effects on public transport systems.

The all-or-nothing approach, despite its simplicity, may be useful in refining a transit network, often a subtler task than debugging a road network, and a task that benefits from many good existing data sources.

There are many versions of multipath approaches implemented in current software, some better than others at handling the issues discussed above. An interesting one allows travellers to adopt, as they do in many cases, a flexible strategy to reach their final destination. A *strategy* is a set of rules that allow the traveller to reach his destination.

Example 10.7 Consider the public-transport network of Figure 10.13.

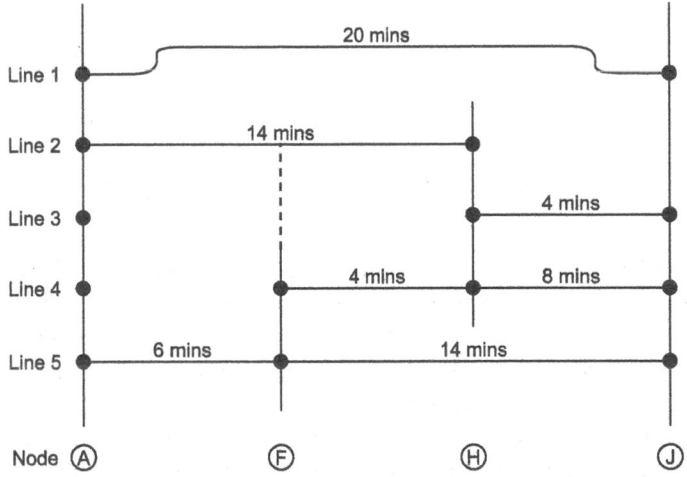

Figure 10.13 A simple public-transport network with transfers

A simple strategy could be:

- take line 2 to stop H; transfer to line 3 and then exit at stop J;

A more complex one may take the form:

- wait up to three minutes for a line 5 vehicle or up to four minutes for a line 2 vehicle; otherwise, take line 1; if line 5 is taken and you see a line 4 vehicle at stop F then board it and alight at J; if there is no line 4 vehicle at F, continue to J; if line 2 vehicle was taken then transfer at H to line 4 if it is about to depart; otherwise, wait for line 3 to reach J; etc.

In general terms, a good flexible strategy will produce shorter expected travel times than the choice of the single path that minimises travel time; the choice of this single minimum path has been for many years the conventional approach to the problem.

In contrast, a more realistic, flexible strategy allows the passenger to take advantage of the variability of waiting times and the opportunistic choice of a good, but low-frequency, service. This is well illustrated by Spiess and Florian (1989).

One can then define, for each node, the set of attractive lines that would be part of a good strategy to reach a given destination j. Given a strategy, an actual trip is then carried out according to a mechanism like:

1) Set i to origin node;
2) Board the first arriving vehicle from the set of attractive lines at i;
3) Alight at a predetermined node;
4) If not yet at destination, set i to the current node and return to step 2; otherwise, the trip is completed.

Note that although this mechanism has a well-defined destination node, the origin is not part of the strategy. A strategy is the set of rules that enables travellers to reach their destination starting from any node in the network. This treatment is helped by the following additional notation:

S_{jk} = set of line sections connecting directly nodes j and k;
L_j^+ = set of outgoing (ingoing if − instead of + is used) line sections from node j;
v_s = flow on line section s;
t_s = in-vehicle travel time on line section s;
f_s = frequency associated to line section s;
g_j = number of trips going to destination node j;
V_{jk} = total flow on route section jk.

We can now identify the set of attractive routes emanating from node j using the dummy variable X_s which takes the value 1 if the line section s, belonging to the set of sections from j to k, is attractive and zero otherwise. Then, for a given pair of nodes j, k the associated values X_s ($s \in S_{jk}$) define the optimum or attractive set of lines towards k.

The total waiting time for users travelling from j to k can be written as:

$$w_{jk} = \frac{V_{jk}}{\sum\limits_{s \in S_{jk}} f_s X_s} \qquad (10.14)$$

The problem of finding an optimum strategy for travelling from all origins to a destination can now be written as:

$$\text{Minimise} \sum_s v_s \cdot t_s + \sum_{jk} w_{jk} \qquad (10.15)$$

subject to:

$$\sum_{s \in L_j^+} v_s + g_j = \sum_{s \in L_j^-} V \qquad (10.16)$$

$$v_s = \frac{X_s f_s V_{jk}}{\sum\limits_{s \in S_{ij}} f_s X_s} = X_s f_s w_{jk} \qquad (10.17)$$

The first term of the objective function (10.15) represents the in-vehicle travel time while the second is the total waiting time. This objective function is linear in v_s and w_{jk}, and the main problem seems to be generated by the non-linear constraints (10.17). Spiess (1983) has shown that these constraints can be relaxed as follows:

$$v_s \leq f_s w_{jk} \tag{10.18}$$

We can further introduce constraints (10.17) into the objective function:

$$\text{Minimise} \sum_{jk} \frac{V_{jk} \left\{ \sum_s t_s \cdot X_s \cdot f_s + 1 \right\}}{\sum_{s \in S_{ij}} f_s \cdot X_s} \tag{10.19}$$

subject to (10.16). This is a (0,1) hyperbolic programming problem.

Two different approaches can be followed here. The one proposed by Spiess and Florian (1989) is based on the linear programming version of this problem, whilst that proposed by De Cea and Fernández (1989) uses a hyperbolic programming (non-linear) formulation. If there are no congestion or capacity problems, the tasks above can be simplified, as the set of optimal strategies will not depend on the actual flows. The Florian–Spiess algorithm is used in EMME (inrosoftware.com/en/products/emme/) and the De Cea–Fernández algorithm in ESTRAUS (De Cea et al. 2005). Tests showed that the De Cea–Fernández approach was faster than the Florian–Spiess method and nearly 50 times faster than the best conventional approach. This improvement in performance, which is crucial to modelling realistic size problems, was achieved at the cost of additional memory requirements, which are not a significant requirement today.

10.6.4 Assignment of Public Transport Trips

Once the best set of line segments to join origin and destination have been identified, one needs to consider the assignment of trips to them. Most programs seek to obtain a reasonable and realistic spread of trips among feasible routes. Conventional approaches, not dealing with the common lines problem explicitly, adopted a number of measures to generate this wider spread of trips. For example, to explicitly distinguish the different access points (bus stops, stations) for each zone and to build trees from each of them (and not just from the centroids) to all destinations. In this way, several alternative routes are identified, one via each different access point. Passengers can then be assigned to these routes using a MNL function of the costs of joining origin and destination via each path.

Spiess and Florian (1989) perform the assignment stage following the identified optimal strategies. This is achieved by assigning to each link the proportion of the volume accumulated at the upstream node that corresponds to the frequency served by the link. De Cea and Fernández (1989) follow a similar approach, but in two stages:

1) First, once the set of common lines for all (i, j) pairs has been identified, a new network is built on the basis of *nodes* and *route sections*. Note that route sections contain only the lines that minimise the total expected travel time for the section; they have an associated travel time (t_r) and a frequency (f_r) corresponding to the sum of the attractive frequencies (i.e. those in the common lines). With these two elements it is possible to obtain a composite cost of travelling along this route section and therefore an efficient private-transport tree-building algorithm can be used to find the best paths. Loading onto these trees results in a set of 'route section flows' v_r.

2) Second, we can decompose the route section flows into their 'line section' components:

$$v_s = \frac{f_s v_r}{f_r} \tag{10.20}$$

The treatment so far has not discussed the problems associated with special fare systems. If the fare system is proportional to the distance travelled, this is not a major problem, as it is normally possible to convert it to time units and add them to the travel time on each link. However, this type of fare structure is hardly common. A flat fare system could also be accommodated, but the treatment of more complex schemes (from a modelling point of view) may pose additional problems for algorithm design.

In most practical cases, it will not be possible to model the whole complexity of fare systems, and some approximate shortcuts will have to be taken in accordance with the most common type of ticket used. For example, in the case of a zonal fare system, assignment may be performed based on time alone and the fare cost added at the end. This may still ignore the importance of special pass holders, but is probably good enough for places like London.

10.6.5 Discrete Route Choice Modelling

It seems natural to apply the significant advances in discrete choice modelling discussed in earlier chapters to the task of representing the decision to select the best combination of services to reach a particular destination. The use of the discrete choice framework casts the problem as maximising the random utility of each traveller. It allows a richer selection of attributes and the representation of a wider range of person types.

The approaches followed may differ by the source of the behavioural data (stated or revealed preference), the number of public transport models included, the choice model tested (mostly MNL, NL, and Cross Nested Logit) and how the alternatives are identified. When stated preference (SP) data are used, the modeller defines the alternatives to consider. For example, Schmid et al. (2019) used SP data from a panel of travellers to estimate a MNL model between car, bus, tram, metro, rail, and active modes. The utility functions included all times, costs, and transfer attributes.

Most efforts are based on revealed preference (RP) data, that is, the actual choices made by travellers. In this case, the alternatives are either observed routes or routes extracted from an analysis of the network, limiting the reasonable routes to a manageable number. Some focus only on one main mode, for example, metro. This is the case of Raveau et al. (2014), using data from the London and Santiago undergrounds and testing MNL structures with a full range of attributes, including the probability of finding a seat. Anderson et al. (2017) considered bus, metro, and rail in Copenhagen and tested several alternative models including a MNL. Meyer de Freitas et al. (2019) used data from the Swiss national household travel survey to analyse the socioeconomic determinants of inter-modal travel in Switzerland and estimate a large-scale multimodal recursive Logit route choice model for urban trip making. They show that intermodal travel is mostly associated with ownership of transit subscriptions, which allow free public transport at the point-of-use.

The work of Domarchi et al. (2019b) is interesting as they tested different configurations for Nested Logit models for multi-modal transit route choice in addition to a Cross Nested Logit (CNL) model. In all cases, they used an extensive set of level-of-service attributes. The configurations tested are illustrated in Figure 10.14.

They found better results when nesting as a function of the first mode as, in NL3; for example, choices when bus was followed by metro were different than for metro followed by bus. In their data-set of observed trip choices, the CNL model performed only marginally better than the NL3 structure.

Passive data collection has also been used to explore these issues. An example of this approach is Janosikova et al. (2014), who used individual data from public transport smartcards and open-source data on routes and frequencies in Slovakia, to model route choice between bus and trolley-bus. Kim et al. (2014) used similar data sources in South Korea, to look at route choice in their metro system. Both tested MNL structures with simplified utility functions.

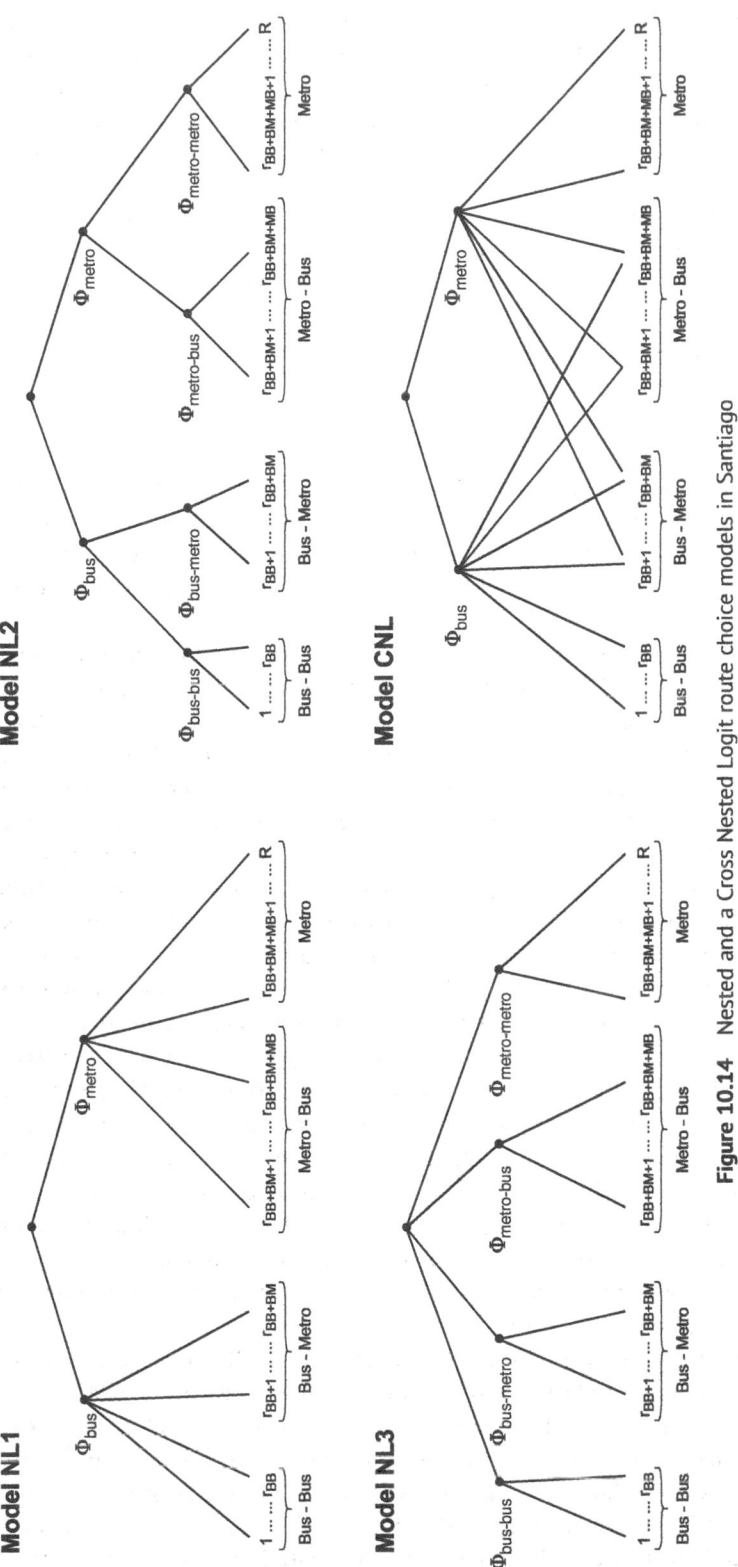

Figure 10.14 Nested and a Cross Nested Logit route choice models in Santiago

Finally, we must stress that public-transport assignment suffers, in general, from similar weaknesses to those identified for road networks. Furthermore, it is fair to say that congested assignment is less well developed for transit networks. There are two effects in play here: first, the limited capacity of the units (buses, trains) may prevent some travellers from implementing their optimal strategies, thus increasing their travel times; second, there is interaction between public transport and private cars sharing the same road network; therefore, increased traffic on one mode will affect travel times on the other as well. We will consider some approaches to dealing with these issues in the next chapter.

10.7 Limitations of the Classic Methods

In previous sections, we have described the most important classic methods for traffic assignment. Before considering more detailed and, to some extent, advanced methods, it is worthwhile reviewing what are seen as the main limitations of these approaches (Willumsen 2007). These deficiencies may come from different sources.

10.7.1 The Assumption of Perfect Information about Costs in all Parts of the Network

Although this is common to all models, having perfect information is essentially overoptimistic, at least until the widespread use of road transport informatics makes more realistic modelling a possibility. Drivers have only partial information about traffic conditions on the same route the last time they used it, and on problems in other parts of the network depending on their own experience, disposition to explore new routes and the use of traffic information services. Moreover, there is evidence that many drivers are heavily influenced by road signs in their choice of route and that signed routes are not always the cheapest (Wootton et al. 1981). Current methods ignore these effects. The influence of variable message signs and more advanced route guidance technology over part of the vehicle fleet is likely to place new requirements for traffic assignment methods.

10.7.2 The Assumption that all Movements can be Represented by a Trip Matrix

Most data collection methods focus on identifying trips and tours, and most models seek to assign trips to the respective networks. All tours can be processed, and their trips are represented in trip matrices. However, not all real movements fit neatly into that description. Taxis cruising for passengers, delivery drivers, tourists, and even people who simply missed the right exit of a motorway or misinterpreted a map. All of them contribute to traffic counts but are very difficult to represent by means of trip matrices. Passive data collection methods, like the use of mobile phone or app data, can identify these diversions, tours, and trips, and, if desired, they can be added to trip matrices.

10.7.3 Limitations in the Node-link Model of the Road Network

These include the fact that not all real road links are considered in the network (incomplete networks), 'end effects' due to the aggregation of trip ends into zones represented by single centroids, banned and penalised turning movements not specified in the network, and the fact that intrazonal trips are not fully treated.

The main problem with incomplete networks arises in heavily congested areas where some of the medium- and long-distance trips will use minor roads as 'rat runs'; a new road scheme may relieve congestion and attract some of these rat-run trips, which will seem to be 'generated journeys' when they are not. Even when great care is taken in connecting the network to zone centroids, end effects are inevitable. These will make estimated link volumes less reliable in the vicinity of centroid connectors, probably overestimating the flows.

It is possible to expand simple nodes to represent all turning movements at a junction and then penalise or remove those links representing banned manoeuvres. An example of a fully expanded junction is given in Figure 10.15; any particularly difficult manoeuvre (e.g. an opposed turn) can then be penalised by associating a longer delay to it. Good software provides efficient ways of automatically expanding junction representations and banning or penalising movements; alternatively, this must be done by hand in the network-building stage itself. In either case, it is likely that some turning movements will not be properly treated.

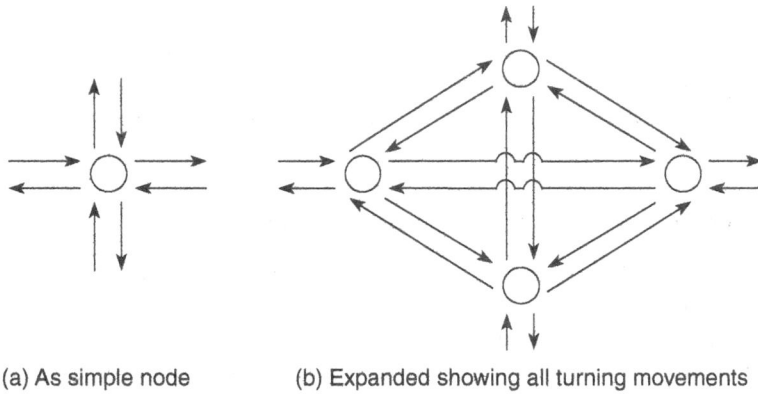

(a) As simple node   (b) Expanded showing all turning movements

Figure 10.15   Representations of a junction

It is also possible to employ models to represent the delays of each movement at a junction as a function of their configuration, geometric characteristics, signal timings, and, of course, flows on each approach. Experience shows that it is better to restrict this treatment to key junctions to retain the convergence properties of good models, as discussed in Chapter 12.

The treatment of intra-zonal movements is also a source of problems; some of them could make use of main links in the road network, but they will not appear in the network model. It is difficult to devise a good method to account for them in assignments.

All of these problems are more difficult to handle when the zones are large and the network representation is sparse. As usual, greater resolution in network and zonal definition will increase realism, but at the cost of data collection, processing, and interpretation.

10.7.4   Errors in Defining Average Perceived Costs

We do not have enough evidence about how these costs are likely to change with time, journey purpose, length of journey, income, predictability, and the environment. Moreover, when we wish to forecast components of cost, for example, fuel consumption, we rely on simplifying assumptions, which may give rise to additional errors.

10.7.5   Not all Trip Makers Perceive Costs in the Same Way

Our stochastic methods are an approximation of this phenomenon, but even they must limit the number of randomisations for reasons of economy. Another possibility is to consider several different user classes, each with its own set of perceived costs.

It is possible to express the deterministic equilibrium assignment problem with multiple user classes, each with its own set of parameters defining perceived link costs; see, for example, the work of Leurent (1998). Convergence to a unique solution is achieved under analogous conditions to those required for the single-user problem. Moreover, the problem and the solution can also be extended to elastic demand modelling (combined mode choice and assignment, for example). Most modern software packages offer this type of facility.

The modelling of multiple user classes (each with different willingness to pay for a better service) is often quite critical in demand studies for private sector facilities and services like high-speed rail links or toll roads. In the context of toll roads and rail journeys, some users may have high willingness-to-pay for services because their costs are covered by their employers; others may be very price sensitive because of personal income or cash constraints. These different user classes can be well represented in these cases, although good stated preference/revealed preference studies will be required to determine the best parameters for each model.

It is not unusual to have different fares for different types of users of public transport: special discounts for students and pensioners, peak and off-peak fares, and so on. Particularly difficult is the treatment of period or season tickets, where marginal trips have an effective zero cost.

A good example of this variety of perceptions is the nature of money. Fares and tolls are increasingly being paid by electronic means: smart cards, tags, and video tolling. It is generally recognised that this type of transaction brings some distance between use and charge, and is perceived as less onerous than a cash payment.

10.7.6 Day-to-Day Variations in Demand

These probably prevent true equilibrium from ever being reached in practice. In that sense, Wardrop's equilibrium represents 'average' behaviour if all travellers think alike and have perfect information. Its solution, however, has enough desirable properties of stability and interpretation to warrant its use in practice; however, it is still only an approximation to the traffic conditions on any one day.

In the same vein, there are time variations in demand and flow within each day. This makes 24-hour models very poor in terms of traffic assignment and, therefore, travel times and costs. The use of peak and off-peak periods for modelling and assignment is essential in congested urban areas but even then, we know that the build-up of congestion produces important changes in travel time in very short time frames. Moreover, a ten-minute delay in departure for the same journey may produce a much greater delay on arrival at the destination because of increased congestion in the network. The costs on links change dynamically in response to traffic: some drivers understand this well and plan their journeys accordingly; others may lack the necessary experience.

10.7.7 Imperfect Estimation of Travel Time Changes with Link Flow Changes

This is partly due to the nature of the cost–flow relationships used. As stated in Section 10.1.3, it is normally assumed that the travel time on a link depends only on the flow on the link itself. At least in urban areas, the delay on a link depends in general on flow on other links too, for example at a priority junction, thus creating interaction effects. This assumption will be discussed later, as it requires better delay models than those assumed in conventional cost–flow relationships.

10.7.8 The Dynamic Nature of Traffic

Most classic assignment methods assume the existence of a trip matrix that is valid over a modelling period, say one hour in the peak. Traffic is then assigned to the network under the assumption of steady-state conditions over that period. In practice, however, traffic behaviour is dynamic, and *steady state* is only a useful simplification. This assumption is relaxed in Chapter 11.

10.7.9 Input Errors

The accuracy of an assignment model also depends on the accuracy of other elements in the transport model, in particular the trip matrix to be loaded. This matrix will inevitably contain many errors and discrepancies, whether it is a synthetic one obtained from a Gravity model or a carefully observed one using extensive surveys. Errors in the conversions from passenger to vehicle trip matrices also limit the accuracy of traffic assignment. This conversion is usually assumed to be a uniform (and constant over time) occupancy rate for each type of vehicle and perhaps journey purpose. Simple observations will show that this is only an average with significant variations across regions.

To some extent, most of these difficulties can be overcome, at least partially, with appropriate tools, but at a cost in terms of data collection, analysis, and running time. Also, in some cases, it may be more difficult to interpret results. Moreover, sometimes these improvements may not provide the reassurance that we have finally reached true equilibrium conditions so that results do not depend on some arbitrary decision on the number of iterations or a similar measure. We will also discuss a more rigorous approach in the next chapter.

10.8 Practical Considerations

The assignment sub-model is critical to the implementation of the whole transport modelling package. However, in contrast with the other classic sub-models, there is no standard calibration procedure to make sure the assignment stage reproduces observations as closely as possible. The most likely candidate for external validation of the model is the use of traffic or cordon counts. The following procedure seems applicable to all kinds of assignment packages, including public-transport and equilibrium methods as discussed in the next chapter.

Goodness of fit for assignment. The assignment stage is critical in that it is relatively simple to cast doubts on the quality of a model because it does not reproduce a particular observation, perhaps flows on a link that is well-known to the decision-maker. There are many ways to present the quality of an assignment run for a particular time period. Most of them are based on comparing modelled with observed flows, either at link level or on one or more screen lines. It is good practice to start by plotting observed versus modelled link flows and fitting the best straight line to them (Figure 10.16).

One would also show the corresponding R^2, the slope, and intercept. The closer the slope is to 1 the better (here it is good at 0.97); and the closer the intercept is to zero, the better. The cloud of points and the parameters above will help identify any bias in the results.

Transport authorities in different countries adopt different indicators and thresholds to judge the overall fitness of an assignment model. These often take the form of measures of differences between observed and modelled flows; for example, the root mean squared error (RMSE) in absolute or percentage terms. A difficult issue is always how to consider variations in flows in a network

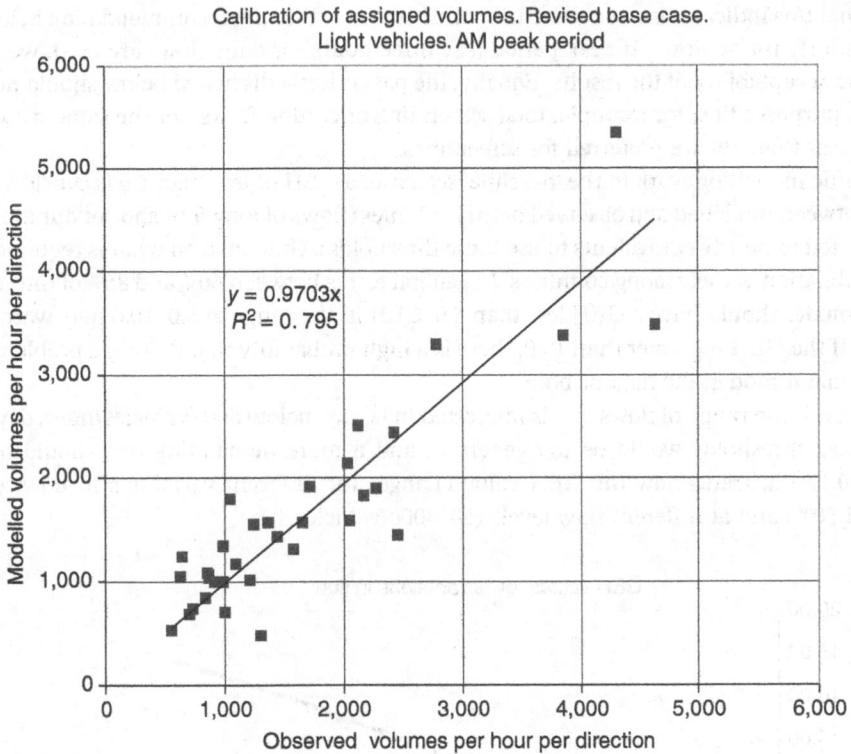

Figure 10.16 Observed versus modelled link flows and best fit straight line

when some of them are very large (say on a motorway) and some offer lower flows, for example on local links. The GEH 'statistic' which gets its name from Geoffrey E. Havers (who proposed it in the 1970s while working as a transport planner in London), has been suggested to overcome this difficulty. Although its mathematical form is similar to a chi-squared test, it is not a true statistical test. Rather, it is an empirical formula that has proven useful for a variety of traffic analysis purposes.

The GEH measure is defined as:

$$\text{GEH} = \sqrt{\frac{(O_i - E_i)^2}{0.5 \cdot (O_i + E_i)}} \tag{10.21}$$

where O_i are observed values and E_i modelled or estimated values for one variable i.

This may be seen as the square root of the product of the absolute difference $(O-E)$ and the relative difference $(O-E)/0.5\,(O+E)$. The reason for using this statistic is the inability of both (absolute and relative differences) to cope with a wide range of flows. For example, an absolute difference of 100 pcu/h may be considered large if the flows are of the order of 200 pcu/h but unimportant for flows of the order of several thousand vehicles an hour. Equally, a 10% error in 100 pcu/h may not be important whereas a 10% error in, say, 6000 pcu/h might mean the difference between building an extra lane or not.

Generally speaking, the GEH statistic will be less sensitive to these problems, as a modeller would probably feel that an error of 20 in 100 would be roughly as bad as an error of 90 in 2000, and both would have a GEH of around 2.

Note that this indicator is not a-dimensional. This means that the recommendation below applies only to hourly traffic flows. If peak period (say three hours) or daily flows are used, we will exaggerate the acceptability of the results. Equally, the pass criteria discussed below should not be used for other purposes like, for example, total screen-line or cordon flows, for the same reason; differences of less than 5% are preferred for screenlines.

For traffic modelling work in the 'baseline' scenario, a GEH of less than 5.0 is considered a good match between modelled and observed hourly volumes (flows of longer or shorter durations should be converted to hourly equivalents to use these thresholds). Guidance on what is required for good model validation varies among countries. In general terms between 60% and 85% of the volumes in a traffic model should have a GEH less than 5.0. GEH in the range of 5.0–10.0 may warrant investigation. If the GEH is greater than 10.0, there is a high probability that there is a problem with the travel demand model, the data or both.

However, if the range of flows one is interested in is, say, below 500 (vehicles/hour/day or whatever), these thresholds would be too generous, and a more demanding one should be sought. Figure 10.17 illustrates how the GEH value changes for different variations in flows (5%, 10%, 20%, and 30%) and at different flow levels (50–4000 vehicles/h).

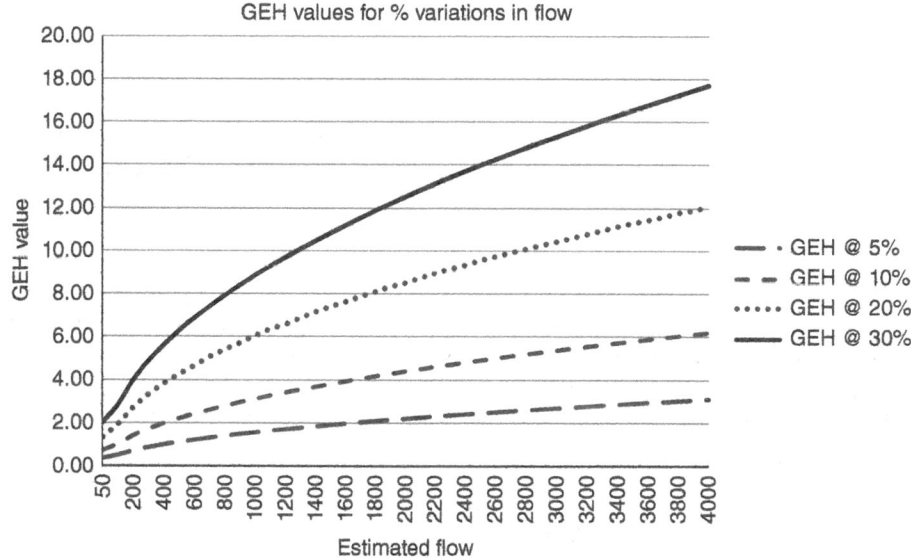

Figure 10.17 GEH indicator values for different flow levels and percentage variations

Another indicator that must be checked is the model's ability to reproduce the travel times observed during the travel time surveys. The best way of presenting these are to plot observed and modelled cumulative times along the routes travelled during the survey.

Check and double-check the network. This is the most important source of error in traffic assignment. There are numerous potential errors in coding a network: the omission of links and nodes previously thought irrelevant, miscoding of distances, use of wrong directions, missing turning-movement penalties, specification of incorrect capacities and time-flow curves, etc. Good software packages will flag many of these errors on input; the use of graphic displays of the network and, even better, graphic editing of networks, is very important. It is easy to underestimate the time taken to input and check a network for a particular study. Any facility likely to speed up and increase the accuracy of the process is worth many professional days.

An additional method for checking a network consists in loading a unit trip matrix (i.e. with a single trip per cell) and then checking modelled flows. This will facilitate the identification of unused links (perhaps because they were coded with too slow speeds, or too long distances) and also heavily used ones; these serve as pointers for coding errors. The plotting of minimum path trees is also a useful aid for network checking. Odd shortest routes and unreachable nodes will also help to identify sources of problems. Improve the connection of centroids to the network if some routes look too strange. Keep in mind, however, that under congested conditions other routes will become attractive and be used. In the case of detailed (microsimulation) assignment models, there will be additional sources of problems as more local data are needed. The same applies to public-transport assignment, where the connection to bus stops or stations is critical for good route choice representations; the same is true of interchange facilities, frequencies, and speeds. The basic rule is: before going to the next step in model fitting, make sure all the observable (measurable) data are correctly represented in the network. Check connectivity first, then link attributes, and then detailed data like saturation flows, signal timings, and so on.

Fit the generalised cost function. Assign weights to time, distance, and any other variables included in it (link status, scenic quality, etc.). Use the GEH measure to assess goodness of fit. This can be applied to cordon counts or to groups of traffic counts on parts of the network thought to be most critical, say primary and secondary roads. The value of the statistic for the whole network also provides an indication of overall fit.

Usually, a good starting point is to assume that time alone explains route choice: use this assumption, run a complete assignment, and then calculate the statistics above. Then begin increasing the weight attached to distance (or other factors) and recalculate the statistics so that the choice of parameters that produces the best fit can be made. One must resist the temptation of improving the fit in one step by trivial alteration of link speeds or turning penalties at this stage, as this reduces the value of the model for forecasting purposes. True errors discovered at this stage must, of course, be corrected; the model should then be re-run for other generalised cost coefficients as well.

Note that the statistic proposed above gives greater weight to a given absolute difference at low flow levels than at high ones. If this is undesirable, collect it for different flow ranges. The percentage of over- and under-estimations of flows can give some indication of bias, which, if present, should be investigated more thoroughly. Note too that if the link capacities are well identified and coded and there is considerable congestion, then equilibrium assignment will tend to produce a good fit with observed flows, even to the extent of masking a few errors in other sub-models.

There may be evidence suggesting that different weights should be applied to different user classes, for example, that heavy lorries are more sensitive to distance and gradient than cars. In that case, the classes should be assigned separately onto the network using their best coefficients in each case.

In the case of public-transport assignment, the relative weights of walking, waiting, and in-vehicle time are part of this calibration process. Interchange penalties play a similar role and provide an additional element for making the model more realistic. Passenger counts at interchanges and stops should be considered separately for the calibration of these weights. An approach similar to that of Suh et al. (1990) may well prove advantageous in fitting generalised cost functions once all other errors have been reduced to a minimum.

Fine-tune the assignment model. This involves finding the best dispersion parameters for stochastic assignment models. Particular care should be exercised at this stage, as depending on the implementation, these parameters may have different interpretations and even dimensions. The documentation of the programs should be examined in detail to guide us in this task.

Detailed urban assignment models like those described in the next chapter offer additional opportunities for fine-tuning. These make them powerful but may also inadvertently hide more fundamental errors in coding. Examples of this type are the fine-tuning of gap acceptance parameters at some junctions, the representation of opposed turning movements at traffic signals, and so on. Particular care should be taken to make sure these modifications correspond to actual traffic engineering conditions on the ground and not to fudge factors included simply to improve the fit of the model.

It must be recognised that no assignment model will ever reproduce the observations exactly. There will always be variability in the traffic counts themselves, errors in the trip matrices used, and a proportion of the actual route choice behaviour that will remain unexplained. What matters, however, is that the resulting costs are as accurate as possible and that the model rests on a sound basis to compare alternative tactical or strategic schemes as required.

Exercises

10.1 The road network represented in Figure 10.18 links two residential areas A and B with two major shopping centres L and M. Travel times between nodes are depicted in minutes, and all links are two-way. Assume first that the costs of these links do not depend on traffic levels.

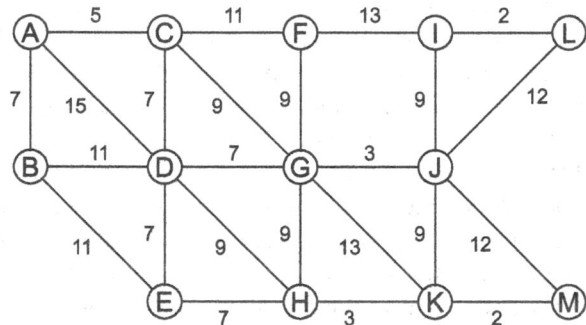

Figure 10.18 Simple network for Exercise 10.1

a) Use a systematic procedure to find the quickest routes between origins A and B and destinations L and M; calculate the corresponding travel times.

b) During a Saturday morning peak hour, the numbers of vehicle movements from A and B to L and M are as follows:

$$A - L = 600 \qquad A - M = 400$$
$$B - L = 300 \qquad B - M = 400$$

Estimate the traffic flow on each link during this period.

c) Consider now that travel time on each link increases by 0.02 of a minute for each vehicle/ hour of flow. Use an incremental loading technique to obtain a capacity-restrained set of flows. Calculate final travel times for each O–D pair.

d) Use an iterative loading procedure to obtain flows and costs under the conditions (c) above.

10.2 A study area contains two residential zones A and B and three workplace zones J, K, and L. The zones are connected by a road network as shown in Figure 10.19, which also depicts travel costs in either direction; these are independent of the traffic flows.

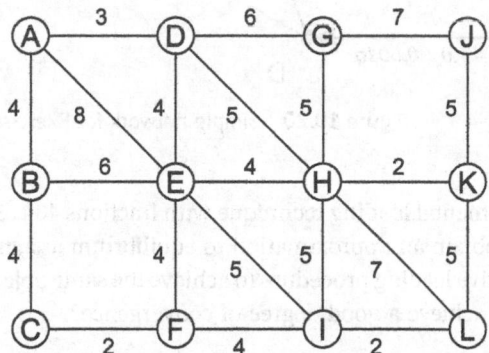

Figure 10.19 Simple network for Exercise 10.2

a) Use a systematic procedure to find the cheapest routes from nodes A and B to destinations J, K, and L and obtain the matrix of travel costs **C**.

b) The total number of trips originating and terminating in each zone during the morning peak are shown in Table 10.4:

Table 10.4 Total trip origins and destinations for Exercise 10.2

Origin	Trips	Destination	Trips
A	1000	J	700
B	2000	K	1000
		L	1300

Run an origin-constrained Gravity model in which the deterrence function is proportional to exp $(-0.1\ C_{ij})$ and obtain a trip matrix. Use this matrix to calculate flows on all the links of the network.

c) Run a doubly constrained Gravity model with the same type of deterrence function and obtain a new trip matrix and link flows. Compare your results of (b) and (c).

10.3 Consider the simple network in Figure 10.20 where there are 100 vehicles/hour travelling from A to X and 500 from B to X. The travel time versus flow relationships are depicted in minutes and the flow q in vehicles/hour.

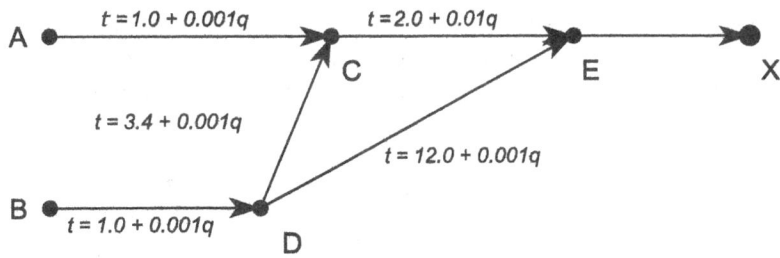

Figure 10.20 Simple network for Exercise 10.3

a) Use an incremental loading technique with fractions 40%, 30%, 20%, and 10% of the total demand to obtain an approximation to equilibrium assignment.
b) Use an iterative loading procedure to achieve the same objective. How many iterations do you need to achieve a good degree of convergence?

11

Dynamic Assignment

11.1 Introduction

In Chapter 10, we introduced static assignment techniques for both private vehicles and public transport. We also identified some of their limitations, among them the fact that static assignment ignores the dynamic nature of traffic. This chapter discusses some ways of overcoming some limitations of classic assignment; these include treatment of travel time reliability, improvements to the estimation of junction delays, and the more radical replacement of static assignment with more realistic dynamic traffic assignment (DTA).

11.2 Travel Time Reliability

The variability and unreliability of travel time in congested urban areas have become significant issues for many types of trips. As traffic increases in heavily loaded networks, the travel time required to perform a particular journey becomes particularly difficult to estimate. Under these circumstances, users may have to make large time allowances to avoid missing a plane or a business meeting; for other activities, they may just accept the penalty associated with unpredictable delays. It has been argued that one of the key benefits of pricing road space is to increase the reliability of journey times and, therefore, produce significant resource savings.

It is possible to use stated preference (SP)/revealed preference (RP) surveys and data to develop appropriate generalised cost functions incorporating these effects. In this way, one could develop a subjective value of travel time reliability. This requires a measure of reliability, for example, the expected standard deviation of travel time σ_t or the expected coefficient of variation of travel time. Note the emphasis on *expected* or *subjective* measures of travel time variability.

An equally important requirement is to develop models that link travel time variability to congestion and supply conditions (i.e. incident management facilities, redundancy in the network, etc.). This is a less well-researched area for a number of reasons. First, data collection is often expensive in this field, as one would require repeated journeys (i.e. same departure time, same origin, and destination) over a large number of days to pick up systematic and random variations; this has to be repeated several times of the day for several origin–destination (O–D) pairs. Second, the supply models must be reasonably straightforward for their use in large strategic models or sensitive to key policy instruments (e.g. new traffic control measures, variable message signs/route guidance) for detailed tactical modelling.

Modelling Transport, Fifth Edition. Juan de Dios Ortúzar and Luis G. Willumsen.
© 2024 John Wiley & Sons Ltd. Published 2024 by John Wiley & Sons Ltd.
Companion website: www.wiley.com/go/ortuzar5e

Willumsen and Hounsell (1998) reported a general study for use in strategic models in the context of road pricing. They used a significant number of observations in a congested network (London) and extended their value by doing simulation runs for over 2000 O–D pairs and over a large number of 'days'. As independent variables, they selected *actual journey time* (JT), *free flow travel time* (FFTT), and a *congestion index*, defined as their ratio (i.e. CI = JT/FFTT).

Several models were calibrated to estimate the standard deviation of travel time under different congested conditions. The authors recommended the following formulation as offering a good compromise between simplicity and realism:

$$\sigma_t = 0.9\, \text{FFTT}^{0.87}(\text{CI} - 1) \tag{11.1}$$

In practical terms, this equation simply relates the standard deviation of travel time with network conditions and is relatively insensitive to trip length; therefore, it offers a promise of adaptation to environments different from London. One advantage of this treatment is that journey time variability can be estimated after assignment and then incorporated into other choice models (e.g. time-of-day, mode, and destination choice).

Work for the UK Department of Transport used distance d and the coefficient of variation ($CV = \sigma/\text{mean}$) of travel time and recommended the following formulation for a particular modelling period (Department of Transport 2007):

$$CV = \alpha\, CI^{\beta} d^{-0.39} \tag{11.2}$$

where α was in the range 0.1–0.19 (expected value 0.16), and the range for β was 0.50–1.37 (expected value 1.02).

This variation is due to the nature of the links and junctions on each route. Note that both formulations were originally developed on the basis of route travel time variability. In practice, it is easier to handle link travel time variability as a component of generalised costs in assignment.

Interest in travel time variability has led to research into the distribution of travel times. Statistical analyses have found that the travel time distributions are asymmetric and significantly skewed to the right, with the Log-normal model being the most recommended for its good fit and relative simplicity (Clark and Watling 2005; Hollander and Liu 2008). Taylor and Susilawati (2012) suggested later that the Burr distribution offers some advantages because of its ability to describe the long tails found in observed distributions of travel time in urban areas. Ma et al. (2016) used this distribution to model travel time variability of bus services.

11.3 Junction Interaction Methods

Classic assignment methods often use the simplification that delays on a link depend only on flow of the link itself; this is useful to set a straightforward traffic assignment problem convergent to a unique solution. However, this assumption may not be realistic enough for congested urban areas. If one considers the route choice and assignment problems in greater detail, one should search for better delay models and a better treatment of dynamic problems. In addition, there is a need to consider the interaction between traffic control and route choice and to treat different vehicle classes separately.

Different software packages offer facilities to represent delays at different types of junctions, from simple 'give way' to signal-controlled junctions and more complex roundabouts, signalised or not. Modelling them requires a more thorough understanding of traffic engineering, and this is often restricted to the most critical junctions in a model. Junction models may be just simple fixed delay penalties, including bans, applied to opposed turning movements, or more realistic implementations that consider the interaction between the different traffic streams at a junction; from this treatment, the models will normally generate an updated volume delay formulation for each turning movement.

So far, we have considered traffic as a continuous variable operating under steady-state conditions. But in reality, traffic is made up of discrete entities (vehicles) that form queues at junctions and bottlenecks in urban areas. If a particular assignment model puts more traffic on a junction than its capacity, the flows downstream will probably be overestimated; this happens because the junction will actually put an effective upper limit, not recognised by the model, and the modelled flows downstream will be greater than the actual flows. Therefore, potential routes using these links may well be ignored by the model. Double counting of delays and missing of potential routes are perverse effects of this simplistic treatment of traffic delays.

Two types of improvement are needed here: first, to consider the physical nature of queues at junctions and their effects in limiting traffic downstream; and second, the need to model the time-dependent nature of queues at junctions as demand builds up and decays before, during, and after the peak period. The second problem can be treated using time-dependent queueing models as proposed by Kimber and Hollis (1979). These approaches model the way in which queues and delays change over time, as traffic demand evolves, and even allow for the presence of queues at the start of a time period of interest.

The first problem requires a physical model of queues, and this can be undertaken through a simple conversion of vehicles queued into queue length or, in more detailed models, through the simulation of the actual queues. A critical issue is the ability of these models to represent the situation where a queue begins to block back an upstream junction and the additional delays this generates to other streams.

Although it is recognised that these developments, from junction models to more advanced queue modelling, are likely to improve assignment accuracy, it is still possible to find models that retain just a node to represent each junction and ignore turning movement prohibitions. An in-depth review by Caliper Corporation (2015) focussed on five different transport models with zoning systems ranging from 3800 to almost 6000 traffic zones and from 30 000 to 74 000 links. They found that two of these models used neither turn prohibitions nor penalties, even in an area where over 9000 turning movement bans were in operation. Nowadays, it is much easier to code networks and identify turning movement bans using digital maps from different sources.

11.4 The Dynamic Nature of Traffic

11.4.1 Delays over Time and Space

There are three common assumptions used in all assignment models that have proved helpful in devising more rigorous mathematical programming formulations and determining the conditions for a unique equilibrium, as discussed in Chapter 12:

- travellers have full knowledge of the generalised costs of travelling on every link and route in the network (perfect information assumption);
- delays on links can be described using a function of flows on that link alone (separability assumption);
- the demand and flows during a modelled period do not change over time (steady-state assumption).

In congested real-world networks, none of these assumptions is actually realistic. Even with the best GPS based guidance, knowledge about travel costs on any network requires perfect foresight about the future costs when the traveller eventually reaches more distant parts of the network. Stochastic assignment goes some way to address this issue, but the introduction of time-dependent delays makes it more difficult to handle.

In a static assignment model, inflow to a link is always equal to the outflow: the travel time simply increases as the flow increases. The volume on such links may increase indefinitely and exceed the physical capacity (in pcu per hour) of the road, as represented by a volume-to-capacity (V/C) ratio greater than one. This is a useful approximation to provide fast assignment routines but in practice it is not possible to exceed the physical capacity of a stretch of road. Thus, any flow with V/C > 1 can only be considered as a demand volume and not as an effective flow that could be contrasted with traffic counts.

Outflow from a link may be less than its inflow for various reasons, for example:

- oversaturation of the traffic signal-controlled junction at the link exit;
- oversaturation of other types of exit junctions: roundabouts, stop signs, and so on;
- merging of two lanes into one or when entering a motorway or freeway;
- weaving of two flow streams near junctions.

Traffic becomes congested and queueing takes place at the end of a link because link inflow is greater than link outflow. Whenever this happens, the risk of queues blocking back the upstream junctions becomes real; unfortunately, this congestion spill-back is rarely represented in static assignment models.

A more subtle limitation of static assignment is that it respects the FIFO (first-in, first-out) rule. This means that all vehicles travelling on the link experience the same travel time. This implies that there is no overtaking between vehicles, in particular if they exit the link as part of different turning movements. This does not happen in reality when one of the movements is opposed (or facing a blocked exit) and the others are not.

As demand varies over time, peak flows propagate during the peak period. One can try to handle this by modelling a short time period, say just the peak hour, where demand can be considered to be more or less uniform. But even then, real capacity constraints in the network create dynamic conditions that standard speed–flow curves cannot handle correctly. Conventional flow-delay curves, as those discussed in Section 10.1.3, allow flows to exceed capacity, and normal equilibrium assignment assumes that all the demand in a time period reaches its final destination. Reality is different; real capacity constraints generate dynamic queues at bottlenecks that prevent all traffic from reaching their destinations during the modelled period. Moreover, these queues remain and grow until demand declines below capacity when they start to clear.

Consider, for example, a road that provides access to a town centre, and that most drivers will like to reach it around 9:00 AM. Figure 11.1 represents an idealised diagram of traffic along this road, starting from a place 60 minutes away from the town centre.

As can be seen, traffic at each 30-minute time slot is different from the average conditions assumed in any classic assignment model. In a real network, with more entry and exit points, real traffic is more like a series of *surges* or *waves* that interact at junctions and at bottlenecks, generating a different set of optimal routes depending on the time of day and how far ahead the user is able to estimate delays on alternative routes.

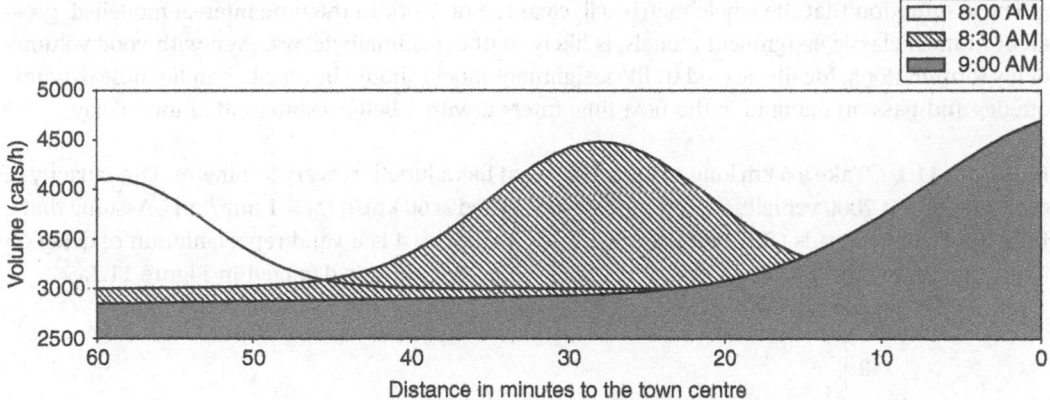

Figure 11.1 Simplified traffic volumes on a link

However, Figure 11.1 is an oversimplification for illustration purposes. In reality, the waves may be fatter and, as more traffic joins the main road to the town centre, the volumes will increase faster than suggested in the figure. The same phenomenon takes place in public transport systems. The most congested section of an underground will be closer to the most desirable destination. One consequence of this is that care must be taken when allocating trips to a particular time interval (say, 8–9 AM peak): a different assignment will be achieved if trips are allocated according to the time they start, the time they arrive at a destination, or an average of the two. When public transport congestion is an issue, it is advisable to allocate trips according to the time of arrival at their destination, as this is where the most severe congestion usually takes place.

Moreover, a different set of conditions will be generated if, for example, there is a bottleneck limiting capacity to some 4000 cars/hour at a distance of 30 minutes from the town centre. In this case, the time profile of traffic would look more like that in Figure 11.2. Not all traffic will be able to get through the bottleneck in one go; queues will build up that will be cleared once demand falls below the 4000 cars/hour limit.

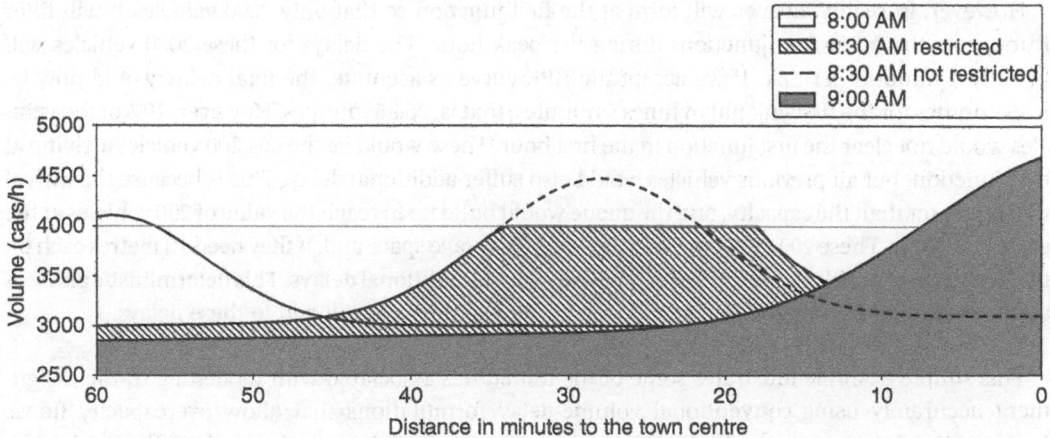

Figure 11.2 Simplified traffic volumes on a link with a capacity bottleneck

The assumption that the whole matrix will clear the network in the time interval modelled, prevalent in most classic assignment models, is likely to underestimate delays, even with good volume-delay formulations. Ideally, a good traffic assignment model should be able to handle these dynamic queues and pass on demand to the next time interval with a better estimation of total delay.

Example 11.1 Take a 5 km long road corridor that has a junction every kilometre. The capacity of each junction is 2000 vehicles/h, and the free flow speed is 60 km/h ($t_0 = 1$ min/km). Assume that a Bureau of Public Roads (BPR) function with $\alpha = 4$ and $\beta = 4$ is a valid representation of delay on these five links. The flow-travel time relationship for each link is depicted in Figure 11.3.

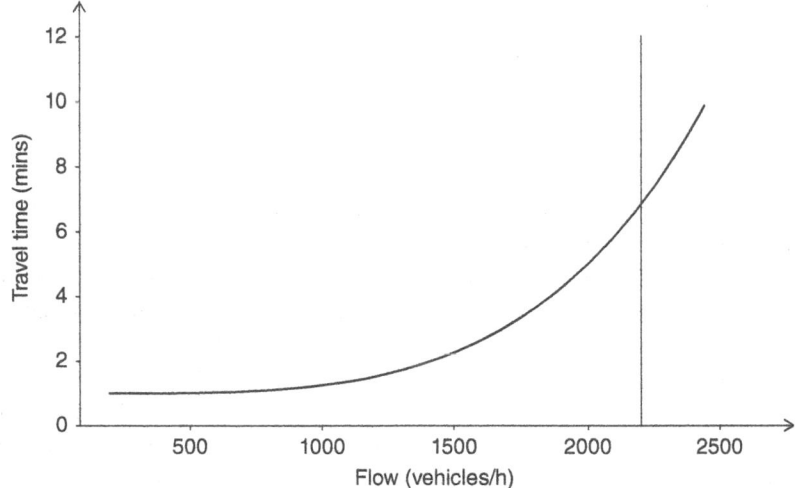

Figure 11.3 BPR curve for 1 km links in Example 11.1

Consider now that a greater capacity road feeds onto this corridor with 2200 vehicles during the peak hour. Using a conventional traffic model based on BPR curves, the time at each link (junction) would be 6.86 minutes. As the BPR curves accept flows above capacity, the total time spent on that 5 km would be 34.30 minutes.

However, in reality, queues will form at the first junction so that only 2000 vehicles/h will filter through to the other four junctions during the peak hour. The delays for these 2000 vehicles will be only 5 minutes per link. If we accept the BPR curve as accurate, the total delay would now be 6.86 minutes for the first link plus 4 times 5 minutes, that is, 26.86 minutes. However, 10% of the vehicles would not clear the first junction in the first hour. These would be the last 200 vehicles arriving at that junction, but all previous vehicles would also suffer additional delay. This is because the arrival rate is greater than the capacity, and the queue would build up to reach the value of 200 vehicles at the end of the hour. These 200 queueing vehicles take up storage space and, if they need 10 metres each on two lanes, they will block the upstream junction causing additional delays. This deterministic analysis assumes regular arrivals and departures; randomness in arrivals will add to these delays.

This simple example illustrates some of the difficulties associated with modelling traffic assignment accurately using conventional volume-delay formulations that allow overcapacity flows. Improved assignment methods should consider the physical characteristics of traffic and handle issues like real capacity constraints, the storage capacity of links to handle queues, and queues remaining at the end of a modelling period and spilling over into the next time slice. Another problem that is gaining importance is the role of reliability in the estimation of travel times. This is, of course, central to time-critical journeys like those to catch a flight or attend an important meeting. How best to model this feature is also important in improving assignment models.

11.4.2 Average and Experienced Travel Times

The fact that delays occur at different times and places during a modelling period, say the AM peak, poses a significant modelling challenge. One can assume that drivers performing a recurrent journey will adapt to these conditions and choose routes in response to their experienced travel times and not the average times estimated using static assignment.

The differences between average and experienced travel times are illustrated in Figure 11.4, adapted from Chiu et al. (2011). In this example, we have a series of three sections of road with traffic joining them at each node. Six groups of drivers (A–F) join the first link at 10 minutes intervals. The values on the cells under each link number reflect the experienced travel time, in minutes, during each 10 minutes period.

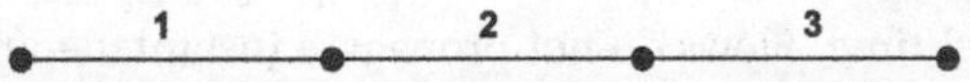

Group	Departure time	Travel time at link (mins)			Total time
		1	**2**	**3**	
A	07:00	10	10	10	50
B	07:10	10	20	20	60
C	07:20	10	20	20	65
D	07:30	20	20	20	?
E	07:40	20	20	30	?
F	07:50	20	20	35	?
Average		15	18	23	56

Figure 11.4 Example of a single link with departures every 10 minutes

Group A joins link 1 at 7:00 and takes 10 minutes to reach link 2. By then, the travel time in link 2 is 20 minutes; therefore, 20 minutes later, group A joins link 3 and takes 20 additional minutes to reach the final destination. The 'Total Time' column records the time taken for each of the first three groups to reach the end of link 3; the rest get there after 8:00. The 'Average' row shows the mean travel time as it would presumably be estimated in a static assignment model.

This difference between experienced and average travel times may well generate different routes for each of these groups because of their different departure times and the way delays take place in different locations and at different times; see, for example, the first exercise in this chapter.

As soon as one recognises that delays occur at different times in different parts of the network, two issues become critical:

- are the routes identified with a static assignment realistic enough? It may well be that other routes are preferred at certain times during the static assignment modelling period;
- some of the difficulties in calibrating a model for a large area may be the result of forcing a less than realistic static assignment on a set of counts that reflect these variations of delay in time and space over the modelling period.

The answer to these questions depends on the objectives of the model. Most strategic models, which are used to develop transport plans and assess the value of key projects, may accept the limited realism of static assignment to achieve the stable system equilibrium necessary to compare schemes. In other cases, a classic model is used before refining the design of an intervention using a more realistic dynamic assignment.

11.5 Dynamic Traffic Assignment (DTA)

11.5.1 General Requirements

There are some basic requirements for a truly DTA model. These have been identified by, among others, Heydecker and Addison (2005) as:

- Positivity. We are only interested in non-negative flows on links, paths, trip matrices, and costs;
- Conservation. The model must satisfy flow conservation requirements;
- FIFO. In real traffic, the first-in-first-out behaviour generally prevails (except at some junctions), and this must be maintained in the model if proper delays are to be estimated;
- Minimum travel time. Flows do not propagate instantaneously;
- Finite clearing time. There are no queues left at the end of the modelling period; infinite delays do not occur (as a standard queuing model might suggest);
- Capacity. There is a strict capacity constraint in the sense that actual flows cannot exceed it even for a short period of time;
- Causality. Delays are affected by what other vehicles do or have done in the past, not in the future.

These requirements lead to correct flow propagation and a consistent interrelationship between travel time and link outflow. Finite clearing time ensures that no travellers remain in the network indefinitely and that it returns to free-flow conditions after the study period. The causality requirement ensures that response follows stimulus.

Wardrop's user equilibrium (UE) principle of route choice can be extended to the dynamic problem as follows:

Under equilibrium conditions in networks where congestion varies over time, traffic arranges itself so that at each instant the costs incurred by drivers on those routes that are used are equal and no greater than those on any unused route.

If travellers choose not only route but also departure time, Wardrop's equilibrium expression can be further extended:

Under equilibrium conditions in networks where congestion varies over time and travellers can choose their time of travel, traffic arranges itself so that the total cost associated with travel on those routes that are used by travellers at the time when they are used, are equal and no greater than those on any route at a time when it is not used.

These two principles assume that travellers know and can anticipate future travel conditions along the journey.

It is possible to present the dynamic user equilibrium (DUE) problem, with or without choice of time of travel, in a closed form. However, practical methods for its solution often rely on modelling discrete time slices or time intervals. Therefore, a key element in the development of numerical methods for the solution of dynamic assignments is the transition from the continuous time formulation of the equilibrium conditions to a discrete-time formulation.

The numerical solution method will typically assign a calculated flow at time s, $T_{ij}(s)$, to a path p throughout a time increment $[s, s + \Delta s]$. It is important to use the flows and costs of that time interval to model equilibrium conditions. If the previous costs are used for assignment, the result will not represent the new traffic conditions.

This approach can be applied to a mathematical programming formulation discussed by Han and Heydecker (2006). In this case, the objective $Z(s)$ to be minimised is calculated at each incremental time interval using the flow assigned throughout that increment together with the costs $c(s + \Delta s)$ at the end of it. Although somewhat outside the scope of this book, we must mention that the variational inequality formulation developed initially by M.J. Smith (1979) and Dafermos (1980) provides an approach to modelling dynamic traffic assignments within the present framework. This was introduced for DTA by Friesz et al. (1993) and has been adopted by others since then.

11.5.2 Discretising Time in DTA

Finding a Dynamic User Equilibrium (DUE) solution for a set of time-varying link and route volumes, and travel times that satisfy the Wardrop's equilibrium for a given network and time-varying O–D demand pattern, is non-trivial. Each traveller's best route choice depends on congestion levels throughout the trip, and these in turn depend on the route choices and progress through the network of other trip makers who leave at different times. This interdependence means that solutions are found through an iterative process, starting from some initial set of route choices, and gradually improving them. A practical goal of many current DTA models is to find something close to equilibrium within a reasonable length of time.

Practical DTA models need to handle time at a finer level than static assignment models. Rather than one hour period with uniform flow levels across the network, a DTA model must consider the variability of demand over time and how flows and congestion evolve in different parts of the network. This is usually handled with some form of discretisation of time, that is, separate consideration of sub-hour time slices, often 10 or 15 minutes each. This, in turn, requires subdividing the trip matrices for these same time slices. To at least approach the traffic dynamics, any model should also be able to pass the queues from one time slice to the next at each link.

The issue of anticipating future traffic conditions beyond the time slice at the start of the trip becomes both important and difficult. An initial estimate of travel times on all links based on the departure time, what may be called 'instantaneous' travel times, may differ significantly from the travel times experienced by the users later in their journey, their 'experienced' travel time. A good anticipation of future travel conditions may be the result of experience from repeated trips or the use of a well-updated GPS navigation application. The DTA model should try to replicate this intelligence either fully (i.e. perfect information assumption) or at least partially.

Convergence to the equilibrium principles outlined above is desirable if one wants to compare performances. It is also a demanding task as there are multiple interactions in the model and across time. The typical definition of *total relative gap*, as we will see in Chapter 12, must be adapted to consider these separate time slices.

11.5.3 Micro- and Meso-Simulation

The discussion above suggests that to achieve a more realistic DTA model, it is necessary to have a better representation of traffic than hourly flows and volume delay formulations. Several approaches have been tried over the years to deliver practical ways to solve these problems. As computer power has increased, new and better software has led to more interesting and persuasive solutions. The approaches could be classified under the labels *meso-* and *micro-simulation*.

Meso-simulation models came first. One approach was to represent traffic using packets of vehicles released sequentially during a time period, as treating them one by one was not possible at the time. Each packet would then select the optimal route according to travel times obtained from a pre-loaded network in the first instance and from previous iterations later on. This was the approach followed by CONTRAM (Leonard and Gower 1982); the cost of using each path was calculated from cost–flow and queueing formulae, and the path costs were then updated. This process was iterated until a degree of convergence was achieved. A similar approach is followed in several commercial implementations of DTA.

Another approach was to use *platoon dispersion* formulations, such as those successfully used in TRANSYT (Robertson 1969), to represent the movement of vehicles and their interaction at different types of junctions. This is the approach used in SATURN (Hall et al. 1980) by dividing the period of interest into shorter time intervals, typically 10 or 15 minutes long. Each time interval is then treated as a steady-state assignment problem. This captures some of the effects of the build-up of congestion but still assumes that all vehicles in the same time interval are faced with the same set of costs.

SATURN combines a platoon-dispersion simulation module with a good equilibrium assignment module. The simulation module is based on the use of cyclic flow profiles to represent the movement of platoons of vehicles over a network, taking good account of the interaction of different flows at roundabouts, signal-controlled and priority junctions. It needs information about the volume of each movement (represented by a link) on the network to estimate capacity, queues, and delays. Therefore, an assignment model is required to load a trip matrix onto the network and to obtain an estimate of these flows. This is achieved through a separate assignment model that can perform Wardrop's selfish and stochastic user equilibrium assignments. The link between the two is through link volumes (from assignment to simulation) and through speed–flow relationships (from simulation to assignment), as depicted in Figure 11.5.

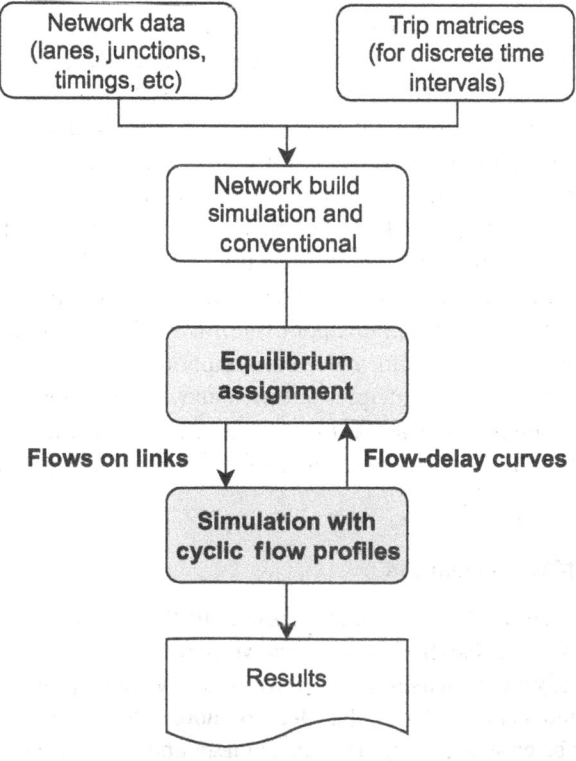

Figure 11.5 The simplified structure of SATURN

The simulation model is used, therefore, to generate suitable cost–flow relationships for the assignment problem. The cost–flow relationships for each link are produced in terms of the flow on the link itself and take the form of a polynomial:

$$C(V_a) = a_0 + a_1 V_a^n \tag{11.3}$$

However, as these relationships are calculated from the current simulation model, they consider the interaction and constraints generated by the flows on the other links in the network. In fact, several iterations of the simulation–assignment cycle must be performed before the whole process converges to a self-consistent set of flows and costs.

Improved computer power has meant that it is now possible to simulate the movements of vehicles individually, and this has generated a group of micro-simulation models. These models are based on a combination of traffic engineering relationships, car following, lane choice, and gap acceptance/merging models, including the treatment of pedestrians, motorcycles, and trams. Micro-simulation models are very powerful, and most of them include a visualisation module that produces good animations of traffic on the network. Micro-simulation models offer a large number of parameters for calibration, including some that relate to the driving culture of a city or country, for example, parameters for *aggression, anticipation,* and the variability of gap acceptance with queue length.

The visualisation/animation modules are useful in two main areas. First, they provide a useful environment to verify the reasonableness of the modelled traffic behaviour and, therefore, assist model calibration. Second, and this has been most valued, they are very persuasive tools to demonstrate problems and solutions to decision-makers, but herein lies a risk. Sometimes a poorly calibrated model may produce very persuasive animations and support solutions that may not be appropriate. Some of the best-known micro-simulation models include AIMSUN (www.aimsun. com), PARAMICS (www.paramics.com), and VISSIM (www.ptv.de). In general terms, these packages have a detailed simulation of traffic dynamics and delays but a less rigorous treatment of equilibration.

11.5.4 Equilibrium and Simulation

There seems to be a degree of conflict between a detailed and accurate treatment of the dynamics of traffic delays and equilibrium. In a congested and well-connected network, like those in urban areas, the cost of a link depends not just on the flow on that link but on all other flows in the network (albeit especially on those joining the same junction). The flow delay functions are, therefore, *non-separable* in the sense that they cannot be written as a function of the flow on the link alone, so we get:

$$C_a = C_a(V_1, V_2, ..., V_a, ..., V_n) \tag{11.4}$$

The strict condition for the convergence of this type of scheme requires that the delay on a link depends mainly on the flow on the link itself and less strongly on flows on the other links (Sheffi 1985). In practice, however, this condition is not satisfied, as delays at, for example, priority junctions and roundabouts, depend primarily on the flows on the links having priority (i.e. circulating and main-road flows, respectively).

For example, SATURN attempts to *diagonalise* the flow-delay formulations after simulation. If we fix all flows except that on link a and vary V_a between, say, zero and the capacity at a, then we can *calibrate* a cost–flow relationship that, in this iteration, depends only on V_a. We can then perform a conventional Wardrop's equilibrium assignment using, for example, the Frank–Wolfe algorithm, obtain a new set of flows on all links, and run the simulation program again.

Example 11.2 Consider the simple network in Figure 11.6 corresponding to two routes from an origin to a destination merging into one. The total flow is 100 vehicles from A to Z.

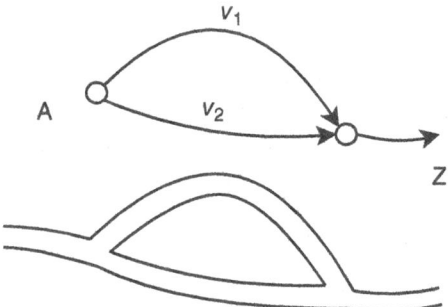

Figure 11.6 A simple network with a merge or give-way junction

Consider first the case in which both streams perform a merge operation; therefore, delays on each link also depend on flow on the other link. Assume now that the cost–flow relationships are as follows:

$$C_1(V_1, V_2) = 8 + 0.3\,V_1 + 0.2\,V_2$$
$$C_2(V_2, V_1) = 13 + 0.4\,V_2 + 0.2\,V_1$$

This can be solved to find a single equilibrium point at $V_1 = 83.5$ and $V_2 = 16.5$ with a minimum cost of 36.35. However, it is illustrative to show a range of values for V_1 and the corresponding link and total expenditure, as shown in Table 11.1:

Table 11.1 Variation of costs and expenditure with V_1

V_1	C_1	C_2	Expenditure
0	28.0	53.0	5300
10	29.0	51.0	4880
20	30.0	49.0	4520
30	31.0	47.0	4220
40	32.0	45.0	3980
50	33.0	43.0	3800
60	34.0	41.0	3680
70	35.0	39.0	3620
80	36.0	37.0	3620
83	36.3	36.4	3632
84	36.4	36.2	3636
90	37.0	35.0	3680
100	38.0	33.0	3800

As can be seen, the solution is a unique, stable equilibrium point. If some flow switches to link 2, then that link has increased delay and drivers will come back to the original route. The same is true if more traffic switches to link 1. The fact that the total expenditure is minimal at another point, approximately $V_1 = 75$, is another example of the difference between social and selfish user equilibrium.

Now consider a slightly different problem with the same type of network. Now the junction is of a give-way type for link 1; link 2 has right of way, and therefore its travel time does not depend on the flow on link 1. The new relationships are now:

$$C_1(V_1, V_2) = 8 + 0.1\,V_1 + 0.2\,V_2$$
$$C_2(V_2, V_1) = 20 + 0.05\,V_2$$

The same type of table can be used to illustrate possible solutions to this assignment problem, as shown in Table 11.2:

Table 11.2 New variation of costs and expenditure with V_1

V_1	C_1	C_2	Expenditure
0	28	25.0	2500
10	27	24.5	2475
20	26	24.0	2440
30	25	23.5	2395
40	24	23.0	2340
50	23	22.5	2275
60	22	22.0	2200
70	21	21.5	2115
80	20	21.0	2020
90	19	20.5	1915
100	18	20.0	1800

In this case, the solution $V_1 = 60$ and $V_2 = 40$ is not stable. A switch to link 1 will decrease costs on that link faster than on link 2, therefore precipitating the solution $V_1 = 100$ and $V_2 = 0$. However, a switch in the other direction, that is to link 2, has the opposite effect and increases costs on link 2 slower than on link 1, therefore, leading to another solution: $V_1 = 0$ and $V_2 = 100$. These two extreme solutions are stable, albeit not with equal costs by each route; however, these two are user equilibrium (UE) solutions as the costs of the paths not used are greater than the costs of the paths used. Any departure from these extreme points will result in new costs, pulling the solution back to the starting point. Note that the equations chosen are simple but not unreasonable. Observe too, that the equation for the non-priority flow shows that delay depends mainly on the flow on the priority link, therefore violating the requirement for a unique solution.

The fact that the solution $V_1 = 100$ is preferable because of its lower overall expenditure is only relevant in terms of network design. For example, we may wish to direct drivers to choose link 1 and ignore link 2. Without this advice, drivers may find either of the two extremes or even one on a particular occasion and the other the following day. Reality may be non-convergent to a stable equilibrium solution; good assignment models may fail to converge simply because they represent well this feature of reality.

SATURN and models like it, therefore, can only be said to provide a reasonable practical approximation to the ideal of Wardrop's equilibrium in congested urban areas. They normally offer practical indicators to estimate how close to a possible equilibrium the iterative process has been able to reach at any one stage. Meso- and micro-simulation models do represent however, the state-of-the-art in detailed traffic assignment for the design of traffic management and other schemes in urban areas.

Exercises

11.1 Consider the simple network in Figure 11.7 together with the travel times on each link at different times during the AM peak. Identify the shortest path from the Origin (O) to Destination (D) using static assumptions, that is, the average travel times on each link. Find also the minimum path for the group that departs at 07:00 using the experienced times.

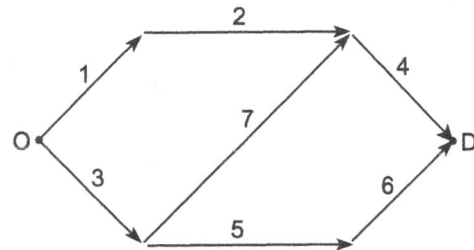

Group	Departure time	Travel time at link (mins)						
		1	2	3	4	5	6	7
A	07:00	10	10	20	10	20	10	20
B	07:10	10	20	10	30	20	20	30
C	07:20	20	30	10	50	10	30	30
D	07:30	20	30	10	60	10	10	30
E	07:40	30	40	10	40	10	10	10
F	07:50	20	40	10	40	10	10	10
Average		18	28	12	38	13	15	22

Figure 11.7 Simple network with times on links during AM peak

11.2 Identify the optimal path for vehicles leaving at 07:10. Are they different from the minimum paths at other times or using average travel times?

11.3 What would you recommend as the next step if difficulties in calibrating an assignment model for a large area were thought to be due to the dynamic nature of traffic?

12

Equilibrium

12.1 Introduction

In Chapter 10, we introduced assignment techniques for both private vehicles and public transport, and we extended these to consider the dynamic nature of traffic in Chapter 11. We identified three main reasons for the spread of routes between each origin-destination (O-D) pair that can be observed in practice. The first was the different objectives of drivers, for example, time or cost minimisers. The second was their imperfect perceptions about travel and link costs. The third was associated with congestion effects, and we used Wardrop's principles as a general framework to discuss this issue. Wardrop's first principle states that under congested conditions, drivers will choose routes until no one can reduce their costs by switching to another path; if all drivers perceive costs in the same way, this produces equilibrium conditions where all the routes used between two points have the same and minimum cost, and all those not used have equal or greater cost.

Congested assignment techniques, as discussed in Chapters 10 and 11, try to approximate to this type of equilibrium. But we saw that the proposed heuristic methods often failed to achieve true Wardrop's equilibrium; therefore, the problem deserved a better treatment. In section 12.2 we cast equilibrium assignment in a more rigorous mathematical programming framework. Although the section is restricted to problems where the delay on a link depends only on flows on the link itself, extensions to social and stochastic user equilibrium, and to congested public transport assignment are also discussed there. Section 12.3 extends the treatment of equilibrium to mode choice and trip distribution modelling; the objective is to make sure that the travel times implied in the generalised costs used to run these *demand models* are consistent with those generated during *assignment*. The naive iteration or feedback of these three sub-models does not lead naturally to equilibrium conditions, as it is somewhat akin to the hard speed-change methods used in congested assignment. Improved methods and practical considerations are included in this section.

12.2 Equilibrium

Here we discuss methods specifically designed to achieve traffic assignment solutions satisfying Wardrop's first principle. We shall follow a combination of intuitive and analytical arguments, but we shall not pursue the latter beyond what is necessary to understand and use equilibrium assignment techniques; readers interested in the more theoretical aspects of equilibrium assignment are directed to the seminal book by Sheffi (1985) or the more recent text by Bell and Iida (1997).

Modelling Transport, Fifth Edition. Juan de Dios Ortúzar and Luis G. Willumsen.
© 2024 John Wiley & Sons Ltd. Published 2024 by John Wiley & Sons Ltd.
Companion website: www.wiley.com/go/ortuzar5e

In what follows, we seek first to establish a more formal formulation of the assignment problem, using mathematical programming, and then we explore its properties and the solution methods that can be used to solve it; this often involves some kind of iterative method and, therefore, the issue of *degree of convergence* to the right solution is important. Finally, we look at some practical issues and extensions to the problems we have considered.

12.2.1 A Mathematical Programming Approach

Consider first some of the properties of Wardrop's selfish equilibrium, in particular, that all routes used (for an O-D pair) should have the same (minimum) travel cost and that all unused routes should have greater (or at most equal) costs. This can be written as:

$$c_{ijr} \begin{cases} = c_{ij}^* & T_{ijr}^* > 0 \\ \geq c_{ij}^* & T_{ijr}^* = 0 \end{cases}$$

where $\{T_{ijr}^*\}$ is a set of path flows that satisfies Wardrop's first principle, and all the costs have been calculated after the $\{T_{ijr}^*\}$ have been loaded. In this case, the flows result from:

$$V_a = \sum_{ijr} T_{ijr} \cdot \delta_{ijr}^a \tag{12.1}$$

where δ_{ijr}^a is 1 if path r between i and j uses link a and zero otherwise. The cost along a path can be calculated as:

$$C_{ijr} = \sum_a \delta_{ijr}^a \cdot c_a(V_a^*) \tag{12.2}$$

Although Wardrop presented his principles in 1952, it was not until four years later that Beckman et al. (1956) proposed a rigorous framework to express them as a mathematical programme; then, it took several more years before suitable algorithms for practical implementations were proposed and tested. Readers interested in the history of these developments should read the excellent book by Boyce and Williams (2015).

The mathematical programming approach expresses the problem of generating a Wardrop assignment as one of minimising an objective function subject to constraints representing properties of the flows. The problem can be written as:

$$\text{Minimise } Z\{T_{ijr}\} = \sum_a \int_0^{V_a} C_a(v)dv \tag{12.3}$$

subject to

$$\sum_r T_{ijr} = T_{ij} \tag{12.4}$$

and

$$T_{ijr} \geq 0 \tag{12.5}$$

The objective function corresponds to the sum of the areas under the cost–flow curves for all links in the network. We will attempt to show below why this is a sensible objective to minimise to obtain Wardrop's equilibrium, but first, we must consider the general properties of this mathematical programme.

The two constraints (12.4 and 12.5) were introduced to make sure that we work only on the space of solutions of interest, that is, non-negative path flows T_{ijr} making up the trip matrix. The role of the second constraint (non-negative trips) is important but not essential at this level of discussion of the problem. The interested reader is referred to Sheffi's book or to some of the classic papers on the topic like Fernández and Friesz (1983) and Florian and Spiess (1982).

It can be shown that the objective function Z is convex as its first and second derivatives are non-negative:

$$\frac{\partial Z}{\partial T_{ijr}} = \frac{\partial}{\partial T_{ijr}} \sum_a \int_0^{V_a} C_a(v) \, dv$$

$$= \sum_a \frac{d}{dV_a} \left(\int_0^{V_a} C_a(v) \, dv \right) \frac{\partial V_a}{\partial T_{ijr}}$$

but from (12.1)

$$\frac{\partial V_a}{\partial T_{ijr}} = \delta_{ijr}^a$$

Now, as V_a only depends on T_{ijr} if the path goes through that link,

$$\frac{d}{dV_a} \int_0^{V_a} C_a(v) \, dv = C_a(V_a)$$

therefore,

$$\frac{\partial Z}{\partial T_{ijr}} = \sum_a C_a(V_a) \cdot \delta_{ijr}^a = c_{ijr} \tag{12.6}$$

and the second derivative of Z with respect to the path flows is:

$$\frac{\partial^2 Z}{\partial T_{ijr}^2} = \frac{\partial}{\partial T_{ijr}} \sum_a C_a(V_a) \cdot \delta_{ijr}^a$$

$$= \sum_a \frac{dC_a(V_a)}{dV_a} \cdot \frac{\partial V_a}{\partial T_{ijr}} \delta_{ijr}^a \tag{12.7}$$

$$= \sum_a \frac{dC_a(V_a)}{dV_a} \delta_{ijr}^a \cdot \delta_{ijr}^a$$

This expression is greater than or equal to zero only if the derivative of the cost–flow relationship is positive or zero. This is a general requirement for convergence of Wardrop's equilibrium to a unique solution. The meaning of this condition is that the cost–flow curve should not have sections where costs decrease when flows increase.

As the problem identified in (12.3)–(12.5) is a constrained optimisation problem, its solution may be found using a Lagrangian method. The Lagrangian can be written as:

$$L\left(\{ T_{ijr}, \phi_{ij} \} \right) = Z(\{ T_{ijr} \}) + \sum_{ij} \phi_{ij} \left[T_{ij} - \sum_r T_{ijr} \right] \tag{12.8}$$

where the ϕ_{ij} are the Lagrange multipliers corresponding to constraints (12.4).

Taking the first derivative of (12.8) with respect to ϕ_{ij}, one obtains, of course, the corresponding constraints. Taking the derivative with respect to T_{ijr} and equating it to zero (for optimisation), one has:

$$\frac{\partial L}{\partial T_{ijr}} = \frac{\partial Z}{\partial T_{ijr}} - \phi_{ij} = c_{ijr} - \phi_{ij}$$

Here, we have two possibilities with respect to the value of T^*_{ijr} at the optimum. If $T^*_{ijr} = 0$, then:

$$\frac{\partial L}{\partial T_{ijr}} \geq 0 \quad \text{as the function is convex}$$

If $T^*_{ijr} \geq 0$, then:

$$\frac{\partial L}{\partial T_{ijr}} = 0$$

This can be translated into the following conditions at the optimum:

$$\phi^*_{ij} \leq c_{ijr} \text{ for all } ijr \text{ where } T^*_{ijr} = 0$$
$$\phi^*_{ij} = c_{ijr} \text{ for all } ijr \text{ where } T^*_{ijr} > 0$$

In other words, ϕ^*_{ij} must be equal to the costs along the routes with positive T_{ijr} and must be less than (or equal to) the costs along the other routes (i.e. where $T_{ijr} = 0$). Therefore, ϕ^*_{ij} is equal to the minimum cost of travelling from i to j: $\phi^*_{ij} = c^*_{ij}$.

In this way, the set of T^*_{ijr} that minimises (12.7) has the following properties:

$$c_{ijr} \geq c^*_{ij} \text{ for all } T^*_{ijr} = 0$$

$$c_{ijr} = c^*_{ij} \text{ for all } T^*_{ijr} > 0$$

Therefore, the solution satisfies Wardrop's first principle.

Example 12.1 Consider, again, the town-centre/bypass problem of Example 10.4. Figure 12.1 shows the cost–flow relationships, and the shaded area is the objective function that we want to minimise. Of course, one way to minimise this area is to have no flow, that is: $V_b = V_t = 0$, but this solution is not only trivial but also of little interest.

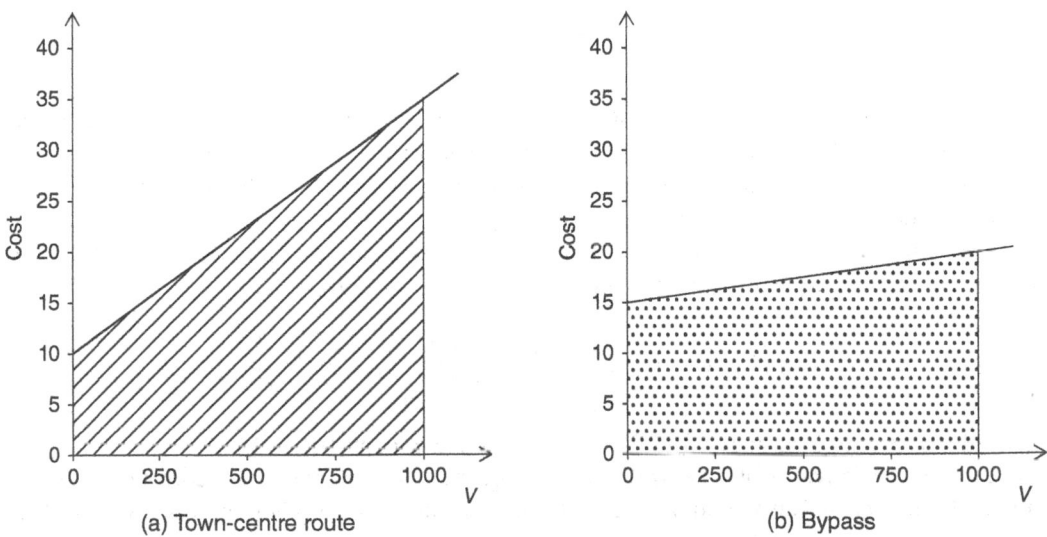

Figure 12.1 Two cost–flow relationships for bypass–town centre problem

What we want is a solution that satisfies the total demand (2000 vehicles), and this is shown in Figure 12.2, where the two cost–flow functions are now displayed with the X-axis running in opposite directions and separated by the total flow that must be split between the two routes.

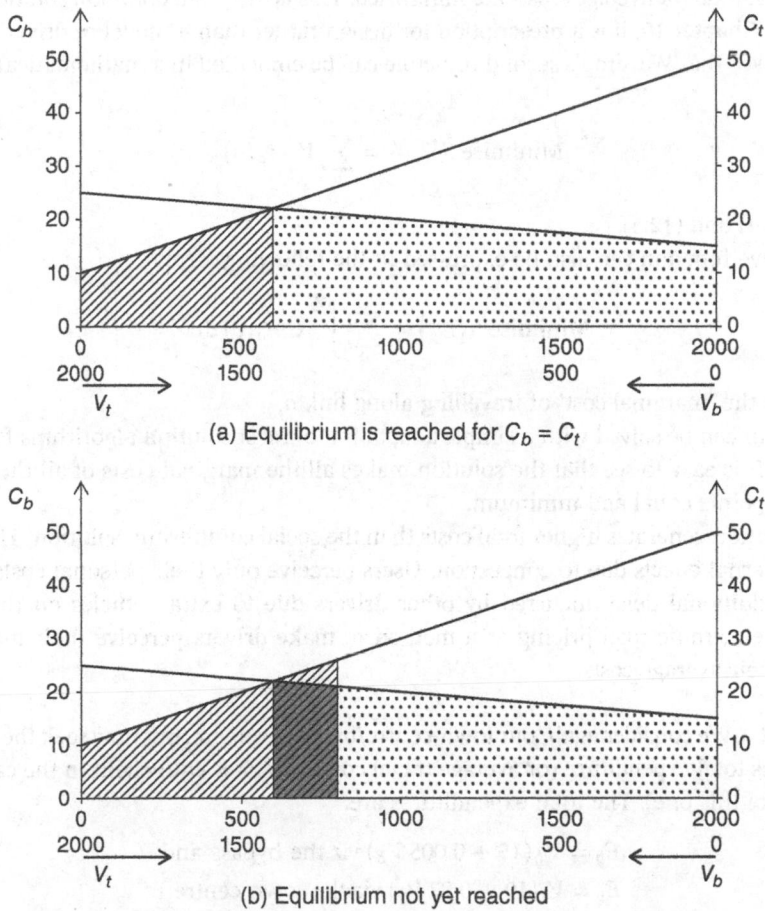

(a) Equilibrium is reached for $C_b = C_t$

(b) Equilibrium not yet reached

Figure 12.2 Equilibrium in simple network

It can easily be seen, in Figure 12.2a, that the sum of areas under the cost–flow curves is minimised for $C_b = C_t$; any departure from this point will simply add a new section to the area, as illustrated in Figure 12.2b. Thus, the equilibrium solution involves a flow via the town centre of 600 vehicles and 1400 via the bypass. It is worth noting that the cost via each route is 22 minutes and the total expenditure in the network is then 44 000 vehicle-min.

In this treatment of equilibrium assignment, we have omitted a number of issues; for example, that of *uniqueness* of the solution. It can be shown that only the link costs c_a^*, inter-zonal costs c_{ij}^*, and link flows V_a^* are unique in the optimum. The path flows T_{ijr}^*, however, are in general not unique. What this means is that there may be several combinations of paths and trips using them, which result in the same link flows and costs; as all used routes (for an O-D pair) have the same minimum cost, the total inter-zonal costs are the same. This can be easily seen if one thinks of several external zones of origin feeding trips into junction A and then exiting to different destinations at junction B in Figure 10.4; although these trips can be distributed in many ways between town-centre and bypass routes under equilibrium conditions, the link flows and costs will remain the same.

12.2.2 Social Equilibrium

Most of what has been discussed so far applies to Wardrop's first principle or *user equilibrium* (UE) problems. Wardrop's second principle specifies that drivers should be persuaded to choose routes in such a way that total (or average) costs are minimised. This is the *social optimum* solution and, as we mentioned in Chapter 10, it is a prescription for design rather than a model of driver's behaviour.

It is easy to see that Wardrop's second principle can be embodied in a mathematical programme of the form:

$$\text{Minimise } S\{T_{ijr}\} = \sum_a V_a \cdot c_a(v) \tag{12.9}$$

subject to (12.4) and (12.5).

This objective function can also be expressed in the following form:

$$\text{Minimise } S\{T_{ijr}\} = \sum_a \int_0^{V_a} Cm_a(v)\, dv \tag{12.10}$$

where Cm_a is the 'marginal cost' of travelling along link a.

This problem can be solved with a simple adaptation of most solution algorithms for the selfish UE problem. It is easy to see that the solution makes all the marginal costs of all the routes used between two points equal and minimum.

The UE solution generates higher total costs than the social equilibrium solution. The difference lies in the external effects due to congestion. Users perceive only their personal costs and do not discern the additional delay incurred by other drivers due to extra vehicles on the road. One can envisage electronic road pricing as a method to make drivers perceive their marginal costs rather than their average costs.

Example 12.2 We consider again our town-centre/bypass problem, but now seek the flow pattern that minimises total expenditure (or average travel costs, which is equivalent in the case of a fixed trip matrix like this one). The total expenditures are:

$$E_b = V_b(15 + 0.005\, V_b) \text{ via the bypass and}$$
$$E_t = V_t(10 + 0.02\, V_t) \text{ via the town centre}$$

The respective marginal costs are:

$$\frac{\partial E_b}{\partial V_b} = 15 + 0.01 V_b$$

$$\frac{\partial E_t}{\partial V_t} = 10 + 0.04 V_t$$

Equating the two and taking advantage of the fact that $V_b + V_t = 2000$, one can find that for social equilibrium conditions, we get the values in Table 12.1:

Table 12.1 Social equilibrium values for Example 12.2

	Town centre	Bypass	Total
Flow	500	1500	2000
Marginal cost	30	30	
Average cost	20	22.5	
Expenditure	10 000	33 750	43 750

Note that the total network expenditure is now 250 vehicle-min less than the user equilibrium solution found in Example 12.1. Of course, one cannot expect drivers to choose the bypass in these numbers, as at least some could reduce their travel costs by choosing the town-centre route. To achieve this social optimum, one would need to increase user costs by 2.5 minutes via the town centre, for example, by charging the equivalent as a town-centre toll. This would represent simply a transfer from private to social consumption resulting in a saving in the use of resources (time, fuel).

12.2.3 Solution Methods

We have described a mathematical programme and shown its relevance in solving the traffic assignment equilibrium problem. The mathematical programme is non-linear and can be solved by a number of methods. Although understanding the theory of equilibrium assignment requires some mathematical background, the actual application of the principles and solution algorithms is much less demanding.

A key consideration when looking into solution algorithms is how quickly (and well) they converge to the correct solution of Wardrop's equilibrium. It is important to select a good convergence criterion to ensure that the solution reached is stable and suitable for project or strategic evaluation. Without this guarantee, small and localised changes in some links may be reflected all over the network and an arbitrary stop in the iterations may result in unreliable results.

Rose et al. (1988) researched a variety of convergence criteria and looked into to the correct solution. They recommend the Relative Gap (RG) as the most reliable measure of convergence:

$$RG = \frac{\sum_a V_a^* \cdot c_a - \sum_a V_a^{AON} \cdot c_a}{\sum_a V_a^* \cdot c_a} \tag{12.11}$$

where c_a is the cost (time) at the current flow on link a, V_a^{AON} is the all-or-nothing (AON) flow on link a, and V_a^* is the current flow on link a.

RG is an estimate of the distance between the current solution and the optimal equilibrium solution. This is because the all-or-nothing solution can be seen as a lower bound for the traffic assignment problem. At true equilibrium, RG would be zero, but as true equilibrium may be too onerous to achieve, a number of tests have been proposed to determine how close is 'close enough'. This depends on the relative size of the user benefits being estimated. The general guideline is to make sure that user benefits, in terms of percentage time savings, are at least 10 times the RG (in %). Boyce et al. (2004) investigated this issue in some practical cases and recommended that RG should be at most 0.1% (0.001) for satisfactory convergence. This is an exacting requirement, probably too demanding for early stages in the model development process. However, it is good and solid advice for the final stages of model calibration and, in particular, for strategic project evaluation.

Patriksson (1994) developed a systematic way of looking at the various algorithms that can be used to solve the mathematical programme for user equilibrium (12.3–12.5). Those found in practice (i.e. implemented in commercial software) can be grouped into:

- a linear approximation (Frank–Wolfe),
- route or path-based assignment, and
- origin-based assignment.

The most commonly used algorithms are a variation of that put forward by Frank and Wolfe (1956). This algorithm can be seen as an improvement on the standard iterative method discussed in Section 10.5.4.

12.2.3.1 The Frank–Wolfe Algorithm

This is presented in both conventional and pseudo code format:

---

1) Select a suitable initial set of current link costs, usually free-flow travel times C_a (0). Initialise all flows $V_a^0 = 0$; make $n = 0$.

2) Build the set of minimum-cost trees with the current costs; make $n = n + 1$.

3) Load the complete matrix **T** of these trees all-or-nothing (AON), obtaining a set of auxiliary flows F_a.

4) Calculate the current flows as:

$$V_a^n = (1 - \phi)V_a^{n-1} + \phi F_a$$

choosing ϕ such that the value of the objective function Z is minimised.

5) Calculate a new set of current link costs based on the flows V_a^n; use a good convergence indicator (say Relative Gap < 0.001) to decide whether to stop or to proceed to step 2

Initialisation
for every link in the network
    Let $c_a = C_a$ (0) for all a
    Let $V_a^n = 0$ for all a and $n = 0$;
end for

Main loop
for $n = 1$ to number of iterations
    build minimum path trees with V_a^{n-1}
    flow costs
    load **T** AON and obtain flows F_a
    estimate ϕ to minimise Z
    make $V_a^n = (1 - \phi)V_a^{n-1} + \phi F_a$
    update $c_a = C_a(V_a^n)$
    if Relative Gap < 0.0001 stop
end for

The main improvement over the iterative method is in step 4, where ϕ is calculated using the mathematical programming formulation instead of a fixed rule. In essence, Frank–Wolfe solves a linearised sub-problem to get a good descent direction and finds a new solution using a linear search. This is enough to guarantee reasonable convergence to Wardrop's equilibrium.

The Frank–Wolfe algorithm can be visualised as a descent approach to the problem of minimising the objective function. The problem is similar to the establishment of the rules to find the lowest point of an enclosed valley in thick fog. A suitable set of rules, in this case, would be:

1) Choose what looks like a good downhill direction; in thick fog, this will depend essentially on local topography.
2) Walk in that direction until you start to go uphill again.
3) Stop at that point and choose another good downhill direction and proceed to step 2, unless you have found a point with no downhill directions (i.e. the bottom of the valley).

This is essentially what the Frank–Wolfe algorithm does, albeit in a space with many more dimensions. At each step in the iterations, we have a current feasible solution (a location in the valley) and the algorithm uses the latest all-or-nothing assignment to provide a descent direction. This can be seen as a local approximation to minimising the objective function Z. Given that the current feasible solution is specified by the path flows $\{T_{ijr}\}$, Frank–Wolfe seeks a second attractive feasible direction $\{W_{ijr}\}$ using a linear (Taylor series expansion) approximation to Z:

$$Z'(\{W_{ijr}\}) = Z(\{T_{ijr}\}) + \sum_{ijr} \frac{\partial Z}{\partial T_{ijr}} (W_{ijr} - T_{ijr})$$

$$= Z(\{T_{ijr}\}) + \sum_{ijr} C_{ijr} \cdot W_{ijr} - \sum_{ijr} C_{ijr} \cdot T_{ijr} \tag{12.12}$$

Here, the only term that is not fixed by the feasible solution $\{T_{ijr}\}$ is $C_{ijr} W_{ijr}$; so, if we wish to minimise a local approximation to Z, we must choose routes W_{ijr} such that the corresponding multipliers C_{ijr} are minimised. A way to do this is to choose routes that are currently and locally minimum cost (i.e. all-or-nothing assignment on trees from current costs).

In general terms, the Frank–Wolfe algorithm tends to converge rapidly over early iterations but less so as it starts to approach the optimum. Related to this is the problem that link flows tend to oscillate during iterations making it more difficult to achieve the necessary precision in the final solution. Notwithstanding, it has the advantage of requiring little computer memory as only link variables need to be stored. However, as modern computers offer plenty of memory, the original advantage is less of a constraint. The slow convergence of Frank–Wolfe is a well-known problem, and a number of improvements have been suggested to speed up convergence, see, for example, the work of Weintraub et al. (1985) and Arezki (1986). Alternative solution methods, such as those discussed below, offer better convergence properties, especially for large and congested networks. It is interesting to note that better solutions are often helped by the adoption of a new framework to cast the problem in.

12.2.3.2 Route-Based Assignment

There are at least two important algorithms that work on the path-flow (rather than link flow) space. We will present here one due to Jayakrishnan et al. (1994) as a *gradient projection* algorithm. It uses a transformed objective function that incorporates the flow conservation constraints into the objective.

The formulation of the algorithm is based on the traffic demand constraints:

$$\sum_r T_{ijr} = T_{ij}$$

The shortest path flows can be expressed as:

$$T_{ij\bar{r}} = T_{ij} - \sum_{r \notin \bar{r}} T_{ijr} \tag{12.13}$$

Now, the optimisation problem can be re-stated as:

$$\min \bar{Z}(T_{ij\bar{r}})$$
$$\text{subject to } T_{ijr} \geq 0 \quad \forall T_{ijr} \in \bar{T}_{ijr}$$

where $\bar{Z}$ is the new objective function, and $\bar{T}_{ijr}$ is the set of non-shortest path flows.

At any (non-optimal) stage in the algorithm, a better solution can be found by moving in the negative gradient direction. This is calculated with respect to the flows on the non-shortest paths, and a move size is found using second derivatives with respect to these path flows. For a fuller description of the algorithm, see Jayakrishnan et al. (1994). Larsson and Patriksson (1992) have developed a related algorithm called Disaggregate Simplicial Decomposition.

12.2.3.3 Origin-Based Assignment

Origin-based assignment represents, in fact, a family of solution methods (Bar-Gera 2002). The basic idea is to define the solution variables in an intermediate way between links and routes. The main solution variables in this algorithm are *origin-based approach proportions*, α_{ia}, for every origin i and every link a, such that for every origin i and node p, the sum of origin-based approach proportions over all links ending at node p is equal to one. Using origin-based approach proportions, *route proportions* are determined by the product of the approach proportions of all the links along the route, that is:

$$\gamma_{ijr} = \prod_{a \subseteq r} \alpha_{ia}$$

Route flows are determined by the product of O-D flow and route proportion, that is:

$$h_{ijr} = T_{ij} \cdot \gamma_{ijr}$$

Bar-Gera (2002) shows that if link a goes from node p to node q and if the total flow from origin i to node q is g_{iq}, then the total flow from origin I arriving at node q through link a is $f_{ia} = \alpha_{ia} \, g_{iq}$.

This representation of the solution allows for an efficient storage of route flows. A key element in this solution method is that for every origin, an *a-cyclic* restricting sub-network, A_i, is chosen such that for origin i the approach proportions of links not included in A_i are restricted to zero.

The following outline of the algorithm is based on Boyce (2007). Start with trees of minimum-cost routes as restricting sub-networks, leading to an all-or-nothing assignment. Next, consider all origins in a sequential order. For each origin, the restricting sub-network is updated, and the origin-based approach proportions are adjusted within the given restricting sub-network.

To update a restricting sub-network, unused links are removed; the maximum cost from the origin to node q (v_q), within the restricting sub-network, is calculated for all nodes and links p–q, where $v_p < v_q$ is added to the restricting sub-network. Once a new restricting sub-network is found, several computationally intensive steps are needed, including reorganisation of the data structure.

As the restricting sub-networks tend to stabilise quickly, it is useful to update origin-based approach proportions while keeping the restricting sub-networks fixed. This is done by introducing *inner iterations*. To update origin-based approach proportions within a given restricting sub-network, a search direction based on shifting flow from high-cost alternatives to low-cost alternatives is used. In addition to current costs, estimates of cost derivatives are used to improve the search direction in a quasi-Newton fashion.

The full algorithm can be displayed in pseudo-code as (Boyce 2007):

<u>Initialization:</u>
  for every origin i
    let A_i be a tree of minimum-cost routes under free-flow conditions from i
    let α_{ia} equal 1 for all links in A_i and 0 otherwise (all-or-nothing assignment)
  end for

Main loop:
for $n = 1$ to number of main iterations
      for every origin i
          update restricting subnetwork A_i
          update origin-based approach proportions α_{ia}
      end for
      for $m = 1$ to number of inner iterations
          for every origin i
              update origin-based approach proportions α_{ia}
          end for
      end for
end for

Update restricting sub-network for origin i:
   remove unused links from A_i
   for every node p compute the maximum cost v_p from i to p
   for every link $a = [p,q]$
      if $v_p < v_q$ add link a to A_i
   find new topological order for new A_i
   update data structures

Update origin-based approach proportions for origin i:
   compute average costs and Hessian approximations
   for step size 1, 1/2, 1/4, 1/8, ...
      compute flow shifts and scale by step size
      project and aggregate flow shifts
      if convergence criteria is met then stop
   end for
apply flow shifts
update total link flows and link costs

Dial (2006) and Bar-Gera and Luzon (2007) have developed variations on this approach that represent improvements on the original algorithm. Table 12.2 compares some of the features of the three general approaches:

Table 12.2 Comparison of algorithmic approaches

	Link based	Path based	Origin based
Decision space	Link flows	Path flows	Origin-based approach proportions and link flows
Memory requirements	Minimum	Greatest	Medium
Speed of convergence	Fast early, slower close to optimum	Fast	Fast

Either the Path- or the Origin-based approaches are preferable over the traditional Frank–Wolfe in most cases, depending on what is available in a particular software package.

12.2.4 Stochastic Equilibrium Assignment

We have discussed pure stochastic and pure user-optimised equilibrium traffic assignment models. In the first case, a spread of routes between two points is produced because of variability in the perceived route costs, and in the second, because of capacity-restraint effects. One would expect that, in reality, both types of effects should play a role in route choice. Models that try to include both effects are called Stochastic User Equilibrium (SUE) models, and they seek an equilibrium condition where:

> *Each user chooses the route with the minimum 'perceived' travel cost; in other words, under SUE no user has a route with lower 'perceived' costs and therefore all stay with their current routes.*

The difference between stochastic and Wardrop's user equilibrium is that in SUE models, each driver is meant to define 'travel costs' individually instead of using a single definition of costs applicable to all drivers.

In theory, models incorporating stochastic and equilibrium properties look particularly attractive; there are, however, operational and practical difficulties in applying them. From a practical point of view, the most important difficulty lies on the convergence properties of these algorithms. To examine this problem, let us define *convergence* here in the following way: an assignment algorithm is said to be convergent if:

- one starts with a particular set of link costs C_a, for example, free-flow costs in the first iteration but calculated costs as a function of flows in subsequent ones; and
- one assigns a matrix using specific rules, say Dial's, and produces new link flows $\{V_a\}$, and then one finds that:

$$C_a = C_a(V_a)$$

In other words, the costs resulting from the new flows are practically the same as those used to find routes and assign traffic. If an algorithm is not convergent, the solution (flows and costs) will depend on when the iterative process was stopped (i.e. an arbitrary decision). For example, the next planner dealing with exactly the same problem but specifying a different number of iterations would find different costs; this is obviously not a desirable property for the assessment of transport projects.

It can be shown that under specific circumstances, it is possible to formulate convergent SUE algorithms (Sheffi 1985). In fact, a practical algorithm to perform SUE assignments is just an extension of the iterative loading methods (MSA algorithm) described in Section 10.5.4. Such an algorithm can be described as follows:

1) Set current costs $C_a = C_a(0)$, that is, free-flow travel costs; initialise $V_a = 0$ for all a, and make $n = 0$.
2) Make $n = n + 1$; build a set of minimum-cost trees with the current costs.
3) Assign the trip matrix to the network using the current trees and a suitable stochastic method (e.g. Burrell's); obtain a set of auxiliary flows F_a.
4) Calculate current flows as:

$$V_a^n = (1 - \phi)V_a^{n-1} + \phi F_a$$

with $\phi = 1/n$.

5) Calculate a new set of current link costs based on the flows V_a^n; if the flows (or current link costs) have not changed significantly in two consecutive iterations, stop, otherwise proceed to step 2.

This algorithm will always tend to produce small changes in flows and costs as ϕ is small for large n. However, it is important to prove that it converges to the right SUE solution. Sheffi (1985) has shown that it does in the long run, that is, for a large number of iterations, perhaps 50 or more. The convergence of the algorithm is not monotonic because the search direction is only a descent direction on average. The speed of convergence depends on the level of network congestion and on the dispersion parameter.

Sheffi (1985) has also shown that for very congested networks, UE provides a good approximation to SUE and is faster in convergence. This suggests that using SUE would only be advantageous in low- to medium-congested assignment problems.

12.2.5 Congested Public Transport Assignment

In Chapter 10, we looked at public transport assignment with fixed costs. This means that the link costs do not depend on the number of passengers on the bus or train and that they do not depend on the carrying capacity of each unit. It is a reasonable approach in cases where the goal of the planning process is to provide enough capacity for all public transport passengers on the routes of their choice. It also has the advantage of facilitating the solution to the public transport assignment problem.

There are, however, situations where it is not feasible to provide enough public transport capacity to preclude congestion, or when that capacity is not present in the base year. In these cases, the route choice of public transport passengers is likely to be influenced by the congestion onboard the vehicles; some travellers will switch from congested to less congested routes, even if the less congested routes are not as attractive in terms of travel time or cost.

We now turn our attention to the assignment problem, where link travel times are no longer constant. The dependency of link costs on passenger flows may take different forms, but from a solution viewpoint, the simplest and most convenient are continuous, non-decreasing functions of the corresponding link flows. This dependence of the link cost on the public transport volume may represent an actual slowing down of the vehicle due to the number of passengers boarding and alighting, or it may be interpreted as a generalised cost that includes a *discomfort* term that increases as the vehicles get crowded.

In this context, the transit assignment problem is no longer separable by destination node, since the link costs depend on the total flow of passengers. The total transit volumes are the sum of the volumes bound for each of the destinations. As the expected cost of any given strategy is no longer fixed but depends on the total volumes, only strategies with minimal expected cost will be used by the travellers (Wardrop 1952).

Spiess (1983) has shown how the problem above can be formalised and solved by applying the successive linear approximation method (Frank and Wolfe 1956). An important advantage of this method is that only total volumes need to be computed and stored since the destination-dependent volumes are dealt with implicitly. The approach is easy to implement in packages like EMME.

A variant of the macro outlined above is being used at London Transport for modelling crowding in the London underground. Instead of using one of the default congestion functions based on nominal capacity, the macro has been modified to include the actual profile of train density and passenger load during peak periods (Abraham and Kavanagh 1992).

However, there are conditions where it is not reasonable to assume that link costs depend only on passenger levels on that link. For example, the delay at a stop may depend significantly on the number of passengers already on the public transport unit (bus or train/metro), as some people may not be able to board the first vehicle that comes along. In this case, delays or generalised costs on a link will depend significantly on flows on other links as well; the situation is not entirely dissimilar to junction delays.

In these cases, congestion modelling should be done using asymmetric generalised cost functions. Here, the perceived waiting time for a service (line) for a boarding passenger depends on the number of passengers already on board; alternatively, the dwell time of a line at a node depends on the number of boarding and alighting passengers there. Although such phenomena occur in reality, their inclusion into assignment models leads to models without the guarantee of a unique solution.

De Cea and Fernández (2000) developed a multimodal/multiple user classes' equilibrium model that incorporates asymmetric generalised cost functions for public assignment (but symmetric functions for road assignment). The model combines destination, mode choice, and assignment in an equilibrium framework. Destination and mode choice are treated in a Nested Logit formulation (destination at the top) combined with equilibrium assignment with capacity constraints. The problem is formulated as a variational inequality, and a diagonalisation algorithm is used for its solution. It is recognised that there is no guarantee of convergence to a unique solution. However, the authors state that they have achieved convergence in all their model applications (De Cea et al. 2005).

12.3 Transport System Equilibrium

12.3.1 Equilibrium and Feedback

The types of equilibrium problems discussed so far concern just a single mode in a network. Wardrop's first principle models this type of behaviour, and a suitable algorithm permits the identification of the routes and flows that will generate consistent costs for all users. As stated before, a similar principle applies to congestion or capacity problems in public transport networks.

The problem becomes more complex when one considers interactions between two or more modes. These may take the following forms:

1) Congestion generated by cars will affect bus travel times on certain routes and therefore change assignment strategies for public transport users. The congestion generated by buses (and street-running LRT systems) and bus stop operations will affect capacities and speeds for cars, and therefore their route choices.
2) Interactions due to park-and-ride and kiss-and-ride operations for buses and segregated track systems. The attractiveness of these mixed-mode operations will depend on road congestion, service frequency, and fares (mode and parking), and all of these are, in general, mutually related.

Pragmatic approaches to this problem are usually of the hard or soft speed-change nature discussed in Section 10.5.2: assume bus times and flows are fixed and known, assign cars to the network to equilibrium, assign passengers to the transit network, obtain new speeds and travel times and fix them, re-assign, obtain new speeds, re-assign, etc. Of course, if one is not prepared to change the bus frequencies in accordance with demand, the problem may converge soon at this level.

In the case of mixed-mode users, the problem is more difficult because they may decide to change their park-and-ride station as a result of congestion on the road network and, therefore, change link flows and levels of congestion when they do so. Even if mixed-mode movements are few at present, not including them in the equilibrium procedure may cause severe problems for design-year forecasts in heavily congested networks.

In all the cases above, we have kept the assumption of a fixed trip matrix (inelastic demand) for each mode. However, what we have seen in earlier chapters must lead us to treat the assumption of a fixed matrix with caution, at least when considering major changes to the transport network or longer timescales. Indeed, the whole point of distribution, mode, and time-of-travel choice models is that demand is elastic, in particular, to travel and route generalised costs. This leads us to consider the influence of congestion and delay at least on mode and destination choice: the issue of *system equilibrium* or at least model system consistency.

Looking at the whole modelling system in forecasting mode, the generalised costs of travel will be affected by congestion and future interventions like new links and modes. Any assumption about travel costs must be revised after assignment, and the system of models should be run again to obtain demand consistent with future costs.

What we have now is a nested set of models, and we need to make sure that the travel costs used by all of them are consistent. A naive (in the sense of simple, not pejorative) iterative strategy would be ... 'run all the models first, obtain new travel times, feedback the new travel times to the models above and repeat until convergence'. This naive feedback strategy is similar to either hard or soft-speed adjustment methods for assignment, as discussed briefly in Chapter 10. Similarly, it has all the makings of a *non-convergent* approach. Oscillations are likely to be a feature of this type of technique unless special conditions are met, or we pay considerably more attention to the development and use of better algorithms.

Example 12.3 Consider again the town-centre/bypass problem of Example 10.4, but add a rail service that links A–B in 12 minutes, with a frequency of ten trains per hour. For simplicity, let us assume that car occupancy is just one person per car and that there are no fares and no fuel or access costs; time is the only cost element. The cost of using rail is then 18 generalised min $(12 + 2$ times $3)$. The total demand V_T is still 2000 passengers an hour. The choice between rail and car is estimated using a Logit model with only one parameter λ. In this case, it is very easy (and fast) to reach equilibrium on the roadside using the fact that:

$$V_b = 0.8\, V_C - 200$$

and

$$t_b = 15 + 0.005 V_b$$

where the car demand V_c is the total demand minus rail demand:

$$V_c = V_T - V_r$$

The quality of convergence to equilibrium can be measured by the proportion of total demand that is displaced from one mode to the other at each iteration. Convergence is reached when the displaced demand is zero. One would expect that the speed of convergence would depend on the parameter λ as higher values (giving greater weight to cost differences) will make the Logit result get closer to all-or-nothing mode choice.

Figure 12.3 shows the number of iterations needed to reach particular levels of convergence, as a function of λ for this simple example. As can be seen, low values of λ enable reasonable convergence for some 15–20 iterations. The reader can verify that 10 iterations are enough if λ is 0.05, for example.

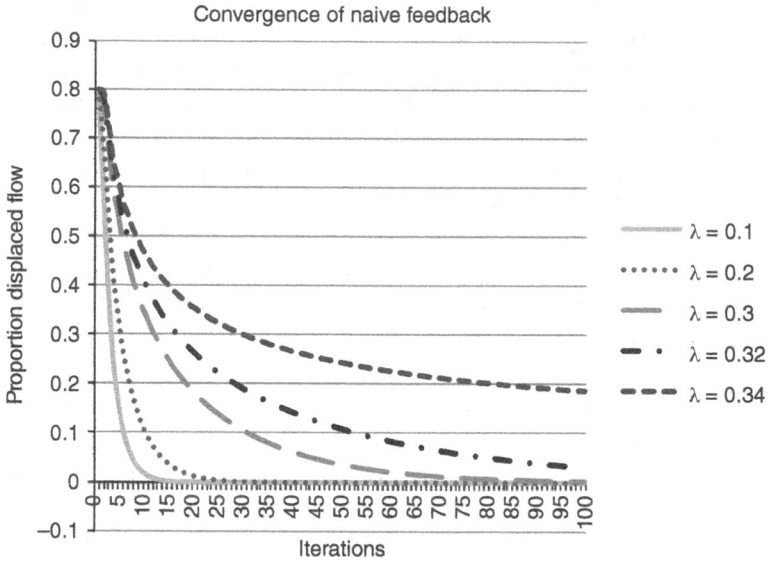

Figure 12.3 Convergence of naive feedback

For larger values of λ, convergence requires solving the whole model 100 times or more. Indeed, for λ greater than 0.34, convergence is never achieved in this example, making it unsound to compare any two schemes after an arbitrary number of iterations. For instance, for $\lambda = 0.34$, the number of trips by rail after 100 iterations oscillate between 650 and 950 each time. Of course, one should not generalise from this simple example to real networks and problems. However, at least it shows that great care must be placed in organising the interaction between different sub-models.

This is an important issue that, in our experience, is sometimes ignored or handled incorrectly in practice. Moreover, in the United States of America at least, running the models with *feedback* is required by Congressional and judicial mandates. Therefore, we attempt here to approach the issue from two complementary perspectives. We will explain the key components of the problems both intuitively and using a more formal mathematical framework.

First, let us mention that the problem is not even a recent one. It was researched in the seventies by John Murchland, a British applied mathematician, and Suzanne Evans, a graduate student at University College London (Evans 1976). She successfully solved the combined distribution and assignment problem in her PhD thesis, but after publishing a couple of papers, she changed her field of inquiry. They introduced the terminology of *combined models* that has been applied ever since.

One early review of the state of the art is by Fernández and Friesz (1983). More recent developments have focussed on developing improved and more practical algorithms and recommendations. The next sections are inspired by the work of Professor David Boyce, whose efforts to convey the importance of the problem and the rigour required to tackle it correctly are exemplary.

12.3.2 Formulation of the Combined Model System

A useful way of tackling this problem is to frame it as a mathematical programme. As we have already done this separately for distribution, mode choice, and equilibrium assignment, framing a combined mathematical programme seems a natural next step. For convenience, we start first with the combined mode choice and assignment problem.

A reasonable start is to collapse as many sub-models as possible into one, in particular, if one can include assignment in the same process. What may be important, however, is not to compromise too much on the realism of the modelling process for the sake of expedience in equilibrium, particularly in short-term tactical decision-making.

Consider first the problem in general terms where a typical demand curve may be inverted to give travel costs as a function of number of trips $C_{ij} = g_{ij}(T_{ij})$. We then have a combined problem described in terms of relationships between flow levels and costs; some of these flows are trips on real links a and others are flows on O-D pairs (hyperlinks).

Now consider the following objective function:

$$\text{Minimise } Z = \sum_a \int_0^{V_a} c_a(\upsilon)\, d\upsilon - \sum_{ij} \int_0^{T_{ij}} g_{ij}(t)\, dt \tag{12.14}$$

subject to

$$T_{ijr} \geq 0$$

$$T_{ij} = \sum_r T_{ijr} \tag{12.15}$$

$$V_a = \sum_{ijr} T_{ijr} \cdot \delta_{ijr}^a \tag{12.16}$$

The derivative of Z with respect to T_{ijr} is:

$$\frac{\partial Z}{\partial T_{ijr}} = c_{ijr} - g_{ij}$$

We can now consider the behaviour of Z at T_{ijr}^* directly:

$$\text{If } T_{ijr}^* = 0 \quad \text{then} \frac{\partial Z}{\partial T_{ijr}} \geq 0 \quad \text{and } c_{ijr} \geq g_{ij} \tag{12.17a}$$

$$\text{If } T_{ijr}^* > 0 \quad \text{then} \frac{\partial Z}{\partial T_{ijr}} = 0 \quad \text{and } c_{ijr} = g_{ij} \tag{12.17b}$$

Therefore, if a particular path is used, then the path cost specifies a value for the demand curve, so we must have:

$$g_{ij}(T_{ij}) = c_{ij}^*$$

A couple of issues are worth noting here. First, for assignment, one usually deals with vehicular flows, whereas in distribution and mode choice, the main variables are person trips. There is a need to account for vehicle occupancy in combined models, although this factor has been omitted for simplicity. The inverted demand function could be of a very general form provided the problem remains convex. However, in many practical problems, it may not be possible to develop suitable closed analytical forms.

Consider a slightly more general case where we add a transit mode b to the system; assume first that the travel times on this transit mode are independent from road speed. The function can be written as follows:

$$\text{Minimise } Z = \eta \sum_a \int_0^{V_a} c_a(\upsilon)\, d\upsilon + \sum_{ij} c_{ij}^b \cdot T_{ij}^b \tag{12.18}$$

Subject to

$$\sum_{ijr} \eta \cdot T_{ij}^c \cdot \delta_{ijr}^a = V_a \qquad \text{a flow conservation constraint} \tag{12.19}$$

where η is vehicle occupancy, b indicates the public transport mode, and c *is* the car; m, below, is the index for mode, either b or c.

We now add a constraint stating that total O-D flows are split between the two modes and add also a dispersion constraint to ensure that this split is not all-or-nothing, as we cannot account for all factors explaining mode choice.

$$T_{ij} = T_{ij}^c + T_{ij}^b \qquad \text{for all } i,j \tag{12.20}$$

$$\sum_{jm} T_{ij}^m \cdot \log T_{ij}^m = -S_0 \tag{12.21}$$

$$T_{ij}^m \geq 0 \tag{12.22}$$

In this combined problem, we have relaxed slightly the user equilibrium conditions to allow trips to choose a different mode (hyper-route) whilst retaining the Logit formulation for mode choice.

If we further add an origin–destination choice element to the problem, to relax the assumption that T_{ij} is fixed, we have:

$$\sum_{jm} T_{ij}^m = O_i \tag{12.23a}$$

$$\sum_{im} T_{ij}^m = D_j \tag{12.23b}$$

The optimality conditions for this model are the same as for user equilibrium plus:

$$T_{ij}^m = \frac{T_{ij} \exp\left(-\lambda C_{ij}^{m*}\right)}{\sum_n \exp\left(-\lambda C_{ij}^{n*}\right)} \tag{12.24}$$

and

$$T_{ij}^m = A_i O_i B_j D_j \exp\left(-\lambda C_{ij}^{m*}\right) \tag{12.25}$$

where C_{ij}^{m*} is the user equilibrium cost of travelling from i to j by mode m.

This solution has the same dispersion coefficient λ for mode and destination choice. A second dispersion constraint (on ij) can be added to convert it to a problem where these dispersion constraints are allowed to be different.

The constraint (12.21) can be integrated into the objective function using the Lagrange multiplier λ and retain only the linear constraints:

$$\text{Minimise } Z = \eta \sum_a \int_0^{V_a} c_a(\upsilon)\, d\upsilon + \sum_{ij} c_{ij}^b \cdot T_{ij}^b + \frac{1}{\lambda} \sum_{ijm} T_{ij}^m \cdot \log T_{ij}^m \tag{12.26}$$

subject to constraints 12.19, 12.20, 12.22, 12.23a and 12.23b.

These are all linear constraints, and the objective function is the sum of convex functions. The solution algorithm proposed by Evans (1976) can be generalised as follows (Boyce 2007):

1) Initialisation. Make iteration counter $n = 0$; compute an initial solution for $\left(T_{ijm}^0\right), \left(V_a^0\right)$. This normally involves using free-flow costs to estimate Gravity and mode choice models and assign trips to the networks

2) Update link costs $C_a = C_a\left(V_a^n\right)$; increment n by 1.

3) Compute new shortest routes from each origin i to destination j and obtain new car costs $\left(C_{ijc}^n\right)$.

4) Solve the O-D and mode choice models and obtain the sub-problem flows $\left(e_{ijm}^n\right)$ sometimes called auxiliary flows, in this case at the Origin-Destination-Mode level.

5) Perform all-or-nothing assignment of car flows $\left(e_{ijc}^n\right)$ to the shortest path from i to j obtaining car flows $\left(g_a^n\right)$.

6) Compute the Relative Gap and test for convergence.

7) Perform a line search to determine the optimal step length (weight) λ^n.

$$\text{Minimise } Z = \eta \sum_a \int_0^{V_a^n} c_a(\upsilon)\, d\upsilon + \sum_{ij} c_{ij}^b \left(T_{ij}^b\right)^n + \frac{1}{\lambda} \sum_{ijm} \left(T_{ijm}\right)^n \log \left(T_{ijm}\right)^n \qquad (12.27)$$

where

$$V_a^n = (1 - \lambda^n) V_a^{n-1} + \lambda^n g_a^n \quad \text{and} \quad T_{ijm}^n = (1 - \lambda^n) T_{ijm}^{n-1} + \lambda^n e_{ijm}^n$$

8) Update the O-D mode and link flows

$$T_{ijm}^n = (1 - \lambda^n) T_{ijm}^{n-1} + \lambda^n e_{ijm}^n$$
$$V_a^n = (1 - \lambda^n) V_a^{n-1} + \lambda^n g_a^n$$

9) Retest the updated value of the objective function for convergence; if not achieved, go back to Step 2

This approach has a critical difference with the naive feedback treatment of the problem. Evans' solution is to average flows (on links and trips by mode and O-D pair) rather than just feedback costs, averaged or otherwise.

Similar formulations have been produced by, among others, Gartner (1980) and Sheffi (1985). This type of approach has been extended further by De Cea et al. (2008), who presented very general combined models with hierarchical demand choices using a multi-objective entropy maximisation approach. This is an important and valuable generalisation as it allows for a general hierarchy of choice models in combination with what they correctly call 'demand-performance' models. The choices may include destination, mode, time of travel, and modal transfer point. The main characteristics of their approach are:

1) Demand choices are assumed to have a hierarchical structure where entropy must be maximised to produce the most likely arrangement subject to the corresponding constraint at each level of the nested tree.

2) A combined demand model incorporating these choices can be posed as a multi-objective programming problem; they put the destination choice model at the top of this tree but other arrangements are possible.

3) All users by class and mode behave according to Wardrop's first principle; the link flow-cost functions are separable and convex.
4) The combined performance–demand equilibrium models are also formulated as multi-objective programming problems.
5) A convex optimisation problem cannot be formulated if the network cost functions are asymmetric, but the problem may be specified as a variational inequality.

With these conditions, the set of choice models turns out to be a Nested Logit model where the scaling parameters must comply (as we saw in Chapter 7) with the requirements for their relative values decreasing from the top to the bottom of the hierarchy.

12.3.3 Solving General Combined Models

The considerations above are particularly useful when it is possible to formulate an appropriate mathematical programming problem with the necessary conditions of convexity and separability or symmetry of the performance–demand functions. The solution algorithm will depend, in general, on the specific formulation in each case. One of the attractions of the naive feedback approach is that it is general enough and does not require assumptions about the characteristics of the model formulation. As we have seen, however, there is no guarantee of convergence to a unique solution that would allow us to compare strategies or projects.

It is generally accepted that the weights in the generalise Evans' solution method could be replaced by pre-determined weights, for example, the rules of the method of successive averages (MSA). The sequence of sub-problem weights or step-sizes λ^n applied in the MSA are 1, 1/2, 1/3, ..., 1/n. The use of this sequence is known to converge to equilibrium, albeit at a fairly slow rate.

An alternative approach is to use relatively arbitrary constant weights instead of the MSA sequence. Perhaps surprisingly, the Constant Weight (CW) approach has been found in practice to converge faster to equilibrium than MSA (see Bar-Gera and Boyce 2006). A more general version of this algorithm could be presented in an intuitive form as follows:

Step 1 Input data, the road and public transport networks, and trip end constraints O_i, D_j.
Step 2 Compute an initial solution for iteration $n = 1$ using free-flow costs or another suitable starting point.
   Initialise travel costs C_{ijm}^n
   Solve the demand model $e_{ijm}^n = T_{ijm}^n$ (a provisional solution for O-D-mode)
   Assign T_{ijc}^n to the road network, where the sub-index c indicates cars
Step 3 Compute a solution for $n = n + 1$ the next iteration
   Calculate costs on used route-mode combinations C_{ijm}^n
   Solve the demand model e_{ijm}^n (auxiliary demand volumes)
Step 4 Average trip matrices T_{ijm}^{n-1} and e_{ijm}^n

   For MSA $T_{ijm}^n = \left(\dfrac{n-1}{n}\right)T_{ijm}^{n-1} + \left(\dfrac{1}{n}\right)e_{ijm}^n$

   For CW $T_{ijm}^n = wT_{ijm}^{n-1} + (1-w)e_{ijm}^n$
Step 5 Assign T_{ijc}^n to the road network to the desired degree of convergence and produce V_a^n
Step 6 Check for convergence of e_{ijm}^n to T_{ijm}^{n-1}

$$\text{Total Misplaced Flow} \quad \text{TMF} = \sum_{ijm} \left| T_{ijm}^n - e_{ijm}^n \right| \leq \text{E} \tag{12.28}$$

or

$$\text{Root Squared Error} \quad \text{RSE} = \sqrt{\sum_{ijm} \left(T_{ijm}^{n-1} - e_{ijm}^n \right)^2} \leq \text{E} \tag{12.29}$$

if converged, stop, otherwise continue to Step 3.

This is a general formulation that can be applied with the MSA and CW methodologies. Bar-Gera and Boyce (2006) and Boyce et al. (2008) report that the use of CW performs better than MSA on real networks in terms of speed and consistency of convergence. They recommend their adoption with w in the range of 0.2–0.5. In particular, the same weight $w = 0.25$ performed well for three cases with very different congestion levels. A general (provisional) rule seems to be to use the CW method, testing a few weights w around 0.25 to find what works best for a particular network and matrices. Naive feedback performed poorly in all cases and, therefore, should be always avoided.

Equilibrium in transport systems and markets is not an end in itself. There are good reasons to suspect that equilibrium does not happen in practice, not even at the simplest network level. Real systems are in a permanent state of change, with travellers experimenting new routes, modes, and destinations. Families change residences, jobs, shopping, and social patterns and lifestyles. However, the state-of-the-art in dynamic modelling of these phenomena is still many years behind that of equilibrium modelling.

The main reason to use models is to provide advice on transport decisions, and this requires comparing alternative ways of intervening in the transport system. Consistency in the use of models to estimate the performance of these interventions is then of capital importance, as we wish to compare 'like with like'. Casting the transport modelling effort into a general equilibrium framework seems a prerequisite for ensuring this consistency. It is not, of course, a sufficient condition: there will be cases where partial modelling of the system will be enough to discriminate a good scheme from one that is not so good. However, the state-of-the-art of equilibrium modelling is such that one seldom has to sacrifice too much realism to achieve it.

Computer memory and speed constraints are mostly a thing of the past. Most modern software now provides all the facilities required to seek equilibrium solutions involving route, mode, destination, and time-of-departure choice. There seems to be little reason not to use these facilities, at the very least for the final runs used to compare two alternative strategies or schemes.

12.3.4 Monitoring Convergence

The combined problem-solving methods discussed above rely on a valid estimate of the degree of convergence to an equilibrium solution. In general terms, two convergence criteria are needed, one for the trip matrix (destination, mode, and time of day) and one for the link flow arrays. These are sometimes combined under the banner of the relative gap (RG).

Feasible solutions to convex optimisation problems have a lower bound associated with them; this is defined in terms of auxiliary demand matrices e_{ijm}^n and flows V_a^{n-1} and g_a^n. The RG is obtained from the lowest objective function (LOF) value, which is always a result of the current iteration, and the best (or highest) lower bound (BLB), which may be the result of an earlier iteration:

$$\text{RG} = \frac{(\text{LOF} - \text{BLB})}{\text{BLB}}$$

The measure is useful for global convergence but is not that easy to interpret intuitively, and this makes it difficult to establish a desired level for the tolerance E in (12.28). When dealing with O-D (plus mode and time-of-travel) volumes, it seems natural to compare the current solution with that resulting from the generalised costs of travel under current conditions. The preferred measure is the total misplaced flows as defined in (12.28) and measured in trips per unit time (hour). This lends itself to an easier intuitive interpretation.

For example, consider a scheme involving the introduction of a new metro station that is expected to attract/generate 2000 trips per hour during the peak. It will be desirable to know that the solution found to the combined problem is misplacing less than, say, 200 trips. Depending on the problem and model, it may be acceptable to have 2000 misplaced trips over a larger area, but figures above, say, 5000 trips are likely to cast some doubt about comparisons with alternatives using this particular model.

Assignment accuracy can be ascertained using the distribution of excess costs (EC) among all used routes:

$$EC_{ijr} = C_{ijr} - C_{ij}^* \tag{12.30}$$

where C_{ijr} is the current cost from origin i to destination j via route r, and C_{ij}^* is the minimum cost between those O-D pairs

A good measure is the average excess cost (AEC), defined as:

$$AEC^n = \frac{\sum_{ijr} T_{ijr} \cdot EC_{ijr}}{\sum_{ij} T_{ij}} = \frac{\sum_{ijr} T_{ijr}\left(C_{ijr} - C_{ij}^*\right)}{\sum_{ij} T_{ij}}$$

which is equivalent to a normalised gap for the fixed demand problem.

Exercises

12.1 A 12-km expressway connects two urban areas. The supply function for each of the three lanes per direction of the link may be approximated by

$$t = 20 + q/200$$

where t is the travel time in min and q is the flow per lane in passenger car units (PCU) per hour. The road is normally used by cars and express (non-stop) buses only; the corresponding vehicle travel times are t_c and t_b. The bus service has a peak-hour frequency of one bus per minute. The demand function for car travel has been estimated to be:

$$V_c = 3480 - 60t_c$$

where V_c is the total car flow per hour and direction. In a similar way, the demand function for bus trips is thought to be:

$$V_b = 4200 - 75t_b$$

where V_b is the number of passengers per hour and direction. You may assume that both t_c and t_b can be calculated from the above supply functions and that a bus is equivalent to 2 PCUs.

a) What is the initial equilibrium state? If a bus has 60 seats, what is its load factor (occupancy divided by capacity)?

b) One of the lanes is now taken for exclusive use by buses. What are the new equilibrium state and the new load factor for buses?

c) Discuss the assumptions implicit in the demand functions used above.

12.2 Two cities 60 km apart are connected by a two-way road over which cars operate throughout the day. The peak-hour demand for travel by car between the two cities is thought to be well described by the following function:

$$q = 6000 - 1500t$$

where q is the demand in vehicles per hour and t is the travel time in hours. The travel times versus flow relationship for the road is:

$$t = 0.90 \exp (0.0003q)$$

a) Estimate how many vehicular and person trips per hour are made under equilibrium conditions if each car carries 1.5 passengers on average.

b) A frequent (but slow) rail service is now implemented between the cities, where each train has a nominal capacity of 300 passengers. During the peak hour, the rail company is prepared to run a train every 10 minutes with an estimated travel time of 90 minutes. If passengers are assumed to use the fastest mode available, is this a sensible level of service?

12.3 Consider the network and conditions described in Exercise 10.3.

a) Express the objective function of the mathematical programme corresponding to Wardrop's selfish equilibrium in terms of the flows and travel time–flow relationships in the figure.

b) Calculate the equilibrium flows on each link and the travel time for each group of travellers. Calculate the value of the objective function above under equilibrium conditions and the total expenditure in travel time in the system.

c) Local traffic engineers have decided to install speed restrictions on link C–D so that the new travel time versus flow function is:

$$t = 5.2 + 0.001q$$

Calculate the new equilibrium conditions in terms of flows and travel times and show that, under these conditions, the total expenditure in travel time in the system is less than in (b).

12.4 The network in Figure 12.4 is loaded during the peak hour with 100 vehicles travelling from A to D. The equations in the network show the travel time on each link in minutes as a function of the flow q on the link in vehicles per hour. All links are unidirectional, as shown.

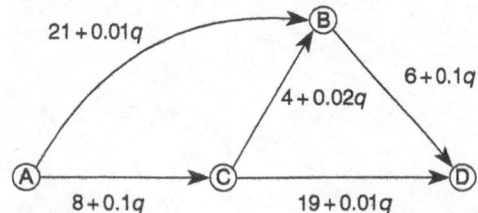

Figure 12.4 Simple network for Exercise 12.4

a) Identify the minimum-cost routes used, their flows and their corresponding equilibrium costs. What is the total expenditure in travel time in the network?

b) Assume that link CB is pedestrianised and therefore unavailable to vehicular traffic. Identify the new equilibrium flows, costs, and total expenditure in travel time in the network.

c) Discuss your results.

13
Departure Time Choice

13.1 Introduction

Peak spreading is a phenomenon widely observed in most large urban areas. As congestion increases, drivers change their departure times to avoid the worst delays and, therefore, the duration of the peak is increased. In large and congested urban areas, it is not uncommon to observe extended peaks (morning and evening) and an inter-peak period with high levels of flow and delay.

The change of departure time is probably the second most likely response to changes in travel conditions, the first being the change of route. This is mostly due to efforts to avoid the worst of congestion, but it should increasingly reflect greater flexibility in working hours, better pricing structures for toll roads, road user charging in general, as well as more sophisticated public transport fare systems.

Traditional approaches to modelling this phenomenon have been rather simple. It is always possible to adopt pragmatic assumptions about the duration of the peak in the future and how the expected demand will spread over this period. This requires only simple factoring of demand, perhaps assuming an elasticity, for future peak periods to generate reasonable levels of congestion and delay. However, these pragmatic approaches lend themselves to arbitrary decisions that will affect the evaluation of schemes and policies, and ignore the fact that travelling at a less desirable time increases the disutility of travel.

In this chapter, we outline current thinking in modelling time-of-day (TOD) choice. Because of its close relationship with assignment and delays, this theme integrates both assignment and choice modelling. The chapter first considers the distinction between macro and micro departure time choice approaches. We then provide a summary of the underlying principles behind the micro modelling of time-of-day choice and the seminal work of Small (1982). In Section 13.4, we introduce the idea of supply–demand equilibrium in this context, and in the next section we discuss the complexities of equilibrium assignment when the model includes a time-of-day choice component. In Section 13.6, we consider modelling time-of-day choice in a discrete choice model framework, and in the final section we add the complexity of incorporating mode choice into the model.

13.2 Macro and Micro Departure Time Choice

It is useful to distinguish between macro and micro time-of-travel choices. Macro time choice involves the selection of how to travel between broad time periods (say, 2–3 hours), for example, the decision to go shopping at an off-peak period instead of at 05:00 PM. Micro departure time

Modelling Transport, Fifth Edition. Juan de Dios Ortúzar and Luis G. Willumsen.
© 2024 John Wiley & Sons Ltd. Published 2024 by John Wiley & Sons Ltd.
Companion website: www.wiley.com/go/ortuzar5e

choice is related to the phenomenon of peak-spreading. As urban congestion increases, some travellers will choose to depart a bit earlier or later than originally desired to avoid the worst delays.

In principle, macro departure time choice can be approached using a Logit model to consider travelling at different periods. Each period will offer some advantages in terms of desirability and some disadvantages in terms of travel times and costs (parking and/or congestion charging). However, if the demand model uses the typical division of time into two peak periods and an inter-peak, the freedom of most trips to transfer between them will be constrained: only a few work trips, for example, could move outside the three-hour peak periods entirely; therefore, such a mechanism might be applied predominantly for shopping as opposed to the journey to work or education.

To model macro level choices, it is necessary to know what proportion of each type of trip takes place in each period. This information is best collected from household survey data that contains complete tours. At a macro level, trips must be allocated to discrete time periods, even those that start and finish in different periods. An incremental Logit model (see Chapter 14) can then be used to modify the total number of trips of each type in each time period according to changes in the mean generalised costs for each period. In these cases, it will be important to apply different sensitivity parameters to different trip purposes, since work/education and business travellers, for example, are less likely to reschedule their activities than shoppers.

Peak spreading is a case of micro departure time choice; it is a normal result of increased congestion, and many transport models treat it using simpler approaches. These try to avoid the problem of ending with some links beyond capacity (V/C > 1) by means of allocating some trips to the shoulders of the peak hour.

A common approach is to model a peak period, typically three hours, rather than a peak hour, and assume that all peak spreading takes place within that period, with the expectation that this peak period will not be loaded to a V/C > 1 in the future. A similarly drastic approach is to identify the links where V/C > 1 and to adjust, in effect capping, the elements of the trip matrix that use them, so that traffic does not exceed capacity; this will result in some 'lost trips' that can be allocated to other time periods, usually pre or post peak hour shoulders.

13.3 Underlying Principles of Micro Departure TIME Choice

A basic concept in micro departure time choice is that travellers have a 'preferred time' of travel and any shift away from it incurs disutility, known as *schedule disutility*. The preferred time of travel may be defined as the preferred departure time or preferred arrival time, the second one being more important for certain activities (e.g. work with a fixed starting time, business meetings, and theatre). The schedule disutility can be added to the travel time disutility to express a combined utility function for travel with variable departure time.

The work of Small (1982) inspires most applications of these ideas. If we focus on arrival time, Small's function takes the following form:

$$U(\tau) = -\alpha \cdot C(\tau) - \beta \cdot \text{SDE}(\tau) - \gamma \cdot \text{SDL}(\tau) - \delta \cdot d_L(\tau) \tag{13.1}$$

where τ is the arrival time and C is the travel duration, expressed as a function of the arrival time, since traffic conditions vary by time of day. SDE and SDL are called the *early schedule delay* and the *late schedule delay*. SDE and SDL express the difference between the chosen time of arrival and the preferred arrival time (PAT), in the case of early and late arrival, respectively. Therefore, SDE and SDL can be defined as:

$$SDE = \max (PAT - \tau, 0)$$
$$SDL = \max (\tau - PAT, 0)$$

(13.2)

The parameters α, β, γ, and δ are positive. α, β, and γ measure the disutility associated with a unit of increase in C, SDE, and SDL, respectively; δ is a fixed penalty for late arrival, and d_L is a dummy variable for late arrival (equal to 1 if τ > PAT and 0 otherwise). The fixed penalty is often omitted from the function and subsumed within the utility parameter for late schedule delay γ.

The utility function defined above can be regarded as the sum of a travel duration term $(-\alpha C(\tau))$ and a 'schedule utility' term expressing the variation in utility associated with the arrival time per se $(-\beta SDE - SDL - \delta d_L)$. This term is maximised when travellers arrive at their preferred arrival time (PAT) (i.e. τ = PAT, making the schedule utility equal to zero). Therefore, when travel duration is constant and no trade-off is possible between it and the schedule utility, the distribution of actual arrival times is identical to the distribution of PATs. However, when travel duration varies by time of day, travellers will shift from their preferred arrival time if the schedule disutility is outweighed by the gain from reduced travel time, resulting in a distribution of actual arrival times wider than the distribution of PATs.

The parameters for these combined utility functions can be estimated by stated preference (SP)/revealed preference (RP) techniques; see, for example, Small (1982), Bates (1996), or more recently Arellana et al. (2012). An example is shown in Figure 13.1, where the y-axis is the schedule utility in travel duration units (hours) and the x-axis is the arrival time, either earlier or later than PAT (hours).

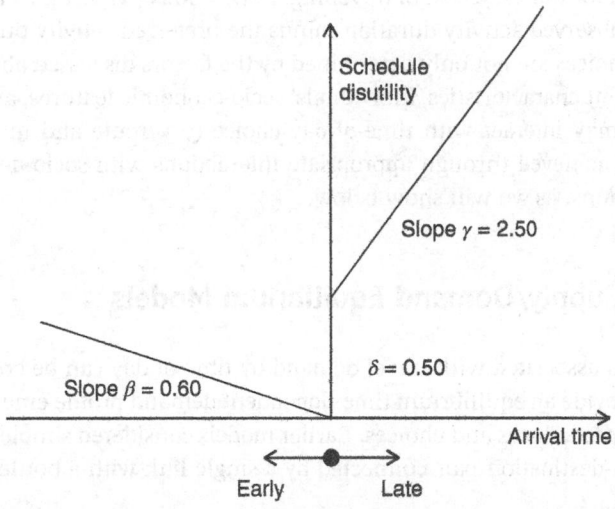

Figure 13.1 Idealised schedule disutility (equivalent minimum travel time)

The coefficients in the figure reflect an idealised disutility function. The asymmetry of the function is something observed in many SP/RP studies: a five minute delay in arrival is generally perceived as worse than arriving five minutes too early. In this idealised case, an arrival 30 minute earlier than PAT would be justified if the individual could achieve a travel time saving of more than

18 minutes (0.3 hours). An arrival 30 minutes later than PAT would incur a fixed penalty equivalent to 30 minutes of travel time plus an additional penalty of 90 minutes (1.25 hours). Therefore, this 30 minutes late arrival would only be justified if the individual could save more than two hours of travel time, an unlikely event. Of course, different (groups of) individuals will have different values for α, β, γ, and δ, and also different preferred arrival times.

The basic formulation, where the penalty term is omitted, was extended by Hyman (1997) to include an 'indifference band' around the PAT, during which arrivals incur no schedule delay. Hendrickson and Plank (1984) proposed a quadratic form of the utility functions, and Polak et al. (1991) proposed a piecewise linear model. Addison and Heydecker (1999) have also examined three classes of smooth functions, namely the sheared hyperbola, the super hyperbola, and a simple non-convex function, the Witch of Agnesi. Despite these efforts, most practical applications have relied on linear functions, with or without a fixed disutility for late arrival δ.

Good departure time models must consider travel time variability (Ettema et al. 2004) and the duration of activities, along with scheduling and associated levels of service information. Daily activity participation time is relevant too, due to its influence on trip making, the order of activity participation, and trip departure time choice. Performing other activities during the day could impose restrictions on departure time choices; therefore, both travel and the time required for each activity should be considered when modelling tours.

De Jong et al. (2003) and Hess et al. (2007b) reported schedule models (SM) including explicit penalties for decreased and increased activity participation time (PTD_i and PTI_i), using a generic utility function such as:

$$V_i = \theta_t\, t_i + \theta_C\, c_i + \beta\, SDE_i + \gamma\, SDL_i + \theta_{PTD}\, PTD_i + \theta_{PTI}\, PTI_i + \delta_L\, d_L \qquad (13.3)$$

where t_i and c_i stand for time and cost of travelling, $PTD_i = \text{Max}\{-PT_i, 0\}$, $PTI_i = \text{Max}\{0, PT_i\}$, and PT_i is equal to the observed activity duration minus the preferred activity duration.

Departure time choices are not only determined by the factors discussed above, but should consider also employment characteristics, individuals' socio-economic features, and information from other choices that may interact with time-of-day choice (e.g. route and mode choices), among others. This can be achieved through appropriate interactions with socio-demographic terms in the above specifications, as we will show below.

13.4 Simple Supply/Demand Equilibrium Models

The utility functions associated with travel demand by time of day can be combined with supply characteristics to provide an equilibrium time-dependent demand profile emerging from the interaction of travellers' preferences and choices. Earlier models considered simple network types, consisting of one origin-destination pair connected by a single link with a bottleneck in between, as depicted in Figure 13.2.

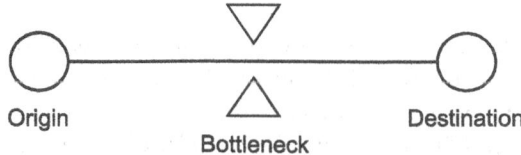

Figure 13.2 Simple network type

Vickrey (1969) examined equilibrium with a fixed number of identical commuters travelling through a single link, where flow is uncongested (travel time is constant and equal to zero for simplicity), except at a bottleneck with fixed capacity that causes delay directly proportional to the length of the queue. Applying the basic principle of equilibrium, namely that no commuter can increase their overall utility by altering their departure time, Arnott et al. (1993, 1994) extended Vickrey's model, calculating the resulting departure profile of the commuters.

Figure 13.3 illustrates the Arnott et al. (1994) equilibrium departure profile for homogeneous commuters. This is fully defined by closed-form expressions for the departure rates (q_1, q_2), the arrival times of the first and last arrival (τ_1, τ_2) and the switching time τ_p at which the departure rate changes from q_1 to q_2 and the maximum queue occurs.

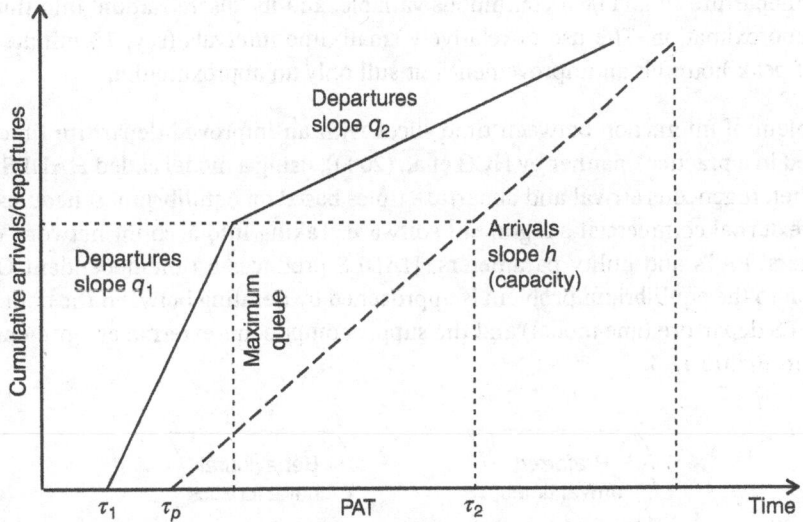

Figure 13.3 Equilibrium departure profile for homogeneous commuters

Several authors have extended Vickrey's model to account for heterogeneity among travellers in their PATs and/or in the parameters associated with travel duration and schedule delay. Hendrickson and Kocur (1981) considered Vickrey's problem when travellers have a distribution of PATs. Arnott et al. (1994) approached the issue of heterogeneity by segmenting the population into homogeneous subgroups according to the values of their PATs and utility parameters.

13.5 Time of Travel Choice and Equilibrium Assignment

Practical applications of the above principles require casting the problem in the context of variable demand equilibrium assignment modelling, since both the flows and the level of demand for every time period (slice) need to be determined. Given the progress discussed earlier in combining equilibrium assignment formulations with Logit choice models, it appears natural to use a similar approach to include time of travel choice as well. Willumsen et al. (1993) assumed that, for each O–D pair, $C(\tau)$ was variable in the peak (and computed through equilibrium assignment) but constant in time periods outside it; this, and the use of linear-in-the-parameters Small-like utility functions, enabled them to cast the problem in a simple combined Logit choice and equilibrium assignment formulation. The time of travel choice was then discrete: travel 'now' or travel during an 'earlier time slice' or during a 'later time slice'. Similar approaches have been put forward by Hendrickson and Plank (1984) and Cascetta et al. (1992).

These approaches, although superior to ignoring time-of-travel choices altogether, have limitations:

1) They ignore any interaction between time periods. Trips displaced from the peak to other time periods will increase their travel times, therefore, new calculations for $C(\tau)$ will be required. As it is generally not possible to estimate a function for $C(\tau)$ the full treatment of time-of-day choice is not possible, and, therefore, this approach is an approximation of the dynamics between different time periods.
2) Time of departure should be a continuous variable, and its 'discretisation' into time slices is a coarse approximation. The use of relatively small-time intervals (say, 15 minutes instead of peak/off-peak hours) is an improvement but still only an approximation.

The problem of interaction between time slices with an improved departure time model has been tackled in a practical manner by HCG et al. (2000), using a model called HADES; this model allows for heterogeneous arrival and departure times based on equilibrium schedules, and interfaces with external commercial assignment software. Taking into account network travel times and travellers' PATs and utility parameters, HADES produces a time-dependent O–D matrix. The solution to the equilibrium problem is approached by iterating between the demand component (HADES departure time model) and the supply component (external assignment model), as illustrated in Figure 13.4.

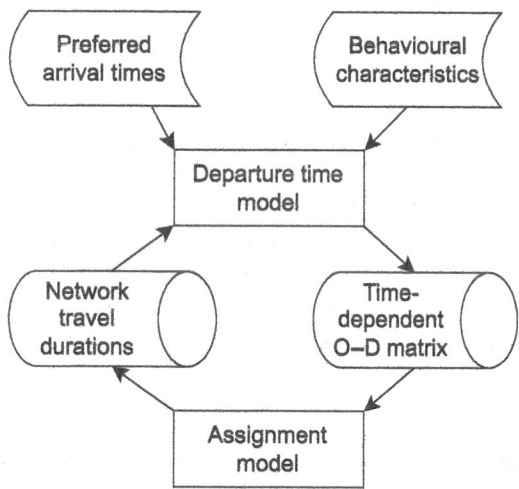

Figure 13.4 Operation of HADES in conjunction with an assignment model

The assumption of independent random utility components of the MNL model is restrictive and unrealistic in this case, as adjacent intervals are likely to be correlated because the unobserved attributes (i.e. random component of the utility function) affect the desirability of the alternatives in a similar way. Small (1987) also stressed the fact that correlation usually arises when the dependent variable is only a discrete representation of an underlying continuous variable; this is the case for the time variable. However, choice models have been extended and improved in a variety of ways to accommodate various patterns of stochastic correlation among alternatives, as discussed in Chapters 7 and 8, allowing to relax the assumption of independence of random components across alternatives.

Several researchers have pointed out that a richer representation of time-of-travel choice behaviour could be achieved within the context of all the activities undertaken by trip makers in each sojourn. This is certainly correct, but as indicated by Mahmassani (2000), it is not a simple problem, and perhaps we must wait until improved passive data collection methods become more common in practice. Novel measuring techniques using global positioning systems (GPS) and the now ubiquitous mobile phones significantly improve data collection in this field.

On the other hand, the discrete choice framework has some drawbacks that arise from the need to discretise time, a variable that is essentially continuous. Firstly, it is likely that the random errors of the trip departure time alternatives are correlated (de Jong et al. 2003), and that correlation increases as the intervals get more precise (e.g. 15 minutes). This issue can be addressed by using model structures such as the Nested Logit (NL) model (Williams 1977; Daly and Zachary 1978), the Ordered Generalized Extreme Value (OGEV) model (Small 1987; Bhat 1998), or the highly flexible Mixed Logit (ML) model, which can approximate any discrete choice model based on random utility theory (McFadden and Train 2000), as in recent studies using mobile-phone data (Bwambale et al. 2019). Finally, new studies have also estimated Hybrid Choice models, incorporating latent constructs to account for individual attitudes or perceptions (Haustein et al. 2018; Thorhauge et al. 2016, 2020), and Latent Class models (Cheng et al. 2020; Thorhauge et al. 2021).

Another issue, which cannot be mitigated even by the most flexible model structures, is that the length of the time intervals is generally determined arbitrarily by the modeller (Habib 2012), and models with different time intervals will likely lead to different results (Bhat and Steed 2002). In the literature, the length of the TOD alternatives has varied from 5 minutes (Small 1982) to fairly aggregate time periods (Tringides et al. 2004).

Although the joint mode-time-of-day choice has received the most attention; the choice decision, mode or time of day, and which individuals are more likely to change when level-of-service (LOS) variables vary, should be analysed specifically for each dataset. This is an important result to look for because it would indicate how travellers might respond to a travel demand strategy. For commuting trips, Hendrickson and Plank (1984) and Hess et al. (2007b) found that travellers were more sensitive to changing time of day than to change mode in their datasets. On the other hand, de Jong et al. (2003) found that travellers appeared to be more sensitive to mode changing than to modifying the times they travel. We will consider this issue in more detail in Section 13.7.

13.6 Modelling Disaggregate Time of Day Choice

There are useful opportunities for refining the treatment of time-of-day choice when modelling it at a disaggregate level. The choices available to the modeller include:

- use SP or RP data, or a combination of both;
- tackle the full schedule modelling (SM) of trips over time or deal only with TOD choice of particular trips;
- model the timing of trips on their own, or jointly with the choice of mode.

Consider first the use of SP experiments as the main source of information for time-of-day choice modelling: a wide range of different procedures have been used to obtain SP data for departure time modelling. Orthogonal in differences (Börjesson 2008) and fractional factorial designs (Tseng and

Verhoef 2008) are examples of standard design techniques used previously. Studies including a tour-based approach have based their SP surveys on more complex designs, combining orthogonal and manual designs to account for a large number of attributes and levels (de Jong et al. 2003; Ettema et al. 2007). However, except for the simulation step within the design procedure by Small et al. (1999), these design techniques have not used prior information about parameters. This absence of efficiency criteria in selecting attribute combinations potentially leads to larger sample size requirements. Koster and Tseng (2010) developed a procedure that includes efficiency criteria in the design generation to address one of the most important difficulties associated with generating SM-based choice experiments, namely that the variables used in the model are functions of the attributes shown to respondents in the survey rather than their actual values.

To achieve realistic choice experiments, the design procedure must also deal with, first, the potential dependency among different attribute levels of the same alternative (for example, travel time reliability and trip length), and, second, the fact that choice situations should be personalised to each respondent's circumstances. Both issues can lead to difficulties in producing a design that has good statistical qualities for the entire sample. For example, 'dependency', where attribute levels of alternative *j* are generated based on those of a reference alternative *i*, can be accommodated in pivot designs (Rose et al. 2008). However, additional complications arise when an attribute level within alternative *j* depends on another attribute level of the same alternative *j*, which in turn is also part of the design. This latter type of dependency is usually present in SM work. Not accounting for it (e.g. that travel time depends on departure time) can give rise to unrealistic choice situations, as can a failure to align scenarios with actual perceived possibilities in terms of realistic combinations from the respondent's perspective.

Note that while pivoting around current values can help in this context, customised levels must be carefully checked before applying the survey to avoid presenting unfeasible or irrelevant trade-offs to respondents. Occasionally, certain variation levels may not work well for the entire sample, as the differences postulated are too big or too small. For these reasons, Arellana et al. (2012) proposed the inclusion of additional constraints to give even more realism to the choice situations and avoid presenting 'meaningless' trade-offs from a respondent perspective. They used a Bayesian efficient SP-off-RP step design that accommodated interdependence among attribute levels and coped with the other difficulties. A necessary condition for developing these designs is to have prior information and reference point schedule data for each respondent. This is commonly the case when collecting a specific sample for the sake of conducting an SP survey. The procedure works as follows:

1) **Definition of preliminary design features**. This includes all activities prior to developing an efficient design, such as steps to:
   - define the context of the experiment and the attributes to be presented;
   - identify constraints and dependencies among attribute levels;
   - in the case of SM, optimise first the shifts in departure/arrival time;
   - identify a priori coefficients;
   - define the number of choice situations and, if necessary, block.
2) **Optimisation to obtain SP generic designs**. Since individuals face different choice situations, a generic design containing attribute levels expressed as relative changes from a reference point is required. This design may be common for all respondents, but it is also possible to create different designs for several predefined segments within the population. This stage will optimise

attribute levels without dependency relations and attribute levels that condition other attribute levels within the design. Within this phase, we need to:

- define efficiency and stop selection criteria;
- select a candidate SP design, including constraints to avoid dominance among alternatives;
- calculate probabilities and the asymptotic covariance matrix based on design attributes and *a priori* coefficients;
- calculate design efficiency;
- choose another SP candidate design until the stop criterion has been reached.

3) **Customisation of choice situations**. Here we move from a generic to a customised design for each respondent. The following activities should be performed:
   - customise choice situations using prior information (i.e. the actual choice) and reference point schedule data;
   - define non-optimised attribute base levels (e.g. based on measured travel times, observed cost, etc.);
   - include dependency constraints among attributes;
   - include other constraints if necessary (e.g. thresholds for differences between attribute levels).

4) **Selection of the final SP design**. This step is similar to the second, but with two fundamental differences: (i) the attribute levels optimised at this stage are different from those optimised at step 2; (ii) at this stage, a full covariance matrix is computed from the total sample data, considering the customised attributes presented to respondents. There is no common design for all respondents, but a tailored design containing as many rows as the number of participants times the number of choice situations per respondent.

5) **Simulation**: The purpose of this stage is to test if the best design obtained above can recover a wide range of 'true' coefficient values. The simulation must be done for the full sample.

6) Return to step 2 if the design does not allow recovering a wide enough range of 'true' coefficient values.

Example 13.1 To develop a departure time choice model for the city of Santiago, a three-step RP–SP-attitudinal survey was designed and applied to some 500 workers in the city. The first stage of the survey was a CAPI at the workplace, which focussed on collecting demographic and employment data, factors influencing scheduling decisions, and information about the schedule of planned activities for the following working day. Respondents did not have to report any trips done on the day of the first CAPI. The second stage, two days afterward, involved filling in a web page travel diary registering all trips completed before and after work during the previous working day. Finally, the third stage involved another CAPI to collect responses to a SP experiment based on recent experience (RP) along with an attitudinal questionnaire (both focussed on work-based trips). Information about respondents' income was also collected during this third stage.

Two sets of SP experiments were presented sequentially to each respondent for evaluating re-timing and/or mode switching behaviour, considering work hour flexibility and the implementation of congestion charging. The first focussed on trips to work in the AM peak (as shown in Figure 13.5a), while the second looked at complete work tours comprising outbound and return legs (Figure 13.5b). In each scenario, respondents faced a choice between four alternatives, of which the first three were for journeys on the current mode departing at different times (namely, travelling at early/current/late time), while the fourth offered the possibility of travelling by a different mode but around the same time as the originally reported trip. Public transport was the alternative mode for car users; if available, car was the primary alternative for transit users; if not, they were offered a new shared taxi service. To minimise the impacts of *inertia* or 'reading left-to-right' effects, the position of the re-timing alternatives was randomised across tasks for each respondent.

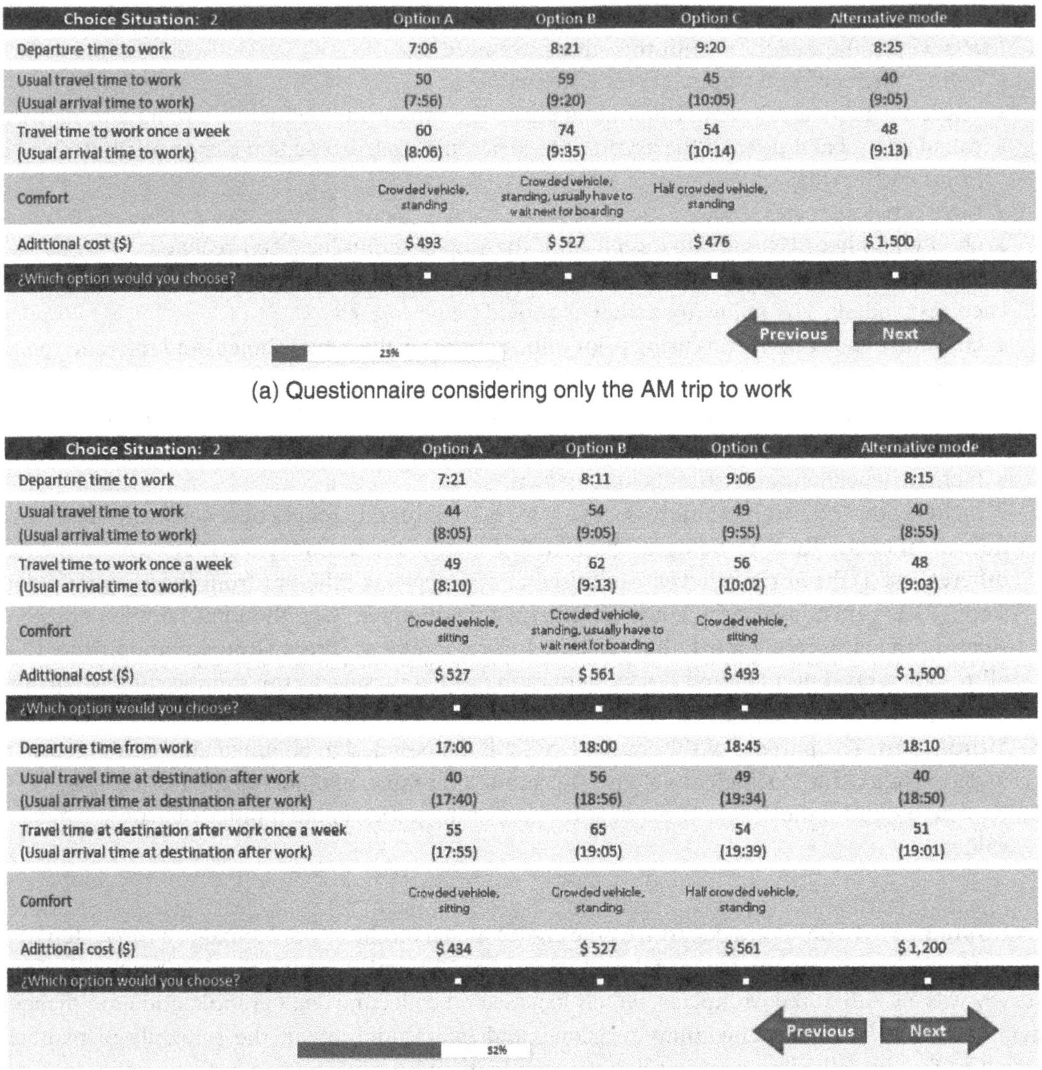

Choice Situation: 2	Option A	Option B	Option C	Alternative mode
Departure time to work	7:06	8:21	9:20	8:25
Usual travel time to work (Usual arrival time to work)	50 (7:56)	59 (9:20)	45 (10:05)	40 (9:05)
Travel time to work once a week (Usual arrival time to work)	60 (8:06)	74 (9:35)	54 (10:14)	48 (9:13)
Comfort	Crowded vehicle, standing	Crowded vehicle, standing, usually have to wait next for boarding	Half crowded vehicle, standing	
Adittional cost ($)	$ 493	$ 527	$ 476	$ 1,500
¿Which option would you choose?	▪	▪	▪	▪

25%

◄ Previous | Next ►

(a) Questionnaire considering only the AM trip to work

Choice Situation: 2	Option A	Option B	Option C	Alternative mode
Departure time to work	7:21	8:11	9:06	8:15
Usual travel time to work (Usual arrival time to work)	44 (8:05)	54 (9:05)	49 (9:55)	40 (8:55)
Travel time to work once a week (Usual arrival time to work)	49 (8:10)	62 (9:13)	56 (10:02)	48 (9:03)
Comfort	Crowded vehicle, sitting	Crowded vehicle, standing, usually have to wait next for boarding	Crowded vehicle, sitting	
Adittional cost ($)	$ 527	$ 561	$ 493	$ 1,500
¿Which option would you choose?	▪	▪	▪	
Departure time from work	17:00	18:00	18:45	18:10
Usual travel time at destination after work (Usual arrival time at destination after work)	40 (17:40)	56 (18:56)	49 (19:34)	40 (18:50)
Travel time at destination after work once a week (Usual arrival time at destination after work)	55 (17:55)	65 (19:05)	54 (19:39)	51 (19:01)
Comfort	Crowded vehicle, sitting	Crowded vehicle, standing	Half crowded vehicle, standing	
Adittional cost ($)	$ 434	$ 527	$ 561	$ 1,200
¿Which option would you choose?	▪	▪	▪	▪

52%

◄ Previous | Next ►

(b) Questionnaire considering the complete tour to and from work

Figure 13.5 Illustrative SP choice screens for both questionnaires

While the first experiment simply involved the choice between the four options, in the second, respondents had to make choices for both the outbound and return legs of their tours. This means that unless respondents decided to change mode, they had the possibility of choosing different alternatives for the outbound and return trips, generating a 10-alternative choice set as illustrated in Figure 13.6.

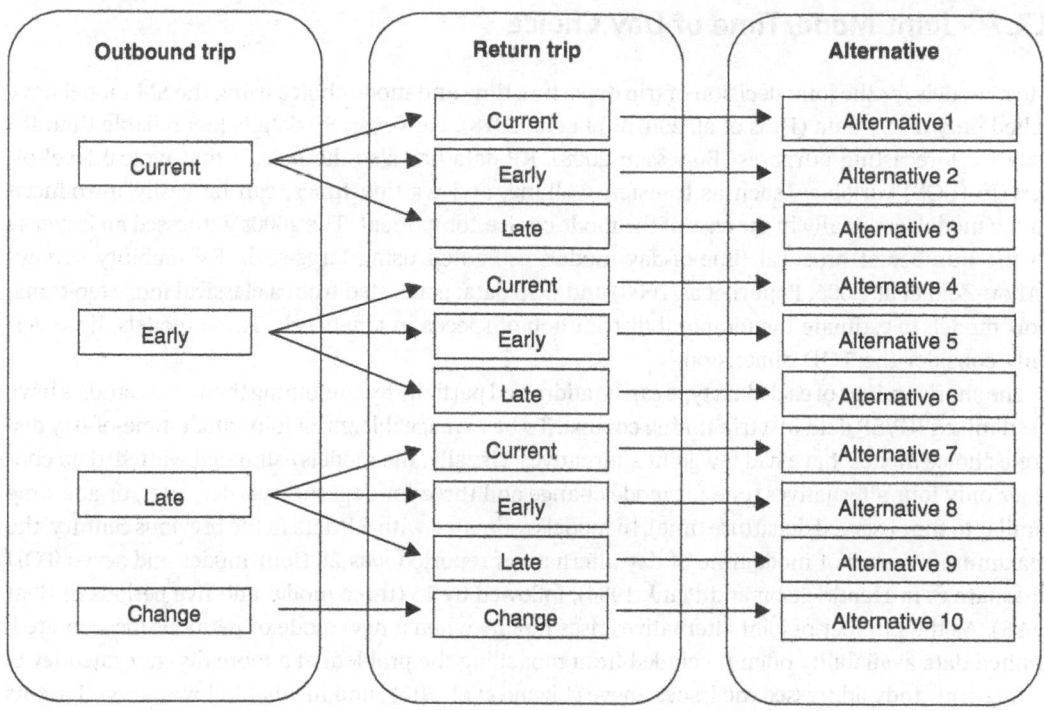

Figure 13.6 Available alternatives in tour models

Example 13.2 Using the data from the experiment described in Example 13.1, two cases were considered (i.e. trip data and tour data). For trip data, the following generic model was used:

$$V_i = \text{ASC}_i + \theta_t\, t_i + \theta_{\text{Time\_diff}}\, \text{Time\_diff}_i + \theta_C\, c_i + \beta\, \text{SDE}_i + \gamma\, \text{SDL}_i$$

where ASC is the alternative specific constant, Time_diff stands for the difference between 'worst' and 'best' possible travel times normalised by the 'best' travel time presented in each alternative. All remaining variables were as defined previously, with subscript i indexing the four alternatives in the trip data. Note that as the d_L constant was not significant, it was removed from the model.

For tour data modelling, to link trips before and after work, an *activity participation* penalty was introduced as proposed by Hess et al. (2007b). Here, the generic model can be written as follows:

$$V_i = \text{ASC}_i^{\text{outbound}} + \text{ASC}_i^{\text{return}} + \theta_t t_i + \theta_{\text{Time\_diff}} \text{Time\_diff}_i + \theta_C c_i$$
$$+ \beta^{\text{outbound}} \text{SDE}_i^{\text{outbound}} + \gamma^{\text{outbound}} \text{SDL}_i^{\text{outbound}} + \theta_{\text{PTD}} \text{PTD}_i$$
$$+ \theta_{\text{PTI}} \text{PTI}_i$$

where PTD_i and PTI_i are participation time, decreased, and increased.

Separate constants for outbound and return legs can be used across alternatives to capture general preferences for departing at specific times or on specific modes for either leg. The attributes time (t), time difference (Time_diff) and cost (c) refer to both legs, while SDE and SDL are outbound-specific (i.e. return-specific values cannot be included in a model that also has activity duration values). Subscripts i in this model represent the ten available alternatives.

In addition to the trip and tour models, a joint model was also estimated, allowing for scale differences between both SP games. A likelihood ratio test confirmed that the null hypothesis required by the joint model (i.e. that it was equivalent to the two separate trip and tour models) could not be rejected.

13.7 Joint Mode/Time of Day Choice

Most models for the joint decision of trip departure time and mode choice using the SM model have relied only on SP data (Hess et al. 2007b; Li et al. 2018). However, SP data is less reliable than RP data for forecasting purposes (Börjesson 2008). RP data has also the benefit that typical level-of-service (LOS) variables, such as transfer, walking, and waiting times, can be easily introduced in the model, especially in the case of the mode choice component. The 2000s witnessed an increase in the number of practical time-of-day models estimated using large-scale RP mobility surveys (Abou-Zeid et al. 2006; Popuri et al. 2008) and LOS data, generated from a classical four-step transport model, to estimate the temporal distribution of speeds in a network. These models, however, only consider the TOD dimension.

The shortcomings of each data type can be addressed partially by combining them. Few studies have used mixed RP/SP data in a trip timing context. To be manageable, most joint mode-time-of-day discrete choice models have had few joint alternatives. Usually, the models estimated with SP data consider only four alternatives (one for mode change and three for departing earlier, later, or at a time similar to the observed departure time). In models estimated with RP data in the previous century, the maximum number of mode-time of day alternatives reported was 28 (four modes and seven TOD alternatives in Hendrickson and Plank 1984), followed by 15 (three modes and five periods in Bhat 1998). As the number of joint alternatives rises rapidly when a new mode or period is incorporated, limited data availability often precluded from modelling the problem at a more disaggregated level.

A recent study addressed the issues above (Lizana et al. 2021) and in what follows, we will use its methodology to illustrate some important points.

13.7.1 Data Collection

A RP databank was built from a departure time survey of nearly 500 employees (Arellana et al. 2012, 2014), which included trip diary information for a specific day, as well as sociodemographic data and employment characteristics. Actual, rather than 'usual trips' were used because in the latter case people tend to report rougher information and data tends to be more approximate when dealing with alternatives that are not used regularly (i.e. times and costs at other times of the day).

In particular, the PAT was defined as the stated time that respondents aimed to arrive at work on the day of the survey, as this definition had been found superior to using the official work entry time as PAT (and, of course, it is more theoretically appealing). Furthermore, as daily variations in traffic in the short run were low due to the city's mild weather conditions, endogeneity issues in this sense (see Peer et al. 2013) were not expected.

The final RP databank considered nine modes: (i) car-driver, (ii) car-passenger, (iii) taxi, (iv) walk, (v) bus, (vi) metro, (vii) bus-metro, (viii) shared taxi, and (ix) a composite mode, called combination-metro, that grouped together trips transferring to/from metro by taxi, shared taxi, car-driver and car-passenger (this was done to reduce the number of joint mode-time of day alternatives, and took advantage that the number of choices of the four individual combined modes was low). One of the study objectives was to evaluate the effect of using different time period lengths; for this reason, three lengths were proposed: 15, 30, and 60 minutes (hereafter referred to as 15-minute, 30-minute and 1-hour databanks).

In the survey, *working flexibility* was measured by the number of minutes respondents could arrive early or late with respect to their official work starting times. Individuals were assumed to have all the time periods available and work flexibility was incorporated by interacting the schedule delay terms with dummy variables for the degree of flexibility. In the survey, the official work starting times were explicitly requested, as well as the existence of entry or exit work constraints. Therefore, in those cases where there was no official work starting time and no restrictions regarding entry work hours, respondents were requested to report their usual arrival time to work, and this value was used as the official work starting time. The survey also collected data regarding the route choices made by every respondent for their chosen mode.

13.7.1.1 Alternatives and Their Attributes

The trip departure time alternatives may be determined using the chosen (or observed) departure time as a reference point; 15 minutes (or 30 minutes/1 hour for the other two databanks) were added and subtracted from the observed departure time until the entire spectrum of time-of-day alternatives was covered. For example, Table 13.1 highlights the trip starting times of the time-of-day alternatives for an individual who declared having departed for work at 8:50.

Table 13.1 Alternative trip starting times for observed departure time at 8:50

15 minutes		30 minutes		1 hour	
5:00–7:00	6:50	5:00–7:00	6:50	5:00–7:00	6:50
7:00–7:15	7:05	7:00–7:30	7:20	7:00–8:00	7:50
7:15–7:30	7:20	7:30–8:00	7:50	**8:00–9:00**	**8:50**
7:30–7:45	7:35	8:00–8:30	8:20	9:00–11:00	9:50
7:45–8:00	7:50	**8:30–9:00**	**8:50**	11:00–14:00	11:50
8:00–8:15	8:05	9:00–11:00	9:20		
8:15–8:30	8:20	11:00–14:00	11:20		
8:30–8:45	8:35				
8:45–9:00	**8:50**				
9:00–11:00	9:05				
11:00–14:00	11:05				

As some individuals may report intermediate stops during their trip to work, the travel time variable has to be taken as the in-vehicle time from origin to destination, discounting the duration of intermediate stops. The duration of the intermediate activities is not considered in model estimation.

The resolution of the LOS data used in each databank was consistent with the interval length in each case (i.e. 15, 30, and 60 minutes). This was done with the aim of evaluating if LOS data aggregation had a negative impact on model results.

13.7.1.2 Stated Preference Data

SP data was consistent with that shown in Example 13.1. Every respondent had to answer five choice situations, where each comprised four alternatives with their respective attributes (see Figure 13.1 for the case of a public transport user). Three of these alternatives consisted of travelling in the same mode but at different times:

- start trip at a time similar to that observed in the RP component;
- start trip earlier; and
- start trip later than the observed starting time.

The fourth alternative entailed the option of switching modes and departing at a time close to the actual departure time. These four alternatives do not completely cover the range of options that commuters may actually have; for example, it is not possible to switch to public transport and leave earlier in order to arrive at work the same time. This was done to simplify the respondent's task. By limiting the choice set in this way, the number of people changing mode to public transport could be underestimated because the only option given to individuals switching to public transport was arriving later than usual, as transit has often longer travel times than car.

The attributes of each alternative are typically travel time (similar to that reported by the respondent), trip starting time, arrival time at work, monetary cost, a measure of travel time variability in the network (e.g. a larger travel time implied to occur once a week), and 'comfort' of the trip. This last attribute will normally appear for public transport alternatives and may consider vehicle occupation and whether the trip was made standing or sitting.

Table 13.2 shows the distribution of choices obtained in the SP experiment of Lizana et al. (2021). As can be seen, respondents tended to prefer their actual departure times and modes.

Table 13.2 Distribution of choices in SP experiment

Alternative	Choice (%)
Start trip at actual time	56
Start trip before	20
Start trip after	10
Switch mode	14

13.7.2 Model Estimation

The variables used in the models were divided into four groups: (i) level-of-service, (ii) scheduling, (iii) socioeconomic or interactions between socioeconomic variables and LOS or scheduling variables, and (iv) mode-specific constants (ASC). While the mode choice preferences are captured by the LOS variables, the TOD preferences are explained both by the LOS variables (mainly travel time) and the scheduling variables. The utility functions of the various alternatives, using an MNL structure as an example, were as follows:

$$
\begin{aligned}
V_{im}^{RP} = {} & ASC_m + \beta_{TT} \text{ Travel time}_{im} + \beta_{AT} \text{ Access time}_{im} \\
& + \beta_{Transfer} \text{ Transfer time}_{im}'' \\
& + \beta_{Transfer_{metro\_comb}} \text{ Transfer time}_{i,metro\_comb} \\
& + \left(\beta_C + \beta_{cw}/_{wage}\right) \text{ cost}_{im} \\
& + \left(\beta_{SDE} + \beta_{SDE-HF} \text{ High flexibility} \right. \\
& \left. + \beta_{SDE-MF} \text{ Medium flexibility}\right) SDE_{im} \\
& + \left(\beta_{SDL} + \beta_{SDL-HF} \text{ High flexibility} \right. \\
& \left. + \beta_{SDL-MF} \text{ Medium flexibility}\right) SDL_{im} + \beta_{Late} \text{ Late to work}_{im} \\
& + \beta_{IS} \text{ Intermediate stops}_{i,Car\ Driver} + \beta_{LC} \text{ License car ratio}_{im*}
\end{aligned}
$$

$$
\begin{aligned}
V_j^{SP} = {} & \mu\left(ASC_j + \beta_{TT} \text{ Travel time}_j + \left(\beta_C + \beta_{cw}/_{wage}\right) \text{cost}_j \right. \\
& + \beta_{TU} \text{ Travel time uncertainty}_j + \beta_{SDE} SDE_j + \beta_{SDL} SDL_j \\
& \left. + \beta_{Late} \text{ Late to work}_j + \beta_{Comfort} \text{ Seated}_j\right)
\end{aligned}
$$

With this notation, the subscript i refers to time period alternatives (i.e. 11, 7, and 5 for the 15-minute, 30-minute and 1-hour databanks, respectively), m indicates modes (i.e. nine modes), and j represents current, early, late, and switch mode alternatives in the SP environment. Subscript m'' refers to public transport alternatives, while m^* stands for both car-driver and metro combination alternatives; the coefficients β are to be estimated.

Two models were estimated with satisfactory statistical tests by Lizana et al. (2021). The first was a MNL model (both in the RP and SP utility functions), whereas the second had an error component Mixed Logit (ECL) structure in the SP utilities. The latter, more flexible structure, allowed to treat adequately the SP data *pseudo panel effect* (i.e. multiple answers made by each respondent, as we saw in Chapter 8).

Regarding the LOS variables, the in-vehicle travel time and cost coefficients were the same in the RP and SP utilities (i.e. generic). Access time (i.e. a combination of waiting and walking times) was only included in the RP utilities. The transfer time, separated into two coefficients – one for the mass transit modes (combination of bus and metro) and another for the combination-metro mode – was only used in the RP utilities. LOS variables included only in the SP functions were a dummy for *sitting comfort* level (expected to have a positive effect) and a travel time *uncertainty* measure. This last variable was defined as the percentage difference between the usual travel time and the once-a-week travel time shown in the experiment; it was expected to cause disutility because the bigger the percentage difference, the more uncertain the travel times were.

The model included the schedule delay early (SDE) and late (SDL) terms, and a dummy variable that took the value of one if the individual arrived late to work (i.e. after the official work starting time) for that specific time-of-day alternative. This dummy is slightly different from that proposed by Small (1982), because the *lateness penalty* in this case was defined with respect to the official work starting time and not with respect to the preferred arrival time.

Even though the same PAT definition was used to compute the schedule delay variables, these could be different for the RP and SP utilities because the models were estimated separately (note that the RP values were consistently higher than the SP values). Notwithstanding, the dummy for being *late to work* was considered generic across the RP and SP choices. Another scheduling variable was the *number of intermediate stops* on the journey to work. The most common intermediate stops were dropping off or picking up someone on the way to work; this variable was added only in the car-driver RP utility function, because driving should give more flexibility than other modes when making multiple stops on a journey.

The socioeconomic variables included the ratio between the number of cars and the number of driving licenses in the household with a ceiling of one, as is usual practice (Ben-Akiva and Lerman 1985); therefore, the availability of cars in the household was limited by the number of licenses. This variable was added only in the car-driver and combination-metro modes, as higher car availability should make these modes more attractive.

A variable cost divided by the hourly wage rate (w) was added to the utility function in addition to cost (and its parameter was the same for the RP and SP choices). Although it might appear theoretically and/or statistically mistaken to include it in the model at first sight, Jara-Díaz et al. (2012) have shown that in samples with relatively large coefficients of variation (CV) for income, an inordinate increase in the values of time (VOT) tends to appear when using the traditional *cost/w* variable, leading to misleading policy evaluations. In fact, in this sample, the income CV was 0.5, so its standard deviation was half the value of the average income, and it was found that the VOT indeed increased by 50% when only the *cost/w* variable was used. On the other hand, the VOT remained constant if only the cost parameter was used, independent of the income CV. Jara-Díaz et al. (2012) proposed using a model specification incorporating simultaneously both the *cost* and *cost/w* parameters to solve this problem, and this worked well in this application.

The models also considered a set of interactions between the schedule delay variables and dummies for the degree of work flexibility in the RP functions to capture the different temporal preferences of individuals with different levels of work schedule flexibility. Six categories were defined for the degree of flexibility: less than 15 minutes early and late (low flexibility early and low flexibility late), up to 30 minutes (medium) and more than 30 minutes (high). The early flexibility dummies interacted with the SDE term, while the late ones interacted with the SDL variable. These interactions were statistically significant, suggesting that the level of work flexibility influenced the temporal choices.

To mix data from different sources, a scale factor is needed (typically, the coefficient μ multiplying the SP utilities) to equate the random error variances associated with each data type, as we saw in Chapter 7. Lizana et al. (2021) found that the scale factor was statistically equal to one, suggesting that the variances of the RP and SP errors were equivalent in their case. This is an interesting finding, since in joint RP/SP analysis, it is mostly found that the SP data tends to have more variance than the RP data (Börjesson 2008). Finally, the variance of the error components was statistically different from zero, confirming the presence of unobserved heterogeneity regarding the multiple SP responses by each individual. When compared with the corresponding MNL model, the ECL model was clearly superior in terms of fit; however, the time uncertainty coefficient was less significant and the variance of the RP data was slightly bigger than that of the SP data (although the scale factor remained not significantly different from one).

13.7.2.1 Analysis of Results

Lizana et al. (2021) found that their 60-minute model followed a NL structure with mode choice located in the root and time-of-day choice in the lower nest of the hierarchy; this suggests that the time-of-day alternatives had a higher degree of substitution than the mode alternatives. However, the only nests with structural parameter ϕ smaller than one (as required by theory, see Section 7.4.2) were for car-passenger, car-driver, bus, bus-metro, walk and metro. This implies that only for these modes do the time-of-day alternatives appear to be correlated, while for the remaining modes they would appear to be independent.

Several NL structures (for example, first time-of-day choice followed by mode choice) were tested for the 15-minute and 30-minute databanks, but none was successful, leaving the MNL structure as the preferred one. This was unexpected, since time-of-day alternatives with a higher time resolution (i.e. 15 minutes versus 30 minutes and 60 minutes) should be more correlated, as there are more time alternatives and they are more similar between them (i.e. the times of departure are closer). A possible explanation could be that, as in the 15-minute and 30-minute databanks, the correlation structure was more complex (i.e. the number of mode/time-of-day alternatives were 99 and 63, respectively) than in the 60-minute base databank, the NL model could not handle it.

Lizana et al. (2021) also found that the values of in-vehicle travel, access, and transfer time progressively decreased when using more aggregate trip departure time alternatives. This change was more pronounced for access and transfer times, with reductions of up to 44% when comparing the 15-minute and 60-minute models. This difference could be attributed to the different number of alternatives in each model and because the correlation between alternatives was not replicated adequately in the 15- and 30-minute models (i.e. using a MNL structure in both the 15-minute and 30-minute models implies that the main difference between both is just the number of alternatives).

Furthermore, every minute of transfer time between a private mode (car, taxi, and shared taxi) and metro causes considerably more disutility than a minute of transfer time between mass transit modes. Also, access time had a higher value than travel time but was less valuable than transfer time.

This means that although transfer and access time are technically the same (i.e. a sum of walking and waiting time), people tend to dislike the mere fact of having to transfer; this has also been found by Navarrete and Ortúzar (2013) for public transport users.

13.7.2.2 Model Valuations

An important feature of the Scheduling Model is the trade-off ratio between the SDE and SDL parameters, and that of travel time. These are the relative values given to a minute of travel time compared to a minute early or late to work. In the RP dataset, the ratios of SDE/travel time and SDL/travel time were larger for people with low work schedule flexibility, and smaller for individuals with high flexibility. This is reasonable since individuals with higher work schedule flexibility are expected to put a lower value on being late or early to work. On the other hand, people with low flexibility levels must be willing to pay a premium to meet their work schedule commitments. For these people, arriving late can even carry monetary penalties.

Also, in RP, people appeared to value arriving a minute early or late to work more than one minute of travel time (in some cases, up to three times more), while in SP, the contrary occurred (the SDE/travel time and SDL/travel time ratios were lower than one). These differences could be attributed to several reasons. In the first place, there are different temporal perspectives between the RP and SP choices (something similar was found by Börjesson 2008). The RP choice is most likely the result of a long-term adaptation process; therefore, changing the trip departure time should imply less time at home or less time available for other activities early in the morning or later in the afternoon. On the other hand, it is probable that in a SP experiment, some people may be willing to change sporadically their trips, but not permanently (given the trip conditions of the SP choice situation); therefore, yielding lower coefficients (in absolute value) for SDE (and SDL) in SP than in RP.

A second and complementary explanation is that the context of the SP experiment entailed the implementation of a flexible working schedule scheme in the respondent's job (and the implementation of a congestion charge in the city as well). SP respondents were told that they could change their trip departure time, ignoring their previously stated work schedule flexibility, with the condition that they had to work the same number of hours in a week and consider their personal restrictions (e.g. activities with the family) as well. This is reaffirmed by the fact that the interactions between SDE (SDL) and the degree of work schedule flexibility were not significant in SP, while in RP they were highly significant. Also, the RP SDE(SDL)/travel time ratios of the 60-minute model for people with high work flexibility levels were more similar to the same ratios in SP. This suggests that individuals with a high capacity to change their trips were penalized for arriving a minute early or late to work in RP in a similar fashion to how all the SP respondents did.

Finally, a minute late was valued more than a minute early in RP (i.e. the ratio of SDL/SDE was higher than 1), while in the SP dataset people were indifferent to being early or late (the ratio was equal to 1). Reasons for this finding could be the same as those explained above.

13.8 Conclusion

The literature indicates a diversity of adopted approaches in modelling departure time choice, as well as scarce consensus. There is still much work to be carried out, both theoretically and in practice, to bring this important area of travel behaviour more into the mainstream of transport modelling.

Dynamic user equilibrium approaches seem to offer possibilities for incorporating departure time choice, possibly within a stochastic context, and a robust network performance sub-model predicting travel times on a continuous basis. However, such an explicit treatment of time requires a detailed description of flow through the network as well as robust solution algorithms, which makes it both analytically and computationally demanding. Its practical implementation in this form must await further developments in these fields.

A better avenue has been to try and overcome the limitations of the models proposed in software like HADES, both in terms of their internal consistency and of the time of travel choice model. We have shown some recent practical examples, but further work is needed to explore practical ways of incorporating these effects into strategic transport models for their use in congested urban areas.

14
Complementary Techniques

14.1 Introduction

This chapter covers a few complementary techniques to the use of a full strategic model as needed to develop a Transport Master Plan. Modellers are asked, many times, to provide advice on questions that do not require implementing a full conventional transport model. For example, which could be the impact of reducing public transport fares by $x\%$ or whether it would be appropriate to increase the tax on fuel to reduce greenhouse gas emissions. A quick response may be necessary; at least to have a 'first cut' estimate leading, perhaps, to a more detailed modelling effort.

The interest in quick response and simplified modelling techniques has spanned more than 40 years (see, for example, the compilation in Ortúzar 1992). As consultants and local authority modellers are often asked to study transport proposals in very short time spans, the development of better and sounder simplified methods will always be welcome. These quick-response approaches range from simpler *sketch modelling* tools to rigorous *incremental* approaches and techniques for adjusting trip matrices using multiple data sources. This chapter covers their range.

The idea of using quick-response models is not new. The practice of not using any formal model for transport project assessment is much more prevalent than what official documents and the technical literature would lead one to believe. Of course, the idea of not using any formal model simply means that decision makers are using their own, mental models, to make decisions. These may be quite powerful and certainly more sensitive to political and social variables than any formal mathematical effort.

Mental models are formed and refined through observation, analogies, discussions, experimentation, and mistakes; human rather than machine learning. Mental models are indeed essential to make use of formal ones, interpret their results, and add considerations normally outside their scope. For this end, the limited numerical processing ability of mental models is not a major limitation.

However, mental models have three major weaknesses: (i) sometimes they fail to identify all the impacts of decisions thus resulting in unintended consequences, more often than not negative ones; (ii) they cannot normally be 'opened up' to discuss them and qualify the recommendations resulting from their use; personal circumstances (e.g. car user, cyclist, etc.) and preferences influence problem definition and bias the analysis.

There is a whole range of modelling approaches in between the extremes of using no formal models at all and employing the most advanced and complex simulation techniques. One way of looking at these is to consider the manner in which different approaches represent space and, hence, distance, the key element in transport. Some models ignore space completely. These are usually of the kind that concentrate on the financial implications of subsidies, taxation, and so on. They may be

Modelling Transport, Fifth Edition. Juan de Dios Ortúzar and Luis G. Willumsen.
© 2024 John Wiley & Sons Ltd. Published 2024 by John Wiley & Sons Ltd.
Companion website: www.wiley.com/go/ortuzar5e

simple elasticity models, sometimes used to discuss fare increases or changes to petrol prices and car taxes. In other cases, they may include more complex interactions, for example, between road, petrol, and car taxes and car ownership and use.

Some authors have advocated the use of *structural modelling* techniques; see, for example, the interesting work of Roberts (1975) in respect of fuel consumption. In this case, a directed graph is often used to connect elements in the transport system, for example, the number of cars, fuel tax, improved fuel consumption, pollution emissions, and costs. Weights could be attached to these linkages to represent the relative strength of each relationship.

If weights are replaced with formal equations, calibrated from actual observations, one ends up with a *non-spatial interaction* model. Khan and Willumsen (1986) developed a model of this kind to enhance the study of car ownership in less developed countries; the thinking behind it was that in developing countries, car ownership should not just be forecast but examined together with its implications for resource allocation to roads and fuel consumption. The model included, in addition to the variables above, functions representing fuel consumption and the need for additional expenditure on road maintenance and new construction. Some of these, in particular construction and the importation of new cars, have severe implications for the balance of payment in these countries and should be explored before deciding on policies that may relax restrictions to car ownership and use.

A better representation of space can be obtained with *idealised models* of the type first proposed by Smeed (1968) and also used by Wardrop (1968) to study, among other policy issues, the limits of car commuting in urban areas. As more people use cars for the journey to work, more space needs to be devoted to roads and parking until radical changes are needed to the nature of the urban area. These models have seldom been used for decision-making but have served to illustrate important policy issues.

The next stage in space modelling involves simplifications to more conventional modelling approaches, as addressed in previous chapters of this book. Sketch planning models have been developed specifically to provide quick response and limited data collection requirements; they are discussed in Section 14.2. Increasing the degree of realism, we then discuss the idea of using simplified incremental split models in Section 14.3. Section 14.4 covers an important group of models that make use of readily available data, in particular traffic counts. Finally, the interpretation of model output and the use of models can also benefit from special training techniques; gaming simulation has been put forward as assisting in this area and is discussed in the last section of this chapter.

14.2 Sketch Planning Methods

Sketch planning models have been proposed as tools for long-range planning by many authors, as reported in OECD (1974) and Sosslau et al. (1978). These are models with a greater level of detail than the idealised network approaches mentioned in the previous section, but they are much simpler than conventional computer suites. This feature facilitates the analysis of broad transport and land-use strategies at a coarse level of resolution without requiring large amounts of data or the rigid assumptions of ideal space models. Their practical implementation ranges from scaled-down conventional aggregate modelling suites of programs to *ad hoc* approaches developed from some simple ideas and assumptions.

Most sketch planning methods rely considerably on the transfer of parameters and relationships from one area or country to another; only certain aspects of the models are made location dependent, usually network characteristics, population, income levels, and so on. At one extreme of sketch planning models are those relying heavily on assumed regularities in human behaviour in the transport field. A typical example was the Unified Mechanism of Travel (UMOT) model proposed by Zahavi (1979), which was based on the assumption that the following relationships are transferable over time and space (regions, countries):

- the average daily travel time per traveller (i.e. an assumption of constant travel time budgets);
- the average daily travel expenditure (money) as a function of income and car ownership (i.e. a money budget relationship);
- the average number of travellers per household as a function of household size and car ownership;
- the unit cost of owning and running a car;
- the speed–flow relationship by road type;
- the threshold of daily travel distance that justifies owning a car.

These relationships were developed following an extensive compilation of databanks from all over the world. UMOT only required as location-specific input the following:

- the number of households and their sizes in the study area;
- the income distribution of households;
- the unit cost of travel by mode;
- the length of the road network in the study area.

An interesting feature of UMOT was that it produced the following results as output:

- car ownership per household by income group;
- aggregate modal choice for the whole city;
- average travel times and speeds;
- other performance indicators like total expenditure and travel times.

UMOT gained some support as a tool for testing broad policy options, for example on fiscal policy (taxation), fuel and car ownership, public transport pricing, and even broad infrastructure investment programmes. However, the model was tested by Downes and Emmerson (1983), Ortúzar (1988), Willumsen and Radovanać (1988), among others. It was found that, in general, UMOT did not represent situations in other countries well, not even at a very high level of aggregation. In fact, the transferability of relationships and budgets was not found to be consistent enough to warrant the use of UMOT, even after improvements to the models were implemented by the authors.

Sketch planning techniques seem to offer advantages in terms of simplicity, fast response, and low data requirements. However, very often, they rely too heavily on the transfer of relationships and parameters from one context to another. This detracts from the analysis unless it is performed only as an initial coarse sketch to select possible solutions for more detailed consideration.

14.3 Incremental Demand Models

A number of approaches have been put forward to perform quick demand analysis of the impact of changes in fares, levels of service (LOS), or other attributes of a particular mode. The best-known methods fall under the headings of *incremental elasticity analysis* and *pivot-point modelling*. In both cases, the aim is to estimate small changes in demand as a result of (small) changes in one (seldom more) of the LOS attributes at a given point in time.

14.3.1 Incremental Elasticity Analysis

Consider an initial situation where the level of demand for a mode is T_0, its level of service S_0 (probably a vector including attributes like travel time, fare, and waiting time). The elasticity of demand with respect to LOS (at a given level of demand and LOS) is given by:

$$E_s = \frac{S_0}{T_0} \frac{\partial T}{\partial S} \approx \frac{S_0}{T_0} \frac{T - T_0}{S - S_0} = \frac{S_0}{T_0} \frac{\Delta T}{\Delta S} \qquad (14.1)$$

There is an initial distinction between *arc* and *point* elasticities. The right-hand side of (14.1) is an expression for the arc elasticity. As S approximates S_0, the elasticity will approach the exact 'point' value $\frac{\partial T}{\partial S}$. In general, point elasticities are more often estimated from demand models and arc elasticities from time series data. The above definition leads to:

$$T - T_0 = \frac{E_s T_0 (S - S_0)}{S_0} \qquad (14.2)$$

The left-hand side of this equation is the estimated change in demand for the mode due to a relative change in the level of service of size $(S - S_0)/S_0$. This type of calculation is often used during fare or frequency reviews for public-transport services. This is, of course, an approximation that assumes that we have calculated E_s beforehand (perhaps from time series data), that this elasticity is constant (or that the demand function is linear–not very likely), and that everything else remains the same. This result is a reasonable approximation for small changes in the LOS variables.

Example 14.1 The fare/demand elasticity of public transport is often taken to be -0.30. If a public-transport system carries 200 000 passengers in the peak period at an average fare of 80 pence/trip:

- estimate the fall in the demand if the average fare increases by 2.5%;
- find out how sensitive is the result to the elasticity value.

In this case $T_0 = 200\,000$; $E_s = -0.30$, and $(S - S_0)/S_0 = 0.025$, so using (14.2) we get:

$$T - T_0 = -0.30 \cdot 200\,000 \cdot 0.025 = -1500 \text{ passengers}$$

If $E_s = -0.2$, the expected reduction in patronage would be 1000 passengers; if it is -0.4, it would then be 2000 passengers.

It is also possible to define a *cross elasticity*, that is, the change in demand for one alternative (mode, destination, route) when the LOS of another alternative changes; say, the change in demand for inter-city rail when air travel fares increase.

We define cross-elasticities of demand for mode i with respect to attributes of mode j as:

$$E_{ij} = \frac{S_j}{T_i} \frac{\partial T_i}{\partial S_j} \approx \frac{S_j}{T_i} \frac{\Delta T_i}{\Delta S_j}$$

Elasticities for some demand functions with respect to changes in one attribute S of the LOS are given in Table 14.1:

Table 14.1 Elasticities for various demand functions

Type	Functional form	Elasticity
Linear	$T = \alpha + \beta S$	$E = \dfrac{\beta S}{T} = \dfrac{1}{1 + \alpha/\beta S}$
Product	$T = \alpha S^{\beta}$	$E = \beta$
Exponential	$T = \alpha \exp(\beta S)$	$E = \beta S$
Share	$p_i = \dfrac{T_i}{\sum_j T_j}$	$E_{S_i}(p_i) = 1 - p_i$ $E_{S_j}(p_i) = -p_j$

There are plenty of compilations of elasticities from around the world. A useful one is that from the Victoria Transport Policy Institute (www.vtpi.org) compiled by Todd Litman (http://www.vtpi.org/tdm/tdm11.htm). One would expect all elasticities with respect of components of generalised cost of travel to be negative (an increase in cost results in a reduction in demand).

Note that point elasticities, estimated from an analytical function, are symmetric: the absolute value of a positive change is the same as that of a negative change. However, we know from experience that this is not the case. The impact of a 10% increase in fares is greater than that of a 10% reduction; people place greater value on a loss than on a gain of the same magnitude, an issue that we also revisit in other chapters. Economists also distinguish between short- and long-term elasticities based on the fact that it may be difficult to adapt instantly to some changes in costs. For example, a moderate change in fuel costs may have little impact on travel in the short term as people will continue to travel to work. However, in the long run, people will change jobs and/or places of residence, and in considering these choices, they will also take into account travel costs and the availability of public transport, something they could not do in the short run. We would expect, therefore, long-term elasticities to be larger than short-term ones: demand should be more elastic in the long term.

It should also be noticed that if the change in costs is large, say a doubling of fuel prices, the additional expenditure incurred by travellers will affect consumption of other goods and services as incomes and budgets are fixed in the short run. This is an *income effect* equal to the change in consumption resulting from changes in one or more prices.

Finally, one can also estimate elasticities of travel demand to changes in attributes of the traveller (for example, income levels) or the region (for example, gross domestic product, GDP). We would expect these to be positive and most likely declining with per capita income levels. Evidence suggests that transport demand elasticities to GDP are greater than one in emerging countries but less than one in post-industrial ones. This is important, as it will help to de-couple economic development and traffic growth.

14.3.2 Incremental or Pivot-Point Modelling

Approaches seeking to model change in trip-making rather than absolute values for some year in the future have always been attractive. They tend to place emphasis on establishing a good reference point, often a solid and well-estimated base year, and then focus on the expected change in key level-of-service (LOS) attributes. We consider different methods in this category.

This classic *incremental* approach estimates future travel demand on the basis of knowledge of the current levels of demand and changes in the LOS variables for each alternative. In this case, we require knowing the demand function but not the specific values of the LOS variables that are not to change, for example, parking charges in different parts of a city. The only data needed are the current market shares of each mode and the proposed changes in the LOS variables; then, an incremental form of the demand model is used to 'pivot' around the current situation.

The incremental form of the Multinomial Logit (MNL) mode choice model was first given by Kumar (1980):

$$p'_k = \frac{p^0_k \exp\left(V_k - V^0_k\right)}{\sum_j p^0_j \exp\left(V_j - V^0_j\right)} = \frac{p^0_k \exp(\Delta V_k)}{\sum_j p^0_j \exp(\Delta V_j)} \qquad (14.3)$$

where p'_k is the new proportion of trips using mode k; p^0_k is the original proportion of trips by mode k; and $(\Delta V_k = V_k - V^0_k)$ is the change in the utility of using mode k, in our case generated by changes to the LOS attributes of mode k.

It is also possible to develop incremental forms for the Nested Logit model (Bates et al. 1987; Martínez 1987). In this case we will have a change in utility at the lower nest because: $\Delta V_i = \beta\,(V_i - V^0_i)$ and for choices above the lower nest the change in utility is the 'composite change' over the alternatives at the level below:

$$\Delta V^* = \ln \sum_i p^0_i \exp(\Delta V_i)$$

Example 14.2 Consider a transport system with three modes: car, bus, and rail with market shares 0.4, 0.45, and 0.15. Assume that the utility function has the following linear form:

$$V_k = -0.10 t_k - 0.20 w_k - 0.05 C_k / I + \delta_k$$

where t_k stands for in-vehicle travel time, w_k for waiting time and C_k / I, cost divided by income; δ_k is a modal penalty.

Assume also that we are only interested in changes in frequency that would reduce expected waiting time by rail from 10 to 7.5 minutes and increase that of bus from 3 to 4 minutes; therefore, we would have for rail:

$$V_r - V^0_r = -0.2(7.5 - 10) = 0.5$$

and for bus:

$$V_b - V^0_b = -0.2(4 - 3) = -0.2$$

The change in modal share would then be:

$$p'_r = \{0.15 \exp(0.5)\}/\{0.15 \exp(0.5) + 0.45 \exp(-0.2) + 0.4\}$$

the reader can verify that this produces:

$$p'_r = 0.24 \quad \text{and} \quad p'_b = 0.36$$

In the same vein, the singly constrained incremental Gravity model can be written as:

$$T_{ij} = \frac{G_i T^0_{ij}\, a_j \exp(-\beta \Delta\, GC_{ij})}{\sum_l T^0_{lj}\, a_j \exp(-\beta \Delta\, GC_{lj})} \qquad (14.4)$$

where G_i represents the total trips generated at zone i, ΔGC_{ij} the difference in generalised cost between the base and design years, and a_j growth factors reflecting changes in the destinations j.

Incremental forms for most travel choice models are not, in general, difficult to develop or implement. For example, Abraham and Kavanagh (1992) report on an incremental model for the whole of London, handling both mode and doubly constrained Gravity models for different person types and modes. This was implemented in EMME, taking advantage of its macro facilities. Other pieces of software have similar modules to implement incremental mode, distribution, and other Logit choice models (see Willumsen et al. 1993).

Incremental or pivot-point model formulations are helpful as we only need to account for changes in the generalised costs or utility functions, not their complete values. Therefore, if we are not introducing new modes, modal penalties can be ignored as they cancel out in ΔGC or ΔV. An additional advantage is that the model preserves the current (or base) matrices, therefore retaining any special associations detected in the data but never completely accounted for in a model; this is particularly valuable when dealing with destination choice where the Gravity model has never performed sufficiently well. The incremental Gravity model is expected to represent changes in the trip pattern resulting from changes in travel costs and generations and attractions.

The way pivot point or incremental models have been described is in accordance with the underlying principles of Logit and Gravity model development. A similar, and practical, idea is to focus on changes in demand as a result of changes in certain attributes, but using absolute models incrementally instead of pivot point models. The main motivation behind this approach lies in the difficulties in calibrating a distribution, and sometimes a mode choice, model that fits observations sufficiently well.

It is common practice in many countries, like the UK, to spend considerable resources in collecting origin–destination (O–D) data and developing one or more robust O–D matrices (by trip purpose and time of day). It is very difficult indeed to adopt any type of distribution or destination choice model that would not distort these matrices significantly. It is highly desirable, in these cases, to use the rich information in the 'observed trip matrix' $\left[T_{ij}^0\right]$ fully and attempt to model only 'changes' in trip patterns as a function of cost and trip-end future states.

In this case, modellers would use absolute models but apply them incrementally. To this end, an absolute (usually Gravity) general model is estimated for the base year $\left[GM_{ij}^0\right]$ and then used for a future year $\left[GM_{ij}^1\right]$. One approach would be to estimate the future matrix as:

$$T_{ij}^1 = \frac{T_{ij}^0}{GM_{ij}^0} GM_{ij}^1 \qquad \text{for all } ij$$

Note that this multiplicative approach is equivalent to adopting a full set of k factors in a Gravity model. The problem is that those cells in the base year matrix T^0 that are zero will remain zero in the future; this may be unrealistic for zones that are fairly empty in the base year but are expected to have increased activity in future years. An alternative approach that avoids this problem is to employ an additive form:

$$T_{ij}^1 = T_{ij}^0 + \left(GM_{ij}^1 - GM_{ij}^0\right)$$

This has the potential danger that some cells may turn out to have negative values that should be rounded up to zero. The essential feature of these two approaches is capturing any significant difference between the base year output from a calibrated model and the observations and passing on these differences to future forecasts.

These potential limitations have led to the development of rules to deal with cases of zero trips in the base year or when the growth predicted by the model is considered to be extreme and judgments must be applied to moderate it. Daly et al. (2012a) proposed a detailed analysis of each cell in the model and adopted two moderating parameters, X_1 and X_2, to deal with cases of extreme growth. They produced a table similar to Table 14.2, to process each of eight different cases:

Table 14.2 Rules to deal with extreme growth cases

Base	Modelled base	Modelled future	Predicted	Cell type
0	0	0	0	1
0	0	>0	T_{ij}^0	2
0	>0	0	0	3
0	>0	>0	Normal growth = 0	4
0	>0	>0	Extreme growth = $GM_{ij}^1 - X_1$	
>0	0	0	GM_{ij}^0	5
>0	0	>0	$T_{ij}^0 + GM_{ij}^1$	6
>0	>0	0	0	7
>0	>0	>0	Normal growth = $T_{ij}^0 \left(GM_{ij}^1 / GM_{ij}^0 \right)$	8
>0	>0	>0	Extreme growth = $T_{ij}^0 \left(X_2 / GM_{ij}^0 \right) + \left(GM_{ij}^1 - X_2 \right)$	

Source: Adapted from Daly et al. (2012a)

A common value for both X_1 and X_2 is 0.5, but the paper by Daly et al. (2012a) suggests further ways to refine this whenever necessary. This general approach has been widely used in incremental models, as empty cells are a common feature of trip matrices obtained by conventional methods.

14.4   Model Estimation From Traffic Counts

14.4.1   Introduction

Conventional methods for collecting O–D information from, for example, home or roadside interviews tend to be costly, labour intensive, and time disruptive to the trip makers. The problem is even more acute in developing countries, where rapid changes in land use and population shorten the 'shelf-life' of data. The need for developing low-cost methods to estimate the present and future O–D matrices is apparent.

Traffic counts can be seen as the result of combining a trip matrix and a route choice pattern. As such, they provide direct information about the sum of all O–D pairs that use the counted links. Traffic counts are very attractive as a data source because they are non-disruptive to travellers, they are generally available, they are relatively inexpensive to collect, and their automatic collection is well advanced. The idea of estimating trip matrices or demand models from traffic counts deserves serious consideration, and a number of approaches have been developed.

Consider a study area that is divided into N zones inter-connected by a road network that consists of a series of links and nodes. The trip matrix for this study area consists of N^2 cells, or $(N^2 - N)$ cells if intra-zonal trips can be disregarded. The most important stage for the estimation of a transport demand model from traffic counts is to identify the paths followed by the trips from each origin to each destination. The variable p_{ij}^a is used to define the proportion of trips from zone i to zone j travelling through link a. Thus, the flow (V_a) in a particular link a is the summation of the contributions of all trips between zones to that link. Mathematically, it can be expressed as follows:

$$V_a = \sum_{ij} T_{ij} \cdot p_{ij}^a, \quad 0 \le p_{ij}^a \le 1 \tag{14.5}$$

The variable p_{ij}^a can be obtained using various trip assignment techniques ranging from a simple all-or-nothing to a more complicated equilibrium assignment. Given all the p_{ij}^a and all the observed traffic counts ($\hat{V}_a$), there will be N^2 unknown T_{ij} values to be estimated from a set of L simultaneous linear equations such as (14.5), where L is the total number of traffic counts.

In principle, N^2 independent and consistent traffic counts would be required to determine uniquely the trip matrix **T**. In practice, the number of observed traffic counts is much less than the number of unknown T_{ij} values. Therefore, it is impossible to determine a unique solution to the matrix estimation (ME) problem. In general, there will be more than one trip matrix that, when loaded onto the network, will reproduce (satisfy) the traffic counts. Two basic approaches have been proposed to resolve this problem: *structured* and *unstructured* methods. In the first case, the modeller restricts the feasible space for the estimated matrix by imposing a particular structure, which is usually provided by an existing travel demand model, for example, a Gravity or Direct-Demand model. The unstructured approach relies on general principles, like maximum likelihood or entropy maximisation, to provide the minimum amount of additional information required to estimate the matrix. These two general approaches will be discussed below, but first we must consider the relationship between route choice and matrix estimation.

14.4.2 Route Choice and Matrix Estimation

Robillard (1975) classified assignment methods for trip matrix estimation from counts under two main groups: *proportional* and *non-proportional* assignment. The former methods make the proportion of drivers choosing each route independent from flow levels. The most common example is all-or-nothing assignment and, in this case, p_{ij}^a is defined as:

$$p_{jq}^a = \begin{cases} 1 & \text{if trips from origin } i \text{ to destination } j \text{ use link } a \\ 0 & \text{otherwise} \end{cases}$$

Pure stochastic assignment methods such as Burrell's and Dial's also fall into this group but in these cases p_{ij}^a can also take intermediate values between 0 and 1.

Non-proportional assignment techniques take explicit account of congestion effects and, therefore, the proportion of travellers using each link depends on the link flows. Equilibrium and stochastic user equilibrium assignment methods are members of this group. These techniques are thought to be more realistic for congested conditions.

However, the advantage of proportional assignment methods is that they permit the separation of the route choice and matrix estimation problems; the proportion of trips using each link p_{ij}^a can be assumed to be independent of the trip matrix to be estimated. In contrast, non-proportional route choice requires the joint or iterative estimation of route choice and trip matrices so that both are consistent. In what follows, we shall assume that proportional assignment methods are a reasonable approximation to route choice; we shall discuss later the extensions needed to cover non-proportional methods.

14.4.3 Transport Model Estimation from Traffic Counts

The calibration of a Gravity model was one of the first methods put forward for estimating trip matrices from traffic counts. The basic idea was to postulate a particular form of Gravity model and examine what happens when it is assigned to the network. For example, in the case of inter-urban travel the trip matrix could be:

$$T_{ij} = \frac{\alpha P_i P_j}{d_{ij}^2}$$

where P_j is the population of urban area j, d_{ij} is the distance between both areas, and α is a constant for calibration, in this case the only one. If a matrix of this kind is assigned onto the network we get:

$$V_a = \sum_{ij} \frac{P_{ij}^a \alpha P_i P_j}{(d_{ij})^2} = \alpha \sum_{ij} \frac{P_{ij}^a P_i P_j}{(d_{ij})^2} \tag{14.6}$$

Note that on the right-hand side of this equation the only unknown is α, the other variables are provided by external data and a reasonable route choice model. One can generalise this model slightly and include other trip generation/attraction factors like employment, industrial production, shopping floor space, and so on. If we denote the gravity part of this model by:

$$G_{ij} = \frac{O_i D_j}{d_{ij}^2}$$

and allow several journey purposes k (or commodities, if one is dealing with freight movements), one can write:

$$V_a = \sum_k \sum_{ij} P_{ij}^a \alpha_k O_i^k D_j^k / (d_{ij})^2 = \sum_k \alpha_k \sum_{ij} P_{ij}^a G_{ij}^k \tag{14.7}$$

Here, the α_k are parameters for calibration but the rest of the data are, once more, assumed to be available. It is relatively simple to see that the parameters α_k may be estimated using least squares. In this case, we postulate that $V_a' = V_a + \varepsilon_a$, where ε_a is an error term. A change of variable:

$$X_k = \sum_{ij} P_{ij}^a G_{ij}^k$$

permits writing:

$$V_a' = \alpha_0 + \sum_k \alpha_k X_k \tag{14.8}$$

where α_0 is the intercept and may be deemed to depict the part of the flow not represented by the Gravity model, for example, local or intra-zonal traffic. This type of approach was followed by the first researchers in this area, Low (1972) for urban areas and Holm et al. (1976) for planning inter-urban networks in Denmark.

Equation (14.7) has at least one obvious deficiency. If a particular O_i and a particular D_j are each doubled, then the number of trips between these zones would quadruple, when it would obviously be more likely that they should also double. To improve on this, the following more conventional model can be used:

$$T_{ij} = \sum_k \left[\alpha_k O_i^k D_j^k A_i^k B_j^k f_{ij}^k \right] \tag{14.9}$$

where α_k is a scale parameter that allows for different units for T_{ij} and O_i^k, D_j^k. A_i^k and B_j^k are the balancing factors expressed as:

$$A_i^k = \left[\sum_j \left(B_j^k D_j^k f_{ij}^k \right) \right]^{-1}$$

$$B_j^k = \left[\sum_i \left(A_i^k O_i^k f_{ij}^k \right) \right]^{-1}$$

and f_{ij}^k is a deterrence function, for example $\exp(-\beta_k C_{ij})$.

Estimating this more conventional model from traffic counts represents a greater effort as the parameters for calibration are now $A_i^k, B_j^k, \beta_k,$ and α_k. This calls for alternative calibration methods, for example, non-linear regression as used by Högberg (1976) or Robillard (1975).

Tamin and Willumsen (1989) generalised this approach following suggestions from Wills (1986) to combine features of the Gravity and the Intervening Opportunities (OP) models in a single function. Wills proposed a flexible Gravity-Opportunity (GO) model for trip distribution in which standard forms of the Gravity and OP models were obtained as special cases. The choice between Gravity and OP approaches is decided empirically by allowing the estimation of parameters that control the global functional form of the trip distribution mechanism.

We can define a transformation δ_{dj}^i such that it equals 1 if destination j is the dth position in ascending order of distance away from i, and zero otherwise; then, the ordered (opportunities) trip matrix can be obtained by the following transformation:

$$Z_{id} = \sum_j \left[\delta_{dj}^i T_{ij} \right] \tag{14.10}$$

Note that while the ordering transformation δ_{dj}^i produces an ordered trip matrix, its inverse $\left(\delta_{dj}^i \right)^{-1}$ allows the observed trip matrix to be recovered by

$$T_{ij} = \sum_d \left[\left(\delta_{dj}^i \right)^{-1} Z_{id} \right] \tag{14.11}$$

It should also be noted that this class of transformation is applicable to any variable based on the O–D matrix, notably the cost matrix and the proportionality factor, in addition to the trip matrix. We can also define a direct Box–Cox transformation such as (8.2) on a variable y as:

$$y^\tau = \begin{cases} (y^\tau - 1)/\tau & \tau \neq 0 \\ \log y & \tau = 0 \end{cases}$$

and an inverse Box–Cox transformation as

$$y^{(1/\tau)} = \begin{cases} (y\tau + 1)^{1/\tau} & \tau \neq 0 \\ \exp y & \tau = 0 \end{cases}$$

These transformations may be combined into a new function, which we introduce as a convex combination in μ,

$$y^{(\tau,\mu)} = \mu y^{(\tau)} + (1-\mu) y^{(1/\tau)}, \quad 0 \leq \mu \leq 1 \tag{14.12}$$

The proposed model can finally be written then as:

$$T_{ij} = \sum_k \left[\alpha_k O_i^k D_j^k A_i^k B_j^k f_{ij}^k \right] \tag{14.13}$$

where:

$$f_{ij}^k = \sum_d \left[\left(\delta_{dj}^i \right)^{-1} F_{id}^k \right] \tag{14.14}$$

$$F_{id}^k = \left(\sum_p^d U_{ip}^k \right)^{(\tau,\mu)} - \left(\sum_p^{d-1} U_{ip}^k \right)^{(\tau,\mu)} \tag{14.15}$$

$$U_{ip}^k = \exp \left[(1-\tau)\gamma_m \log D_{pk}^i - \beta_m C_{ip} \right] \tag{14.16}$$

and

$$D_{dk}^i = \sum_j \left[\delta_{dj}^i D_j^k \right] \tag{14.17}$$

The (τ, μ) transformation is defined by equation (14.12).

From this general form several special cases may be derived by setting τ and μ to particular values. Three extreme cases generating specific models are easily identified: the Gravity (GR), the pure Logarithmic-Opportunity (LO) and the pure Exponential-Opportunity (EO) models.

Three estimation methods were implemented by Tamin and Willumsen (1989) to calibrate the general form from traffic counts, namely: non-linear least squares (NLLS), weighted non-linear least squares (WNLLS), and maximum likelihood estimation (MLE). The general model was tested for both freight transport in Bali, Indonesia (Tamin and Willumsen 1992) and passenger traffic in Ripon, UK (Tamin and Willumsen 1989). In the first case, even if the traffic counts were not classified by lorry type it was possible to discriminate up to nine different commodity types, one of them empty trucks. In this case, proxy data for the O_i^k and D_j^k were required, for example, production levels of certain commodities. The parameter α_k then played the double role of converting these proxies first to tonnes and then to lorries.

The main conclusions from this research were:

1) The GO and OP models were more time consuming than the GR model since they required more complicated algebra and procedures, which took longer to solve.
2) Good fit at the traffic count level produced a general good fit at the trip matrix level as well.
3) Although Burrell's stochastic assignment was also used to estimate the p_{ij}^a, it gave no better fit to the traffic counts than all-or-nothing assignment.
4) Although the GO was the best model in terms of matching the observed traffic counts, the authors could not guarantee that it would also produce the best fit to an independently observed trip matrix. In fact, it was found that the model that gave the best fit at the trip matrix level was the GR model with the NLLS method and Burrell assignment.

Holm et al. (1976) extended the gravity model approach to include some features of equilibrium assignment. They made use of an iterative loading with $\phi = 1/n$ (see Section 10.5.4) to obtain the proportion of trips using each link. However, this is only a heuristic approximation as under strict equilibrium conditions the proportions are not, in general, unique.

Other, perhaps Direct-Demand models, could also be used in this type of estimation method. One interesting advantage of the approach is that once a demand model is calibrated it may be used for forecasting purposes too, provided future values for parameters like O_i and D_j are available or estimable.

14.4.4 Matrix Adjustments Using Traffic Counts

Entropy-maximising and information-minimising techniques have been used as model-building tools in urban, regional, and transport planning for many years, particularly after the seminal work of Wilson (1970). For example, we discussed the derivation of the conventional Gravity model from an entropy-maximising formalism in Chapter 5. In this context, entropy-maximising provides a naive, least-biased trip matrix that is consistent with the information available as constraints to a maximisation (of an entropy function) problem. In the case of the Gravity model, the constraints represent trip-end and total cost information.

This idea was used by Willumsen (1978) to derive a model to estimate trip matrices from traffic counts. The problem can be written as:

$$\text{Maximise } S(T_{ij}) = -\sum_{ij} \left(T_{ij}\log T_{ij} - T_{ij}\right) \tag{14.18}$$

subject to:

$$\hat{V}_a - \sum_{ij} T_{ij}p_{ij}^a = 0 \tag{14.19}$$

for each counted link a, and:

$$T_{ij} \geq 0$$

Constraints (14.19) replace the trip-end and cost constraints of the Gravity model derivation. The use of Lagrangian methods permits the formal solution to this problem to be found as:

$$T_{ij} = \exp \sum_a \left(-\tau_a p_{ij}^a\right) = \prod_a X_a^{p_{ij}^a} \tag{14.20}$$

where τ_a are the Lagrange multipliers corresponding the constraints (traffic counts) and,

$$X_a = \exp\left(-\tau_a\right)$$

The availability of an old matrix, or simply a matrix estimated (or cordoned off) from another study, could be accommodated to some advantage. Let **t** be this prior matrix, sometimes called a 'reference trip matrix'; in this case, the new objective function becomes:

$$\text{Maximise } S_1\left(T_{ij}/t_{ij}\right) = -\sum_{ij} \left(T_{ij}\log T_{ij}/t_{ij} - T_{ij} + t_{ij}\right) \tag{14.21}$$

subject to the same constraints (14.19) and non-negativity. This objective function is, of course, convex and the term t_{ij}, being a constant, is only there for convenience; it can actually be eliminated from the derivation of the model.

Using the same methodology and change of variables, the formal solution can be seen to be:

$$T_{ij} = t_{ij} \exp \sum_a \left(-\tau_a p_{ij}^a\right) = t_{ij}\prod_a X_a^{p_{ij}^a} \tag{14.22}$$

Example 14.3 Consider the simple network depicted in Figure 14.1. This network has two origins (1 and 2) and two destinations (3 and 4). The flows on all links are also shown in this figure.

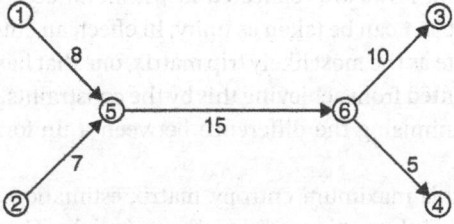

Figure 14.1 Simple network with traffic counts

There are only six (integer) trip matrices that can reproduce the observed flows as shown in Table 14.3:

Table 14.3 Integer trip matrices for Example 14.3

Matrix	First		Second		Third		Fourth		Fifth		Sixth	
$\begin{array}{c}\quad j\\ i\quad\end{array}$	3	4	3	4	3	4	3	4	3	4	3	4
1	8	0	7	1	6	2	5	3	4	4	3	5
2	2	5	3	4	4	3	5	2	6	1	7	0
$S(T_{ij})$	−11.07		−7.46		−5.98		−5.78		−6.84		−9.96	
$S_1(T_{ij}/t_{ij})$	−5.79		−3.69		−3.70		−5.07		−7.22		−12.20	

The entropy-maximising formalism seeks to identify the most probable trip matrix consistent with the information available, in this case five traffic counts. Incidentally, the reader can verify that only three of these counts are independent (see Section 14.4.5); therefore, the problem is, indeed, underspecified.

The values of the objective function $S(T_{ij})$ are also shown in this table. According to this, the most probable trip matrix would be the fourth, {5, 3, 5, 2}, as it has maximum entropy value. If a prior matrix was available then a second objective function (14.21) should be used. Assume the prior matrix {3, 2, 1, 3} is available; the new values from the entropy function are also depicted above. The most probable trip matrix in these circumstances would be the second one, {7, 1, 3, 4}.

Of course, in more practical problems, we cannot hope to directly calculate the entropy values of all possible matrices. Note, for instance, that reducing the number of counts increases the number of feasible trip matrices. More importantly, flows of the order of hundreds or thousands increase the number of possible (integer) trip matrices enormously. What is needed is an effective solution method that does not require matrix identification.

There are several possible methods to solve model (14.22). The most widely used is the multi-proportional approach. This is, in essence, an extension of the bi-proportional and tri-proportional methods discussed in Chapter 5. In this case, instead of balancing the trip matrix and trying to match trip-end totals (and cost-bin totals in the tri-proportional case), we undertake successive corrections to the prior trip matrix in order to reproduce the observed traffic counts. There is one correction factor X_a for each traffic count, and its calculation involves the iterative estimation of these factors until the observed link flows are replicated to within an acceptable tolerance.

If no prior matrix is available, **t** can be taken as unity; in effect, an entropy-maximising formalism may be considered to generate as the most likely trip matrix, one that has the same number of trips in each cell, unless being prevented from achieving this by the constraints. In other words, maximising entropy is equivalent to minimising the difference between a uniform target and the estimated matrix.

The detailed analysis of this maximum entropy matrix estimation (ME2) model and that of a related approach, based on information-minimising principles, is given by Van Zuylen and Willumsen (1980). Both models are practically equivalent and share most of their properties. The ME2 model will always reproduce the observations V'_a to within a given tolerance provided the constraints define a feasible space, that is, Eq. (14.19) must have at least one solution in non-negative T_{ij}. An additional condition for the prior matrix **t** is discussed below.

It can be shown that minimising the negative of the objective function (14.21) is approximately equivalent to minimising:

$$S_2(T_{ij}/t_{ij}) = \frac{0.5(T_{ij} - t_{ij})^2}{T_{ij}} \tag{14.23}$$

This is an error-like measure of the difference between the values of t_{ij} and T_{ij}. In effect, the negative of $S_1(T_{ij}/t_{ij})$ is also a natural measure of the difference between these cell values: it is zero when $t_{ij} = T_{ij}$ and increasingly positive as the difference increases. In this sense, the estimated matrix is that closest to the prior matrix, which when loaded onto the network, can reproduce the traffic counts.

The model can accommodate other sources of data, provided they can be incorporated as linear constraints. An example of this type may be information about the trip length distribution (TLD) thought to be realistic for the study area. This type of information can be translated into constraints equivalent to those of cost bins, as discussed in Chapter 5; for example:

$$\frac{1}{T}\sum_{ij} T_{ij}\delta_{ij}^k = P_k \tag{14.24}$$

where T is the total number of trips, P_k is the proportion of trips in cost (length) range (bin) k, and δ_{ij}^k is 1 if trips between i and j have cost in range k, and zero otherwise.

Public-transport systems with a zonal or other variable fare system permit the introduction of constraints of this type to help estimate the corresponding trip matrices using passenger counts and ticket sales data (see de Cea and Cruz 1986).

Moreover, the mathematical program can also be written with a combination of equality and inequality constraints, thus enhancing the value of this type of approach. For example, the planner may know that the capacity of a link is Q_a but not have a traffic count for it; or that no more than D_j' vehicles can go to a particular destination because of parking capacity there. This type of information can be incorporated as inequality constraints, for example:

$$\sum_{ij} T_{ij}p_{ij}^a \leq Q_a \quad \text{for some links } a \tag{14.25}$$

$$\sum_i T_{ij} \leq D_j' \quad \text{for some destinations } j \tag{14.26}$$

The solution to this program is still a multiplicative model; Lamond and Stewart (1981) have shown how the multi-proportional algorithm can be extended to handle inequality constraints; therefore, the same solution method may be used for this expanded model.

One of the features of the (extended) ME2 model is its multiplicative nature. This means that if a cell in the prior matrix is zero it will remain zero in the solution as well. This may be a source of problems if the cell in the prior matrix was zero by chance (i.e. because of the sampling rate adopted in the study) instead of representing an O–D pair with no trips at all. One pragmatic solution to this problem, for very sparse prior matrices, is to 'seed' the empty cells with a small value, for example 0.5 trips. The constraints, through the multi-proportional or other solution algorithm, will then ensure that some of these trips 'grow' to one or more full trips while others regain a zero value.

Example 14.4 Consider the same network of Example 14.3, but assume that now we only have two traffic counts, on links 5–6 and 2–5 (15 and 7). Table 14.4 shows the multi-proportional algorithm as applied to this problem. First, the full solution for the case of uniform (no) prior matrix, Case A.

Table 14.4 Multiproportional solution for two traffic counts

		Traffic count	Modelled flow	Ratio	Trips per O–D pair			
					1–3	1–4	2–3	2–4
A	*Prior matrix*				*1.00*	*1.00*	*1.00*	*1.00*
	Iteration	15	4.00	3.750	3.75	3.75	3.75	3.75
	1	7	7.50	0.933			3.50	3.50
	Iteration	15	14.50	1.034	3.88	3.88	3.62	3.62
	2	7	7.24	0.967			3.50	3.50
	Iteration	15	14.76	1.016	3.94	3.94	3.56	3.56
	3	7	7.11	0.984			3.50	3.50
	Iteration	15	14.89	1.008	3.97	3.97	3.53	3.53
	4	7	7.05	0.992			3.50	3.50
	Iteration	15	14.95	1.004	3.99	3.99	3.51	3.51
	5	7	7.03	0.996			3.50	3.50
B	*Prior matrix*				*3.00*	*2.00*	*1.00*	*3.00*
	Iteration	15	15.03	0.998	4.81	3.21	1.75	5.24
	5	7	6.98	1.002			1.75	5.25
C	*Prior matrix*				*3.00*	*2.00*	*0.00*	*3.00*
	Iteration	15	15.06	0.996	4.82	3.21	0.00	6.97
	6	7	6.97	1.004			0.00	7.00
D	*Prior matrix*				*3.00*	*2.00*	*0.50*	*3.00*
	Iteration	15	15.04	0.998	4.81	3.21	1.00	5.99
	6	7	6.98	1.002			1.00	6.00

As can be seen, it takes only five iterations to reach convergence within 5% tolerance. The solution {3.99, 3.99, 3.5, and 3.5} does not coincide with the maximum-entropy solution of Example 14.3, because the number of traffic counts is not the same. Case B shows the problem with the prior matrix {3, 2, 1, 3}; again, it takes only five iterations to reach satisfactory convergence. The solution {4.81, 3.21, 1.75, and 5.25} is indeed different, thus showing how the information contained in an outdated trip matrix can be used to advantage in matrix estimation; there is something of value in past information worth making use of.

Case C illustrates what happens when there is a zero entry in the trip matrix. There is still a solution but the zero is preserved in it. Finally, Case D shows the effect of 'seeding' the zero in the prior matrix with 0.5. The solution this time, {4.81, 3.21, 1.0, and 6.0} affects only trips from the origin previously containing the zero.

Consider now the effect of increasing the number of counts to three by including link 6–3. The corresponding results are depicted in Table 14.5.

First, note that the number of iterations required has now increased. This seems to depend not so much on the actual number of counts used but on how close they are to removing all flexibility in the matrix. In this case, three out of four degrees of freedom are removed by these counts. The solution in case A, {5.33, 2.68, 4.67, and 2.35}, is the one that maximises $S(T_{ij})$ and if rounded to integers coincides with the solution in Example 14.3.

The solution for case B, {6.55, 1.51, 3.45, and 3.58}, has the same properties in respect of $S_1(T_{ij})$. Case C is interesting as it shows that with the inclusion of a zero in the prior matrix, the algorithm

Table 14.5 Multiproportional solution for three traffic counts

		Traffic count	Modelled flow	Ratio	Trips per O–D pair			
					1–3	1–4	2–3	2–4
A	*Prior matrix*				*1.00*	*1.00*	*1.00*	*1.00*
	Iteration	15	4.00	3.750	3.75	3.75	3.75	3.75
	1	7	7.50	0.933			3.50	3.50
		10	7.25	1.379	5.17		4.83	
	Iteration	15	15.05	0.997	5.32	2.68	4.65	2.35
	10	7	7.00	1.000			4.65	2.35
		10	9.97	1.003	5.33		4.67	
B	*Prior matrix*				*3.00*	*2.00*	*1.00*	*3.00*
	Iteration	15	15.11	0.992	6.51	1.51	3.41	3.56
	14	7	6.97	1.004			3.42	3.58
		10	9.94	1.006	6.55		3.45	
C	*Prior matrix*				*3.00*	*2.00*	*0.00*	*3.00*
	Iteration	15	17.15	0.875	8.75	0.13	0.00	6.12
	20	7	6.12	1.143			0.00	7.00
		10	8.75	1.143	10.00		0.00	
D	*Prior matrix*				*3.00*	*2.00*	*0.50*	*3.00*
	Iteration	15	15.10	0.994	6.98	1.05	2.96	4.01
	19	7	6.97	1.004			2.97	4.03
		10	9.95	1.005	7.01		2.99	

fails to converge, even after 20 iterations. The reader may verify that forcing cell 2–3 to zero makes the problem unfeasible: there are seven trips out of node 2 but only five are permitted to reach their destination. Case D illustrates the effect of seeding the empty cell with 0.5 trips; the algorithm now converges to a reasonable solution.

14.4.5 Traffic Counts and Matrix Estimation

One can ask at this stage whether any set of counts is suitable for trip matrix estimation. For example, is it possible that certain combinations of counts make it impossible to estimate a matrix that satisfies them? These problems will be discussed under the headings of *independence* and *inconsistency* in traffic counts.

14.4.5.1 Independence

Not all traffic counts contain the same amount of 'information'. For example, in Figure 14.2 traffic link c is made up of the sum of traffic on links a and b. Counting traffic on link c is then redundant and only two counts there can be said to be independent.

Wherever a flow continuity equation of the type 'flows into' a node equals 'flows out of' the node can be written, its counts will be linearly dependent. In this case, it will always be possible to describe one link flow as a linear combination of the rest. Note that a centroid connector attached to node 5 would remove the dependency in Figure 14.2.

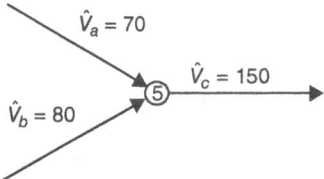

Figure 14.2 Dependent counts

14.4.5.2 Inconsistency

Counting errors and the fact that traffic counts are often obtained on different occasions (hours, days, or weeks) are likely to lead to inconsistencies in the flows. In other words, the expected flow continuity relationships will not be met. If the count V_c in Figure 14.2 had been 160 instead of 150, the corresponding equations would be inconsistent, and no trip matrix could possibly reproduce these flows. One way of reducing this problem is to allow an error term in the equations or to remove the inconsistencies beforehand.

It is possible to identify two sources for inconsistencies in the link flows. The first one is simply the fact that errors in the counts may lead to situations in which the 'total flow into' a node does not equal the 'total flow out of' the same node, thus not meeting link flow *continuity* conditions. The second source is a mismatch between the assumed traffic assignment model and observed flows. For example, an assignment model may allocate no trips on a link having an observed (perhaps small) flow. In these conditions, there will be no trip matrix capable of reproducing the observed link flows using that route choice model.

Example 14.5   It is useful to distinguish between these two types of inconsistency, first at *flow level* and then at *path flow level*. Assume, we have observations about flows of four links (identified by the pair of nodes delimiting them) and we would like to find non-negative trip matrices satisfying these and a route choice model as depicted in Figure 14.3.

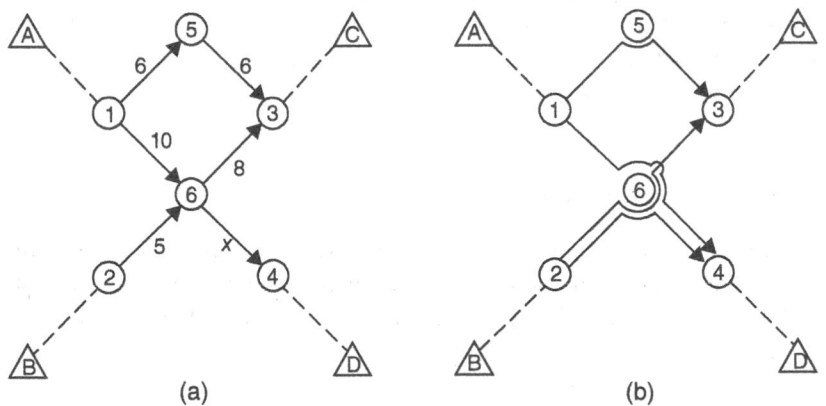

Figure 14.3 An example of path flow inconsistencies with counts

Consider first the case where the count x has been found to be 8, thus making the total flow into node 6 equal to 15, and the flow out of this node equal to 16. These counts are then inconsistent, perhaps because they were taken on different days or simply because of counting errors. We can remove this inconsistency by arbitrarily increasing the flows on links $(1, 6)$ or $(2, 6)$ by one, or by reducing the flows on links $(6, 3)$ or $(6, 4)$ by one. We can be more systematic and make the least adjustments necessary to preserve flow continuity conditions. For example, if we want to minimise the sum of squares of the increments/reductions, then the optimum change would be 0.25 on each link.

An alternative approach is to seek a maximum-likelihood solution to this problem, as proposed by Van Zuylen and Willumsen (1980). This assumes that link flows are Poisson distributed and that the observations available are samples from this distribution. Maximum likelihood is then used to generate a model for producing improved and consistent estimates of the flows. On the other hand, model calibration from traffic counts, as discussed in the previous section, makes an explicit allowance for errors in the observed link flows. These methods are not limited, therefore, by independence and consistency problems.

Consider now the case when the count x is 7. It can be seen that now the link flow continuity conditions are met. However, the assumed assignment depicted in Figure 14.3b is incompatible with the flows shown in Figure 14.3a. No feasible trip matrix can reproduce the count of 8 at link $(6, 3)$ because the only path using it, B–C, is limited to a maximum of 5 by link $(2, 6)$.

The set of linear equations corresponding to this example is given by:

$$\text{Link } (1,5) \quad T_{AC} \qquad\quad = 6 \tag{14.27}$$
$$\text{Link } (5,3) \quad T_{AC} \qquad\quad = 6 \tag{14.28}$$
$$\text{Link } (1,6) \quad T_{AD} \qquad\quad = 10 \tag{14.29}$$
$$\text{Link } (2,6) \quad T_{BC} + T_{BD} = 5 \tag{14.30}$$
$$\text{Link } (6,3) \quad T_{BC} \qquad\quad = 8 \tag{14.31}$$
$$\text{Link } (6,4) \quad T_{AD} + T_{BD} = 7 \tag{14.32}$$

Clearly Eqs. (14.30) and (14.31) are incompatible with the non-negativity of T_{BC}. The same applies to Eqs. (14.29) and (14.32), making it impossible to solve this set of equations. In simple problems like this, inconsistencies can be ascertained by inspection, but in more complex networks they can only be identified by means of row and column operations on the linear equations. For large systems, these operations are likely to be expensive in terms of computer requirements.

In this simplistic example, it is not difficult to see that the problem originates in the assumed single route between A and C. If two paths were allowed, one via node 5 and the other via node 6, the inconsistency could be removed. Furthermore, the value of the resulting variable p_{AC}^6 cannot be arbitrarily chosen; in effect, a feasible solution requires

$$0.2 \le p_{AC}^6 \le 0.5$$

The fact that the *path flow* continuity conditions are not met seems to reflect errors in assignment, whereas the *link flow* discontinuities are a reflection of errors in the traffic counts alone. Thus, it seems reasonable to develop a technique for removing the link flow inconsistencies in the counts to ensure that the link flow continuity conditions are met. On the other hand, a reasonable approach to dealing with the lack of consistency at the path flow level seems to be the adoption of a better route choice model. In general terms, consistency at the link flow level is a necessary but not sufficient condition for consistency at path flow level. Consistency at path flow level is, however, a sufficient condition for link flow consistency.

The interested reader may verify that there are only seven different (integer) trip matrices that can satisfy the observed flows in the example above.

14.4.6 Limitations of ME2

ME2, probably because of its simplicity, relative efficiency, and ease of programming, has been widely implemented and used, particularly in the UK. However, the model has some known limitations, and it is worth exploring them before discussing opportunities to improve it.

One of the limitations arises when traffic has grown (or declined) markedly between the prior (or old) trip matrix and the present. The model estimates the matrix closest to the prior, which, when loaded onto the network, reproduces the traffic counts, but this may lead to distortions. In these cases, it is probably better to consider the structure of the prior matrix, say through the proportion of total trips, which appear in each cell, and not the absolute number of trips in each O–D pair. One would then try to find a matrix with the closest structure to that of the prior matrix that reproduces the traffic counts when loaded onto the network. This can be approximated by means of a general growth factor first, for example:

$$\tau = \frac{\sum\limits_a \hat{V}_a}{\sum\limits_a \sum\limits_{ij} t_{ij} p_{ij}^a} \tag{14.33}$$

which is then applied to the prior matrix before using the ME2 model. In this way, the structure of the prior matrix is preserved as much as possible. The estimation of τ above is only an approximation; for a more rigorous approach see Bell (1983).

A second limitation of ME2 is the fact that it considers traffic counts as error-free observations of non-stochastic variables. In effect, the model gives complete credence to the traffic counts and uses the prior matrix only to compensate for the fact that they do not contain sufficient information for estimation purposes. However, this may not be very appropriate in practice. For a start, one must acknowledge that traffic counts are certainly not error-free. Apart from counting errors, there is the problem of time variations (hourly, seasonal, etc.). Traffic counts obtained on different days or at different times can hardly be considered observations of a non-stochastic variable.

Willumsen (1984) suggested an approach to compensate for this second difficulty. It starts from the idea that functions of the type $\{X \log X/Y - X + Y\}$ can be seen as useful measures of the difference between X and Y. Then he constructed a composite objective function to satisfy the following:

$$\text{Minimise } S_3 = \sum_{ij} \left(T_{ij} \log T_{ij}/t_{ij} - T_j + t_{ij} \right) + \sum_a \phi_a (V_a \log V_a/v_a - V_a + v_a) \tag{14.34}$$

where, V_a is now the *true* value of the traffic count at a, v_a is the value of one observation of the flow made at a, and ϕ_a is a weighting factor that depends on the confidence attached to the observation v_a.

The use of the Lagrangian method now leads to the solution:

$$T_{ij} = t_{ij} \prod_a X_a^{p_{ij}^a} \tag{14.22}$$

$$V_a = v_a X_a^{1/\phi_a} \tag{14.35}$$

Again, this model can be solved using the multi-proportional algorithm, but now we also need to correct the observations to obtain a better estimation of the true value of the link flows. Note that if ϕ_a is large (i.e. we assign a high weight to the counts as we believe them to be very accurate), V_a tends to v_a; in the limit, with $\phi_a = \infty$, we revert to the original model as $V_a = v_a$. Note that the smaller the value of ϕ_a, the greater the credence given to the prior matrix **t**.

One would expect that the weights ϕ_a depend on the variability of the observations. Brenninger-Gothe et al. (1989) have discussed this model in detail, showing that a natural value for the weights ϕ_a is the standard deviation associated with the observations. If these are not available, they can be estimated using some assumptions about the distribution of the error terms. These authors have further extended the model to consider weights attached to both the prior matrix (μ_{ij}) and the traffic counts (ϕ_a); thus, the new objective function becomes:

$$\text{Minimise } S_3 = \sum_{ij} \mu_{ij} \left(T_{ij} \log T_{ij}/t_{ij} - T_{ij} + t_{ij} \right) + \sum_a \phi_a (V_a \log V_a/v_a - V_a + v_a) \tag{14.36}$$

The main limitations of ME2 can therefore be reduced using reasonably simple methods. However, other authors have proposed alternative approaches to solving the matrix estimation problem, some of which start from a different basic framework.

14.4.7 Improved Matrix Estimation Models

Bell (1983) formulated a model that tries to preserve the structure of the prior matrix, in the sense described in the previous section, by adding a new constraint and thus modifying the mathematical programme as follows:

Minimise $-S_2$ subject to:

$$\hat{V}_a - \sum_{ij} T_{ij} p_{ij}^a = 0 \text{ for each counted link } a \qquad (14.19)$$

$$\tau = \sum_{ij} T_{ij} \bigg/ \sum_{ij} t_{ij} \qquad (14.37)$$

and

$$T_{ij} \geq 0$$

In addition, Bell suggests the use of the Newton–Raphson method to solve this model with an iterative estimation for τ. Alternatively, one may assume an initial value for τ, solve the standard model using a multi-proportional method, and then check if it is consistent with Eq. (14.37). The cycle should be repeated until the value of τ converges.

The use of the Newton–Raphson algorithm has advantages in terms of computer time and is also useful in tracing the effect of errors in the traffic counts through to the estimated trip matrix (Bell 1983); this type of sensitivity analysis is an alternative to the treatment of errors in the traffic counts suggested above. However, the Newton–Raphson method requires more memory and is therefore restricted to small and medium-sized networks.

A variant on the standard objective function (S_1) is either to linearise it using Taylor's expansion or to construct a generalised least squares formulation. In both cases, we still try to minimise the difference between prior and estimated matrices subject to the same constraint (14.19). Bell (1984) suggested the Taylor series expansion solution, whereas McNeil and Hendrickson (1985) and Cascetta (1984) have proposed versions involving generalised least squares approaches. One problem is that under certain circumstances, these models may produce negative entries in the estimated trip matrix, in particular when the original prior matrix has small values. As this is not an uncommon occurrence, this feature is undesirable.

Maher (1983) proposed a Bayesian approach for the trip matrix estimation problem, which results in functional forms equivalent to the generalised least squares method. A prior estimate of the trip matrix is updated in light of a set of traffic counts; both are assumed to be multivariate Normal distributed variables with known covariance.

Spiess (1987) proposed a maximum likelihood model to solve the problem. He considered a specific formulation where, for each O–D pair t_{ij} is obtained by observing an independent Poisson process with mean $\Omega_{ij} T_{ij}$. This corresponds to the problem of taking a sample of an existing trip matrix with a sampling rate of $\Omega_{ij} < 1$. The probability of observing t_{ij} is:

$$\text{Prob}\big[\text{Poisson}(\Omega_{ij} T_{ij}) = t_{ij}\big] = (\Omega_{ij} T_{ij})^{t_{ij}} \exp(-\Omega_{ij} T_{ij}) / t_{ij}! \qquad (14.38)$$

The joint probability of observing the sample matrix $\{t_{ij}\}$ is therefore:

$$\text{Prob}\big[\{t_{ij}\}\big] = \prod_{ij} \text{Prob}\big[t_{ij}\big] = \prod_{ij} (\Omega_{ij} T_{ij})^{t_{ij}} \exp(-\Omega_{ij} T_{ij}) / t_{ij}! \qquad (14.39)$$

Applying the MLE technique to this problem requires finding the matrix $\{T_{ij}^*\}$, which satisfies the constraints and yields the maximum probability (14.39) of observing $\{t_{ij}\}$. By taking logarithm of Eq. (14.39) and adopting the usual convention that $0 \log 0 = 0$, we can formulate the maximum likelihood model as:

$$\text{Max} \sum_{ij} \left(t_{ij} \log \left(\Omega_{ij} T_{ij} \right) - \Omega_{ij} T_{ij} - \log t_{ij}! \right) \tag{14.40}$$

subject to the usual non-negativity constraints and to Eq. (14.19). Separating the logarithm into the sum and discarding constant terms, one can rewrite (14.40) as:

$$\text{Min} \sum_{ij} \left(\Omega_{ij} T_{ij} - t_{ij} \log T_{ij} \right) \tag{14.41}$$

This objective function is convex in T_{ij}; provided the set of constraints is consistent and the flows are feasible then, the existence of an optimal solution is assured. The solution may be obtained by any standard method for convex programming problems. However, Spiess (1987) has developed an algorithm that exploits some of the specific properties of this problem.

For further comments on this problem and possibilities for extensions see Cascetta and Nguyen (1988) and Willumsen (1991).

14.4.8 Treatment of Non-Proportional Assignment

The ME2 model discussed in the preceding sections is based on the assumption that it is possible to obtain the route choice proportions $\left\{ p_{ij}^a \right\}$ independently from the matrix estimation process. Wherever congestion plays an important role in route choice this assumption becomes questionable as the route choice, proportions and the trip matrix become interdependent. Because of its theoretical and practical advantages, equilibrium assignment is the natural framework for extending the ME2 model for the congested network case.

The main problem in incorporating Wardrop's equilibrium into trip matrix estimation is that now the route choice proportions and the trip matrix to be estimated are interdependent. One way of tackling this problem is to adopt an iterative approach: assume a set of route choice proportions $\left\{ p_{ij}^a \right\}$, estimate a matrix $\mathbf{T}$, load it onto the network and obtain a new set of route choice proportions; repeat the process until route choice proportions and estimated matrices are mutually consistent.

This general scheme can be implemented in different ways. For example, in SATURN (Hall et al. 1980), the route choice proportions are estimated using the value ϕ in the Frank–Wolfe algorithm (the optimum linear combination of accumulated and auxiliary flows; see Section 12.2.3). It is recognised that, in general, the path-flows under equilibrium conditions are not unique. However, this method assumes that they are unique.

An alternative approach requires restating the original problem in terms of a three-dimensional matrix (origin, destination, and route) as follows:

$$\text{Maximise } S_4 = - \sum_{ijr} T_{ijr} \left(\log T_{ijr} / t_{ijr} - 1 \right) \tag{14.42}$$

subject to

$$\sum_{ijr} T_{ijr} \delta_{ijr}^a - \hat{V}_a = 0 \tag{14.43}$$

and

$$T_{ijr} \geq 0$$

where the index r indicates the route or path chosen; δ_{ijr}^a is 1 if route r between i and j uses link a, and zero otherwise.

It is always possible, of course, to reconstruct the O–D matrix $\{T_{ij}\}$ by aggregating the path flow matrices $\{T_{ijr}\}$. Again, the solution to this new program is:

$$T_{ijr} = t_{ijr} \prod_a X_a^{\delta_{ijr}^a} \tag{14.44}$$

and

$$T_{ij} = \sum_r T_{ijr} \tag{14.45}$$

The prior path flow may be calculated from the prior trip matrix as $t_{ijr} = t_{ij}/R_{ij}$, where R_{ij} is the number of paths between i and j. In this case, the path flows can take any value as they are not assumed to be unique. The Frank–Wolfe algorithm for equilibrium assignment is used to identify attractive paths (those selected at each all-or-nothing step) but not to define the strict proportions of the trip matrix using them. This is only a heuristic scheme and a suitable algorithm for its solution is as follows:

1) Assign, using equilibrium assignment methods, a base-year matrix $\{t_{ij}\}$ to the network and save the corresponding routes (trees). Set the cycle counter n to 1.
2) Estimate a trip matrix $\{T_{ij}\}^n$ for iteration n, using independent routes $\{\delta_{ijr}^a\}$ and observed flows $\{\hat{V}_a\}$.
3) Assign $\{T_{ij}\}^n$ to equilibrium, saving the routes (trees) used in the process.
4) Increment n by 1 and return to step 2 unless the changes in routes $\{\delta_{ijr}^a\}$ or estimated matrices have been sufficiently small.

For a test of this approach and a comparison with proportional assignment techniques in the case of a comprehensive data set for Reading in the UK, see Willumsen (1982).

A more general approach has been put forward by Fisk (1988) and Oh (1989), where maximum-entropy matrix estimation and user equilibrium assignment are combined as a single mathematical program.

14.4.9 Quality of Matrix Estimation Results

Matrix adjustment from traffic counts includes a powerful group of techniques that provide significant help in developing useful and robust trip matrices. However, to use the approach in a sound and reliable manner, a number of points require careful attention. One particular aspect to bear in mind is that matrix estimation techniques may try to force an adjusted trip matrix to reproduce traffic counts even if there are significant errors in the network, the assignment method, or the counts themselves. The following recommendations reflect our views on pitfalls to avoid when using this type of technique:

- make sure the network is fully debugged and that all relevant turning movements are well represented;
- use an assignment method appropriate to the context; this usually means equilibrium assignment;
- ensure that any prior matrix is reasonable and do not over-rely on one that is not;
- set aside some 10–15% of the traffic counts for validation of the adjusted trip matrix;
- ensure all traffic counts are adjusted using seasonal and daily factors to a common representative day and that only relevant vehicle types are included (i.e. do not use pcus when car trips are needed);

- if possible, assign a level of confidence to each count and allow greater tolerance for those that are less reliable;
- bear in mind that some bottlenecks may restrict actual traffic on the network to levels below demand (metering effect); it may be better to ignore counts affected by this constraint;
- apply matrix estimation techniques in small increments and obtain network and matrix statistics at the end of each run: compare number of trips and travel speeds and trust only the matrices that do not change these indicators by more than 10%; monitor the trip length distribution before and after matrix estimation, as significant changes probably indicate that the trip matrix is being distorted by the procedure;
- use only the validation counts above to report fitness for purpose;
- never accept a post-matrix estimation trip table without thorough checks on its validity; these methods are powerful and generally easy to use but may distort a perfectly good prior matrix too much and render the results of any scheme test unreliable.

14.4.10 Estimation of Trip Matrix and Mode Choice

The idea of extending this type of approach to matrix and mode choice estimation is attractive. Let us consider a singly constrained destination/mode choice model of the following Logit form:

$$T_{ij} = O_i \frac{S_j \sum_k \exp\left(\sum_p \theta_p X_{ijk}^p\right)}{\sum_d S_d \sum_k \exp\left(\sum_p \theta_p X_{idk}^p\right)} \tag{14.46}$$

where the mode choice component of the model is given by:

$$P_{ij}^k = \frac{\sum_p \exp\left(\sum_p \theta_p X_{ijk}^p\right)}{\sum_m \exp\left(\sum_p \theta_p X_{ijm}^p\right)} \tag{14.47}$$

T_{ij} are trips between zones i and j, O_i is the total number of trips originating at zone i, S_j is a measure of the attractiveness of zone j, P_{ij}^k is the proportion of trips using mode k between zones i and j, X_{ijk}^p is the pth explanatory variable for mode k (for example, in-vehicle travel time) and θ are model parameters.

Although the derivations we will present below are for the simpler MNL case, they can easily be extended to consider the simultaneous estimation of more general Nested Logit forms (Ortúzar and Willumsen 1991).

14.4.10.1 Simple Unimodal Case

Let us first consider a single mode case with just one scale parameter μ, multiplying a 'generalised cost' variable X_{ij}, to be estimated. In this simple case (14.46), reduces to:

$$T_{ij} = O_i \frac{S_j \exp(\mu X_{ij})}{\sum_d S_d \exp(\mu X_{id})} \tag{14.48}$$

Now, assume we possess observations on a set of link flows $\hat{V}_a$, and also that we know, from an assignment model, the proportions p_{ij}^a for all links with observed flows. In such case, we can postulate that Eq. (14.19) holds, and to estimate the value of μ we can, for example, seek to minimise the following normalised non-linear (generalised) least squares function:

$$S = \sum_a \left[\left(\hat{V}_a - \sum_{ij} T_{ij} p_{ij}^a \right) \Big/ \hat{V}_a^2 \right]^2 \tag{14.49}$$

To find the minimum, we usually require first and second derivatives of S with respect to μ. These are provided by Ortúzar and Willumsen (1991); unfortunately, even in this simple case, the derivatives look rather intractable, so a unique solution to the problem may be difficult to establish.

14.4.10.2 *Updating with Aggregate Modal Shares*

Let us consider the transference of models (14.46) and (14.47) with parameters θ estimated in another context; we ignore the original mode-specific constants as they ensure reproduction of the aggregate market shares in that context. Define a transfer utility function as:

$$V_{ijk} = \mu \left(\sum_p \theta_p X_{ijk}^p \right) + M_k \tag{14.50}$$

where X_{ijk}^p are zonal values for the level-of-service and socioeconomic variables in the new context, μ is a scale parameter as before and $\mathbf{M}$ a set of $(K-1)$ mode-specific constants to be estimated; K is the total number of modes.

In this case it is possible to find maximum likelihood estimators for μ and $\mathbf{M}$ but it is only possible to guarantee a unique optimum for fixed μ (i.e. when only the constants are updated).

14.4.10.3 *Updating with Traffic Counts*

The main problems arise in this case if we are interested in mixed-mode combinations but only have counts for the 'pure' modes. For example, consider the case of choice between car, bus, underground, and combinations of the latter with the first two. It is obvious that even if we have separate counts for each pure mode, these include observations corresponding to mixed-mode movements. If we settle for a mode aggregation and are interested in estimating the scale parameter μ and a set of constants for the pure modes, the problem can be solved using a generalised least squares formulation similar to (14.49), as shown by Ortúzar and Willumsen (1991).

14.4.10.4 *Updating with Combined Information*

Assume we wish to update μ and $\mathbf{M}$ of (14.50) and have available observed aggregate shares P_k and sets of observed passenger counts $\hat{\mathbf{V}}$ for each competing mode. The problem can be formulated either as a maximum likelihood or generalised least squares one.

In the first case, we will get different functions to maximise and hence different first-order conditions and optima, depending on the assumptions made about the distribution of count errors. The favourite assumptions have been multinomial, independent Poisson and independent Normal (see Tamin and Willumsen 1992). As data on counts can be assumed to be independent of data on aggregate shares, the log-likelihood function takes the form of a sum of two expressions. If it is assumed that the counts have no error, a final case of interest results, which requires maximising a much simpler function subject to (14.19). Expressions for each of these cases are given by Ortúzar and Willumsen (1991); there is no guarantee, however, that either of them leads to a unique optimum.

> The generalised least squares formulation has two advantages: the first is that no distributional assumptions are needed on the data set; the second is the possibility of incorporating explicitly differences in the accuracy of each data item prior to estimation. A need for normalising, which is also a feature of this approach, is very evident here given the different order of magnitude of the differences between observed and modelled values for both types of data. For example, the maximum difference in the case of aggregate shares is just 1, while differences in count data may easily run into the hundreds or thousands.
>
> The range of methodologies available in principle to solve this important problem is difficult to evaluate without recourse to experimentation; by the end of 2022, such an exercise had not been reported.

14.5 Gaming Simulation

Mathematical models do not solve real-life transport problems, but the interpretation of mathematical solutions is useful to aid decision-making. Simplified models may help in reducing the effort required to find a mathematical answer and in facilitating the subsequent interpretation of this solution in relation to the real problem. We use conceptual or mental models to understand, interpret, and act in our professional life. Mental models are, in effect, a prerequisite for the development and application of mathematical ones run on a computer.

Despite their significance and because of their character, it is difficult to examine mental models, and this often leads to quite unmanageable communication problems. Better and richer mental models in the minds of planners and decision-makers are probably as important as the use of rigorous and sound behavioural models in the computer if transport planning is to be improved. Given the key role played by mental models in the use and application of mathematical ones, it seems sensible to investigate techniques for improving the first to get better solutions through the second.

But how are mental models acquired, revised, rejected, or enhanced? The main factors seem to be formal and informal education, discussions, and, above all, practical experience. One of the main problems facing planning education and training is how to provide realistic experiences. This is particularly acute in the transport field, where the most important consequences of a policy measure or infrastructure project may follow only after considerable time. Besides, it is surprisingly easy to become too involved in the details of particular techniques and lose sight of the wider process into which they must fit.

The need for methods to develop a general comprehension of a system rather than detailed information about its parts has been recognised in several fields, particularly in management and business training. Several educational techniques have been developed to this end: case studies, role playing, and different types of exercises. Gaming simulation is a particularly attractive technique in this field. It was originally developed for military purposes, in the form of *war games*, but since computers became widely available, it has spread successfully into management science, politics, sociology, and regional and transport planning.

Educational games are sequential decision-making exercises structured around an artificial environment acting as surrogate for the real world. This artificial environment may be just a set of instructions and graphical material or may involve an elaborate simulation exercise using computer programs, physical models, and animated displays. As in real life, games usually have a competitive dimension. This feature can be incorporated in at least two forms: by dividing the players into teams with partially conflicting objectives (e.g. car owners, environmental protection officers, local residents, etc.) or by facing each player with a computer model of a complex system plus a common set of initial conditions and final objective. Key indicators can then be used to assess the performance of each player in achieving these objectives. The first approach stresses the need for negotiation and compromise whilst the second emphasises efficiency in pursuing objectives. Both methods enhance understanding of complex systems and support the development of learning skills. In both cases, the success of players depends on their ability to learn from the outcome of their own decisions, those of others and the effects of unexpected events like a strike or fuel price increases. The final objective of any gaming-simulation exercise is to augment the ability to learn through the enrichment of the conceptual model every player has of a system. For a good background on gaming-simulation design and experience, the reader is directed to Greenblat and Duke (1975) or Taylor (1971), and in the transport field to Ortúzar and Willumsen (1978).

Several gaming simulations have been developed specifically for the transport field. Some of these cover problems like negotiating the alignment for a new road or planning new public-transport services. Probably the most widely used game in the urban transport management field is GUTS (Willumsen and Ortúzar 1985). The original objectives for this computer-based game were:

1) The game should treat the transport sector of an urban area as a system, that is, it should highlight the interrelations between modes, traffic management and investment decisions, and financial constraints; therefore, the computer program contains relationships conveying these interactions.
2) The game should be realistic but manageable; the most common types of investment and traffic management decisions should be included and key financial and resource constraints should be simulated.
3) The model should allow for a range of alternative and even conflicting objectives to be pursued and consequently, the program should produce not a single but multiple performance indicators. At the same time, the information available to players should not be too different from that commonly available to real decision makers.
4) The game should stress the importance of continually monitoring the performance of a transport system.
5) The model should allow the representation of different types of urban areas in terms of residence, employment, car ownership, income distributions and growth rates, public-transport patronage, and related indicators.

GUTS is available now as a friendly web-based application (www.MicroGuts.net). The model is based on a simplified urban area with circular symmetry. Two main modes, car and bus, operate freely and in competition; in addition, trips can be done using a combination of car and bus (park-and-ride) and also by bicycle; the user can make decisions about public-transport fares and levels of service, the introduction of bus lanes, road pricing, parking provision, and charges, as well as major investment projects. The program checks these decisions and runs its inner model to represent one year of operation of the whole transport system. At the end of the run indicators on flow levels, speeds, modal split, travel time, and expenditure by person type are produced; the financial performance of the public transport company is also reported. Changes in accessibility levels and the impact of new investment are also simulated, as are unexpected events inducing changes to the cost structure of the transport modes operating in the city. The symmetry condition imposed on the city simplifies the model, which has advantages in terms of speeding up the learning curve of the user and enhancing running time of the computer.

Games like GUTS can enhance transport planning in a number of ways. First, in their normal training-tool mode, they can be used to educate new recruits to a team and to develop a common language throughout an office. Second, a model of this type may be seen as a simple *sketch planning* tool valuable in discussing broad policy options and particular conceptions of decision makers. Although no substitute for full-scale models, games like GUTS may help to bridge the gap between broad strategies and specific modelling studies. A third use of tools of this kind is in demonstrating the advantages and limitations of mathematical models. The extremes of total rejection of a transport model or its blind acceptance are still present in some political and planning quarters. The evident simplicity of a gaming-simulation exercise combined with its capacity to represent interactions between modes, decisions, and decision- makers, provide a good example of what the formal modelling approach can offer. The use and subsequent critique of the game by politicians and planners help them to understand each other's activities and interests better.

Exercises

14.1 The network in Figure 14.4 represents a small area with two origins A and B and two destinations Y and Z.

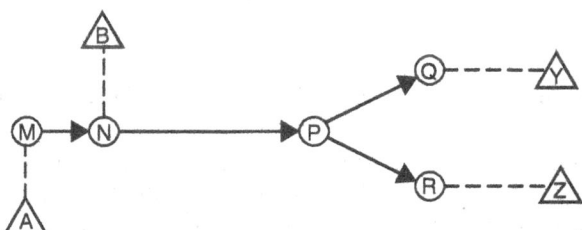

Figure 14.4 Simple network for Exercise 14.1

Traffic counts have been made of the car flows using the network with the results shown in Table 14.6:

Table 14.6 Traffic counts for Exercise 14.1

Link	Flow
M–N	400
N–P	700
P–Q	500

a) Use an entropy-maximising model to estimate a trip matrix from the information above. Assume a suitable prior matrix for this problem if necessary. A 3% error in the modelled flow is considered acceptable for this question.
b) Repeat the calculations assuming the prior matrix is given (Table 14.7):

Table 14.7 Prior matrix for Exercise 14.1

	Y	Z
A	100	50
B	80	200

14.2 The network in Figure 14.5 represents links connecting two origins, A and B to two destinations, C and D, in a developing country. The populations of the two origins are 10 000 and 20 000 inhabitants, respectively, and the markets held at C and D are equally attractive in terms of size and prices. The link distances (in km) are indicated in the figure.

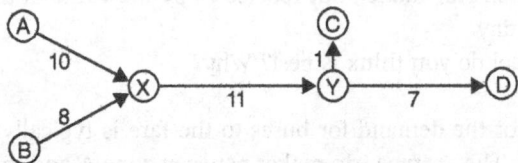

Figure 14.5 Simple network for Exercise 14.2

Person counts have been obtained for three links, as shown in Table 14.8:

Table 14.8 Person counts for Example 14.2

Link	Persons/day
A–X	3400
X–Y	11 900
Y–D	4100

Calibrate a model of the type:

$$T_{ij} = \frac{bP_iD_jd_{ij}^{-n}}{\sum\limits_{j} D_jd_{ij}^{-n}}$$

where P_i is the population of zone i, D_j is the attractiveness index for the market in zone j, and d_{ij} is the travel distance between i and j. Try at least two values for the power n, including $n = 2$ and $n = 2.5$.

14.3 Three villages, A, B, and C, are connected by a navigable river in an underdeveloped country. Village A has a population of 1000 inhabitants; village B is 30 km downstream of A and has a population of 2000; village C is 10 km downstream of B and has a population of 300 inhabitants. The value of the goods exchanged in each village per day is 500, 600, and 600 pesos respectively.

Two observers have spent some time making directional counts of passengers travelling in boats along the river with the results shown in Table 14.9:

Table 14.9 Directional boat counts for Example 14.3

River section	Passengers per half day
A–B	45
B–A	60
B–C	360
C–B	560

a) Calibrate a Gravity model of the form suggested in Exercise 14.2, where D_j is replaced by the population of village j. Use $n = 2.0$.

b) Calibrate a similar model, but replace D_j by the value of the goods exchanged in each village per day.

c) Which model do you think is best? Why?

14.4 The elasticity of the demand for buses to the fare is typically acknowledged to be in the region of -0.3. The average trip maker between zone A and the centre of town (CBD) currently faces a bus fare of $2 per trip; the bus share of all trips between A and the CBD is 60%, other trips use either car or underground.

If the total number of trips between both zones is 2000 estimate the loss in patronage of the buses if the fare is raised to $3 per trip, all other things being equal, using the incremental Logit method. Compare your result with the cruder elasticity calculation; discuss your findings

Hint: Estimate the parameter θ_c from the data given the simple expression for the Logit direct elasticity.

15
Freight Demand Models

15.1 A Subject of Increasing Importance

Most of this book concentrates on modelling passenger movements, with a strong emphasis on urban problems. However, freight movements, and in particular road haulage, are an important source of congestion and other traffic problems. The noise and nuisance generated by heavy lorries, the problems created by on-street loading and unloading of goods vehicles to serve shops and premises, and the usual complaint about lorries taking up a good deal of the capacity of interurban roads are only some of the problems associated with this type of traffic.

Unfortunately, in urban areas the policy options available to influence road haulage are limited. They are mainly controls on loading/unloading, on the size of vehicles allowed in certain areas (lorry routeing), the provision of major freight interchanges, the encouragement of rear access to premises, and improved layouts at new developments.

Freight demand modelling may play a particularly important role in developing countries, where the efforts to increase exports and gain access to underdeveloped areas are more urgent. Facilitating freight movements there is likely to impact economic development significantly. Moreover, the competition between road and rail in some of these countries is a key issue in resource allocation for investment and maintenance.

In the case of interurban and international movements, there is more scope for policies to influence freight mode choice and to regulate competition between rail and road. Improved allocation of road user charges and targeting subsidies to key rail or road services are also important policy options. The design of these policies may require more refined modelling efforts than those used in urban studies.

One might expect that the choices made for freight movements would follow economic rationality alone, for example, to minimise a combination of travel times and costs appropriate to the value of the goods being transported. In this case, the 'value of the goods' is not only how much they cost but what are the implications of their delayed or early arrival in terms of storage costs and downstream manufacturing/sales delayed. However, an observation of real flows shows many examples where this economic rationality seems difficult to interpret or it is more subtle and complex than we would expect: moving bottled water all across the world, for example.

One can envisage the complexity of freight movements as the result of four layers of decisions and activities:

1) Decisions on productions, destinations, type of product, volume, and trade relationships: who produces what, in what quantities, and for what intermediate or final consumer.
2) Logistics decisions, like the use and location of inventories and supply chain management, for example, 'just-in-time' contracts and lean manufacturing.

Modelling Transport, Fifth Edition. Juan de Dios Ortúzar and Luis G. Willumsen.
© 2024 John Wiley & Sons Ltd. Published 2024 by John Wiley & Sons Ltd.
Companion website: www.wiley.com/go/ortuzar5e

3) Choice of transport modes, vehicles and multi-modal facilities to deliver the good according to the previous decisions.

4) Decisions on the multi-modal transport routes required to reach each particular destination. The actual route chosen may be relatively simple in the case of road haulage. However, it will become more complex when there is a need to route containers or intermodal terminal units (ITU) on a rail network unless there are enough to make up a full trainload.

Attention to each of these layers depends on the scope and geography of the study. Urban studies will probably focus mostly on the fourth layer and consider the upper three more or less given and identifiable through relatively conventional data collection surveys. Future conditions may imply changes in both route availability and the location of the origin and destination of movements, thus focusing on the first and fourth layers.

Regional and international studies will tend to cover the four layers with, perhaps, a simplified approach to modelling changes in logistic decision-making. Trade flows, mode, and route choices will likely be a significant focus of attention; data collection and processing will look deeper into these issues.

Given these facts, it appears surprising that much less research has been undertaken on modelling this type of movement than on the effort allocated to passenger demand. Why would this be the case? We believe there are several reasons for this:

1) Many aspects of freight demand make it more challenging to model than passenger movements; we discuss some of these below.

2) For some time, urban congestion has been high in most industrialised countries' political agendas; on this issue, passenger movements play a more critical role than freight.

3) The movement of freight involves more actors than the movement of passengers; we have the industrial *firms* sending and receiving the goods, the *shippers* organising the consignment and modes, the *carriers* undertaking the movement, and several other agents running transshipment, storage, and customs facilities. In some cases, two or more of these may coincide, for example in *own-account* operations, but there is always scope for conflicting objectives which are difficult to model in detail in practice.

4) Recent trends in freight research have emphasised the role that inventory control and management of stocks play in the overall production process. These trends are a departure from more traditional passenger modelling techniques and share little in common (see Regan and Garrido 2002).

This chapter summarises approaches to freight demand modelling. It starts with a discussion of the main difficulties associated with modelling freight movements. It then presents what is probably the most traditional approach to the problem, that is, to adapt the classic aggregate transport model to the case of commodities. Extensions of the disaggregate approach to freight demand are also outlined. The section closes with some practical considerations for the implementation of these ideas. The interested readers are directed to the classic book by Harker (1987) for further details.

15.2 Factors Affecting Goods Movements

As in the case of passenger demand, it is useful to consider first the factors that one would expect to influence freight movements. The following is not an exhaustive list but covers the most important ones:

1) *Location factors*. Freight is always a derived demand and usually part of an industrial process. Therefore, the location of sources for raw materials and other inputs to a production process,

plus the location of intermediate and final markets for their products, will determine the level of freight movements involved and their origins and destinations.

2) The *diversity of products* needed and produced is high, greater than the most detailed segmentation of travel demand by person types and journey purposes. A demand for bolts cannot be satisfied by providing cashew nuts. There will be a significant number of commodity matrices in any freight demand study.

3) *Physical factors.* The characteristics and nature of raw materials and end products influence the way in which they can be transported: in bulk, packaged in light vans, in secure vehicles if the products are of high value, and in refrigerated containers if they are perishable. There is a greater variety, therefore, of vehicle types to match commodity classes than in the case of passenger transport.

4) *Operational factors.* The size of the firm, its policy for distribution channels, its geographical dispersion and so on, strongly influence the possible use of different modes and shipping strategies.

5) *Geographical factors.* The location and density of the population may influence the distribution of end products.

6) *Dynamic factors.* Seasonal variations in demand and changes in consumers' tastes play an important role in changing goods' movement patterns.

7) *Pricing factors.* Compared to passenger demand, prices are not generally published because they are more flexible and subject to negotiations and bargaining power.

15.3 Pricing Freight Services

It is usually difficult for the analyst to obtain reliable data about freight charges. For example, in Europe, both transport firms and users try to keep them confidential to strengthen their positions when it comes to negotiations. The factors affecting charges or cost imputations, and therefore mode choices, are thought to be:

1) *The length of the supply contracts.* A better price can be obtained if the shipper guarantees demand for one or more years rather than just for one single shipment. The existence of price adjustment clauses helps to extend the lengths of contracts.

2) *The extent of volume discounts.* Following the above, a contract guaranteeing steady high-volume shipments is likely to benefit from a lower price.

3) *The importance of terminal facilities.* The availability of a rail terminal nearby, or even at the firm, would certainly reduce the cost of shipping by rail. Its absence would increase the likelihood of using road transport all the way, without even considering rail or water transport.

4) *The use of own-account operations, especially in road haulage.* Some firms prefer this type of operation for reasons other than transport (image, reliability, integration). These firms will tend to extend the use of own-account operations for marginal products rather than consider a completely new mode.

5) *Some modes are better suited to transport certain commodities.* For example, *pipelines* are ideal for bulk liquids and some suspensions, and *merry-go-round* (non-stop) trains are ideal for movements from mines to processing centres. This closer fit of supply characteristics to demand would certainly influence the charges made for those products.

6) *Hierarchical transport systems.* For example, in the case of petroleum products, use of large tankers to refineries, then small tankers and pipelines to major terminals, rail to other terminals, and lorries to petrol stations and final users. These structures are difficult to modify in the short run as they have evolved over a long period and are well established; thus, their pricing mechanisms may be difficult to change.

15.4   Data Collection for Freight Studies

As we have seen, the business of moving freight is more complex than moving passengers (see Figure 15.1). Data collection must, therefore, be planned considering the key features of goods transport to be captured in a particular region or city. What follows is a simplified version of the key participants influencing choices in the movement of goods (Friedrich et al. 2003):

1) The *sender* (shipper, consignor) who requires delivery of its goods to a particular destination and puts these goods-units in the care of others (freight forwarder, carrier) to be delivered to a consignee. The sender will decide on a freight forwarder based on reliability, speed of delivery, price, and other factors.

2) The *freight forwarder* who organises the shipping process. This firm will provide and schedule uni- or inter-modal transport chains for transporting the goods. To deliver these services, the firm may subcontract carriers or provide its own carrier service.

3) The *carriers* are responsible for transporting the goods. The carriers will provide the ships, trains, and/or vehicles (sometimes in combination) required for the transport operation along a section of the transport chain. The vehicle units operate on a network connecting origin, hubs, and destination. The carriers may specify a route to be followed by vehicle units.

4) The *drivers* guide the vehicle/transport units along a predefined route. In the case of road transport, a driver may decide to change the route between two points of the journey during the trip.

5) The *consignee* is entitled to take delivery of the goods.

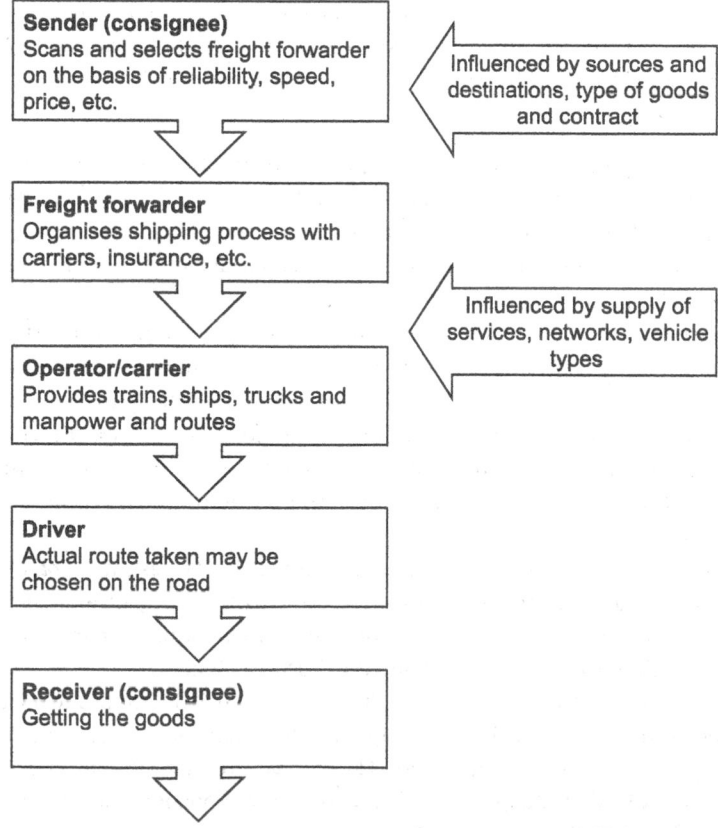

Figure 15.1   Actors and processes in freight movements

Additional actors appear at different stages in this process, for example, insurance companies, quality assurance inspectors, customs officials, facilities, and intermediate storage units. Some of these services are sometimes provided by shippers or carriers. These complexities are some of the reasons why transport modelling for freight is so drawn out. It is often difficult to identify exactly who takes actual decisions on mode combinations or routes, and, therefore, it is difficult to collect data and develop choice models, aggregate or otherwise.

Whenever goods are transferred from one of these participants to another, a small amount of data is generated and stored in some way. As more of this data is collected and stored electronically; it should be easier, in principle, to access and process it. Commercial confidentiality, however, continues to make this task very difficult.

For most urban transport studies, it may be enough to collect information at roadside interview sites on the type of vehicle used, the goods transported and their origin, and destination. The delivery of goods in urban areas is often made more complex by the use of distribution/collection tours with multiple stops. This information is difficult to collect at the roadside because of time constraints. This explains why it is customary to collect additional information from carriers, and from major generators and attractors of goods movements.

An important source of data on truck movements results from the use of Global Positioning System (GPS) devices in the vehicle. They are increasingly used to monitor movements and traffic conditions as well as to enable real-time tracking of deliveries. But again, it is generally difficult to access these data that are considered confidential.

An even more valuable source of truck movements may be available in cases where GPS data is used to charge commercial vehicles for the use of the roads. This provides a much larger sample and may be available in an anonymised form for use in large-scale strategic models.

In summary, for urban studies, the main sources of information would be:

- asking drivers on the road;
- identifying specific carriers (couriers, delivery companies, or refuse collectors) and interviewing them about tours and schedules;
- sometimes a mail-back survey to senders or freight forwarders may offer a moderate degree of success.

On the other hand, one must remember that in rapidly growing cities, construction work provides a significant source and destination for freight movements, including empty vehicles. This issue is problematic for modelling purposes as future construction activity will almost certainly occur in different locations, and it is more challenging to predict these locations than on the base year.

For regional and international freight studies, the movements of interest are somewhat simpler. However, the logistic and multi-modal aspects of decision making are taking a more important role in defining modes, routes, and timings and these may be more difficult to model both in the base year and in the future. Additional information is available from waybills and other instruments that accompany consignments and this may be accessible under favourable circumstances; alas, this is seldom the case.

15.5 Aggregate Freight Modelling

The great majority of freight demand models applied in practice have been of the aggregate kind (see for example Van Es 1982; Friesz et al. 1983; Harker 1985). These applications followed the classic four-stage model with some adaptations specific to freight. A typical example of this approach was the work of Kim and Hinkle (1982), who used the American urban transport planning suite (UTPS) with some adaptations to model state-wide freight movements. In outline, this approach involves:

- estimation of freight generations and attractions by zone;
- distribution of generated volumes to satisfy 'trip-end' generation and attraction constraints; the usual methods for this task are linear programming or Gravity models;
- assignment of origin-destination movements to modes and routes.

We shall look at these and other factors in some detail below; note that this aggregate approach implicitly assumes that the supply of freight services is more or less fixed and recurrent, and it is only a matter of establishing it using reliable data collection methods. In reality, this is unlikely to be the case; most truck and van movements in urban areas are demand-responsive and influenced by a dynamic pattern of deliveries. It would be more natural to handle these characteristics in a disaggregate and, ideally, agent-based model.

15.5.1 Freight Generations and Attractions

The techniques used to obtain total trip ends depend on the scope of the study, the level of aggregation envisaged initially, and the type of products considered:

1) A direct survey of demand and supply may be undertaken for major flows in the case of homogeneous products: sugar, petroleum products, iron ore, coal, cement, fertilizers, grains, etc. These may be forecast using industry or sector studies. This approach may be used for interurban movements but is not recommended for urban problems.
2) The use of macroeconomic models, for example, of the input-output nature, based on regional rather than national data.
3) Growth-factor methods, such as those discussed in Chapter 4, are often used in forecasting future trip ends.
4) Zonal multiple linear regression is often used to obtain more aggregate measures of freight generations and attractions, in particular in urban areas.
5) Demand may be associated with warehouse capacity or with the total shopping area at each zone (in urban studies) rather than with industrial development.

15.5.2 Distribution Models

Many urban studies simply apply growth-factor methods to observed goods movement matrices, as discussed in Chapter 5. However, many inter-urban freight transport studies have used synthetic aggregate models, even of the Direct Demand type. The two aggregate techniques most used in this area are briefly discussed here: the Gravity model and the Linear Programming approach.

In the case of the Gravity model, it is relatively simple to re-interpret its functional form as:

$$T_{ij}^k = A_i^k B_j^k O_i^k D_j^k \exp\left(-\beta^k C_{ij}^k\right) \tag{15.1}$$

where k is a commodity type index; T_{ij}^k are tonnes of product k moved from i to j; A_i^k, B_j^k are balancing factors with their usual interpretation; O_i^k, D_j^k are supply and demand for product k at zone

i (or j); β^k are calibration parameters, one per product k; and C_{ij}^k are generalised transport costs per tonne of product k between zones i and j.

The idea of using a generalised cost function formulation for freight demand is apparently due to Kresge and Roberts (1971). This can be interpreted as follows (omitting superscript k for simplicity):

$$C_{ij} = f_{ij} + b_1 s_{ij} + b_2 \sigma s_{ij} + b_3 w_{ij} + b_4 p_{ij} \qquad (15.2)$$

where f_{ij} is the out-of-pocket charge for using a service from i to j; s_{ij} is door-to-door travel time between i and j; σs_{ij} is the variability of travel time s; w_{ij} is the waiting time or delay from request for service to actual delivery (i.e. it may be a long time for maritime transport), and p_{ij} is the probability of loss or damage to goods in transit.

All of these depend on the mode used and to some extent on the commodity being transported. The constants b_n are, in general, proportional to the value of the goods. For example, in the case of the probability of loss, the cost is at least the goods' value, but probably more, due to penalties for delays in delivery. In the cases of delay, variability of delay and transit times, the values of b_n are at least proportional to those of the goods, essentially through increased inventory costs. Modern industrial production techniques, such as those emphasising 'just-in-time' deliveries, try to minimise these elements together with stocking costs. The minimum for b_1 to b_3 is the cost of the interest rate applied to the value of the goods during the time period considered.

In general terms, it is important to consider the relative contribution of transport (generalised) costs to the final cost of a commodity. For example, in the case of wheat, coal, cement, and bricks, transport costs are a key element in their final price; however, in the case of convenience foods, consumer goods, chocolates, or electronics, transport costs have a low (direct) contribution to price.

A second approach to the distribution modelling is Linear Programming (LP). This usually takes the form of a minimisation program: minimise total haulage costs (in money terms, very rarely in terms of generalised costs), subject to supply, and demand constraints.

$$\text{Minimise } Z = \sum_{ij} T_{ij} \cdot C_{ij} \qquad (15.3)$$

subject to:

$$\sum_i T_{ij} = D_j \qquad (15.4)$$

$$\sum_j T_{ij} = O_i \qquad (15.5)$$

This is the well-known Hitchcock's transportation problem, which can be solved efficiently in a very simple way (Dantzig 1951). More advanced formulations may involve non-linear costs and perhaps more elaborate constraints involving a time element and minimum shipment sizes.

This minimisation problem makes some sense from the point of view of a large firm trying to satisfy its customers at a minimum cost. Alternatively, if an industry has several plants with different production capacities and costs, the objective function may be to maximise profits or to minimise total cost at the marketplace. From the point of view of modelling, the LP approach has a better chance of being realistic when:

- the industry is concentrated in a few firms;
- there are low-value goods and relatively high transport costs;
- there are few demand points (zones), perhaps a monopsony (a single buyer).

However, it must be recognised that although LP may be a reasonable model for the behaviour of a single client or industrial firm, it cannot hope to represent aggregate behaviour for various commodities. The LP solution will tend to be too sparse, with particular destinations being served only by certain origins. On the other hand, the Gravity model is quite flexible. By changing the value of β, it is possible to vary the relative importance of cost compared with supply and demand constraints.

The formal relationship between LP and Gravity models has been explored by Evans (1973). She showed that in the limit, $\beta = 0$ in (15.1) will produce a matrix of movements where transport costs play no role (in fact this is Furness's solution to the growth-factor problem); whereas a large value for β will generate a solution closer to an LP model, that is, where transport costs are dominant (in the limit $\beta = \infty$ will reproduce the LP solution). Therefore, it is possible to use the Gravity model formulation to represent the whole range of client behaviour for destination choice, from that almost indifferent to transport costs (electronics?) to the behaviour expected in the case of low-cost, high-bulk commodities such as cement or sand, where transport costs are paramount.

15.5.3 Mode Choice

This is essentially a shipper's decision as to which carrier should be used to deliver the goods to their destination. When modelled at this aggregate level, modal choice is often treated using a Multinomial Logit (MNL) formulation based on generalised costs, as described above. This may turn out to be coarse, because the information can only capture those elements of mode choice incorporated in the generalised costs concept.

Shippers' decisions are, of course, dependent on the rates charged by carriers, which in turn depend on the volumes they move between each O– D pair. As the size of many consignments is significant in terms of the impact on carriers' rates, there are interactions inside mode choice which go beyond that encountered between passengers and public-transport operators. This problem is often ignored at high levels of aggregation.

In the case of urban freight movements, the problem of mode choice is trivial; the coverage provided by non-road modes is extremely limited.

15.5.4 Assignment

In the case of road haulage, this is now a carrier's decision, sometimes modulated by the driver of each vehicle: the choice of the best route to take the goods from origin to destination. To some extent, this is the easiest problem. The use of capacity restraint is probably relevant to most urban situations. In the case of interurban movements, on the other hand, it may be sufficient to use a stochastic assignment model. However, it may be argued that different types of vehicles must be modelled in different ways; for example, light vans may be less sensitive to the hilliness of routes than heavy lorries; also, vehicles carrying perishable goods might give greater priority to minimising time than those carrying, say, bulk cement. The use of multi-class assignment methods appears warranted to cope with this variety of cost concepts.

Investigations into road haulage have sometimes revealed unexpected influences on route choice. For example, newly built toll roads sometimes lack rest, food, and refuelling facilities, thus making them unattractive routes for long-distance drivers. Lorry drivers often prefer to drive at night to avoid the worst of congestion but they are sometimes limited in their choice by deliveries on narrow time windows. Notwithstanding, and particularly in less developed countries, safety and security may turn out to be the paramount factors for route choice.

In the case of rail, trains are sometimes scheduled according to a semi-variable timetable (roll-on roll-off trains, mail). In these cases, the algorithms from timetable-based public transport assignment can also be applied to rail freight assignment. More often, freight trains operate in response to demand, and a timetable does not exist, not even a line network with headways or frequencies. This requires a train formation algorithm to build the train journeys and their implied timetable. As this may be an inappropriate level of detail in a regional study, a short- or multi-path search algorithm may be appropriate in that case.

The use of a shortest path algorithm is likely to require incremental path searches, where the links already used in the previous steps are penalised, to favour routes using other links. This could be important when it is necessary to distinguish different train types (slow local trains, faster direct trains) and implies realistic penalties for shunting operations at transfer locations. It is also important to remember that marshalling yards and flat crossings impose capacity constraints on route choice and assignment models.

Intermodal terminals have gained significant market shares in recent times. They are usually conceived as interconnected by rail corridors although shipping and road haulage also provide services. Containers are transferred from one mode to another using gantry cranes and front lifters. Intermodal terminals are often perceived as a set of platforms with special equipment that serve a user catchment area via road and rail networks.

Intermodal assignment requires a multi-modal network model where many routes may be used for a specific O–D pair. A multi-modal route tree concatenates uni-modal route legs into intermodal routes. A route leg describes the part of a journey between two transfer points which does not require a transfer between vehicles. An intermodal freight assignment based on a route tree would consist of the following steps (Friedrich et al. 2003):

- generation of direct route legs between all origins and destinations using a uni-modal search;
- generation of route legs between transfer points using a uni-modal search;
- construction of route trees;
- calculation of generalised costs for all routes including transfer costs, and
- distribution of demand onto routes.

15.5.5 Equilibrium

As in the case of passenger demand, the problem of system or market equilibrium pervades the whole modelling exercise, but the techniques to achieve it are still under development. One early formulation of this problem is due to Friesz et al. (1983), who developed a freight network equilibrium model (FNEM). This model considers explicitly the decisions of both shippers and carriers for an inter-modal freight network with non-linear costs and delay functions that vary with commodity volumes.

FNEM treats shippers and carriers sequentially; shippers are assumed to be user optimisers trying to minimise the delivered price of the commodities they send and, therefore, Wardrop's first principle is used to replicate their behaviour. This sub-model is an elastic transport demand model expressed as a mathematical programming problem, solvable by the usual extension to the Frank–Wolfe algorithm discussed in Chapter 11. The assignment to carriers is performed through a *perceived* network including only the O–D pairs, transhipment nodes, and associated links considered by shippers in their decisions.

The carrier sub-model uses a full description of the actual transport networks. Carriers are assumed to be operating-cost minimisers and are modelled using Wardrop's second principle. The flow patterns of individual carriers are aggregated to obtain global network flows.

A similar approach was formulated by Moavenzadeh et al. (1983) for planning intercity transport demand in Egypt. In this case, the approach was based on the simultaneous transportation equilibrium model (STEM) (Safwat and Magnanti 1988), discussed in Chapter 11.

At a higher level of analysis, it may well be that the macroeconomic models used to generate the total demand and supply levels, and in some cases, the matrix of movements, use transport costs that are inconsistent with those generated by other parts of the model. Consequently, when such models are employed sequentially with a detailed freight network model, the two may fail to converge to stable solutions.

Harker (1985) formulated a generalised spatial price equilibrium model (GSPEM), which ties together the concepts of spatial process and shipper-carrier equilibrium to simultaneously predict:

- the production and consumption of goods;
- the shippers' routeing of freight traffic, and
- the freight rates.

A variant of the Frank–Wolfe algorithm was developed to solve a particular implementation of this problem and it was applied in a large-scale study (with approximately 3600 nodes and 14 600 arcs) concerning the USA's coal economy.

Example 15.1 Three types of aggregate freight models were estimated by Tamin and Willumsen (1992) for the island of Bali, Indonesia: a Gravity model, an Intervening Opportunities model, and a combined Gravity–Opportunities (GO) model. They considered five different types of commodities but used only traffic counts. The resulting freight matrices were compared with those observed in a major survey of the island. It was found that although the GO model performed slightly better than the pure Gravity model, its gain in accuracy did not compensate for the greater computational effort involved. The Gravity model was capable of discriminating between the five groups of commodities yielding a different β value for each. It was also far superior to the simple application of the Furness growth factor method.

15.5.6 Freight and Service Trips

Recent research has established the importance of estimating the number of vehicular trips associated with the pick-up or delivery of supplies, and service activities such as air conditioning repair technicians, to both households and commercial establishments (Holguín-Veras et al. 2017). Although these trips are a relatively small portion of the total transport demand in metropolitan areas, they are key to the economy, the environment, and the quality of life. Freight trips perform a crucial service to the economy delivering the supplies needed by commercial establishments and, since the advent of e-commerce, to households as well. In fact, internet deliveries to households represent the fastest-growing segment of transport demand in the world; in most developed countries, the number of deliveries to households is larger than the number of deliveries to commercial establishments. Moreover, the service activities that prompt the service trips typically take long periods of time, frequently in the range of hours. As a result, the percentage of commercial parking spaces used by service vehicles is significantly larger than the percentage of private vehicles. This, in turn, creates a situation where the more numerous freight vehicles cannot find suitable parking forcing them to double-park (Holguín-Veras et al. 2021). Thus, it is important to accurately forecast freight and service trips in urban and metropolitan areas.

The research on freight and service trip generation clearly indicates that traditional explanatory variables such as roofed area or the total area of the establishment are poor predictors of freight and service trip generation (Holguín-Veras et al. 2017). Instead, it has been shown that it is better to model the number of shipments sent and deliveries received (in the case of freight activity), and the number of service visits (in the case of services); and then estimate the corresponding number of trips dividing these numbers by the average number of shipments and deliveries that are made on a single trip, and the average number of services that can be performed in a single trip. Among other benefits, decoupling the demand and the vehicle trips enable analysts to consider the effects of novel vehicular technologies. Research has also shown that establishment-level models are better than zone-level models. In terms of explanatory variables, the most important one is the establishment's employment classified by industry sector.

Another important issue found is that using the correct model specification is critical, as some industry sectors receive and produce a constant number of deliveries and shipments, regardless of the number of employees (Holguín-Veras et al. 2011). In these cases, using a generation rate per employee will overestimate the freight trips for large establishments, and underestimate those for small establishments. Although freight and service activities are influenced by spatial factors (Sánchez-Díaz et al. 2016), these factors are difficult to consider in practical aggregate model applications; they could be treated more naturally in agent-based disaggregate models.

A good review of the state of practice of aggregate freight modelling is available in Chow et al. (2010).

15.6 Disaggregate Approaches

Since discrete choice models were developed and applied to model passenger demand, the idea of extending them to cover freight movements also gained currency; see for example Gray (1982) and Van Es (1982). The demand for freight transport is seen as that for a number of individual consignments, each with its own characteristics, for which the individual shipper has to take certain transport-related decisions. Every decision is seen as a choice made from a discrete set of alternatives. There is a number of related choices required in each case, for example, to transport x tonnes at time t of commodity k by transport mode m from origin i to destination j. The carrier would then have to choose the route to perform this task.

The flexibility of discrete choice modelling permits the construction of general utility functions for these types of choices. They can include, for example:

- the characteristics of the transport services, such as tariffs, times, reliability, damage and loss, minimum consignment, among others;
- the attributes of the goods to be transported, such as type of product, volume/weight ratio, and value/weight ratio; if the good is perishable, inventory system, and ownership;
- the characteristics of the market, such as its relative prices, firm size, availability of loading/ unloading facilities, and general infrastructure facilities;
- the attributes of the shipping firm, such as its production level, sales prices, plant location, available infrastructure facilities, storage policy, and so on.

However, the approach has found scarce applications at a national scale, due mainly to the limited understanding of all the elements involved in formulating the utility functions and the demanding data-collection efforts required to estimate this type of models. Notwithstanding, its application to particular sub-markets or commodities may provide valuable insights for policy formulation. For example, Ortúzar (1989) was able to use stated preference data to examine the question of offering a new service (refrigerated containers) for international maritime cargo. A similar application was also done by Fowkes and Tweddle (2000).

An important example of disaggregate freight modelling is provided by de Jong and Ben-Akiva (2007), who considered a set of commodity distribution chain choices. Their model was specified at the level of the decision-maker, from one sender to one receiver. Several choices were modelled, including frequency/shipment size, number of legs or stops, location and use of consolidation and distribution centres, and mode and vessel type for each leg. The aggregate production flows were first disaggregated into annual firm-to-firm flows using the number of employees per firm by zone. Once the annual flows were estimated the model calculated the shipment size at the destination of a chain such that total logistics costs were minimised. Transport costs were estimated only after transport chains were identified.

Empty truck movements were included in this model as carrying *empties* and using exogenously determined return loads. The method is based on the empty truck model by Holguín-Veras and Thorson (2003). Service delivery truck tours were not included since the approach focused on the movement of commodities. Like the other models in this subclass, integrated shipper-carrier operations are assumed.

Other examples of disaggregate freight modelling can be found in Chow et al. (2010).

15.7 Conclusions

Despite efforts in recent years, freight demand modelling is still less advanced than passenger demand modelling. The leading edge of research and development has been passenger demand forecasting, with freight following its footsteps trying to adapt models to its particular needs.

The problems of data collection may be compounded in the case of freight. For example, data collection for disaggregate approaches suffers from confidentiality and reliability problems. Even collecting data for aggregate modelling represents a much greater effort than that for passenger movements due to the great dispersion of firms, important daily and seasonal variations, among other factors.

Opportunities for extensive roadside interviews are limited, except at points where long delays are inevitable (for example, waiting for a ferry). In some cases, such as international travel, it may be advantageous to collate data from customs or a collection of waybills.

Because simplified models use low-cost and regularly collected data (traffic counts), it may be possible to run them often enough to update forecasts and provide corrective measures for plans, that is, they offer opportunities for implementing a continuous planning approach.

In the case of urban freight modelling, simple approaches are normally followed. They are usually based on models of vehicle movements, disregarding the commodities shifted, the type of locations served, and the underlying economic activities that originate this demand. It is often considered sufficient to obtain a commercial-vehicle matrix using roadside interviews (at cordon and screen-line points) and then to gross it up to the planning horizon by means of growth-factor methods.

Some software packages offer specialised facilities to solve relatively simple logistic problems like the planning of tours and routes. Others offer more sophisticated modules to optimise the formation of trains, routeing via hubs, and using multi-modal networks taking advantage of intermediate storage facilities at different costs.

Efforts to improve aggregate and disaggregate freight modelling are likely to prove fruitful from both research and practical viewpoints.

16

Activity-Based Models

16.1 Introduction

Travel has always been seen as *derived demand*. We rarely travel just for the sake of travelling. We do it to satisfy a particular need or requirement at a different location. We can perceive life as a sequence of *activities* undertaken at different *locations,* over a period of time, for example, a day or even a week. To perform these activities, we need to make *trips*; these, in turn, form tours linking a sequence of activities over time.

The conventional trip-based approach, exemplified in the classical transport model, has produced some sound transport systems analyses, with travel demand and network performance procedures estimating flows that tend towards equilibrium, with input from land use and transport supply. These models can be entirely trip-based, or more likely today, based on the estimation of *Productions* and *Attractions* and modelled as such until the trip-based assignment stage. This approach can be seen as a simplified way of handling the link between *travel* and *activities* (i.e. the reason why we travel between two points).

Mitchell and Rapkin (1954) established the link between travel and activities and called for a comprehensive framework and inquiries into travel behaviour. For several reasons, these ideas were not taken forward at the time, at least partially because it was challenging to operationalise them for practical planning purposes.

Many authors have contributed to the basic thinking of *activity analysis*. Among them, one must mention two contributions from the 1970s. Hägerstrand (1970), proposed a *time–geographic* approach that delineated systems of constraints on activity participation in time and space. Jones (1979), led the first comprehensive study of activities and travel behaviour at the Transport Studies Unit at Oxford, where the approach was defined and empirically tested, and where initial attempts to model complex travel behaviour were first completed.

Activities take place in space and to reach their desired locations people must travel. Looking at trips independently misses some of the behavioural richness of linking activities in different locations and with different time windows or constraints. Some activities can be rescheduled in time (i.e. postponing a trip to the gym) but only within constraints (gym opening hours). Others, like work or school attendance, are more difficult if not impossible to shift in time. Moreover, some activities may be rescheduled and reassigned to different individuals in the household and then to a different day of a week; for example, undertaking a main shopping trip for groceries. It is clear that, at least in principle, getting a better understanding of how people organise activities and the tours associated with them, should provide a more solid basis for travel demand modelling. This chapter explores how much of that understanding of activities and their schedules can actually be incorporated into operational models and what approaches can be followed to achieve this.

Modelling Transport, Fifth Edition. Juan de Dios Ortúzar and Luis G. Willumsen.
© 2024 John Wiley & Sons Ltd. Published 2024 by John Wiley & Sons Ltd.
Companion website: www.wiley.com/go/ortuzar5e

The increasing popularity of remote working (or 'tele-working'), and internet-procured shopping and entertainment, favours the generation of more flexible and less recurrent trip patterns; these become more difficult to represent in an aggregate model. Moreover, the successful introduction of new, often demand responsive, modes of travel facilitate these flexible schedules and represent a new challenge to modelling their availability and the level of service provided to potential users. These factors contribute to an interest in modelling travel demand, and supply, at a more granular level.

In what follows, we address first the issue of *tours* in greater depth looking also at the activities they link. We then look at the activities and how we can model the complex interactions within a household that help schedule them and, therefore, generate trips. The next section identifies the econometric structures that can be used to represent these scheduling processes. We introduce agent and activity based modelling. A key element in Activity-Based Models (ABM) is to synthesise detailed populations for both the base and future years. We then discuss approaches to model the daily schedule of activities of members of that population and the downstream tours and trips that result. Finally, we discuss some general points about the approach.

16.2   Activities, Tours, and Trips

It is useful to define the key terms in this discussion before advancing further. In this chapter we will consider the following concepts:

- an *activity* is a continuous interaction with the physical environment, a service or a person, within the same socio-spatial environment, that is relevant to the sample/observation unit; it includes any pure idle times before or during the activity (e.g. waiting at a clinic);
- a *stage* is a continuous movement using one mode of transport, more precisely one vehicle; it includes any pure waiting (idle) times immediately before or during that movement (e.g. waiting for a bus, searching for a parking space and making parking manoeuvres);
- a *trip* is a continuous sequence of stages between two activities (note that a trip can have one stage, for example, a car trip, or more, in a multi-mode trip);
- a *tour* is a sequence of trips starting and ending at the same location; a *trip chain* is the same as a tour, but it may not end at the same location;
- a *trip purpose* is defined by the most important activity undertaken at one of the ends of the trip.

Tours may be classified by length and by their 'most relevant' activity, for example, Home-Based Tours, Business-Based Tours, etc. Consider, for example, an urban area where a classic Production-Attraction model is using the following trip purposes:

- Home-Based Work (HBW) that includes the journey from work back home;
- Home-Based Education (HBEd), including the journey back home;
- Home-Based Other (HBO), including shopping, leisure, etc.;
- Non-Home-Based Business (NHB);
- Non-Home-Based Other (NHO).

Figure 16.1 illustrates the concepts and distinctions between activities, trips, tours, and purposes. The diagram identifies four typical individuals (A–D) that can undertake six different (aggregated) activities at home, work (i.e. factory and office), education (school), and shopping and leisure (i.e. a restaurant).

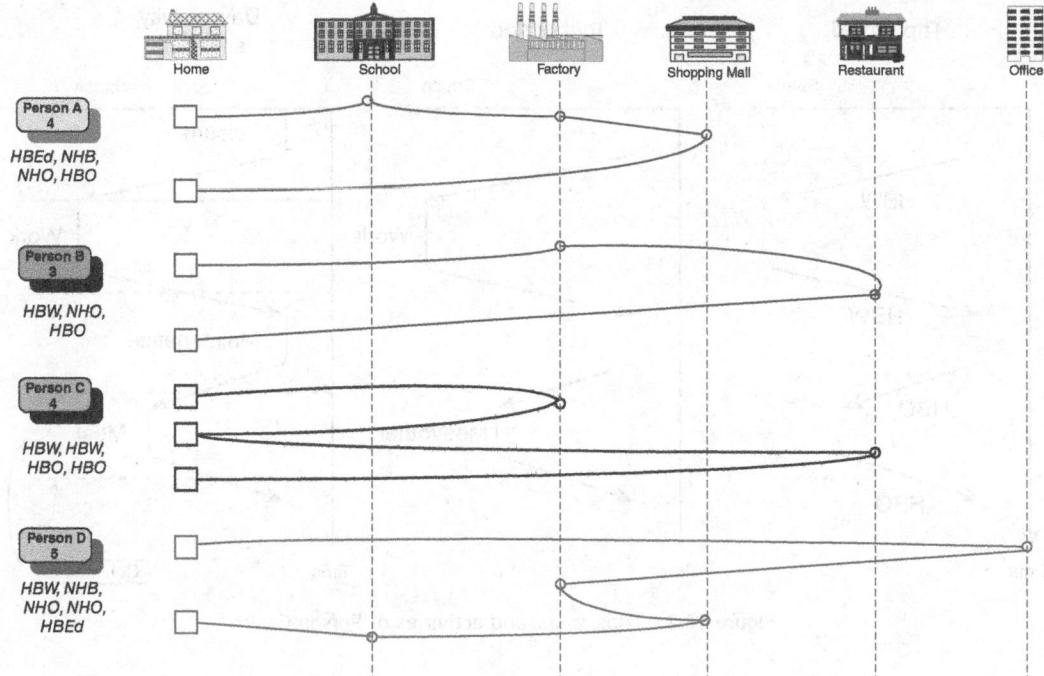

Figure 16.1 Daily activities, tours, trips, and purposes

In this diagram, Person A undertakes one tour visiting the school, factory, and shopping mall and then back home. Person A may have taken a child (Person E, not shown) to school and then proceeded to work. In a classic model, this tour would appear as four trips, one HBEd (or Escort), one NHB, and two NHOs. Depending on how the data were collected and processed, the first two might have been condensed into one HBW trip.

Person B has also undertaken one tour with three trips, HBW, NHO, and HBO. A classic model would have captured this sequence a bit better in practice. Person C undertakes two tours. The first is a simple one to work and back and the second is also a simple tour with two HBO trips. These two tours would be perfectly picked up in a classical model as two HBW and two HBO trips.

The longest tour is made by Person D who goes to work in an office, visits a factory, goes shopping, and finally attends an evening course before returning home. Despite the complexity of the tour, a classic model would have picked up these trips more or less accurately; however, the motivation and time constraints for some trips would normally be missed.

The choice of mode is, of course, also related to tour structure and length. If a person chooses to drive a car for the first trip in a tour, it is almost certain that this will remain the choice for the other legs. A possible exception is for short tours from work (not shown above), where public transport could be chosen for convenience, speed or to avoid parking problems. Similarly, if public transport is chosen for the first leg of a tour, this is likely to remain the choice for the rest of the outing, including taxis.

The description above is appropriate to compare tours and trips but does not give enough information about activities. This is explored further in Figure 16.2 constructed on the basis of the data for Person C.

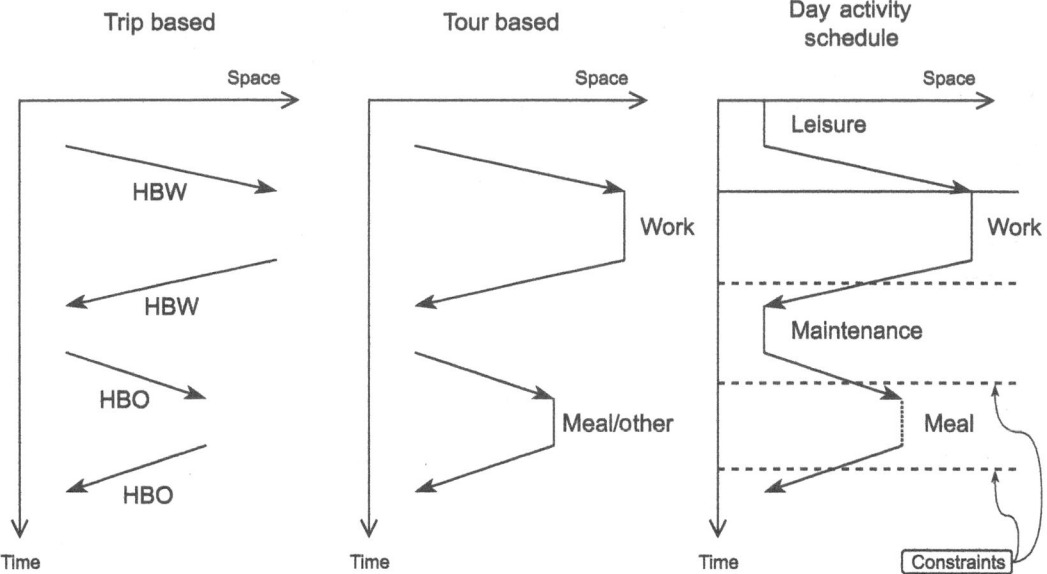

Figure 16.2 Trips, tours, and activities of Person C

Person C starts from leisure at home (although sleep could also be considered *essential mainte-nance* at home, say, seven hours minimum plus breakfast) and then travels to work; this activity has a strict constraint given by its starting time but is more flexible on leaving the job. Person C returns home for some maintenance (rest) and then goes out for a meal at a restaurant that, in this case, has a flexible start and end as an activity (i.e. no strict booking). The role of different time constraints for the activities just illustrated is made more complex if Person C does not wish to dine alone. He may prefer to coordinate with the spouse for this meal, starting either at home or from a different location.

To capture the behavioural richness of activities we must look at the household and the mutual interaction between trips, tours, and activities. We should also consider the following essential aspects of activities and behaviour:

- travel is derived from the need to change locations between successive activities;
- individual activities may be the components of more significant personal projects (i.e. shopping for paint contributes to a plan to redecorate your home), reflect longer-term commitments (work, religious attendance), or satisfy some basic physiological or psychological demands, such as sleeping or enjoying the company of friends;
- scheduling activities involves the choice of time, duration, location, and access/egress mode for the preferred activities;
- some activities are *mandatory* (e.g. work, education attendance) and offer limited flexibility in terms of location and duration; others are required to *maintain* other activities (e.g. eating, sleeping at home, shopping, and personal business away from home); finally, some activities, although *discretionary*, are still essential for a fulfilling life (e.g. social, recreational, entertainment);
- individuals have monetary and time constraints (i.e. money and time budgets);
- individuals schedule their activities in coordination with other members of the household, or of their social network, to maximise satisfaction, considering short- and long-term aspirations;

- individuals are constrained in their scheduling of activities by the resources available to them, in particular vehicles of various kinds, or the availability of reasonable public transport services;
- individuals are further constrained by the need to be available to others at particular times and locations, either face-to-face (e.g. presentation to client) or by phone or videoconference;
- longer term commitments to other household members, residential locations, and work/educational places provide additional constraints to individual choices.

The task of converting these issues into a workable and reliable activity scheduling process, that can be formalised using closed-form formulations or more general computer codes, is a demanding task but the advent of widely available computer power has made possible a number of alternative ways of tackling this difficulty.

Before we look into that problem, we must consider how to model the individuals that will participate in the selection of activities and tours.

16.3 Tours, Individuals, and Representative Individuals

In this section, we consider the unit of application of ABM: households and individuals or representative individuals. To address this question, we look first at tours and their complexity. If the majority of trips in a metropolitan area is of the Person C and D type above, a simpler treatment for tours may be sufficient. But if the type of tours represented by Persons A and B is significant, say over 15%, then it would be important to address this issue in practice.

A recent study of travel in the Auckland (NZ) region, established that overall, 70% of tours have outward and return trips of the same purpose (and corresponding to that of the tour); 57% of tours comprise only two (out and back) legs. For HB work and business multi-leg tours, the average is around two extra legs per tour; for HBO purposes it is 1.3 extra legs and for education, it is just one extra leg (these extra legs being NHB trips).

Inevitably, longer and more complex tours require additional research, time, and resources. It is difficult to make a prior decision about how many different types of tours one should attempt to model. Longer tours are less frequently found but may be more important in the future if policies to manage congestion are implemented: a four-leg tour satisfying three different activities contributes less to congestion than three 'there-and-back' tours doing the same. Of course, not all possible tours would be included in a model, only the most important and frequent; for example, if 95% of the tours have four trips or less, this should probably be the maximum number to model.

The next task is to identify the individuals that would be modelled to undertake these representative tours. In a fully disaggregated approach, these individuals will result from an expansion of a random sample representing the whole universe of travellers. Each of these individuals will have a specific set of characteristics: age, gender, income, type of work, type of family, car availability, etc. In a disaggregate approach, sample enumeration techniques, as discussed in Chapter 9, will be used to model individuals' choices and to apply these results to the entire population of the study area. This is a major task for the base year (when, typically, a household survey is available) but it becomes an even more demanding task for future forecasting years where the population has to be synthesised. The task of fully specifying present and future individuals is termed population synthesis and is discussed later in this chapter. Population synthesis can also be used to replace individual addresses for home or work with better spread addresses and even at a more detailed level of resolution than traffic analysis zones (TAZ).

An intermediate approach is to identify a number of homogeneous behavioural groups or representative individuals, say some 20 segments of the travelling universe. Each group will have a set of characteristics pertaining to travel behaviour but they will still be represented as based on the centroid of a given zone and travelling to other zone centroids. Figure 16.3 illustrates idealised synthetic populations in a zone of a study area.

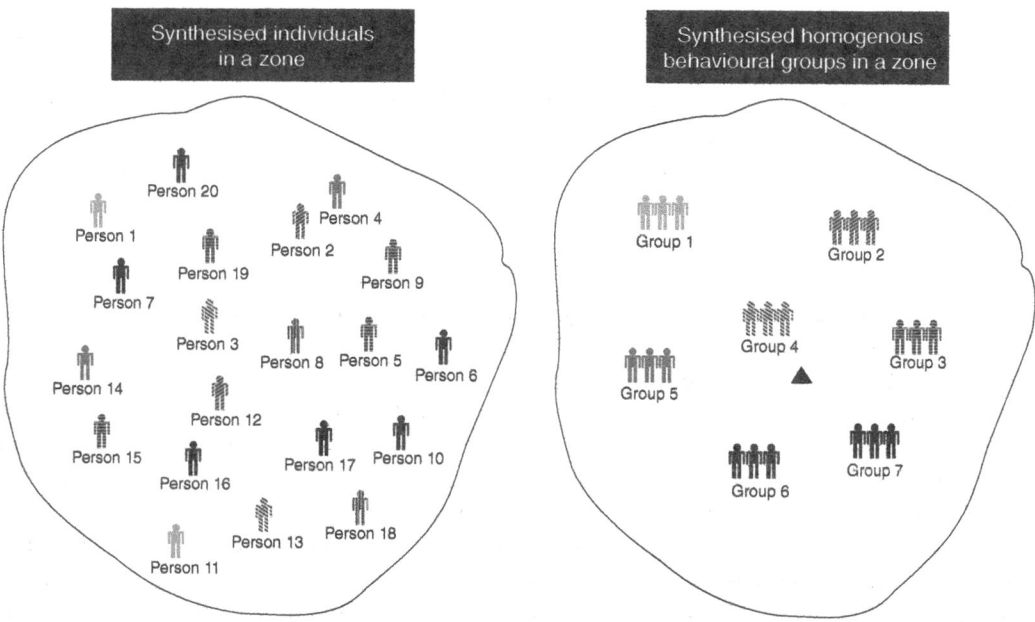

Figure 16.3 Individuals and homogenous behavioural groups in a zone

Both individuals or behavioural groups will require a population synthesis procedure to generate their number and characteristics in each zone (or address coordinates) of the study area in the future, on the basis of known land uses. If the key modelling focus is on tours, then homogenous behavioural groups may provide sufficient disaggregation. If the interest is in the activities and processes that generate those tours, it is difficult to envisage the use of homogenous behavioural groups as capturing the complexity of these interactions. In fact, one would need to model not just individuals but all the members of a household that take part in these decisions.

16.4 Agent-Based Modelling

The task of representing activities and tours at a sufficiently detailed granular level requires a different type of disaggregate model: an Agent-Based Model (AgBM). In this case, systems are modelled as a collection of autonomous decision-making entities called agents. Each agent, individually, assesses its context and opportunities and makes decisions based on a set of rules, many of them probabilistic in nature. Agents may adopt different behaviours appropriate for the system they represent, for example working remote or face-to-face, travelling or resting. An AgBM often represents competitive interactions between agents, for example deciding who goes first at a junction or who does the weekly shopping in a household.

These models are particularly useful when the interactions between agents become easier to describe than transition rules for the complete system. Agents may be capable of evolving, allowing unanticipated behaviours to emerge. An advanced AgBM may incorporate neural networks, evolutionary algorithms, or other learning techniques to allow realistic learning and adaptation.

Agent-based modelling offers several advantages in the context of activities and tours (Vovsha 2017). For example, it may be possible to:

- make individual agents (households and persons) intelligent by having individual memory;
- give individual agents the capacity to adapt to the changing environment, for example, by changing parameters such as willingness to pay as a function of situational time pressures;
- allow direct interactions between individual agents, for example, to allocate tasks within a household;
- explore the competition for activities, for example, in the context of workplace location choice;
- explicitly model the supply of vehicles for demand-responsive transit; for example, the number of autonomous cars available for hire in a particular area and at a particular time, and the car relocation tasks required to make them available where demanded.

In this context, most Activity-Based Models (ABM) discussed in the next section are implemented as Agent-Based Models.

16.5 Activity-Based Modelling

16.5.1 Introduction

Activity-based demand models can be classified into two main groups (Tajaddini et al. 2021): Utility maximization-based econometric models and rule-based computational process models (CPM). Utility maximization models apply the type of structures discussed in Chapters 7–9 in this book, for example, Logit and Probit, to estimate probabilities of certain events. Rule-based computational process models apply different sets of condition-action rules and focus on the implementation of daily travel and scheduling activities. Most recent approaches combine ABM with utility maximisation modelling, sometimes exploiting the opportunity for agents to learn, modify, and improve their interactions with other agents.

It is important to recognise from the outset, that an ABM is only part, albeit a core one, of the complete modelling system. For a start, an ABM covers only residents in the study area. A good deal of the model system is still aggregated, zonal based, and produces the traditional outputs that are needed for the appraisal of projects and policies. However, because of its finer treatment of activities, long- and short-term decision-making, tours and mode choice, an ABM is able to address, at least in principle, a wider range of policy instruments and behavioural responses. Figure 16.4, adapted from Bowman and Bradley (2008), shows an ABM core and the additional components of a complete transport modelling system.

The ABM core contains long- and medium-term choice simulators, a person-day simulator, and covers the full day. The former model the long-term choices, like a normal place of work, car ownership, and season ticket commitments; medium-term issues involve household tasks allocated to individuals (e.g. escorting children to school, convenience shopping, etc.). The person-day simulator searches for the most appropriate set of activities and tours required to satisfy these tasks. The main output is a list of household and person day-tours (including destination, time of travel, and

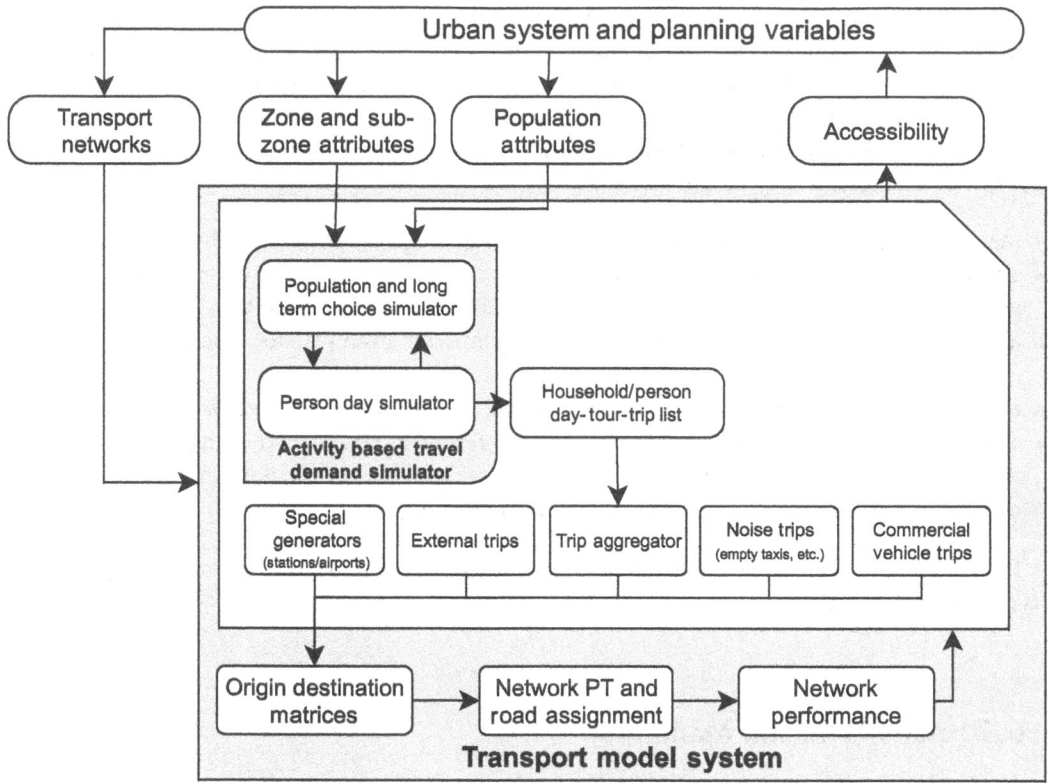

Figure 16.4 A modelling system with an ABM core

mode choices) that, in turn, result in a trip list that goes into a *trip aggregator* where all trips are consolidated. To these residents' trips, one needs to add:

- external trips from outside the study area;
- special generator trips, in particular from an airport or some other large or long-distance travel stations;
- commercial vehicle trips, including courier deliveries and rubbish collection;
- *noise trips*, that is, trips that are seldom modelled but do exist in the network, such as empty taxis cruising for passengers, drivers looking for a parking space, emergency services/police vehicles, people out for a 'drive', or lost drivers.

The combination of all these trips (except normally the *noise* group) is used to build time-dependent trip matrices, which are assigned to the network until equilibrium; this results in network performance indicators (travel times, etc.) which are fed back into the ABM core and trip consolidator modules. Accessibility information is also fed back onto the urban system to influence land use and population attributes.

We will look now at the components of the ABM core itself.

16.5.2 Population Synthesis

The task of population synthesis is not exclusive to ABM. Indeed, aggregate and disaggregate models have a population sample with most of the relevant characteristics available for the base year when Household Travel Survey (HTS) data are collected. Moreover, this population needs to be projected for future years based on a few properties that are actually forecast by planners, such as the number of people per zone, perhaps income, and with some luck car ownership. Other attributes, like distribution of household sizes, age distribution, school and university attendance, or multiple vehicle ownership need to be estimated, more often than not at the level of the representative households in each zone. Population synthesis is an excellent way to expand the base year sample to the full population and project this for future years, retaining key characteristics of the individuals.

An ABM that forecast the activities and travel of urban populations require this task at a higher degree of disaggregation; the most important developments in population synthesisers have been attained by seeking this more recent goal. The first task is to create a synthetic population and then simulate the behaviour of the households and persons in that population.

Population synthesis involves generating an artificial population by expanding the disaggregate sample data to mirror known aggregate distributions of household and person variables of interest (recall the discussion in Chapter 9). The process normally starts by creating a base year synthetic population from Census and HTS data and then using aggregate demographic and land use forecasts to create a synthetic population for each future year. The synthesis procedure involves two main steps. First, a demographic distribution of households is estimated for each transport zone or small Census area (zone), and then a matching sample of households is drawn from a set of household records for which nearly complete Census information is available.

The demographic distribution is defined discretely by the cartesian product of several categorical control variables (dimensions), where each multidimensional category (or cell) is defined as a unique value combination of the one-dimensional control variables. The number of households in each cell is often estimated through an iterative proportional fitting procedure analogous to the Furness method. The procedure starts with an initial joint distribution available for (aggregate) Census geographical units. It then cycles iteratively through a set of control totals, one for each category of each control variable.

Example 16.1 Consider sample data as shown in Table 16.1. There are three household sizes and two income levels. We know from, say Census data, that there are 55 households (HH) with low income and 35 HH with high income in that zone, and that there is a total of 20, 40, and 30 HH of each size. The sample data are shown in the 3 × 2 (say from an HTS) rectangle in the middle.

Table 16.1 Sample data and marginal distributions

| HH size | Adjustment | Income | | Total | HH size marginals |
		Low	High		
1		3.00	1.00	4.00	20
2		2.00	4.00	6.00	40
3+		4.00	2.00	6.00	30
Total		9.00	7.00		
Income marginals		55.00	35.00		

The application of a bi-proportional adjustment, in this case, will solve this population synthesis problem; after only three row and column iterations we would get the nearly perfect fit shown in Table 16.2. What remains is to convert these figures into discrete individuals residing at specific location coordinates.

Table 16.2 Adjusted synthesised data after three iterations

3		Income		Total	HH size marginals
		Low	High		
HH size	Adjustment				
1	0.997	16.19	3.81	20.0	20
2	1.003	16.59	23.41	40.0	40
3+	0.998	22.17	7.83	30.0	30
Total		54.95	35.05		
income marginals		55.00	35.00		

This approach can be extended to cover other dimensions like car ownership, the number of students, or the number of persons of different types. The adjustments will then be multi-proportional. As we saw in the case of matrix adjustments in Chapter 5, a requirement for this procedure to work is to have consistent control of marginal totals. In this case, the iterative procedure will converge so that all control totals are satisfied and the correlation structure of the initial joint distribution is preserved. Control totals are taken from Census tables for the base year. For the forecast years, they will come from demographic and land use forecasts, which may be less detailed. Note that for some model applications, the number of households in each cell may need to be rounded to an integer number.

It is also useful to note that the problem of zero cells or zero marginals, that affected trip matrix expansion or matrix estimation, may apply also to the population synthesisers. Similar corrections would need to be applied. In estimating a year distribution, all population synthesizers control for household income, household size, and the number of workers. Additional household characteristics used as controls in some cases include age and gender of the head of household, presence of children, and family vs. non-family household members.

From an ABM perspective, the process of population synthesis needs entering into a second phase. We need to identify the personal attributes from within each household, while retaining the marginal totals for each zone. However, it is known that the derivation of personal attributes can severely affect the accuracy of the subsequent modelling.

This second phase typically includes three or four steps. The first is to convert into integer the non-integer values for households in zones resulting from the first phase. Second, a Monte Carlo procedure is typically employed to draw the correct number of households of each type from the HTS or an available Census sample. Note that as some of the desired data may not be available in the Census, or it may not be accessible to the modeller, it is often inevitable to sample from the HTS and any activity diary dataset available. Third, the useful household and person variables are extracted from the drawn households and retained for use by the model system.

The fourth step is optional. Many implementations of ABM have sought to use a finer geographical level than that offered by conventional TAZ. This optional fourth procedure is used to assign each household to a more precise location (sub-zone or just coordinates) within its geographic unit. The final output from these processes is a synthetic population where each synthesised household and its members have many clearly defined characteristics of interest for use in the model system and, together, they match the estimated demographic distribution within each zone.

16.5.3 Monte Carlo and Probabilistic Processes

Most ABMs use a Monte Carlo process to represent individuals (and vehicles) and their behaviour in a transport system. The name comes from the use of random numbers (as in a roulette) to sample from a population with a known distribution of attribute or characteristics as we saw in Chapter 8. Pseudo-random numbers between 0 and 1 can be generated very easily, for example in Excel using the RAND( ) function, and these values can be used to sample from any distribution. To create a particular individual, one may sample from a Uniform (0–1) distribution for sex, from a Log-Normal distribution for income, and from special distributions for age, family size, employment, etc., including sampling from a set of possible locations for sub-zone. This is repeated for each individual and then samples are taken for tour characteristics and length, including time of trip making.

Monte Carlo simulations are powerful; it is possible to represent almost any population, both present and future, and include all characteristics believed to be relevant to identify activities and desirable tours. This flexibility comes at two prices. First, that it is often difficult to have full confidence that the resulting model is rigorously *calibrated* and representative of an external reality that may be different from the ideas of the modeller. Second, significant computing power and time are required to obtain reliable results; this limitation has been more or less removed by the recent developments in computing. As random numbers are used to represent individual characteristics and travel behaviour, it is not enough to simulate one day (or one hour in a traffic microsimulation project). It is necessary to repeat the process several times, with different initial random numbers, to gain confidence in the stability of the results.

Monte Carlo simulations can be used to model individual choices within a well-structured family of Nested Logit (NL) models for the choice of, for example, activity patterns, tour length, tour characteristics, choice of destination, time of day, and mode.

16.5.4 Structuring, Activities, and Tours

An ABM is, in fact, an integrated system or combination of several, mostly sequential, econometric sub-models. In principle, these would be structured to cover:

- long-term commitments of the household and its members, including the amount and location of work, residential location relative to work, education and friends, preferred types and locations for shopping and leisure, etc.;
- medium-term schedules for each individual in the household reflecting the tasks allocated to them and their specific activity demands, including projects like getting a degree or learning a new skill;
- daily personal schedule of activities formulated by the individual, although some activities might change during the day in response to changing conditions; this flexibility has been helped by the use of mobile phones.

In practice, most applications start at the Person Day-Activity model and from this activity pattern tours are selected and disaggregated into key components. It is possible to structure these components as a set of NL models, as proposed in the seminal paper by Bowman and Ben-Akiva (2001) that has influenced a number of ABM efforts, in particular TRANSIMS (https://sourceforge.net/projects/transims/). An idealised example of this nested structure is presented in Figure 16.5.

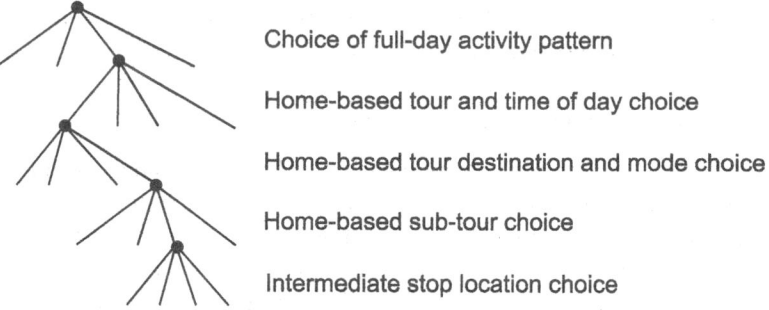

Choice of full-day activity pattern

Home-based tour and time of day choice

Home-based tour destination and mode choice

Home-based sub-tour choice

Intermediate stop location choice

Figure 16.5 Example of nested choice structure for the activity schedule

The lower-tier models are conditional on decisions at the higher tiers. As one would expect, the conditional model probabilities flow down the hierarchy whilst *logsums*, or expected utility values, pass the information from low-tier choices up to the top-tier ones. Going down, one has to calculate probabilities for all alternatives in each nest representing a large number of combinations, as each branch above has many sub-branches below.

Moreover, using logsums in this way, means that we need to calculate the utilities of every combination of alternatives going up the tiers before calculating probabilities from the top to the low-tier nests. As these alternatives include all destinations, times of travel, mode, tour and sub-tour types, and intermediate stops, this is computationally very intensive.

The advantage of using a random utility framework and Logit formulations is their solid theoretical background, sound model estimation techniques, and user's familiarity with the strengths and limitations of the approach.

Inevitably, the top-tier choices are quite critical and, at the same time, complex. The choice of a person's full-day activity pattern may involve selecting among many (i.e. 50 or more) preidentified activity patterns with their respective utility functions estimated from the HTS and travel diary surveys. One of these possible activity patterns is, of course, staying at home; the others require travel in different combinations. It is customary to distinguish between primary activities and tours (work, education) and secondary ones having more flexibility in location, timing, and mode.

Taking as an example the seminal ABM for Portland (Oregon, USA), we can appreciate better the scope and complexity of the task (Bowman and Ben Akiva 2001). The Portland Day Activity Pattern contains 114 alternatives differing in terms of the activities involved and their sequence. The choice set covers primary activities and tours, and secondary tours associated (extensions) with the primary ones:

Primary Activities:

- subsistence (work or education) on tour;
- subsistence (work or education) at home;
- maintenance (shopping, personal business, etc.) on tour;
- maintenance at home;
- discretionary (social, recreational, leisure, etc.) on tour;
- discretionary at home.

If the primary activity is on a tour, the daily activity pattern model also estimates the trip chain type. There are eight possible subsistence tours, four for maintenance and four for discretionary tours. The trip chain type is defined by the number and sequence of stops in the tour (i.e. a simple tour, one or more activities on the way to the primary destination, one or more activities on the way home, and intermediate activities in both directions). For the subsistence tours, there is also a work-based sub-tour addition to each of the four tour types above.

Secondary Tours: At the same time as the primary activity and tour type, the daily activity pattern model estimates the number and purposes of secondary tours with the following alternatives:

- no secondary tours;
- one secondary tour for work or maintenance;
- two or more secondary tours for work or maintenance;
- one secondary tour for discretionary activities;
- two or more tours for discretionary activities;
- two or more tours, at least one for work/maintenance and one for discretionary activities.

As not all tour types apply to all primary activities, there are only 19 possible combinations of primary tour types; these, times six secondary tour types make up the 114 alternatives.

Besides these choices, one has to consider the timings, duration, modes, and destinations/stops for these tours and trips, which makes the set of nested models rather demanding in terms of computational power. Therefore, the methods for solving these large-scale nested models become quite critical.

16.5.5 Solving ABM

In the case of the aggregate classic approach, the model is applied using enumeration at the zonal level. For each travel zone, the number of trips by each mode to each destination zone at each time of the day is computed. This is generally the result of several sub-models: trip generation, the proportion of trips going to each destination, modal shares and so on. Model probabilities/shares are used to distribute demand across all feasible alternatives. We may call this approach *zonal enumeration*.

In the case of ABM, we need to use a different approach, either sample enumeration or Monte Carlo microsimulations. Sample enumeration, as we saw in Chapter 9, also follows an approach of multiplying conditional probabilities. In this case, however, instead of applying the models separately for each travel zone, we apply them to each household and/or person in a representative sample. Thus, sample enumeration tends to work on a less aggregate scale than zonal enumeration, but that is not necessarily the case. Zonal enumeration can be applied also to many different segments of the population in each zone so that we can essentially work with an expanded sample of representative household/person types.

Sample enumeration, however, enables the retention of more complete information about each person and household in the sample – not just those characteristics that are used to define market segments. As a result, sample enumeration allows a wider range of variables to be included in the models. Moreover, if models are estimated at the level of the person or household, then sample enumeration is applied at the same level, avoiding aggregation bias.

Bradley et al. (1999) looked into the use of sample enumeration and Monte Carlo microsimulation as two alternative methods for solving ABMs using the Portland case as a test bed. Despite using a number of shortcuts for the sample enumeration approach, they concluded that it was faster and more practical to use Monte Carlo simulation.

In the case of stochastic Monte Carlo microsimulation, one still needs to use the same choice trees and analytical structure, but just solve the hierarchical model in a different way. The logsum linkages are also used to calculate utilities up the tiers for each individual up to the full activity/tour pattern. But instead of calculating probabilities for all combinations of alternatives down the tree, samples of activity lists from the survey data are taken to replace the information with choice data from the models. For example, in the case of the Portland model (Bradley et al. 1999) the process involved the following steps:

- draw a random sample of a single full-day activity/tour pattern from the top model probabilities;
- if the primary activity of the day is out-of-home, draw the times of day for the primary tour from the tour time-of-day model probabilities;
- use these synthetic choices to sample a corresponding day-long sequence chain of observed activities from the HTS;
- for each tour in the pattern, sample from the destination and mode choice model probabilities to replace the observed destinations and modes in the activity list;
- for any intermediate stops in any tours, apply the intermediate stop location models stochastically to assign locations to those activities.

It should be noted that the only details retained from the HTS activity records are the more precise timings and sequencing of activities since the time-of-day models usually deal only with a discrete number of different time periods (e.g. five in the Portland model). All the other observed choices are replaced by synthesised ones sampled from the models.

The use of Monte Carlo microsimulations has become the preferred method for solving this type of ABM. Of course, some questions remain:

- as we are using random numbers, do the starting point and their sequence influence results?
- do we need to run the model several times with different sequences of random numbers to ensure we obtain reliable and repeatable results?

Bradley et al. (1999) also investigated these issues offering the following conclusions and recommendations:

1) Always run the model simulating the full population of interest.
2) The differences in results when using a different random number sequence at an aggregate level are minor (1% or 2%).
3) When looking at more focused results, bear in mind that if the values are small, for example, the number of trips made by a population segment between one group of zones and another group, then the percentage variations are likely to be large.
4) These variations are healthy reminders that all models inevitably contain errors; notwithstanding, stochastic sampling errors are likely to be small compared to other sources of error present in any model (measurement, specification, input forecast errors, recall the discussion in Chapter 2).

16.5.6 Integration with Assignment

The complete model system requires a multi-modal assignment process. This is often static assignment using period origin-destination trip matrices and producing period static flows as outputs. Because of their detailed macroscopic treatment of demand, ABMs lend themselves to more realistic Dynamic Traffic Assignment (DTA). The combination of ABM and DTA can better represent the interactions between human activity, their scheduling decision, and the underlying congested networks.

There are different approaches to the integration of ABM and DTA, the most common of which is sequential integration. In this case, exchanging data between two major model components (ABM and DTA) happens at the end of a full iteration. To generate daily activity patterns for all the synthetic population, the ABM is run for the whole day. The outputs of the ABM, namely lists of activities, trips, and plans, are then fed into the DTA model obtaining flows. The DTA model then generates a new set of time-dependent skim matrices as inputs to the ABM for the next iteration. This process is continued until convergence is reached (Tajaddini et al. 2021). The sequential nature of this approach means that it cannot be responsive to short-term dynamics of congestion nor represent the use of real-time, enroute, information that is increasingly available. However, achieving a tighter and more efficient integration is a demanding exercise (Rieser et al. 2007).

Most ABM—DTA model integrations are based on sequential integration. Efforts to develop a comprehensive simulation model that can account for all components of dynamic mobility and management strategies is still a work in progress. As Miller (2020) points out, the ability of these models to reflect behavioural realism is still limited. Improvements will have to implement an integrated ABM—DTA platform on a large network including multi-modal dynamic assignment while reducing computational effort via tighter data exchange procedures.

16.6 Refining Activity or Tour-Based Models

The description above has focussed on the Person Day Activity model only; however, as discussed before, an ABM also includes components for long- and medium-term decisions. This section provides some additional information on the handling of these and other issues in a complete ABM system, which will have modules for:

- population synthesis for the geographic allocation of households;
- longer-term decisions: mostly car ownership but in some models also the choice of place of work and education;
- person/household-daily scheduling, including the choices of activity patterns that span the whole day for a household or person;
- tour level choices as discussed in the previous section;
- trip level choices: intermediate stop locations, mode, and timing;
- consolidation and assignment of trips to their respective networks.

Moreover, the ABM should also provide interrelationships between many of these decisions as, for example, the choice of car as a mode by one person in a household will affect the choices of the other members.

16.6.1 Choice of Usual Place of Work and Education

It is recognised that these are long-term decisions that are not adjusted on a day-to-day basis. The choice of a place to live is implicitly modelled in the population synthesiser, so it is included at the top level. The choice of the usual place of work, and school or university education, is better modelled at the top of the hierarchy as well and not as part of the person's day-schedule; most models today include it at this level. Note that some workers (construction, salespersons) may not have a *usual* place of work. It is important to ask, in the HTS exercise, about the usual place of work even if it was not accessed on that survey day.

16.6.2 Car Ownership

This is usually modelled also at the top level using a disaggregate model based on household and person types as discussed in Chapter 18.

16.6.3 In and Out of Home Activities

The ABM approach recognises that some activities (work, study, maintenance, and discretionary) are undertaken at home and we may wish to identify and include them. Most models, however, focus mainly on out-of-home activities and recognise only the probability that a person will not make any external tours in one day.

The number of out-of-home activities is relatively large, at least seven are usually considered: work, school, escort, shopping, meals, personal business, and social-recreational. Additional distinctions are also possible.

16.6.4 Person Day-Patterns Linked Across Household Members

Originally, ABM treated linkages across members of the household implicitly through person type and household composition variables. The use of microsimulation makes it easier to treat these linkages more explicitly. For example, it is possible to simulate the children of a household first and then the adults conditional to what the children do, in particular, their educational activities. This will result in escort activities being correctly allocated to adults and children.

Joint activities, as going together for a meal out, should also be modelled consistently as they are likely to have a significant impact on mode choice, for example. This will require a module for joint activity generation and participation; this additional effort must be traded-off against the greater accuracy achievable for trip choices.

16.6.5 Activities Allocated Explicitly Among Members of the Household

In principle, certain activities are undertaken on behalf of the household rather than individually; for example, shopping and escort trips. Modelling how these activities are allocated to different members of the household and on different days of the week has been limited. This task is further hampered by the limitations of survey methods currently in use; these are less useful to identify which activities are more likely to be allocated to different members at different times. With a few exceptions, most ABMs do not have a module to allocate these activities to members and assume that they depend only on general household and personal characteristics. It is argued that who actually undertakes them is less important than the fact that they are carried out at certain times and destinations.

16.6.6 Number of Zones Used

In most cases, the zones used for developing an ABM are similar in size to those of a trip or tour-based aggregate model. Ultimately, the car and public transport assignment modules are exactly the same. However, the use of microsimulation facilitates the implementation of finer geographical resolutions. Several models use a finer sub-zone system below that of TAZ. For example, the Portland model uses 20 000 *block faces* and a model in Sacramento 700 000 *parcels* (Bowman 2009). This fine level of disaggregation is useful to define more accurate destination choice alternatives, and estimate mode choice using detailed access to public transport information and level of service data. Note that in this way intra-zonal trips practically disappear from the model.

16.6.7 Time Periods and Time Constraints

Most ABM applications contain tour time-of-day models that reflect some sensitivity of time of travel choice to network conditions. However, there is usually only a limited number (3–5) of assignment periods, thus blurring some of the time sensitivity. There is a tendency to use more precise time windows to schedule each tour and trip consistently during the day (Bowman 2009).

This requires keeping track of the available time window after blocking off the time required by each activity and associated trips.

This tendency is converging towards half-hour periods, the main constraint being the ability of people to report times accurately; in fact, there seems to be a tendency to report times rounded to 10 or 15 min intervals. At this level of detail, there would be good reasons to move to dynamic assignment.

16.6.8 Network Equilibrium

Given the level of detail of the microsimulation approach for solving an ABM, it could be argued that there is no role for network equilibrium, as it does not happen in reality and trying to achieve it would distort results. Nevertheless, as indicated earlier, the reason to seek equilibrium solutions is to obtain modelling results that enable the consistent comparison of alternatives. In the case of an ABM, we compound the problem of iterative processing with the use of Monte Carlo simulations based on computer-generated random numbers.

Vovsha et al. (2008) investigated this issue. They looked into a number of alternative methods for ensuring, or at least approaching, convergence of the whole model system. They concluded that the application of the method of successive averages (MSA) to trip consolidated matrices and link flows, led to reasonable results after some 8–9 global (feedback) iterations. Further research is necessary in this field.

16.7 Challenges of Activity-Based Models

The implementation and use of ABM still face challenges that must be addressed in any practical implementation. Kagho et al. (2020) have identified the main ones as:

1) Input data limitations: the aspiration to model all aspects of human behaviour and the transport system is limited by the availability of data and knowledge at a reasonable cost.
2) Cost of computation: as many simulation runs are needed to obtain a well-calibrated model and stable results, the computing requirements are more demanding than those of a classic aggregate model.
3) Transparency: although an ABM is easier to explain in intuitive terms than a classic transport model, the results depend on a large number of interactions and calculations that may subtly depend on input data.
4) Calibration and validation: these two important aspects of model development are less well-established than in the case of classic aggregate models. The requirements are less well-developed and it may not be possible to achieve the same standards that apply to classic aggregate models.

Miller (2023) recently reached similar conclusions and identifies areas for further research including better understanding and models for:

- Activity episode generation to understand the logic of remote work or internet procurement.
- Activity location choice, especially for non-work/education activities.
- Sequencing of location and trip choice.
- In-home versus out-of-home activities, important to understand the 'need to travel'.
- Use of passive location tracking data, mobile phone network data, and other sources.
- Intra-household interactions and negotiations.

Addressing these issues requires greater experience in the development and application of an ABM. Software is available to support ABM development. In particular, MATSim (Horni et al. 2016) provides an open access and fully integrated Agent-Based Model of traffic flow and the resulting congestion, by tracing synthetic travellers' daily schedules and decisions. Further, SimMobility (Adnan et al. 2016) is an agent-based integration of a fully econometric activity-based demand model system with a simulation-based dynamic traffic assignment system. These and other efforts are facilitating the introduction of ABMs in different contexts and providing the basis for the transfer of models from one context to another. The use of a 'donor model', to be at least partially re-estimated in the new context using the same software, reduces the cost of implementing an ABM.

16.8 Extending Random Utility Approaches

Despite their emphasis on activities, the majority of ABMs are essentially microsimulation tour-based models that use a random utility choice-modelling framework; this has limitations. Current ABMs offer only a little in the way real activity re-scheduling happens, or of negotiations within the household on task allocations, and even less in terms of postponing tours to a later day of the week.

There are some experimental models that attempt to go further into treating these issues with greater realism. The most promising approaches depart from the econometric methods that are solved by sample enumeration or microsimulation. Econometric methods are ultimately based on the idea that individuals seek to optimise their utilities choosing the best among available alternatives. In practice, human behaviour actually recognises the costs of information acquisition, information representation, information processing, and decision-making. The new methods seek to represent behaviour and negotiations within this framework and are grouped under the name of Computational Process Models (CPM).

CPM are also microsimulations due to their disaggregate nature, the sequential decision process and the use of heuristics. However, the heuristics employed by CPM involve 'if-then' rules rather than utility-maximizing decision criteria. Models in this line of research are ALBATROSS (Arentze and Timmermans 2004) and PlanomatX (Feil et al. 2009).

Although these models have seen a number of applications, they will remain experimental for a while because of their nature. There are significant differences in the way these models handle the search for improved activity schedules and these rely on assumptions about behaviour and the nature of intra-household negotiations that are difficult to transfer from one context to another. The area of behavioural science is progressing very fast and we are likely to see better models implemented first in a research environment before general adoption for transport *decision-making* and policy development.

17

Model Design

17.1 Introduction

The main role of a transport modelling system, as an aid to decision-making, is to provide forecasts supporting insights about the impacts, positive and negative, of potential transport interventions. The model outputs are then useful to:

- establish a structured understanding of the performance of the transport system, especially regarding changed conditions, today and in the future, to help diagnose problems and identify the need to consider interventions; and
- compare the performance of future transport systems, with and without interventions, to understand their mix of costs and benefits.

Then, the transport model must be designed to deliver these broad aims. So, modellers face three particular challenges when specifying the model. The first is to balance the complexity and costs of both the modelling and forecasting tasks, given the decision-making needs. Too much complexity is a potential source of human errors and may be a waste of resources; and too much simplification may not provide the confidence in the model outputs required to support decision-making. The second challenge is to ensure that the necessary model inputs can, in turn, be forecast with confidence as they will underpin any travel projections. This is a difficult challenge, as there are few input variables that can be forecast 30 or even 10 years ahead with any degree of confidence. The uncertainty associated with any forecast is an integral part of the use of transport models to assist decision-making. This brings in the third challenge: to explain this forecasting uncertainty to decision-makers, so that they understand the reliance that can be placed on particular outputs and their inherent risks. This last challenge is associated with the modeller's recognition of the limitations of the model and its input data.

The design of a model system is a demanding task, and different approaches and preferences may result in very different models. Transportation authorities in many countries have produced guidelines on how to develop transport models to satisfy local requirements and support good and consistent decision-making. Most of them are available on the internet; for example, the Transport Analysis Guidance (TAG https://www.gov.uk/guidance/transport-analysis-guidance-tag) in the UK, the Australian Transport Assessment and Planning Guidelines (ATAP https://www.atap.gov.au/) or the New Zealand Transport Model Development Guidelines (TMDG https://www.nzta.govt.nz/assets/resources/transport-model-development-guidelines/docs/tmd.pdf); other authorities have also developed and adopted their own guidelines with different

Modelling Transport, Fifth Edition. Juan de Dios Ortúzar and Luis G. Willumsen.
© 2024 John Wiley & Sons Ltd. Published 2024 by John Wiley & Sons Ltd.
Companion website: www.wiley.com/go/ortuzar5e

degrees of specificity and flexibility; and these should be adhered to in each case. This chapter focuses only on the key principles for good model design behind any such guidelines. It starts with a consideration of the difference between accuracy and precision, as pursuing the second may undermine the first, and then considers the different aspects of the specification of a transport modelling system.

17.2   Accuracy and Precision

In the case of a measurement instrument (say a particular type of survey), *repeatability* is the degree to which several measurements under unchanged conditions show the same results. For example, the repeatability of stated preference surveys requires extreme care in sample selection, the delivery of the survey instrument, and the subsequent analysis of the data. Given this, there is a useful distinction between precision and accuracy, concepts that are sometimes confused in practice.

Precision is the measurement resolution, that is, the level of detail the instrument is capable of delivering. It can be interpreted as the number of significant figures in a measurement. We could say that a model based on more and smaller zones, or greater user segmentation, should be more precise.

The *accuracy* of a measurement tool is the degree of closeness of the measured quantity with its actual (true) value. A typical calibration process will try to ensure that the model is as accurate as possible. Accuracy requires repeatability of the model output. Different runs with the same inputs and parameters should produce exactly the same outputs.

Model error refers to the disparity between the actual values and the values predicted by a model for a specific variable. Typically, the primary concern lies in the predictive accuracy of a model, gauging its ability to correctly represent future conditions that remain uncertain and yet to unfold. The accuracy of these forecasts can only be evaluated after the events have transpired, as an *ex-post* assessment.

It is risky to pursue precision when we should be aiming for accuracy. A common mistake is to assume that because a model is more detailed and precise, it will also be more accurate and reliable.

The problem is that we cannot eliminate future uncertainty. Mathematicians and engineers, because of our education, tend to fall into the Laplacian illusion: 'Given perfect knowledge of present conditions and perfect knowledge of the laws that govern the universe, we ought to be able to make "perfect" predictions'. Sadly, this is indeed an illusion, as we have only partial knowledge of present conditions; further, these are driven by influences and behavioural traits that we do not fully understand, and probably never will. Nevertheless, our role as modellers is to provide the best advice to decision-makers about the likely impacts of any proposed interventions.

Following Willumsen (2014), in this book, we distinguish between projecting and forecasting. A *projection* is a conditional ('if, then') statement about the future. Projections are estimates of future conditions that may exist (the 'then') as a result of adopting a set of assumptions (the 'if'). For example, '*If* current economic trends continue and no new mobility technology disrupts the demand for public transport, *then* transit patronage will continue to increase at a rate similar to the recent past'. This is not predicting that it *will* happen, only stating what would likely happen *if* our assumptions remain valid.

A *forecast* is a judgmental statement of what the modeller believes to be the most likely future. Unlike analysts, who only state what would happen if certain assumptions are satisfied (a projection), forecasters accept the responsibility of evaluating the 'ifs', adopting only those they believe are most likely to occur. This is, of course, a riskier position to take, but it may be the one requested in a particular case.

This book does not advocate the production of forecasts in all cases. In most situations, decision-makers should be able to accept conditional projections or, preferably, a range of projections under different future scenarios. They will also consider the opportunities available to adapt the intervention to changing conditions or to influence them through policies and regulations.

17.3 Model Specification

The process of designing a model system to support transport planning can be broken up into five stages (Willumsen and Ortúzar 2016):

1) Set out the objectives of the model.
2) Identify potential interventions to be assessed using the model.
3) Identify the behavioural responses that are likely to result from these interventions.
4) Select the model structures and functional forms aiming to capture the primary behavioural responses identified above.
5) Select the level of detail (granularity), required to implement these models.

These steps are not entirely sequential, as there are several trade-offs to be resolved between the objectives of realism, level of detail, data and budget availability, and our current understanding of all subtle aspect of travel behaviour. Nevertheless, it is useful to have these decisions formalised in a set of documents, so that others can judge, later on, whether the model addressed the right objectives even if the context was changing.

17.3.1 Model Objectives

It is useful to formalise the model design process into written documentation including:

- a statement of requirements;
- a functional specification of the transport model;
- a technical specification of the transport model; and
- a set of criteria to confirm the goodness-of-fit of the model: the calibration, validation and audit process.

The *statement of requirements* must consider how the model will be used. For example, this could be a strategic model to assist in planning a set of interventions in the future and help assess their performance; this will usually be a multi-modal transport model, either aggregate or disaggregate, or even an ABM. Alternatively, it may be a model developed only to assess a particular project or a limited package of measures, for example, a new road link or the expansion of an airport. A transport model can also be developed to explore ways of improving the operational performance of a system, for example, re-allocating road space to cyclist and active modes. Each of these categories of models will have a different specification.

17.3.2 Identify Possible Interventions

The statement of requirements helps identifying a modelling tool sensitive to the relevant issues at hand. In the case of a transport project, the proposed intervention will have already been identified. In the case of a strategic model to support plan making, a long list of possible interventions must be prepared. This should be more than the list of possible projects in the mind of decision-makers. It should include also other potential interventions that may be considered politically unfeasible, but perhaps necessary in the future, such as the introduction of road user charges (RUC). This list of possible interventions will drive some of the characteristics of the model, for example, the type of networks to code and what attributes to include in their links, for example, a toll in the case of RUC.

The statement of requirements should also recognise what aspects of the performance of a transport system are of interest and, therefore, define the type of outputs required from the model. For example, emissions of greenhouse gases (GHG) have become an important performance output this century but were widely ignored in the previous one. In the same vein, there is now a requirement to look at the equity impacts of any intervention in the transport system, when in the past it was enough to estimate costs and benefits to society in general.

17.3.3 Identify Relevant Behavioural Responses

When defining the scope of a model, it is important to consider the behavioural responses that are likely to arise from the proposed interventions. Will these be only route choice changes or more complex mode and time of day choices? The potential behavioural responses help determining the type of demand model to be specified.

The combination of the identified interventions and likely behavioural responses are essential to define the *functional specifications* of the model; in other words, what modes and interventions should be possible to analyse with the model and what type of demand model should be considered as well.

These considerations will also result in a decision about the minimum demand segmentation required. Usual segmentations combine person type and journey (or activity) purpose. Person type is often a function of income and vehicle ownership/availability, but increasingly considers the type of job or profession to account for the potential for remote work. The main rationality for having person-type segmentations is the different response expected to certain interventions, for example, parking charges or a new public transport service. Journey purposes will at least include to work, to study (both assumed to be recurrent), home based and non-home based other, and in the course of work. The main rationality behind segmenting by journey purpose is, again, that different responses are expected, in particular in relation to destination and time of travel choices.

It is tempting to select a detailed segmentation of demand on the grounds of greater realism. However, this would place onerous demands on data collection and the development of different models for each segment. A thorough analysis of the data collection results and the model estimation tasks should help in deciding whether this detailed segmentation can, or must, be simplified to obtain more robust models and faster execution times.

The functional specification will also include the time periods to be modelled, that is, whether a full day, just one or two peak periods, or a greater number of time and days of the week modelling periods to facilitate the estimation of average daily and annual performance. As it is not practical to model each future year in a 30-year strategic plan, it is customary to select a few planning horizons, say every 5 or 10 years, where the full model will be run and alternative interventions tested.

17.3.4 Technical Specification and Data Requirements

Selecting the most appropriate transport modelling technique is a nuanced and complex task. It needs to balance the requirement for rigorous analysis, which depends on the significance of the intervention, against the time and cost required for modelling and appraisal.

The *technical specification* of a model should translate the functional requirements into specific model forms. It will specify, for example, the type of assignment required for both private and public transport trips, the characteristics of models for each step in the classic approach, or each module in an Activity Based Model.

When preparing this technical specification, other aspects must also be considered:

- the data that is available and the data that must be collected to calibrate and validate the proposed model form;
- the local knowledge and level of expertise of future model users; and
- the time, computing resources, and funds available to prepare and use the model in the future.

We have known for over 50 years that data collection can consume a large proportion of the budget available for any given study (Boyce et al. 1970), thus restricting the time and resources available for model development and use. Therefore, data collection must be planned carefully, making as much use as possible of existing data including passive sources like mobile phone and smartcard data. It should also consider the opportunities and validity of transferring parameters or donor models from other jurisdictions.

The expertise of the end users of the future model system should also be considered. Of course, this can be supplemented with the help of consultants with valuable experience obtained in other model development projects, and with a training programme associated with the model-building task. Experience takes time to develop but is essential for the valid interpretation of model outputs.

The time and money available for developing a new modelling system, or to update an existing one, is a real and unavoidable constraint. Decision-makers tend to underestimate the time needed to develop a new modelling system; this usually takes years rather than months, especially if a new mobility survey is needed to obtain up-to-date behavioural data.

Another dimension is the expected run time of the model. Large transport models ideally should run overnight (8–10 hours), so that there is enough time for preparing and checking the inputs and outputs, and consider the realism of the results and their possible interpretation. Especially during model development, multiple runs are needed to refine and calibrate the model, identifying and correcting modelling and data errors, with the objective of delivering a useful and reliable tool.

All these aspects should be considered when deciding the level of detail, that is, the granularity to be adopted for the model. This will consider the additional data costs, and the availability and reliability of future data required to achieve the desired level of detail.

Willumsen and Ortúzar (2016) suggest the implementation of a multi-criteria granularity appraisal to discuss these issues more transparently, as shown in Table 17.1.

Table 17.1 Multi-criteria appraisal of model granularity

Dimension	Additional realism and accuracy in base year	Ability to forecast inputs and parameter stability in future years	Changes in data collection costs	Changes in model processing cost (runtimes)
Number of zones	Improved geography of trip generation/attraction and more accurate level-of-service for mode choice and assignment	Increased errors in forecast of population, employment and student enrolment	Small increase for trip generation, larger cost increase for trip matrix development	Runtimes roughly proportional to square of number of zones
Person types (age gender, income, HH size, car ownership)	Improved realism in all sub-models	Increased uncertainty in the projections of the number of each segment in each zone	Significant increase in data collection costs to achieve required sample size	Multiple sub-models for each segment increase runtimes
Journey purposes				
Network connectivity				
Junction details, etc.				

Such a table would need to be populated with the available options for each dimension (we only exemplify some of them). It will not provide a quantified answer to the questions associated with model design, but should help in deciding on the best approach, level of detail, and data collection needed. In practice, budget and timescale constraints will play a critical role in making these decisions.

17.3.5 Quality Assurance

Most strategic transport models involve thousands of lines of code, millions of data points, and hundreds of parameter values; the opportunity for inadvertently introducing, or not spotting, errors is high, so much so that it is impossible to be 100% certain that a model is error-free. It is important, therefore, to have a solid *quality assurance* (QA) process as part of the project, but independent from the model development team. Guidelines have been developed for this QA process. For example, the UK Department of Transport has published a very useful one (Department of Transport 2014).

The overarching principle is that the quality assurance of analytical models should be taken by professionals other than those in the modelling team and the effort should be proportionate to the risk and impact of errors slipping through. The QA process should go beyond checking that the model perform the correct calculations. It should cover model governance and version control. If external consultants are employed when developing the model, they should have in place their own rigorous quality assurance process in addition to that of the end-user.

17.4 Model Calibration and Validation

All useful models need calibration and validation stages. *Calibration* is the task of finding the set of model parameters that allow to represent reality best; for example, finding the best values for the parameters in a utility function. It requires choosing a defined set of parameter values to optimise one or more goodness-of-fit measures that are a function of the observed data. The concept is closely related to that of model *estimation*, but the latter is more general and may include examining some specification issues (e.g. Multinomial or Nested Logit) and the best functional form and parameter set (e.g. eliminating non-significant variables).

Overfitting may occur when a model is excessively complex, for example, having too many parameters relative to the quality and number of observations; such a model will tend to describe noise and imperfections in the data rather than a true underlying relationship (Willumsen and Ortúzar 2016). An overfitted model is likely to have poor predictive performance, as it can exaggerate minor fluctuations in the data. Transport models offer plenty of opportunities for overfitting, ranging from poorly designed data collection efforts to overly detailed elements of the model when neither the base year nor future data support such level of detail.

Overfitting, or over calibration, is particularly dangerous if matrix estimation using traffic counts is employed. Not only it may give a false impression of the model's ability to reproduce reality, but it may also distort the trip matrix in excess, making it unreliable and unusable in an incremental model.

The *validation* process, on the other hand, attempts to quantify how well a transport model reproduces a set of reference (base) year conditions (such as traffic volumes or patronage estimates). Validation should use data separate to that used in the model calibration as we discussed earlier in the book.

Section 10.8 offers some examples of the validation of traffic assignment contrasting modelled volumes with traffic counts on links and screenlines. The network validation should also include a comparison of modelled against observed travel times and mode patronage against known statistics such as ticket sales. Moreover, it is also useful to confirm that key paths through the networks are realistic; this task sometimes identifies errors in network coding that must be corrected.

Because most transport models have been built on the basis of cross-sectional data (i.e. a snapshot in time when the data was collected), there has been a tendency to interpret model validation exclusively in terms of the goodness-of-fit achieved between observed behaviour and base-year model results. However, this is a necessary but by no means a sufficient condition for model validation.

A more rigorous version of validation is to use the model to predict a previous state of the transport system, perhaps 10 years ago (*backcasting*), and compare the results with known conditions at the time. Unfortunately, it is rare to have the data and resources to undertake this exercise. Therefore, it is necessary to find other ways to confirm that the model is 'fit for purpose'. One way of attempting this is to test the response of the model to particular interventions; in essence, to perform sensitivity tests of the model. For example, one can test the response of the model to a 20% increase in public transport fares and contrast results with known or accepted elasticities of demand to fare changes. As the model will be used for forecasting, an additional test is to increase the 'all modes' trip matrices by a reasonable percentage and use the model with the base-year networks to explore the reasonableness of the demand model responses and resulting flows. These sense checks are essential to gain confidence in the model.

A transport model validation report should cover:

- a description of the data used in estimating, calibrating, and validating the model;
- a report on the 'fit' achieved to the estimation and calibration data;
- a report on the validation outcomes for a reference (base) year; and
- a report on the sensitivity tests undertaken, and their realism.

17.5   Model Review

The reliability of a model to support decision-making cannot be determined as being *good* simply by looking at forecasts. Confidence in the processes used to derive the forecasts should be sought via the structure of the transport model and its calibration and validation; this normally involves the task of an independent model review and auditing.

As we have seen, the development of a transport model involves four broad processes:

- data collection;
- model specification;
- model estimation and calibration; and
- model application to the scheme appraisal.

The model review task should confirm that these processes have been carried out according to best practice or specific local guidelines. A model review report would normally cover:

- a statement of the modelling objectives and the elements of the model specification that aim to satisfy them;
- a specification of the base year data, including a description of the surveys, sample sizes, bias assessment, and validation;
- a description of the transport networks, their structure, data sources, and validation checks;
- a description of the demographic and employment data available;
- a document reporting on model specification and model estimation, model structures, variables and coefficients, statistical estimation procedures, and model fit to data;
- evidence of validation tests, including fit to independent data, comparison with other models, sensitivity tests, and elasticities;
- a description of the forecast year inputs, networks, demographic data, and economic assumptions;
- documented validation of the forecasts, comparison with other forecasts, where available, comparison with historic trends, and reasoned explanations of the forecasts (such as the sources of the diverted traffic, amount of time savings, and so on).

As the risk of model errors increases with the complexity or degree of innovation of the modelling approach, the level of quality assurance and model review to be deployed should be stricter.

17.6   Plan Making

Transport modellers do not design strategic transport plans or specific interventions, they have a supporting role in aiding decision-makers. The design of a transport plan is not an easy task. There are two main approaches and two main problems.

In terms of approach, the traditional view has been that one must identify how much demand for future mobility will grow and then design a plan or intervention to cope with it. This approach has been characterised as *Predict and Provide*. The problem is that it has been shown to lead to the provision of additional road capacity that inevitably has resulted in even more congestion and delays. An alternative view is gaining support; it poses that one must decide what type of future is most desirable and design a set of interventions to ensure that it will come to pass. This approach has been characterised as *Vision and Validate* or *Decide and Provide*; it is justified as having a better fit with the current objectives of reducing the transport carbon footprint and achieving a more sustainable future. Nevertheless, the success of the approach in achieving these goals still needs to be tested over time.

The two main problems in plan making are how to deal with uncertain futures, and how to design a plan that is flexible enough to cope with these uncertainties. These are some of the most critical challenges facing the planning profession.

When planning for future mobility, it is useful to consider the Triple Access model formalised by Lyons and Davidson (2016). This mental model invites planners to think about alternative ways to provide the accessibility demanded by people: physical mobility, virtual or digital connectivity, and spatial proximity, via appropriate better land use. The model is illustrated in Figure 17.1.

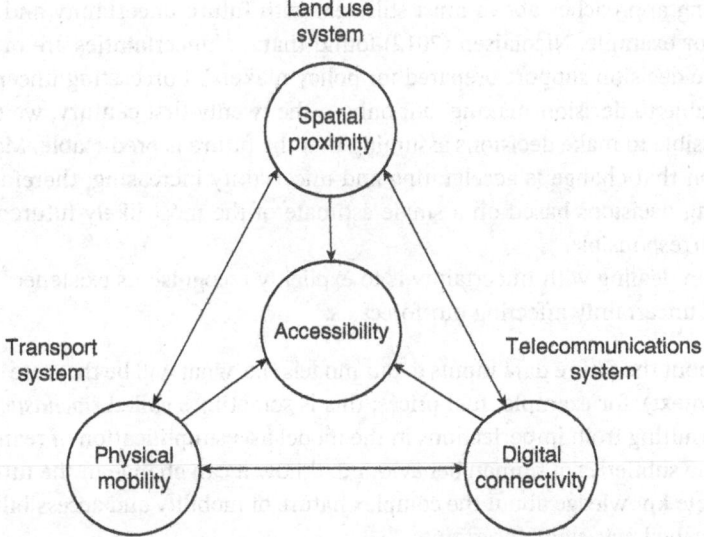

Figure 17.1 The Triple Access model

The Triple Access model can be used in both the Predict and Provide and Vision and Validate modes. It has implications for transport and land use modelling and decision-making, as well as for the development of better digital connectivity infrastructure.

Developing a good package of measures for a transport plan can be complex. There will be a large number of candidate projects and interventions, and it is not practical to test all possible combinations of projects in the model therefore, a simplified approach is needed.

A possible approach is to include all sensible interventions (and their discounted capital and operating costs, CAPEX and OPEX) and run the model estimating total benefits against the 'Do Nothing' (or 'Do Minimum') alternative. This set, which considers all the projects, may probably satisfy the vision but is likely to exceed the budget. The next step is to start removing projects, one at a time, and compute the resulting cost savings and potential loss of benefits, providing an estimate of the incremental costs and benefits of adding that project. The ratio of Benefits/Costs would be an approximation of the 'value for money' offered by each particular project. The list of key projects and their Benefit/Cost ratios should help in deciding which projects to include and also the ideal sequence over time for each intervention. Then, attractive package combinations that meet the budget constraints can be tested for each planning horizon, ascertaining how close to the vision can the system get in a particular time horizon. This would be followed by adjustments and fine tuning of policies to realistically achieve the objectives.

Of course, there will be many other considerations to bear in mind when preparing a transport plan. Some of them could be simply the need to generate benefits that will be appropriately distributed among the population. It will also be desirable to estimate whether different land use arrangements or facilities for remote work, for example, could generate alternative benefits at a lower cost in the long run. This approach could be adapted and applied, in a slightly different way, in the Predict and Provide mode.

17.7 Dealing with Uncertainty

Both plan making approaches above must still deal with future uncertainty and risk, but this is often ignored; for example, Nicolaisen (2012) found that: ... 'uncertainties are often toned down or ignored in the decision support prepared for policy makers'. Forecasting uncertainty has been always unavoidable to decision-making, but only in the twenty-first century, we have recognised that it is not possible to make decisions assuming that the future is predictable. Moreover, there is now a perception that change is accelerating and uncertainty increasing; therefore, ignoring the issue and making decisions based on a single estimate of the most likely future (i.e. the base or central case) is irresponsible.

The first step in dealing with uncertainty is to explicitly recognise its existence. There are three main sources of uncertainty affecting our forecasts:

- uncertainty about the future data inputs to our models (i.e. what will be the state of the world and immediate context), for example, fuel prices; this is sometimes called *stochastic uncertainty*.
- uncertainty resulting from imperfections in the model as a simplification of reality (we have not captured all the subtleties of human behaviour and how it can change in the future); this comes from incomplete knowledge about the complex nature of mobility and accessibility systems, and is sometimes called *epistemic uncertainty*.
- uncertainty from unforeseen events or developments, even if they are foreseen (e.g. autonomous vehicles). These may be events we cannot yet estimate their impact on the mobility spectrum or events and situations that may be entirely unpredictable because we have not experienced them before; this is sometimes called *ontological uncertainty*.

Good practice in dealing with these issues is to prepare an *uncertainty register* or Log, where all known sources of uncertainty, their likelihood, and the importance of their impacts are registered, hopefully identifying measures to mitigate them. An example may be found in the UK Department of Transport's TAG uncertainty kit (https://www.gov.uk/government/publications/tag-uncertainty-toolkit). It should also include sources of uncertainty stemming from weaknesses in the base year model as a result of, for example, limitations in calibration and validation.

There are two main treatments for dealing with uncertainty in travel forecasts. The simplest is *stochastic risk analysis* (Willumsen 2014; Yu and Prevedouros 2017), which has been used for some time to deal with the stochastic uncertainties of any forecast.

The first step of this approach is to identify the main drivers of future demand for a particular project or plan, and estimate the elasticities of demand to variations in these drivers; this requires using sensitivity tests. This is followed by a consideration of the variability of these drivers in the past, for example, GDP or population growth; a statistical distribution with mean and standard deviations is fitted to these past variabilities. A simpler model, usually based on a spreadsheet, is then developed showing the links between the future values of these key drivers and the future demand (or welfare impacts) for the project using the estimated elasticities. Then, a number of simulated variations on each of the input variables is run using Monte Carlo methods and the range of outcomes is recorded. A more detailed discussion of this approach is presented in Section 19.5.

However, despite its attractions, this approach deals only with one type of uncertainty, the variability of a future around 'business as usual' plus some 'random noise'. The future may be more different in several dimensions and may be subject to unexpected events or new technologies.

Dealing with these more radical dimensions of uncertainty may require the development of a range of possible future scenarios, agreed by all key stakeholders, to test the robustness of any transport plan. Issues, like (i) the future role and importance of new mobility technologies like autonomous vehicles or Mobility as a Service, (ii) the uncertainty surrounding energy costs, and (iii) future preferences for virtual or presential work and leisure, would influence the nature of these scenarios.

Scenario planning has been used for many years in fields other than transport but is increasingly used in forecasting. For example, the UK Department of Transport has been producing road traffic forecasts under multiple scenarios since 2018 (Department of Transport 2018).

The key elements of this approach can be summarised as follows (Willumsen 2014):

1) Decide drivers for change/assumptions. Identify which are the main ones affecting future traffic and revenue in a specific transport situation. For example, it could be GDP growth, population location, self-driving cars, or the preference of people for digital rather than physical experiences.
2) Bring these drivers together into a viable framework. The longer list of drivers must be prioritised and, ideally, linked together: how may GDP growth affect future willingness to pay and the share of automated vehicles.
3) Produce a limited number (say 5–10) of initial mini-scenarios. These should be consistent; for example, high economic growth linked to higher values of time. These scenarios should be described and their elements quantified, as much as possible, in a 'first cut' approach.
4) Reduce to 2–4 scenarios that can be described in detail and quantified, and that are sufficiently different to teach us something about future challenges. Identify mitigation measures that could be taken to make this future a bit malleable.
5) Draft the scenarios, in other words, convert them into quantitative inputs to the model plus qualitative considerations for the interpretation of outputs.
6) Run the model under each scenario and produce key outputs, for example, revenue profiles in the case of a road concession.
7) Identify the issues arising and how they affect decisions. This could help define whether the overall uncertainty makes the project too risky, or whether potential upsides are sufficiently attractive to go for it with confidence.

It is useful to bring different disciplines and backgrounds into this task to avoid the trap that considers the unfamiliar improbable. For sure, the scenarios should be agreed in advance by all stakeholders.

Note that there are different ways to develop and handle scenarios. For example, the Rand Corporation has proposed an approach that can be used at a high level to underpin more specific local scenarios (Zmud et al. 2013). A similar approach has been followed by the UK Department of Transport, which has designed a set of 'Common Analytical Scenarios' and an 'Uncertainty Toolkit' that may be adapted to local conditions and uncertainties (Department of Transport 2022). The identification of key attributes that could shape alternative futures should drive the development of a few internally consistent scenarios. Then, the model system and different transport plans should be run for each scenario to identify the strengths and weaknesses of each package, and key decision points identified when plans may have to change and adapt.

The fact that we must deal with multiple scenarios makes the design of coherent packages for future interventions in a transport system particularly difficult. It is not practical to test all possible combinations of transport interventions under each different scenario and select the package that performs best under all conditions. A more realistic approach is needed.

A possible way out is to develop a 'Meta Model' based on the model runs for each scenario and the elasticities implied in their sensitivity tests. This could be implemented even in Excel and is illustrated in Figure 17.2, with reference to an example involving the study of the potential impact of new technologies like connected and autonomous vehicles (CAVs, see Chapter 20) and energy prices to traffic in the access to an airport. The individual impact of each of these disruptions was obtained from the detailed modelling as well as the elasticities of demand to interventions like parking and access pricing. This Meta Model can then be run under different assumptions about income growth and business air travel and produces outputs in terms of traffic levels, GHG, social welfare, and equity (who benefits and who loses).

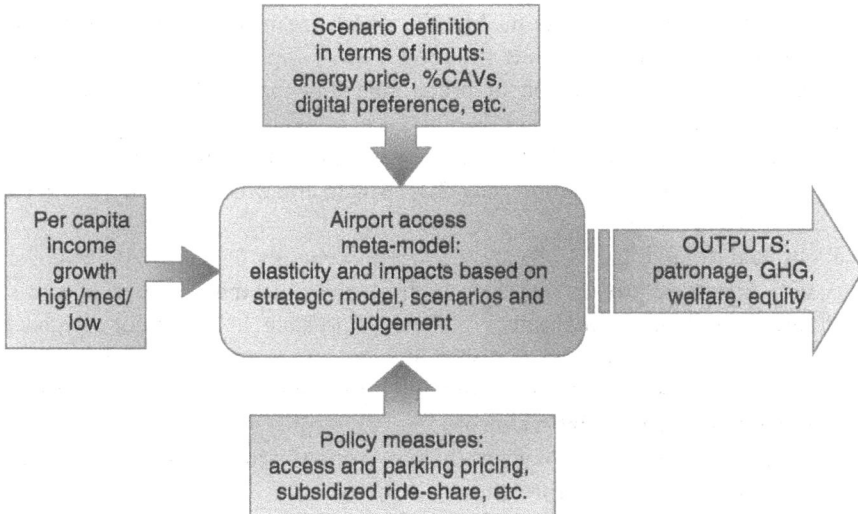

Figure 17.2 A meta model to consolidate scenario outcomes

Rejecting the idea of using only a single scenario to prepare a plan as misguided, the Travel Model Improvement Program (TMIP) in the USA has adopted a similar approach with an *ad hoc* application: the TMIP-Exploratory Modelling and Analysis Tool (Milkovits et al. 2019). TMIP-EMAT has been developed to facilitate the use of exploratory techniques with travel forecasting models. The aim is to explore the deep uncertainties that exist in future land use, demographics, and transport systems inputs, plus the uncertainty that exists in the model itself.

It must be recognised that these are only initial steps in dealing with uncertainty. Significant challenges remain in incorporating these issues into cost-benefit analysis, which is still usually based on a central or 'most likely' case. Dealing with multiple scenarios and cases and making sense of many benefit/cost ratios to assist decision-making is a demanding task.

18

Key Parameters, Planning Variables, and Value Functions

This chapter covers some essential aspects of transport modelling. The first is the important issue of forecasting planning variables; for example, future population, employment, school places, shopping areas, and income distribution. These are variables that are needed to make predictions with transport planning models. Sometimes they are provided externally to the study; in others, they must be estimated as part of the planning exercise. In either case, they play a key role in determining the forecasting ability of the models discussed in this book.

General approaches to obtaining planning variables for aggregate models are discussed in Section 18.1; these are also key inputs to a more disaggregate approach to synthesising populations, as discussed in Chapter 16. A particular way to develop these estimates is using Land Use Transport Interaction models that aim to capture the mutual influence between changes in accessibility and land use allocation; these are outlined briefly in Section 18.2.

One of the most important planning variables is car ownership, and this is the subject of Section 18.3. Both time-series and econometric models to forecast car ownership are discussed, together with some more recent approaches.

Many issues surrounding the concept, estimation, and application of the *value of time* are presented in Section 18.4. Finally, Section 18.5 discusses the concept and methods used to value external effects of transport, such as accidents and pollution. The book would not have been complete without this discussion.

18.1 Forecasting Planning Variables

18.1.1 Introduction

As discussed in Chapter 1, modellers distinguish between *endogenous* variables (i.e. to be forecast as part of the modelling exercise) such as flows, and *exogenous* or *independent* variables; the latter are required to run the models and are supposed to originate externally from them. Typical examples in the transport field are population, employment, car ownership, and income. Values for these variables should be provided for the base year and for each of the years for which forecasts are needed from the transport model.

The level of detail and disaggregation required for these variables depends on the type of model being used. In general terms, an aggregate demand model has fewer requirements than a disaggregate one in this sense. For example, at the trip generation level, an aggregate, zone-based, linear

Modelling Transport, Fifth Edition. Juan de Dios Ortúzar and Luis G. Willumsen.
© 2024 John Wiley & Sons Ltd. Published 2024 by John Wiley & Sons Ltd.
Companion website: www.wiley.com/go/ortuzar5e

regression model should only require population, car ownership, and average income by zone; a cross-classification or category analysis model, on the other hand, will need the number of households in each of the categories used, say 108 per zone, as we saw in Chapter 4, if the model is stratified by income (6 levels), household structure (6 levels), and car ownership (3 levels).

The importance of these variables in influencing the accuracy of the whole modelling exercise is high, as established by Mackinder and Evans (1981) in a study of 44 British urban transport studies. They found that all models overestimated key indicators of performance, but that the most important element in explaining this overestimation was the error in the values used for the planning variables. Model specification errors played a much lesser role in the overall inaccuracies. The planning variables were often wrong even though they used official global forecasts, which could have been more accurate in the first place.

There are good reasons why forecasting planning variables is such a difficult exercise. The values of many of them in the future depend on complex interactions with other actors and influences that are difficult to predict. This is certainly the case with the allocation of population and employment to geographical areas, as these are influenced by interactions among factors such as:

- population, income, and car ownership;
- levels of employment and their type;
- land use master plans, zoning, and building regulations that affect what can be done, where, and at what densities;
- parking standards (minimum or maximum) for new developments;
- land parcels available for development (green and brown fields) and their cost;
- developers' activity regarding new and second-hand properties and the evolution of their 'land banks';
- local politicians and decision-makers adapting plans and regulations to changing conditions;
- changing views about desirable lifestyles and work practices;
- international and local trends on how best to tailor retail and services to a changing population.

The question then arises: how can we reduce, as much as possible, the errors in these planning variables? This is a difficult problem with no simple or single answer. A full discussion of the techniques available for forecasting these variables is outside the scope of this book; for practical methods, the reader may consult England et al. (1985). However, we will discuss some of the ideas behind these techniques to appraise their strengths and weaknesses.

18.1.2 Use of Official Forecasts

The apparently simplest option in dealing with planning variables is to use official forecasts. In the United Kingdom, for example, there are estimates at District Council (and London Borough) level, of:

- population, households, employed residents, and employment;
- number of households owning 0, 1, and 2 or more cars;
- private-vehicle trip ends by different journey purposes.

The UK Department of Transport also produces, from time to time, forecasts of future demand expressed as expected vehicle kilometres for different types of vehicles. Other official institutions provide other types of forecasts for planning variables, at least at a highly aggregate level. Of course, these forecasts are seldom at a sufficient level of disaggregation to be directly usable in a detailed modelling exercise; however, they do reduce the amount of work needed to generate the required values for the planning variables at zone level. Some techniques to achieve this are discussed in the next section.

To some extent, the problem with using official forecasts is that they sometimes reflect the expected effect of economic and regional policies, the success of which may actually depend on other, uncontrollable factors such as international trade and cooperation. Mackinder and Evans (1981) found that errors in forecasting these global indicators were at the root of the problem of inaccurate planning variables at the local level.

We will come back to this problem again. How can we accurately forecast transport activity if there are significant errors in some of the key inputs used in our transport models?

18.1.3 Forecasting Population and Employment

If no forecasts of planning variables are available for cities or districts, the planning team will need to develop methods for their estimation. There are several potential methods to this end, and some may be more appropriate than others for each application.

18.1.3.1 Trend Extrapolation

The direct extrapolation of current trends is the simplest but least satisfactory procedure, even if it is only applied at the level of the whole study area. Trend extrapolation does not consider decisions already made about the availability of land for future development; it does not value new regional development policies nor does it consider the expected growth in employment in the study area. In addition, it does not provide any information about the age structure of the population, an important element in, for example, trip generation modelling.

18.1.3.2 Cohort Survival

A more detailed technique considers deaths, births, and immigration in and out of a study area to forecast future population:

$$P_{t_1} = P_{t_0} + B_{t_0 t_1} - D_{t_0 t_1} + NI_{t_0 t_1} \tag{18.1}$$

where P_{t_1} is population at time t_1; P_{t_0} is population at time t_0; $B_{t_0 t_1}$ are surviving births in the period t_0 to t_1; $D_{t_0 t_1}$ are deaths in the same period, and $NI_{t_0 t_1}$ is the net migration in the same period.

Used in this highly aggregate fashion, Eq. (18.1) ignores the age structure of the population and could under or overestimate, for example, the corresponding fertility rates. For this reason, the method is usually applied to subgroups of the population, or *cohorts*, and becomes a *cohort survival* approach. This involves the following stages:

1) The population is separated into cohorts; males are separated from females, and each sex group is divided into age strata (usually of five years) to give a population structure for the base year.
2) Fertility rates are then applied to females of child-bearing age.
3) The new-borns are added up and 'sexed' in known proportions.
4) The female and male babies make up the first cohort in the next round of calculations.
5) Survival rates are applied to females and males in all cohorts, starting from the youngest generation; survivors are then 'aged', that is, moved forward to the next cohort.
6) The process is repeated, re-starting from stage 2, until the forecasting period has been reached.

Migration is an increasingly relevant phenomenon requiring estimates of the sex and age structure of migrants. It is easy to see how the method may be adapted to include this new input and the uncertainties involved.

The information demanded by this technique includes the initial number, age/sex structure of the population, and its associated survival, fertility, and migration rates. The main source of uncertainty lies in the prediction of these rates, in particular fertility and migration rates.

18.1.3.3 Transitional Probabilities

An interesting alternative approach to cohort survival methods is to follow *family cycles* and use *transitional probabilities* reflecting the chances of moving from one stage in the cycle to another; for example, from married couple with no children to married couple with one child under school age, and from there to married couple with two children, and so on. A whole matrix of transitional probabilities is then built and processed to obtain the household population at different stages in the family cycle in the forecast years. This approach certainly offers the potential of providing a detailed account of population growth at the level required for trip generation modelling. However, the uncertainty in the estimation and stability of the transitional probabilities is likely to be greater than that associated with fertility and migration rates in cohort survival methods.

Both cohort survival and transitional probability approaches can be usefully adapted to a continuous planning framework where periodically collected data about fertility, migration, and survival rates, and/or probabilities of changing family cycle status, permit the updating of previous estimates of population in the future and, hence, the changing of trip generation rates, and so on.

When forecasting employment change, we are faced with similar problems. General trends in employment depend on economic policy, international trade, and regional incentives. At a more local level, aspects such as the availability of land and qualified labour force in the study area play an important role, as well as the type of economic activity prevailing there. Moreover, the type and levels of employment also play a key role in determining the levels of income available to the households in the study area, which in turn influence car ownership and trip-making behaviour.

18.1.3.4 Economic Base

A useful distinction in employment forecasting is that of *basic* and *non-basic* activities. The latter are created in response to local demands, whereas the former require an external stimulus. Basic activities produce goods or services that are exported to other areas and regions. Non-basic activities produce goods and services to attend the needs of the local population. The growth of basic activities creates additional non-basic ones (shops, banks, services, and so on) to satisfy the needs of additional population. Thus, the basic activities of a region constitute its 'economic base', and strengthening it should result in economic, employment, and population growth.

18.1.3.5 Input–Output Analysis

Finally, in forecasting the growth of a particular activity one should also follow the concomitant growth it generates in other industries providing inputs to it. Some of these will be based outside the study area whilst others may be located inside it. The use of an 'input–output matrix' is the traditional method to follow these linkages at the national or regional levels. Such a matrix depicts how much input from other sectors of the economy is needed to increase output from one particular activity. However, the availability of such matrices at the local level is rare, since the lowest level of disaggregation is usually a regional one.

18.1.4 The Spatial Location of Population and Employment

Having estimated population and employment (in different subgroups) for the study area, it becomes necessary to allocate these forecasts to specific zones previous to the application of our transport models. This work is usually carried out in conjunction with local planning authorities, who usually have established plans for future development and re-allocation of land uses to different zones in the study area. The use of age- or life-cycle-specific forecasts is helpful in this process, as different types of housing development are more likely to attract different types of families.

The location of employment depends on its nature; for example, industrial development, commercial services, consumer services, and so on. Major changes in the location of economic activities should probably be discussed with those involved in carrying them out. Industrial development may require special sites, good availability of water and energy services, and access to major roads and railway/port terminals. In the absence of restrictive planning controls, office employment tends to be located close to good communication facilities and as close as possible to other office developments.

These two examples show that, in the final analysis, the location of population and employment is not independent of the transport system. Changes in accessibility are likely to affect the potential for development of different parts of a study area. This can be considered in the discussions with planning authorities, or more formally, in a more comprehensive model, as outlined in the next section.

In summary, the allocation of population and employment to zones usually requires a combination of formal models and discussions with planning authorities. The practical ways in which these tasks are carried out owe much to heuristic approaches and context-dependent choices. It seems difficult to eliminate current uncertainties about national, regional, and local forecasts for these planning variables, and this has important implications for the whole planning process. Despite these difficulties, land-use models and their interaction with transport systems have been developed and used as discussed in the next section.

The issue of disaggregating these allocations at the individual or household level or to generate a synthetic populations has already been discussed in Chapter 16.

18.2 Land-Use Transport Interaction Modelling

There is an almost universal recognition that transport, in particular accessibility, and land use are interrelated. One attractive approach to forecasting population and employment and allocating them to zones is, therefore, to internalise these exogenous planning variables in an integrated model of land use and transport. This has been an active area of research since the early 1970s; see, for example Wilson et al. (1977), Foot (1981), de la Barra (1989), Echeñique et al. (1990), and Simmonds (2001). However, after an initial period of optimistic claims about the success of such models, researchers became more modest in their aspirations (see Mackett 1985).

The importance of the interaction between transport and land use is twofold. First, if transport strategies significantly change accessibility, this will change demand for land and generate new developments in some areas; these will in turn affect the pattern of trips (i.e. the trip matrices) and, therefore, have an impact on the performance of the transport system. These interactions are illustrated in Figure 18.1.

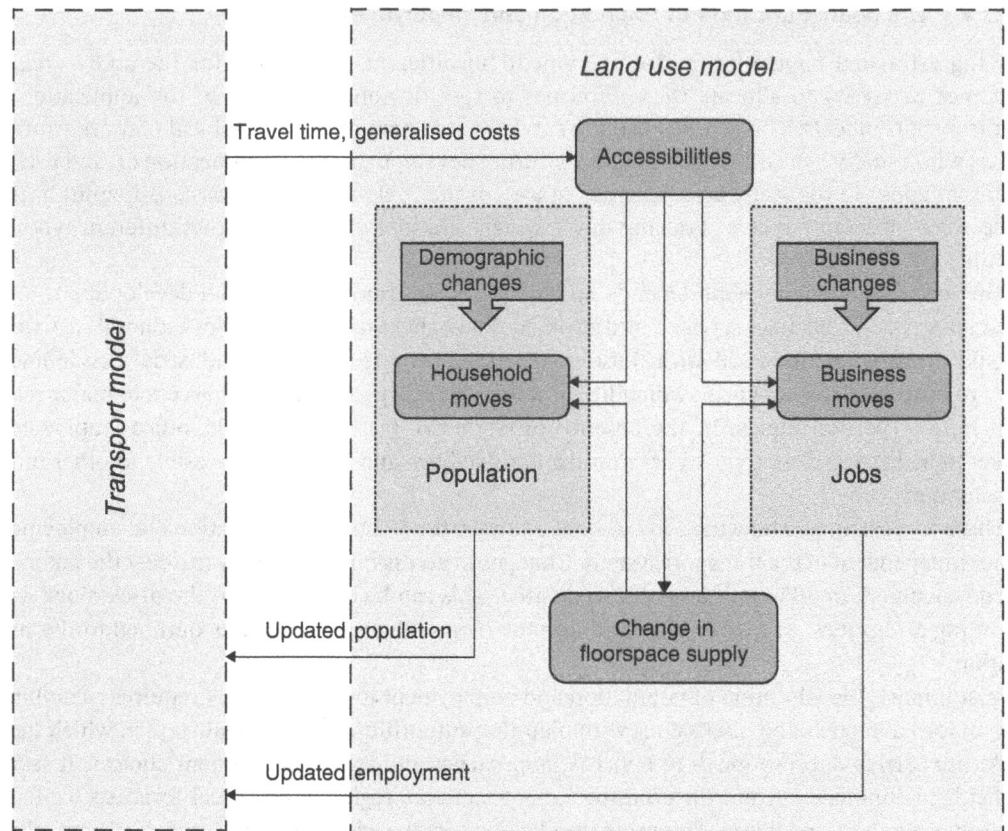

Figure 18.1 Transport and land use interactions

Second, changes in the attractiveness of some areas will affect their land prices; this can be interpreted as the capitalisation of user benefits into land prices and implies a transfer of benefits to land owners. This capitalisation issue raises the question of who gains and who loses as a result of a transport scheme, and how a local authority can recover some of the increase in land prices from the land owners.

The complexity of the relationships involved and their still fluid theoretical underpinnings have led to a situation where models and software are almost inseparable, an indication, perhaps, of the lack of consensus on what constitutes a good approach. It is impossible to do justice to this specialist modelling area in a book like this; the reader is directed to reviews like those of Wegener (2004) and Zöllig-Renner et al. (2015).

It is interesting to note that the basic design structure of most land use models is similar; however, there are at least four different fundamental design perspectives:

- behavioural or structure-explaining approach;
- bid-rent or discrete choice approach;
- aggregate or microsimulation approach; and
- emphasis in equilibrium or change dynamics.

Behavioural approaches treat relevant behaviour explicitly, for example, birth, marriage, job change, or relocation. Structure-explaining approaches attempt to model the outcome directly;

for example, distribution of jobs, without dealing with the processes that lead to that particular distribution. In practice, many models are somewhere between these two approaches.

The bid-rent theory assumes that every actor on the land use market makes bids for a piece of land, and the bidder with the highest offer gets it. Because of transport costs, everybody is willing to bid more for a location with good accessibility, and whoever values this more highly, often businesses, gets the most accessible land, usually city centres. In the discrete choice approach, households, firms, and developers make choices among a finite set of alternatives for locations, jobs, and land, for example. It has been suggested (see NCHRP 2010) that bid-rent approaches work best in markets that are highly competitive and transparent and discrete-choice approaches in markets that react with some time lag and where users decide with imperfect information.

Aggregate models aggregate actors into certain groups (for example, households by household type or firms by industry type), and these are assumed to have homogenous preferences. As we have seen in the case of Activity-Based Models, microsimulation approaches offer advantages in terms of model development and in treating interactions explicitly.

Finally, some modelling approaches are underpinned by general equilibrium considerations, whereas others emphasise the fact that change, and the rate at which it happens are inherent features of land use, business, and transport markets.

We try to identify here some of the distinct theoretical components behind this type of models.

18.2.1 The Lowry Model

Many practical applications in the past have followed the lines put forward nearly 60 years ago by Lowry (1965). His model considered the spatial characteristics of an urban area in terms of three broad sectors of activity: employment in basic industries, employment in population-serving industries, and the household or population sector.

The Lowry model starts by allocating exogenously specified basic employment to zones, and then the spatial distribution of households and non-basic employment is assigned using endogenous relationships. In addition, there are constraints on the maximum number of households for each zone (according to local regulations) and on the service employment thresholds for any zone; different types of service employment are assumed to have different minimum thresholds for their viability in any one zone.

The basic equations of the Lowry model can be written as:

$$\mathbf{P} = \mathbf{EA} \tag{18.2}$$

$$\mathbf{E}^{S} = \mathbf{PB} \tag{18.3}$$

$$\mathbf{E} = \mathbf{E}^{b} + \mathbf{E}^{S} \tag{18.4}$$

where $\mathbf{P}$ is a zonal population vector; $\mathbf{E}$ is a row vector for total employment in each zone j, $\mathbf{E}^{b}$ and $\mathbf{E}^{S}$ are row vectors for basic and non-basic (service) employment in each zone j; and $\mathbf{A}$ and $\mathbf{B}$ are zone-to-zone matrices of workplace-to-household and household-to-service-centre accessibilities.

The accessibility variables have two components: one corresponding to the participation rate in each zone (households per employee for $\mathbf{A}$ and service employment per household for $\mathbf{B}$) and a second corresponding to proper accessibility indices. These are normally calculated as:

$$A'_{ij} = \frac{E_j \exp\left(-\beta C_{ij}\right)}{\sum_{ij} E_j \exp\left(-\beta C_{ij}\right)} \tag{18.5}$$

$$B'_{ij} = E_j^{S} \exp\left(-\alpha C_{ij}\right) \sum_{ij} E_j^{S} \exp\left(-\alpha C_{ij}\right) \tag{18.6}$$

and can be derived directly from the Gravity model as shown in Chapter 5.

Lowry (1965) proposed a sequential solution to this problem, including the constraints and thresholds mentioned above. More recent research efforts have emphasised the simultaneous solution of the same model and its extensions. Most have dealt with additional disaggregation into different person and household types, and their treatment over space. For example, certain types of people may be more willing (or capable) to pay for increased accessibility than others, thus influencing land prices and the type of development to be undertaken in different zones.

Integrated land-use and transport models have been implemented in a number of computer suites. To keep the problem tractable, some compromise in the level of detail of the transport part of the model is necessary; the hope is that what is lost in richness in the representation of the transport sector is more than compensated by gains in the forecasting of employment, population, and household development in the study area. An important problem with these models, however, is that they may suffer greatly from convergence problems in their extremely complex equilibration mechanisms. For a comparison of different implementations and extensions to this approach, the reader should consult Webster et al. (1988).

18.2.2 The Bid-Choice Model

This approach was put forward by Martínez (1992) and implemented in a software package called CUBE (https://www.bentley.com/software/cube/) with applications in several countries. The approach follows two modelling streams; the first one, originally proposed by Alonso (1964), is a *bid-auction* location model where land is assigned to the highest bidder. The proportion $P_{h/i}$ of customers type h making a successful bid for a given location i depends on whether h's willingness-to-pay (WP_{hi}) is the highest among the bidders $g \in \mathbf{H}$. The assumption that WP_{hi} is a function of real estate and neighbourhood attributes of the location and socio-economic characteristics of the bidder plus an IID EVI distributed error term, leads to a MNL expression:

$$P_{h/i} = \frac{H_h \exp{(\mu WP_{hi})}}{\sum_g H_g \exp{(\mu WP_{gi})}} \tag{18.7}$$

where μ is the usual scale factor of the distribution of error terms. The expected market price p_i is equal to the expected maximum bid from potential buyers, given by

$$p_i = (1/\mu) \log \left\{ \sum_g \left(H_g \exp{(\mu WP_{gi})} \right) \right\} \tag{18.8}$$

The probability of an agent being the best bidder for a location in (18.7) can be combined with a latent class approach (as we saw in Section 8.6.3.2), where class membership is a function of location attributes (Cox and Hurtubia 2021). This introduces spatial heterogeneity into the model (i.e. the individual valuation of attributes may change across locations), while also allowing for behaviour-based classifications for urban areas.

18.2.2.1 Elasticities in Bid-Auction Location Choice Models

The traditional formulation of elasticity in Logit models considers the variation of the probability of choosing an alternative, with respect to the variation of a specific attribute, as we discussed in Chapter 7. But it is calculated for attributes that vary only for one of the alternatives (for example, bus travel time).

In bid-auction models, the attributes included in the WP functions are not specific to households (i.e. the *alternatives*), but refer to the location (accessibility, for example). Therefore, a variation in location attributes affects not only one but every alternative (i.e. the bidding households). This condition makes the traditional formula for direct elasticities in (choice) Logit models not valid for bid-auction location models.

Cox and Hurtubia (2022) derive an equivalent logit function for a bid-auction model, finding that the elasticity ($E_{P_{h/i},z_{ki}}$) for the location probability with respect to the k-th location attribute, for a household type h is given by:

$$E_{P_{h/i},z_{ki}} = \beta_{hk} z_{ki} \left(1 - P_{h/i}\right) - \sum_{g \neq h} \left[P_{g/i} \beta_{gk} z_{ki}\right]$$

where $P_{h/i}$ is the location probability of household h in location i, β_{hk} is the preference parameter of the k-th location attribute, for household h, and z_{ki} is the value of the k-th attribute at location i.

18.2.2.2 Consumer Surplus

The second modelling stream is a maximum consumer surplus model or *choice* model, equivalent to Anas (1982)'s maximum utility model. The consumer surplus CS_{hi} of individual h from choosing location i is given by the difference between its willingness-to-pay (WTP) and the price of the location:

$$CS_{hi} = WP_{hi} - p_i$$

Under some simplifying assumptions the proportion $P_{h/i}$ of consumers h choosing location i is given by:

$$P_{h/i} = \frac{S_i \exp[\mu(WP_{hi} - p_i)]}{\sum_j S_j \exp[\mu(WP_{hj} - p_i)]} \tag{18.9}$$

Martínez (1992) proved that the distribution of households and firms obtained from the bid-auction model in Eqs. (18.7) and (18.8) is identical to that obtained from the choice version in Eq. (18.9). His *bid-choice* model is then summarised in these equations. These in turn can be simplified further when used at an aggregate level.

The transport system is represented by suitable accessibility (to destinations) and attractiveness (with respect to origins) indices, which are included as location attributes in the WTP functions. The next task is to specify these functions; this must be done more or less on a case-by-case basis as the best function will depend on the availability of data and consumers' behaviour. Real estate developers are assumed to maximise profit, calculated as the price (p_i) minus the development costs (c_i), with profit assumed IID EVI distributed with scale parameter λ, such that the proportion of development allocated to a given zone is a MNL model:

$$P_i = \frac{\exp \lambda(p_i - c_i)}{\sum_i \exp \lambda(p_i - c_i)} \tag{18.10}$$

CUBE calculates the random bidding and supply market equilibrium (Martínez and Henríquez 2007), where the total number of consumers (households and firms) equals the total number of location units (for residential and non-residential use). Each consumer is allocated to one location; consumer behaviour is affected by other consumers' choices (i.e. social externalities and agglomeration economies), and both suppliers and consumers are constrained by external regulation or zoning schemes using a 'constrained' MNL model (Martínez et al. 2009).

18.2.3 Systems Dynamics Approach

Models based on the two formulations above seek, at least to some extent, to achieve equilibrium conditions. The Systems Dynamics (SD) approach places more emphasis on the rate of change and the processes of positive and negative feedback. The approach is based on the pioneering work of Forrester (1969), updated thanks to the availability of low-cost software capable of implementing it on a wider scale and with refined resolution. We follow here the ideas of Swanson (2003), whose Urban Dynamics Model (UDM) has seen many practical applications, in particular relating transport investment for urban regeneration.

In common with previous models, SD focuses on *accessibility* as a key driver of the attractiveness of a location to businesses and residents. A good location provides people access to the activities outside the home (including work) they wish to take part in; it provides businesses access to customers, a workforce, and markets. A SD land use transport interaction model will focus on a few features that make locations attractive, adding markets for jobs, transport, building residences and business premises.

From the point of view of residents, a location will be more attractive if it provides good access to suitable employment and offers adequate housing. From the point of view of business, the main features would be access to a suitable workforce, the availability of adequate premises, and access to markets and suppliers. The combined effect is illustrated in Figure 18.2, which shows the feedback between households and businesses: as the number of households rises, the accessible workforce also rises, increasing the attractiveness of a location for businesses. This will tend to attract more businesses and increase the number of accessible jobs, making areas with good accessibility more attractive to live in. This is an example of positive feedback, as increases in population lead to more business activity, which in turn attracts additional households. Of course, other constraints would start to apply, as the supply of premises runs out, accessible land becomes fully utilised and/or congestion becomes severe.

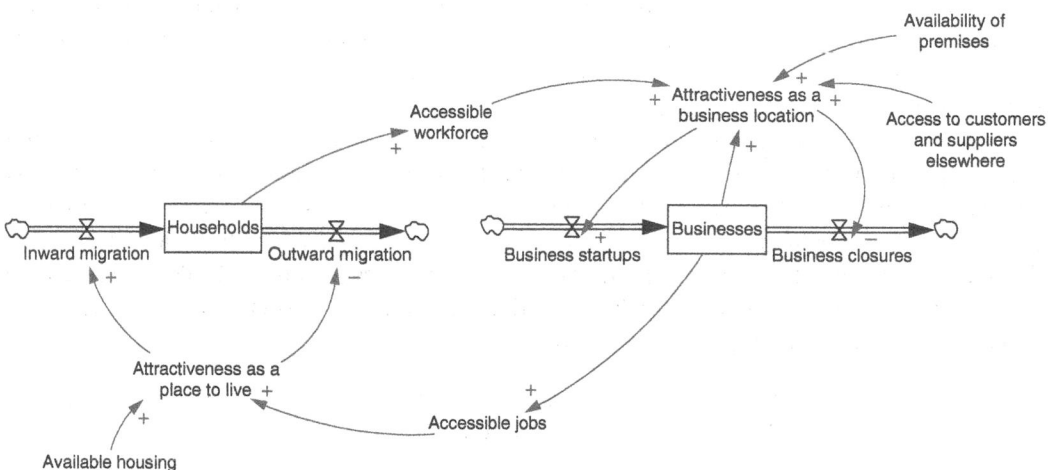

Figure 18.2 Access to business and residences

Houses are seen as infrastructure *stock*; they are built and remain in place for many years, occupying land and providing accommodation either to owners or renters. Houses are built by developers who consider how attractive a location is and whether it will lead to a reasonable financial return; this in turn will depend on their assessment of current and future demand for residences. A similar process applies to the construction or refurbishment of business premises, whether they are built by developers or by companies seeking places to expand or relocate.

Construction will tend to rise as an area becomes more attractive, but there will be delays in the response by developers. Builders need planning permission, land must be prepared, and houses and business premises take time to build. There will be lags, counted in years, between attractive conditions arising and new premises and houses becoming available in the marketplace.

Such processes of change, their causes and effects, are what SD models are designed to address. Using modern software, they are able to distinguish between different person, households and job types, different businesses, and their land requirements and different modes of transport. The transport component of the UDM, for example, can handle most responses provided in the classic aggregate transport model, including hierarchical mode choice and congested assignment.

Figure 18.3 illustrates some of the sequences of cause and effect that the full UDM recognises. The focus on dynamic change provides some valuable insights into policy development and implementation and provides a useful tool to track the evolution of markets, businesses, and residential location.

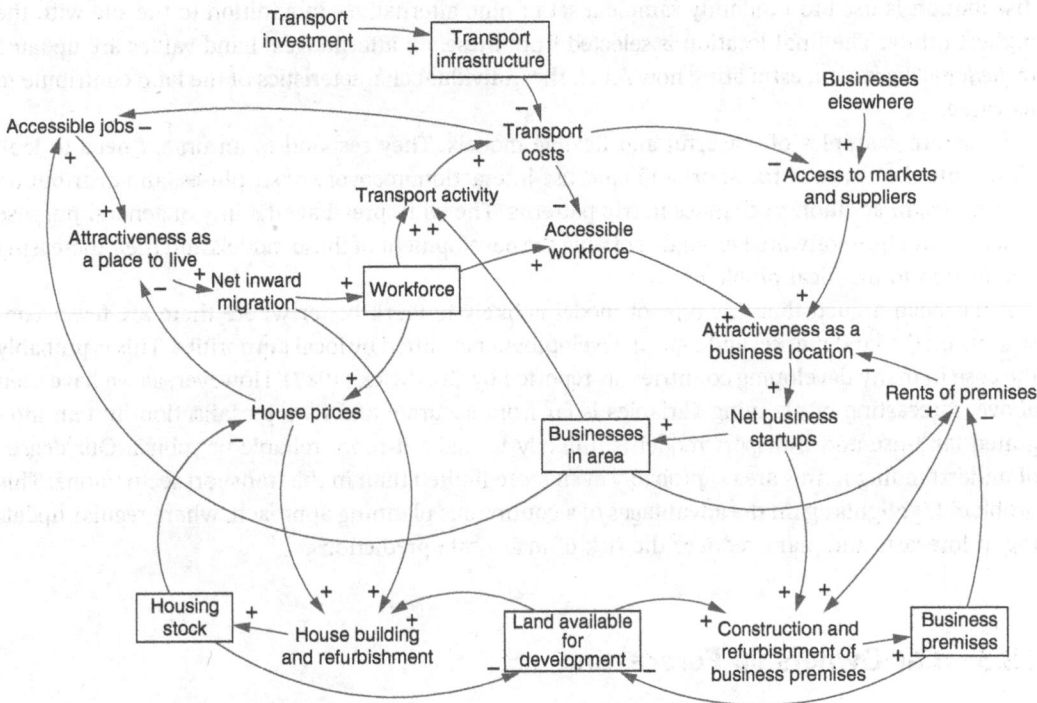

Figure 18.3 Relationships in a dynamic urban model

18.2.4 Urban Simulation

The advent of powerful computers and low-cost software has led to the development of many modelling approaches based on microsimulation for dynamic assignment and activity-based modelling. In the case of land-use transport interactions, the responses of interest with pre-specified probability distributions are analogous to those described above. What microsimulation can offer is the power to incorporate a number of dimensions of both individuals and their choice processes, which would otherwise require an excessive amount of disaggregation in classic model-based accounts.

Microsimulation models are relatively easy to understand and implement, and permit tracking of individuals, households, businesses, and parcels of land. Nevertheless, even a detailed microsimulation model will always leave out many variables that are necessary (but even then, never sufficient) to explain individual behaviour.

The MASTER model, developed in the UK by Mackett (1990) and UrbanSim, developed in the USA by a team led by Paul Waddell (Waddell 2002; Waddell et al. 2003) are two examples of this type of approach. MASTER, for example, considers households one at a time, allowing first for demographic processes including aging, giving birth, dying, divorce, and marriage. Marriage and divorce lead to the creation of new households with divorcees becoming one class of *forced* movers. Voluntary movers include newly married couples, individuals leaving the parental home, and wholly-moving households influenced by changes in life cycle. Both public and private housing markets are recognized, and dwelling occupancies are tracked from one period to the next. The choice of residence zone is based on a weighted function of generalized travel to work costs for the head of household.

UrbanSim simulates households, employees, developers, and real estate prices with a similar degree of refinement. Location decisions are based on MNL models. To select a location, a uniform distribution is used to randomly sample a set of nine alternatives in addition to the site with the highest utility. The final location is selected from these ten alternatives. Land values are updated by *hedonic regression*, estimating how much the individual characteristics of the land contribute to its value.

These are examples of powerful and flexible models. They respond to an urgent need to look closer into the issues of transport and land-use interaction, recovery of surpluses, and distribution of benefits, in addition to changes in trip patterns. The widespread availability of general-purpose model estimation software has made possible the development of these models and their increasing application to practical problems.

It has been argued that this type of model is likely to work better where there are fewer constraints on the land market and type of development permitted by local authorities. This is probably the case in many developing countries, as reported by Chadwick (1987). However, as we have seen above, forecasting of planning variables is far from accurate and its internalisation into an integrated land-use and transport model is unlikely to make it more reliable or robust. Our degree of understanding in this area is probably even more limited than in the transport sector alone. This problem highlights again the advantages of a continuous planning approach, where regular updating of forecasts and plans reduces the risk of inaccurate predictions.

18.3 Car-Ownership Forecasting

18.3.1 Background

Although the total number of passenger cars active on the road in industrialised countries almost doubled at the end of the last century (de Jong 1989), the rate of growth was dramatically higher in developing countries and has continued increasing well into the new century. For example, the fall in import duties for small cars in Chile (from 120% to only 10%) in 1977 meant that average car ownership in Santiago went up by more than 100% in only five years (see Fernández et al. 1983); later, a comparison of 1991 and 2001 data for Santiago revealed that car ownership continued growing at more than 3% per year (DICTUC 2003). Cars bring benefits to owners but the total increase in passenger car-kilometres represents a high cost to society in terms of accidents, fuel, pollution, increased traffic congestion, and additional road construction and maintenance costs.

A problem faced by the planners of any nation is that forecasts of the number of cars and/or vehicle kilometres for, say, the year 2050, imply that the above adverse effects may become unmanageable. In fact, by the end of the 1980s there were already cities like Athens, Los Angeles, Mexico, Sao Paulo, Seoul, and Tokyo, which had become notorious for their congestion and pollution problems; today their number has increased tenfold. A more recent challenge for car-ownership forecasting is the potential success of vehicle sharing arrangements with or without connected and autonomous vehicles (CAVs). If car sharing and Mobility as a Service (MaaS) become as successful as optimists expect, this could decouple car ownership from income growth and the provision of travel services.

Models to predict changes in car ownership have been under development since the early 1940s. It can be said, in general, that these efforts have been made with the following three different purposes in mind:

- market research studies for vehicle manufacturers and petrol companies, which are of limited interest to transport modellers as they are more concerned with vehicle attributes like size, engine capacity, electric range and so on;
- government-sponsored studies seeking to determine the need for new infrastructure (basically highways) at a national level; until the end of the 1970s, simple time-series models were used for this task;
- local studies, which are usually part of strategic transport studies and which have made use of more advanced econometric methods with either cross-sectional and/or longitudinal data.

We will not attempt to cover all aspects of the car-ownership forecasting problem here, as whole books and theses have been devoted to the subject (see, for example, Mogridge 1983; Train 1986; de Jong 1989). We will only briefly discuss two basic methods:

- time-series extrapolations using aggregate data at a national or regional level (basically the seminal work of John Tanner at the British Transport and Road Research Laboratory), and
- econometric methods using disaggregated data at the household level, since it has been argued that the decision to acquire a car cannot be modelled correctly at the strictly individual level or at the zone level (see, for example Bates et al. 1978).

Modern methods sometimes incorporate features of both approaches and extend estimates to car usage as well. Critical reviews of these and other methods have been given by Button et al. (1982) and de Jong (1989); forecasting sales and ownership of electric vehicles has become important in recent years, see Ding and Li (2021).

18.3.2 Time-Series Extrapolations

It seems clear that car ownership rates (e.g. cars/head of population) should not increase indefinitely in time (i.e. in general, people with a driving licence are not going to indulge in several cars each); for this reason, the curves that are usually used to model this phenomenon are S-shaped. For example, if the number of cars/person in the United States and in the United Kingdom were plotted against time, one could find approximately the shapes depicted in Figure 18.4.

One curve that proved popular in this field was the Logistic, pioneered by Tanner (1978). The following three parameters were needed to adjust it:

- C_0, the car-ownership rate in the base year (cars/person);
- g_0, the rate of increase of the car-ownership rate in the base year given by $\frac{1}{C}\frac{dc}{dt}$ evaluated at $t = 0$, and
- S, the saturation level of car ownership.

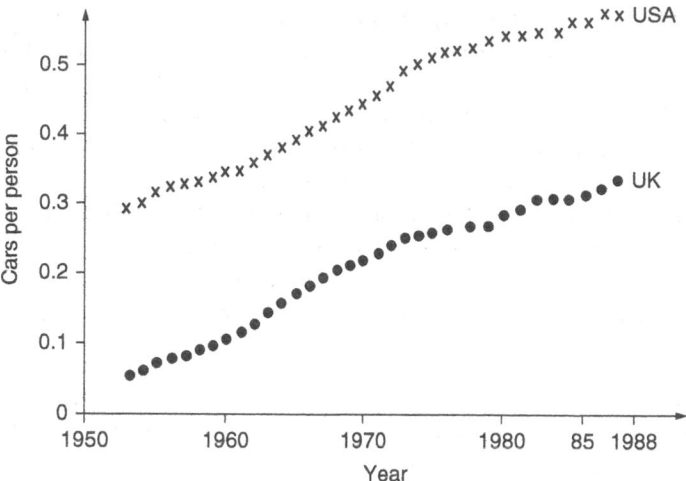

Figure 18.4 Shape of car-ownership increase

In Logistic curves we have that:

$$\frac{dC}{dt} = aC_t(S - C_t) \tag{18.11}$$

where a is a constant. Solving this differential equation yields:

$$C_t = \frac{S}{1 + b\exp(-aSt)} \tag{18.12}$$

where b is an integration constant. To find the values of a and b we can resort to the boundary conditions at $t = 0$; from (18.11 and 18.12) we obtain respectively:

$$g_0 = a(S - C_0) \text{ and } C_0 = \frac{S}{1 + b}$$

and replacing these values in (18.12) we finally get:

$$C_t = \frac{S}{1 + [(S - C_0)/C_0]\exp[-g_0 S \cdot t/(S - C_0)]} \tag{18.13}$$

Therefore, knowledge of C_0 and g_0 for one year taken as a base allows us to extrapolate C_t for any future year if S is known; however, S is not known but must be estimated. Tanner's method consisted of fitting a regression line:

$$g = \alpha + \beta C_t$$

Saturation corresponds, by definition, to that instant when the rate of change in the number of cars per capita (g) is zero; in this case, we get $S = -\alpha/\beta$, and as we would expect α to be positive and β less than zero, we can deduce that $S > 0$ (Figure 18.5):

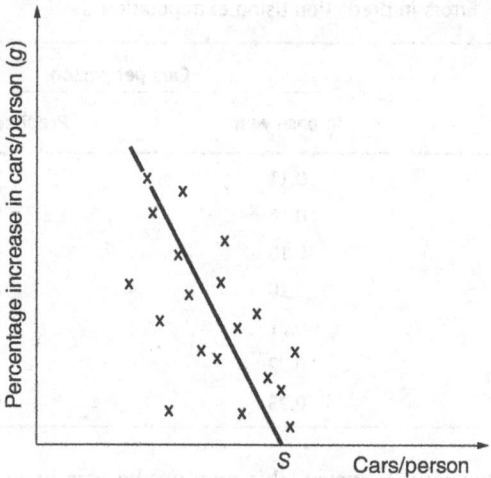

Figure 18.5 Determining the saturation level

Unfortunately constructing the graph in Figure 18.5 with data for the USA and the UK gave rise to Figure 18.6, implying that the method could work in the latter case but was doubtful in the former. For this and other reasons, the method was heavily criticised by Button et al. (1982).

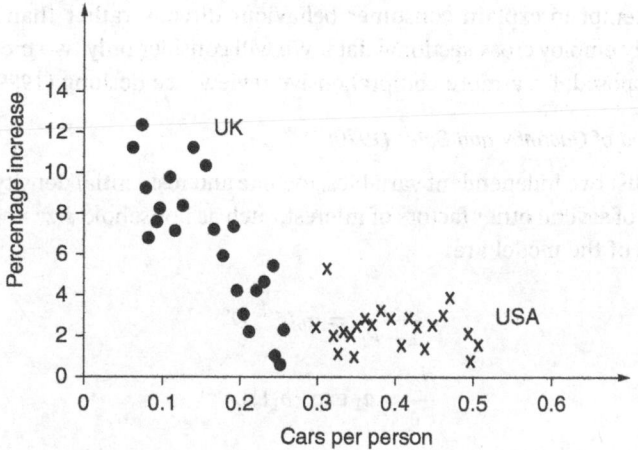

Figure 18.6 Saturation rates for UK and the USA

With the above data, Tanner (1974) estimated S as 0.45 for Great Britain. Table 18.1 compares predictions for 1975 made in different years with the observed figure of 0.25 cars/head in that year. As can be seen, the method was not very reliable.

In summary, the main objections to the Logistic extrapolation method are:

1) The model is not sensitive to policy variables. It is impossible to study the effects on car ownership of changes in car prices, road tax and import duties, fuel costs, and so on. Also, as it does not consider the influence of economic variables, if the correlation among them changes in time, perverse results may be obtained (i.e. consider the effect on car ownership brought about by the petrol crisis in 1973, or the aforementioned effect of reducing import duties in Chile in 1977).

Table 18.1 Errors in prediction using extrapolation

Base year	Cars per person	
	In base year	Predicted for 1975
1960	0.11	0.28
1964	0.16	0.32
1966	0.18	0.31
1968	0.20	0.30
1969	0.21	0.28
1971	0.22	0.27
1972	0.23	0.26

2) S is assumed to be a constant; however, this may not be true in practice as attitudes tend to change with time.
3) The model does not yield information about different types of cars or, more importantly in the case of planning purposes, the proportion of people belonging to households with 0, 1, and 2 or more cars.

18.3.3 Econometric Methods

These methods attempt to explain consumer behaviour directly rather than looking at general trends and normally employ cross-sectional data. We will consider only two methods out of several that have been proposed; for a more comprehensive review, see de Jong (1989).

18.3.3.1 The Method of Quarmby and Bates (1970)

This method uses just two independent variables, income and residential density, although it recognises the existence of several other factors of interest, such as household size and vehicle price. The basic relationships of the model are:

$$\frac{P_0}{1 - P_0} = a_0 I^{-b_0} D^{c_0} \tag{18.14}$$

$$\frac{P_2}{P_1} = a_1 \exp{(b_1 I)} D^{-c_1} \tag{18.15}$$

$$P_0 + P_1 + P_2 = 1 \tag{18.16}$$

where I is annual family income (thousands of \$), D is the number of residents per acre, and P_i is the probability of owning 0, 1, and 2 or more cars; a_i, b_i, and c_i are parameters to be estimated.

Substituting P_1 from (18.16) into (18.15) and taking logarithms we get:

$$\log\{P_2/(1 - P_0 - P_2)\} = \log(a_1) + b_1 I - c_1 \log(D) \tag{18.17}$$

Now, because D is a discrete variable for any given segment it may be considered a constant and (18.17) reduces to:

$$\log\{P_2/(1 - P_0 - P_2)\} = b_1 I + \text{constant}$$

It is instructive to consider that as income (I) increases, so does the left-hand side term of Eq. (18.17); therefore, one can deduce that $(1 - P_0 - P_2)$ tends to zero or, what comes out to be the same, P_2 tends to $(1 - P_0)$. However, as P_0 is nearly zero for high income, that would mean that

P_2 would tend to 1 and this is obviously incorrect as one would expect a lower limit for it. This upper bound, or saturation level (S) of P_2, may be incorporated to the model by adjusting (18.17), yielding:

$$\log\{P_2/[S(1 - P_0) - P_2]\} = \log(a_1) + b_1 I - c_1 \log(D) \qquad (18.18)$$

where S must be determined empirically; now, as this may be difficult in practice, the usual procedure involves trying different values in a sensitivity analysis. The types of curves obtained by this method are illustrated in Figure 18.7.

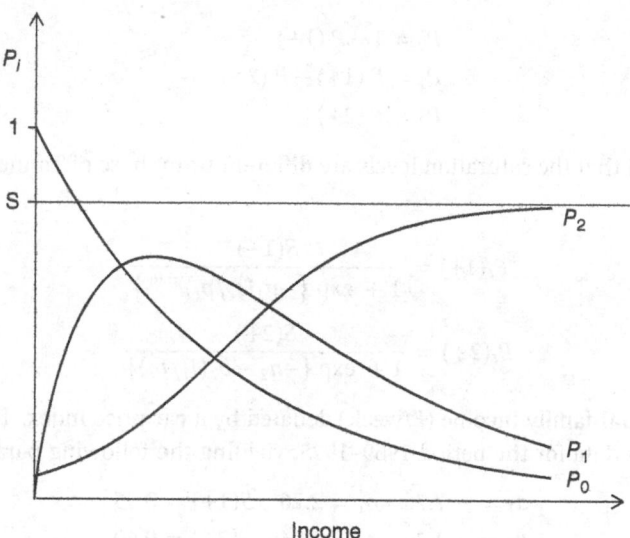

Figure 18.7 Car ownership versus income

Example 18.1 Consider the data in Table 18.2 and assume a value of $S = 0.78$; the problem is to estimate the parameters of the Quarmby and Bates's model for a fixed residential density value.

Table 18.2 Car ownership proportions by income

Income	P_0	P_1	P_2
1	0.61	0.34	0.05
2	0.35	0.47	0.18
3	0.22	0.44	0.34
4	0.16	0.37	0.47
5	0.10	0.30	0.60
6	0.08	0.24	0.68

If we take the logarithm of (18.14) for fixed D (i.e. c_0 is of no interest) we get:

$$\log\{P_0/(1 - P_0)\} = \log(a_0) - b_0 \log(I)$$

and fitting a regression line to the data we obtain $a_0 = 1.74$ and $b_0 = 1.60$. On the other hand, if we replace the value of S in Eq. (18.18) for constant D, we get:

$$\log\{P_2/[0.78 (1 - P_0) - P_2]\} = \log(a_1) + b_1 I$$

and fitting another regression line to the data we finally obtain $a_1 = 0.10$ and $b_1 = 0.84$.

18.3.3.2 The Regional Highway Transport Model (RHTM) Method

This method (Bates et al. 1978) combined the best features of the previous two approaches. First, it is necessary to define the following variables:

- $P(1+)$ = percentage of households with one or more cars, with a saturation level of $S(1+)$
- $P(2+)$ = percentage of households with two or more cars, with a saturation level of $S(2+)$

Therefore, the previous method's values can be derived as:

$$P_0 = 1 - P(1+)$$
$$P_1 = P(1+) - P(2+)$$
$$P_2 = P(2+)$$

but it must be noted that the saturation levels are different from those of Tanner. The model takes the following form:

$$P_t(1+) = \frac{S(1+)}{1 + \exp\left\{-a_1(I_t/p_t)^{-b_1}\right\}} \tag{18.19}$$

$$P_t(2+) = \frac{S(2+)}{1 + \exp\left\{-a_2 - b_2(I_t/p_t)\right\}} \tag{18.20}$$

where (I_t/p_t) is annual family income (£/week) deflated by a car price index. The model was calibrated using British data for the period 1969–1975, yielding the following parameter values:

$$a_1 = -7.76 \quad b_1 = 2.26 \quad S(1+) = 0.95$$
$$a_2 = -3.76 \quad b_2 = 0.04 \quad S(2+) = 0.60$$

To forecast, it was necessary to assume a certain distribution of income (for example, one of the Gamma type); also, to convert the modelled results to cars/person (C_p) it was necessary to use Census data. For example, Bates et al. (1978) postulated the following conversion rule:

$$C_p = P(1+) + 2.17 \cdot P(2+)$$

To obtain cars/household the analyst finally required information about the future average number of persons per household.

18.3.3.3 Models of Car Ownership and Use

Some authors have argued that car ownership should not be considered in isolation of other processes like motorcycle ownership, mode choice or at least car usage, as the latter is more critical than ownership. Train (1980) developed a Structured Logit Model of car ownership and mode choice. The work of de Jong (1990, 1996) has always emphasised the need to model jointly car ownership and use (kilometrage), and he has used different approaches, including indirect utility and discrete choice.

In a different context, Khan and Willumsen (1986) argued that in developing countries, growth in car ownership (and use) commits future resources to additional investment in roads and road maintenance. They insisted that car ownership should be considered a policy variable rather than an exogenous factor; to support these ideas, they developed policy-sensitive models of car ownership and use, and calibrated those using data from different countries and time periods. They also developed an 'analysis' model where the total number of cars, car-kilometres, fuel consumption, tax revenues, and road maintenance and investment costs were calculated for one or more years in the future. Alternative policies regarding taxation, import duties, and road construction could be compared in terms of their implied costs to the country.

18.3.3.4 Models of Motorcycle Ownership

It is surprising how little work has been carried out in the modelling of motorcycle ownership. Motorcycles are a much-maligned mode of transport despite their importance in many countries, including some in Europe. They have a poor safety record but offer a low-cost mode of transport with smaller requirements for roads and parking spaces than cars. The use of four-stroke engines should make them less polluting than cars, and in their modern incarnation as electric two-wheelers, they could deserve more attention.

There is a certain degree of substitution between motorcycle and car ownership, especially as income increases. Motorcycles are present in all countries, even in those where car ownership has reached saturation levels. This suggests that an interesting way to model motorcycle ownership could be jointly with car ownership. A simple model would assume that the saturation level of motorcycle ownership must be related to the actual level of car ownership: if the current level of car ownership is high, the saturation level of motorcycle ownership should decrease. However, Iglesias et al. (2022) studied this issue in depth as part of an Inter-American Development Bank study to understand car and motorcycle use in two Latin American metropolises, Bogotá and Santiago. They estimated hybrid choice models considering sociodemographic variables (e.g. income and household structure), built environment variables (e.g. population density, number of bus stops and whether the dwelling was located in the periphery), and two latent constructs: *Functionality,* which in Santiago varied quadratically with age and linearly with trip frequency, and *Style,* which in Bogotá was associated with indicators such as: 'I like to drive', 'having a car goes with my life style', and 'my car is my personal space'. They found that, in Bogotá, car ownership was almost exclusively associated with the two higher income strata, while for the rest it was very important to own and use motorcycles. In Santiago the use of motorcycles is still limited to delivery services and has not yet become a significant element in traffic.

Mesa-García et al. (2023) recently studied the factors affecting motorcycle ownership and use using data from Bogotá. They estimated hybrid choice models for the decision to own a motorcycle and the decision to use it more or less frequently. They found that both models depended on socio-demographic and built environment variables, but also that attributes and perceptions played an important role. For example, having a negative perception of public transport, having a pro-motorcycle personality, or having developed habit for using a motorcycle, were key determinants for deciding to own and use motorcycles.

18.3.4 International Comparisons

Energy use in the transport sector grows faster than in any other sector of the global economy. Of that growth, an increasing proportion originates in emerging countries. This is a reflection of the low levels of car ownership in these countries and the near saturation levels achieved in nations like the USA. Therefore, it is important to understand better how increases in wealth affect car ownership and use, and how these in turn will affect the choice of powertrain (ICE or EV) and emissions.

Dargay and Gately (1999) comprehensive study into the effect of income level on car ownership included international comparisons as part of this process. They used income and car ownership data for the period 1960–1992 from 26 countries, ranging from the USA to India and China (data was not available for all these years in all countries). Then they searched for suitable functional forms to model car ownership as a function of income level. After experimenting with a number of functional forms, they chose a Gompertz model, where the equation for long-run vehicle ownership V^* as a function of per capita income I can be written as:

$$V^* = \gamma \exp\left(\alpha e^{\beta I}\right) \tag{18.21}$$

where α and β are negative values. The parameter γ defines the saturation level since for $\beta > 0$:

$$\lim_{I \to \infty} V^* = \gamma$$

The parameter α specifies the value of the function at $I = 0$, that is:

$$V^*_{I=0} = \gamma e^{\alpha}$$

and since the saturation level γ cannot be equal to 0, the value of the Gompertz function approaches 0 as α increases negatively.

The Gompertz function has a long-run elasticity that can be calculated by appropriate differentiation:

$$\eta^{LR} = \alpha \beta I e^{\beta I} \tag{18.22}$$

The income level that produces the maximum elasticity is obtained by setting the derivative of the elasticity to 0:

$$I_{ME} = -1/\beta \tag{18.23}$$

and the maximum elasticity is defined by:

$$\eta^M = -\alpha e^{-1} = -0.3678\alpha \tag{18.24}$$

Dargay and Gately (1999) recognised that vehicle ownership cannot change instantly; there are lags and inertia effects that must be considered. They postulated a simple partial adjustment mechanism to account for these lags:

$$V_t = V_{t-1} + \theta\left(V^*_t - V_{t-1}\right)$$

where θ is the speed of adjustment ($0 < \theta < 1$) and V_t is vehicle ownership at time t. That converts into:

$$V_t = \gamma\theta \exp\left(\alpha e^{\beta I_t}\right) + (1-\theta)V_{t-1} \tag{18.25}$$

For a number of theoretical and practical reasons, the authors restricted the values of α, θ, and γ to be the same for all countries but allowed β to be country specific, leading to the following model:

$$V_{jt} = \gamma\theta \exp\left(\alpha e^{\beta_j I_{jt}}\right) + (1-\theta)V_{jt-1} \tag{18.26}$$

where the subscript j represents a given country.

Using their datasets, they found a common saturation level $\gamma = 0.85$ vehicles per person (and 0.65 cars per person) and a value of $\alpha = -5.9$. They also found the value of $\theta = 0.09$, suggesting that 9% of the total response to income takes place within one year. The values of β ranged from -0.3 to -0.2 in different countries.

Given the range of countries in their databank, the models developed by Dargay and Gately (1999) are quite useful for application in different countries where there are limited time series available for car ownership forecasting.

18.4 The Value of Travel Time

18.4.1 Introduction

The question 'has time a value?' is answered in the affirmative by most people. A more complex question is 'what value?' and this can be extended to consider under what circumstances it can or must be measured.

This theme has generated an enormous debate in the literature for more than 40 years (see, for example, Bruzelius 1979) simply because time savings continue to be the single most important benefit of transport improvement projects all over the world. However, in spite of its importance, a consensus has not been reached about the size and nature of the values to be used in project evaluation. We will not attempt to review the subject in great detail here, but refer the reader to Jara-Díaz (2007) for a deeper discussion.

For example, in Great Britain (and other countries, such as Chile), social values of time corresponding to a fixed proportion of the average hourly rate are recommended. On the other hand, increasing values for three ranges of time savings (0–5 minutes, 5–15 minutes, and 15 or more minutes) have been recommended in the USA (AASHTO 1977). Clearly the use of linear or non-linear valuation functions should lead to different benefits and, hence, different investment priorities. For example, the British norm tends to favour schemes generating small time savings, while the norm above would favour schemes generating more substantive time savings.

Most studies distinguish between *subjective* (or behavioural) and *evaluation* (or social) values of time. The first corresponds to, for example, the value of the parameter associated with in-vehicle travel time in the generalised cost functions we studied in Chapter 5, which should have been derived by estimating, typically, a discrete choice demand model with real disaggregate data. The evaluation value is that used, as the name implies, to compare alternative schemes that produce different levels of time and other resource savings. It is argued, therefore, that the behavioural value of time reflects mostly the ability of the traveller to pay and not the intrinsic value of a particular time saving. This is why, often, the value of time used for evaluation purposes is an *equity* value, taken as being the same for all travellers, independently from their age or socioeconomic group, as we will see below.

On the other hand, it may be argued that the use of different 'values of time' for evaluation and demand modelling purposes introduces inconsistencies of approach at different stages of the same exercise. This was, for example, one of the criticisms levelled at the controversial implementation of the *Transantiago* public transport system in 2007 (see Muñoz et al. 2009) in Chile, as the low social-equity values of waiting time were confronted with normal operators' earnings in a complex optimisation programme used as part of the system design. There is little dispute, however, that the subjective values of time are heavily dependent on model specification and data (see Gaudry et al. 1989); this is an undesirable property because a consistent evaluation of projects is sought over a wide range of models and areas.

18.4.2 Subjective and Social Values of Time

The utility function estimated from discrete travel choice models can be used to calculate the subjective value of time saving (SVT) or, equivalently, the willingness-to-pay (WTP) to reduce travel time (in-vehicle, walking, or waiting) by one unit. As shown by Jara-Díaz (2000), because travel utility is, in fact, a conditional indirect utility function, the SVT can be given a microeconomic interpretation that depends upon the arguments that are assumed to enter the utility function and the type of constraints considered; see also Bates (1987).

Time valuation analysis comes from three sources: the pure time allocation theories, the home production framework, and the literature on travel demand. Everything started with Becker's (1965) approach, based upon the idea of utility depending on the amount of *final goods* (i.e. a prepared meal) consumed, each of which requires market goods and time as inputs. This was the origin of a time value equal to hourly income, because 'time can be converted into money' by spending more time at work and less in consumption. This elementary result was soon proved limited, after the successive analysis by Johnson (1966), Oort (1969), DeSerpa (1971), and Evans (1972), because work time should enter utility as an argument.

Later on, the fixed-income approach to mode choice models, introduced earlier as the *expenditure rate* approach (Jara-Díaz and Farah 1987; Jara-Díaz and Ortúzar 1989), also supported a travel time value that is not necessarily related to the wage rate. The result of this stream of papers was a framework in which the economic actions of individuals were looked at as if they maximised a utility function that depends upon all types of activities undertaken and, on all goods consumed, subject to three types of constraints: a money budget, a time constraint, and a set of technical relations between goods and time (Jara-Díaz 1998, 2007).

Up until now, the SVT has been shown to reflect the value of relaxing the minimum time requirement for travel. Analytically, this is the ratio of the multiplier on that constraint over the marginal utility of income (MUI) and can be shown to be equal to the resource value of time (or, equivalently, the value of leisure) minus the value of the marginal utility of travel. The former represents the value of reassigning the travel time saved to other activities, and is analytically given by the ratio of the multiplier of the time constraint over the MUI. The latter is the lost value, in 'direct utility' terms, because of less travel, and is expected to be negative. Thus, the SVT adds up the value of a gain in leisure plus the value of a reduction in an unpleasant activity (see the discussion and the extra references provided by Jara-Díaz 2007).

It is important to note that if individuals choose the work schedule (hours at work) at a given wage rate, they will adjust that schedule until the value of leisure equals the value of work; this is given by the addition of the money earned (the wage rate) plus the value of the marginal utility of work (which can be positive or negative). Jara-Díaz and Guevara (2000), and Munizaga et al. (2006), among others, have managed to estimate simultaneous models of travel and activities, obtaining not only the SVT but also its component elements.

Finally, a word on the price of travel time that should be used for social appraisal of projects (the evaluation or *social value of time*). There is no reason for society to value an individual's reassignment of travel time at her SVT. For the analysis of discretionary travel, the state of the art is the work by Gálvez and Jara-Díaz (1998), who showed that a proper social price of time (SPT), consistent with a social appraisal framework within the field of welfare economics, should be equal to the ratio of the marginal utility of time over what they call 'social utility of money'. This is given by a weighted sum of individuals' MUI, with the weights given by the proportion of marginal taxes paid by the corresponding group. This approach advocates for a potentially different SPT by group, which is generally different from each group's SVT. It is important to note that these authors show, analytically, that accepting the SVT as SPT is equivalent to assigning to each group a social weight that increases with income. This has important and generally undesired policy implications, but sadly, it reflects the approach sometimes taken in practice.

18.4.3 Some Practical Results

Heggie (1983) argued that the value of time debate was more empirical than theoretical. The practical difficulties associated with measuring values of time encouraged the use of indirect methods such as the discrete choice approach mentioned above. However, this method generates the usual empirical problems such as:

- how to choose an appropriate sample, that is, one that basically contains people with a real choice among clearly defined alternatives in terms of time and cost of travel;
- how to correctly measure the travel attributes, that is, avoiding aggregation, perception, and other sources of bias; and
- which demand function to use that is consistent with the situation under study.

All these problems suggest that values derived from models estimated with revealed preference data (the large majority of cases) may be suspect for evaluation purposes.

Perhaps the most complete study ever undertaken about the value of travel time savings was performed between 1981 and 1986 by a consortium of consultants and academic experts in Britain, using a series of models estimated with revealed preference and stated preference (SP) data for various choice situations in several areas of Great Britain (Bates and Roberts 1986). Its principal recommendations (Department of Transport 1987) were:

1) The value of working time (i.e. trips made during or as part of work) is equal to the gross hourly income of the traveller, including all additional costs to the employer.
2) The trips for all other purposes, including trips to work, increased their valuation from 27% to 43% of the average hourly income of full-time employed adults (this is an increment of 85%).
3) For the majority of cases a single 'equity' value of time should be used; however, in cases where the proportion of children, pensioners, or employed adults is judged to differ significantly from the national average, an *ad hoc* equity value of time should be estimated using the individual values for each of these groups.
4) To update these values, information about real hourly incomes for each year should be used; to forecast, these incomes should be estimated as a function of the domestic per capita product.
5) The values of waiting and walking time should be taken as twice the value of in-vehicle travel time; bicycle users should be treated as pedestrians in this sense.
6) Small time savings should be valued equally as more significant savings. Additional research contradicted this recommendation and the current advice in the UK is to apply a different valuation to travel time depending on trip length (ARUP et al. 2015 and Department of Transport, 2022a).

In 1994, the UK Department of Transport commissioned a new value of time study (Accent and HCG 1996). In what follows, we summarise some of their most interesting conclusions, which are broadly in line with the findings of an earlier study done in Holland (HCG 1990) using the same methodology:

1) For any level of variation around the original journey time, travel time gains are valued less than losses. For non-work-related journeys, variations up to five minutes in journey times are generally ignored. Business travellers are more sensitive to gains and losses than commuters, who in turn are more sensitive than those on non-work-related journeys.
2) There is a clear relationship between income and SVT (as was found in 1986), which is monotonically increasing but not directly proportional. At the same income levels, the 1994 SVT

values were significantly lower than those recorded in 1986. This may have been caused by changes in the composition of the car-using population (those with high SVT were earlier to acquire and use cars), with the growth in usage then biased towards market segments with lower SVT.

3) SVT values under congested conditions are significantly higher than for trips done under free-flow conditions. However, the types of road mix (i.e. percentage of time travelling on motorways, trunk roads, and other roads) were not significantly different. Regular users of motorways were relatively indifferent to the number of lanes, but appeared to be very sensitive to travelling with lorries in the traffic, and clearly disliked roads with no shoulders (the strongest effect of all).

4) In relation to peak shifting, it was found that the disutility of departing earlier increased linearly with the time difference. This was also true for later departures up to one hour as, curiously, they found that the burden did not increase much beyond that hour; see also the discussion in Bianchi et al. (1998).

Further national value of time studies had been conducted in several countries more recently, for example, Denmark (Fosgerau et al. 2007), Germany (Dubernet and Axhausen 2020), Holland (Significance et al. 2013), Sweden (Börjesson and Eliasson 2014), Switzerland (Axhausen et al. 2008), and the UK (ARUP et al. 2015), and these should be consulted by interested readers as they offer interesting new insights.

Finally, other issues related with time valuations that we are not considering here, are the following:

- the value of time can be a function of both the duration of the trip and the magnitude of the time saved (see the discussion in DICTUC 2021); it may also depend on traffic congestion (Wardman and Ibañez 2012);
- aversion to losses, or 'sign effect'; that is, savings in time may be valued differently than losses (de Borger and Fosgerau 2008);
- variability of travel time, where it is possible to estimate a value associated to reducing it (Dubernet and Axhausen 2020), and
- effect of crowding in the value of time (Wardman and Whelan 2011; Li and Hensher 2013; Tirachini et al. 2013; Batarce et al. 2016).

18.4.4 Methods of Analysis

18.4.4.1 Estimation of Subjective Values of Time

To estimate the WTP for savings in travel time (i.e. the SVT) in the classic transport microeconomic literature, modellers need to measure the trade-offs between travel time and cost faced by a target population represented by a statistical sample (e.g. individuals commuting from certain suburbs to the CBD). The SVT corresponds to the marginal rate of substitution between perceived times t_i (in-vehicle, walking, or waiting time) and costs c_i of travel at constant utility (Gaudry et al. 1989), yielding the following expression:

$$\text{SVT} = -\frac{dC_i}{dt_i}\bigg|_v = \frac{\partial V_i/\partial t_i}{\partial V_i/\partial c_i} \tag{18.27}$$

The representative utility function in a classical discrete choice model is assumed to be linear and additive in the (fixed) marginal utility parameters. Under this assumption, the SVT corresponds to the ratio between the estimated parameters, θ_t and θ_c, of the attributes travel time and cost[1]; for example, in the case of the popular wage rate (w) specification (Train and McFadden 1978), this simply yields:

$$\text{SVT} = \frac{w\theta_t}{\theta_c} \tag{18.28}$$

and one can easily see that the ratio θ_t/θ_c represents SVT as a percentage of income. On another hand, for the linear-in-parameters expenditure rate (g) specification (Jara-Díaz and Farah 1987), where g is given by (8.7), Eq. (18.27) yields:

$$\text{SVT} = \frac{g\theta_t}{\theta_c} \tag{18.29}$$

Further, in the non-linear Box–Cox case (8.3), if we use a wage rate specification we would get:

$$SVT = \frac{w\theta_i t_i^{\tau_i - 1}}{\theta_c (C_i/w)^{\tau_c - 1}} \tag{18.30}$$

which will clearly vary across alternatives if τ_k is not equal to 1. This latter formula implies that if both τ's are equal and they are less than one (i.e. as required by their microeconomic conditions), the model will necessarily yield higher value of time estimates for modes which are more expensive per minute; however, this may not be the case if the τ's differ (Gaudry et al. 1989).

Román et al. (2014) provide a good application of value of travel time savings estimation using revealed preference data in the context of intercity travel in Spain.

18.4.4.2 Confidence Intervals for the Value of Time

As θ_t and θ_c above are maximum likelihood estimates (MLE) of the *true* model parameters, they are not really constants but random variables that distribute asymptotically Normal. For this reason, the 'SVT point estimate' (i.e. θ_t/θ_c) is also a random variable with an unknown probability density function (PDF), and it is appropriate to examine the consequences of replacing this single value by the construction of a confidence interval. A simpler way out consists in judging the significance of the SVT by means of a pseudo t-ratio test. Making a first- order expansion of a Taylor series for the random variable θ_t/θ_c around its mean value (the ratio of the estimated coefficients), the following t-ratio may be constructed:

$$t_{tc} = \left(\frac{\sigma_t^2}{\theta_t^2} + \frac{\sigma_c^2}{\theta_c^2} - \frac{2\text{Cov}(\theta_t, \theta_c)}{\theta_t \theta_c} \right)^{-1/2} \tag{18.31}$$

where σ_t and σ_c are the standard errors of the estimated coefficients. Daly et al. (2012b) have given credentials to this formula, arguing that due to the asymptotic properties of the maximum likelihood estimator, it should be an exact measure in the immediate vicinity of the maximum (i.e. the Delta method we briefly discussed in Chapter 9).

If a vector of random variables (in our case the parameter estimates) converges asymptotically to a joint distribution (in our case the multivariate Normal), then any continuous function of the parameters, such as the ratio, converges in distribution (to the ratio of two Normal variables), according to the 'continuous mapping' theorem (see theorem 5 in Mann and Wald 1943).

---

[1] Note that sometimes the cost coefficients can be different (i.e. for operating cost and for tolls in the case of cars) and it is not obvious which yields the correct SVT estimate (Hensher 2011).

Consequently, the SVT point estimate would be a random variable governed by a probability distribution (i.e. for large samples that for the ratio between two Normal distributed variables) without an explicit form (Fieller 1933; Hinkley 1969; Shanmugalingam 1982) and may turn out to be unstable (Meijer and Rouwendal 2000). Note that in the special case of two Normal variables with mean zero, its ratio follows a Cauchy PDF (Arnold and Brockett 1992), which has an indefinite variance and its mean does not have an analytical expression.

Given these facts it is possible that for large samples the ratio between the parameters θ_t and θ_c, which are components of a general multivariate Normal population, will be governed by an unyielding PDF (the only exception being when the coefficient of variation of θ_c approaches zero, in which case the ratio approximates the Normal distribution). It would then be necessary to find an econometric procedure to make statistical inference on this ratio without resorting to the direct use of the associated PDF (but see the discussion in section 18.4.4.3).

To solve this problem, several methods are available in the literature. For example, Ettema et al. (1997) discuss a general method to construct confidence intervals for the SVT even in cases where the parameters of travel time and cost are allowed to interact with other segmentation variables. Simulation is used to simultaneously calculate the parameters from a multivariate Normal distribution defined by the covariance matrix of the estimated travel time and cost parameters. Values for these are generated a sufficiently large number of times, and the confidence interval is constructed on the basis of the mean and variance estimates of the generated sample; it is possible to simulate values for the parameters of travel time, waiting time, walking time, and cost simultaneously. Finally, by simply calculating the 0.025 and 0.975 percentiles, the limits of the confidence interval at the 95% level are obtained.

This method does not need to introduce additional assumptions (other than normality for the MLE). In addition to being applicable to any type of utility function specification, it considers the variance of the parameters and the correlation between them. The results of Ettema et al. (1997) suggest that when the correlation increases, the size of the intervals decreases, indicating that we may obtain extreme results when correlation is not considered.

Further advances on this method and an application to an RP/SC model including interactions in the utility specification and the introduction of intangible variables, such as comfort, were done by Espino et al. (2006). Their results indicate that the size of the confidence interval is affected by the outliers of the simulation as well as by the magnitude of the simulated parameters. The estimated parameters should be consistent in relation to all the microeconomic principles underpinning the model, that is, the marginal utilities of the different attributes must have a correct sign for every individual in the sample (i.e. before applying, in their case, sample enumeration to obtain the corresponding SVT); otherwise, such individuals should be removed from the calculation. They found that elimination of outliers in two steps (first, from the simulated multivariate Normal distribution and second, from the simulated distribution of the SVT), as well as the removal of individuals with inconsistent marginal utilities, was the strategy that provided narrower confidence intervals. Furthermore, in this case the amplitude of the intervals remained constant as the number of simulations (up to 100 000) was increased (but, again, see the discussion in 18.4.4.3).

Armstrong et al. (2001) discuss two further methods. They called the first *asymptotic t-test* method and is based on the following null hypothesis:

$$H_0 : \theta_t - VT \cdot \theta_c = 0 \tag{18.32}$$

where *VT* represents the SVT point estimate (i.e. the ratio between the parameters of time and cost in a linear utility). The confidence interval is given by the set of *VT* values for which it is not possible to reject H_0 at a given level of significance. The corresponding statistic is:

$$t = \frac{\theta_t - VT \cdot \theta_c}{\sqrt{\text{Var}(\theta_t - VT \cdot \theta_c)}}$$

This expression distributes Normal for linear models and asymptotically Normal for non-linear models like the MNL (Ben-Akiva and Lerman 1985). Armstrong et al. (2001) also derived the upper and lower bounds for the interval as follows:

$$V_{U,L} = VT\left(\frac{t_c}{t_t}\right)\frac{(t_t t_c - \rho t^2)}{(t_c^2 - t^2)} \pm VT\left(\frac{t_c}{t_t}\right)\frac{\sqrt{(\rho t^2 - t_t t_c)^2 - (t_t^2 - t^2)(t_c^2 - t^2)}}{(t_c^2 - t^2)} \quad (18.33)$$

where t_t and t_c correspond to the t-statistics for θ_t and θ_c, respectively, and ρ is the coefficient of correlation between both parameter estimates. Eq. (18.33) is a real number only if the radical argument is non-negative; it can be shown that this condition is met when the parameters θ_t and θ_c are statistically significant (so that t_c and t_t are greater than t). This condition assures positive upper and lower bounds.

The confidence interval derived from this formulation is not symmetrical with respect to the SVT point estimate (VT), and the interval's midpoint is greater than VT as well. Another feature is that the value of ρ has a strong influence; for example, the interval size decreases with the value of ρ and vice versa. The size of the interval also narrows as the t-statistics get more significant.

Finally, note that for large samples the following equality holds:

$$\lim_{\substack{N \to \infty \\ t_t, t_c \to \infty}} V_{U,L} = VT \quad (18.34)$$

which agrees with the intuition that the larger the sample size, the smaller should be the interval size.

Armstrong et al. (2001) called their second approach *likelihood ratio* (LR) *test* method. It is based on imposing the linear restriction (18.32) to the maximum likelihood estimation process and comparing the statistical efficiency of the estimation with respect to the unrestricted case. The procedure is to search for values of VT for which the linear restriction is valid given a certain significance level. The null hypothesis is still the same as in the previous case, but the test is performed according to the following statistic:

$$\text{LR} = -2[l(\theta_r) - l(\theta)] \quad (18.35)$$

where $l(\theta_r)$ and $l(\theta)$ represent the logarithm of the likelihood function for the restricted and unrestricted models, respectively. LR is distributed χ^2 with one degree of freedom (corresponding to the single restriction imposed). In section 18.4.4.3 we discuss these further.

Example 18.2 Let us consider the following systematic utility function to be estimated:

$$V_{iq} = \theta_t t_{iq} + \theta_C C_{iq} + \sum_k \theta_k x_{kiq}$$

where t_{iq} and C_{iq} are the travel time and cost for individual q; x_{kiq} are attributes (different from travel time and cost) for individual q, and θ_k are their corresponding parameters. Replacing the ratio of θ_t and θ_C by VT in the above equation, the following utility function is obtained:

$$V_{iq} = \theta_C(VT \cdot t_{iq} + C_{iq}) + \sum_k \theta_k x_{kiq}$$

These two equations allow us to compute the unrestricted and restricted log-likelihood functions, $l(\theta)$ and $l(\theta_r/VT)$. Clearly, if the SVT is equal to VT then we get that $l(\theta) = l(\theta_r/VT)$, but different values of VT will yield different values of the restricted log-likelihood function. This method requires a search for the maximum and minimum values of VT for which the following inequality holds:

$$-2[l(\theta_r/VT) - l(\theta)] \leq \chi^2_{1,1-\alpha}$$

An advantage of this method over the asymptotic t-test method (18.33) is that it is not restricted to linear utility functions. However, the process of constructing the intervals is more tedious because it requires an iterative procedure to obtain each limit. Armstrong et al. (2001) present the results of using all the above methods for various cases of interest.

The subjective values of time and their confidence intervals (both bounds and size) vary strongly with model specification (i.e. they are strongly dependent on the functional form assumed for the representative utility and on the model structure). But with cross-sectional data it is not easy to give a clear rejection of any reasonable model form; see the discussion in Jara-Díaz and Ortúzar (1989).

18.4.4.3 *A Deeper Look at Computing Measures of Uncertainty for WTP*

This section draws heavily on the definitive paper by Daly et al. (2023). Let us consider a reparameterisation of the choice model by an invertible one-to-one vector function to obtain a vector η with the same dimension as θ:

$$\eta = f(\theta) \text{ and } \theta = f^{-1}(\eta) \tag{18.36}$$

With these conditions, Cramer (1986) shows that the essential MLE properties are not affected by the transformation, so that $\hat{\eta} = f(\hat{\theta})$ is a maximum likelihood estimate (MLE) of η. Greene (2008) similarly indicates that, if θ is asymptotically Normally distributed and $f(\hat{\theta})$ is continuous, then it is also asymptotically Normally distributed.

This highlights the distinction between Normality and asymptotic Normality. If we have that a given parameter θ_k follows a Normal distribution in a population, then the distribution of its inverse does not have finite moments. However, if $\hat{\theta}_k$ is an asymptotically Normal coefficient estimate, then the estimate of its inverse maintains the property of asymptotic Normality. As stated by Daly et al. (2012b) ... 'the estimate of the reciprocal has just as much status as the initial estimate'.

A simple method for calculating the standard deviation of a function of random parameters is the Delta method, which is generally presented as an approximation, based on a Taylor series expansion, as we saw in Chapter 9. On the other hand, given appropriate conditions, a first-order approximation to the error in $\eta = f(\theta)$ induced by the error in θ is given by (e.g. Greene 2008):

$$cov(\eta) = f'^T \Omega f \tag{18.37}$$

where f' is the matrix of first derivatives of the function f with respect to θ, and Ω is the covariance matrix of the estimates of θ.

However, in the case of MLE, these formulae can be given a different status. Cramer (1986) shows that, in the context of (18.36), the covariance of η as given in (18.37) is the Cramer–Rao lower bound, so that the estimate of η has the full MLE properties. Daly et al. (2012b) used this result to show that a likelihood function, which has been maximised with respect to one set of parameters, can also be considered to have been maximised with respect to parameters derived by transformations of the first set. The optimum values of the new coefficients are the transformed values of the old ones, and their estimation errors are given exactly by the formulae of the Delta method, which in this context is not an approximation. Therefore, a model can be estimated with a specification that is convenient and then transformed to a different specification as required, without losing the maximum likelihood properties of the estimates.

It should also be noted that the use of the Delta method does not, in any way, imply that the cost coefficient needs to be fixed. If θ_c and possibly also θ_k are distributed across individuals, the analyst needs to first use the results in Daly et al. (2012b) to determine the existence of moments for the distribution of WTP. If these moments can be expressed as a function of estimated parameters, then the analyst can again use the Delta method.

The above discussion has explained how standard errors can be calculated accurately for WTP measures. Our attention now turns to confidence intervals around the estimates. Care is again required given the asymptotic nature of the distribution of coefficient estimates. The natural tendency to calculate a C% confidence interval using $\hat{\theta} \pm z^*\sigma_\theta$, where z^* is the upper $\frac{1-C}{2}$ critical value for a $N(0, 1)$ distribution, relies on the assumption of Normality, which applies only in the close neighbourhood of $\hat{\theta}$. In the case of coefficients (or functions thereof, such as WTP) with small standard errors relative to their estimates (i.e. high t-ratios), the above calculation can be acceptable. However, it is far from clear what level of statistical significance is required to make the $\hat{\theta} \pm z^*\sigma_\theta$ calculation acceptable. Therefore, other approaches for calculating confidence intervals deserve attention.

A commonly used method in environmental economics is that of Fieller (1954), which calculates the confidence limits as:

$$V_{\min}, V_{\max} = \frac{S_{cx} \pm \sqrt{S_{cx}^2 - S_{cc}S_{xx}}}{S_{cc}} \tag{18.38}$$

where $S_{cc} = \hat{\theta}_c^2 - t^2\sigma_c^2$; $S_{xx} = \hat{\theta}_x^2 - t^2\sigma_x^2$; $S_{cx} = \hat{\theta}_c\hat{\theta}_t - t^2\sigma_{cx}$, and t is the critical t-value. This formulation assumes exact Normality rather than asymptotic Normality and is exactly equivalent to the first approach of Armstrong et al. (2001); although their formula (18.33) looks different it is in fact the same.

Another approach is the *likelihood ratio* (LR) method of Armstrong et al. (2001), which we saw in the previous section. This approach is ingenious and avoids the issue caused by the denominator approaching zero, but it relies heavily on knowledge of the likelihood function; also, as it is a two-stage procedure, it could bring a relative loss of efficiency and potentially introduce bias (Daly et al. 2023). For example, if the function contains local optima, it is quite possible that spurious results can be obtained.

Finally, an approach that has received less attention than expected in this context is bootstrapping[2]. It operates by sampling, with replacement, N observations from the original sample; this

---

2 An alternative resampling technique that could also be used is the Jackknife (Shao and Tu 1995).

is repeated a number of times S, leading to samples $D_1, ..., D_S$. Individual models are then estimated yielding S sets of coefficient values (e.g. the vector $\hat{\theta}^{(s)}$ in run s). The approach is based on the concept that if the original data is a representative sample from the population being studied, then the bootstrap samples also resemble samples that might be drawn if sampling were done again. For that reason, they give the expected sampling variance.

In the context of data with multiple observations per individual, it is necessary to sample at the level of individuals rather than observations. The covariance matrix of the coefficient estimates $cov\left(\hat{\theta}^{(1)}, ..., \hat{\theta}^{(S)}\right)$ is calculated as their covariance over the bootstrap samples and an empirical confidence interval for WTP can be obtained from the distribution of

$$WTP^{(s)} = \frac{\hat{\theta}_x^{(S)}}{\hat{\theta}_c^{(S)}}, \forall s.$$ Although this process is computationally expensive for large values of S,

it does not rely on any assumptions about Normality.

18.4.4.4 Special Problems Brought in by the Use of More Flexible Models

If tastes are assumed to be homogeneous, as in the classical MNL or NL models, it is possible to derive a single WTP value for a fictitious average individual. In this case, it is also straightforward to examine if the model satisfies the required micro-economic conditions. But this assumption can be too restrictive, as WTP may vary from one person to another depending not only on observable social and economic characteristics, but also on unobserved variables or attributes that are difficult to measure. For this reason, it is important to study the distribution of preferences in the population to obtain more accurate measurements.

As we saw in Chapters 7 and 8, advances in the field have enabled analysts to use increasingly more flexible models, such as Mixed Logit (ML), that allow one to define broader behavioural patterns (Train 2009). However, these models have been infrequently applied in evaluation studies and a consensus on the correct way to interpret their results has not yet been reached (Hensher and Greene 2003; Sillano and Ortúzar 2005). Furthermore, most applications have been limited to estimating just the mean and spread of the distribution of population parameters and not individual parameter values. Now, the estimation of WTP values involves taking ratios of stochastic variables, and in this case, the problem we discussed in the previous section is compounded by the fact that not only the estimates, but the parameters themselves, are random variables, and this is not a trivial issue (Meijer and Rouwendal 2000).

Amador et al. (2005) analysed individual preference heterogeneity with different methods and compared their benefit measures. To capture heterogeneity, they used two approaches discussed in Chapter 8. First, *systematic taste variations* as in Eq. (18.17) where each level-of-service parameter is allowed to be a function of observed socio-economic characteristics (i.e. age, sex, income, and vehicle ownership). Second, capturing *random taste variations* through the specification of a ML model (see Section 8.6). Both approaches can also be used in a single model allowing us to incorporate non-observed heterogeneity as well as systematic variations in preferences.

Amador et al. (2005) compared subjective values of time (SVT) computed from a MNL imposing preference homogeneity and from various specifications allowing for taste variations (see

Table 18.3). They found that the values derived from a model with homogeneous preferences (MNL-1) were similar to those obtained when systematic variations in tastes were considered (MNL-2); however, if travel time tastes were allowed to vary randomly, significant differences appeared (i.e. up to 40% increase in SVT) even when a systematic variation for gender was allowed for as in model ML-2. This suggests that using a restrictive specification may lead to an underestimation of the value of travel time savings.

Table 18.3 Subjective values of travel time[a]

	Men	Women	Mean
MNL-1	—	—	14.9
			(14.3–15.6)
MNL-2	10.4	18.7	15.3[b]
	(10.0–10.8)	(17.9–19.4)	
ML-1	—	—	21.4
			(20.4–22.4)
ML-2	17.0	24.7	21.5[b]
	(16.4–17.6)	(23.7–25.9)	

[a] Confidence intervals for SVT at the 95% level are presented in parenthesis following Armstrong et al. (2001).
[b] Weighted averages considering that the sample was composed of 204 men and 290 women (Amador et al. 2005).

However, previous experience suggests that conclusions actually depend on the nature of the data and specifications used in each study. For example, Hensher (2001a,b) also found that more restrictive models tend to underestimate the value of time; notwithstanding, other authors have found no significant differences between values produced by different models (Train 1998; Carlsson 2003), and in some cases even lower SVT values have been obtained when more flexible models than the MNL are specified (i.e. Box–Cox Logit, see Gaudry et al. 1989).

One possible explanation for the empirically observed discrepancies is the re-scaling that all parameters undergo when moving from a fixed specification to one where some parameters are allowed to vary randomly (see Example 8.8). But if all parameters were re-scaled in the same proportion, the SVT should not be affected by changing the specification. However, empirical evidence shows that not all parameters are re-scaled by the same magnitude. Sillano and Ortúzar (2005) suggest that an intuitive explanation for this would be that the explicit treatment of parameter variation over the population into the systematic utility component is equivalent to the incorporation of an explanatory variable previously left out in the original (MNL) model. This is analogous to one of the misspecification problems discussed in Section 2.2.1.4 and would lead to the restructuring of the utility parameters to compensate for the extra explanation accounted for. Thus, depending on the variables included in the model, the functional form chosen for the indirect utility function, and the nature of the data, a fixed-parameters model may lead to over/underestimates of the true values of time.

In what follows, we will discuss some econometric aspects of four different methods to achieve WTP estimates from parameter distributions. These methods can be applied to jointly distributed parameters, but we will assume independent distributions for simplicity. However, in many cases, results are coincidental (Sillano and Ortúzar 2005).

Ratios of Population Means. The simplest way to derive WTP values is to take the ratio of the means of the parameter distributions involved. In other words, if

$$\theta_t \sim f(\mu_t, \sigma_t) \wedge \theta_c \sim g(\mu_c, \sigma_c) \quad \text{then} \quad \frac{\theta_t}{\theta_c} \rightarrow \frac{\mu_t}{\mu_c} \tag{18.39}$$

This is not the mean value of the WTP, but a WTP value derived from the coefficients of the 'average individual' for each parameter. Therefore, this interpretation should not be used in cost-benefit analysis, and the calculation of this index may only be used as a means of testing model specification. Also, as the method disregards the rest of the distribution, it considers a unique value for the parameters neglecting all information about heterogeneity in the population. So, at the end, the model is treated almost as an MNL, making in some sense the extra estimation effort worthless.

Simulation. This method has been applied to construct confidence intervals (Ettema et al. 1997; Armstrong et al. 2001) as we saw in the previous section, and has been also used to derive WTP values from ML models by Hensher and Greene (2003) and Espino et al. (2006). It is a first approach to constructing a WTP distribution over the population using information neglected by the previous method. An important feature of this method is that no assumptions are needed about the resulting distribution of parameter ratios.

However, one problem with the method is that it can yield rather large spreads for the distributions, as the simulation process may involve drawing values that are close to zero. Hensher and Greene (2003) discuss the effect of removing parts of the simulated distributions of WTP, and compare this action with constraining the distributions. But in relation to the validity of this method, the real issue is not whether or how to constrain the distribution to make it theoretically correct. Hensher and Greene (2003) acknowledge that the mere fact of applying statistic distributions – which are already analytical constructs – to behavioural parameters governed by an unknown logic, makes constraining (or removing parts of) the parameters or WTP distributions neither better nor worse than an unconstrained distribution, unless there is a theoretical rationale behind.

A consistent rationale for cutting off the tails of the distributions is that there are no *real* people with such extreme values to fill in the tails we are cutting. So, when applying this method, the analyst must remember that the final goal is to estimate WTP values for the sampled population, and for sample sizes smaller than infinity, this is a finite set of values. Therefore, the real problem with simulating WTP distributions from sampled values is not how to constrain them in the right way, but the fact that we are simulating countless numbers of values for people who do not even exist.

Log-Normal Distribution for WTP. The use of Log-Normal distributions for the parameters over the population in ML models has been proposed by many authors, as this constrains their signs to be consistent; further, it yields an analytical expression for the resulting WTP distribution since the ratio of two Log-Normal distributed variables is also Log-Normal.

Consider a random variable x such that $x{\sim}N(\mu_x, \sigma_x)$. Then, a variable defined as $X = \exp(x)$, has a Log-Normal distribution with mean $\exp(\mu_x + \sigma_x^2/2)$, and standard deviation given by $\exp(\mu_x + \sigma_x^2/2)\cdot\sqrt{(\exp(\sigma_x^2) - 1)}$. Now consider the ratio of two Log-Normal variables, say X/Y, then:

$$\frac{X}{Y} = \frac{\exp(x)}{\exp(y)} = \exp(x - y) = \text{WTP}$$

where

$$\text{WTP} \sim \log N\left(\exp\left(\mu_{wtp} + \frac{\sigma_{wtp}^2}{2}\right), \exp\left(\mu_{wtp} + \frac{\sigma_{wtp}^2}{2}\right) \cdot \sqrt{\exp(\sigma_{wtp}^2) - 1}\right) \tag{18.40}$$

As x and y are Normal variables, their difference is also Normal with:

$$(x - y) \sim N\left(\mu_x - \mu_y, \sigma_x^2 + \sigma_y^2 - 2\sigma_{xy}\right)$$

Since we are assuming independent parameters, the covariance term disappears. Then, replacing the above expression in (18.40), we get that an expression for the Log-Normal WTP distribution is:

$$\text{WTP} \sim \log N\left(\exp\left((\mu_x - \mu_y) + \frac{\left(\sigma_x^2 + \sigma_y^2\right)}{2}\right),\right.$$

$$\left.\exp\left((\mu_x - \mu_y) + \frac{\left(\sigma_x^2 + \sigma_y^2\right)}{2}\right) \cdot \sqrt{\exp\left(\sigma_x^2 + \sigma_y^2\right) - 1}\right) \tag{18.41}$$

This expression can be used to calculate cumulative proportions and confidence intervals. However, both Hensher and Greene (2003), and Sillano and Ortúzar (2005) found that in the case of this distribution, there are considerable differences between taking the ratios of the means and the means of the ratios; this brings new evidence to the discussion. The ratios of the means do not yield the WTP for the mean individual household, but for a virtual one who perceives the mean marginal utility of the population for each attribute (i.e. an *individual household* who has the mean parameter for, say, travel time and also the mean parameter for cost). The existence of this household is not a fact but a mere coincidence, and even if it existed, its WTP value would not be representative.

An analytical explanation for this difference can be easily derived. Consider two independently distributed Log-Normal structural parameters θ_t and θ_c with associated Normal means b and c and variances s_t^2 and s_c^2, respectively. The ratio of their means can be expressed as a function of the coefficients of the underlying Normal distributions:

$$\left.\begin{array}{l}\bar{\theta}_t = \exp\left(b + \dfrac{s_t^2}{2}\right) \\[2mm] \bar{\theta}_c = \exp\left(c + \dfrac{s_c^2}{2}\right)\end{array}\right\} \quad \frac{\bar{\theta}_t}{\bar{\theta}_c} = \exp\left(b - c + \frac{s_t^2 - s_c^2}{2}\right) \tag{18.42}$$

And from (18.41) we can express the mean of the WTP log-Normal distribution in terms of the same coefficients:

$$\overline{wtp} = \exp\left(b - c + \frac{s_t^2 + s_c^2}{2}\right) \tag{18.43}$$

From here we can derive the relation:

$$\overline{wtp} = \left(\frac{\overline{\theta_t}}{\overline{\theta_c}}\right)\exp(s_c^2) \tag{18.44}$$

Thus, the ratio of the means of Log-Normal parameters is equal to the mean WTP value deflated by the exponential of the variance of the Normal distribution underlying the Log-Normal cost coefficient (i.e. the parameter in the denominator of the WTP ratio). In other words, the WTP mean and the ratio of parameter means are scaled by a proportionality factor which, by the way, is fixed for the model (i.e. the three attributes considered in this example are scaled by the same factor). The logic of this effect is the following: the larger the variance of the cost coefficient, the larger the portion of the denominators' mass that will be near to zero, and hence the mean WTP will grow larger.

In conclusion, the use of Log-Normal distributions for valuation purposes is not recommended. Their wide tail tends to give extremely large WTP values with high probabilities, yielding large portions of cumulative mass close to zero and distorting the analysis. Its main appeal is that it allows constraining the parameters to be strictly positive (for negative coefficients, they enter with a negative sign in the utility formulation). However, as we saw in Example 8.8, the relative ease of estimation with Normal distributions may also lead to structural parameters with correct theoretical signs. Thus, it is not worthwhile to undergo the effort to estimate the model with Log-Normal distributed parameters, since even if the individual values show a large portion of incorrectly signed people, the right course of action should be to investigate them for consistency and perhaps remove them from the sample.

Fixing the cost coefficient. Another method that has been used considers fixing the cost coefficient and thus letting the WTP distribution follow the distribution of the numerator; if the parameter in the numerator follows a Normal distribution the resulting WTP distribution would be simply given by:

$$\left. \begin{array}{c} \theta_t \sim N(\mu_t, \sigma_t) \\ \theta_c \quad \textit{fixed} \end{array} \right\} \quad \frac{\theta_t}{\theta_c} \sim N\left(\frac{\mu_t}{\theta_c}, \frac{\sigma_t}{\theta_c}\right) \tag{18.45}$$

Revelt and Train (2000) cite three reasons for fixing the cost coefficient:

- it effectively solves the problem under discussion;
- the ML model tends to be unstable when all coefficients vary over the population, and identification issues arise (Ruud 1996); and
- the choice of an appropriate distribution for the cost coefficient is not straightforward since the Normal and other distributions allow for positive values, and the Log-Normal is both hard to estimate and gives values close to zero, as discussed above.

Notwithstanding, there is one key drawback to this method that needs attention.

Example 18.3 Table 18.4 compares estimates of WTP derived from a MNL with those of a ML model with a *fixed* cost coefficient (Sillano and Ortúzar 2005). As can be seen, the means of the resulting WTP distributions (for travel time to work, travel time to study, and an environmental attribute, in a residential location choice experiment) are considerably higher than the MNL point estimates, a result that has also been reported by Revelt and Train (1998).

Hensher (2001a,b,c) have also found higher mean WTP values for heteroskedastic and autoregressive specifications; this could indicate that ML models (with any error structure) tend to overestimate WTP values. But these works did not explore the possibility that by constraining only part of the error structure they could be causing an unbalanced growth in the model coefficients, hence producing higher welfare estimates.

Table 18.4 Mean WTP estimates for fixed cost coefficient ML and MNL

		Willingness-to-pay	
Attributes		MNL	ML
Travel time work	Mean	36.0	51.0
(Ch$/min)	Std. dev.		54.8
Travel time study	Mean	22.0	31.0
(Ch$/min)	Std. dev.		47.5
Days of alert	Mean	124 362	126 160
(Ch$/DA per year)	Std. dev.		107 430

In Example 8.8, we explained why larger means for ML parameters, in relation to the MNL, should be expected because of the extra variance explained by the random parameters; we have also discussed above possible reasons for obtaining uneven enlargement factors. The fact is that constraining a taste coefficient to be fixed over the population, may make it grow in a less-than-average proportion (i.e. the parameters that are allowed to vary grow more than the parameters that should vary over the population, but are constrained to be fixed). Note that this is not the case of parameters that are eventually fixed because their standard deviations were originally estimated and found not significant (see the discussion by Sillano and Ortúzar 2005 on this issue).

Willingness-to-pay estimation from individual level parameters. In Section 8.6, we discussed two forms to estimate individual-level parameters for ML models, both of which involved the use of Bayesian statistics. The estimation of individual taste parameters eliminates the issue of analysing the WTP distribution resulting from the division of two random variables over the population. Instead, individual-level WTP point estimates can be computed, along with their individual confidence intervals.

Example 18.4   Figure 18.8 present frequency charts for the valuation of the two attributes, whose distribution was shown in Figure 8.5. The charts show high concentrations on each edge of the axis, accounting for extremely large positive and negative WTP values. It is important to mention that notwithstanding the sign of the WTP value, all implausibly large values belong to individual households in the sample with non-significant *cost* (i.e. the rent in the case of this location choice example) parameters. That is, the denominator of the WTP ratio is statistically close to zero, yielding an inordinately large value.

It is also important to mention that in Figure 18.8a, the only negative WTP values are also associated with extreme cases. In fact, they correspond to the few observations with an incorrect sign for the *rent* parameter, but as it was also not significant in those cases, it caused the ratio to grow disproportionably.

This suggests paying special attention to observations with a cost parameter statistically equal to zero. In these cases, the WTP ratio grows to implausibly large monetary valuations for reductions in the corresponding attribute. On the other hand, as the individual household

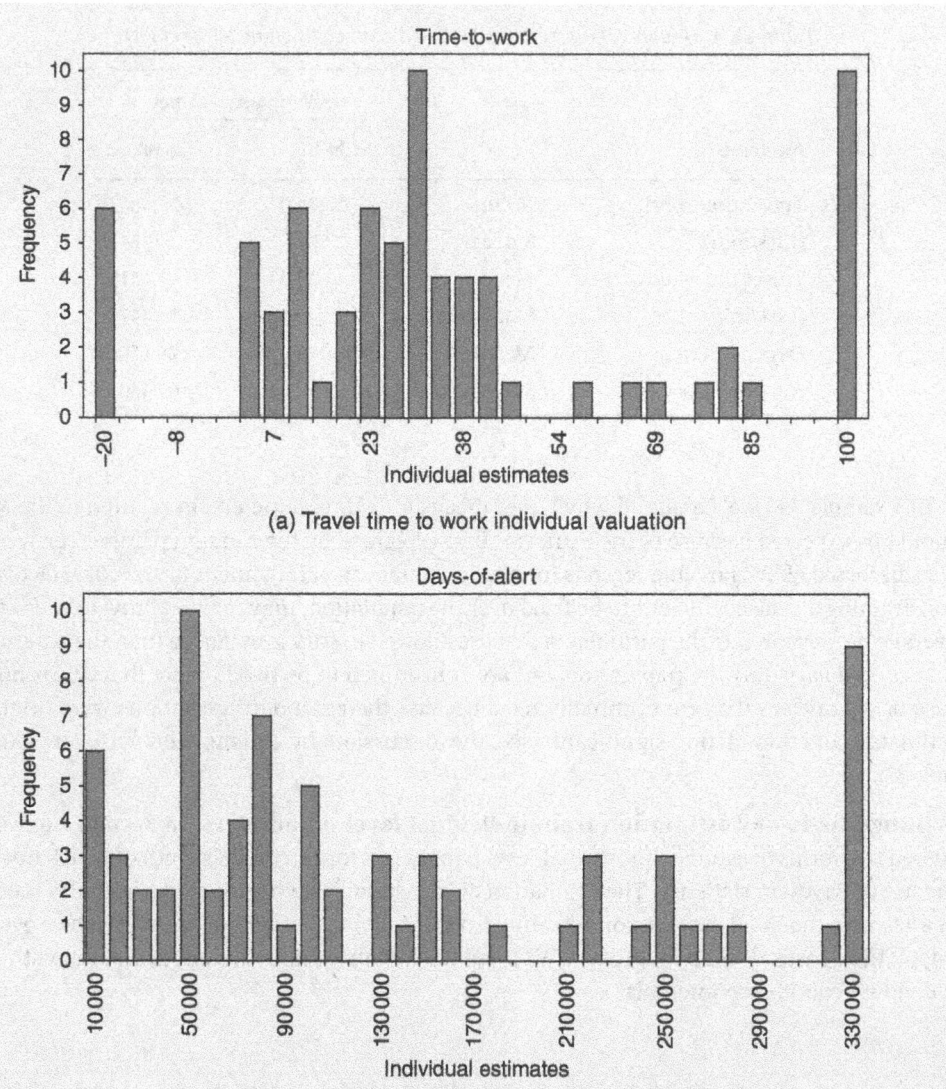

(a) Travel time to work individual valuation

(b) Days of alert individual valuation

Figure 18.8 Individual level WTP point estimates

does not place *any* weight on the cost attribute, we can debate whether those observations do not consider the cost attribute at all (i.e. similar to *attribute non-attendance* as we saw in Section 7.7.6), or whether the weight they place on it is negligible in relation to the rest of the attributes. If the latter is the case, the interpretation of an extremely large WTP value would be correct. If not, monetary valuations cannot be computed for these observations. Further theoretical developments are necessary to define criteria to help answer this question, but note that this is case specific (i.e. it depends on the survey design, the underlying microeconomic model, and the characteristics of the valued attributes).

The estimation of individual-level WTP values is as close as we can get to the correct method of valuation inference from ML models. However, for project evaluation and cost-benefit analysis, we usually need data for different groups or strata in the population. One beauty of individual-level data is that an analysis at the level of a given stratification can simply be performed by averaging the WTP values of those individuals present in each stratum along with their cluster variance. In fact, thresholds (or strata boundaries) can even be defined *ex-post* in order to minimize the variance of the WTP values across the group, and hence be able to define more homogeneous segments for project evaluation and detailed analysis. Sillano and Ortúzar (2005) discuss this and other points in more detail.

18.4.4.5 The Transfer Price Approach

In the context of travel demand analysis, *transfer price* has been understood as the amount by which the cost of one option would have to be varied to equalise its overall attractiveness with that of another predefined option (see Bonsall 1983).

A typical application of the method involves asking individuals, for example, how much the fare of their currently preferred option should increase to persuade them to switch to another alternative. It is clear that an important problem of the technique (in common with other forms of SP analyses, and in particular contingent valuation, which is the closest one) has to do with the reliability that the analyst can associate with such a dataset. On the other hand, a strong advantage of the method, if it works, is that it allows one to ascertain not only the direction of individual preferences but also the difference (in preference terms) among the various available options. Thus, in theory, and in common with other SP studies, less data are required than for an RP study to obtain a model of similar accuracy. We will not attempt to discuss the method in detail here, but interested readers are referred to Gunn (1984) for a good discussion of its advantages and limitations, in particular its general inconsistency with conventional random utility theory.

Example 18.5 Consider a random utility model such as (7.2) in a binary-choice situation and assume that the transfer price (TP) corresponds to the difference between the utility of the chosen alternative (U_c) and the other (U_r), that is, it represents the increment in the cost of the chosen alternative that would made the traveller indifferent. Thus, we have:

$$TP = U_c - U_r$$

However, the expected value of ($U_c - U_r$) is precisely the difference in representative utilities ($V_c - V_r$); so assuming these to be linear in the parameters, as usual, we can form the following linear regression system:

$$TP(observed) = \theta_1(X_{1c} - X_{1r}) + \theta_2(X_{2c} - X_{2r}) + \dots$$

which should allow us to estimate the unknown parameters θ knowing the attributes $\mathbf{X}$ for both options. Furthermore, different values of time for *time savers* and *cost savers* may be calculated with this method (see Lee and Dalvi 1969).

One important problem, first noted by Hensher (1976), concerns the treatment of *habit* in TP models. Gunn (1984) shows that specifications using TP as a dependent variable but restricting its sign (i.e. by modelling the options separately or by switching the observable characteristics to reflect the difference between chosen and rejected option), cannot easily be made consistent with conventional random utility theory (see also the discussion in Chapter 8).

18.4.4.6 The Stated Preference Approach

Stated preference (SP) methods, as discussed in depth in Chapter 8, have become the most used method to estimate subjective values of time in recent years. For example, in their final report to the Department of Transport, the consultants for the 1994 UK value of time study noted that:

> *evidence has amassed during the last ten years sufficient to have confidence that a well-mounted SP survey with a well-designed questionnaire and proper analysis can yield reliable results, though this is preferable with a supporting base in RP data if actual forecasts of levels of demand are to be made* (Accent and HCG 1996).

An interesting discussion related to the use of SP methods in location choice and the implications for the value of time is given by Pérez et al. (2003).

We do not attempt to review the large number of SP-based value of time studies reported in the literature during the last few years, but only remind readers of the recent European reports on national studies discussed above, as some brought in a lot of innovations. Other studies have considered important issues such as the estimation of randomly distributed values of time (Gopinath and Ben-Akiva 1995), or estimating time values using SP data allowing for interaction effects (Ortúzar et al. 2000c; Rizzi and Ortúzar 2003). These studies cover new areas and use state-of-the-art models and specifications, as discussed in Chapters 7 and 8.

18.5 Valuing External Effects of Transport

18.5.1 Introduction

In many countries of the developed world, WTP methods have been used for the monetary valuation of a range of external effects of transport such as accidents, pollution, noise, visual intrusion, and amenity loss. Several examples have been compiled by Hansson and Markham (1992), OECD (1994a), Mauch and Rothengatter (1995), Maddison et al. (1996), Friedrich et al. (1998), ECMT (1998), and Rizzi and Ortúzar (2015). The thrust of this work has been to establish the full social costs of transport as a basis for efficient pricing in this sector and to extend the scope of social cost-benefit analysis (SCBA) for improved project appraisal.

Although there has been much academic enthusiasm for the monetary valuation of these non-market goods, this has been challenged on principle and practical grounds; a good expression of the nature of this dissent can be found in Adams (1992) and Whitelegg (1993). However, there is considerable force in the argument that, while empirically well-founded monetary values are difficult to achieve and may be valid only in specific contexts, their expression will help to ensure that externalities are not marginalised or understated in project and programme planning. This is particularly important in the context of road investment appraisal and resource allocation for accident countermeasures and pollution control strategies. Indeed, at the end of the previous century, the attribution of monetary values to accidents of different severity was an important stimulus to increasing resources towards road safety and establishing priorities over different safety measures in many countries around the world (Allsop 1999).

Also, as part of the expectation to respond to increasingly challenging environmental standards and targets, many national and local governments have been establishing, extending, or refining databases relating to accidents, noise, and a variety of gaseous pollutants. These are intended for use in monitoring changes over time and evaluating fiscal, regulatory, and investment policies.

In many developed countries, this process is already established, while for most developing countries, it is currently at a relatively early stage of development, and the scope and quality of such data vary considerably (Chesnut et al. 1997).

Now, although sufficient evidence was amassed over the 1990s, the economic costs of accidents, noise, and pollution were all subject to considerable variation, partly due to the different sources of data and methods of measurement. For example, Quinet (1994) noted that '*for all forms of transport pollution*', estimates based on WTP provided the highest numerical values of a statistical life (VOSL), a feature long known in the case of accident costing. In fact, the United Kingdom government replaced the human capital approach to fatality costing by the WTP approach in 1988, and this was extended to non-fatal accidents in 1994, drawing on the national studies of Jones-Lee et al. (1985, 1992). However, in the case of fatalities, the government was not persuaded to accept the considerably higher values emerging from the former WTP study and instead it implemented a compromise value (Dalvi 1988), thereby exercising an element of caution in the face of a radical change of methodology (Department of Health 1999).

Until the mid-1990s, monetary valuation of environmental externalities was seldom given official support (OECD 1994b; Lee and Kirkpatrick 1996). However, the situation changed rapidly afterwards, and surveys of *official* transport appraisals (Bristow et al. 1998; DETR 1998) suggested that monetary values for noise, air pollution, and (to a lesser extent) barrier effects, were increasingly used in many European countries by the new century. However, the appraisal of road investments undertaken or supported by national authorities still involves a limited cost-benefit analysis (with unit monetary values confined to savings in time, accidents, and operating costs), applied in conjunction with an environmental and socio-economic impact assessment. For example, in a survey of United States governmental agencies responsible for highway developments, Waters (1992) noted that relatively few embraced a sophisticated SCBA, preferring rather simpler needs-based or cost-effectiveness approaches.

Although several academic studies urged the extension of the SCBA framework to embrace a wider range of impacts (Bateman et al. 1993; Willis et al. 1998), governments remained cautious about its formal extension to pollution, noise, visual intrusion, amenity loss, and ecosystem damage. This is partly because of gaps in knowledge, both in impact assessment and economic valuation (Mullen 1997), and partly because the site-specific nature of some of the impacts inhibited the use of standardised unit values. These are universal concerns.

It remains a considerable research challenge to integrate environmental impact assessment, cost-benefit and multi-criteria analysis traditions (Commission of the European Communities 1994; OECD 1994c; Lee and Kirkpatrick 1996; Nardini 1997) in a context in which environmental objectives are assuming increasing importance. Efforts include new ways of assembling and presenting qualitative and quantitative information to minimise bias against non-monetary valuation items and the construction of appraisal frameworks that establish a 'level playing field' between different modes and allow transport problems to be addressed with less emphasis on highway solutions (Price 1999; Glaister 1999). Monetisation will increasingly be applied in multimodal settings with heavier demands on data.

18.5.2 Methods of Analysis

There are several taxonomies for valuation methods available in the literature and a much larger economic discussion than we could attempt here (ECMT 1996; Mauch and Rothengatter 1995, Nash 1997; Verhoef 1994). In this section, we will just quickly review, for the sake of completeness, two methods, the Human Capital approach and the Contingent Valuation method, as it is fair to say that currently the method which clearly dominates the field is our old acquaintance, the Stated Preference approach, reviewed in Chapters 3 and 8 (see Rizzi and Ortúzar 2003 for a well-designed methodology that has been used already as far as Australia and Norway).

18.5.2.1.1 Human Capital Approach It is based on the assumption that the value of an individual is what she produces, and this is usually measured by the gross salary perceived at work (i.e. before taxes in order to include the government and hence society). If the person dies, this production is lost. This, 40-year-old approach (Landefeld and Seskin 1982), postulates that the value of preventing the death of an individual aged t is equal to the net present value (PV_t) of her expected earnings for the rest of her life:

$$PV_t = \sum_{i=1}^{T-t} \frac{\pi_{t+i} E_{t+i}}{(1+r)^i} \tag{18.46}$$

where π_{t+i} is the probability that the individual will survive from age t to age $t+i$, E_{t+i} are the expected earnings of the individual at age $t+i$, r is the discount rate, and T is the retirement age.

The method has been heavily criticised as being the antithesis of the conventional premises of welfare economics. Discussion has also touched on how to value production of individuals that are not in the labour market (i.e. housewives), or what discount rate should be used to calculate PV (a sensitive issue in the case of children and young adults); classical rates ranged from 6% to 10%, but nowadays values below 5% are preferred to avoid punishing any age stratum in excess. Table 18.5, taken from Landefeld and Seskin (1982), shows the effects of age and discount rate on the human capital value of life.

Table 18.5 Net present value by age and discount rate

Age group	Net present value (US$)		
	Discount = 2.5%	Discount = 6.0%	Discount = 10.0%
1–4 years	761 047	205 101	59 859
20–24 years	967 221	534 799	320 114
40–44 years	625 508	454 972	338 232
65–69 years	47 506	40 886	35 304

Due to difference in wages, if applied strictly, the human capital approach would yield smaller values for the life of women than for the life of men, and smaller values may also have to be assigned to non-Caucasians. Zero values would be assigned to retired individuals or to those incapacitated either by illness or for any other reason. For this reason, as in the case of the value of time, the proper methodology should be to estimate a single equity value to be used in project evaluation. Notwithstanding, it is widely accepted that, as this approach does not consider pain and suffering by the victim and their relatives, its values constitute an underestimation of the true value of the social loss, and, therefore, its use should just allow us to establish a lower bound for the value of life.

18.5.2.2   Contingent Valuation

As mentioned in Section 2.4.1, this is a technique for eliciting values for goods that are not or cannot be bought and sold in a normal market. People are asked for their value of a good, contingent on a market existing for it. A hypothetical market is created and described to the respondent, who is then asked to make a market (purchase) decision. Contingent markets define the good or amenity of interest, the existing level of provision, possible increments or decrements, the institutional structure under which the good is to be provided, and the method of payment. Mitchell and Carson (1989) provide a comprehensive explanation of the theoretical foundations of the contingent valuation (CV) technique, methodological issues and practical application. Overviews are provided by Bateman and Turner (1993) and Hanemann (1994).

CV questions can ask for people's WTP values or for their willingness-to-accept (WTA) compensation values. The WTP value is the income an individual would forego to achieve an increase in the level of a good and remain at the same level of utility, and WTA is the inverse. A problem here is property rights; WTP assumes that these belong to the consumer and WTA the contrary. However, WTP is most commonly used because it resembles familiar consumer purchase decisions (although in cases of environmental deterioration, for example, WTA should be the correct theoretical value to obtain). Thus, CV attempts to measure the change in income necessary to offset a change in amenity, while leaving utility unchanged.

There are three main methods of eliciting CV values:

- open-ended questions, where respondents are just asked how much they would be willing to pay for a good;
- iterative questions, where respondents are asked first whether they would be willing to pay a specified amount; if they answer yes, the question is repeated with small increments in the cost until they say no, then the cost is reduced by smaller decrements until a final figure is reached (and vice versa, if they start by saying no to the original figure);
- referendum questions, also known as dichotomous choice questions, where respondents answer yes or no to a WTP question with a specified payment; the double-bounded dichotomous choice question has an extra question after the first.

The referendum approach is the most attractive because it presents scenarios similar to those that respondents, as consumers, encounter in day-to-day market transactions. The payment mechanisms for actually buying or selling the good can include property taxes, income or sale taxes, utility bills, community charges, fares, entry fees, subscription schemes, or even an abstract instrument. Since its early application in the 1970s the CV approach has been used to value a wide range of non-market goods. Carson et al. (1995) provides a bibliography of CV studies containing 1,400 references, indicating the wide applicability of the method.

On the other hand, a strong critical assessment is provided in a collection of conference papers edited by Hausman (1993) and Diamond and Hausman (1994), who believe that the evidence suggests that CV surveys do not measure the preferences they attempt to measure, and that changes in survey methods are unlikely to alter this. However, the method is still popular and has certainly been used in many important studies related to valuing externalities in the transport sector (e.g. Jones-Lee et al. 1992; Feitelson et al. 1996).

In the case of road accidents, the two most feared outcomes are to die or to become a severely injured victim. Not surprisingly, road project appraisal practice in most industrialised countries has given those two outcomes the highest economic values, with fatalities being more valued than severe injuries.

Conventional practice until the end of the 1990s was to elicit WTP values for preventing both fatalities and severe injuries using contingent valuation (CV) and risk-risk trade-offs (or standard gambling) methods (Jones Lee et al. 1993, 1995). But CV basically involves a trade-off between money and risk expressed as a tiny probability. Usually, a question is posed to respondents asking for their willingness to pay to buy some special safety device designed to reduce *only* the likelihood of a particular outcome of a road crash; for example, the likelihood of becoming a fatal victim or the likelihood of suffering – say – a head concussion.

The risk–risk trade-off, on the other hand, demanded respondents to exchange the risk of one likely trauma outcome of a road crash for another one. Usually, respondents had to assume they were already a road accident victim suffering a particular trauma; then they were offered the alternative of a medical intervention that, with probability p, would return them to their health state before the crash and, with probability $1 - p$, they would end up in a health state worse than the current hypothetical one – this state was usually death. Respondents had to state the value of p that would make them undertake the medical intervention. Hence, it was possible to *chain* different risks with the risk considered in the CV survey, allowing the researcher to monetise risks other than those considered in the CV exercise. The reader may ask, why not use the CV to put a monetary value on all types of risk. The reason was that money-risk trade-offs were deemed unstable, so researchers would rather avoid the overuse of CV.

18.5.2.2.1 The Stated Choice Approach Although the above methods may work as a first empirical approximation, they do not address the issue under analysis (i.e. risk of a road accident) in its proper dimension. First, the road safety schemes an authority wants to evaluate are of a public-good nature. It is about reducing a public risk; that is, a risk that displays no rivalry in consumption since the benefits of the scheme accrue to all drivers on that particular stretch of road. The safety device considered in the CV approach is a private good, not a public one (but this could be corrected by substituting a public good for the private good, and this critique would lose substance). Second, and more important, a road safety scheme is about decisions on *ex ante* risk management, in the sense of what can be done to prevent road crashes or mitigate the impact of a road crash. However, the risk–risk trade-off is akin to a post-trauma alternative medical treatment, associated with decisions to be taken after the accident has occurred. This information should be more relevant for health insurance companies than for public road agencies.

So, if WTP values are required for appraising road safety projects stated choice (SC) methods are a superior elicitation approach (Rizzi and Ortúzar 2003, 2006; Iragüen and Ortúzar 2004; Hojman et al. 2005; Hensher et al. 2009; Veisten et al. 2013; Flügel et al. 2015; González et al. 2018). This technique places the respondent in the correct context, for example, having to choose between two routes with different levels-of-service (i.e. travel time, toll, number of fatalities, and number of severely injured victims). This way, people implicitly reveal WTP not only for safety improvements, but also for travel time savings, probably the most important trip attribute. The quota of increased realism afforded by the SC approach is necessary to uncover the value people actually place on safer roads. It also avoids the problem of *embedding* (Sælensminde 2001), since both the reduction of fatalities and severely injured victims are valued, together with travel time, in an integrated framework where the individual is always conscious of her budget constraint.

As a caveat, SC methods are not without problems. As with CV, the hypothetical nature of the choice scenarios is the main disadvantage of any stated preference survey, as we have discussed in Chapters 2 and 8. However, we strongly believe that SC surveys outdo conventional CV surveys with respect to increasing realism.

Example 18.6 Hojman et al. (2005) designed a route SC survey for car trips between the cities of Santiago and Valparaíso, and another for car trips between Santiago and Rancagua (i.e., the capital and two important Chilean cities, respectively). The distance between Santiago and these two cities is around 120 km via Class A roads (Routes 68 and 5-South, respectively), which are fairly safe for Chilean standards.

After a detailed experimental design phase, including focus groups and two pilots, the final survey instrument contained five parts. The first asked for driving experience on interurban roads and on Routes 68 and 5 in particular. A question was included about the last time the respondent drove on any of these routes; if the answer was more than a year ago, the survey ended. The second part included the choice experiment itself (which varied according to the purpose of the trip and the route where the driver had more experience), and the third part included different types of questions, some related to the choices themselves and others to road crash experience and attitudes. The fourth part enquired about socio-economic data, and the fifth allowed respondents to give their personal definition of what constitutes a severe injury.

When respondents are asked to examine a series of choice situations, it is important to set up a realistic context. According to the answer given in the first part of the survey, people were asked to consider that they had to travel from Santiago to Valparaíso or to Rancagua. Invoking a recent trip to either destination was a way to reduce to a minimum the problem of not including as a third alternative the option of not doing the trip at all (see Section 2.4.2.6). The trip to either city had the following characteristics (the underlined parts varied across contexts):

➤ you drive your car;
➤ you travel during a *regular weekend (without extra holiday days)*;
➤ you pay for the total cost of the trip, including the toll;
➤ you start the *trip in the morning and is raining*;
➤ you have to choose between two routes (both are similar to Route *68*), taking into account the following four elements: (i) toll charge, (ii) travel time, (iii) number of fatal victims per year and (iv) number of severely injured victims per year'.

A short explanation was also given on what was considered a fatal victim and a severely injured victim. In explaining the latter, several road-crash-related severe traumas were mentioned, so respondents focused their attention on these types of traumas. These definitions were analysed in the focus groups and pre-test surveys. Finally, statistical data was also given about the number of fatalities, severely injured victims, and total annual flows on Routes 68 and 5 during the previous year. Nothing was said about any accompanying member within the car; hence, a question was asked whether or not the driver was considering travelling alone or with someone else.

As can be seen, the context was clearly defined: the day, the time of day, and the purpose of the trip were all specified; it was assumed that those who answered the questionnaire were the drivers, and they were also assumed to pay for the toll. As many highways operate under a private toll system in Chile people are already familiar with toll charges. In particular, as safety is related to a particular trip taken by the respondent, there was little room for an altruistic choice.

So, as it was in the best interest of respondents to give a truthful answer, this way they managed to increase the *realism* of the hypothetical choice context to a plausible maximum, reducing the possibility of strategic bias.

The statistical design used allowed, in principle, to estimate different parameters for the safety variables of each alternative route. One parameter was considered for a lower number of crashes and another for higher values. The aim was to test the *prospect theory* hypothesis (Kahneman and Tversky 1979) that increases in the level of danger are valued differently (once the sign is taken into account) than improvements in the level of safety; thus, it was expected that the modulus for higher numbers of crashes would be greater than that for lower numbers. However, as this result did not show up in the pilot study phase, they considered only one parameter in the final survey.

The survey was programmed on a web page following the good results obtained in previous experiences (Iragüen and Ortúzar 2004). To recruit respondents, key officials at several institutions (both public and private) who accepted to cooperate with the study were contacted. Then, these officials sent emails to employees, enticing them to participate. Hojman et al. (2005) obtained approximately 500 answers, 250 for each route, but did not calculate the response rate since (for confidentiality reasons, they did not register the e-mail of respondents and did not enquire how many people were contacted at each institution). Most individuals in the survey were middle- to high-income people by Chilean standards, as car ownership is low compared to European or US levels.

Using this data, a variety of models were estimated – ranging from MNL models allowing for systematic taste variations to ML. Hojman et al. (2005) concluded that the WTP values estimated from their data were between 10 and 15 times higher than the values used in social project evaluation in Chile at the time (computed from the human capital approach). Their values were also compared with values obtained in other countries using both similar and different methods, finding that – in general – the Chilean values differed from the others in more ways than could be accounted for by income differences. In fact, they concluded that:

> our values should also be more relevant for road planners in developing nations than transferring values from industrialised nations (i.e. Miller 2000 derived a 'value of risk reductions' for Chile in a range of two to three times higher), since accounting for differences in risk aversion is by no means an easy task.

Exercises

18.1 Consider the following simple econometric model to determine car ownership as a function of income:

$$P_0/(1 - P_0) = \alpha I^{\beta}$$
$$P_2/[0.8 (1 - P_0) - P_2] = 0.09 \exp (0.751)$$
$$P_0 + P_1 + P_2 = 1$$

a) Calibrate the model using the data in Table 18.6 (*Hint*: do it graphically)

Table 18.6 Data for Exercise 18.1

I	P_0	P_1	P_2
1	0.60	0.35	0.05
2	0.40	0.50	0.10
3	0.25	0.55	0.20
4	0.20	0.45	0.35
5	0.15	0.35	0.50

b) Indicate what proportions of 0, 1, and 2 or more cars the model predicts for an annual income of six monetary units.

18.2 Table 18.7 presents the results of a TP survey made on the sample of eight individuals in Exercise 9.3. TP indicates the reported increment in the monetary cost (expressed in time units after deflating by income) of the currently chosen mode that would leave each individual indifferent to both alternatives. The study assumed that only time (t) and cost/income (c) were relevant variables.

Table 18.7 Transfer price data for Exercise 18.2

Individual	Chosen option	TP	t_1 (min)	t_2 (min)	c_1 (min)	c_2 (min)
1	1	8.0	47.5	83.2	14.8	7.0
2	1	6.5	30.2	45.0	10.4	5.0
3	1	2.5	22.0	30.4	12.6	4.0
4	2	0.5	45.0	50.6	8.2	5.0
5	2	1.5	15.3	20.5	50.0	17.0
6	1	8.5	34.8	50.2	55.0	35.0
7	2	130.0	65.5	100.5	200.3	53.5
8	2	6.0	12.0	14.0	44.6	17.0

a) Use the data to estimate the individuals' subjective value of time. Discuss the role, size, and sign of the intercept of the TP linear regression equation (*Hint*: if you do not have available a calculator with a linear regression facility, do it graphically assuming the coefficient of time, θ_t is known and equal to -0.03).

b) If the revealed preferences parameter for the time variable is indeed -0.03 and the mode specific constant of option 1 is 1.35, estimate the SVT using another method. Compare your results and discuss.

19

Pricing and Revenue

19.1 Pricing and Welfare

The overarching objective of transport planning should be to improve the welfare of the residents and travellers in a region sustainably. This goal typically requires reducing emissions and accidents and managing congestion whilst supporting a more equitable accessibility pattern. Authorities have many tools available to achieve these broad objectives: transport and land use policies, investment in infrastructure, and measures to influence the price of travel, such as subsidies and taxes.

Price and cost are terms used interchangeably in our normal day-to-day conversations, but when it comes to economics, each one takes on a separate meaning. *Cost* is the amount of resources incurred in producing a particular good or service, in our case, the resources involved in providing and managing the infrastructure and modes of transport required to deliver a travel service; this should also include any costs imposed on others when using the service. *Price* refers to the money charged to prospective customers for the provision of goods and services; users must pay an agreed amount to get them. In the case of transport systems, the traveller directly perceives prices, but the costs incurred to provide the service may not be fully understood and internalised. Travel service prices may be fixed in advance or dynamically change over time to manage demand; a typical example of variable pricing is the adjustment of airline fares over time by most companies to ensure high occupancy of their planes.

Price is one of the key attributes that travellers consider when selecting a mode of transport or the best route to reach their destination. This chapter explores the impact of price in seeking to maximise welfare and draws some lessons from the involvement of the private sector in the delivery of transport infrastructure and services.

Free market economists postulate that social welfare should be maximised through the operation of the market if all goods and services are properly priced, consumers have perfect information and are unconstrained in their choices. But this is a tall order, as prices are often distorted and perfect information is an impossible aspiration in most cases.

There are two main problems with transport prices and costs: (i) they do not represent well the cost of delivering services and therefore are sub-optimal; (ii) there are costs that are not perceived and, therefore, are not relevant to a particular travel choice. Both these limitations mean that consumer choices cannot be aligned with the broad objectives above.

Mobility provides many benefits, but it also incurs significant costs that are not perceived by the traveller; examples include the pollution, accidents, and congestion delays forced on other drivers, and in the case of greenhouse gases (GHG), the whole world and future generations. These *externalities* distort the market and result in over-consumption of under-priced resources and lower use of modes that contribute more to welfare (Bonsall and Willumsen 2013). The problem is compounded because transport is co-produced by consumers (who buy vehicles and fuel, maintain them, etc.), businesses (that provide parking, operate transport services, and develop land), and governments (that supply infrastructure, parking facilities, and regulate modes and land use development).

Authorities would like to influence the pricing of transport services to correct for negative externalities and support the adoption of modes of travel that minimise them and offer some positive benefits, like better health in the case of walking and cycling.

19.2 Correcting Prices for Externalities

As we saw in Section 1.3.2, externalities are generated when the production of a good or service causes costs or benefits to others. In the field of transport, all modes generate some negative externalities; for example, air travel generates noise and air pollution which impacts people living near airports and under flight paths, not to mention the greenhouse effect of emissions in the upper atmosphere. Similarly, each additional vehicle on a crowded road imposes a quantum of delay on all other vehicles and also adds to the noise and emissions generated. More generally, the negative externalities of transport services are wide-ranging and include environmental impacts such as local air pollution, GHG emissions, noise and light pollution, safety hazards, community severance, and congestion.

According to the International Energy Agency (https://www.iea.org/), the transport sector generates between 20% and 25% of the GHG (mostly CO_2 in this case) and within the sector, road traffic represents around 74% of the emissions; rail contributes about 1% and aviation and shipping generate about 12% each; pipelines and other modes contribute the rest. Reducing these emissions to reach carbon neutrality by 2050 is a key objective in most countries.

Focussing on road users, part of the problem is that they do not perceive the full costs of their journeys. They only perceive direct costs and charges (fuel, tolls, and parking) ignoring not just the unpriced externalities but also the cost and management of infrastructure and, in most cases, semi-fixed costs like the depreciation, maintenance, and insurance of their vehicles. The incomplete perception of the full costs of their choices leads to overuse of modes, such as the private car, that generate significant externalities. To correct this distortion, it is necessary to expose travellers to the true (social) marginal costs of their journeys; however, calculating these costs, translating them into an efficient price, and imposing the charge in a cost-effective manner represents a considerable challenge.

Optimal pricing requires consideration of equity as well as efficiency and must not ignore transaction costs (the costs of designing, implementing, and enforcing a pricing regime). Externality cost calculations require estimates about the use of transport facilities (e.g. how many vehicles of what type are and will be used on which roads and particular times of day) and of relevant prices (e.g. future price of fuel, time lost in congestion, materials, carbon emissions, etc.). Moreover, an accurate allocation and charging of these prices may imply important transaction costs with a negative impact on disadvantaged groups. These issues have discouraged the adoption of such charges, the best known of which is the congestion or road user charge (RUC), which has been implemented in several places, and has been particularly successful in Singapore, London, and Stockholm (Metz 2018).

Implementing any kind of correcting charges requires consideration of all other taxes and subsidies in the make-up of transport costs and how well these are perceived and internalised by travellers in their decision-making.

19.3 The Perception of Travel Costs

The use of prices to support more sustainable travel choices requires that they are appropriate and, perceived and understood by travellers when making their choices. It is not enough to establish a price to cover the relevant costs; it is also essential that the price signal is known and perceived directly by the users when they decide to use a facility (Bonsall and Willumsen 2013).

Car users are already subject to several pricing signals relating to the costs of running their cars and meeting the legal requirements to keep a car on the road. However, few people have a complete and accurate idea of the cost of running their vehicles (i.e. maintenance, depreciation, insurance and annual road trax). Estimating these costs and linking them to a car journey is complex and most drivers would see no reason to make the effort.

In contrast, out-of-pocket costs like tolls or parking charges are more directly perceived and therefore influence choice. In the case of externalities, without a clear signal as to their value, travellers will have little idea of their magnitude and no means of considering them in their choices; so, they have no way of judging the extent to which they ought to change their behaviour to reduce the burden on others and even on future generations.

One must accept that socially optimal pricing, taking full account of all relevant externalities, is often politically difficult to pursue, technically difficult to achieve, and may be expensive to implement, thus reducing its value. Moreover, car users already face a set of prices that is only partially known and understood, often not perceived at the time of making decisions and, therefore, their responses to complex prices are difficult to predict. When considering the introduction of correcting prices, one must focus on their appropriate values, the cost of implementing the charges, and how they will be communicated to travellers.

The use of electronic means of payment tends to distance the act of consumption from that of payment and is likely to blur the perception of these charges. Information technology offers the possibility of delivering the required information at the appropriate time, but it must be provided with minimum effort on the part of the traveller to facilitate its use.

19.4 Pricing Tools

This section considers different tools to introduce correcting prices and their likely effectiveness in influencing travel choices. The main tools currently available are:

- car ownership taxes;
- fuel taxes;
- parking charges;
- pay-as-you-go insurance;
- tolls (inter-urban and urban), and
- congestion charges/road user charges

It is worth considering two aspects of each one. First, how effective is the proposed measure at correcting distortions and omissions at the location and time when the decision is taken (i.e. to use a car, or another mode). Second, how efficient is it as a means of collecting and enforcing payment (i.e. what is the level of transaction costs); this is not a secondary issue, as transaction costs may outweigh, or at least reduce, the benefits of deploying a particular pricing mechanism.

19.4.1 Car Ownership Taxes

Taxation of vehicle ownership through purchase taxes often combined with an annual charge may influence vehicle acquisition. Most countries impose some form of an annual charge on car owners. There is usually a notion that the revenue generated will be used to provide and maintain road infrastructure but the link is often less direct. Although some countries levy a fixed charge irrespective of vehicle size, many have a sliding scale with larger, or more luxurious, vehicles attracting a higher charge.

Certain nations have sought to use high purchase taxes or annual charges to limit the number of cars owned. Singapore has decades of experience with a transportation policy based on balanced development of road and transit infrastructure and restraint of traffic. In addition to its taxes on vehicle ownership, Singapore requires each vehicle to have a certificate of entitlement (COE) before it can be driven on the road and is currently valid for 10 years. Its cost may be higher than that of a new car. Singapore's COE has succeeded in restraining growth in its car fleet, but this type of car ownership restraint clearly raises issues of equity and may only be feasible under special geographical and political conditions. Furthermore, it has the problem that once people manage to obtain a COE, they will tend to make as much use as possible of the vehicle to make the investment worthwhile.

Purchase taxes, annual ownership taxes, and other annual charges can ensure that revenue from vehicle ownership covers the costs of providing (and maintaining) roads and their complementary services but, crucially, they are not related to car usage and are simply perceived as part of the unavoidable cost of owning a car. They may even be an incentive to use the vehicle to get maximum benefits out of an expensive asset, as mentioned above. A charge more closely related to vehicle miles travelled would get closer to the ideal of internalising infrastructure costs and emissions.

19.4.2 Fuel Taxes

Fuel taxes are an efficient way of collecting revenue, because the collection costs are low and largely borne by fuel retailers rather than the government. As fuel consumption is closely related to kilometres travelled, it has often been suggested that fuel taxes should be increased to replace some or all of the annual taxes discussed above. The argument being that, without adding to the net costs of the average motorist, the annual taxes could be more effectively targeted at vehicle use. The fact that fuel consumption is higher in congested conditions and that high-emitting vehicles tend to consume more fuel, adds justification for this approach.

Despite its attraction, rising fuel taxes is politically risky and therefore unpopular. Also, fuel taxes cannot be used to target vehicle use at specific locations or times of day to internalise congestion impacts. Moreover, the potential role of fuel taxes to influence behaviour is now being eroded by the introduction of electric vehicles, as it is not possible to use tax on electricity for this purpose. The growth in electric vehicles is creating a gap in public finances as fewer vehicles use taxed fuel. This conundrum will force exploring other means of collecting the funds required to provide infrastructure and internalise emissions and congestion impacts.

19.4.3 Parking Charges

Cars spend most of their life stationary in parking spaces, and car use is not possible without parking. The provision of parking facilities is a significant cost to society; not only do the spaces consume valuable urban land but on-street parking may cause delay to moving vehicles and accident risk to pedestrians (Bonsall and Willumsen 2013). It is not surprising, therefore, that parking policy has sought to limit the availability of parking spaces in city centres and to price them accordingly.

Two approaches to pricing parking can be identified: parking management and mobility management (Shoup 2005). In the first case, one seeks to price parking such that space is used efficiently and there is always one spot available for somebody valuing it highly. This strategy requires variable pricing so that some 10–15% of the spaces are always available, reducing parking search time and the associated congestion.

From the perspective of mobility management, parking policy could seek to charge or tax private non-residential parking such that average parking charges are higher than the price of a return journey by public transport. Charging a premium for all-day parking can similarly help to influence the timing of trips. Transaction costs for parking pricing are relatively low, particularly if enforcement costs can be funded from fines, and generally leave a significant positive balance to the authorities.

Notwithstanding, parking pricing is not an efficient enough way to price externalities because it only covers the static part of car usage and is poorly related to the movement part of the journey, which generates congestion and emissions.

19.4.4 Tolled Facilities

Toll roads have existed for over 300 years despite objections to them. Extensive networks of toll roads have been developed in the USA, Canada, France, Spain, Italy, and Australia. Some emerging countries have also adopted the tolling of roads as a means to provide much-needed good quality infrastructure.

Excluding the case of shadow tolls, where the authority pays users tolls, all toll roads require facilities to collect money from travellers. A toll plaza requires considerable space, facilities, and personnel thus increasing the transaction costs and often generating queueing delays. The use of electronic tags either linked to a bank account or charged with pre-paid money, speeds up tolling transactions as it is not necessary to stop. The same advantage can be obtained with an account linked to a vehicle and using an automatic number plate recognition (ANPR) system.

The use of free-flow or open-road electronic toll collection is now common in many countries. In this case, a toll plaza is replaced by a gantry with sensors for detection and transaction recording and video cameras for enforcement. Open road tolling (ORT) offers important advantages to toll road operators, but it is particularly attractive for implementation in urban areas where there is no space for toll plazas. Moreover, ORT generates no delays, reduces emissions, and removes the capacity constraints of a toll plaza.

The availability of non-stop payment technologies has facilitated another variant of priced road space: managed lanes, also known as high occupancy and tolled (HOT) lanes; these additional lanes are available free to high occupancy vehicles (usually cars with two or more passengers) and to any vehicle on paying a toll. The lanes are segregated from the rest of the traffic on a motorway and a key feature of many of them is that their pricing is dynamic and usually set to ensure free-flowing conditions in the priced lanes. The availability of an adjacent free, but usually congested, set of general-purpose lanes (GPL) makes this approach more acceptable to drivers but it means that congestion externalities are not charged to all the drivers who generate them.

Urban ORT roads can have complex pricing schemes. The urban toll roads in Santiago de Chile, for example, have three levels of charging per kilometre depending on time-of-day, direction of movement, and speed achieved, as measured over a longer period. This is a semi-dynamic approach to toll pricing and is similar to the criteria used in Singapore's congestion charge (see below). The prices are displayed on relevant websites and on many of the gantries where the transactions are recorded. The use of a complex pricing structure is theoretically sound but, if it becomes too difficult for the driver to predict the charge in advance, its efficiency as a means of influencing behaviour is undermined. Research into responses to complex pricing structures (see Bonsall et al. 2007) has suggested that, beyond a certain point, additional complexity causes the user to misperceive the price signal or to ignore it altogether.

The extensive use of geopositioning systems (GPS) suggests an alternative mean for collecting tolls. Vehicles equipped with a GPS device can be charged for their effective use of roads. There is no reason in principle why this approach could not be used to charge for road use by private cars. Modern vehicles are often connected remotely to the manufacturer and report on their movements, engine conditions, and other data perhaps facilitating such a charging scheme in the future.

19.4.5 Pay-As-You-Drive Insurance

Most governments require drivers to have some form of third-party insurance before they are allowed to drive on a public road. The premiums are generally in the form of annual or monthly payments which reflect an assessment of the driver's (and/or car) risks of being involved in an accident.

Under pay-as-you-drive (PAYD) insurance, which is being offered to motorists in an increasing number of countries, the insurance premium is calculated dynamically and reflects the distance driven. Also, depending on the technology adopted, it may also consider the times at which driving occurs, the type of roads used, and even the driving style adopted. Such systems obviously provide an incentive to drive safely and to reduce the overall distance driven.

19.4.6 Congestion and Road User Charging

The economic theory underpinning congestion charging was put forward by Pigou in 1920, with Walters (1961) relating it specifically to road traffic. The specific case for congestion charging was made some 60 years ago in the Smeed Report (1964) in the UK. The report observed that the structure of motoring taxation did not reflect the real journey costs and was particularly oblivious to costs imposed on others. Since then, transport economists and planners have advocated the adoption of congestion charging, provided it is targeted to the locations and times where excessive delays take place.

Singapore was the first country to adopt congestion charging (Metz 2018). Its Area Pricing Scheme, introduced in 1975, comprised a cordon that vehicles could cross during the morning peak only if they displayed a pre-purchased paper license. The immediate impact was a significant reduction in peak period car traffic and congestion and an increase in public transport patronage. In 1998, the paper-based system was replaced by a better-targeted electronic system using a combination of cordon and hotspot electronic charging. The level of the charges at each gantry is reviewed regularly and is set to maintain speeds close to a target level. The charges are displayed prominently at each gantry and on a website. Menon (2000) reports that the introduction of electronic road pricing brought about a further 15% reduction in traffic levels and that there is considerable sensitivity to the timing of the different charge levels.

The introduction of the London congestion pricing scheme in 2003 represented a breakthrough both technically and politically. It involved charging a daily fee to drivers who wished to drive in the defined central area. Payment of the charge entitled users to drive as much as they wished within the area on that day. The charge was initially set at £5 with a 90% discount for residents of the charged area and was very effective in changing behaviour; there was a 30% reduction in congestion, most of which was attributed to a shift to public transport or to diversion around the charge area (Transport for London 2008). In later years, it has been necessary to increase the charge to retain the congestion reduction benefits but the scheme retains its popularity among Londoners.

The implementation of congestion charging in Stockholm was technically more advanced than London's. Under the Stockholm scheme, users pay a toll that depends on the time of the day when crossing a cordon around the city centre. Although the charge is significantly lower than London's, it is more directly related to usage and has resulted in a 22% reduction in traffic crossing the cordon during the charge period and a 6% increase in use of public transport (Börjesson et al. 2012).

Despite these early examples, two main considerations have delayed the implementation of congestion charging elsewhere. The first is the concern about its possible regressive nature: the rich will pay and benefit from faster travel whilst the poor will have to use 'inferior' modes of transport. The second objection is that voters may find congestion charging an unacceptable form of taxation (see the discussion about this in Ortúzar et al. 2021). Experience has demonstrated that in the right conditions these objections can be overcome.

19.5 The Experience of Private Sector Projects

The twenty-first century has seen an increase in the involvement of the private sector in the design and delivery of transport infrastructure and services. We do not discuss here the reasons for this involvement; instead, we seek to draw lessons for both future projects involving the private sector and for the design, structuring, modelling, and operations of congestion charging or RUC projects by the public sector.

19.5.1 Involvement of Private Sector in Transport Projects

Many transport projects today are implemented through some sort of concession where the private sector invest in constructing and operating a facility and then transfers it to the public sector at the end of the contract period (usually between 20 and 40 years). Willumsen (2014) provides a description of the issues involved in the design, bidding, and financing of this type of projects, together with the implications for modelling. We provide here only a summary of these issues.

Many agents, professionals, and advisors play key roles in the process of developing a transport project from conception to successful implementation by the private sector. We simplify these here into three main participants:

- the sponsor, usually the government, who identifies a project, develops and takes it to the marketplace;
- bidders, often a consortium of construction companies, operators, and their advisors, who prepare offers for a concession to build, operate and eventually transfer the asset back to the sponsor; and
- financial institutions, often a combination of banks, infrastructure, and other funds, who would either invest in the concession or lend money under different forms of debt to the concessionaire.

There may be also other agents like insurance companies, rating agencies, and pension funds, who may take some degree of risk and/or influence the outcome of the transaction.

Figure 19.1 provides a simplification of the process and the role and concerns of the three main agents: banks and financial institutions, bidders/consortia, and sponsor. We look at each stage in turn.

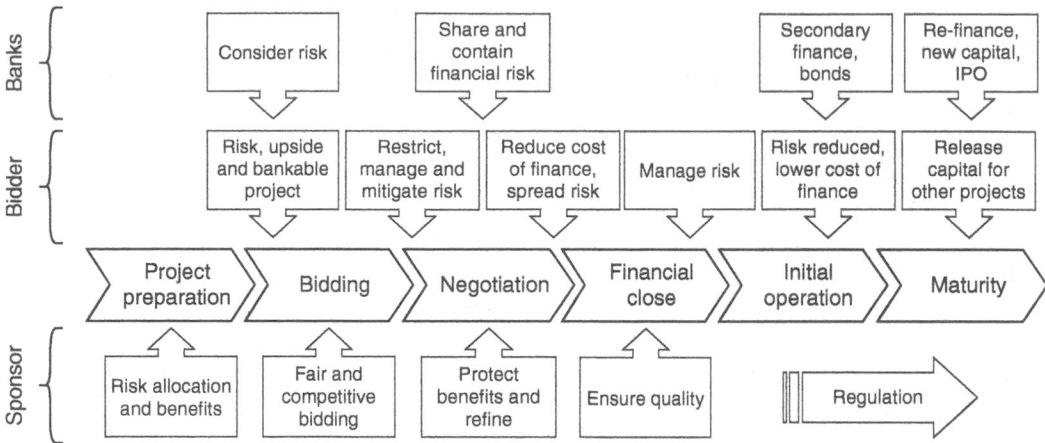

Figure 19.1 Simplified project development process

Project preparation. The Sponsor/Government aims to design a concession that will provide significant benefits to society whilst offering an attractive role for the private sector. In doing this, the Sponsor will identify and assess the risks involved and allocate each to whoever has a greater capacity to manage it (i.e. to do something about it). Although the acceptance of a risk costs money, those who can do something to manage and mitigate it are likely to charge less for accepting responsibility. Traffic and revenue risks are often transferred to the concessionaire because it is in a position, through the provision of a good service, to manage it. At this stage, the Sponsor will try to provide a clear and transparent view of risks in order to get good competitive and comparable bids.

Bidding process. At this stage, each consortium studies the project and its terms of reference to decide how much to ask for accepting the risks, obligations, and compensating revenue streams on offer. The nature of the revenue risk will depend on a number of factors including the conditions of the concessions and the award criteria (on toll levels, duration, minimum subsidy/maximum payment, and discounted value of revenue stream). The consortia will try to get a clear view of the risks and determine whether they have a special competitive advantage (faster construction, better finance) that could be exploited in the bid.

Negotiations. Sometimes these are very short as the conditions of the bid would have specified the project fully. More often there is a period of negotiation once a preferred bidder has been selected; this period is used to refine the Concession Contract and its conditions considering variations that may have not been envisaged originally. It usually deals mostly with risks other than traffic but it may involve obtaining stronger guarantees from the Government that unexpected alternative routes/services will not be provided in the future. At this stage, the concession is assigned to a Special Purpose Vehicle (SPV), set up by the consortium to build and operate the project until it is transferred back to the Government. The financial institutions, in turn, will try to share and spread the risk among different banks and to press the sponsor and consortium for guarantees. Rating agencies may play a key role here in assessing project risks.

Financial close. Here, all the funds needed to implement the project are finally secured and made available. This often involves obtaining a loan to cover the construction/rehabilitation period plus one or two years into operation. The partners of the SPV provide the rest of the finance as equity. The repayment of this loan is sometimes structured around lower-cost longer-term finance once the project is well into operation. This may mean a longer-term loan or a bond issue. Therefore, financial institutions would like to be confident that this second stage finance is assured. There may be a grant provided by the Sponsor to strengthen the financial viability of the project. In the case of existing assets that need rehabilitation and operation over many years, the consortium may offer a payment to the government in compensation.

Second stage finance. Once the project is in operation, all risks are much reduced and, therefore, it should be possible to obtain finance at lower rates. Often the main remaining risk is associated with the future revenue stream. A review of previous traffic projections may be needed to offer additional confidence in future revenue streams.

Third stage finance. Once the project is operating well and its long-term financial structure is in place, it is possible to offer equity participation in the market, totally or in part, depending on the conditions of the concession and the strategy of the concessionaire. This may take the form of an Initial Public Offering (IPO) or just a privately agreed opportunity to invest in the SPV holding the concession. This injection of capital will enable the release of some capital of the original investors that they could use in another concession.

The process just described is fairly different from the usual consideration of projects and strategic transport planning for the public sector. The number of agents or stakeholders is significant and each is concerned with risks but from different perspectives. The forecasting of traffic, patronage, and revenue is central to these concerns and becomes the most important risk as the project matures.

Traffic and revenue forecasting is produced for each of these key stakeholders and in each case their different perspectives are brought to bear. It is not surprising, therefore, that what is considered important and included in the model may vary and so would the traffic and revenue projections.

Given the variety of agents and the sums of money at stake, the need for transparency becomes paramount; model black boxes are not just unacceptable but are seen as a source of additional risk. The ability to explain a traffic model and deliver a clear narrative relying on several complementary sources of evidence to support revenue projections becomes essential.

19.5.2 Uncertainty and Risk

Uncertainty is a result of our failure to ascertain a future event with certainty. It is a reflection of our lack of knowledge and it is, in principle, impossible to quantify. Pure uncertainty is not helpful in deciding whether to invest in a particular scheme or not; but it may be inevitable. On the other hand, we may be able to assign probabilities to the level of economic growth in the future for a particular region and from this infer future levels of traffic in a specific road section. Risk is, in this sense, quantifiable uncertainty.

Uncertainty has always been a feature in our modelling and forecasting work, but the involvement of private investors and financial institutions has given a clear monetary value to the issue of risk. In terms of demand modelling, private investors and financial institutions are interested in a revenue stream, year after year for the duration of a concession, and the degree of confidence that can be associated with these figures. These risks change over time and to understand this we need to start by considering the different actors and processes that are central to private sector projects. Figure 19.2 shows a nominal and simplified profile of risks over the time of a project.

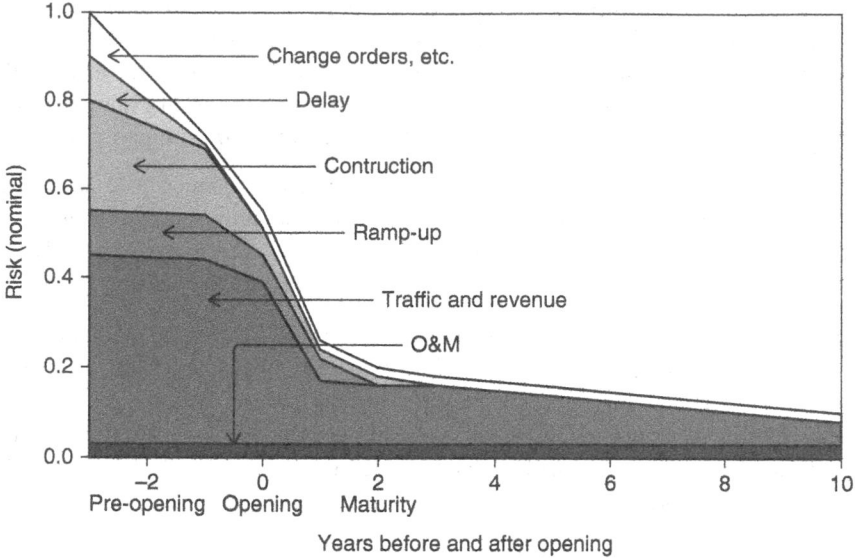

Figure 19.2 Nominal risk profile for a private sector transport project

The main sources of risks during the bid preparations are:

- Will all the 'right of way' required be released to the concession in time for construction?
- Are all construction costs sufficiently well-defined and known?
- Are the ground conditions sufficiently investigated, including the possible need to displace utilities?
- Are the costs of operating and maintaining (O&M) the future asset well-known and quantifiable?
- How confident can one be about the future traffic and revenue (T&Revenue) streams?
- How long and steep will be the period of transition between starting operations and the time when stable traffic levels materialise? This is known as the *ramp-up* period.

As can be seen from Figure 19.2, these risks are higher before construction starts. During construction, most of these risks are reduced so when the project starts operating the only remaining risks will be some hidden faults in construction, traffic and revenue, residual O&M, and ramp-up. Finally, the project will reach maturity when traffic levels stabilise and the only remaining risk is associated with the level of growth and the possibility that a competing facility is provided some time in the future.

The importance of traffic risk and the mismatch between traffic projections and traffic outturn has been recognised for some time. JP Morgan studied 14 pre-opening toll road studies in the USA and compared them with the traffic achieved after opening (Muller 1996). They found that in two cases the original studies underestimated traffic and revenue by between 10% and 30%. In four cases, there were moderate over-estimations of revenue of between 12% and 25%. There were, however, eight cases (57%) where the over-estimation of revenue was from 45% to 75%.

The main reasons for this over-estimation of revenue were considered to be:

- poor analysis of alternatives;
- poor or no analysis of willingness to pay tolls to save time; and
- overoptimistic growth rates, mostly based on over-estimation of generated traffic through new developments.

Similar results have been found in other markets and countries as documented by Bain (2009) in the case of toll roads and by Flyvbjerg et al. (2005) for public works. In fact, Bent Flyvbjerg and colleagues have amassed considerable data and cases about these issues; for a good summary see Flyvbjerg et al. (2018).

There are many reasons for this, some of them outside the control of the traffic forecaster (e.g. unforeseen economic recessions that escaped even the banks financing these facilities). However, many of the criticisms levied to the craft of demand forecasting remain valid. Too often, the modelling approach adopted mirrors the classic approach employed on behalf of the public sector for the past thirty years and fails to identify and isolate the key drivers of traffic (and revenue) in a way that recognizes the associated risks.

These risks relate to four main sources:

- the size of the relevant market that can possibly be attracted to the new facility (in-scope traffic);
- the estimation of capture rates for that market;
- the development of reliable growth models (including where appropriate induced traffic); and
- consideration of future new alternatives to the new facility that may affect the effective traffic and revenue capture.

RUC projects face similar modelling challenges identifying who will pay the charge, who will change route, mode, or time of travel together with the estimation of the impacts on congestion and emissions. RUC projects require an additional focus on the complementary measures required to make it work and the equity impacts of the pricing measures.

19.5.3 Risk Management and Mitigation

Before discussing the implications of all of the above for modelling, it is useful to consider what concessionaires can do to manage and mitigate risks. Bidders can, of course, request better guarantees from sponsors. This may take the form of explicit, or sometimes implicit, minimum revenue guarantees over the life of the concession. Other assurances include automatic adjustment of tolls or fares in line with inflation or other formula and a commitment that no competing facilities will be provided, at least for the initial years of the project.

During construction, the consortium can ensure minimum opposition from locals by a good communications strategy that should also help to smooth the transition into paying for the use of the new facility. The concessionaire can ensure that a good service is provided at all times and that a good relationship is developed with users and clients. The provision of complementary services (fuelling stations, food and rest facilities at toll roads, and newsstands and refreshments with public transport services) is also important to attract and support customers.

Rapid response to incidents and emergencies and quick restoration of services after a *force majeur* event are meant to be defining characteristics of private sector involvement in transport projects. Marketing can play a useful role in identifying those users that could start using the new facility given the right information or encouragement. This is important for road haulage companies that sometimes may not be fully aware of the reduction in risks and operating costs that result from using a better, if paid, road.

19.6 Demand Modelling

19.6.1 Willingness to Pay

Willingness to pay (WTP) plays an obvious key role in the estimation of the impact of tolls/prices on travel behaviour. In the case of a concession, any changes in behaviour translate into patronage and revenue collection. WTP is usually represented through the subjective value of time (SVT), ascertained through stated preference (SP) or revealed preference (RP) surveys.

In this respect, segmentation is very important as the use of a single SVT is not reliable enough and will tend to exaggerate, one way or another, the real capture rate of any facility. Segmentation can be done on the joint basis of trip purpose and income levels; this is particularly important if differential growth is expected in the future. An additional and important segmentation is to distinguish travellers who have their costs, including tolls and fares, covered by somebody else, usually their employer; they tend to have a high but not unlimited WTP consistent with the travel policies of their companies.

In the case of trucks, WTP depends on a number of factors including the size and type of goods hauled, the type of contract for each shipment (e.g. 'just in time' arrangements), company policy and legal requirements (i.e. dangerous goods are often required to use the safest road, normally a tolled one) and, in some cases, opportunistic decisions by the driver. Some road haulage companies, in particular one-man operations, are not fully aware of their operating costs and may be more cash sensitive; these will display a lower WTP for tolls to save time and operating costs than an objective evaluation would suggest.

In the case of urban schemes, one must also consider that many trips will be made day after day and the impact of charges or fares over the monthly budget may be significant. Income effect may have to be considered for such toll roads and RUC schemes. The attractiveness of new public transport facilities is also influenced by WTP, especially if a new mode is better than existing services and is required to pay a premium fare.

WTP may also vary with the quality of the service or the road provided. For example, one should be willing to pay more to reduce the time spent under less comfortable conditions, say heavy congestion or merely the need to stop at junctions as opposed to enjoying free flow conditions on a good highway. This is sometimes referred as a *motorway premium* or a *standard road malus* and may be up to a difference of between 20% and 40% of the SVT depending on each case.

It is generally recommended to use at least 10 WTP user classes for toll roads, including at least four for trucks. WTP is likely to grow in line with the income level of the relevant population. This may be just the car-owning population that will turn to be the wealthier proportion in an emerging country. The rate of growth of SVT with per capita income is somewhat uncertain and in dispute. In the largest meta-study to date, Wardman et al. (2016) suggest that SVT will grow at approximately 0.7 times the rate of growth of income per capita of the relevant population.

19.6.2 Simple Projects

A small number of toll road/bridge projects can be handled using simple models on a spreadsheet. This may be the case of some new estuarial crossings where there is only one or at most two alternatives. If the alternatives are clear and limited, it is possible to use a Logit formulation to consider them and the effect of introducing a new one.

Depending on the nature of the new alternative, this can be incorporated as another choice in a MNL, or in a NL structure if it is, for example, a new shorter bridge to compete with a longer road and ferry crossings. Whatever the case, the alternative specific constants are always going to be contentious and should be supported by evidence complementary to the SP extracted values (Cherchi and Ortúzar 2006). The need to incorporate as many service and personal attributes as possible in the choice structure should reduce the importance of these constants.

Note also that segmentation remains important even if a simple Logit formulation is used. Simple cases will tend to be predominantly inter-urban ones and therefore choices might be modelled on a full-day basis. The exception will be cases with significant variations in travel times during the day, either because of congestion or the availability of some alternative only at certain times.

19.6.3 Complex Projects

Most transport concessions will require developing and implementing a network model and, in many cases, a multi-modal one. It is hard to separate and distinguish the sources of revenue risk in a conventional model: the size of in-scope traffic, growth, traffic capture, and the true elements of choice: willingness to pay and the relative advantages of each alternative. These issues are confounded in a large-scale model with less relevant material and a full range of behavioural responses. To de-construct the components of future demand, it is useful to adopt an approach as depicted in Figure 19.3. Here, we extract the main components of in-scope trips, benefits of the new facility compared to alternatives, WTP for these benefits, and growth from a conventional transport model. Each of these will have risks associated with them, and it is the task of the modeller to identify and reduce them to provide more reliable forecasts.

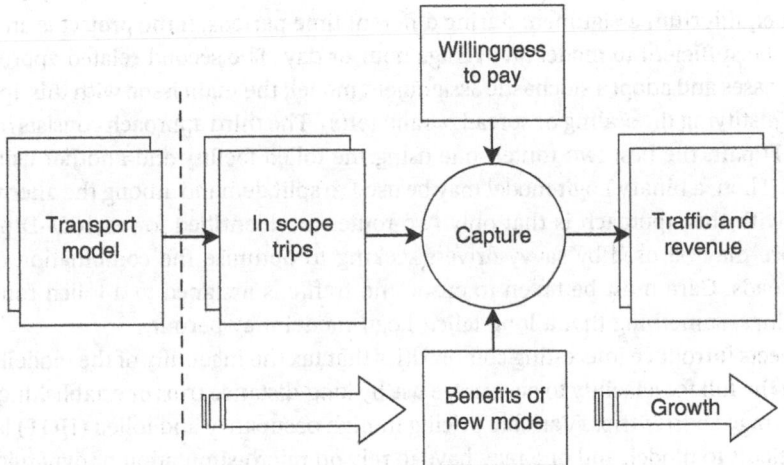

Figure 19.3 Estimation of traffic and revenue from a new facility

Traffic revenues depend on the size of the relevant travel market, its future growth, the choice mechanisms available to users, and their own preferences. When faced with a new facility, users can have the following main responses:

- change their route;
- change mode;
- change their destination to one easily reachable using the new system;

- change their trip-making frequency (generated/suppressed traffic), and
- change the time of travel as a result of price and congestion profiles.

How many of these responses will be important depends on the new facility and on the alternatives available now and in the future. Experience with similar schemes and a bit of experimental modelling can help in deciding how many of these responses should be included in revenue projections. These will also help in identifying the scope of the traffic that could be captured by the new facility.

In-scope trips refer to the traffic that *might* be attracted to the project. As such, it can be considered the target market for the concession. With a toll road or bridge, it will be the traffic that would use the route if no charge was implemented. For public transport services, it represents an initial judgement on the patronage that could be captured both from other competing public transport services and potentially attracted from other modes under the most favourable circumstances.

The most reliable way to estimate this potential demand is to undertake a battery of transport surveys (passengers and freight). Investors are more convinced by actual 'on the ground' data than by any outputs from elegant and sophisticated models. Origin–destination (O–D) surveys, traffic counts, and travel time surveys, undertaken probably at different times of the year and days of the week, and for at least 16 hours per day, would be ideal. These should be combined with some permanent traffic counting methods to obtain a suitable profile of demand throughout a year.

Most toll roads can be modelled using just the assignment stage in a commercial package capable of handling multiple user classes. It is important that these models are handled in terms of generalised costs of travel and because of their final use, it is convenient to quantify these in monetary units (see the discussion by Hensher 2011).

Three alternative approaches can be used here. The most common is to employ 10 or more user classes with equilibrium assignment during different time periods; if the project is an uncongested area, it may be sufficient to model an average hour or day. The second related approach is to use fewer user classes and adopt a stochastic assignment model; the main issue with this approach is the difficulty in justifying the scaling or spread parameter(s). The third approach consists in identifying, for each O–D pair, the best two routes, one using the tolled facility and another using only non-tolled roads. Then, a binary Logit model may be used to split demand among the alternative routes. A problem with this approach is that only two routes are identified for each O–D pair, when in practice more may be used by savvy drivers seeking to optimise the combination of tolled and non-tolled roads. Care must be taken to ensure no traffic is assigned to a tolled route if it offers no time savings, something that a long-tailed Logit model may permit.

Some projects introduce interesting complexities that tax the ingenuity of the modeller, for example, capping the toll for a facility to encourage use by long-distance trips or establishing a minimum toll to discourage shorter trips. Variable pricing in high occupancy and tolled (HOT) lanes are particularly difficult to model, and one may have to rely on micro-simulation or dynamic assignment techniques.

Public transport projects, new metro, LRT, rail, or BRT schemes, are also more complex to model as they inevitably involve mode choice and other behavioural responses. Nevertheless, the same principles apply: the identification of in-scope trips, transparent representation of choices, in-depth WTP analysis, and consideration of present and future alternatives.

The viewpoints of the bidder and financial institutions are similar, although the second focuses on default risk and the first on the probability of achieving a significant surplus after debt coverage. Both benefit from looking at revenue projections in the context of the risks associated with each contribution. One way of handling this is to build different scenarios for the future: Optimistic, Base Case, and Downside are commonly used and must be clearly defined.

The financial strength of a project of this nature will depend on several factors. An often critical one is the Debt Service Coverage Ratio (DSCR). This is the ratio of revenue from operations to principal and interest obligations; that is, payments due to lenders at each period. The most critical stage will be the earlier years of the project. When a project is implemented in stages, it will be important to model them separately and to provide estimates that incorporate ramp-up effects from the outset.

The ability of the traffic advisor to explain the workings of the model to non-specialists and to demonstrate in-depth understanding of the underlying drivers of its financial success is critical to a successful bid preparation.

19.6.4 Road User Charge Projects

Although similar considerations apply, the modelling of RUC projects is more complex than for transport concessions. RUC modelling must represent all common responses to a charging system, not just route choice. The objective of RUC is maximising welfare, not revenue collection and the funds collected can be used to improve other aspects of the transport system. The model should pay particular attention to who benefits and who experiences a loss of welfare as a result of the charging system. Equity considerations are paramount.

The model should be able to generate all outputs of interest in the evaluation of the costs and benefits of a RUC scheme. These include delays and congestion, emissions and local pollution, and route and mode shifts. As the target is often to reduce congestion and emissions, the RUC tolls will have to depend on location and time of day. As different pricing levels and charging points are tested during design, the location of congestion will also change requiring an iterative approach.

There will be close interaction between the technology used for charging and enforcement and the modelling task. As RUC is usually considered for urban areas, the location of charging and enforcement points can vary widely, unless the city has a configuration that makes charging simpler as is the case of Stockholm.

RUC schemes are likely to involve discounts on residents and a cap on maximum charges per day (as in London); these require additional considerations in the model, for example, modelling the full-day tours of each traveller. Furthermore, RUC schemes are always planned with complementary measures to ensure their acceptance and smooth operation. These measures may include junction improvements to avoid problems with the new routes for drivers, improved public transport services and new cycle lanes to promote mode shift, and exceptions to the charge for certain types of vehicles, sometimes taxis and always emergency units.

The RUC schemes for Singapore, London, and Stockholm were designed and refined using classic but sophisticated aggregate models. This required the estimation of some of these features outside the main model; however, this did not prevent their fairly accurate estimation of impacts and successful implementation.

19.6.5 Scheme Design

The private sector has a long experience in the collection of tolls and ticket revenues and their enforcement. The comments below although focussed on transport concessions, apply also to cases where the collection of RUC tolls is allocated to the private sector, perhaps as the result of a competitive bidding process.

The government is usually interested in offering a concession that transfers a significant element of risk to the private sector. It is also interested in tapping into the creativity and good management of the private sector to secure intelligent design, innovative financial packages, and to offer a high level of service throughout the concession.

To achieve this, the sponsor requires a competitive and transparent tendering process over a well-designed Concession Package. This will be assisted by low bidding costs and wide international promotion of the concession programme, if appropriate. The government would normally retain those elements of risk that it is best equipped to handle, for example, securing the right of way in a timely manner.

Revenue risk is often transferred to the concessionaire as it can influence it through good service and pricing. Even then, to facilitate financial closure, the government may be persuaded to offer some measures to reduce revenue risk. The main instruments available to provide manageable revenue risk are the containment of future competition, minimum revenue guarantees (MRG), the choice of decision rules for awarding the concession, and the provision of a well-documented database.

MRG are sometimes offered over the first few years of the concession and at a level below that of the Sponsor's base case scenario. The level of this guaranteed revenue stream is important in determining de debt/equity ratio for the concession.

If the future is very uncertain, for example, when the government does not want to commit to not building alternatives in the future, some concessions have been awarded to the bidder requesting the *lowest present value* (LPV) of the revenue stream discounted at a pre-determined rate. If the revenue stream is below expectations, the result is just an extension of the concession up to a pre-determined limit. Revenue risk is therefore reduced. Revenue projections are still needed in order to secure financial backing for the project, but they become less important than in concessions awarded on the basis of the lowest toll level or maximum duration.

The costs associated with the production of concession bids are generally high and naturally, the bidder consortia would like to recover them through successes in their bidding programmes. The sponsors are, therefore, interested in reducing bidding costs as much as feasible without compromising the quality of the concession agreements. Traffic and revenue studies are an expensive element of bidding for a concession. There are significant advantages for the sponsor to conduct a good Reference Study:

- undertaken to international standards; this means either by an international company or at least a technical audit by one;
- transparent and well documented, where data should be collected with good quality assurance and provided both processed and in raw (e.g. interview records) form;
- data should cover the relevant periods and be segmented generously; at least some traffic/person counts should be continuous over a whole year;
- if software packages are used to process the data and model demand, they should be internationally and commercially available; and
- the provision of geo-coded data and the whole database on electronic form is highly desirable.

Travel surveys are expensive and time-consuming. They do not fit well within the timescales and budgets available for bidding for a concession. Therefore, it is highly desirable that these are undertaken as part of the Reference Study for the sponsor in preparation for the concession. To be of use, they should be well documented and made available to all bidders in a transparent format including the processed and raw data. Geocoding these data provides an added benefit of allowing consortia to develop their own zoning systems. The bidder would like to confirm this information with its own mobile phone trip matrices, traffic counts and other observations, seeking, at the same time, to identify opportunities to obtain a competitive advantage over other consortia.

19.6.6 Ramp-Up, Leakage, and Discounts

This section discusses some issues that do not figure significantly in public work projects but have great importance in private sector projects and will be relevant in the case of RUC schemes.

The ramp-up, or transitional period following the opening of a facility, was never considered particularly important in the case of public work projects. However, revenue collections of the first few years of a concession play a significant role in their financial viability. This is why quick implementation and good estimation of this transitional period are essential. During ramp-up, potential users learn about the new facility and the advantages it may offer to their journeys. There is often strong resistance to the introduction of a new tolled facility instead of an equivalent free one. This resistance may result in a slow adoption rate even if the advantages more than compensate the imposed toll. The adoption of good communication and marketing strategies should help in reducing the length of this transitional period.

Nobody has come up yet with a good theory to support the estimation of ramp-up durations. Notwithstanding, we know that this will depend on issues like:

- the frequency of trip-making in the area, as the more frequent repeated trips are, the shorter the ramp-up period will be;
- the significance of the advantages offered by the new facility; for example, a major time saving will result in shorter ramp-ups;
- information on the new facility and the advantages it will offer;
- the local tolling culture; if people are used to toll roads, then it will be easier to adopt a new one; and
- the provision of a short period when the new mode or facility is provided without a charge may facilitate its appreciation, but may generate a backlash when a toll is introduced; these periods should be short and well communicated.

In the absence of a good theory, one must rely on benchmarking transitional periods with other similar facilities and contexts. Anything between six months and several years is possible depending on the characteristics above (Willumsen 2014).

Expansion from the modelled time periods to the year is usually undertaken based on traffic counts or ticket sales that are assumed to capture daily, weekly, and seasonal variations. Whenever congestion plays a role in the capture rate of a new facility or service, it will be necessary to model different periods of the day (and sometimes of the week) to ascertain their corresponding different capture rates. The expansion task is now dependent on the number of hours a year that is represented by each modelled period.

It is generally not practical to model every year of operation using a full transport model. Common practice is to model only those years when significant changes in the network, or in prices, are expected and interpolate the other years. Years that are far in the future are sometimes extrapolated from the last year modelled with confidence. Later years bear little influence on the financial strength of a project.

Not every penny that is collected at the toll plaza or fare box reaches the coffers of the concessionaire. There are inevitable losses in the trail from transaction to bank account, even when electronic fare and toll collection are dominant. Some losses are the result of straight avoidance on the part of users, others may be due to technical failures, misclassification of vehicles, and human error. In most projects, the expected loss rates are reasonably well-known. This is now the case for electronic toll collection as well.

Enforcement is an important component of the transaction costs of a concession or RUC scheme. In urban areas, toll collection will be free-flowing and therefore enforcement will be necessary to secure revenue and ensure that the decongestion benefits accrue. This is usually achieved with video recording and automatic number plate recognition (ANPR) to validate payments; in this case, cameras are placed at or near the gantries used for charging. ANPR can also be used to collect the tolls as is the case of the London Congestion Charge scheme. The use of GPS signals to collect tolls does not remove the need for some type of enforcement and this is a particular challenge for that technology.

Discounts may be offered to certain users, for example, on the basis that they live close by and are practically forced to use a paid facility; in this case, discounts are used to help acceptance of the new concession or RUC. In other cases, discounts are offered to frequent users to reduce the cost of recurrent use of the facility. Both types of discounts may be required by the concession contract or offered to attract users that otherwise would avoid the new concession. It is essential to fix these discounts to avoid cannibalising users who would have paid a full fare.

Finally, some projects will offer fares or toll rates that are shared among different suppliers of services, for example metro and feeder buses. In this case, it will be necessary to perform additional calculations to correctly allocate revenues to these different agents and concessions. This is also the opportunity to consider most discounts, period tickets, and concessions (free passes, exempt users) that will influence the final revenue stream figures.

As stated in previous chapters, modelling is mostly useful when benefiting from good interpretation of results. In the case of private sector projects, sound interpretation of results is of paramount importance. The ability to understand and communicate modelling results is based on the capacity to track influences from inputs to outputs. This is where good understanding of the theories underpinning the models is essential. Explanations should be delivered in the language and conceptions of the interested parties, not those of the modeller.

19.7 Risk Analysis

Although the Reference Study prepared by the Sponsor should identify the main revenue risks, urgent consideration of these will only start in preparation for the bidding process. A traffic and revenue study for a bidding consortium will normally consider first the production of a Risk Register. This will contain also the revenue risks and they will serve to focus the attention and the data collection effort for the traffic study.

It is difficult, even undesirable, to generalise on these risks but they are likely to include:

- over-estimation of willingness to pay;
- over-estimation of growth prospects;
- ignoring future changes to the network; and
- underestimation of the importance of technology or trend changes.

Some risks are inherently difficult to identify. These are the *black swans* of Taleb (2007), events that are almost impossible to foresee like pandemics or the impact of oil prices on the price of tortillas in Mexico via biofuel production in the USA. Oil prices are indeed difficult to forecast and they may have a strong influence on travel behaviour, in particular when they take the form of a significant shock. Experience has shown that forecasts of economic growth and recessions are less reliable than one would like.

There are three basic ways of handling the issue of risk in traffic and revenue projections: sensitivity analysis, stochastic simulations, and scenarios. We discuss these below.

19.7.1 Sensitivity and Sources of Risk

Sensitivity analysis is performed to identify the dependence of model outputs on small changes in model parameters and inputs. It is used for two reasons: first, to ensure that the model responses are reasonable and explainable; second, to identify what are the key risk sources that are most likely to affect the financial strength of a project.

Sensitivity tests are usually undertaken at least for: SVT, growth rates usually linked to GDP, timing of competing projects, and toll or fare levels. Variations of +/−10% or 20% on SVT are useful to assess how dependent are the estimated revenues on our evaluation of these parameters. Financial institutions linked to the project should be able to provide estimates of possible variations of future GDP growth. These will affect incomes and therefore car ownership, traffic, and revenues.

Toll and fare level sensitivity tests are also important, even if these are fixed in the concession contract, because one would like to be confident that increasing them will increase revenue. These sensitivity tests may suggest, for example, that toll or fares have been set too high and that more revenue (and benefits) would be collected with lower rates. An example of this type of toll sensitivity tests is shown in Figure 19.4.

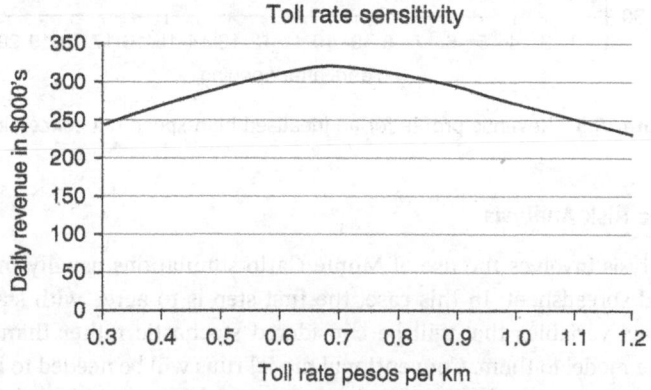

Figure 19.4 Toll sensitivity tests

The figure shows that the optimal toll rate, in terms of maximising revenue, is around 0.70 pesos/km. The Sponsor would have the toll fixed at some 0.5 or 0.6 pesos/km to protect user benefits.

Different aspects of our transport models generate different confidence levels in their outputs. We believe more in the results of an assignment model because it is straightforward to diagnose why it fails on the base year. On the other hand, we have less confidence in the forecasts of mode, destination, or frequency of travel choices because of the difficulties in performing such a diagnostic test quickly and effectively enough.

Moreover, the drivers for some components of demand that may be captured by the new service may differ from those by other contributors. For example, in analysing future patronage of a high-speed rail concession, shares from air travel may depend on the pricing policies of low-cost airlines, but these are difficult to predict; and shares from other rail services or car users may be more certain as their pricing policies are better understood and, therefore, easier to forecast.

A useful way of presenting these results is to deconstruct the outputs of a traffic model in a manner that enables the interested party to assign their own risk indices to different components of future demand. This is illustrated in Figure 19.5 for a hypothetical high-speed rail link. The figure shows the different contributions of 'induced' and 'redistributed' traffic plus the traffic captured from alternative modes. A bidder that has good information about long-distance bus/coach operations will be more confident of this particular component of future demand capture.

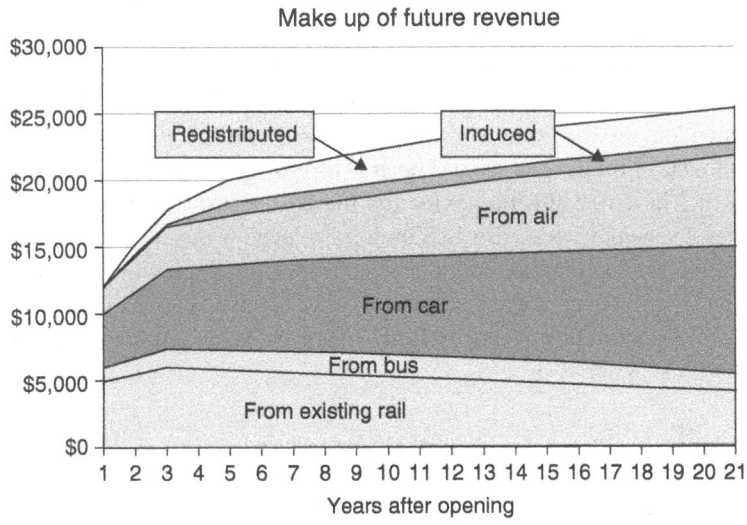

Figure 19.5 Revenue profile for an idealised high-speed rail concession

19.7.2 Stochastic Risk Analysis

Stochastic risk analysis involves the use of Monte Carlo simulations usually implemented as an ad-on to a standard spreadsheet. In this case, the first step is to agree with stakeholders on the (few) input or model variables that will be considered stochastic rather than fixed, and relate the outputs from the model to them. Conventional model runs will be needed to identify, for example, how variations in GDP growth affect revenues. Most of these would have been undertaken as part of the sensitivity tests mentioned previously.

The next step would be to adopt some probability distributions around the mean expected values of these variables, for example, SVT. It is tempting to assume that these would be Normal distributions. However, we should be warned that the probabilistic distributions of some key variables (GDP is a good example) usually have *fat tails*; that is, they tend to display extreme values more often than a Normal distribution.

The next step is to construct a model where this handful of variables influences revenue outcomes and where their probability distributions are sampled repeatedly in a Monte Carlo simulation. Note that in most cases these distributions are assumed to be independent. This is convenient but may be more difficult to accept in cases such as the relationship between GDP and SVT. Each run of the Monte Carlo simulation reflects one possible revenue path diverging from the base case. This is illustrated in Figure 19.6, where each path represents a diversion from the expected base case assumed to be unity; a revenue factor value of 0.95 in one year implies that collections in that case would be only 95% of the base case for that year.

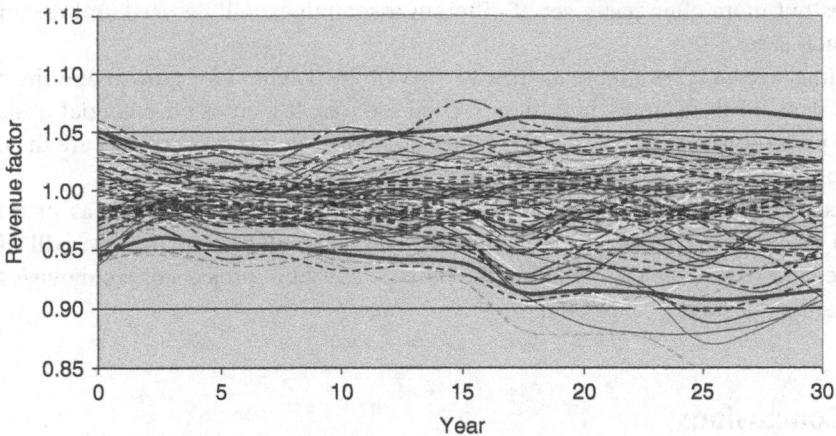

Figure 19.6 Monte Carlo revenue paths on a toll road

The end result is a distribution of revenue outcomes over the life of the project. Of these ranges, lenders would be more interested in the so called P90 or P75 revenue streams, which are the revenues that will be exceeded 90% or 75% of the time. P50 is the expected or base case revenue stream. Equity investors might be interested in P40, that is, revenues that have only a 40% probability to materialise, but represent a significant upside of the project. Figure 19.7 illustrates the distribution of values for a toll road with extremes of P90 (the lower band) and P10 (the upper one).

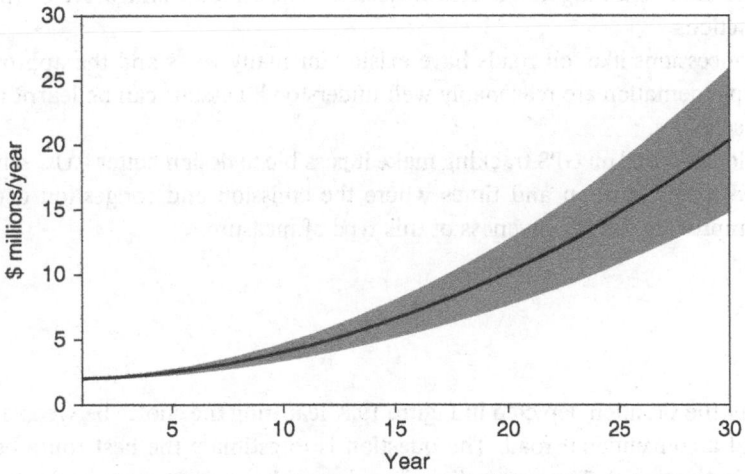

Figure 19.7 P90 revenues after Monte Carlo risk analysis

19.7.3 Scenarios

The scenario planning method discussed in Chapter 17 is seldom applied to private sector projects but it may be employed when considering a RUC scheme. The main approach is usually to base most decisions on a base (or most likely) case and to build relatively simple Optimistic/Upside and Pessimistic/Downside scenarios. Stochastic risk analysis may be used to develop these cases but more often just a set of different assumptions will be used and documented to develop each case.

The optimistic case takes a more aggressive view of the potential for growth and the capture of additional demand through the provision of good services. It focuses on potential upsides to the patronage projections and this is why it is sometimes called 'equity case' as they are the main beneficiaries of upsides.

The pessimistic case usually reflects the perception of risk of the lender and as such it is more concerned with potential downsides of the project. This, often called, 'lenders case' will take a more pessimistic view of economic growth and the possible advent of projects or technologies that will subtract demand from the concession.

19.8 Conclusions

There is a good case for introducing pricing mechanisms in the transport system in an effort to correct some of the biases and imperfections of market prices. The internalisation of externalities, in particular GHG emissions and congestion, should play a key role in justifying this type of intervention. The task of estimating these externalities and devising a charging system to internalise them is difficult. Moreover, the charging system should be easy to understand so that it is accepted and, therefore, influences behaviour.

Road user charges have been proposed as the best way to internalise the significant externalities of road traffic. Despite its economic case, there have been few implementations of RUC schemes and therefore limited experience in their design, modelling, and implementation. There are good reasons to expect that, following the success of electric vehicles, some kind of RUC will be necessary in most jurisdictions.

Transport concessions like toll roads have existed for many years and the approaches to their design and implementation are reasonably well understood. Lessons can be learnt that are applicable to RUC schemes.

New technologies based on GPS tracking make it possible to design better RUC schemes, as they can be tailored to the location and times where the emission and congestion externalities are greater, thus improving the effectiveness of this type of measure.

Exercises

19.1 Consider the situation depicted in Figure 19.8, featuring the choice between a tolled motorway and a conventional road. The question is to estimate the best route between origin O and destination D. The conventional road provides a direct route and takes 45 minutes. The alternative is to go first from O to X to access the tolled motorway in 7 minutes, then travel through it for 20 minutes paying a $1.50 toll, to finally connect with the destination in 5 minutes via link Y-D.

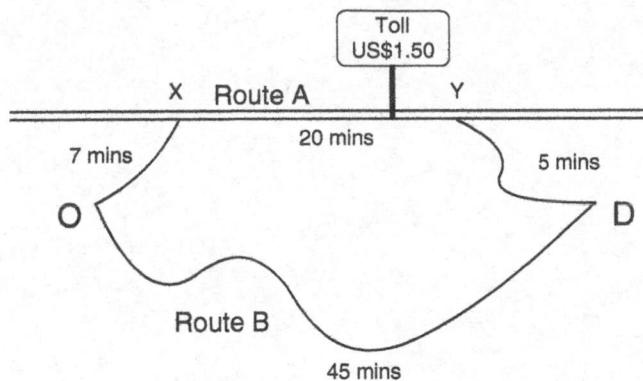

Figure 19.8 Two alternative routes between O and D

Assume now that the SVT is 0.10 $/min (6 $/h) and ignoring any vehicle operating costs:

a) determine the best route to follow; and
b) assume now that there is a motorway premium that reduces the SVT by 20% when using the tolled road but not conventional ones; what is the best route now?

19.2 A managed, or HOT, lane is planned with dynamic pricing. The system will have two general purpose lanes (GPL) and one tolled lane. It is estimated that the capacity of the tolled lane will be 2000 pcu/h and for the GPL 4000 pcu/h. The lanes are 20 km long and have both a free-flow speed of 120 km/h. Both HOT and GPL have a travel time that varies with flow as follows:

$$T = 10 + 10 \cdot (\text{Flow/Capacity})^Z (\min)$$

It is estimated that the average SVT is 0.20 $/min (12 $/h) and over the period some 4000 cars per hour wish to use the lanes of which 400 are HOVs.

a) Find the toll that maximises revenue, the revenue generated per hour and the travel times on each lane.
b) If total demand is 5000 vehicles/h but 100 of those are trucks and they are not allowed to use the HOT, find the optimal toll, the revenue generated and the travel times on each lane.
c) Provide a critical view of the design of this particular managed lane.

Consider the same arrangement for the managed lane as above with a demand of 5400 vehicles per hour, 100 of them trucks that cannot use the HOT lane and 400 are HOVs. Assume now a distribution of SVT and find the toll that maximises revenues, the revenue per hour, and the travel times on each lane.

20

Modelling the Less Common

20.1 Introduction

As simplifications of reality, transport models do not generally attempt to represent all the available transport alternatives. For example, strategic transport models usually ignore active modes, like walking and cycling, because they are normally not loaded onto a network. On the other hand, taxis and delivery vehicles are often omitted from strategic models due to their low market shares and limited impact on congestion; finally, motorcycles, moto-taxis and rickshaws may be ignored too in many locations and for similar reasons.

Ignoring certain transport alternatives is not necessarily wrong; it depends on their relevance in the context of the decisions under consideration. For example, strategic models tend to focus on motorised travel, leaving decisions concerning walking and cycling to approaches less reliant on modelling and simulation. It is also true that modellers have made efforts to include some of these *forgotten modes*; for example, London's strategic models (https://content.tfl.gov.uk/londons-strategic-transport-models.pdf) include taxis and cycling on the grounds that they do represent now a significant share of demand.

In truth, however, another reason to ignore some transport alternatives is that they may pose some difficult modelling issues. In this chapter, we aim to address some of these questions and, where possible, suggest ways forward. This is an area of continuous development, facing the challenge of new technologies and services. For example, up to the 2010s it was reasonable to assume that walking and cycling were restricted to relatively short distance trips (i.e. less than 7 km); today, the popularity of electric bicycles and e-scooters makes it necessary to review this assumption. Moreover, recent years have seen considerable resources allocated to the development of new modes, in particular connected and automated (or autonomous) vehicles (CAVs). Efforts have also been deployed to coordinate transport modes to offer an integrated service often named Transport or Mobility as a Service (MaaS). Litman (2021) reviews the range of these emerging services, identifying their benefits and costs and their implications for transport planning. In turn, Metz (2019) reviews new technologies and explores the business models and policy choices likely to influence the evolution of old and new forms of mobility.

In common with other sections of this book, we consider the issues posed by these old and new technologies from the double viewpoint of supply and demand. Modelling the demand for the abovementioned alternatives poses new challenges; for example, the significant growth in cycling in most countries is only partially explained by improvements in infrastructure (e.g. cycleways and parking facilities). To understand their choice, it is necessary to widen the range of utility function

Modelling Transport, Fifth Edition. Juan de Dios Ortúzar and Luis G. Willumsen.
© 2024 John Wiley & Sons Ltd. Published 2024 by John Wiley & Sons Ltd.
Companion website: www.wiley.com/go/ortuzar5e

variables, incorporating *latent* ones, such as habits, and attitudes towards health and the protection of the environment (Gutierrez et al. 2020). Most new modes do not operate on a fixed route and schedule (e.g. taxi or paratransit services) and, therefore, require the development of a supply model as they are not always available everywhere (Domarchi et al. 2019a).

Some of these modes, like cycling and motorcycling, also have a public version that we will consider together with by-the-minute car hire and e-scooters. These *public services* can be characterised as vehicle-sharing arrangements that when incorporated into a MaaS offer may provide a multi-modal, door-to-door, service. The eventual advent of CAVs will have additional implications and they are, therefore, considered separately. All vehicle-sharing systems have the potential to replace owned cars or motorcycles and, therefore, weaken some of the standard vehicle ownership models discussed in Chapter 18.

For each of these *less common* cases, we identify below some of their main operational characteristics and potential impacts; we also point to some research reports that shed light on them. Then, we outline modelling approaches that could be used to treat these new cases. Given the limitations in our knowledge about these new or forthcoming mobility services, their modelling is likely to be restricted to exploring their potential impact as part of future scenario planning.

We will start our analysis with some *new modes* that offer, or would offer, scheduled services; in particular, cable cars, hyperloop and other systems that operate on a timetable and fixed route, as these are simpler to incorporate in a model.

20.2 New Scheduled Services

Hyperloop has been suggested as a new, high-speed service running on capsules inside a tube with a partial vacuum to reduce friction (Hansen 2020). Although not a completely new idea, it was promoted heavily in the 2010s as a fast mode that would not conflict with surface modes as it would run mostly underground. However, we are not aware of any commercial implementations of hyperloop and its future is uncertain.

Another not quite new mode of transport is the aerial cable car, with several implementations for skiers and tourists, which has also been extensively used as a public transport mode in steep areas of cities, in particular Medellín and Bogotá in Colombia (Escobar et al. 2022) and La Paz, Bolivia. Aerial cable-car public transport systems work particularly well as feeders to metro and bus rapid transit (BRT) systems.

These types of cases can be treated as a classic public transport system with routing independently from the road network, a known capacity, stops and frequency. Capacity restrictions in cable-car systems may create long queues during peak hours or when many passengers alight from an underground train and join the cable-car service. Care should be taken when estimating queueing delays in these cases.

In some countries, shared taxis operate almost as a fixed route and frequency service, or at least the service is so frequent that waiting times are below 5 minutes. Although these services could be seen as a case of ride-sharing (discussed below), in many contexts they can also be treated as a conventional public transport system; in this sense, their defining characteristics would be their high frequency and (almost) fixed route.

20.3 Walking

Walking is usually identified as a mode in Household Mobility Surveys; it is also the most versatile mode of travel. Walk trips are usually short and do not contribute to congestion and emissions; for these reasons, they are often included only at the Trip Generation stage of strategic models and ignored in subsequent sub-models (i.e. they are normally considered intrazonal trips). Local micro-simulation models, on the other hand, often treat them explicitly, as the configuration of junctions and the availability of good sidewalks (pavements in the United Kingdom) influence delays to both pedestrians and traffic (Vallejo-Borda et al. 2020).

Health and environmental considerations have convinced authorities that more should be done to encourage walking and, therefore, policies and projects supporting pedestrians are becoming more strategic in nature, increasing the interest in modelling them. The aim would be to identify interventions, at a large scale, that would best support walking. This requires acquiring better knowledge about which features of infrastructure and policies pedestrians value most, and also which policies/actions may persuade car users to walk rather than drive in the case of shorter trips.

Habib et al. (2014a) researched the generation of walking trips for commuting. They used econometric models to investigate the importance of considering walking distance jointly with walking propensity in the trip generation model. Their data sources were household travel surveys collected in the Greater Toronto and Hamilton areas in three different years. As expected, they found that for walking, travel distance should be considered jointly with the propensity to walk. The relationship between car ownership and walking-trip generation proved to be strong, with high car ownership resulting in fewer walking trips. Their models suggest that the baseline walking propensity and distance have remained unchanged over the years, despite significant efforts to encourage active modes through mixed land-use policies. This finding would need to be confirmed in other contexts with different land use configurations and soft measures to stimulate walking. This is an area deserving additional research and investment.

Note, that a great deal of work in this area has been done by modellers from other disciplines, and some of their work has already permeated the transport profession. We refer here to the seminal work of Bill Hillier and colleagues on the development of the Space Syntax approach (Hillier and Hanson 1984; Koohsari et al. 2016; Penn et al. 1998; Turner 2007).

20.4 Cycling

In the last century, it was possible to say that extensive use of cycling for commuting and non-sport purposes was restricted to a few countries like Denmark, the Netherlands, and China. However, cycling experienced a significant growth during the first two decades of the twenty-first century, and it now represents a mode of travel that is difficult to ignore, with a significant share of trips in many places. The reasons behind this growth in popularity are partly the improvements in cycling infrastructure (connected cycle lanes with a good standard), support at destination, provision of parking facilities, and, in the case of companies, showers and changing rooms for cyclists. But perhaps the most significant driver has been a change in attitudes, an interest in the health and fitness benefits of cycling, and the aspiration to contribute to a reduction of emissions from internal combustion engines (ICE); these are *softer*, attitudinal latent variables.

It can be argued that in congested cities, most able travellers could choose cycling over other modes on the grounds of savings in travel time and costs. But with a few exceptions, this has not been the case. It is important, therefore, to identify deterrents to cycling and factors that make cycling attractive beyond travel time and money.

At the beginning of the century, Ortúzar et al. (2000a) investigated the potential bicycle demand in Santiago, Chile, a city where cycling represented less than 2% of the trips. The reasons for this scant use had been traced back to: (i) a sense of insecurity on the road; (ii) the lack of proper/secure bicycle storage facilities; and (iii) culture, as a low social status was attached locally to riding a bike except as sport. The main objective of their study was to reconsider the use of bicycles as an alternative mode of transport in the context of a future city with considerably better transport and cycling facilities.

They conducted a household survey, including a stated preference experiment for potential bicycle users, and used this data together with information from a large-scale mobility survey, to estimate and apply models for the city as a whole. They found that there were sectors of the city where bikes could capture more than 10% of the trips, and that, on average, the use of bicycles could jump to some 5.8%. Their results also showed that trip length was indeed a fundamental variable. It is interesting to note that 20 years later, the share of bicycle trips in Santiago is now 10%, even if the full improvement in cycling infrastructure has not materialised yet; it seems that the main driver was a change in attitudes.

Wardman et al. (2007) studied the factors influencing the propensity to cycle to work. They concluded that ... 'Time spent cycling is valued almost three times more highly than travel time for the other modes'. Therefore, while cycling can be faster than a bus, it can still cost more in value-of-time terms. This can be explained in terms of exposure to the weather, risk of accidents, and the physical effort involved. The authors commented that it was strange that this high *value of time* spent cycling is often ignored in the appraisal of schemes that protect and facilitate cycling.

The current perception is that cycling offers many advantages over traditional transport modes in highly congested cities. One of them is shorter travel times (Hamilton and Wichman 2017; Rodríguez-Valencia et al. 2021), but this is not enough to explain cycling growth. Other factors include improvements in the physical and mental health of users (Oja et al. 2011; Mueller et al. 2015), economic savings, due to lower costs, and social benefits such as zero emissions or noise, among others (Pucher and Buehler 2012).

Subjective variables, such as attitudes and perceptions, also play a role in the adoption of cycling by different types of users (Maldonado-Hinarejos et al. 2014; Muñoz et al. 2016; Sottile et al. 2019). In particular, it has been found that individuals with a pro-environment attitude find cycling more attractive and that this may be determinant in explaining a shift towards this mode (Handy et al. 2010; Heinen et al. 2011; Maldonado-Hinarejos et al. 2014; Fernández-Heredia et al. 2016; Clark et al. 2021).

In particular, Clark et al. (2021) investigated the impact of bicycle facilities on perceptions of 'bikeability' using data from a large survey. Respondents were asked to rate a series of images of hypothetical roadways in terms of perceived comfort, perceived safety, and willingness to try bicycling. Latent-class regression was used to model these responses, with bicycle facility type and roadway characteristics as explanatory variables and attitudinal factors (car preference, bike enjoyment, risk tolerance, and anti-exercise) as covariates for class membership. They were able to identify two pro-bicycle classes: *risk-embracing* and *risk-cautious*, and one *pro-car* class. The impact of segregated and protected cycling facilities on perceptions was much higher for the pro-bike/risk-cautious group suggesting different approaches to promote cycling to them. Work on the role of perceptions, the built environment, and cycling facilities on the use of bicycles has continued, as shown in the work of Ricardo Hurtubia and colleagues in Chile (Echiburu et al. 2021; Rossetti et al. 2018, 2019).

On the other hand, a strong deterrent to cycling is the subjective perception of risk (safety and security) associated with this mode. This reflects concerns about personal safety when sharing road space with motor vehicles and the risk that their bicycle could be stolen; people who are very sensitive to these risks are less likely to use a bicycle for their trips (Titze et al. 2008). Further, many studies have shown a relationship between the *habit* associated with cycling and greater continuity in cycling, as well as in a positive attitude towards cycling (Heinen et al. 2011; Gatersleben and Haddad 2010; Gutierrez et al. 2020).

Another obvious barrier to cycling is trip distance. For example, Heinen et al. (2011) found that distance was a decisive factor in choosing cycling for trips to work. Different estimates for the feasible range for cycling trips vary. Winters et al. (2011) concluded that trips of less than 5 km were attractive for cycling, while Mandic et al. (2020) estimated the ideal range as 2–4 km; Heinen et al. (2011) found that longer trips, up to 10 km, were feasible. These estimates depend, of course, on context: facilities, gradients, and climate; the use of electric bikes extends the range of cycling trips.

Factors supporting the choice of cycling for regular trips include cycle-inclusive infrastructure in good condition, direct routes between O–D pairs, safe bicycle parking, speed limits for motor vehicles, and showers and lockers at the destination (Habib et al. 2014b; Gutierrez et al. 2021; Márquez et al. 2021). It has also been found that certain land use arrangements are correlated with increased cycling, such as mixed land uses, green areas, and a higher density of commerce (Titze et al. 2008; Handy et al. 2010; Oliva et al. 2018).

Gutierrez et al. (2021) explored several of the above issues in an environment considered aggressive to cycling: the city of Barranquilla, Colombia. They designed a stated choice experiment geared to students and staff living within three kilometres of two college campuses and travelling by bus or bike. They used two latent variables (*Unsafety/insecurity* and *Convenience*) in addition to typical level-of-service (LOS) variables in their choice models. They found that the perception of Unsafety/insecurity had an important negative effect on the probability of using bicycles. Interestingly, in terms of Convenience, their results suggest that it was just as important to have showers on campus as bike lanes on the route.

In summary, there is an understanding that route quality matters for cycling, and that other subtler factors like interest in health and the environment and the perception of accident risk are also critical. Nevertheless, the inclusion of a demand model capturing these influences in strategic modelling and appraisal is still experimental.

20.5 Motorcycling

Motorcycling is a common mode of transport in many countries in Asia and Latin America (Chu et al. 2019; Estupiñan et al. 2015). It provides an efficient way of moving people (and sometimes goods) using a low-cost vehicle that is easy to maintain and park. Two-stroke engines were very polluting, but most motorcycles today use cleaner four-stroke power. The mode's main problem is still high accident risk and exposure to weather conditions; in some places, motorcycles are seen as a 'stepping stone' to car ownership (Iglesias et al. 2022).

The ownership and use of motorcycles, despite their popularity, are much less researched than those of bicycles. We reviewed a joint motorcycle and car ownership model in Section 18.3.3.3, and Willumsen (2014) illustrates such a model for a city in Latin America.

Burge et al. (2007) modelled motorcycle ownership and mode choice in the UK, a country with low levels of motorcycle use. However, the structure of the ownership model is of interest; it is a disaggregated Nested Logit model based on stated preference surveys combining the choice of motorcycle ownership and engine size.

Wen et al. (2012) used panel data to develop motorcycle ownership and usage models based on a large-scale questionnaire panel survey of vehicle owners in Taiwan. They tested Multinomial, Nested, and Mixed Logit (ML) formulations and three types of panel data regression models – ordinary, fixed, and random effects. They found that the fixed effects model provided the best results, indicating strong evidence for the existence of heterogeneity.

Despite the paucity of research in this field, compared to car ownership, the examples above show the feasibility of developing motorcycle ownership and use models using classic data collection methods and functional forms. In that sense, motorcycles can be treated in a similar way as private cars, using the same choice models and assignment techniques. Although motorcycles travel mostly at the speed of the rest of the traffic, their drivers take advantage of the vehicle's narrow profile to slip to the front of the queue at junctions with advantage (i.e. motorcycle queueing delay is different and shorter than that of other road vehicles); however, drivers also tend to exhibit reckless behaviour while lane-changing, leading to many serious accidents. In summary, given the current greater interest in reducing inequalities, this is a surprisingly under-researched field.

20.6 Parking

This is certainly not a new mode but presents some interesting challenges both for aggregate and disaggregate models. There are two conditions when specific parking models may be needed: trips to a central area with limited parking spaces, and trips that may use park-and-ride facilities. In each of these cases, the drivers face several parking alternatives, as illustrated in Figure 20.1: on-street and off-street (lot), public or reserved parking for certain employees or residents, priced by the hour, on subscription, or free, but with limited stay duration.

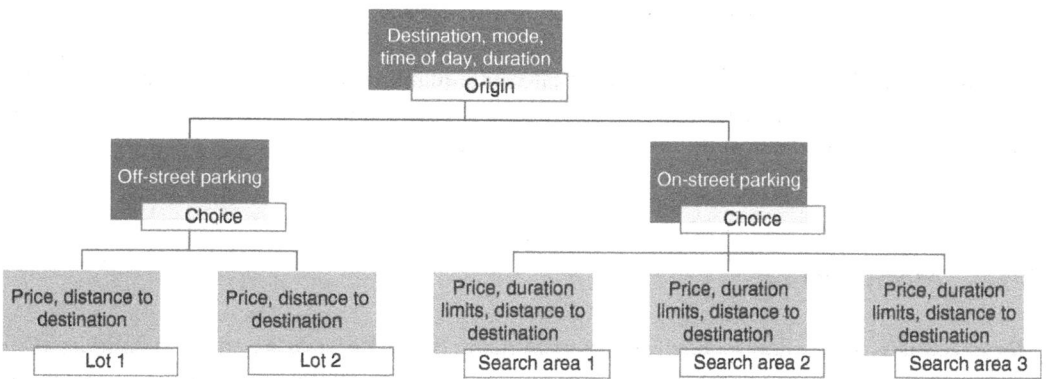

Figure 20.1 Modelling parking alternatives

All these types of parking will have capacity constraints, and, in many cases, this will generate search times to find a suitable parking space. In most instances, there will be a walk time from the parking spot to the final destination. These factors should not only influence the utility functions in a choice of parking location model but also mode choice and, in the case of discretionary trips, destination choice.

Modelling these factors is a difficult task. However, if these parking constraints are ignored, the forecast number of car trips could well be overestimated. Moreover, given the reluctance of decision-makers to adopt road user charges, it may well be that parking policy and pricing are the next best tools to manage demand.

It is important to ensure an appropriate balance between car parking demand and supply in all forecasts made with any type of transport model. In some cases, simple approximations to a fully specified treatment of parking may suffice. In other contexts, it may be necessary to model parking choices with a greater level of detail. We describe here some of the features that may be required when modelling parking choices.

Parking is best modelled using tours rather than trips. The activities that travellers undertake at a destination usually have a minimum duration, and this may preclude the use of certain parking spaces with time limits. Parking duration often determines the parking charge. Moreover, use of tours allows to place constraints on time changes for outbound and return trips.

The availability of a parking space depends on the car park capacity and car park occupancy levels at the time the space is required. This needs considering how many cars have already parked there before the time period and how many could park during it. This means that the day should be divided into a number of time slices or periods, and the number of arrivals and departures from the car park should be recorded for each time slice. The capacity of on-street parking is more difficult to establish with precision; there is a trade-off between an ideal parking location and additional walking time, but some approximation may be sufficient.

The calculation of the parking accumulations requires information about parking durations, which will vary by trip/tour purpose. Car park occupancy levels may also be affected by the search and queuing times required to access a space, the charges levied, and the length of stay permitted. This requires a representation of the trade-off between search and queuing times, charges, and walking distances and times, in order to model the choice of parking location.

It is also desirable to include an increased segmentation of demand to represent the situation where access to certain spaces, such as private non-residential spaces, is restricted to particular users (e.g. some employees and clients). In locations with uneven enforcement, it may be necessary to include illegal parking as an option.

Modelling Park and Ride (P&R) offers some additional challenges. P&R is often seen as a major component of transport plans for urban areas and, therefore, it is necessary to model it with a reasonable degree of realism. P&R facilities are usually incorporated to provide a better access to a rail, metro, or BRT services to a town centre. In these cases, the choices available to travellers include alternative stations for joining the main public transport service that will take them to their destination. This involves changing the length of the access and main segments of the trip with different implications for travel time and costs (Ortúzar 1983).

P&R may attract users from both car all-the-way and public transport all-the-way; this suggests that its choice may be either a sub-mode of car or a sub-mode of public transport. The nesting structure should be determined, in principle, by the fit of the choice model to the data available (Ortúzar 1980).

When modelling P&R, it is important to represent the balance between out-and-return trips across the day, in particular to ensure that P&R is used in both directions. The need to track car park occupancy is as important as in the general urban case; a particular park-and-ride site may be preferred by some drivers as it offers spaces unavailable at the time at a preferable location.

Research has shown that many of the features above can be treated using discrete choice models. For example, Axhausen et al. (1989) proposed an MNL model for the choice among three parking types: illegal parking, on-street parking, and off-street parking. The alternatives were described by their access time, search time, egress time, and parking cost. The study emphasised the importance of distinguishing between different groups of individuals when a parking policy was set.

Teknomo and Hokao (1997) introduced an MNL model for the choice between an on-street parking space, off-street parking lot, and an off-street, multi-storey parking facility using revealed preference data. Hess and Polak (2004) estimated a random coefficients Mixed Logit (ML) model for the choice among five parking types: free on-street, charged on-street, charged off-street, multi-storey parking facility, and illegal parking using stated choice data. More recently, Ibeas et al. (2014) examined car driver's behaviour when choosing a parking place; the alternatives available were free on-street parking, paid on-street parking, and parking in an underground multi-storey car park. A ML model, allowing for correlation between random taste parameters and estimated using stated choice data, was used to infer values of time, both when looking for a parking space and for accessing the final destination. They also found that the perception of the parking charge was fairly heterogeneous, depending both on the drivers' income levels and whether or not they were local residents.

Hilveert et al. (2012) proposed a framework for parking choice and search behaviour for three time-space phases: (i) pre-trip static decision; (ii) during travel passive search; and (iii) in-area search strategy adaptation. Their model estimation confirmed that the choice of parking location was influenced by parking costs, search time, walk time to the destination, facility type, and decision-maker characteristics.

Exploring the advantages of agent-based modelling Rodríguez et al. (2022) used a microsimulation parking model based on two sub-models: choice of parking place and search for a parking place. Their model was able to dynamically simulate various policies based on charging for on-street parking spaces with charge updates at short time intervals of between 5 and 15 minutes.

The full modelling of parking choice and P&R may not be practical in the context of a large-scale strategic model, and some simplifications may be needed. Possible simplifications include:

- use trips rather than tours and independent modelling of time periods;
- geographic restriction of P&R, specifying catchment areas for each site;
- manual allocation of origin and destination pairs to the individual P&R sites;
- use of highway assignment and simple mode choice models;
- base-year P&R representation within a public transport assignment model.

A highway network can be coded so that a particular destination (centroid) may be accessed from different car parks with appropriate walking times; the estimated charge can be added as an extra link cost, and the link will have a capacity equivalent to that available during the modelled period (for example, assume the car park is empty at the beginning of the AM peak). The treatment of capacity restraint is more difficult, but it is possible to assign a delay-occupancy function similar to a travel time-flow relationship; the delay in this case would be associated with the parking search time and/or the extra walking time to the destination.

20.7 Demand-Responsive Transport

20.7.1 Introduction

Taxis are probably the most common form of demand-responsive mobility. They do not operate on a schedule and must be requested either on the street or through an application. Taxis pose difficult modelling questions, in particular their availability over time and space, which is a critical element of the level-of-service they provide. Second, taxis also travel without passengers, empty, when cruising for customers or when responding to a remote request; therefore, they generate more vehicle kilometres travelled than private cars to deliver the same useful trips.

These issues are not exclusive to taxis. Recent years have seen the introduction of new *on-demand* services under a variety of names and with a wide range of characteristics. They include:

1) Shared bicycles (electric or not) and electric scooters for hire by the minute (often called *micro-mobility*) can have fixed stations or be *free floating* within a defined geo-fence. They usually operate on a subscription plus payments on use ('pay-as-you-go').

2) Shared cars and vans hired by the minute; these can be, again, electric vehicles (EV) or internal combustion engine (ICE), back to station (or round-trip rental) or free-floating (sometimes called one-way rental); they often require subscription to a company, institution, or *club*, plus pay-as-you-go for use.
3) Taxis and Uber-like services, either EV or ICE; usually pay-as-you-go.
4) MaaS, an integrated multi-modal service that may have a flat fee for a period, pay-as-you-go, or a combination of these.

Demand-responsive transport (DRT) services can be considered a form of vehicle sharing. They can be provided by a company and, in the case of car/van sharing, they could be offered as peer-to-peer (P2P) or business-to-consumer (B2C). In the case of MaaS, some services could be for a single person or for a small group travelling together from the same origin to the same destination. Alternatively, travellers may share part or all of the services with other persons not known to them; this would be a case of ridesharing. The power source, EV or ICE, is important as a way to reduce emissions and for the need to provide facilities for charging.

The eventual introduction of CAVs adds another degree of modelling complexity. An owned CAV may be shared with other members of the family and friends and, in that case, travel empty to serve them as necessary. A CAV may also operate as owned some of the time but be available to a commercial MaaS fleet when not required by the owner, or it may be shared on a P2P basis.

When facing these new services, the natural inclination is to aim for an Agent-Based Model (AgBM); however, just considering all possible variations is a daunting task. Moreover, the future market share, or even existence, of some of these new modes is uncertain, and they currently represent only a small market penetration in the base year. There are many unknowns that restrict the use of any DRT model just to exploratory research and scenario testing until more is understood about how people would react and use these alternatives.

Nevertheless, there are some key questions that should be addressed, at least to develop some useful scenarios for strategic planning and modelling. For example:

- to what extent does vehicle sharing reduce vehicle ownership, and how this will influence travel?
- what type of vehicle and ride-sharing system would effectively reduce vehicle kilometres travelled (VKT) and, therefore, lower energy consumption, emissions, and time wasted in congestion?
- what type of person and user would be attracted to each of these new services?
- which of these services would be financially sustainable and therefore more likely to survive without subsidies?

It is important to answer these questions to develop the right policies in advance of the introduction of new services. Figure 20.2 illustrates the many dimensions that may characterise the service offered by DRT modes.

DRT may be offered with different forms of sharing: a vehicle that can be hired by the minute, a service that can be requested for a single trip or a ride-sharing mode that serves different passengers and destinations. These can be offered in a P2P or B2C format and with the exception of hailing a taxi in the street, they will all use applications (apps) to request and pay for the service.

The type of booking may also differentiate DRT services: whether the service responds more or less on demand in quasi-real time (short wait) or if it must, or can, be booked in advance (i.e. reservation based). Some suppliers offer only one mode, whereas others offer a door-to-door multi-mode service through an integrator. A high degree of integration with public transport is an important characteristic of a good DRT service.

In all these cases, the role of the integrators and dispatchers is critical. They can receive a service request from a user and have to determine the best way to serve it. In the case of car or bike hire, this

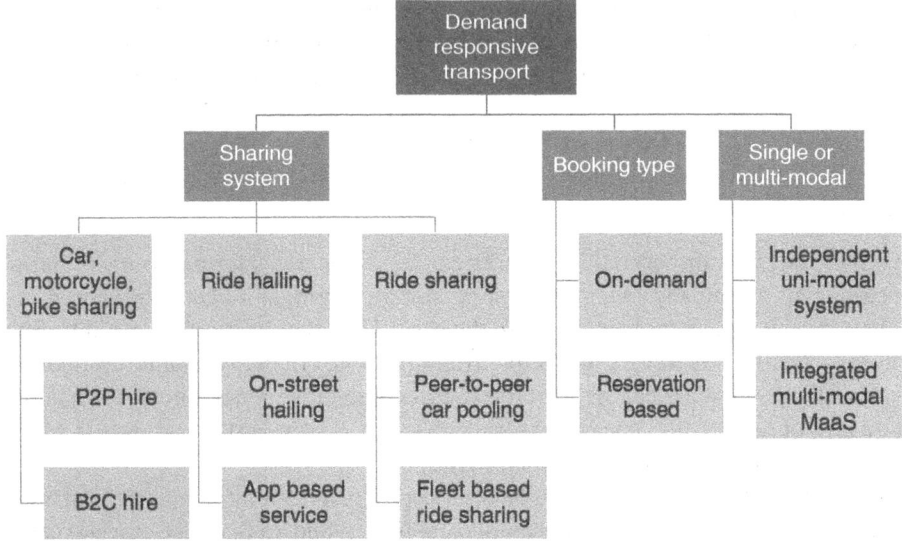

Figure 20.2 Characteristics of demand-responsive services

role is shared with the user, who must identify the nearest unit available (and the level of charge in the case of an EV). In all other cases, the dispatcher must locate available (or soon to be available) vehicles at suitable locations, so that it can offer one or more alternatives to the user. The alternatives may be different in terms of travelling alone or sharing the vehicle with others, the fare, wait and travel time. For some ride-sharing services, there may be a request to walk to a suitable pick-up point to reduce detour time for all passengers. The user can then select the service that suits her best. This type of arrangement is illustrated in Figure 20.3, which shows the importance of the information available to the dispatcher and the task of optimising the use of vehicles to serve customers.

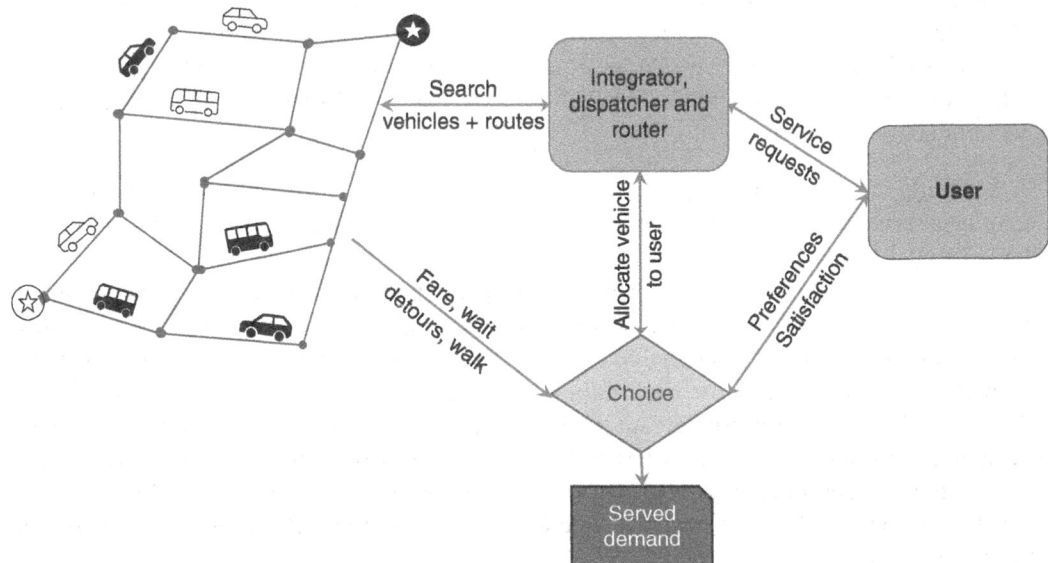

Figure 20.3 The user and dispatcher in a DRT system

We look at some of the individual cases of DRT below, considering the characteristics of the service they can provide and the impact they may have on demand.

20.7.2 Micro-Mobility Sharing

In this group, we include bike and e-scooter sharing. They have become a popular way of moving in cities in their two main versions: back to a station or free-floating. Adoption has increased with the introduction of electric bikes, especially in cities with hilly areas where downhill hire preferences require a constant effort to relocate bikes.

Back-to-station and free-floating services offer slightly different services. Back-to-station has inevitably a limited spatial coverage, as it is not economical to install stations everywhere. This poses the challenge that a given station may be empty when a bike is needed or full when the user is searching for a location to park the bike. From the point of view of the supplier, there is the additional cost of providing the station and optimising its location, something that Mix et al. (2022), among others, have investigated. The supplier may have the additional cost of relocating bikes to areas where they are needed at particular times.

Free-floating services are more flexible in this sense, but the problem of finding a bike or e-scooter when needed remains. The area of legal use is usually defined by a geo-fence, and optimising it to maximise use and reduce idle time is also a challenge. Free-floating services have also been blamed for the parking of units in unsuitable places, creating a nuisance for pedestrians and degrading the built environment. This has resulted in authorities defining specific locations for parking bikes and, in particular, e-scooters.

El-Assi et al. (2017) investigated the factors affecting Toronto's (station-based) bike share ridership. The models developed show the importance of socio-demographic attributes, land use, and built environment, as well as different weather conditions on bike share ridership. The effect of bike infrastructure (bike lanes, paths, etc.) was also found to be critical in increasing bike-sharing demand. Not surprisingly, the study revealed a significant correlation between temperature, land use density and bike share trip activity.

Li et al. (2019) investigated the use of free-floating bike-sharing services in Kunming, China, contrasting them with private bikes and station-based systems. The data was collected through questionnaires, and then a MNL model was applied to identify the key influential factors. The results are not unexpected, but it is helpful to find support for intuitive perceptions. Theft risk and maintenance costs were the main deterrents to owning a bike when bike sharing was available. Recurrent trip makers preferred station-based services, whereas those with more flexible schedules opted for free-floating ones. In Kunming, and presumably elsewhere, private and free-floating bikes are preferred for longer trips.

Mix et al. (2022) looked at the optimal location of bike-sharing stations. Evidence has shown that bike-sharing systems (BSS) can only provide benefits when their network is adequately designed, in order to capture ridership and generate demand. This study proposed an integrated approach to model the demand for bike-sharing services and the optimal location of stations in the system, based on built environment and accessibility-based variables. The methodology had two steps. On the first, trip generation models were estimated using multiple regression for different types of trips and periods of the week. On the second step, maximum demand coverage models were developed to allocate the BSS stations, according to the trip generation models and to different proposed scenarios. To test the proposed methodology, information from the BSS of Santiago, Chile, was used. Results suggest a relationship between the built environment and the use of public bicycles, with a main effect from residential and office land uses and the presence of long bicycle lanes near the stations. In addition, they confirmed the presence of endogeneity, associated with the location of BSS stations and BSS demand generation, which was controlled using accessibility variables. As for the optimal location models, their outcomes differed significantly from the observed spatial distribution of stations in Santiago, with higher density in central areas and along corridors with cycling infrastructure. The predicted demand level for the optimal distribution of stations was 64% higher than the observed demand. This study confirms the benefits of an integrated modelling of trip generation and station location to foster higher public bicycle usage, a relevant point for BSS decision planning and the promotion of more sustainable mobility.

20.7.3 Car, Motorcycle, and Van Sharing

We consider these three modes as a group because they share similar characteristics. They are also offered as station-based or free-floating and present similar problems of availability when needed, parking and the objective of the supplier to maximise their use. It has been argued that the availability of these vehicle-sharing services tends to reduce vehicle ownership, and this impact has been observed, especially in dense urban areas. Reducing the size of the fleet (owned and provided) helps in reducing the emissions generated in manufacturing the vehicles; this is partly counterweighed by the shorter economic life of intensely used shared vehicles.

An early version of car sharing is the taxi, but in this case, one must also cover the cost of the driver, a major element of the fare. Modern car sharing (no driver) also incurs additional costs for refuelling, cleaning, maintenance, and repairs that must be covered by the fee.

Dong et al. (2020) investigated when it was more advantageous, cost-wise, to use taxis or car-sharing services in Beijing. Their results suggest that for very short and very long trips, the taxi offers a cost advantage; this is also the case for evening trips, but car sharing seems to offer a cost advantage the rest of the time. These findings are, of course, specific to Beijing and at that particular time.

One of the arguments in support of car sharing is that it should reduce car ownership. Abandoning car ownership entails an element of risk: what will happen if I need a car and there is no one available? Jain et al. (2022) investigated the potential reduction of car ownership in Melbourne, Australia, using data from a survey of members and non-members of car-sharing clubs (all respondents had a shared car facility within a 10-min walk of their home). One in three households reduced car ownership and, rather puzzling, most reductions occurred in the year prior to joining the car share scheme, suggesting a planned decision. Fleet-based car share members reported a larger reduction in car ownership compared to P2P members; this difference may be due to the perception that in P2P arrangements there may be less certainty of accessing a car when needed. Residents in densely populated inner areas used car share to avoid or delay car ownership, while middle suburb residents used car share to avoid purchasing a second car. Either way, parking spaces are released for other uses, and emissions from manufacturing cars are reduced, even if car use may be unaffected.

Chicco et al. (2022) found similar reductions in car ownership in three German cities and compared station-based and free-floating services. They found that all car-sharing users reported lower levels of car ownership than before joining the service, but station-based members were significantly more likely to reduce car ownership than free-floating members. Again, this may be linked to the need for confidence in finding a vehicle when needed; in modelling terms, this reflects an expectation about the level of service that each type of car-sharing system would offer.

It is also interesting to learn whether car sharing substitutes trips for more sustainable modes like walking, cycling, and public transport. If that is the case, this substitution would undermine the positive impact of reducing car ownership. There is limited research on this impact. Göddeke et al. (2022) investigated this issue using data from 80 German cities. They found that car-sharing members used walking, cycling, or public transport 1.4–1.5 times more often than non-members; also, members used motorised private transport less often than non-members. However, this does not prove causality; it is equally likely that people who preferred these sustainable modes also tended to join car-sharing arrangements. To explore the effect of an increase in the car-sharing fleet, they used MNL models for the years 2013, 2015, and 2017. They found that increasing car-sharing supply and being car-sharing members were not effective measures to increase walking, cycling, and public transport use. Additional research is needed to provide an unambiguous answer to this key question; however, policies are needed now to decide whether support for car sharing is an effective way of encouraging better mode choices.

This dilemma was explored by Nansubuga and Kowalkowski (2021), who reviewed the policy-making literature in the context of vehicle-sharing services. They recognise that EV car sharing could be a good introduction to the use of electric vehicles replacing ICE power. They concluded that policy interventions supporting car sharing are needed but not if they are at the expense of public transport. Policy should make car sharing a more attractive alternative than owning a car and a complement to public transport.

20.7.4 Connected and Automated Vehicles

CAVs have been forecast to revolutionise transport and mobility for many years. Initially, they were predicted for around 2020, but their commercial and widespread availability have been delayed several times. The problem of delivering fully automated and safe CAVs, that is at Level 5, has proved more difficult than originally anticipated.

Level 5 CAVs will drive themselves without assistance anywhere in the network. A less ambitious version, Level 4, should be able to drive unassisted in certain restricted environments or domains. At the time of writing, Level 4 CAVs are only available on an experimental basis.

It has been suggested that CAVs will have a number of impacts, the most notable being a reduction in accidents by eliminating human error; the most relevant impacts to transport modelling could be:

- their ability to be driven at shorter headways depending on whether they are also connected to other vehicles or not; this should increase the capacity of all lanes and perhaps also that of junctions;
- trip induction, at least because some people unable to drive today would be able to use them in the future; this will also depend on whether they are hired or owned and how they are used;
- lowering the value of in-vehicle time as the user will be able to undertake other activities while travelling, for example, working or sleeping;
- the introduction of much lower cost taxi-like services, deployed as automated MaaS;
- CAVs may be cars and also small and large buses and trucks;
- they may have a potential negative impact on public transport, in particular when they are low-frequency services.

Much has been written as to the potential impact of CAVs, either owned or rented by the minute, including how cities will eventually have to adapt to their use, see for example, Metz (2019), Litman (2021), and Zhang (2022).

There are many uncertainties surrounding the time of general introduction of CAVs, their characteristics, and how they will be used – probably both as privately owned and as part of a mobility fleet. Nevertheless, they may still need to be incorporated into different scenarios to explore their potential impact on cities and countries.

20.7.5 Mobility as a Service

MaaS is a type of mobility facility that uses a digital platform or application to enable users to plan, book, monitor, and pay for a service that involves multiple modes. Travel planning typically begins with a journey planner, which is available in most places. The user can then select their preferred trip, based on cost, time, and convenience. It may then book the service, either for immediate use or in advance, and monitor its delivery, for example, by being reassured that the mode is approaching. The charge for the complete service is normally collected when the journey is completed.

It is clear from this definition that DRT and MaaS overlap. Both are demand-responsive and both (in most cases) use applications to request the service. We identify MaaS separately here to refer to the provision of door-to-door multi-modal services, a challenge requiring the integration of different modes to provide a *seamless* service with a single fare.

There are many ways to arrange these services, as shown in Figure 20.2. Of course, many MaaS journeys will involve a mixture of modes, for example, bike sharing to a station, ride sharing on a train, and e-scooter rental for reaching the final destination. Again, it is important to ascertain where the travellers attracted to these new mobility services come from. If they take customers away from public transport, this is likely to increase energy consumption and emissions. Unfortunately, as public transport users are familiar with the concept of sharing a ride and have already accepted the idea of a walk to a pick-up point, they could be more easily attracted to the advantages of MaaS than car owners. This is an important challenge for authorities trying to develop policies that reduce congestion and emissions whilst satisfying travellers' needs. Transport modelling, even if it is only on a scenario basis, should help shape these policies and regulations.

A complete book has been dedicated to the subject, so we refer our readers to it for more details (Hensher et al. 2020).

20.8 Modelling Demand-Responsive Mobility

20.8.1 The Challenge

Modelling demand-responsive mobility presents some new challenges. For a start, it does not serve a well-defined public transport system with routes and schedules; therefore, it is necessary to develop a model for the supply of these new systems. This must include aspects such as the size and characteristics of the fleet, the method for allocating vehicles to customers and the modality of use offered (vehicle or ride sharing), the repositioning of empty vehicles to where they are needed at any one time and the type of pricing (fixed, period, or dynamic pricing with *surges*). Moreover, it is also necessary to estimate the level-of-service (LOS) offered to travellers (walk, wait, transfer, in-vehicle times, fares) and the impact on travellers, operators, and the city.

Models attempting to forecast the future demand for shared mobility usually combine analytical or agent based approaches with assumptions about the type of service that could be delivered in practice and the share of travellers that would use them or be prevented from using private modes.

It is common, even in detailed AgBM, to assume that only a few services will be offered; for example, two types of ride-sharing services in the case of simulations for the International Transport Forum (ITF 2020). It has also been common to assume that total demand will be fixed (i.e. no induction) and, in some cases, that people will be forced to change mode because some alternatives will be banned or highly charged. This is a modelling short-cut to explore the potential for shared mobility under specific, usually favourable, conditions.

A more realistic approach would be to assume that DRT service demand depends on a range of choices open to individuals or groups of travellers. This poses a particular challenge, as it is not clear how to represent requests for services and the role of a dispatcher (human or AI-based) allocating vehicles to meet demand in an aggregate, zonal-based model. This is the case even when modelling car or bike-sharing services, floating or otherwise, as the availability of a unit nearby depends on earlier fulfilled requests. This type of issue is more naturally treated in an AgBM (see, for example, ITF 2020).

On the other hand, any model should be able to consider the size and characteristics of the fleet available and compare it to requirements for particular levels of demand. The fleet will also need 'stabling' (except for P2P), with charging facilities for EVs and a good rationale for repositioning to serve customers more effectively.

Figure 20.4 illustrates the different dimensions that might be considered when modelling demand-responsive mobility either in a strategic or microsimulation model.

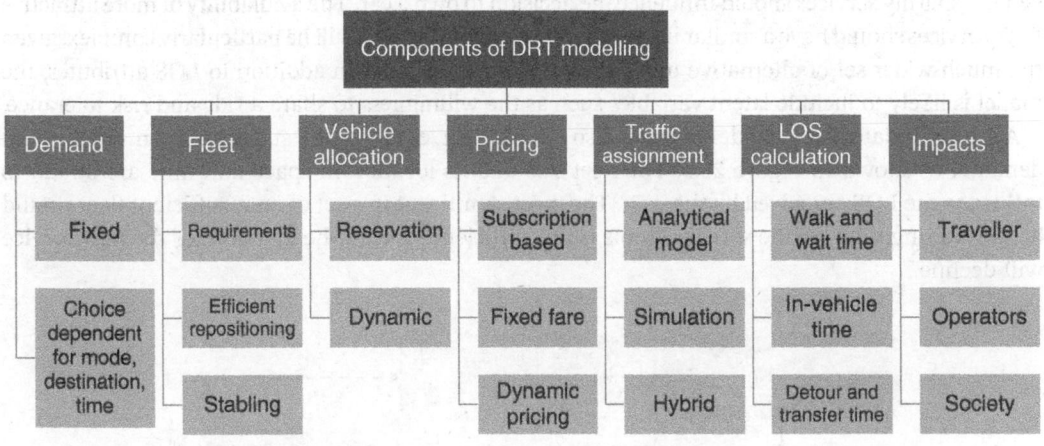

Figure 20.4 Modelling requirements for DRT

The conventional approach to assignment is to distribute trips between O–D pairs on routes and services identified using an aggregate analytical model, including congestion and crowding effects. In the case of DRT, there are two different 'trip matrices', for travellers and for vehicles. The need for repositioning empty vehicles and routing them to serve different customers in a demand-responsive arrangement cannot be represented using conventional minimum-cost assignment techniques. This may require full microsimulation or the adoption of a hybrid approach, where some modes are pre-assigned conventionally and those that are demand responsive follow a different set of rules in an agent-based microsimulation.

20.8.2 Modelling Approaches

Modelling DRT requires new supply and demand models. The LOS provided by the new mobility would be described in terms of cost (fare), in-vehicle, waiting and access times plus detour time in the case of ride sharing. Cost and time attributes will depend on how the new services are organised and also on the level of demand, as these are interrelated; for example, ride-sharing works better when there is enough demand for the service. The supply model would be similar to the collection and delivery of parcels from specific locations with the problem of optimising the allocation of vehicles to serve these requests. The supply model will have to replicate this logistics exercise, allocating vehicles in the fleet to individual requests, finding the optimal route in each case, and minimising empty travel and idle time.

It is simpler to assume that only one operator or integrator will serve the travellers in a city, but there is little reason to believe this will always be the case. There will be inefficiencies resulting from multiple operators that may be counterweighed, at least partially, by increased competition.

Solving this logistic problem also requires two additional specifications: the fleet available, in terms of number and vehicle characteristics (number of seats), and the maximum standards for waiting and detour time. The conventional public transport system and its services are also part of the solution to supplying the best LOS to all users. It is clear that a good supply model cannot be aggregated and must be run as an AgBM at a highly detailed level of spatial and time granularity.

In terms of demand modelling, the impact of new mobility services is likely to be greater in the mode choice, destination choice and car ownership sub-models. As we saw above, just the availability of vehicle-sharing services should influence the decision to own a car. The availability of more attractive DRT services should have a similar impact. Mode choice modelling will be particularly complex, given the much wider set of alternative modes available in the future. In addition to LOS attributes, the model is likely to include latent variables such as the willingness to share a ride and risk tolerance.

An appropriate DRT model should also include an equilibrium stage between supply and demand, as shown in Figure 20.5. The fleet size and its location in space and time are bound to influence the LOS perceived by the user. The most significant impact of an insufficient fleet should be the waiting time for the service. If this is not sufficiently short, the demand for the new service will decline.

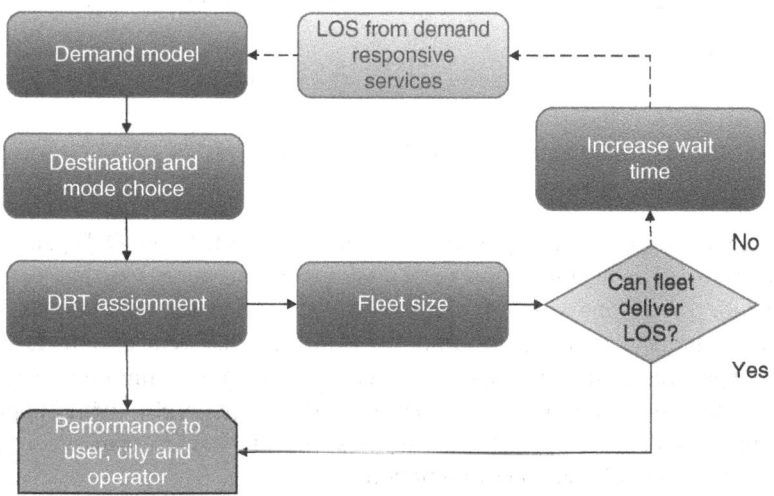

Figure 20.5 Demand, fleet size, and equilibrium

To fully capture the fine spatial and temporal granularity associated with the DRT levels of service, the model is likely to be disaggregated and agent-based. When a well-established aggregate model exists, it may be interesting to explore a hybrid approach where AgBM is only used where it is essential to model the new demand-responsive technologies. Such an approach is illustrated in Figure 20.6, which is adapted from Jones and Willumsen (2021).

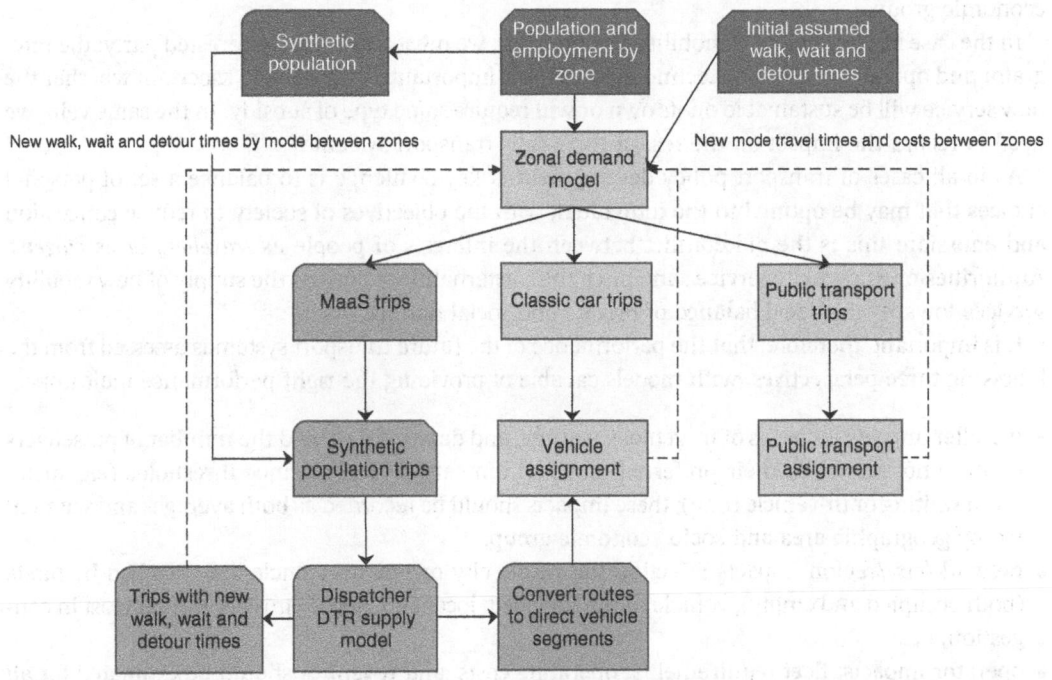

Figure 20.6 Hybrid DRT modelling

The zonal data for population and employment/student enrolment is used to create a synthetic population that is also allocated to specific points (corners) in the network. Assumptions are initially required about the zone-to-zone LOS attributes of all modes. These are fed into a zone-based demand model that generates initial estimates of the trips by mode and time of day.

DRT trips from the aggregate model are then converted into point-to-point trips using the synthetic population. An agent-based supply model will then assign vehicles to these trips and calculate a set of performance indicators. The new estimates for walk, wait, and detour times are then fed back (dotted lines) to the demand model. Point-to-point DRT trips can then be aggregated to zone-to-zone vehicle trips. Ride-share routes can be decomposed into direct segments between stops, again aggregated to zone-to-zone vehicle movements. Note that empty vehicle movements from the supply model must be added and pre-loaded to the network. Normal car and DRT zone-to-zone vehicle movements can then be combined to finally perform a conventional assignment. The resulting direct travel times and costs must then be fed back to the demand model.

This is only an intermediate approach towards a full AgBM. The mixture of point-to-point and zone-to-zone LOS variables is a bit awkward, and an aggregate, zonal-based demand model inevitably loses some of the detailed granularity available in an AGBM, reducing the realism of the exercise.

20.8.3 Model Outputs

Modelling new mobility technologies places new requirements for the model's outputs. Transport models are usually required to produce two types of outputs. One type reflects the impact of interventions on society as a whole, typically related to congestion, energy consumption, and emissions; the most common indicator of this type is vehicle kilometres travelled or VKT. The second group relates the impact to individual travellers, for example, levels of accessibility per zone or socio-economic group.

In the case of assessing new mobility technologies, we must add a third interested party: the integrator and operator of the new technologies. This is important as we need to ascertain whether the new service will be sustainable on its own or will require some type of subsidy. In the same vein, we need to record the impact on the rest of the public transport system.

As in all cases of transport policy development, a key challenge is to balance a set of personal choices that may be optimal to the individual, with the objectives of society to reduce congestion and emission; this is the old conflict between the interests of people as *travellers* or as *citizens*. Authorities must regulate services and price their externalities to orient the supply of new mobility services towards this ideal balance of private and social optima.

It is important, therefore, that the performance of the future transport system is assessed from the following three perspectives, with models capable of providing the right performance indicators:

- traveller impacts, in terms of in-vehicle, waiting, and detour times, and the number of passengers that are not served by their preferred mode within certain performance thresholds (e.g. maximum waiting or in-vehicle time); these impacts should be recorded as both averages and segmented by geographic area and socio-economic group;
- network/city/region impacts affecting the whole city or region: vehicle-km travelled by mode (both occupied and empty), vehicle hours travelled, local and global emissions, hours lost in congestion, etc.;
- operator impacts: fleet requirements, operating costs, and revenues should be estimated for all main operators.

20.9 Deliveries and Collections

As noted in Chapter 15, freight has often received less attention than passenger movements in strategic models for urban areas. It is generally assumed that the movement of lorries can be represented by reasonably stable vehicle matrices, the main problem being that of collecting data to estimate them. There are, however, movements of goods vehicles that are not well represented by vehicle matrices. These are the pick-up and delivery services, including the collection of domestic and commercial rubbish. The movement of these vehicles is best represented by logistic tours.

City logistics focus on the delivery and picking of goods in urban areas, affecting the performance of businesses and quality of life in cities. Research issues like delivery with time windows, traffic congestion and parking, route optimisation, and dynamic routing are of great importance. Gayialis et al. (2022) reviewed research on the topic and put forward an information system to support the efficient delivery of goods within urban areas. This is an area of continuous improvement supported by the increased availability of location and demand data. Although transport modellers would do well to track developments in this field, the full modelling of the logistic arrangement for these services is beyond the scope of strategic transport models. A simplification is needed for these models, as suggested by Hensher et al. (2022); their work develops an aggregate model in the context of a potential distance-based charging system, but the approach has more general applications.

The growing popularity of internet shopping is also increasing the number of delivery vehicles on urban streets; they follow different routes every day, and parking difficulties create new sources of congestion. Recent research has taken an interest in issues associated with these deliveries. For example, Morganti et al. (2014) looked at alternative parcel delivery arrangements in Germany and France. The paper analysed the key drivers for the development of two delivery services (pick-up points and lockers), with reference to the strategies of service providers and e-commerce firms as well as consumer preferences. This theme has been continued by Schnieder et al. (2021) using data from London.

20.10 Digital and Distant Presence

The general availability of good internet connections and video conferencing software has resulted in at least two trends that impact travel and traffic. One of them is the possibility of working remotely, either from home or from a co-working space. The other is the ability to procure goods and services through the internet: e-Commerce. Both of them remove the need to undertake some trips: commuting and shopping; at the same time, they generate some new trips: local personal errands and trips to deliver, and sometimes return, the goods purchased.

Research on the impacts of remote working was active well before the Covid pandemic. For example, O'Keefe et al. (2016) explored full-day and part-day telecommuting in the Greater Dublin Area, seeking to identify the main drivers and constraints to the practice. They found, from a survey, that around 44% of the population of the area telecommuted at least once a month. The main constraint was the need for contact with colleagues, while greater flexibility and avoiding travelling in peak periods were the most important drivers in the propensity to telecommute. They concluded that remote working reduces the commuting burden and affords valuable flexibility to workers; in societal terms, telecommuting reduces greenhouse gas emissions and limits congestion while improving the quality of life.

The Covid pandemic accelerated the adoption of e-Commerce and telecommuting. Kogus et al. (2022) investigated whether the impact of the pandemic was temporary or would accelerate telecommuting in the long term. Their research was based on remote work experience and trends in Israel and Czechia. They used a combination of revealed preferences about work habits before and during the pandemic and stated intentions on expected work patterns in the new 'normal'. Two models were used to analyse the data, one addressing factors that increased/decreased teleworking trends and the other addressing factors that affected the frequency of actual commutes. Their results reveal that most respondents (62% in Israel and 68% in Czechia) expected to retain the same telecommuting/working from home balance. While recognising the potential for bias, they expected a moderate reduction of between 6.5% and 8.7% in the number of commute trips to remain post-pandemic.

Remote work can take one of two forms: working from home or from a co-working or teleworking centre. Bieser et al. (2021) researched the different impacts of these alternatives. They compared the consequences of work locations (employer's office, teleworking centre, home) on time use and travel using travel diaries collected in Stockholm. Some of their results are intuitive. When working from a teleworking centre, commuting trips are reduced and some new local trips are performed, mostly walking and cycling. When working from home, there is some time spent on local trips in addition to the working time.

Obeid et al. (2022) investigated to what extent teleworking reduced trip making using a combination of passively collected 'point of interest' data and five waves of a survey on a panel of participants. The objective was to quantify the effects of changes in the frequency of telecommuting on the total number of daily and weekly trips, and the distance travelled. They found that teleworking resulted in the generation of, on average, one additional non-commute trip on remote workdays relative to work at the office. However, the total distance travelled on telecommute days was significantly shorter than on commute days. They observed that at the weekly level, one additional day of telecommuting resulted in one additional non-commute trip, suggesting that it was a newly generated trip, not one shifted from other days of the week.

On the other hand, the impact of online shopping on trip making was reviewed by Le et al. (2022). They found the expected evidence that online shopping substitutes for shopping trips; they considered all possible effects of e-Commerce: substitution (i.e. e-Commerce replaces travel), complementarity (i.e. it generates other trips), and modification (i.e. it changes travel patterns like destinations, modes, times, or shopping durations). However, they found that as most studies were limited in their reliance on cross-sectional surveys, they could neither capture all these effects nor the difference between the short- and long-term.

Modelling the future in these cases requires an additional segmentation or characterisation of the population and place of work. It is necessary to identify whether the profession and skills of the person are such that they can be deployed remotely. At the same time, it is necessary to identify the type of work that can be undertaken remotely. The trend at the time of writing is to prefer hybrid approaches with some days at the office and some at home or a co-working centre. The proportion of each is not settled and may change in the future.

20.11 Soft Measures, Smarter Choices

For many years, attempts have been made to influence the way we travel through measures that do not imply new infrastructure or services, nor a new charging policy or travel/parking constraints. These attempts to influence travel choices are not new; car manufacturers have used them to associate ownership with success, freedom, and even sex appeal. The attempt to use similar soft measures to influence more sustainable travel choices is more recent; they attempt to change attitudes and preferences for particular modes: walking, cycling, and public transport. In this context, they are considered *soft measures*. They seek to provide better information and opportunities to help people reduce their car use while showing the benefits of alternatives. In essence, soft measures seek to change attitudes and perceptions in relation to travel choices; in modelling terms, they would influence the latent variables in a choice model.

Most of the time soft measures work as intended but it is difficult to quantify the relationship between the measures adopted, the effort involved, and the resulting changes in behaviour. Early advice from the UK Department for Transport in relation to multi-modal studies was that soft measures might achieve a traffic reduction of about 5%. They later published a pioneering review of soft measures and their impact (Cairns et al. 2004), adopting a wide definition of soft measures that included:

- workplace and school travel plans, providing information and encouragement to adopt more sustainable and healthy choices;
- personalised travel planning, travel awareness campaigns, and public transport information and marketing with similar aims;
- car clubs and car-sharing schemes;
- teleworking, teleconferencing, and home shopping.

We have reviewed in earlier sections the potential impact of the last two measures. Cairns et al. (2004) analysed each of these soft factors separately, followed by an assessment of their combined potential impact. Perhaps surprisingly, the report's findings were consistent with the early advice mentioned above. They also found that the cost of achieving reduced car travel using soft measures was about 1.5 pence per car kilometre (i.e. £15 for removing 1000 vehicle kilometres of traffic a year).

A few years later the Department for Transport published an update of this report (Sloman et al. 2010) considering the impact of a programme of soft interventions. They concluded that the effect of these measures was consistent with the expectations, with a reduction in car trips, mostly short ones, and in car distance travelled combined with an increase in walking and cycling and to a lesser extent in bus travel.

Since then, many countries have implemented journey planners helping travellers to make better-judged trip decisions. These have been enhanced by advancements in sensing technologies and artificial intelligence; it is now possible to predict, with great accuracy, the arrival of the next bus and the total time (and fare) required to reach a particular destination. A recent review of research in the field by Pawluk de Toledo et al. (2022) confirmed previous findings, in the sense that soft measures can result in real changes in behaviour.

In terms of new soft measures, social media has opened many opportunities for the positive influence of enlighted opinion leaders and the less positive shaming of those choosing to fly or travel by car; the impact of social media in terms of changing attitudes and behaviour is yet to be ascertained.

However, studies in psychology and human behaviour highlight that neither soft nor hard measures to encourage changes are really effective if they are not accompanied by triggers and policies that encourage a change in individual habits (Verplanken et al. 2008; Janke and Handy 2019). This was clearly demonstrated during the Covid pandemic, when significant changes in travel behaviour were introduced and several of these were retained in subsequent years.

References

AASHTO (1977) *A Manual of User Benefit Analysis of Highway and Bus Transport Improvements*. American Association of State Highway and Transportation Officials, Washington, DC.

Abdel-Aty, M.A., Kitamura, R. and Jovanis, P.P. (1997) Using stated preference data for studying the effect of advanced traffic information on driver's route choice. *Transportation Research* **5C**, 39–50.

Abou-Zeid, M. and Ben-Akiva, M.E. (2009) A model of travel happiness and mode switching. In S. Hess and A. Daly (eds.), *Choice Modelling: The State-of-the-Art and the State-of-Practice*. Emerald, Bingley.

Abou-Zeid, M., Ben-Akiva, M., Bierlaire, M., Choudhury, C. and Hess, S. (2010) Attitudes and value of time heterogeneity. In E. van de Voorde and T. Vanelslander (eds.), *Applied Transport Economics – A Management and Policy Perspective*. De Boeck Publisher, Brussels.

Abou-Zeid, M., Rossi, T. and Gardner, B. (2006) Modelling time-of-day choice in the context of tour and activity based models. *Transportation Research Record* **1981**, 42–49.

Abraham, H. and Kavanagh, C. (1992) Modelling public transport in-vehicle congestion using EMME/2 Release 5. *Proceedings 1st European EMME/2 Users Conference*, London, England.

Accent and HCG (1996) *The Value of Travel Time in UK Roads–1994*. Final Report to the UK Department of Transport, Accent Marketing & Research and Hague Consulting Group, London.

Adams, J. (1992) Horse and rabbit stew. In A. Croker and C. Richards (eds.), *Valuing the Environment: Economic Approaches to Environmental Valuation*. Belhaven Press, London.

Addison, J.D. and Heydecker, B.G. (1999) Dynamic traffic equilibrium with departure time choice. In A. Ceder (ed.), *Transportation and Traffic Theory*. Pergamon, Oxford.

Adnan, M., Pereira, F., Lima, C., Basak, K., Lovric, M., Raveau, F., Zhu, Y., Ferreira, J., Zegras, C. and Ben-Akiva, M. (2016) SimMobility: a multi-scale integrated agent-based simulation platform. *95th Annual TRB Meeting*, Washington, DC.

Akçelik, R. (1991) Travel time functions for transport planning purposes: Davidson's function, its time-dependent form and an alternative travel time function. *Australian Road Research* **21**, 49–59.

Algers, S., Daly, A.J., Kjellman, P. and Widlert, S. (1995) Stockholm model system: application. *7th World Conference on Transport Research*, Sydney, Australia.

Allaman, P.M., Tardiff, T.J. and Dunbar, F.C. (1982) *New Approaches to Understanding Travel Behaviour*. NCHRP Report 250, National Cooperative Highway Research Program, Transportation Research Board, Washington, DC.

Allen, J., Muñoz, J.C. and Ortúzar, J. de D. (2019) Understanding public transport satisfaction: using Maslow's hierarchy of (transit) needs. *Transport Policy* **81**, 75–94.

Allenby, G. (1997) An introduction to hierarchical Bayesian modelling. *Tutorial Notes*, Advanced Research Techniques Forum, American Marketing Association, Carrollton, TX.

Modelling Transport, Fifth Edition. Juan de Dios Ortuzar and Luis G. Willumsen.
© 2024 John Wiley & Sons Ltd. Published 2024 by John Wiley & Sons Ltd.
Companion website: www.wiley.com/go/ortuzar5e

Allenby, G. and Rossi, P. (1999) Marketing models for consumer heterogeneity. *Journal of Econometrics* **89**, 57–78.

Allsop, R.E. (1999) Road safety strategy and targets in Great Britain. *Traffic Engineering* **34**, 72–79.

Alonso, W. (1964) *Location and Land Use.* Harvard University Press, Cambridge, MA.

Alonso, W. (1968) Predicting best with imperfect data. *Journal of the American Institute of Planners* **34**, 248–255.

Amador, F.J., Gonzalez, R.M. and Ortúzar, J. de D. (2005) Preference heterogeneity and willingness-to-pay for travel time savings. *Transportation* **32**, 627–647.

Ampt, E.S. (2003) Respondent burden. In P.R. Stopher and P.M. Jones (eds.), *Transport Survey Quality and Innovation.* Pergamon, Amsterdam.

Ampt, E.S. and Ortúzar, J. de D. (2004) On best practice in continuous large-scale mobility surveys. *Transport Reviews* **24**, 337–363.

Anas, A. (1982) *Residential Location Markets and Urban Transportation.* Academic Press, London.

Anderson, M.K., Nielsen, O.A. and Prato, C.G. (2017) Multimodal route choice models of public transport passengers in the Greater Copenhagen Area. *EURO Journal on Transportation and Logistics* **6**, 221–245.

Andrews, R.L., Ansari, A. and Currim, I.S. (2002) Hierarchical Bayes versus finite mixture conjoint analysis models: a comparison of fit, prediction and partworth recovery. *Journal of Marketing Research* **39**, 87–98.

Arellana, J., Daly, A.J., Hess, S., Ortúzar, J. de D. and Rizzi, L.I. (2012) Development of surveys for study of departure time choice: a two-stage efficient design approach. *Transportation Research Record* **2303**, 9–18.

Arellana, J., Ortúzar, J. de D., Rizzi, L.I. and Zuñiga, F. (2014) Obtaining public transport level-of-service measures using in-vehicle GPS data and freely available GIS web-based tools. In S. Rasouli and H. Timmermans (eds.), *Mobile Technologies for Activity-Travel Data Collection and Analysis.* IGI Global, Hershey, PA.

Arentze, T.A. and Timmermans, H.J.P. (2004) ALBATROSS – a learning-based transportation oriented simulation system. *Transportation Research* **38B**, 613–633.

Arezki, Y. (1986) Comparison of some algorithms for equilibrium traffic assignment with fixed demand. *Proceedings 14th PTRC Summer Annual Meeting*, University of Sussex, England.

Ariely, D. (2009) *Predictably Irrational.* Harper Collins, London.

Armoogum, J. and Madre, J.L. (1998) Weighting or imputations? The example of non responses for daily trips in the French NPTS. *Journal of Transportation and Statistics* **1**, 53–63.

Armstrong, P.M., Garrido, R.A. and Ortúzar, J. de D. (2001) Confidence intervals to bound the value of time. *Transportation Research* **37E**, 143–161.

Arnold, B. and Brockett, P. (1992) On distributions whose component ratios are Cauchy. *American Statistician* **46**, 25–26.

Arnott, R., de Palma, A. and Lindsey, R. (1993) A structural model of peak period congestion: a traffic bottleneck with elastic demand. *American Economic Review* **83**, 161–179.

Arnott, R., de Palma, A. and Lindsey, R. (1994) Welfare effects of congestion tolls and heterogeneous commuters. *Journal of Transport Economics and Policy* **28**, 139–161.

ARUP, ITS and Accent (2015) *Provision of Market Research for Value of Travel Time Savings and Reliability.* Phase 2 Report to UK Department for Transport, London.

Ashley, D.J. (1978) The regional highway traffic model: the home based trip end model. *Proceedings 6th PTRC Summer Annual Meeting*, University of Warwick, England.

Ashok, K., Dillon, W. and Yuan, S. (2002) Extending discrete choice models to incorporate attitudinal and other latent variables. *Journal of Marketing Research* **39**, 31–46.

Atasoy, B., Glerum, A., Hurtubia, R. and Bierlaire, M. (2010) Demand for public transport services: integrating qualitative and quantitative methods. *Proceedings 10th Swiss Transportation Research Conference*, Ascona, Switzerland.

Atherton, T.J. and Ben-Akiva, M.E. (1976) Transferability and updating of disaggregate travel demand models. *Transportation Research Record* **610**, 12–18.

Axhausen, K.W., Beyerle, A. and Schumacher. H. (1989) Choosing the type of parking: a stated preference approach. *68th Annual TRB Meeting*, Washington, DC.

Axhausen, K.W., Hess, S., König, A., Abay, G., Bates, J.J. and Bierlaire, M. (2008) Income and distance elasticities of values of travel time savings: new Swiss results. *Transport Policy* **15**, 173–185.

Axhausen, K.W., Löchl, M., Schlich, R., Buhl, T. and Widmer, P. (2007) Fatigue in long-duration travel diaries. *Transportation* **34**, 143–160.

Axhausen, K.W., Zimmermann, A., Schönfelder, S., Rindsfüser, G. and Haupt, T. (2002) Observing the rhythms of daily life: a six-week travel diary. *Transportation* **29**, 95–124.

Bacharach, M. (1970) *Biproportional Matrices and Input Output Change*. Cambridge University Press, New York.

Badoe, D.A. (2007) Forecasting travel demand with alternatively structured models of trip frequency. *Transportation Planning and Technology* **30**, 455–475.

Bahamonde-Birke, F.J., Kunert, U., Link, H. and Ortúzar, J. de D. (2017a) About attitudes and perceptions – finding the proper way to consider latent variables in discrete choice models. *Transportation* **44**, 475–493.

Bahamonde-Birke, F.J., Navarro, I. and Ortúzar, J. de D. (2017b) If you choose not to decide, you still have made a choice. *Journal of Choice Modelling* **22**, 13–23.

Bahamonde-Birke, F.J. and Ortúzar, J. de D. (2020) How to categorize individuals on the basis of underlying attitudes? A discussion on latent variables, latent classes and hybrid choice models. *Transportmetrica* **17A**, 856–877.

Bain, R. (2009) *Toll Road Traffic and Revenue Forecasts: An Interpreter's Guide*. RBconsult Ltd., West Malling, Kent.

Bain, R. and Plantagie, J.W. (2003) *Fair's Fair? Why Tram Projects are on a Bumpy Road*. Standard and Poor's, London.

Bain, R. and Polakovic, L. (2005) *Traffic Forecasting Risk Study Update 2005: Through Ramp-up and Beyond*. Standard & Poor's, London.

Balbontín, C. and Hensher, D.A. (2020) Identifying the role of stated process strategies in business location decisions. *Transportation Research* **141E**, 102028.

Balbontin, C., Hensher, D.A. and Collins, A.T. (2019) How to better represent preferences in choice models: the contributions to preference heterogeneity attributable to the presence of process heterogeneity. *Transportation Research* **122B**, 218–248.

Balcombe, R., Mackett, R., Paulley, N., Preston, J., Shires, J., Titheridge, H., Wardman, M. and White, P. (2004) *The Demand for Public Transport: A Practical Guide*. Transportation Research Laboratory Report TRL593. Transportation Research Laboratory, Wokingham.

Banister, D. (2003) Critical pragmatism and congestion charging in London. *International Social Science Journal* **176**, 249–264.

Bar-Gera, H. (2002) Origin-based algorithm for the traffic assignment problem. *Transportation Science* **36**, 398–417.

Bar-Gera, H. and Boyce, D.E. (2006) Solving a nonconvex combined travel forecasting model by the method of successive averages with constant step sizes. *Transportation Research* **37B**, 351–367.

Bar-Gera, H. and Luzon, A. (2007) Non-unique route flow solutions for user-equilibrium assignments. *Traffic Engineering and Control* **48**, 408–412.

Bastin, F., Cirillo, C. and Toint, P.L. (2006) Application of an adaptive Monte Carlo algorithm to mixed logit estimation. *Transportation Research* **40B**, 577–593.

Batarce, M., Muñoz, J.C. and Ortúzar, J. de D. (2016) Value crowding in public transport: implications for cost-benefit analysis. *Transportation Research* **91A**, 358–378.

Bateman, I.J. and Turner, R.K. (1993) Valuation of the environment, methods and techniques: the contingent valuation method. In R.K. Turner (ed.), *Sustainable Economics and Management: Principles and Practice*. Belhaven Press, London.

Bateman, I.J., Turner, R.K. and Bateman, S. (1993) Extending cost-benefit analysis of UK highway proposals: environmental evaluation and equity. *Project Appraisal* **8**, 213–224.

Bates, J.J. (1988) Econometric issues in stated preference analysis. *Journal of Transport Economics and Policy* **22**, 59–69.

Bates, J.J. (1996) *Time Period Choice Modelling: A Preliminary Review*. Final Report for UK Department of Transport–HETA Division, John Bates Services, Oxford.

Bates, J.J., Ashley, D.J. and Hyman, G. (1987) The nested incremental logit model: theory and application to modal choice. *Proceedings 15th PTRC Summer Annual Meeting*, University of Bath, England.

Bates, J.J., Gunn, H.F. and Roberts, M. (1978) A model of household car ownership. *Traffic Engineering and Control* **19** (486–491), 562–566.

Bates, J.J. and Roberts, M. (1986) Value of time research: summary of methodology and findings. *Proceedings 14th PTRC Summer Annual Meeting*, University of Sussex, England.

Battellino, H. and Peachman, J. (2003) The joys and tribulations of a continuous survey. In P.R. Stopher and P.M. Jones (eds.), *Transport Survey Quality and Innovation*. Pergamon, Amsterdam.

Becker, G. (1965) A theory of the allocation of time. *Economic Journal* **75**, 493–517.

Beckman, M.J., McGuire, C.B. and Winsten, C.B. (1956) *Studies in the Economics of Transportation*. Yale University Press, New Haven, CT.

Beckman, R.J., Baggerly, K.A. and McKay, M.D. (1995) Creating synthetic baseline populations. *LA-UR-95-1985*, Los Alamos National Laboratory, Los Alamos.

Bekhor, S., Ben-Akiva, M.E. and Ramming, S. (2002) Adaptation of logit kernel to route choice situation. *Transportation Research Record* **1805**, 78–85.

Bekhor, S. and Prashker, J.N. (2001) Stochastic user equilibrium formulation for the generalized nested logit model. *Transportation Research Record* **1752**, 84–90.

Bell, M.G.H. (1983) The estimation of an origin destination matrix from traffic counts. *Transportation Science* **17**, 198–217.

Bell, M.G.H. (1984) Log-linear models for the estimation of origin-destination matrices from traffic counts: an approximation. In J. Volmüller and R. Hamerslag (eds.), *Proceedings of the Ninth International Symposium on Transportation and Traffic Theory*. VNU Science Press, Utrecht.

Bell, M.G.H. and Iida, Y. (1997) *Transportation Network Analysis*. Wiley, Chichester.

Ben-Akiva, M.E. (1974) Structure of passenger travel demand models. *Transportation Research Record* **526**, 26–42.

Ben-Akiva, M.E. (1977) Choice models with simple choice set generating processes. *Working Paper*, Centre for Transportation Studies, MIT.

Ben-Akiva, M.E. (2009) Planning and action in a model of choice. In S. Hess and A. Daly (eds.), *Choice Modelling: the State-of-the-Art and the State-of-Practice*. Bingley, Emerald.

Ben-Akiva, M.E. and Bierlaire, M. (1999) Discrete choice methods and their applications in short term travel decisions. In R. Hall (ed.), *The Handbook of Transportation Science*. Kluwer, Dordrecht.

Ben-Akiva, M.E. and Bolduc, D. (1987) Approaches to model transferability and updating: the combined transfer estimator. *Transportation Research Record* **1139**, 1–7.

Ben-Akiva, M.E. and Bolduc, D. (1996) Multinomial probit with a logit kernel and a general parametric specification of the covariance structure. *Working Paper*, Department d'Economique, Université Laval.

Ben-Akiva, M.E. and Lerman, S.R. (1979) Disaggregate travel and mobility choice models and measures of accessibility. In D.A. Hensher and P.R. Stopher (eds.), *Behavioural Travel Modelling*. Croom Helm, London.

Ben-Akiva, M.E. and Lerman, S.R. (1985) *Discrete Choice Analysis: Theory and Application to Travel Demand*. MIT Press, Cambridge, MA.

Ben-Akiva, M. and Morikawa, T. (1990) Estimation of travel demand models from multiple data sources. *Proceedings 11th International Symposium on Transportation and Traffic Theory*, Yokohama, Japan.

Ben-Akiva, M., Morikawa, T. and Shiroishi, F. (1992) Analysis of the reliability of preference ranking data. *Journal of Business Research* **24**, 149–164.

Ben-Akiva, M.E., Walker, J.L., Bernardino, A.T., Gopinath, D.A., Morikawa, T. and Polydoropoulou, A. (2002) Integration of choice and latent variable models. In H.S. Mahmassani (ed.), *In Perpetual Motion: Travel Behaviour Research Opportunities and Challenges*. Pergamon, Amsterdam.

Ben-Akiva, M.E. and Watanatada, T. (1980) Application of a continuous spatial choice logit model. In C. F. Manski and D. McFadden (eds.), *Structural Analysis of Discrete Data: With Econometric Applications*. MIT Press, Cambridge, MA.

Berndt, E.U., Hall, B.H., Hall, R.E. and Hausman, J.A. (1974) Estimation and inference in non-linear structural models. *Annals of Economic and Social Measurement* **3** (4), 653–655.

Berry, S., Levinsohn, J. and Pakes, A. (1995) Automobile prices in market equilibrium. *Econometrica* **63**, 841–890.

Bhat, C.R. (1995) A heteroskedastic extreme value model of intercity travel mode choice. *Transportation Research* **29B**, 471–483.

Bhat, C.R. (1997) An endogenous segmentation mode choice model with an application to intercity travel. *Transportation Science* **31**, 34–48.

Bhat, C.R. (1998) Accomodating flexible substitution patterns in multidimensional choice modelling: formulation and application to travel mode and departure time choice. *Transportation Research* **32B**, 455–466.

Bhat, C.R. (2001) Quasi-random maximum simulated likelihood estimation of the mixed multinomial logit model. *Transportation Research* **35B**, 677–695.

Bhat, C.R. (2003) Simulation estimation of mixed discrete choice models using randomized and scrambled Halton sequences. *Transportation Research* **37B**, 837–855.

Bhat, C.R. (2015) A new generalized heterogeneous data model (GHDM) to jointly model mixed types of dependent variables. *Transportation Research* **79B**, 50–77.

Bhat, C.R. and Castelar, S. (2002) A unified mixed logit framework for modelling revealed and stated preferences: formulation and application to congestion pricing analysis in the San Francisco Bay Area. *Transportation Research* **36B**, 593–616.

Bhat, C.R. and Guo, J. (2004) A mixed spatially correlated logit model: formulation and application to residential choice modelling. *Transportation Research* **38B**, 147–168.

Bhat, C.R. and Steed, J.L. (2002) A continuous-time model of departure time choice for urban shopping trips. *Transportation Research* **36B**, 207–224.

Bhatta, B. and Larsen, O. (2011) Errors in variables in multinomial choice modelling: a simulation study applied to a multinomial model of travel mode choice. *Transport Policy* **18**, 326–335.

Bianchi, R., Jara-Díaz, S.R. and Ortúzar, J. de D. (1998) Modelling new pricing strategies for the Santiago Metro. *Transport Policy* **5**, 223–232.

Bierlaire, M. (2016) PythonBiogeme: A Short Introduction. *Report TRANSP-OR 160706*, Series on Biogeme. Transport and Mobility Laboratory, Ecole Polytechnique Federal de Laussane.

Bierlaire, M., Bolduc, D. and McFadden, D. (2008) The estimation of generalized extreme value models from choice-based samples. *Transportation Research* **42B**, 381–394.

Bierlaire, M., Hurtubia, R. and Flötteröd, G. (2010) An experimental analysis of the implicit choice set generation using the constrained multinomial logit model. *Transportation Research Record* **2175**, 92–97.

Bieser, J., Vaddadi, B., Kramers, A., Höjer, M. and Hilty, L. (2021) Impacts of telecommuting on time use and travel: a case study of a neighborhood telecommuting center in Stockholm. *Travel Behaviour and Society* **23**, 157–165.

Blase, J.H. (1979) Hysteresis and catastrophe theory: a demonstration of habit and threshold effects in travel behaviour. *Proceedings 7th PTRC Summer Annual Meeting*, University of Warwick, England.

Bliemer, M.C.J. and Rose, J.M. (2009) Efficiency and sample size requirements for stated choice experiments. *88th Annual TRB Meeting*, Washington, DC.

Bliemer, M.C.J. and Rose, J.M. (2010) Construction of experimental designs for mixed logit models allowing for correlation across choice observations. *Transportation Research* **44B**, 720–734.

Bliemer, M.C.J., Rose, J.M. and Hensher, D.A. (2009) Efficient stated choice experiments for estimating nested logit models. *Transportation Research* **43B**, 19–35.

Bolduc, D. and Alvarez-Daziano, R. (2009) On estimation of hybrid choice models. In S. Hess and A. Daly (eds.), *Choice Modelling: The State-of-the-Art and the State-of-Practice*. Emerald, Bingley.

Bolduc, D., Boucher, N. and Alvarez-Daziano, R. (2008) Hybrid choice modelling of new technologies for car choice in Canada. *Transportation Research Record* **2082**, 63–71.

Bolduc, D. and Giroux, A. (2005) *The Integrated Choice and Latent Variable (ICLV) Model: Handout to Accompany the Estimation Software*. Département d'économique, Université Laval, Québec, Canada.

Bollen, K.A. (1989) *Structural Equations with Latent Variables*. Wiley, Chichester.

Bonsall, P.W. (1983) Transfer price data: its use and abuse. *Proceedings 11th PTRC Summer Annual Meeting*, University of Sussex, England.

Bonsall, P.W., Shires, J.D., Matthews, B., Maule, J. and Beale, J. (2007) Responses to complex pricing signals: theory, evidence and implications for road pricing. *Transportation Research* **41A**, 672–683.

Bonsall, P.W. and Willumsen, L.G. (2013) Pricing methods to influence car use. In T. Gärling, D. Ettema and M. Friman (eds.), *Handbook of Sustainable Travel*. Springer, New York.

Börjesson, M. (2008) Joint RP-SP data in a mixed logit analysis of trip timing decisions. *Transportation Research* **44E**, 1025–1038.

Börjesson, M. and Eliasson, J. (2014) Experiences from the Swedish value of time study. *Transportation Research* **59A**, 144–158.

Börjesson, M., Eliasson, J., Hugosson, M. and Brundell-Freij, K. (2012) The Stockholm congestion charges – 5 years on: effects, acceptability and lessons learnt. *Transport Policy* **20**, 1–12.

Börsch-Supan, A. (1990) On the compatibility of the nested logit model with utility maximization. *Journal of Econometrics* **43**, 373–388.

Börsch-Supan, A. and Hajivassiliou, V.A. (1993) Smooth unbiased multivariate probability simulators for maximum likelihood estimation of limited dependent variable models. *Journal of Econometrics* **58**, 347–368.

Bovy, P.H.L. and Fiorenzo-Catalano, S. (2007) Stochastic route choice set generation: behavioural and probabilistic foundations. *Transportmetrica* **3**, 173–189.

Bowman, J.L. (2009) Historical development of activity-based models: theory and practice. *Traffic Engineering and Control* **50**, 314–318.

Bowman, J.L. and Ben-Akiva, M.E. (2001) Activity-based disaggregate travel demand model system with activity schedules. *Transportation Research* **35A**, 1–28.

Bowman, J.L. and Bradley, M.A. (2008) Activity-based models: approaches used to achieve integration among trips and tours throughout the day. *2008 European Transport Conference*, Leeuwenhorst, The Netherlands.

Boyce, D.E. (2007) Forecasting travel on congested urban transportation networks: review and prospects for network equilibrium models. *Networks and Spatial Economics* **7**, 99–128.

Boyce, D.E., Day, N.D. and McDonald, C. (1970) *Metropolitan Plan Making*. Monograph Series No. 4, Regional Science Research Institute, Philadelphia.

Boyce, D.E., O'Neill, C.R. and Scherr, W. (2008) Solving the sequential travel forecasting procedure with feedback. *Transportation Research Record* **2077**, 129–135.

Boyce, D.E., Ralevic-Dekic, B. and Bar-Gera, H. (2004) Convergence of traffic assignments: how much is enough? *Journal of Transportation Engineering of ASCE* **130**, 49–55.

Boyce, D.E. and Williams, H.C.W.L. (2015) *Forecasting Urban Travel: Past, Present and Future*. Edward Elgar, Cheltenham.

Bradley, M.A., Bowman, J.L and Lawton, K. (1999) A comparison of sample enumeration and stochastic micro-simulation for application of tour-based and activity-based travel demand models. *1999 European Transport Conference*, Cambridge, England.

Bradley, M.A. and Daly, A.J. (1994) Use of the logit scaling approach to test for rank-order and fatigue effects in stated preference data. *Transportation* **21**, 167–184.

Bradley, M.A. and Daly, A.J. (1997) Estimation of logit choice models using mixed stated-preference and revealed-preference information. In P. Stopher and M. Lee-Gosselin (eds.), *Understanding Travel Behaviour in an Era of Change*. Pergamon, Oxford.

Bradley, M. and Daly, A.J. (2000) New analysis issues in stated preference research. In J. de D. Ortúzar (ed.), *Stated Preference Modelling Techniques*. Perspectives 4, PTRC, London.

Braess, D., Nagurney, A. and Wakolbinger, T. (2005) On a paradox of traffic planning. *Transportation Science* **39**, 446–450.

Branston, D. (1976) Link capacity functions: a review. *Transportation Research* **10**, 223–236.

Brenninger-Gothe, M., Jornsten, K. and Lundgren, J. (1989) Estimation of origin-destination matrices from traffic counts using multiobjective programming formulations. *Transportation Research* **23B**, 257–269.

Brey, R. and Walker, J.L. (2011) Estimating time of day demand with errors in reported preferred times: an application to airline travel. *Procedia Social and Behavioural Sciences* **17**, 150–168.

Bristow, A.L., Jansen, G., Mackie, P.J. and Nellthorp, J. (1998) Costs, prices and values in the appraisal of transport projects – European principles and practice. *Proceedings 8th World Conference on Transport Research*, Antwerp, Belgium.

Brög, W. and Ampt, E. (1982) State of the art in the collection of travel behaviour data. In *Travel Behaviour for the 1980's*. Special Report 201. National Research Council, Washington, DC.

Brög, W. and Erl, E. (1982) Application of correction and weighting factors to obtain a representative data base. *Proceedings 10th PTRC Summer Annual Meeting*, University of Warwick, England.

Brög, W. and Meyburg, A.H. (1980) The non-response problem in travel surveys: an empirical investigation. *Transportation Research Record* **775**, 34–38.

Brownstone, D. (1998) Multiple imputation methodology for missing data: non-random response and panel attrition. In T. Gärling, T. Laitila and K. Westin (eds.), *Theoretical Foundations of Travel Choice Modelling*. Elsevier, Oxford.

Bruzelius, N. (1979) *The Value of Travel Time*. Croom Helm, London.

Bunch, D.S. (1991) Estimability in the multinomial probit model. *Transportation Research* **25B**, 1–12.

Bureau of Public Roads (1964) *Traffic Assignment Manual*. Urban Planning Division, US Department of Commerce, Washington, DC.

Burge, P., Fox, J., Kouwenhoven, M., Rohr, C. and Wigan, M.R. (2007) Modelling of motorcycle ownership and commuter usage: a UK study. *Transportation Research Record* **2031**, 59–68.

Burgess, L. and Street, S. (2003) Optimal designs for 2k choice experiments. *Communications in Statistics: Theory and Methods* **32**, 2185–2206.

Burgess, L. and Street, S. (2005) Optimal designs for choice experiments with asymmetric attributes. *Journal of Statistical Planning and Inference* **134**, 288–301.

Burrell, J.E. (1968) Multiple route assignment and its application to capacity restraint. In W. Leutzbach and P. Baron (eds.), *Beiträge zur Theorie des Verkehrsflusses*. Strassenbau und Strassenverkehrstechnik Heft, Karlsruhe.

Button, K.J., Pearman, A.D. and Fowkes, A.S. (1982) *Car Ownership Modelling and Forecasting*. Gower, Aldershot.

Bwambale, A., Choudhury, C.F. and Hess, S. (2019) Modelling departure time choice using mobile phone data. *Transportation Research* **130A**, 424–439.

Cairns, S., Sloman, L., Newson, C., Anable, J., Kirkbride, A. and Goodwin, P. (2004) *Smarter Choices – Changing the Way We Travel*. Final Report to the UK Department for Transport, London.

Caldwell, L.C. and Demetski, M.J. (1980) Transferability of trip generation models. *Transportation Research Record* **751**, 56–62.

Caliper Corporation (2015) *Traffic Assignment and Feedback Research to Support Improved Travel Forecasting*. Final Report to the US Federal Transit Administration, Washington, DC.

Cameron, A.C. and Trivedi, P.K. (1998) *Regression Analysis of Count Data*. Cambridge University Press, Cambridge.

Campbell, D., Hensher, D.A. and Scarpa, R. (2014) Bounding WTP distributions to reflect the 'actual'; consideration set. *Journal of Choice Modelling* **11**, 4–15.

Cantillo, V., Amaya, J. and Ortúzar, J. de D. (2010) Thresholds and indifference in stated choice surveys. *Transportation Research* **44B**, 753–763.

Cantillo, V., Heydecker, B.G. and Ortúzar, J. de D. (2006) A discrete choice model incorporating thresholds for perception in attribute values. *Transportation Research* **40B**, 807–825.

Cantillo, V. and Ortúzar, J. de D. (2005) A semi-compensatory discrete choice model with explicit attribute thresholds of perception. *Transportation Research* **39B**, 641–657.

Cantillo, V. and Ortúzar, J. de D. (2006) Implications of thresholds in discrete choice modelling. *Transport Reviews* **26**, 667–691.

Cantillo, V., Ortúzar, J. de D. and Williams, H.C.W.L. (2007) Modelling discrete choices in the presence of inertia and serial correlation. *Transportation Science* **41**, 195–205.

Cardell, N.S. and Reddy, B. (1977) A multinomial logit model which permits variations in tastes accross individuals. *Working Paper*, Charles River Associates, Boston.

Cardell, N.S. and Dunbar, F. (1980) Measuring the societal impacts of automobile downsizing. *Transportation Research* **14A**, 423–434.

Carlsson, F. (2003) The demand for intercity public transport: the case of business passengers. *Applied Economics* **35**, 41–50.

Carlsson, F. and Martinsson, P. (2002) Design techniques for stated preference methods in health economics. *Health Economics* **12**, 281–294.

Carrasco, J.A. and Ortúzar, J. de D. (2002) A review and assessment of the nested logit model. *Transport Reviews* **22**, 197–218.

Carson, R.T., Louviere, J.J., Anderson, D.A., Arabie, P., Bunch, D.S., Hensher, D.A., Johnson, R.M., Kuhfeld, W.F., Steinberg, D., Swait, J., Timmermans, H. and Wiley, J.B. (1994) Experimental analysis of choice. *Marketing Letters* **5**, 351–368.

Carson, R.T., Wright, J., Carson, N., Alberini, A. and Flores, N. (1995) *A Bibliography of Contingent Valuation Studies and Papers*. Natural Resource Damage Assessment, La Jolla.

Cascetta, E. (1984) Estimation of trip matrices from traffic counts and survey data: a generalised least squares approach estimator. *Transportation Research* **18B**, 289–299.

Cascetta, E. (2009) *Transportation Systems Analysis: Models and Applications*. Second Edition. Springer, New York.

Cascetta, E. and Nguyen, S. (1988) A unified framework for estimating or updating origin/destination matrices from traffic counts. *Transportation Research* **22B**, 437–455.

Cascetta, E., Nuzzolo, A. and Biggiero, L. (1992) Analysis and modelling of commuters' departure time and route choice in urban networks. *Proceedings of the Second International CAPRI Seminar in Urban Traffic Networks*, Capri, Italy.

Casey, H.J. (1955) Applications to traffic engineering of the law of retail gravitation. *Traffic Quarterly* **9**, 23–35.

CASRO (1982) *On the Definition of Response Rates*. Special Report Task Force on Completion Rates, Council of American Survey Research Organisations, New York.

Caussade, S., Ortúzar, J. de D., Rizzi, L.I. and Hensher, D.A. (2005) Assessing the influence of design dimensions on stated choice experiment estimates. *Transportation Research* **39B**, 621–640.

Chadwick, G. (1987) *Models of Urban and Regional Systems in Developing Countries*. Pergamon Press, Oxford.

Chamberlain, G. (1984) Panel data. In Z. Griliches and M. Intriligator (eds.), *Handbook of Econometrics*, Vol. 2. North-Holland, Amsterdam.

Chapman, R.G. and Staelin, R. (1982) Exploiting rank ordered choice set data within the stochastic utility model. *Journal of Marketing Research* **19**, 288–301.

Cheng, Q., Deng, W. and Raza, M.A. (2020) Analysis of the departure time choices of metro passengers during peak hours. *Intelligent Transport Systems* **14**, 866–872.

Cherchi, E., Cirillo, C. and Ortúzar, J. de D. (2017) Modelling correlation patterns in mode choice models estimated on multiday travel data. *Transportation Research* **96A**, 146–153.

Cherchi, E., Meloni, I. and Ortúzar, J. de D. (2013) The latent effect of inertia in the choice of mode. In M. Roorda and E. Miller (eds.), *Travel Behaviour Research: Current Foundations, Future Prospect*. Lulu Press, Raleigh.

Cherchi, E. and Ortúzar, J. de D. (2006) On fitting mode specific constants in the presence of new options in RP/SP models. *Transportation Research* **40A**, 1–18.

Cherchi, E. and Ortúzar, J. de D. (2008a) Predicting best with mixed logit models: understanding some confounding effects. In P.O. Inweldi (ed.), *Transportation Research Trends*. Nova Science Publishers, New York.

Cherchi, E. and Ortúzar, J. de D. (2008b) Empirical identification in the mixed logit model: analysing the effect of data richness. *Networks and Spatial Economics* **8**, 109–124.

Cherchi, E. and Ortúzar, J. de D. (2011) On the use of mixed RP/SP models in prediction: accounting for random taste heterogeneity. *Transportation Science* **45**, 98–108.

Cherrett, T. and McDonald, M. (2002) Traffic composition during the morning peak period. *European Journal of Transport and Infrastructure Research* **2**, 41–55.

Chesnut, L.G., Ostro, B.D. and Vichit-Vadakan, N. (1997) Transferability of air pollution control health benefit estimates from the United States to developing countries: evidence from the Bangkok study. *American Journal of Agricultural Economics* **79**, 1630–1635.

Chib, S. and Greenberg, E. (1995) Understanding the Metropolis-Hastings algorithm. *The American Statistician* **49**, 327–335.

Chicco, A., Diana, M., Loose, W. and Nehrke, G. (2022) Comparing car ownership reduction patterns among members of different car sharing schemes operating in three German inner-city areas. *Transportation Research* **163A**, 370–385.

Chihara, T.S. (1978) *Introduction to Orthogonal Polynomials*. Gordon and Breach, New York.

Chiou, L. and Walker, J.L. (2007) Masking identification of discrete choice models under simulation methods. *Journal of Econometrics* **141**, 683–703.

Chiu, Y., Bottom, J., Mahut, M., Paz, A., Balakrishna, R., Waller, T. and Hicks, J. (2011) Dynamic traffic assignment: a primer. *Transportation Research Circular Number E-C153*, Transportation Research Board, Washington, DC.

Chorus, C.G., Arentze, T.A. and Timmermans, H.J.P. (2008) A random regret-minimization model of travel choice. *Transportation Research* **42B**, 1–18.

Chorus, C. and Kroesen, M. (2014) On the (im-)possibility of deriving transport policy implications from hybrid choice models. *Transport Policy* **36**, 217–222.

Chorus, C.G., van Cranenburgh, S. and Dekker, T. (2014) Random regret minimization for consumer choice modeling: assessment of empirical evidence. *Journal of Business Research* **67**, 2428–2436.

Chow, J.Y.J., Yang, C.H. and Reagan, A.C. (2010) State of the art of freight forecast modeling: lessons learned and the road ahead. *Transportation* **37**, 1011–1030.

Chu, C. (1989) A paired combinational logit model for travel demand analysis. In World Conference on Transport Research (ed.), *Transport Policy, Management and Technology Towards 2001*. Western Periodicals Co., Ventura, CA.

Chu, M.C., Nguyen, L.X., Ton, T.T. and Huynh, N. (2019) Assessment of motorcycle ownership, use, and potential changes due to transportation policies in Ho Chi Minh City, Vietnam. *Journal of Transportation Engineering of ASCE* **145A**. https://doi.org/10.1061/JTEPBS.0000273.

Cirillo, C., Daly, A.J. and Lindveld, K. (2000) Eliminating bias due to the repeated measurements problem in SP data. In J. de D. Ortúzar (ed.), *Stated Preference Modelling Techniques*. Perspectives 4, PTRC, London.

Clark, C., Mokhtarian, P., Circella, G. and Watkins, K. (2021) The role of attitudes in perceptions of bicycle facilities: a latent-class regression approach. *Transportation Research* **77F**, 129–148.

Clark, S. and Watling, D. (2005) Modelling network travel time reliability under stochastic demand. *Transportation Research* **39B**, 119–140.

Collins, A.T. and Hensher, D.A. (2015) The influence of varying information load on inferred attribute non-attendance. In S. Rasouli and H.J.P. Timmermans (eds.), *Bounded Rational Choice Behavior: Applications in Transport*. Emerald, Bingley.

Commission of the European Communities (1994) *Cost-Benefit and Multi-Criteria Analysis for New Road Construction*. DGVII, Research and Development Unit, Brussels.

Cook, R.D. and Nachtsheim, C.J. (1980) A comparison of algorithms for constructing exact *D*-optimal designs. *Technometrics* **22**, 315–324.

Coombs, C.H., Dawes, R.M. and Tversky, A. (1970) *Mathematical Psychology: An Elementary Introduction*. Prentice-Hall, Englewood Cliffs, NJ.

Copley, G. and Lowe, S.R. (1981) The temporal stability of trip rates: some findings and implications. *Proceedings 9th PTRC Summer Annual Meeting*, University of Warwick, England.

Cosslett, S.R. (1981) Efficient estimation of discrete choice models. In C.F. Manski and D. McFadden (eds.), *Structural Analysis of Discrete Data: With Econometric Applications*. MIT Press, Cambridge, MA.

Cox, T. and Hurtubia, R. (2021) Latent segmentation of urban space through residential location choice. *Networks and Spatial Economics* **21**, 199–228.

Cox, T. and Hurtubia, R. (2022) Compact development and preferences for social mixing in location choices: results from revealed preferences in Santiago, Chile. *Journal of Regional Science* **62**, 246–269.

Cramer, J.S. (1986) *Econometric Applications of Maximum Likelihood Methods.* Cambridge University Press, Cambridge.

Dafermos, S.C. (1980) Traffic equilibrium and variational inequalities. *Transportation Science* **14**, 42–54.

Daganzo, C.F. (1979) *Multinomial Probit: The Theory and its Applications to Demand Forecasting.* Academic Press, New York.

Daganzo, C.F. (1980) Optimal sampling strategies for statistical models with discrete dependent variables. *Transportation Science* **14**, 324–345.

Daganzo, C.F. and Kusnic, M. (1993) Two properties of the nested logit model. *Transportation Science* **27**, 395–400.

Daganzo, C.F. and Sheffi, Y. (1979) Estimation of choice models from panel data. *26th Annual Meeting of the Regional Science Association*, Los Angeles, CA, USA.

Dalal, S.R. and Klein, R.W. (1988) A flexible class of discrete choice models. *Marketing Science* **7**, 232–251.

Dalvi, Q.M. (1988) *The Value of Life and Safety: A Search for a Consensus Estimate.* Department of Transport, London.

Daly, A.J. (1982a) Estimating choice models containing attraction variables. *Transportation Research* **16B**, 5–15.

Daly, A.J. (1982b) Applicability of disaggregate models of behaviour: a question of methodology. *Transportation Research* **16A**, 363–370.

Daly, A.J. (1987) Estimating 'tree' logit models. *Transportation Research* **21B**, 251–268.

Daly, A.J. (1997) Improved methods for trip generation. *Proceedings 25th European Transport Forum*, Brunel University, England.

Daly, A.J. (1998) Prototypical sample enumeration as a basis for forecasting with disaggregate models. *Proceedings 26th European Transport Conference*, University of Loughborough, England.

Daly, A.J. (2001) Alternative tree logit models: comments on a paper by Koppelman and Wen. *Transportation Research* **35B**, 717–724.

Daly, A.J., Fox, J., Patruni, B. and Milthorpe, F. (2012a) Pivoting in travel demand models. *2012 Australasian Transport Research Forum*, Perth.

Daly, A.J. and Gunn, H.F. (1986) Cost effective methods for national level demand forecasting. In A. Ruhl (ed.), *Behavioural Research for Transport Policy.* VNU Science Press, Utrecht.

Daly, A.J., Hess, S. and de Jong, G. (2012b) Calculating errors for measures derived from choice modelling estimates. *Transportation Research* **46B**, 333–341.

Daly, A., Hess, S. and Ortúzar, J. de D. (2023) Estimating willingness-to-pay from discrete choice models: setting the record straight. *Transportation Research* **176A**. https://doi.org/10.1016/j.tra.2023.103828.

Daly, A.J. and Ortúzar, J. de D. (1990) Forecasting and data aggregation: theory and practice. *Traffic Engineering and Control* **31**, 632–643.

Daly, A.J., van der Valk, J. and van Zwam, H.P.H. (1983) Application of disaggregate models for a regional transportation study in the Netherlands. In P. Baron and H. Nuppnau (eds.), *Research for Transport Policies in a Changing World.* SNV Studiengesellschaft Nahverkehr, Hamburg.

Daly, A.J. and Zachary, S. (1978) Improved multiple choice models. In D.A. Hensher and M.Q. Dalvi (eds.), *Determinants of Travel Choice.* Saxon House, Westmead.

Dantzig, G.B. (1951) Application of the Simplex Method to a transportation problem. In T.C. Koopmans (ed.), *Activity Analysis of Production and Allocation.* Wiley, New York.

Daor, E. (1981) The transferability of independent variables in trip generation models. *Proceedings 9th PTRC Summer Annual Meeting*, University of Warwick, England.

Dargay, J. and Gately, D. (1999) Income's effect on car and vehicle ownership, worldwide: 1960–2015. *Transportation Research* **33A**, 101–138.

De Borger, B. and Fosgerau, M. (2008) The trade-off between money and travel time: a test of the theory of reference-dependent preferences. *Journal of Urban Economics* **64**, 101–115.

De Cea, J. and Cruz, G. (1986) ESMATUC: un modelo de estimación de matrices de viajes en transporte urbano colectivo. *Apuntes de Ingeniería* **24**, 109–125 (in Spanish).

De Cea, J. and Fernández, J.E. (1989) Transit assignment to minimal routes: an efficient new algorithm. *Traffic Engineering and Control* **30**, 491–494.

De Cea, J. and Fernández, J.E. (2000) ESTRAUS: un modelo de equilibrio oferta-demanda para redes multi-modales de transporte urbano con múltiples clases de usuarios. In J. Colomer and A. García (eds.), *Calidad e Innovación en los Transportes*. Universidad de Valencia, Valencia (in Spanish).

De Cea, J., Fernández, J.E., Dekock, V. and Soto, A. (2005) Solving network equilibrium problems on multimodal urban transportation networks with multiple user classes. *Transport Reviews* **35**, 293–317.

De Cea, J., Fernández, J.E. and de Grange, L. (2008) Combined models with hierarchical demand choices: a multi-objective entropy maximisation approach. *Transport Reviews* **28**, 415–438.

De Grange, L., Fernández, J.E. and De Cea, J. (2010) A consolidated model of trip distribution. *Transportation Research* **46E**, 61–75.

Dehghani, Y. and Talvitie, A.P. (1980) Model specification, model aggregation and market segmentation in mode choice models: some empirical evidence. *Transportation Research Record* **775**, 28–34.

Dehghani, Y. and Talvitie, A.P. (1983) Forecasting accuracy, transferability and updating of modal constants in disaggregate mode choice models with simple and complex specifications. *Proceedings 11th PTRC Summer Annual Meeting*, University of Sussex, England.

De Jong, G.C. (1989) Some Joint Models of Car Ownership and Use. PhD Thesis, Faculteit der Economishe, Universiteit van Amsterdam.

De Jong, G.C. (1990) An indirect utility model of car ownership and private car use. *European Economic Review* **34**, 971–985.

De Jong, G.C. (1996) A disaggregate model system of vehicle holding duration, type choice and use. *Transportation Research* **30B**, 263–276.

De Jong, G. and Ben-Akiva, M. (2007) A micro-simulation model of shipment size and transport chain choice. *Transportation Research* **41B**, 950–965.

De Jong, G., Daly, A.J., Pieters, M., Miller, S., Plasmeijer, R. and Hofman, F. (2007) Uncertainty in traffic forecasts: literature review and new results for The Netherlands. *Transportation* **34**, 375–395.

De Jong, G., Daly, A.J., Pieters, M., Vellay, C., Bradley, M. and Hofman, F. (2003) A model for time of day and mode choice using error components logit. *Transportation Research* **39E**, 245–268.

De la Barra, T. (1989) *Integrated Land Use and Transport Modelling: Decision Chains and Hierarchies*. Cambridge University Press, Cambridge.

Del Mistro, R. and Behrens, R. (2008) How variable is variability in traffic? How can TDM succeed? *27th South African Transport Conference*, Pretoria, South Africa.

Denstadli, J.M., Lines, R. and Ortúzar, J. de D. (2012) Information processing in choice-based conjoint studies: a process-tracing study. *European Journal of Marketing* **46**, 422–446.

Department of Health (1999) Economic appraisal of the health effects of air pollution. In *Ad Hoc Group on Economic Appraisal of the Health Effects of Air Pollution*. HMSO, London.

Department of Transport (1985) *Traffic Appraisal Manual (TAM)*. HMSO, London.

Department of Transport (1987) *Values for Journey Time Savings and Accident Prevention*. HMSO, London.

Department of Transport (1997) *Traffic Appraisal of Road Schemes: Design Manual for Roads and Bridges*. HMSO, London.

Department of Transport (2007) *Forecasting Travel Time Variability in Urban Areas. Deliverable D1. Data Analysis and Model Development*. UK Department of Transport, London.

Department of Transport (2014) *Quality Assurance of Analytical Modelling*. UK Department of Transport, London.

Department of Transport (2018) *Road Traffic Forecasts*. UK Department of Transport, London.

Department of Transport (2022) *TAG: Uncertainty Toolkit*. UK Department of Transport, London.

Department of Transport (2022a) *TAG Unit A1.3 User and Provider Impacts*. Department of Transport, London.

DeSerpa, A. (1971) A theory of the economics of time. *Economic Journal* **81**, 828–846.

DeShazo, J. (2002) Designing transactions without framing effects in iterative question formats. *Journal of Environmental Economics and Management* **43**, 360–385.

DETR (1998) *A New Deal for Highways in England. Guidance to the New Appraisal Framework*. Department of the Environment, Transport and the Regions, London.

Deville, J.C., Särndal, C.E. and Sautory, O. (1993) Generalised ranking procedures in survey sampling. *Journal of the American Statistical Association* **88**, 1013–1020.

Dhar, R. and Simonson, I. (2003) The effect of forced choice on choice. *Journal of Marketing Research* **40**, 146–160.

Dial, R.B. (1971) A probabilistic multipath traffic assignment model which obviates path enumeration. *Transportation Research* **5**, 83–111.

Dial, R.B. (2006) A path-based user-equilibrium traffic assignment algorithm that obviates path storage and enumeration. *Transportation Research* **40B**, 917–936.

Diamond, P. and Hausman, J.A. (1994) Contingent valuation: is some number better than no number? *Journal of Economic Perspectives* **8**, 45–64.

Díaz, F., Cantillo, V., Arellana, J. and Ortúzar, J. de D. (2015) Accounting for stochastic variables in discrete choice models. *Transportation Research* **78B**, 222–237.

DICTUC (1978) *Encuesta Origen y Destino de Viajes para el Gran Santiago*. Final Report to the Ministry of Public Works, Department of Transport Engineering, Universidad Católica de Chile, Santiago (in Spanish).

DICTUC (1998) *Actualización de Encuestas Origen-Destino de Viajes*. Final Report to the Ministry of Planning, Department of Transport Engineering, Pontificia Universidad Católica de Chile, Santiago (in Spanish).

DICTUC (2003) *Encuesta Origen-Destino de Viajes del Gran Santiago 2001*. Final Report to the Ministry of Planning, Department of Transport Engineering and Logistics, Pontificia Universidad Católica de Chile, Santiago (in Spanish).

DICTUC (2021) *Estudio Valor Social del Tiempo de Viaje en el Contexto del Sistema Nacional de Inversiones*. Report to the Ministry of Planning, Department of Transport Engineering and Logistics, Pontificia Universidad Católica de Chile, Santiago (in Spanish).

Dijkstra, E.W. (1959) Note on two problems in connection with graphs (spanning tree, shortest path). *Numerical Mathematics* **1**, 269–271.

Ding, S. and Li, R. (2021) Forecasting the sales and stock of electric vehicles using a novel self-adaptive optimized grey model. *Engineering Applications of Artificial Intelligence* **100**, 104–148.

Domarchi, C., Coeymans, J.E. and Ortúzar, J. de D. (2019a) Shared taxis: modelling the choice of a paratransit mode in Santiago de Chile. *Transportation* **46**, 2243–2268.

Domarchi, C., Raveau, S., Iglesias, P., Muñoz, J.C. and Ortúzar, J. de D. (2019b) Modelos de elección de ruta de transporte público en Santiago: una aplicación con la herramienta STEP y datos de la encuesta origen-destino. *XIX Chilean Conference on Transportation Engineering*, Santiago, Chile (in Spanish).

Domencich, T. and McFadden, D. (1975) *Urban Travel Demand: A Behavioural Analysis*. North-Holland, Amsterdam.

Dong, X., Cai, Y., Cheng, J., Hu, B. and Sun, H. (2020) Understanding the competitive advantages of car sharing from the travel-cost perspective. *International Journal of Environmental Research and Public Health* **17**, 4666.

Dong, X. and Koppelman, F.S. (2004) Comparison of continuous and discrete representations of unobserved heterogeneity in logit models. *83th Annual TRB Meeting*, Washington, DC, USA.

Douglas, A.A. and Lewis, R.J. (1970) Trip generation techniques: (1) Introduction; (2) Zonal least squares regression analysis. *Traffic Engineering and Control* **12**, 362–365, 428–431.

Douglas, A.A. and Lewis, R.J. (1971) Trip generation techniques: (3) Household least squares regression analysis; (4) Category analysis and sumary of trip generation techniques. *Traffic Engineering and Control* **12**, 477–479, 532–535.

Downes, J.D. and Emmerson, P. (1983) Urban transport modelling with fixed travel budgets; an evaluation of the UMOT process. *TRRL Supplementary Report SR 799*, Transport and Road Research Laboratory, Crowthorne.

Downes, J.D. and Gyenes, L. (1976) Temporal stability and forecasting ability of trip generation models in reading. *TRRL Report LR 726*, Transport and Road Research Laboratory, Crowthorne.

DRCOG (2000) *Describing and Reaching Non-Responding Populations*. Denver Regional Council of Governments. http://www.drcog.org/pub_news/about_pub_news.htm.

Dubernet, I. and Axhausen, K.W. (2020) The German value of time and value of reliability study: the survey work. *Transportation* **47**, 1477–1513.

Duffus, L.N., Sule Alfa, A. and Soliman, A.H. (1987) The reliability of using the gravity model for forecasting trip distribution. *Transportation* **14**, 175–192.

Duncan, G.J., Juster, F.T. and Morgan, J.N. (1987) The role of panel studies in research on economic behaviour. *Transportation Research* **21A**, 249–263.

Dunphy, R.T. (1979) Workplace interviews as an efficient source of travel survey data. *Transportation Research Record* **701**, 26–29.

Ebbes, P., Papies, D. and Van Heerde, H.J. (2011) The sense and non-sense of holdout sample validation in the presence of endogeneity. *Marketing Science* **30**, 1115–1122.

Echeñique, M.H., Flowerdew, A.D.J., Hunt, J.D., Mayo, T.R., Skidmore, I.J. and Simmonds, D.C. (1990) The MEPLAN models of Bilbao, Leeds and Dortmund. *Transport Reviews* **10**, 309–322.

Echiburu, T., Hurtubia, R. and Muñoz, J.C. (2021) The role of perceived satisfaction and the built environment on the frequency of cycle-commuting. *Journal of Transport and Land Use* **14**, 171–196.

ECMT (1996) *The Valuation of Environmental Externalities*. European Conference of Ministers of Transport, Paris.

ECMT (1998) *Efficient Transport for Europe: Policies for Internalisation of External Costs*. European Conference of Ministers of Transport, Paris.

Eilon, S. (1972) Goals and constraints in decision making. *Operations Research Quarterly* **23**, 3–15.

El-Assi, W., Mahmoud, S. and Habib, K.M.N. (2017) Effects of built environment and weather on bike sharing demand: a station level analysis of commercial bike sharing in Toronto. *Transportation* **44**, 589–613.

Elldér, E. (2020) Telework and daily travel: new evidence from Sweden. *Journal of Transport Geography* **86**. https://doi.org/10.1016/j.jtrangeo.2020.102777.

England, J., Hudson, K., Masters, R., Powell, K. and Shortridge, J. (eds.) (1985) *Information Systems for Policy Planning in Local Government*. Longman, Harlow.

Escobar, D., Sarache, W. and Jiménez-Riaño, E. (2022) The impact of a new aerial cable-car project on accessibility and CO_2 emissions considering socioeconomic stratum: a case study in Colombia. *Journal of Cleaner Production* **340**, https://doi.org/10.1016/j.jclepro.2022.130802.

Espino, R., Ortúzar, J. de D. and Román, C. (2006) Confidence interval for willingness-to-pay measures in mode choice models. *Networks and Spatial Economics* **6**, 81–96.

ESTRAUS (1989) *Estudio Estratégico de Transporte del Gran Santiago*. Final Report to the Executive Secretariat of the Urban Transport Commission, Consorcio SIGDO-KOPPERS/CIS, Santiago (in Spanish).

Estupiñan, N., Santana, M., Palacios, A. and Rodríguez, D.A. (2015) *Motorcycle Ownership and Use: The Case of Latin America*. CAF-Development Bank of Latin America, Caracas.

Ettema, D., Ashiru, O. and Polak, J. (2004) Modelling timing and duration of activities and trips in response to road-pricing policies. *Transportation Research Record* **1894**, 1–10.

Ettema, D., Bastin, F., Polak, J. and Ashiru, O. (2007) Modelling the joint choice of activity timing and duration. *Transportation Research* **41A**, 827–841.

Ettema, D., Gunn, H., De Jong, G. and Lindveld, K. (1997) A simulation method for determining the confidence interval of a weighted group average value of time. *Proceedings 25th European Transport Forum*, Brunel University, England.

Evans, A. (1972) On the theory of the valuation and allocation of time. *Scottish Journal of Political Economy* **19**, 1–17.

Evans, S.P. (1973) A relationship between the gravity model for trip distribution and the transportation problem in linear programming. *Transportation Research* **7**, 39–61.

Evans, S.P. (1976) Derivation and analysis of some models for combining trip distribution and assignment. *Transportation Research* **10**, 37–57.

Evans, S.P. and Kirby, H.R. (1974) A three dimensional Furness procedure for calibrating gravity models. *Transportation Research* **8**, 105–122.

Fang, S.C. and Tsao, S.J. (1995) Linearly-constrained entropy maximization problem with quadratic cost and its applications to transportation planning problems. *Transportation Science* **29**, 353–365.

Feil, M., Balmer, M. and Axhausen, K.W. (2009) Generating comprehensive all-day schedules: expanding activity-based travel demand modelling. *2009 European Transport Conference*, Leeuwenhorst, The Netherlands.

Feitelson, E., Hurd, R. and Mudge, R. (1996) The impact of airport noise on willingness-to-pay for residences. *Transportation Research* **1D**, 1–14.

Fenichel, E.P., Lupi, F., Hoehn, J.P. and Kaplowitz, M.D. (2009) Split-sample tests of 'no opinion' responses in an attribute-based choice model. *Land Economics* **85**, 348–362.

Fernández-Antolín, A., Guevara, C.A., de Lapparent, M. and Bierlaire, M. (2016) Correcting for endogeneity due to omitted attitudes: empirical assessment of a modified MIS method using RP mode choice data. *Journal of Choice Modelling* **20**, 1–15.

Fernández-Heredia, A., Jara-Díaz, S. and Monzón, A. (2016) Modelling bicycle use intention: the role of perceptions. *Transportation* **43**, 1–23.

Fernández, J.E., Coeymans, J.E. and Ortúzar, J. de D. (1983) Evaluating extensions to the Santiago underground system. *Proceedings 11th PTRC Summer Annual Meeting*, University of Sussex, England.

Fernández, J.E. and Friesz, T.L. (1983) Equilibrium predictions in transportation markets: the state of the art. *Transportation Research* **17B**, 155–172.

Ferrini, S. and Scarpa, R. (2007) Designs with a-priori information for nonmarket valuation with choice-experiments: a Monte Carlo study. *Journal of Environmental Economics and Management* **53**, 342–363.

Fieller, E. (1933) The distribution of the index in a Normal bivariate population. *Biometrika* **24**, 428–440.

Fieller, E.C. (1954) Some problems in interval estimation. *Journal of the Royal Statistical Society* **16B**, 175–185.

Fisk, C.S. (1988) On combining maximum entropy trip matrix estimation with user optimal assignment. *Transportation Research* **22B**, 69–73.

Florian, M. and Spiess, H. (1982) The convergence of diagonalization algorithms for asymmetric network equilibrium problems. *Transportation Research* **16B**, 477–484.

Flügel, S., Elvik, R., Veisten, K., Rizzi, L.I., Meyer, S.F., Ramjerdi, F. and Ortúzar, J. de D. (2015) Asymmetric preferences for road safety: evidence from a stated choice experiment among car drivers. *Transportation Research* **31F**, 112–123.

Flyvbjerg, B., Ansar, A., Budzier, A., Buhl, S., Cantarelli, C.C., Garbuio, M., Glenting, C., Holm, M., Lovallo, D., Lunn, D., Molin, E.J.E., Rønnest, A., Stewart, A. and van Wee, B. (2018) Five things you should know about cost overrun. *Transportation Research* **118A**, 174–190.

Flyvbjerg, B., Skamris Holm, M.K. and Buhl, S.L. (2005) Inaccuracy in traffic forecasts. *Transport Reviews* **26**, 1–24.

Foerster, J.F. (1979) Mode choice decision process models: a comparison of compensatory and non-compensatory structures. *Transportation Research* **13A**, 17–28.

Foerster, J.F. (1981) Nonlinear and non-compensatory perceptual functions of evaluations and choice. In P.R. Stopher, A.H. Meyburg and W. Brög (eds.), *New Horizons in Travel Behaviour Research*. D.C. Heath and Co., Lexington, MA.

Foot, D. (1981) *Operational Urban Models*. Methuen, London.

Forrester, J. (1969) *Urban Dynamics*. Productivity Press, Portland.

Fosgerau, M. and Bierlaire, M. (2007) A practical test for the choice of mixing distribution in discrete choice models. *Transportation Research* **41B**, 784–794.

Fosgerau, M. and Hess, S. (2010) A comparison of methods for representing random taste heterogeneity in discrete choice models. *European Transport* **42**, 1–25.

Fosgerau, M., Hjorth, K. and Lyk-Jensen, S.V. (2007) *The Danish Value of Time Study: Final Report*. Danish Transport Research Institute, Copenhagen.

Fowkes, A.S. (2000) Recent developments in stated preference techniques in transport research. In J. de D. Ortúzar (ed.), *Stated Preference Modelling Techniques*. Perspectives 4, PTRC, London.

Fowkes, A.S. and Tweddle, G. (2000) Validation of stated preference forecasting: a case study involving anglo-continental freight. In J. de D. Ortúzar (ed.), *Stated Preference Modelling Techniques*, London.

Fowkes, A.S. and Wardman, M. (1988) The design of stated preference travel choice experiments, with special reference to interpersonal taste variations. *Journal of Transport Economics and Policy* **22**, 27–44.

Frank, M. and Wolfe, P. (1956) An algorithm for quadratic programming. *Naval Research Logistics Quarterly* **3**, 95–110.

Friedrich, M., Haupt, T. and Nökel, K. (2003) Freight modelling: data issues, survey methods, demand and network models. *10th International Conference on Travel Behaviour Research*, Lucerne, Switzerland.

Friedrich, R., Bickel, P. and Krewitt, W. (eds.) (1998) *External Costs of Transport. Institute for Energy Economics and Rational Energy Use*. Universität Stuttgart, Stuttgart.

Friesz, T.L., Bernstein, D., Smith, T.E., Tobin, R.L. and Wie, B.V. (1993) A variational inequality formulation of the dynamic network user equilibrium problem. *Operations Research* **41**, 179–191.

Friesz, T.L., Tobin, R. and Harker, P. (1983) Predictive intercity freight network models: the state of the art. *Transportation Research* **17A**, 409–417.

Furness, K.P. (1965) Time function iteration. *Traffic Engineering and Control* **7**, 458–460.

Galbraith, R.A. and Hensher, D.A. (1982) Intra-metropolitan transferability of mode choice models. *Journal of Transport Economics and Policy* **16**, 7–29.

Galilea, P. and Ortúzar, J. de D. (2005) Valuing noise level reductions in a residential location context. *Transportation Research* **10D**, 305–322.

Gálvez, T. and Jara-Díaz, S.R. (1998) On the social valuation of travel time savings. *International Journal of Transport Economics* **25**, 205–219.

Gardner, B. (2009) Modelling motivation and habit in stable travel mode contexts. *Transportation Research* **12F**, 68–76.

Gartner, N.H. (1980) Optimal traffic assignment with elastic demands: a review. *Transportation Science* **14**, 192–208.

Gatersleben, B. and Haddad, H. (2010) Who is the typical bicyclist? *Transportation Research* **13F**, 41–48.

Gaudry, M.J.I., Jara-Díaz, S.R. and Ortúzar, J. de D. (1989) Value of time sensitivity to model specification. *Transportation Research* **23B**, 151–158.

Gaudry, M.J.I. and Wills, M.I. (1978) Estimating the functional form of travel demand models. *Transportation Research* **12**, 257–289.

Gayialis, S.P., Kechagias, E.P. and Konstantakopoulos, G.D. (2022) A city logistics system for freight transportation: integrating information technology and operational research. *Operational Research* **22**, 5953–5982.

Geman, S. and Geman, D. (1984) Stochastic relaxation, Gibbs distribution and the Bayesian restoration of images. *IEEE Transactions on Pattern Analysis and Machine Intelligence* **6**, 721–741.

Gibson, J., Baeza, I. and Willumsen, L.G. (1989) Congestion, bus stops and congested bus stops. *Traffic Engineering and Control* **30**, 291–296.

Gilbert, D. (2007) *Stumbling on Happiness*. Harper Perennial, London.

Glaister, S. (1999) Observations on the new approach to the appraisal of road projects. *Journal of Transport Economics and Policy* **33**, 227–234.

Glerum, A., Atasoy, B. and Bierlaire, M. (2014) Using semi-open questions to integrate perceptions in choice models. *Journal of Choice Modelling* **10**, 11–33.

Göddeke, D., Konstantin Krauss, K. and Gnann, T. (2022) What is the role of carsharing toward a more sustainable transport behavior? Analysis of data from 80 major German cities. *International Journal of Sustainable Transportation* **16**, 861–873.

Godoy, G. and Ortúzar, J. de D. (2008) On the estimation of mixed logit models. In P.O. Inweldi (ed.), *Transportation Research Trends*. Nova Science Publishers, New York.

Goldenberg, X. (1996) Choosing a household survey method: results for the Dallas-Fort Worth pretest. *4th International Conference on Survey Methods in Transport*, Oxford, England.

Golob, T.F., Kitamura, R. and Supernak, J. (1997) A panel-based evaluation of the San Diego I-15 carpool lanes project. In T.F. Golob, R. Kitamura and L. Long (eds.), *Panels for Transportation Planning: Methods and Applications*. Kluwer, Boston.

Golob, T.F. and Richardson, A.J. (1981) Non-compensatory and discontinuous constructs in travel behaviour models. In P.R. Stopher, A.H. Meyburg and W. Brög (eds.), *New Horizons in Travel Behaviour Research*. D.C. Heath and Co., Lexington, MA.

González-Valdés, F., Heydecker, B.G. and Ortúzar, J. de D. (2022) Quantifying behavioural difference in latent class models to assess empirical identifiability: analytical development and application to multiple heuristics. *Journal of Choice Modelling* **46**, 100356.

González-Valdés, F. and Ortúzar, J. de D. (2017) The stochastic satisficing model: a bounded rationality discrete choice model. *Journal of Choice Modelling* **27**, 74–87.

González-Valdés, F. and Raveau, S. (2018) Identifying the presence of heterogeneous discrete choice heuristics at an individual level. *Journal of Choice Modelling* **28**, 28–40.

González, R.M., Román, C., Amador, F.J., Rizzi, L.I., Ortúzar, J. de D., Espino, R., Martin, J.C. and Cherchi, E. (2018) Estimating the value of risk reductions for car drivers when pedestrians are involved: a case study in Spain. *Transportation* **45**, 499–521.

Goodman, P.R. (1973) Trip generation: a review of the category analysis and regression models. *Working Paper 9*, Institute of Transport Studies, Leeds University.

Goodwin, P.W. (1977) Habit and hysteresis in mode choice. *Urban Studies* **14**, 95–98.

Goos, P. (2002) *The Optimal Design of Blocked and Split-Plot Experiments*. Lecture Notes in Statistics. Springer-Verlag, New York.

Gopinath, D.A. and Ben-Akiva, M. (1995) Estimation of randomly distributed values of time. *Working Paper*, Department of Civil and Environmental Engineering, MIT.

Gray, R. (1982) Behavioural approaches in freight transport modal choice. *Transport Reviews* **2**, 161–184.

Green, K.C. and Scott Armstrong, J. (2015) Simple versus complex forecasting: the evidence. *Journal of Business Research* **68**, 1678–1685.

Greenblat, C. and Duke, R. (1975) *Gaming-Simulation: Rationale, Design and Applications.* Wiley, New York.

Greene, W.H. (2008) *Econometric Analysis.* Pearson Education Inc, Upper Saddle River, NJ.

Greene, W.H. and Hensher, D.A. (2003) A latent class model for discrete choice analysis: contrasts with mixed logit. *Transportation Research* **37B**, 681–698.

Greene, W.H., Hensher, D.A. and Rose, J.M. (2006) Accounting for heterogeneity in the variance of the unobserved effects in mixed logit models. *Transportation Research* **40B**, 75–92.

Greene, M. and Ortúzar, J. de D. (2002) Willingness-to-pay for social housing attributes: a case study from Chile. *International Planning Studies* **7**, 55–87.

Guerrero, T.E., Guevara, C.A., Cherchi, E. and Ortúzar, J. de D. (2020) Addressing endogeneity in strategic urban mode choice models. *Transportation* **48**, 2081–2102.

Guerrero, T.E., Guevara, C.A., Cherchi, E. and Ortúzar, J. de D. (2021) Forecasting with strategic transport models corrected for endogeneity. *Transportmetrica* **18A**, 708–735.

Guerrero, T.E., Guevara, C.A., Cherchi, E. and Ortúzar, J. de D. (2022) Characterizing the impact of discrete indicators to correct for endogeneity in discrete choice models. *Journal of Choice Modelling* **42**. https://doi.org/10.1016/j.jocm.2021.100342.

Guevara, C.A. (2015) Critical assessment of five methods to correct for endogeneity in discrete-choice models. *Transportation Research* **82A**, 240–254.

Guevara, C.A. (2018) Overidentification tests for the exogeneity of instruments in discrete choice models. *Transportation Research* **114B**, 241–253.

Guevara, C.A. and Ben-Akiva, M. (2006) Endogeneity in residential location choice models. *Transportation Research Record* **1977**, 60–66.

Guevara, C.A. and Polanco, D. (2016) Correcting for endogeneity due to omitted attributes in discrete-choice models: the multiple indicator solution. *Transportmetrica* **12A**, 458–478.

Guevara, C.A. and Thomas, A. (2007) Multiple classification analysis in trip production models. *Transport Policy* **14**, 514–522.

Gunn, H.F. (1984) An analysis of transfer price data. *Proceedings 12th PTRC Summer Annual Meeting,* University of Sussex, England.

Gunn, H.F. (1985a) *Value of Time for Evaluation Purposes: The State of the Art.* Report No. 421–01, Hague Consulting Group, The Hague.

Gunn, H.F. (1985b) Artificial sample applications for spatial interaction models. In *Colloquium Vervoersplanologisch Speurwerk.* The Hague, The Netherlands.

Gunn, H.F. and Bates, J.J. (1982) Statistical aspects of travel demand modelling. *Transportation Research* **16A**, 371–382.

Gunn, H.F., Ben-Akiva, M.E. and Bradley, M.A. (1985) Tests of the scaling approach to transferring disaggregate travel demand models. *Transportation Research Record* **1037**, 21–30.

Gunn, H.F., Fisher, P., Daly, A.J. and Pol, H. (1982) Synthetic samples as a basis for enumerating disaggregate models. *Proceedings 10th PTRC Summer Annual Meeting,* University of Warwick, England.

Gunn, H.F., Kirby, H.R., Murchland, J.D. and Whittaker, J.C. (1980) The RHTM trip distribution investigation. *Proceedings 8th PTRC Summer Annual Meeting,* University of Warwick, England.

Gunn, H.F. and Pol, H. (1986) Model transferability: the potential for increasing cost-effectiveness. In Λ. Ruhl (ed.), *Behavioural Research for Transport Policy.* VNU Science Press, Utrecht.

Guo, J.Y. and Bhat, C.R. (2007) Population synthesis for microsimulating travel behaviour. *Transportation Research Record* **2014**, 92–101.

Gutiérrez, M., Cantillo, V., Arellana, J. and Ortúzar, J. de D. (2021) Estimating bicycle demand in an aggressive environment. *International Journal of Sustainable Transportation* **15**, 259–272.

Gutierrez, M., Hurtubia, R. and Ortúzar, J. de D. (2020) The role of habit and the built environment in the willingness to commute by bicycle. *Travel Behaviour and Society* **20**, 62–73.

Habib, K.M.N. (2012) Modelling commuting mode choice jointly with work start time and duration. *Transportation Research* **46A**, 33–47.

Habib, K.M.N., Han, X. and Lin, W. (2014a) Joint modelling of propensity and distance for walking-trip generation. *Transportmetrica* **10A**, 420–436.

Habib, K.M.N., Mann, J., Mahmoud, M. and Weiss, A. (2014b) Synopsis of bicycle demand in the City of Toronto: investigating the effects of perception, consciousness and comfortability on the purpose of biking and bike ownership. *Transportation Research* **70A**, 67–80.

Hägerstrand, T. (1970) What about people in regional science? *Papers of the Regional Science Association* **24**, 7–21.

Hajivassiliou, V.A. and Ruud, P. (1994) Classical estimation methods for LDV models using simulation. In R. Engle and D. McFadden (eds.), *Handbook of Econometrics*, Vol. **IV**. Elsevier Science, Amsterdam.

Hall, M.D., Daly, A.J., Davies, R.F. and Russell, C.H. (1987) Modelling for an expanding city. *Proceedings 15th PTRC Summer Annual Meeting*, University of Bath, England.

Hall, M.D., Van Vliet, D. and Willumsen, L.G. (1980) SATURN – a simulation assignment model for the evaluation of traffic management schemes. *Traffic Engineering and Control* **21**, 168–176.

Hamilton, T. and Wichman, C. (2017) Bicycle infrastructure and traffic congestion: evidence from DC's Capital Bikeshare. *Journal of Environmental Economics and Management* **87**, 72–93.

Han, S.J. and Heydecker, B.G. (2006) Consistent objective and solutions of dynamic user equilibrium models. *Transportation Research* **40B**, 16–34.

Handy, S.L., Xing, Y. and Buehler, T.J. (2010) Factors associated with bicycle ownership and use: a study of six small U.S. cities. *Transportation* **37**, 967–985.

Hanemann, W.M. (1994) Valuing the environment through contingent valuation. *Journal of Economic Perspectives* **8**, 19–43.

Hansen, I.A. (2020) Hyperloop transport technology assessment and system analysis. *Transportation Planning and Technology* **43**, 803–820.

Hansson, L. and Markham, J. (1992) *Internalisation of External Costs in Transportation*. International Road Transport Union, Paris.

Harker, P.T. (1985) The state of the art in the predictive analysis of freight transport systems. *Transport Reviews* **5**, 143–164.

Harker, P.T. (1987) *Predicting Intercity Freight Flows*. VNU Science Press, Utrecht.

Hartgen, D. (2013) Hubris or humility? Accuracy issues for the next 50 years of travel demand forecasting. *Transportation* **40**, 1133–1157.

Hartley, T.M. and Ortúzar, J. de D. (1980) Aggregate modal split models: is current U.K. practice warranted? *Traffic Engineering and Control* **21**, 7–13.

Hauser, J.R. (1978) Testing the accuracy, usefulness and significance of probabilistic choice models: an information theoretic approach. *Operations Research* **26**, 406–421.

Hausman, J.A. (ed.) (1993) *Contingent Valuation: A Critical Assessment*. North-Holland, Amsterdam.

Haustein, S., Thorhauge, M. and Cherchi, E. (2018) Commuters' attitudes and norms related to travel time and punctuality: a psychographic segmentation to reduce congestion. *Travel Behaviour and Society* **12**, 41–50.

HCG (1990) *The Netherlands Value of Time Study*. Final Report to the Dienst Verkeerskunde, Rijkswaterstaat. Hague Consulting Group, The Hague.

HCG, Halcrow Fox and Imperial College London (2000) *User Manual for HADES 1.0*. Prepared for the Department of the Environment, Transport and the Regions, London.

Heckman, J.J. (1978) Dummy endogenous variables in a simultaneous equation system. *Econometrica* **46**, 931–959.

Heckman, J.J. (1981) Statistical models for discrete panel data. In C.F. Manski and D. McFadden (eds.), *Structural Analysis of Discrete Data: With Econometric Applications*. MIT Press, Cambridge, MA.

Hedayat, A.S., Sloane, N.J.A. and Stufken, J. (1999) *Orthogonal Arrays: Theory and Applications*. Springer-Verlag, New York.

Heggie, I.G. (1983) Valueing savings in non working time: the empirical dilemma. *Transportation Research* **17A**, 13–23.

Heinen, E., Maat, K. and van Wee, B. (2011) The role of attitudes toward characteristics of bicycle commuting on the choice to cycle to work over various distances. *Transportation Research* **16D**, 102–109.

Hendrickson, C. and Kocur, G. (1981) Schedule delay and departure time decisions in a deterministic model. *Transportation Science* **15**, 62–77.

Hendrickson, C. and Plank, E. (1984) The flexibility of departure times for work trips. *Transportation Research* **18A**, 25–36.

Hensher, D.A. (1976) Valuations of commuter travel time savings: an alternative procedure. In I.G. Heggie (ed.), *Modal Choice and the Value of Time*. Clarendon Press, Oxford.

Hensher, D.A. (1987) Issues in the pre-analysis of panel data. *Transportation Research* **21A**, 265–285.

Hensher, D.A. (1999) HEV choice models as a search engine for the specification of nested logit tree structures. *Marketing Letters* **10**, 339–349.

Hensher, D.A. (2001a) The valuation of commuter travel time savings for car drivers: evaluating alternative model specifications. *Transportation* **28**, 101–118.

Hensher, D.A. (2001b) Measurement of the valuation of travel time savings. *Journal of Transport Economics and Policy* **35**, 71–98.

Hensher, D.A. (2001c) The sensitivity of the valuation of travel time savings to the specification of unobserved effects. *Transportation Research* **37E**, 129–142.

Hensher, D.A. (2006a) The signs of times: imposing a globally signed condition on willingness-to-pay distributions. *Transportation* **33**, 205–222.

Hensher, D.A. (2006b) How do respondents process stated choice experiments? Attribute consideration under varying information load. *Journal of Applied Econometrics* **21**, 861–878.

Hensher, D.A. (2008) Joint estimation of process and outcome in choice experiments and implications for willingness to pay. *Journal of Transport Economics and Policy* **42**, 297–322.

Hensher, D.A. (2011) A practical note on calculating behaviourally meaningful generalised cost when there are two cost parameters in a utility expression. *Road and Transport Research* **20**, 74–76.

Hensher, D.A. and Greene, W.H. (2002) Specification and estimation of the nested logit model: alternative normalizations. *Transportation Research* **36B**, 1–17.

Hensher, D.A. and Greene, W.H. (2003) The mixed logit model: the state of practice. *Transportation* **30**, 133–176.

Hensher, D.A., Ho, C.Q., Mulley, C., Nelson, J.D., Smith, G. and Wong, Y.Z. (2020) *Understanding Mobility as a Service (MaaS): Past, Present and Future*. Elsevier, Amsterdam.

Hensher, D.A. and Louviere, J.J. (1998) A comparison of elasticities derived from multinomial logit, nested logit and heteroskedastic extreme value SP-RP discrete choice models. *Proceedings 8th World Conference of Transportation Research*, Antwerp, Belgium.

Hensher, D.A., Rose, J.M. and Greene, W.H. (2015) *Applied Choice Analysis*. Second Edition. Cambridge University Press, Cambridge.

Hensher, D.A., Rose, J.M., Ortúzar, J. de D. and Rizzi, L.I. (2009) Estimating the willingness-to-pay and value of risk reduction for car occupants in the road environment. *Transportation Research* **43A**, 692–707.

Hensher, D., Wei, E., Liu, W., Ho, L. and Ho, C. (2023) Development of a practical aggregate spatial road freight modal demand model system for truck and commodity movements with an application of a distance-based charging regime. *Transportation* **50**, 1031–1071.

Herriges, J.A. and Kling, C.L. (1995) An empirical investigation of the consistency of nested logit models with utility maximization. *Economic Letters* **50**, 33–39.

Herriges, J.A. and Kling, C.L. (1996) Testing the consistency of nested logit models with utility maximization. *American Journal of Agricultural Economics* **77**, 875–884.

Hess, S., Ben-Akiva, M.E., Gopinath, D. and Walker, J.L. (2009) Taste heterogeneity, correlation and elasticities in latent class choice models. *88th Annual TRB Meeting*, Washington, DC.

Hess, S., Bierlaire, M. and Polak, J.W. (2005a) Capturing correlation and taste heterogeneity with mixed GEV models. In R. Scarpa and A. Alberini (eds.), *Application of Simulation Methods in Environmental and Resource Economics*. Springer, Dordrecht.

Hess, S., Bierlaire, M. and Polak, J.W. (2005b) Estimation of values of travel-time savings using mixed logit models. *Transportation Research* **39A**, 221–236.

Hess, S., Bierlaire, M. and Polak, J.W. (2007a) A systematic comparison of continuous and discrete mixtures models. *European Transport* **37**, 35–61.

Hess, S. and Palma, D. (2019) Apollo: a flexible, powerful and customisable freeware package for choice model estimation and application. *Journal of Choice Modelling* **32**, 100170.

Hess, S., and Polak, J. (2004) An analysis of parking behaviour using discrete choice models calibrated on SP datasets. *Conference Papers*, European Regional Science Association, Volos, Greece.

Hess, S., Polak, J., Daly, A.J. and Hyman, G. (2007b) Flexible substitution patterns in models of mode and time of day choice: new evidence from the UK and the Netherlands. *Transportation* **34**, 213–238.

Hess, S. and Rose, J.M. (2009) Allowing for intra-respondent variations in coefficients estimated on repeated choice data. *Transportation Research* **43B**, 708–719.

Hess, S., Shires, J. and Jopson, A. (2013) Accommodating underlying pro-environmental attitudes in a rail travel context: application of a latent variable latent class specification. *Transportation Research* **25D**, 42–48.

Hess, S. and Stathopoulos, A. (2013) A mixed random utility – random regret model linking the choice of decision rule to latent character traits. *Journal of Choice Modelling* **9**, 27–38.

Hess, S., Stathopoulos, A. and Daly, A. (2012) Allowing for heterogeneous decision rules in discrete choice models: an approach and four case studies. *Transportation* **39**, 565–591.

Hess, S., Train, K.E. and Polak, J.W. (2006) On the use of a Modified Latin Hypercube Sampling (MLHS) approach in the estimation of a mixed logit model for vehicle choice. *Transportation Research* **40B**, 147–163.

Heydecker, B.G. (2004) Objectives, stimulus and feedback in signal control of road traffic. *Journal of Intelligent Transportation Systems* **8**, 63–76.

Heydecker, B.G. and Addison, J.G. (2005) Analysis of dynamic traffic equilibrium with departure time choice. *Transportation Science* **39**, 39–57.

Hillier, B. and Hanson, J. (1984) *The Social Logic of Space*. Cambridge University Press, Cambridge.

Hilveert, O., Toledo, T. and Bekhor, S. (2012) Framework and model for parking decisions. *Transportation Research Record* **2319**, 30–38.

Hinkley, D. (1969) On the ratio of two correlated normal random variables. *Biometrika* **56**, 635–639.

Högberg, P. (1976) Estimation of parameters in models for traffic prediction: a non-linear-regression approach. *Transportation Research* **10**, 263–265.

Hojman, P., Ortúzar, J. de D. and Rizzi, L.I. (2005) On the joint valuation of averting fatal and serious injuries in highway accidents. *Journal of Safety Research* **36**, 377–386.

Holden, D., Fowkes, A.S. and Wardman, M. (1992) Automatic stated preference design algorithms. *Proceedings 20th PTRC Summer Annual Meeting.* University of Manchester Institute of Science and Technology, England.

Holguín-Veras, J., Jaller, M., Destro, L., Ban, X., Lawson, C. and Levinson, H.S. (2011) Freight generation, freight trip generation, and perils of using constant trip rates. *Transportation Research Record* **2224**, 68–81.

Holguín-Veras, J., Kalahasthi, L. and Ramirez-Rios, D.G. (2021) Service trip attraction in commercial establishments. *Transportation Research* **149E**, 102301.

Holguín-Veras, J., Lawson, C., Wang, C., Jaller, M., González-Calderón, C., Campbell, S., Kalahasthi, L., Wojtowicz J. and Ramirez-Rios, D. (2017) *Using Commodity Flow Survey Microdata and Other Establishment Data to Estimate the Generation of Freight, Freight Trips and Service Trips: Guidebook.* NCFRC Research Report 37, National Cooperative Freight Research Program, Transportation Research Board, Washington, DC.

Holguín-Veras, J. and Thorson, E. (2003) Modelling commercial vehicle empty trips with a first order trip chain model. *Transportation Research* **37B**, 129–148.

Hollander, Y. (2016) *Transport Modelling for a Complete Beginner.* CTthink! London.

Hollander, Y. and Liu, R. (2008) Estimation of the distribution of travel times by repeated simulation. *Transportation Research* **16C**, 212–231.

Holm, J., Jensen, T., Nielsen, S., Christensen, A., Johnsen, B. and Ronby, G. (1976) Calibrating traffic models on traffic census results only. *Traffic Engineering and Control* **17**, 137–140.

Horni, A., Nagel, K. and Axhausen, K.W. (2016) *The Multi-Agent Transport Simulation MATSim.* Ubiquity Press, London.

Horowitz, J.L. (1980) Confidence intervals for the choice probabilities of the multinomial logit model. *Transportation Research Record* **728**, 23–29.

Horowitz, J.L. (1981) Sources of error and uncertainty in behavioural travel demand models. In P.R. Stopher, A.H. Meyburg and W. Brög (eds.), *New Horizons in Travel Behaviour Research.* D.C. Heath and Co., Lexington, MA.

Horowitz, J.L. (1982) Specification tests for probabilistic choice models. *Transportation Research* **16A**, 383–394.

Horowitz, J.L. and Manski, C.F. (1998) Censoring of outcomes and regressors due to survey nonresponse: identification and estimation using weights and imputations. *Journal of Econometrics* **84**, 37–58.

Huang, G.H. and Bandeen-Roche, K. (2004) Building an identifiable latent class model with covariate effects on underlying and measured variables. *Psychometrika* **69**, 5–32.

Huber, J. and Train, K.E. (2001) On the similarity of classical and Bayesian estimates of individual mean partworths. *Marketing Letters* **12**, 257–267.

Huber, J. and Zwerina, K. (1996) The importance of utility balance and efficient choice designs. *Journal of Marketing Research* **33**, 307–317.

Hunt, G.L. (2000) Alternative nested logit model structures and the special case of partial degeneracy. *Journal of Regional Science* **40**, 89–113.

Hyman, G.M. (1969) The calibration of trip distribution models. *Environment and Planning* **1**, 105–112.

Hyman, G.M. (1997) The development of operational models for time period choice. *Working Paper*, HETA Division, Department of the Environment, Transport and the Regions, London.

Ibeas, A., dell'Olio, L., Bordagaray, M. and Ortúzar, J. de D. (2014) Modelling parking choices considering user heterogeneity. *Transportation Research* **70A**, 41–49.

Iglesias, P., Godoy, F.J., Ivelic, A.M. and Ortúzar, J. de D. (2008) Un modelo de generación, distribución y partición modal conjunta para viajes interurbanos. *Proceedings XIV Panamerican Congress on Traffic and Transportation Engineering*, Cartagena de Indias, Colombia (in Spanish).

Iglesias, P., Greene, M. and Ortúzar, J. de D. (2013) On the perception of safety in low income neighbourhoods: using digital images in a stated choice experiment. In S. Hess and A. Daly (eds.), *Choice Modelling: The State of the Art and the State of Practice*. Edward Elgar Publishing Ltd., Cheltenham.

Iglesias, P., Ortúzar, J. de D., Rodríguez-Valencia, A., Giraldez, F. and Calatayud, A. (2022) Entendiendo la elección modal del automóvil en ciudades de ALC. *Technical Note IDB-TN-2416*, Banco Interamericano de Desarrollo, Washington, DC (in Spanish)

Iragüen, P. and Ortúzar, J. de D. (2004) Willingness-to-pay for reducing fatal accidents risk in urban areas: an internet-based web page stated preference survey. *Accident Analysis and Prevention* **36**, 513–524.

ITF (2020) *Shared Mobility Simulations for Lyon*. OECD-International Transport Forum, Paris.

Jain, T., Rose, G. and Johnson, M. (2022) Changes in private car ownership associated with car sharing: gauging differences by residential location and car share typology. *Transportation* **49**, 503–527.

Jang, T. (2005) Count data models for trip generation. *Journal of Transportation Engineering of ASCE* **131**, 444–450.

Janke, J. and Handy, S. (2019) How life course events trigger changes in bicycling attitudes and behaviour: insights into causality. *Travel Behaviour and Society* **16**, 31–41.

Janosikova, L., Slavík, J. and Koháni, M. (2014) Estimation of a route choice model for urban public transport using smart card data. *Transportation Planning and Technology* **37**, 638–648.

Jansen, G.R.M. and Bovy, P.H.L. (1982) The effect of zone size and network detail on all-or-nothing and equilibrium assignment outcomes. *Traffic Engineering and Control* **23**, 311–317.

Jara-Díaz, S.R. (1998) Time and income in travel choice: towards a microeconomic activity framework. In T. Garling, T. Laitila and K. Westin (eds.), *Theoretical Foundations of Travel Choice Modelling*. Pergamon, Oxford.

Jara-Díaz, S.R. (2000) Allocation and valuation of travel time savings. In D. Hensher and K.J. Button (eds.), *Handbook of Transport Modelling*. Pergamon, Oxford.

Jara-Díaz, S.R. (2007) *Transport Economic Theory*. Elsevier Science, Amsterdam.

Jara-Díaz, S.R. and Farah, M. (1987) Transport demand and user's benefits with fixed income: the goods/leisure trade-off revisited. *Transportation Research* **21B**, 165–170.

Jara-Díaz, S.R. and Guevara, A. (2000) The contribution of work, leisure and travel to the subjective value of travel time savings. *2000 European Transport Conference*, Cambridge, England.

Jara-Díaz, S.R., Munizaga, M. and Guerra, R. (2012) Sensitivity of the value of time to the specification of income in discrete choice models. *International Journal of Transport Economics* **39**, 239–254.

Jara-Díaz, S.R. and Ortúzar, J. de D. (1989) Introducing the expenditure rate in the estimation of mode choice models. *Journal of Transport Economics and Policy* **23**, 293–308.

Jayakrishnan, R., Tsai, W., Prashker, J. and Rajadhyaksha, S. (1994) A faster path-based algorithm for traffic assignment. *Transportation Research Record* **1443**, 75–83.

Jensen, A.F., Cherchi, E. and Ortúzar, J. de D. (2014) A long panel survey to elicit variations in preferences and attitudes in the choice of electric vehicles. *Transportation* **41**, 973–993.

Johnson, L.W. and Hensher, D.A. (1982) Application of multinomial probit to a two-period panel data set. *Transportation Research* **16A**, 457–464.

Johnson, M. (1966) Travel time and the price of leisure. *Western Economic Journal* **8**, 135–145.

Jones, P.M. (1979) New approaches to understanding travel behaviour: the human activity approach. In D.A. Hensher and P.R. Stopher (eds.), *Behavioural Travel Modelling*. Croom Helm, London.

Jones, P.M. and Willumsen, L.G. (2021) Modelling and analysis of shared autonomous mobility. In X. Zhang (ed.), *Cities for Driverless Vehicles*. ICE Publishing, London.

Jones-Lee, M.W., Hammerton, M. and Philips, P.R. (1985) The value of safety: results of a national sample survey. *Economic Journal* **95**, 49–72.

Jones-Lee, M.W., Loomes, G., O'Reilly, D.M. and Philips, P.R. (1992) The value of preventing non-fatal road injuries: findings of a willingness-to-pay national sample survey. *TRL Contractor Report 330*, Transport Research Laboratory, Crowthorne.

Jones-Lee, M., Loomes, G. and Philips, P. (1995) Valuing the prevention of non-fatal road injuries: contingent valuation vs. standard gamble. *Oxford Economic Papers* **47**, 676–695.

Jones Lee, M., O'Reilly, D. and Philips, P. (1993) The value of preventing non-fatal road injuries: findings of a willingness to pay national sample survey. *TRL Working Paper WPSRC2*, Transport Research Laboratory, Crowthorne.

Kagho, G.O., Balac, M. and Axhausen, K.W. (2020) Agent-based models in transport planning: current state, issues, and expectations. *Procedia Computer Science* **170**, 726–732.

Kahneman, D. (2013) *Thinking Fast and Slow*. Farrar, Straus and Giroux, New York.

Kahneman, D. and Tversky, A. (1979) Prospect theory: an analysis of decisions under risk. *Econometrica* **47**, 263–291.

Kam, H.B. and Morris, J. (1999) Response patterns in travel surveys: the VATS experience. *Working Paper*, Transport Research Centre, RMIT.

Kamargianni, M., Ben-Akiva, M. and Polydoropoulou, A. (2014) Integrating social interaction into hybrid choice models. *Transportation* **41**, 1263–1285.

Kannel, E.J. and Heathington, K.W. (1973) Temporal stability of trip generation relations. *Highway Research Record* **472**, 17–27.

Kanninen, B.J. (2002) Optimal design for multinomial choice experiments. *Journal of Marketing Research* **39**, 214–217.

Kass, R., Carlin, B., Gelman, A. and Neal, R. (1998) Markov Chain Monte Carlo in practice: a roundtable discussion. *The American Statistician* **52**, 93–100.

Keeter, S., Miller, A., Kohut, A., Groves, R.M. and Presser, S. (2000) Consequences of reducing non-response in a national telephone survey. *Public Opinion Quarterly* **64**, 125–148.

Kessels, R., Goos, P. and Vandebroek, M. (2006) A comparison of criteria to design efficient choice experiments. *Journal of Marketing Research* **43**, 409–419.

Khan, A. and Willumsen, L.G. (1986) Modelling car ownership and use in developing countries. *Traffic Engineering and Control* **27**, 554–560.

Kim, H., Li, J., Roodman, S., Sen, A., Sööt, S. and Christopher, E. (1993) Factoring household travel surveys. *Transportation Research Record* **1412**, 17–22.

Kim, T.J. and Hinkle, J. (1982) Model for statewide freight transportation planning. *Transportation Research Record* **889**, 15–19.

Kim, K.M., Hong, S., Ko, S. and Kim, D. (2014) Does crowding affect the path choice of metro passengers? *Transportation Research* **77A**, 292–304.

Kimber, R.M. and Hollis, E.M. (1979) Traffic queues and delays at road junctions. *TRRL Report LR 909*, Transport and Road Research Laboratory, Crowthorne.

Kirby, H.R. (1979) Partial matrix techniques. *Traffic Engineering and Control* **20**, 422–428.

Kitamura, R. (1990) Panel analysis in transportation planning: an overview. *Transportation Research* **24A**, 401–415.

Kitamura, R. and Bovy, P.H.L. (1987) Analysis of attrition biases and trip reporting errors for panel data. *Transportation Research* **21A**, 287–302.

Kogus, A., Foltynova, H.B., Ayelet, G.T., Shiftan, Y., Vejchodská, E. and Shiftan, Y. (2022) Will COVID-19 accelerate telecommuting? A cross country evaluation for Israel and Czechia. *Transportation Research* **164A**, 291–309.

Koohsari, M.J., Sugiyama, T., Mavoa, S., Villanueva, K., Badland, H., Giles-Corti, B. and Owen, N. (2016) Street network measures and adults' walking for transport: application of space syntax. *Health & Place* **38**, 89–95.

Koppelman, F.S. (1976) Guidelines for aggregate travel prediction using disaggregate choice models. *Transportation Research Record* **610**, 19–24.

Koppelman, F.S., Kuah, G-K and Rose G. (1985a) Transfer model updating with aggregate data. *64th Annual TRB Meeting*, Washington, DC, USA.

Koppelman, F.S., Kuah, G.-K. and Wilmot, C.G. (1985b) Transfer model updating using disaggregate data. *Transportation Research Record* **1037**, 102–107.

Koppelman, F.S. and Wen, C.H. (1998) Alternative nested logit models: structure, properties and estimation. *Transportation Research* **32A**, 289–298.

Koppelman, F.S. and Wen, C.H. (2000) The paired combination logit model: properties, estimation and application. *Transportation Research* **34B**, 75–89.

Koppelman, F.S. and Wilmot, C.G. (1982) Transferability analysis of disaggregate choice models. *Transportation Research Record* **895**, 18–24.

Koster, P. and Tseng, Y.Y. (2010) Stated choice experimental designs for scheduling models. In S. Hess and A. Daly (eds.), *Choice Modelling: The State-of-the-Art and the State-of-Practice*. Bingley, Emerald.

Kraft, G. (1968) *Demand for Intercity Passenger Travel in the Washington-Boston Corridor*. North-East Corridor Project Report, Systems Analysis and Research Corporation, Boston, MA.

Kresge, D.T. and Roberts, P.O. (1971) *Techniques of Transport Planning: Systems Analysis and Simulation Models*. Brookings Institution, Washington, DC.

Krishnan, K.S. (1977) Incorporating thresholds of indifference in probabilistic choice models. *Management Science* **23**, 1224–1233.

Kroes, E.P., Daly, A.J., Gunn, H.F. and van der Hoorn, A.I.J.M. (1996) The opening of the Amsterdam ring road: a case study on short-term effects of removing a bottleneck. *Transportation* **23**, 71–82.

Kruithof, J. (1937) Calculation of telephone traffic. *Der Ingenieur* **52**, E15–E25.

Kruskal, J.B. (1965) Analysis of factorial experiments by estimating monotone transformations of the data. *Journal of the Royal Statistical Society* **27B**, 251–263.

Kumar, A. (1980) Use of incremental form of logit models in demand analysis. *Transportation Research Record* **775**, 21–27.

Kurth, D.L., Coil, J.L. and Brown, M.J. (2001) Assessment of quick-refusal and no-contact non-response in household travel surveys. *Transportation Research Record* **1768**, 114–124.

Lamond, B. and Stewart, N.F. (1981) Bregman's balancing method. *Transportation Research* **15B**, 239–248.

Lancaster, K.J. (1966) A new approach to consumer theory. *Journal of Political Economy* **14**, 132–157.

Landefeld, J.S. and Seskin, E.P. (1982) The economic value of life: linking theory to practice. *American Journal of Public Health* **72**, 555–566.

Langdon, M.G. (1976) Modal split models for more than two modes. *Proceedings 4th PTRC Summer Annual Meeting*, University of Warwick, England.

Langdon, M.G. (1984) Methods of determining choice probability in utility maximising multiple alternative models. *Transportation Research* **18B**, 209–234.

Lange, K.L., Little, R.J. and Taylor, J.M. (1989) Robust statistical modelling using the t distribution. *Journal of the American Statistical Association* **84**, 881–896.

Lanzendorf, M. (2003) Mobility biographies: a new perspective for understanding travel behaviour. *10th International Conference on Travel Behaviour Research,* Lucern, Switzerland.

Larson, R.C. and Odoni, A.R. (1981) *Urban Operations Research.* Prentice Hall, Englewood Cliffs, NJ.

Larsson, T. and Patriksson, M. (1992) Simplicial decomposition with disaggregated representation for the traffic assignment problem. *Transportation Science* **26**, 4–17.

Lawless, J.F. (1987) Negative Binomial and mixed Poisson regression. *The Canadian Journal of Statistics* **15**, 209–225.

Le, H., Carrel, A. and Shah, H. (2022) Impacts of online shopping on travel demand: a systematic review. *Transport Reviews* **42**, 273–295.

Leamer, E. (1978) *Specification Searches: Ad-Hoc Inference with Nonexperimental Data.* Wiley, New York.

Lee, N. and Dalvi, M.Q. (1969) Variations on the value of travel time. *The Manchester School* **37**, 213–236.

Lee, N. and Kirkpatrick, C. (1996) Relevance and consistency of environmental impact assessment and cost-benefit analysis in project appraisal. *Project Appraisal* **11**, 229–236.

Leonard, D.R. and Gower, P. (1982) User guide to CONTRAM version 4. *TRRL Supplementary Report 735,* Transport and Road Research Laboratory, Crowthorne.

Leonard, D.R. and Tough, J. (1979) Validation work on CONTRAM – a model for use in the design of traffic management schemes. *Proceedings 7th PTRC Summer Annual Meeting,* University of Warwick, England.

Leong, W. and Hensher, D.A. (2012) Embedding multiple heuristics into choice models: an exploratory analysis. *Journal of Choice Modelling* **5**, 131–144.

Leong, W. and Hensher, D.A. (2014) Relative advantage maximisation as a model of context dependence for binary choice data. *Journal of Choice Modelling* **11**, 30–42.

Lerman, S.R. (1984) Recent advances in disaggregate demand modelling. In M. Florian (ed.), *Transportation Planning Models.* North-Holland, Amsterdam.

Lerman, S.R. and Louviere, J.J. (1978) The use of functional measurement to identify the form of utility functions in travel demand models. *Transportation Research Record* **673**, 78–85.

Lerman, S.R. and Manski, C.F. (1979) Sample design for discrete choice analysis of travel behaviour: the state of the art. *Transportation Research* **13B**, 29–44.

Lerman, S.R. and Manski, C.F. (1981) On the use of simulated frequencies to approximate choice probabilities. In C. Manski and D. McFadden (eds.), *Structural Analysis of Discrete Data with Econometric Applications.* MIT Press, Cambridge, MA.

Lerman, S.R., Manski, C.F. and Atherton, T.J. (1976) *Non-Random Sampling in the Calibration of Disaggregate Choice Models.* Final Report to the Urban Planning Division, Federal Highway Administration, US Department of Transportation, Washington, DC.

Leurent, F.M. (1998) Multicriteria assignment modelling: making explicit the determinants of mode or path choice. In P. Marcotte and S. Nguyen (eds.), *Equilibrium and Advanced Transportation Modelling.* Kluwer Academic, Boston.

Li, Z. and Hensher, D.A. (2013) Crowding in public transport: a review of objective and subjective measures. *Journal of Public Transportation* **16**, 107–134.

Li, H., Li, X., Xu, X., Liu, J. and Ran, B. (2018) Modelling departure time choice of metro passengers with a smart corrected mixed logit model – a case study in Beijing. *Transport Policy* **69**, 106–121.

Li, X., Zhang, Y., Du, M. and Yang, J. (2019) Social factors influencing the choice of bicycle: difference analysis among private bike, public bike sharing and free-floating bike sharing in Kunming, China. *KSCE Journal of Civil Engineering* **23**, 2339–2348.

Liem, T.C. and Gaudry, M.J.I. (1987) P-2: a program for the Box–Cox logit model with disaggregate data. *Publication 525*, Centre de Recherche sur les Transports, Université de Montréal.

Likert, R. (1932) A technique for the measurement of attitudes. *Archives of Psychology* **140**, 1–55.

Lim, K.K. and Srinivasan, S. (2011) Comparative analysis of alternate econometric structures for trip generation models. *Transportation Research Record* **2254**, 68–78.

Litman, T. (2009) *Transportation Elasticities How Prices and Other Factors Affect Travel Behavior*. Victoria Transport Policy Institute, Victoria, Canada.

Litman, T. (2021) *New Mobilities: Smart Planning for Emerging Transportation Technologies*. Island Press, Washington.

Lizana, P., Ortúzar, J. de D., Arellana, J. and Rizzi, L.I. (2021) Forecasting with a joint mode/time of day choice model based on combined RP and SC data. *Transportation Research* **150A**, 302–316.

Logan, D.C. (2009) Known knowns, known unknowns, unknown unknowns and the propagation of scientific enquiry. *Journal of Experimental Botany* **60**, 712–714.

Louviere, J.J. (1988a) Conjoint analysis modelling of stated preferences: a review of theory, methods, recent developments and external validity. *Journal of Transport Economics and Policy* **22**, 93–119.

Louviere, J.J. (1988b) *Analysing Decision Making: Metric Conjoint Analysis*. Sage Publications, Newbury Park.

Louviere, J.J., Hensher, D.A. and Swait, J.D. (2000) *Stated Choice Methods: Analysis and Application*. Cambridge University Press, Cambridge.

Louviere, J.J. and Lancsar, E. (2009) Choice experiments in health: the good, the bad, the ugly and toward a brighter future. *Health Economics, Policy, and Law* **4**, 527–546.

Low, D.E. (1972) A new approach to transportation systems modelling. *Traffic Quarterly* **26**, 391–404.

Lowry, I.S. (1965) A model of a metropolis. *Technical Memorandum RM-4035-RC*, The Rand Corporation, California.

Luce, R.D. and Suppes, P. (1965) Preference, utility and subjective probability. In R.D. Luce, R.R. Bush and E. Galanter (eds.), *Handbook of Mathematical Psychology*. Wiley, New York.

Lurkin, V., Garrow, L.A., Higgins, M.J., Newman, J.P. and Schyns, M. (2017) Accounting for price endogeneity in airline itinerary choice models: an application to continental U.S. markets. *Transportation Research* **100A**, 228–246.

Lyons, G. and Davidson, C. (2016) Guidance for transport planning and policymaking in the face of an uncertain future. *Transportation Research* **88A**, 104–116.

Ma, Z., Ferreira, L., Mesbah, M. and Zhu, S. (2016) Modelling distributions of travel time variability for bus operations. *Journal of Advanced Transportation* **50**, 6–24.

Mackett, R.L. (1985) Integrated land use-transport models. *Transport Reviews* **5**, 325–343.

Mackett, R.L. (1990) Comparative analysis of modelling land-use transport interaction at the micro and macro levels. *Environment and Planning* **22A**, 459–475.

Mackinder, I.H. and Evans, S.E. (1981) The predictive accuracy of British transport studies in urban areas. *TRRL Supplementary Report SR 699*, Transport and Road Research Laboratory, Crowthorne.

Maddison, D., Pearce, D., Johansson, O., Calthrop, E., Litman, T. and Verhoef, E. (1996) *The True Cost of Road Transport*. Earthscan Publications, London.

Maher, M.J. (1983) Inferences on trip matrices from observations on link volumes: a Bayesian statistical approach. *Transportation Research* **17B**, 435–447.

Mahmassani, H.S. (2000) Trip timing. In D. Hensher and K.J. Button (eds.), *Handbook of Transport Modelling*. Pergamon, Oxford.

Mahmassani, H.S. and Sinha, K.C. (1981) Bayesian updating of trip generation parameters. *Transportation Engineering Journal* **107**, 581–589.

Maldonado-Hinarejos, R., Sivakumar, A. and Polak, J. (2014) Exploring the role of individual attitudes and perceptions in predicting the demand for cycling: a hybrid choice modelling approach. *Transportation* **41**, 1287–1304.

Mandic, S., Hopkins, D., García Bengoechea, E., Flaherty, C., Coppell, K., Moore, A., Williams, J. and Spence, J.C. (2020) Differences in parental perceptions of walking and cycling to high school according to distance. *Transportation Research* **71F**, 238–249.

Manheim, C.F. (1973) Practical implications of some fundamental properties of travel demand models. *Highway Research Record* **244**, 21–38.

Mann, H.B. and Wald, A. (1943) On stochastic limit and order relationships. *The Annals of Mathematical Statistics* **14**, 217–226.

Manski, C.F. and Lerman, S.R. (1977) The estimation of choice probabilities from choice based samples. *Econometrica* **45**, 1977–1988.

Manski, C.F. and McFadden, D. (1981) Alternative estimators and sample designs for discrete choice analysis. In C.F. Manski and D. McFadden (eds.), *Structural Analysis of Discrete Data: With Econometric Applications*. MIT Press, Cambridge, MA.

Márquez, L., Cantillo, V. and Arellana, J. (2021) How do the characteristics of bike lanes influence safety perception and the intention to use cycling as a feeder mode to BRT? *Travel Behaviour and Society* **24**, 205–217.

Martínez, F.J. (1987) La forma incremental del modelo logit: aplicaciones. *Actas del III Congreso Chileno de Ingeniería de Transporte*, Universidad de Concepción, Chile (in Spanish).

Martínez, F.J. (1992) The bid-choice land-use model: an integrated economic framework. *Environment and Planning* **24A**, 871–875.

Martinez, F.J., Aguila, F. and Hurtubia, R. (2009) The constrained multinomial logit model: a semi-compensatory choice model. *Transportation Research* **43B**, 365–377.

Martínez, F.J. and Henríquez, R. (2007) A random bidding and supply land use equilibrium model. *Transportation Research* **41B**, 632–651.

Maslow, A.H. (1943) A theory of human motivation. *Psychological Review* **50**, 370–396.

Mauch, S.P. and Rothengatter, W. (1995) *External Effects of Transport*. Union International des Chemins de Fer, Paris.

Mayberry, J.P. (1973) Structural requirements for abstract-mode models of passenger transportation. In R.E. Quandt (ed.), *The Demand for Travel: Theory and Measurement*. D.C. Health and Co., Lexington, MA.

McCulloch, J.H. (1985) Miscellanea: on heteros*edasticity. *Econometrica* **53**, 483.

McDonald, K.G. and Stopher, P.R. (1983) Some contrary indications for the use of household structure in trip generation analysis. *Transportation Research Record* **944**, 92–100.

McFadden, D. (1974) Conditional logit analysis of qualitative choice behaviour. In P. Zarembka (ed.), *Frontiers in Econometrics*. Academic Press, New York.

McFadden, D. (1978) Modelling the choice of residential location. In A. Karlquist, L. Lundquist, F. Snickars and J.W. Weibull (eds.), *Spatial Interaction Theory and Planning Models*. North-Holland, Amsterdam.

McFadden, D. (1981) Econometric models of probabilistic choice. In C. Manski and D. McFadden (eds.), *Structural Analysis of Discrete Data with Econometric Applications*. MIT Press, Cambridge, MA.

McFadden, D. (1986) The choice theory approach to market research. *Marketing Science* **5**, 275–297.

McFadden, D. (1989) A method of simulated moments for estimation of discrete response models without numerical integration. *Econometrica* **57**, 995–1026.

McFadden, D. (2001) Economic choices. *American Economic Association* **91**, 351–378.

McFadden, D. (2013) The new science of pleasure. *Working Paper No. 18687*, National Bureau of Economic Research, Cambridge, MA.

McFadden, D. and Reid, F.A. (1975) Aggregate travel demand forecasting from disaggregate behavioural models. *Transportation Research Record* **534**, 24–37.

McFadden, D. and Train, K. (2000) Mixed MNL models for discrete response. *Journal of Applied Econometrics* **15**, 447–470.

McKelvey, R.D. and Zavoina, W. (1975) A statistical model for the analysis of ordinal level dependent variables. *Journal of Mathematical Sociology* **4**, 103–120.

McNeil, S. and Hendrickson, C. (1985) A regression formulation of the matrix estimation problem. *Transportation Science* **19**, 278–292.

Meijer, E. and Rouwendal, J. (2000) *Measuring welfare effects in models with random coefficients. Research Report No. 00F25*, SOM Research School, University of Groningen.

Menon, A.P.G. (2000) ERP in Singapore – a perspective one year on. *Traffic Engineering and Control* **41**, 40–45.

Menon, A.P.G., Lam, S.H. and Fan, H.S.L. (1993) Singapore's road pricing system: its past, present and future. *ITE Journal* **63**, 44–48.

Mesa-García, S.A., Rodríguez-Valencia, A., Ortúzar, J. de D. and Llera-Echeverri, G. (2023) Understanding the factors influencing motorcycle ownership and use. *Travel Behaviour and Society* (in press).

Metz, D. (2018) Tackling urban traffic congestion: the experience of London, Stockholm and Singapore. *Case Studies in Transport Policy* **6**, 494–498.

Metz, D. (2019) *Driving Change: Travel in the Twenty First Century*. Agenda Publishing, Newcastle upon Tyne.

Meyer, R.J., Levin, I.P. and Louviere, J.J. (1978) Functional analysis of mode choice. *Transportation Research Record* **673**, 1–7.

Meyer de Freitas, L., Becker, H., Zimmermann, M. and Axhausen, K.W. (2019) Modelling intermodal travel in Switzerland: a recursive logit approach. *Transportation Research* **119A**, 200–213.

Milkovits, M., Copperman, R., Newman, J., Lemp, J., Rossi, T. and Sun, S. (2019) Exploratory modelling and analysis for transportation: an approach and support tool – TMIP-EMAT. *Transportation Research Record* **2673**, 407–418.

Miller, E. (2023) The current state of activity-based travel demand modelling and some possible next steps. *Transport Reviews* **43**, 565–570.

Miller, E.J. (2020) Travel demand models, the next generation: boldly going where no-one has gone before. In K.G. Goulias and A.W. Davis (eds.), *Mapping the Travel Behavior Genome*. Elsevier, Amsterdam.

Miller, T. (2000) Variations between countries in values of statistical life. *Journal of Transport Economics and Policy* **34**, 169–188.

Mitchell, R.B. and Rapkin, C. (1954) *Urban Traffic: A Function of Land Use*. Columbia University Press, New York.

Mitchell, R.C. and Carson, R.T. (1989) *Using Surveys to Value Public Goods: The Contingent Valuation Method*. Resources for the Future, Washington, DC.

Mix, R., Hurtubia, R. and Raveau, S. (2022) Optimal location of bike-sharing stations: a built environment and accessibility approach. *Transportation* **160A**, 126–142.

Moavenzadeh, F., Markow, M., Brademeyer, B. and Safwat, K. (1983) A methodology for intercity transportation planning in Egypt. *Transportation Research* **17A**, 481–491.

Mogridge, M.J.H. (1983) *The Car Market*. Pion, London.

Moore, E.F. (1957) The shortest path through a maze. In *Proceedings International Symposium on the Theory of Switching*. Harvard University Press, Cambridge, MA.

Morganti, E., Seidel, S., Blanquart, C., Dablanc, L. and Lenz, B. (2014) The impact of e-commerce on final deliveries: alternative parcel delivery services in France and Germany. *Transportation Research Procedia* **4**, 178–190.

Morikawa, T. (1996) A hybrid probabilistic choice set model with compensatory and non-compensatory choice rules. In D.A. Hensher, J. King and T. Oum (eds.), *World Transport Research, World Conference on Transport Research Society*, Sydney.

Morikawa, T., Ben-Akiva, M. and Yamada, K. (1992) Estimation of mode choice models with serially correlated RP and SP data. *Proceedings 6th World Conference on Transport Research*, Lyon, France.

Morikawa, T. and Sasaki, K. (1998) Discrete choice models with latent variables using subjective data. In J. de D. Ortúzar, D.A. Hensher and S.R. Jara-Díaz (eds.), *Travel Behaviour Research: Updating the State of Play*. Pergamon, Oxford.

Moser, C.A. and Kalton, G.K. (1985) *Survey Methods in Social Investigation*. Ashgate, Farnham.

Mueller, N., Rojas-Rueda, D., Cole-Hunter, T., de Nazelle, A., Gerike, R., Gotschi, T., Panis, L., Kahlmeier, S. and Nieuwenhuijsen, M. (2015) Health impact assessment of active transportation: a systematic review. *Preventive Medicine* **76**, 103–114.

Mullen, S. (1997) Determining monetary values of environmental impacts – a DETR perspective. *Presented at Determining Monetary Values of Environmental Impacts*, University of Westminster, London.

Muller, R.H. (1996) Examining toll road feasibility studies. *Municipal Market Monitor*. J.P. Morgan Securities, Inc., New York.

Munizaga, M.A. and Alvarez-Daziano, R. (2005) Testing mixed logit and probit models by simulation. *Transportation Research Record* **1921**, 53–62.

Munizaga, M.A., Correia, R., Jara-Díaz, S.R. and Ortúzar, J. de D. (2006) Valuing time with a joint mode choice-activity model. *International Journal of Transport Economics* **33**, 69–86.

Munizaga, M., Devillaine, F., Navarrete, C. and Silva, D. (2014) Validating travel behavior estimated from smartcard data. *Transportation Research* **44C**, 70–79.

Munizaga, M.A., Heydecker, B.G. and Ortúzar, J. de D. (2000) Representation of heteroskedasticity in discrete choice models. *Transportation Research* **34B**, 219–240.

Munizaga, M.A. and Palma, C. (2012) Estimation of a disaggregate multimodal public transport origin-destination matrix from passive smart card data from Santiago, Chile. *Transportation Research* **24C**, 9–18.

Muñoz, B., Monzón, A. and Daziano, R. (2016) The increasing role of latent variables in modelling bicycle mode choice. *Transport Reviews* **36**, 737–771.

Muñoz, J.C., Ortúzar, J. de D. and Gschwender, A. (2009) Transantiago: the fall and rise of a radical public transport intervention. In W. Saleh and G. Sammer (eds.), *Travel Demand Management and Road User Pricing: Success, Failure and Feasibility*. Ashgate, Farnham.

Murakami, E. and Watterson, W.T. (1990) Developing a household travel survey for the Puget Sound Region. *Transportation Research Record* **1285**, 40–48.

Murchland, J.D. (1977) The multiproportional problem. *TSG Note JDM-263*, Transport Studies Group, University College London.

Nansubuga, B. and Kowalkowski, C. (2021) Carsharing: a systematic literature review and research agenda. *Journal of Service Management* **32**, 55–91.

Nardini, A. (1997) A proposal for integrating environmental impact assessment, cost-benefit analysis and multicriteria analysis in decision making. *Project Appraisal* **12**, 173–184.

Nash, C. (1997) Transport externalities: does monetary valuation make sense? In G. de Rus and C. Nash (eds.), *Recent Developments in Transport Economics*. Ashgate Press, London.

Navarrete, F. and Ortúzar, J. de D. (2013) Subjective valuation of the transit transfer experience: the case of Santiago de Chile. *Transport Policy* **25**, 138–147.

NCHRP (2010) Advanced practices in travel forecasting. *NCHRP Synthesis Report 406*, National Cooperative Highway Research Program, Transportation Research Board, Washington, DC.

Nicolaisen, M.S. (2012) *Forecasts: fact or fiction? Uncertainty and inaccuracy in transport project evaluation*. Department of Development and Planning, Aalborg University, Denmark.

Niederreiter, H. (1992) *Random Number Generation and Quasi Monte Carlo Methods*. Society for Industrial and Applied Mathematics, Philadelphia.

Obeid, H., Anderson, M.L., Bouzaghrane, M.A. and Walker, J.L. (2022) Does telecommuting reduce trip-making? Evidence from a US panel during the COVID-19 impact and recovery periods. Available at: https://doi.org/10.2139/ssrn.4213516.

OECD (1974) *Urban Traffic Models: Possibilities For Simplification*. OECD Road Research Group, Paris.

OECD (1994a) *Internalising the Social Costs of Transport*. Organisation for Economic Cooperation and Development, Paris.

OECD (1994b) *Environmental Impact Assessment of Roads*. Organisation for Economic Cooperation and Development, Paris.

OECD (1994c) *Project and Policy Appraisal: Integrating Economics and Environment*. Organisation for Economic Cooperation and Development, Paris.

Oh, J.H. (1989) Estimation of trip matrices in networks with equilibrium link flows. *Proceedings 17th PTRC Summer Annual Meeting*, University of Sussex, England.

Oi, K.I.Y. and Shuldiner, P.W. (1962) *An Analysis of Urban Travel Demands*. Northwestern University Press, Evanston.

Oja, P., Titze, S., Bauman, A., de Geus, B., Krenn, P., Reger-Nash, B. and Kohlberger, T. (2011) Health benefits of cycling: a systematic review. *Scandinavian Journal of Medicine and Science in Sports* **21**, 496–509.

O'Keefe, P., Caulfield, B., Brazil, B. and White, P. (2016) The impacts of telecommuting in Dublin. *Research in Transportation Economics* **57**, 13–20.

Oliva, I., Galilea, P. and Hurtubia, R. (2018) Identifying cycling-inducing neighborhoods: a latent class approach. *International Journal of Sustainable Transportation* **12**, 701–713.

Olsen, G.D. and Swait, J.D. (1998) Nothing is important. *Working Paper*, Faculty of Management, University of Calgary.

Oort, O. (1969) The evaluation of travelling time. *Journal of Transport Economics and Policy* **3**, 279–286.

Ortúzar, J. de D. (1980) Mixed-mode demand forecasting techniques. *Transportation Planning and Technology* **6**, 81–95.

Ortúzar, J. de D. (1982) Fundamentals of discrete multimodal choice modelling. *Transport Reviews* **2**, 47–78.

Ortúzar, J. de D. (1983) Nested logit models for mixed-mode travel in urban corridors. *Transportation Research* **17A**, 283–299.

Ortúzar, J. de D. (1988) Zahavi's alpha relation: myth or blessing? *International Journal of Transport Economics* **15**, 189–201.

Ortúzar, J. de D. (1989) Determining the preferences for frozen cargo exports. In World Conference on Transport Research (ed.), *Transport Policy, Management and Technology Towards 2001*. Western Periodicals Co, Ventura, CA.

Ortúzar, J. de D. (ed.) (1992) *Simplified Transport Demand Modelling*. Perspectives 2, PTRC, London.

Ortúzar, J. de D. (2001) On the development of the nested logit model. *Transportation Research* **35B**, 213–216.

Ortúzar, J. de D., Achondo, F.J. and Espinosa, A. (1986) On the stability of logit mode choice models. *Proceedings 14th PTRC Summer Annual Meeting*, University of Sussex, England.

Ortúzar, J. de D., Armstrong, P.M., Ivelic, A.M. and Valeze, C. (1998) Tamaño muestral y estabilidad temporal en modelos de generación de viajes. *Actas X Congreso Panamericano de Ingeniería de Tránsito y Transporte*, Santander, Spain (in Spanish).

Ortúzar, J. de D., Bascuñán, R., Rizzi, L.I. and Salata, A. (2021) Assessing the potential acceptability of road pricing in Santiago. *Transportation Research* **144A**, 153–169.

Ortúzar, J. de D. and Donoso, P.C.F. (1983) Survey design, implementation, data coding and evaluation for the estimation of disaggregate choice models in Santiago, Chile. *2nd International Conference on New Survey Methods in Transport*, Sydney, Australia.

Ortúzar, J. de D., Donoso, P.C.F. and Hutt, G.A. (1983) The effects of measurement techniques, variable definition and model specification on demand model functions. *11th PTRC Summer Annual Meeting*, University of Sussex, England.

Ortúzar, J. de D. and Garrido, R.A. (1994a) On the semantic scale problem in stated preference rating experiments. *Transportation* **21**, 185–201.

Ortúzar, J. de D. and Garrido, R.A. (1994b) A practical assessment of stated preference methods. *Transportation* **21**, 289–305.

Ortúzar, J. de D. and Hutt, G.A. (1984) La influencia de elementos subjetivos en funciones desagregadas de elección discreta. *Ingeniería de Sistemas* **4**, 37–54 (in Spanish).

Ortúzar, J. de D., Iacobelli, A. and Valeze, C. (2000a) Estimating demand for a cycle-way network. *Transportation Research* **34A**, 353–373.

Ortúzar, J. de D. and Ivelic, A.M. (1987) Effects of using more accurately measured level-of-service variables on the specification and stability of mode choice models. *Proceedings 15th PTRC Summer Annual Meeting*, University of Bath, England.

Ortúzar, J. de D. and Ivelic, A.M. (1988) Influencia del nivel de agregación de los datos en la estimación de modelos logit de eleción discreta. *Actas del V Congreso Panamericano de Ingeniería de Tránsito y Transporte*, Universidad de Puerto Rico en Mayagüez, Puerto Rico (in Spanish).

Ortúzar, J. de D., Ivelic, A.M., Malbrán, H. and Thomas, A. (1993) The 1991 Great Santiago origin-destination survey: methodological design and main results. *Traffic Engineering and Control* **34**, 362–368.

Ortúzar, J. de D., Martínez, F.J. and Varela, F.J. (2000b) Stated preferences in modelling accessibility. *International Planning Studies* **5**, 65–85.

Ortúzar, J. de D. and Palma, A. (1992) Stated preference in refrigerated and frozen cargo exports. In J. de D. Ortúzar (ed.), *Simplified Transport Demand Modelling*. Perspectives 2, PTRC, London.

Ortúzar, J. de D., Roncagliolo, D.A. and Velarde, U.C. (2000c) Interactions and independence in stated preference modelling. In J. de D. Ortúzar (ed.), *Stated Preference Modelling Techniques*. Perspectives 4, PTRC, London.

Ortúzar, J. de D. and Rodríguez, G. (2002) Valuing reductions in environmental pollution in a residential location context. *Transportation Research* **7D**, 407–427.

Ortúzar, J. de D. and Williams, H.C.W.L. (1982) Una interpretación geométrica de los modelos de elección entre alternativas discretas basados en la teoría de la utilidad aleatoria. *Apuntes de Ingeniería* **7**, 25–50 (in Spanish).

Ortúzar, J. de D. and Willumsen, L.G. (1978) Learning to manage transport systems. *Traffic Engineering and Control* **19**, 239–239.

Ortúzar, J. de D. and Willumsen, L.G. (1991) Flexible long range planning using low cost information. *Transportation* **18**, 151–173.

Outram, V.E. and Thompson, E. (1978) Driver route choice-behavioural and motivational studies. *Proceedings 5th PTRC Summer Annual Meeting*, University of Warwick, England.

Ouwersloot, H. and Rietveld, P. (1996) Stated choice experiments with repeated observations. *Journal of Transport Economics and Policy* **30**, 203–212.

Overgaard, K.R. (1967) Urban transportation planning: traffic estimation. *Traffic Quarterly* **21**, 197–218.

Pakes, A. and Pollard, D. (1989) Simulation and the asymptotics of optimisation estimators. *Econometrica* **57**, 1027–1057.

Pape, U. (1974) Implementation and efficiency of Moore algorithms for the shortest route problem. *Mathematical Programming* **7**, 212–222.

Park, S. and Gupta, S. (2009) Simulated maximum likelihood estimator for the random coefficient logit model using aggregate data. *Journal of Marketing Research* **46**, 531–542.

Patriksson, M. (1994) *The Traffic Assignment Problem: Models and Methods*. VSP, Utrecht.

Pawluk de Toledo, K., O'Hern, S. and Koppel, S. (2022) Travel behaviour change research: a scientometric review and content analysis. *Travel Behaviour and Society* **28**, 141–154.

Peer, S., Knockaert, J., Koster, P., Tseng, Y.Y. and Verhoef, E. (2013) Door-to-door travel times in RP departure time choice models: an approximation method using GPS data. *Transportation Research* **58B**, 134–150.

Pendyala, R.M., Parashar, A. and Muthyalagari, G.R. (2001) Measuring day-to-day variability in travel characteristics using GPS data. *80th Annual TRB Meeting*, Washington, DC, USA.

Penn, A., Hillier, B., Banister, D. and Xu, J. (1998) Configurational modelling of urban movement networks. *Environment and Planning* **25B**, 59–84.

Pérez, P.E., Martínez, F.J. and Ortúzar, J. de D. (2003) Microeconomic formulation and estimation of a residential location choice model: implications for the value of time. *Journal of Regional Science* **43**, 771–789.

Petrin, A. and Train, K. (2010) A control function approach to endogeneity in consumer choice models. *Journal of Marketing Research* **47**, 3–13.

Pickens, J. (2005) *Organizational Behavior in Health Care*. Jones and Bartlett Publishers, Sudbury.

Polak, J.W. (2002) Analysis of non-response in the LATS 2001 pilot household travel diary survey. *81st Annual TRB Meeting*, Washington, DC, USA.

Polak, J.W., Jones, P.M., Vythoulkas, P.C., Meland, S. and Tretvik, T. (1991) *The Trondheim Toll Ring: Results of a Stated Preference Study of Travellers' Responses*. Report to the European Commission DRIVE Programme, Transport Studies Unit, University of Oxford.

Popuri, Y., Ben-Akiva, M. and Proussaloglou, K. (2008) Time-of-day modelling in a tour-based context: the Tel Aviv experience. *Transportation Research Record* **2076**, 88–96.

Prashker, J.N. and Bekhor, S. (2000) Congestion, stochastic and similarity effects in stochastic user equilibrium models. *Transportation Research Record* **1733**, 80–87.

Prato, C.G. (2009) Route choice modeling: past, present and future research directions. *Journal of Choice Modelling* **2**, 65–100.

Prato, C.G. and Bekhor, S. (2006) Applying branch & bound techniques to route choice set generation. *Transportation Research Record* **1985**, 19–28.

Price, A. (1999) A new approach to the appraisal of road projects in England. *Journal of Transport Economics and Policy* **33**, 221–226.

Pucher, J. and Buehler, R. (2012) *City Cycling*. MIT Press, Cambridge, MA.

Puckett, S.M. and Hensher, D.A. (2009) Revealing the extent of process heterogeneity in choice analysis: an empirical assessment. *Transportation Research* **43A**, 117–126.

Purvis, C.L. (1989) Sample design for the 1990 Bay Area Household Travel Survey. *Working Paper 1*, Bay Area Metropolitan Transport Commission, San Francisco.

Quarmby, D.A. and Bates, J.J. (1970) An econometric method of car ownership forecasting in discrete areas. *MAU Note 219*. Department of the Environment, London.

Quinet, E. (1994) The social cost of transport: evaluation and links with internalisation policies. In *Internalising the Social Costs of Transport*. Organisation for Economic Cooperation and Development, Paris.

Raftery, A. and Lewis, S. (1992) How many iterations in the Gibbs Sampler? In J.M. Bernardo, A.F.M. Smith, A.P. David and J.O. Berger (eds.), *Bayesian Statistics 4*. Oxford University Press, New York.

RAND (2004) PRISM West Midlands tour generation modelling. *Report RED-02061-03*, Rand Europe, Cambridge.

Raveau, S., Alvarez-Daziano, R., Yáñez, M.F., Bolduc, D. and Ortúzar, J. de D. (2010) Sequential and simultaneous estimation of hybrid discrete choice models: some new findings. *Transportation Research Record* **2156**, 131–139.

Raveau, S., Guo, Z., Muñoz, J.C. and Wilson, N.H.M. (2014) A behavioural comparison of route choice on metro networks: time, transfers, crowding, topology and socio-demographics. *Transportation Research* **66A**, 185–195.

Raveau, S., Muñoz, J.C. and de Grange, L. (2011) A topological route choice model for metro. *Transportation Research* **45A**, 138–147.

Regan, A.C. and Garrido, R.A. (2002) Modelling freight demand and shipper behaviour: state of the art and future directions. In D.A. Hensher (ed.), *Travel Behaviour Research: The Leading Edge*. Pergamon, Oxford.

Revelt, D. and Train, K.E. (1998) Mixed logit with repeated choices: households' choices of appliance efficiency level. *Review of Economics and Statistics* **80**, 647–657.

Revelt, D. and Train, K.E. (2000) Customer-specific taste parameters and mixed logit. *Working Paper E00-274*, Department of Economics, University of California at Berkeley.

Richardson, A.J. (1982) Search models and choice set generation. *Transportation Research* **16A**, 403–419.

Richardson, A.J., Ampt, E.S. and Meyburg, A. (1995) *Survey Methods for Transport Planning*. Eucalyptus Press, Melbourne.

Richardson, A.J. and Meyburg, A.H. (2003) Definitions of unit non-response in travel surveys. In P.R. Stopher and P.M. Jones (eds.), *Transport Survey Quality and Innovation*. Pergamon, Amsterdam.

Rieser, M., Nagel, K., Beuck, U., Balmer, M. and Rümenapp, J. (2007) Agent-oriented coupling of activity-based demand generation with multiagent traffic simulation. *Transportation Research Record* **2021**, 10–17.

Rivers, D. and Vuong, Q. (1988) Limited information estimators and exogeneity tests for simultaneous probit models. *Journal of Econometrics* **39**, 347–366.

Rizzi, L.I. and Ortúzar, J. de D. (2003) Stated preference in the valuation of interurban road safety. *Accident Analysis and Prevention* **35**, 9–22.

Rizzi, L.I. and Ortúzar, J. de D. (2006) Road safety valuation under a stated choice framework. *Journal of Transport Economics and Policy* **40**, 71–96.

Rizzi, L.I. and Ortúzar, J. de D. (2015) Valuing transport externalities. In C. Nash (ed.), *Handbook of Research Methods in Transport Economics and Policy*. Edward Elgar, Cheltenham.

Roberts, F.S. (1975) Weighted di-graph models for the assessment of energy use and air pollution in transportation systems. *Environment and Planning* **7A**, 703–724.

Robertson, D.I. (1969) *TRANSYT: a traffic network study tool. TRRL Report LR 253*, Transport and Road Research Laboratory, Crowthorne.

Robillard, P. (1975) Estimating the O–D matrix from observed link volumes. *Transportation Research* **9**, 123–128.

Rodríguez, A., Cordera, R., Alonso, B., dell'Olio, L. and Benavente, J. (2022) Microsimulation parking choice and search model to assess dynamic pricing scenarios. *Transportation Research* **156A**, 253–269.

Rodríguez-Valencia, A., Rosas-Satizabal, D., Unda, R. and Handy, S. (2021) The decision to start commuting by bicycle in Bogotá, Colombia: motivations and influences. *Travel Behaviour and Society* **24**, 57–67.

Román, C., Martín, J.C., Espino, R., Cherchi, E., Ortúzar, J. de D., Rizzi, L.I., González, R.M. and Amador, F.J. (2014) Valuation of travel time savings for intercity travel: the Madrid-Barcelona corridor. *Transport Policy* **36**, 105–117.

Rose, G., Daskin, M. and Koppelman, F.S. (1988) An examination of convergence error in equilibrium traffic assignment models. *Transportation Research* **22B**, 261–274.

Rose, G. and Koppelman, F.S. (1984) Transferability of disaggregate trip generation models. In J. Volmüller and R. Hamerslag (eds.), *Proceedings of the Ninth International Symposium on Transportation and Traffic Theory*. VNU Science Press, Utrecht.

Rose, J.M. and Bliemer, M.C.J. (2008) Stated preference experimental design strategies. In D.A. Hensher and K.J. Button (eds.), *Handbook of Transport Modelling*. Elsevier, Oxford.

Rose, J.M. and Bliemer, M.C.J. (2009) Constructing efficient stated choice experimental designs. *Transport Reviews* **29**, 587–617.

Rose, J.M., Bliemer, M.C.J., Hensher, D.A. and Collins, A.T. (2008) Designing efficient stated choice experiments in the presence of reference alternatives. *Transportation Research* **42B**, 395–406.

Rose, J.M., Hensher, D.A., Caussade, S., Ortúzar, J. de D. and Jou, R.C. (2009a) Identifying differences in willingness to pay due to dimensionity in stated choice experiments: a cross country analysis. *Journal of Transport Geography* **17**, 21–29.

Rose, J.M., Scarpa, R. and Bliemer, M.C.J (2009b) Incorporating model uncertainty into the generation of efficient stated choice experiments: a model averaging approach. *International Choice Modelling Conference,* Harrogate, England.

Rossetti, T., Guevara, C.A., Galilea, P. and Hurtubia, R. (2018) Modeling safety as a perceptual latent variable to assess cycling infrastructure. *Transportation Research* **111A**, 252–265.

Rossetti, T., Saud, V. and Hurtubia, R. (2019) I want to ride it where I like: measuring design preferences in cycling infrastructure. *Transportation* **46**, 697–718.

Rothenberg, T. (1971) Identification in parametric models. *Econometrica* **39**, 577–591.

Ruijgrok, C.J. (1979) Disaggregate choice models: an evaluation. In G.R.M. Jansen, P.H.L. Bovy, J.P.J.M. van Est and F. Le Clercq (eds.), *New Developments in Modelling Travel Demand and Urban Systems*. Saxon House, Westmead.

Ruud, P. (1996) Simulation of the multinomial probit model: an analysis of covariance matrix estimation. *Working Paper*, Department of Economics, University of California at Berkeley.

Sælensminde, K. (2001) Inconsistent choices in stated choice data: use of the logit scaling approach to handle resulting variance increases. *Transportation Research* **4D**, 13–27.

Sánchez-Díaz, I., Holguín-Veras, J. and Wang, X. (2016) An exploratory analysis of spatial effects on freight trip attraction. *Transportation* **43**, 177–196.

Sándor, Z. and Wedel, M. (2001) Designing conjoint choice experiments using managers' prior beliefs. *Journal of Marketing Research* **38**, 430–444.

Sándor, Z. and Wedel, M. (2002) Profile construction in experimental choice designs for mixed logit models. *Marketing Science* **21**, 455–475.

Sándor, Z. and Wedel, M. (2005) Heterogeneous conjoint choice designs. *Journal of Marketing Research* **42**, 210–218.

Safwat, K.N.A. and Magnanti, T. (1988) A combined trip generation, trip distribution, modal split and trip assignment model. *Transportation Science* **22**, 14–30.

Scarpa, R. and Rose, J.M. (2008) Design efficiency for non-market valuation with choice modelling: how to measure it, what to report and why. *Australian Journal of Agricultural and Resource Economics* **52**, 253–282.

Schlereth, C. and Skiera, B. (2017) Two new features in discrete choice experiments to improve willingness-to-pay estimation that result in SDR and SADR: separated (adaptive) dual response. *Management Science* **63**, 829–842.

Schmid, B., Jokubauskaite, S., Aschauer, F., Peer, S., Hössinger, R., Gerike, R., Jara-Díaz, S.R. and Axhausen, K.W. (2019) A pooled RP/SP mode, route and destination choice model to investigate mode and user-type effects in the value of travel time savings. *Transportation Research* **124A**, 262–294.

Schneider, M. (1959) Gravity models and trip distribution theory. *Papers and Proceedings of the Regional Science Association* **5**, 51–56.

Schnieder, M., Hinde, C. and West, A. (2021) Combining parcel lockers with staffed collection and delivery points: an optimization case study using real parcel delivery data (London, UK). *Journal of Open Innovation: Technology, Market, and Complexity* **7**, 1–18.

Shanmugalingam, S. (1982) On the analysis of the ratio of two correlated Normal variables. *The Statistician* **31**, 251–258.

Shao, J. and Tu, D. (1995) *The Jackknife and Bootstrap*. Springer, New York.

Sheffi, Y. (1985) *Urban Transportation Networks*. Prentice Hall, Englewood Cliffs, NJ.

Sheffi, Y., Hall, R. and Daganzo, C.F. (1982) On the estimation of the multinomial probit model. *Transportation Research* **16A**, 447–456.

Shoup, D. (2005) *The High Cost of Free Parking*. Planners Press, Chicago.

Significance, VU University Amsterdam, John Bates Services (2013) *Values of Time and Reliability in Passenger and Freight Transport in The Netherlands*. Report for the Ministry of Infrastructure and the Environment, The Hague.

Sikdar, P.K. and Hutchinson, B.G. (1981) Empirical studies of work trip distribution models. *Transportation Research* **15A**, 233–243.

Sillano, M. and Ortúzar, J. de D. (2005) Willingness-to-pay estimation with mixed logit models: some new evidence. *Environment and Planning* **37A**, 525–550.

Silva, M.S. (2002) Secuencias de Baja Discrepancia para Estimación de Modelos Logit Mixto. MSc Thesis, Department of Transport Engineering, Pontificia Universidad Católica de Chile (in Spanish).

Simon, H.A. (1957) *Models of Man: Social and Rational*. Wiley, New York.

Simmonds, D.C. (2001) The objectives and design of a new land-use modelling package: DELTA. In G. Clarke and M. Madden (eds.), *Regional Science in Business*. Springer Verlag, Berlin.

Sloman, L., Cairns, S., Newson, C., Anable, J., Pridmore, A. and Goodwin, P. (2010) *The Effects of Smarter Choice Programmes in the Sustainable Travel Towns: Summary Report*. Report to the UK Department for Transport, London.

Small, K.A. (1982) The scheduling of consumer activities: work trips. *American Economic Review* **72**, 467–479.

Small, K.A. (1987) A discrete choice model for ordered alternatives. *Econometrica* **55**, 409–424.

Small, K.A., Noland, R.B., Chu, X. and Lewis, D. (1999) *Valuation of travel-time savings and predictability in congested conditions for highway user-cost estimation. NCHRP Report 431*, National Cooperative Highway Research Program, Transportation Research Board, Washington, DC.

Smeed Report (1964) *Road pricing: The Economic and Technical Possibilities*. UK Ministry of Transport, HMSO, London.

Smeed, R.J. (1968) Traffic studies and urban congestion. *Journal of Transport Economics and Policy* **2**, 2–38.

Smith, M.D. (2005) State dependence and heterogeneity in fishing location choice. *Journal of Environmental Economics and Management* **50**, 319–340.

Smith, M.E. (1979) Design of small sample home interview travel surveys. *Transportation Research Record* **701**, 29–35.

Smith, M.J. (1979) Existence, uniqueness and stability of traffic equilibria. *Transportation Research* **13B**, 295–304.

Smith, R.L. and Cleveland, D.E. (1976) Time stability analysis of trip generation and predistribution modal choice models. *Transportation Research Record* **569**, 76–86.

Smock, R.J. (1962) An iterative assignment approach to capacity restraint on arterial networks. *Highway Research Board Bulletin* **156**, 1–13.

Sobel, K.L. (1980) Travel demand forecasting by using the nested multinomial logit model. *Transportation Research Record* **775**, 48–55.

Sosslau, A.B., Hassam, A., Carter, M. and Wickstrom, G. (1978) Quick response urban travel estimation techniques and transferable parameters. *NCHRP Report 817*, National Cooperative Highway Research Program, Transportation Research Board, Washington, DC.

Sottile, E., Sanjust di Teulada, B., Meloni, I. and Cherchi, E. (2019) Estimation and validation of hybrid choice models to identify the role of perception in the choice to cycle. *International Journal of Sustainable Transportation* **13**, 543–552.

Spear, B.D. (1977) *Applications of New Travel Demand Forecasting Techniques to Transportation: A Study of Individual Choice Models*. Final Report to the Office of Highway Planning, Federal Highway Administration, US Department of Transportation, Washington, DC.

Spiegelhalter, D., Thomas, A., Best, N. and Lunn, D. (2003) *WinBUGS User Manual*. Institute of Public Health, University of Cambridge, MRC Biostatistics Unit.

Spielberg, F., Weiner, E. and Ernst, U. (1981) The shape of the 1980's: demographic, economic and travel characteristics. *Transportation Research Record* **807**, 27–34.

Spiess, H. (1983) On optimal route choice strategies in transit networks. *Publication 286*, Centre de Recherche sur les Transports, Université de Montréal.

Spiess, H. (1987) A maximum likelihood model for estimating origin-destination matrices. *Transportation Research* **21B**, 395–412.

Spiess, H. and Florian, M. (1989) Optimal strategies: a new assignment model for transit networks. *Transportation Research* **23B**, 82–102.

Steer Davies Gleave (2000) *Diseño Operacional del Sistema Transmilenio: Proyecto de Transporte Urbano para Santa Fe de Bogotá*. BIRF 4021-FONDATT-10, Bogotá (in Spanish).

Steer, J. and Willumsen, L.G. (1983) An investigation of passenger preference structures. In S. Carpenter and P.M. Jones (eds.), *Recent Advances in Travel Demand Analysis*. Gower, Aldershot.

Steinberg, R. and Zangwill, W. (1983) The prevalence of Braess's paradox. *Transportation Science* **17**, 301–318.

Steinmetz, S.S.C. and Brownstone, D. (2005) Estimating commuters 'value of time' with noisy data: a multiple imputation approach. *Transportation Research* **36B**, 865–889.

Stock, J.H. and Yogo, M. (2005) Testing for weak instruments in linear IV regression. In D.W. Andrews and J.H. Stock (eds.), *Identification and Inference for Econometric Models: Essays in Honour of Thomas Rothenberg*. Cambridge University Press, Cambridge.

Stopher, P.R. (1982) Small-sample home-interview travel surveys: application and suggested modifications. *Transportation Research Record* **886**, 41–47.

Stopher, P.R. (1998) Household travel surveys: new perspectives and old problems. In T. Gärling, T. Laitila and K. Westin (eds.), *Theoretical Foundations of Travel Choice Modelling*. Pergamon, Oxford.

Stopher, P.R. and Greaves, S.P. (2004) Sample size requirements for measuring a change in behaviour. *27th Australian Transport Research Forum*, Adelaide, Australia.

Stopher, P.R. and Jones, P.M. (2003) Developing standards of transport survey quality. In P.R. Stopher and P.M. Jones (eds.), *Transport Survey Quality and Innovation*. Pergamon, Amsterdam.

Stopher, P.R. and Metcalf, H.M.A. (1996) Methods for household travel surveys. *NCHRP Synthesis of Highway Practice 236*, Transportation Research Board, Washington, DC.

Stopher, P.R. and Meyburg, A.H. (1979) *Survey Sampling and Multivariate Analysis for Social Scientists and Engineers*. D.C. Heath and Co., Lexington, MA.

Stopher, P.R. and Stecher, C. (1993) Blow-up: expanding a complex random sample travel survey. *Transportation Research Record* **1412**, 10–16.

Stouffer, A. (1940) Intervening opportunities: a theory relating mobility and distance. *American Sociological Review* **5**, 845–867.

Street, D.J. and Burgess, L. (2004) Optimal and near-optimal pairs for the estimation of effects in 2-level choice experiments. *Journal of Statistical Planning and Inference* **118**, 185–199.

Street, D.J. and Burgess, L. (2007) *The Construction of Optimal Stated Choice Experiments: Theory and Methods*. Wiley, Hoboken, NJ.

Street, D.J., Burgess, L. and Louviere, J.J. (2005) Quick and easy choice sets: constructing optimal and nearly optimal stated choice experiments. *International Journal of Research in Marketing* **22**, 459–470.

Suh, S., Park, C. and Kim, T.J. (1990) A highway capacity function in Korea: measurement and calibration. *Transportation Research* **24A**, 177–186.

Supernak, J. (1979) A behavioural approach to trip generation modelling. *Proceedings 7th PTRC Summer Annual Meeting*, University of Warwick, England.

Supernak, J. (1981) Transferability of the person category trip generation model. *Proceedings 9th PTRC Summer Annual Meeting*, University of Warwick, England.

Swait, J.D. and Adamowicz, W. (2001) The influence of task complexity on consumer choice: a latent class model of decision strategy switching. *Journal of Consumer Research* **28**, 135–148.

Swait, J.D., Adamowicz, W. and Buren, M. (2004) Choice and temporal welfare impacts: incorporating history into discrete choice models. *Journal of Environmental Economics and Management* **47**, 94–116.

Swait, J.D. and Bernardino, A. (2000) Distinguishing taste variation from error structure in discrete choice data. *Transportation Research* **34B**, 1–15.

Swait, J.D., Louviere, J.J. and Williams, M. (1994) A sequential approach to exploiting the combined strengths of SP and RP data: application to freight shipper choice. *Transportation* **21**, 135–152.

Swanson, J. (2003) The dynamic urban model: transport and urban development. In R. Eberlein, V. Diker, R. Langer and J. Rowe (eds.), *Proceedings of the 21st International Conference of the Systems Dynamics Society*. Systems Dynamic Society, New York.

Tajaddini, A., Rose, G. and Kockelman, K. (2021) Recent progress in travel demand modelling: rising data and applicability. In S. de Luca, R. Pace and C. Fiori (eds.), *Models and Technologies for Smart, Sustainable and Safe Transportation Systems*. IntechOpen, London.

Taleb, N.N. (2007) *The Black Swan*. Random House, New York.

Tamin, O.Z. and Willumsen, L.G. (1989) Transport demand model estimation from traffic counts. *Transportation* **16**, 3–26.

Tamin, O.Z. and Willumsen, L.G. (1992) Freight demand model estimation from traffic counts. In J. de D. Ortúzar (ed.), *Simplified Transport Demand Modelling*. Perspectives 2, PTRC, London.

Tanner, J.C. (1974) Forecasts of vehicles and traffic in Great Britain: 1974 revision. *TRRL Report LR 650*, Transport and Road Research Laboratory, Crowthorne.

Tanner, J.C. (1978) Long term forecasting of vehicle ownership and road traffic. *Journal of the Royal Statistical Society* **141A**, 14–63.

Tardiff, T.J. (1976) A note on goodness-of-fit statistics for probit and logit models. *Transportation* **5**, 377–388.

Tardiff, T.J. (1979) Specification analysis for quantal choice models. *Transportation Science* **13**, 179–390.

Taylor, T.L. (1971) *Instructional Planning Systems: A Gaming-Simulation Approach to Urban Problems.* Cambridge University Press, New York.

Taylor, M. and Susilawati, S. (2012) Modelling travel time reliability with the Burr distribution. *Procedia-Social and Behavioural Sciences* **54**, 75–83.

Teknomo, K. and Hokao, K. (1997) Parking behaviour in central business districts. *East Asian Science, Technology and Society* **2**, 551–570.

Thorhauge, M., Haustein, S. and Cherchi, E. (2016) Accounting for the theory of planned behaviour in departure time choice. *Transportation Research* **38F**, 94–105.

Thorhauge, M., Swait, J.D. and Cherchi, E. (2020) The habit-driven life: accounting for inertia in departure time choices for commuting trips. *Transportation Research* **133A**, 272–289.

Thorhauge, M., Vij, A. and Cherchi, E. (2021) Heterogeneity in departure time preferences, flexibility and schedule constraints. *Transportation* **48**, 1865–1893.

Timberlake, R.S. (1988) Traffic modelling techniques for the developing world. *67th Annual TRB Meeting*, Washington, DC, USA.

Tirachini, A., Hensher, D.A. and Rose, J.M. (2013) Crowding in public transport systems: effects on users, operation and implications for the estimation of demand. *Transportation Research* **53A**, 36–52.

Titze, S., Stronegger, W.J., Janschitz, S. and Oja, P. (2008) Association of built-environment, social-environment and personal factors with bicycling as a mode of transportation among Austrian city dwellers. *Preventive Medicine* **47**, 252–259.

Toner, J.P., Clark, S.D., Grant-Muller, S.M. and Fowkes, A.S. (1998) Anything you can do, we can do better: a provocative introduction to a new approach to stated preference design. *8th World Conference on Transport Research*, Antwerp, Belgium.

Toubia, O., Hauser, J. and Garcia, R. (2007) Probabilistic polyhedral methods for adaptive choicebased conjoint analysis: theory and application. *Marketing Science* **26**, 596–610.

Train, K.E. (1977) Valuations of modal attributes in urban travel: questions of non-linearity, non-genericity and taste variations. *Working Paper*, Cambridge Systematics Inc. West, San Francisco.

Train, K.E. (1978) The sensitivity of parameter estimates to data specification in mode choice models. *Transportation* **7**, 301–309.

Train, K.E. (1980) A structured logit model of auto ownership and mode choice. *The Review of Economic Studies* **47**, 357–370.

Train, K.E. (1986) *Qualitative Choice Analysis: Theory, Econometrics and an Application to Automobile Demand.* MIT Press, Cambridge, MA.

Train, K.E. (1998) Recreation demand models with taste differences over people. *Land Economics* **74**, 230–239.

Train, K.E. (2001) A comparison of hierarchical Bayes and maximum simulated likelihood for mixed logit. *Working Paper*, Department of Economics, University of California at Berkeley.

Train, K.E. (2009) *Discrete Choice Methods with Simulation.* Second Edition. Cambridge University Press, Cambridge.

Train, K.E. and McFadden, D. (1978) The goods/leisure trade-off and disaggregate work trip mode choice models. *Transportation Research* **12**, 349–353.

Train, K.E., McFadden, D. and Ben-Akiva, M.E. (1987) The demand for local telephone service: a fully discrete model of residential calling patterns and service choices. *RAND Journal of Economics* **18**, 109–123.

Train, K.E. and Sonnier, G. (2005) Mixed logit with bounded distributions of correlated partworths. In R. Scarpa and A. Alberini (eds.), *Application of Simulation Methods in Environmental and Resource Economics*. Springer, Dordrecht.

Train, K.E. and Wilson, W.E. (2008) Estimation of stated-preference experiments constructed from revealed-preference choices. *Transportation Research* **40B**, 191–203.

Transport for London (2008) *Central London Congestion Charging, Impacts Monitoring*. Annual Report. Transport for London, London.

Traugott, M.W. and Katosh, J.P. (1979) Response validity in surveys of voting behaviour. *Public Opinion Quarterly* **42**, 359–377.

Tringides, C., Ye, X. and Pendyala, R.M. (2004) Departure-time choice and mode choice for non-work trips: alternative formulations of joint model systems. *Transportation Research Record* **1898**, 1–9.

Tseng, Y.Y. and Verhoef, E. (2008) Value of time by time of day: a stated-preference study. *Transportation Research* **42B**, 607–618.

Turner, A. (2007) From axial to road-centre lines: a new representation for space syntax and a new model of route choice for transport network analysis. *Environment and Planning* **34B**, 539–555.

Tversky, A. (1972) Elimination by aspects: a theory of choice. *Psychological Review* **79**, 281–299.

Vallejo-Borda, J.A., Ortiz-Ramírez, H.A., Rodríguez-Valencia, A., Hurtubia, R. and Ortúzar, J. de D. (2020) Forecasting the quality of service of Bogota's sidewalks from pedestrian perceptions: an Ordered Probit MIMIC approach. *Transportation Research Record* **2674**, 205–216.

Van Es, J.V. (1982) Freight transport, an evaluation. *ECMT Round Table 58*, European Conference of Ministers of Transport, Paris.

Van Vliet, D. (1977) D'Esopo: a forgotten tree-building algorithm. *Traffic Engineering and Control* **18**, 372–375.

Van Vliet, D. (1978) Improved shortest path algorithms for transport networks. *Transportation Research* **12**, 7–20.

Van Wissen, L.J.G. and Meurs, H.J. (1989) The Dutch mobility panel: experiences and evaluation. *Transportation* **16**, 99–119.

Van Zuylen, H. and Willumsen, L.G. (1980) The most likely trip matrix estimated from traffic counts. *Transportation Research* **14B**, 281–293.

Veisten, K., Flügel, S., Rizzi, L.I., Ortúzar, J. de D. and Elvik, R. (2013) Valuing casualty risk reductions from estimated baseline risk. *Research in Transportation Economics* **43**, 50–61.

Verhoef, E. (1994) External effects and social costs of road transport. *Transportation Research* **28A**, 273–287.

Verplanken, B. and Orbell, S. (2003) Reflections on past behaviour: a self-report index of habit strength. *Journal of Applied Social Psychology* **33**, 1313–1330.

Verplanken, B., Walker, I., Davis, A. and Jurasek, M. (2008) Context change and travel mode choice: combining the habit discontinuity and self-activation hypotheses. *Journal of Environmental Psychology* **28**, 121–127.

Vickrey, W. (1969) Congestion theory and transport investment. *American Economic Review* **59**, 251–261.

Vij, A. and Walker, J. (2014) Hybrid choice models: the identification problem. In S. Hess and A. Daly (eds.), *Handbook of Choice Modelling*. Edward Elgar Publishing, Cheltenham.

Vovsha, P. (1997) The cross-nested logit model: application to mode choice in the Tel Aviv metropolitan area. *76th Annual TRB Meeting*, Washington, DC, USA.

Vovsha, P. (2017) Microsimulation travel models in practice in the US and prospects for agent-based approach. In J. Bajo, Z. Vale, K. Hallenborg, A.P. Rocha, P. Mathieu, P. Pawlewski, E. Del Val, P. Novais, F. Lopes, N.D. Duque Méndez, V. Julián and J. Holmgren (eds.), *Highlights of Practical Applications of Cyber-Physical Multi-Agent Systems*. Springer, Heidelberg.

Vovsha, P., Donnelly, R. and Gupta, S. (2008) Network equilibrium with activity-based microsimulation models: the New York experience. *87th Annual TRB Meeting*, Washington, DC, USA.

Vredin-Johansson, M., Heldt, T. and Johansson, P. (2006) The effects of attitudes and personality traits on mode choice. *Transportation Research* **40A**, 507–525.

Waddell, P. (2002) UrbanSim: modeling urban development for land use, transportation and environmental planning. *Journal of the American Planning Association* **68**, 297–314.

Waddell, P., Borning, A., Noth, M., Freier, N., Becke, M. and Ulfarsson, G. (2003) UrbanSim: a simulation system for land use and transportation. *Networks and Spatial Economics* **3**, 43–67.

Walker, J.L. (2002) Mixed logit (or logit kernel) model: dispelling misconceptions of identification. *Transportation Research Record* **1805**, 86–98.

Walker, J.L. and Ben-Akiva, M. (2002) Generalized random utility model. *Mathematical Social Sciences* **43**, 303–343.

Walker, J.L., Ben-Akiva, M.E. and Bolduc, D. (2007) Identification of parameters in Normal error component logit-mixture (NECLM) models. *Journal of Applied Econometrics* **22**, 1095–1125.

Walker, J.L., Jieping, L., Sirinivasan, S. and Bolduc, D. (2010) Travel demand models in the developing world: correcting for measurement errors. *Transportation Letters* **4**, 231–243.

Walker, J.L. and Li, J. (2007) Latent style preferences and household location decisions. *Journal of Geographical Systems* **9**, 77–101.

Walters, A.A. (1961) The theory and measurement of private and social cost of highway congestion. *Econometrica* **29**, 676–697.

Wardman, M., Chintakayala, V.P.K. and de Jong, G. (2016) Values of travel time in Europe: review and meta-analysis. *Transportation Research* **94A**, 93–111.

Wardman, M. and Ibáñez, J.N. (2012) The congestion multiplier: variations in motorists' valuations of travel time with traffic conditions. *Transportation Research* **46A**, 213–225.

Wardman, M., Tight, M. and Page, M. (2007) Factors influencing the propensity to cycle to work. *Transportation Research* **41A**, 339–350.

Wardman, M. and Whelan, G.A. (2011) Twenty years of rail crowding valuation studies: evidence and lessons from British experience. *Transport Reviews* **31**, 379–398.

Wardrop, J.G. (1952) Some theoretical aspects of road traffic research. *Proceedings of the Institution of Civil Engineers, Part II* **1**, 325–362.

Wardrop, J.G. (1968) Journey speed and flow in central urban areas. *Traffic Engineering and Control* **9** (528–532), 539.

Warner, S.L. (1962) *Strategic Choice of Mode in Urban Travel: A Study of Binary Choice*. Northwestern University Press, Evanston.

Waters, W.G. (1992) *The Value of Time Savings for the Economic Evaluation of Highway Investments in British Columbia*. Report to the Canadian Ministry of Transportation and Highways, Victoria, B.C.

Watson, S.M., Toner, J.P., Fowkes, A.S. and Wardman, M. (2000) Efficiency properties of orthogonal stated preference designs. In J. de D. Ortúzar (ed.), *Stated Preference Modelling Techniques*. Perspectives 4, PTRC, London.

Webster, F.V., Bly, P.H. and Paulley, N.J. (eds.) (1988) *Urban Land-Use and Transport Interaction: Policies and Models*. Gower, Aldershot.

Wegener, M. (2004) Overview of land use transport models. In D.A. Hensher, K.J. Button, K.E. Haynes and P.R. Stopher (eds.), *Handbook of Transport Geography and Spatial Systems*. Elsevier, Amsterdam.

Weintraub, A., Ortiz, C. and Gonzáles, J. (1985) Accelerating convergence of the Frank-Wolfe algorithm. *Transportation Research* **19B**, 113–122.

Wen, C., Chiou, Y. and Huang, W. (2012) A dynamic analysis of motorcycle ownership and usage: a panel data modelling approach. *Accident Analysis and Prevention* **49**, 193–202.

Wermuth, M.J. (1981) Effects of survey methods and measurement techniques on the accuracy of household travel-behaviour surveys. In P.R. Stopher, A.H. Meyburg and W. Brög (eds.), *New Horizons in Travel Behaviour Research*. D.C. Health and Co., Lexington, MA.

Whitelegg, J. (1993) *Transport for a Sustainable Future: The Case for Europe*. Belhaven Press, London.

Williams, H.C.W.L. (1977) On the formation of travel demand models and economic evaluation measures of user benefit. *Environment and Planning* **9A**, 285–344.

Williams, H.C.W.L. (1981) Travel demand forecasting: an overview of theoretical developments. In D.J. Banister and P.G. Hall (eds.), *Transport and Public Policy Planning*. Mansell, London.

Williams, H.C.W.L. and Ortúzar, J. de D. (1982a) Behavioural theories of dispersion and the mis-specification of travel demand models. *Transportation Research* **16B**, 167–219.

Williams, H.C.W.L. and Ortúzar, J. de D. (1982b) Travel demand and response analysis-some integrating themes. *Transportation Research* **16A**, 345–362.

Williams, H.C.W.L. and Senior, M.L. (1977) Model based transport policy assessment: (2) Removing fundamental inconsistencies from the models. *Traffic Engineering and Control* **18**, 464–469.

Williams, I. (1976) A comparison of some calibration techniques for doubly constrained models with an exponential cost function. *Transportation Research* **10**, 91–104.

Willis, K.G., Garrod, G.D. and Harvey, D.R. (1998) A review of cost-benefit analysis as applied to the evaluation of new road proposals in the UK. *Transportation Research* **3D**, 141–156.

Wills, M.J. (1986) A flexible gravity-opportunities model for trip distribution. *Transportation Research* **20B**, 89–111.

Willumsen, L.G. (1978) Estimation of an O–D matrix from traffic counts: a review. *Working Paper 99*, Institute for Transport Studies, University of Leeds.

Willumsen, L.G. (1982) Estimation of trip matrices from volume counts; validation of a model under congested conditions. *Proceedings 10th PTRC Summer Annual Meeting*, University of Warwick, England.

Willumsen, L.G. (1984) Estimating time-dependent trip matrices from traffic counts. In J. Volmüller and R. Hamerslag (eds.), *Proceedings Ninth International Symposium on Transportation and Traffic Theory*. VNU Science Press, Utrecht.

Willumsen, L.G. (1991) Origin-destination matrix: static estimation. In M. Papageorgiou (ed.), *Concise Encyclopedia of Traffic and Transportation Systems*. Pergamon Press, Oxford.

Willumsen, L.G. (2007) Travel networks. In D.A. Hensher and K.J. Button (eds.), *Handbook of Transport Modelling*. Second Edition. Pergamon, Oxford.

Willumsen, L.G. (2014) *Better Traffic and Revenue Forecasting*. Maida Vale Press, London.

Willumsen, L.G. (2021) Use of big data in transport modelling. *Discussion Paper No. 2021/05*, International Transport Forum, OECD Publishing, Paris.

Willumsen, L.G., Bolland, J., Arezki, Y. and Hall, M. (1993) Multi-modal modelling in congested networks: SATURN + SATCHMO. *Traffic Engineering and Control* **34**, 294–301.

Willumsen, L.G. and Hounsell, N.B. (1998) Simple models of highway reliability–supply effects. In J. de D. Ortúzar, D.A. Hensher and S.R. Jara-Díaz (eds.), *Travel Behaviour Research: Updating the State of Play*. Pergamon, Oxford.

Willumsen, L.G. and Ortúzar, J. de D. (1985) Intuition and models in transport management. *Transportation Research* **19A**, 51–58.

Willumsen, L.G. and Ortúzar, J. de D. (2016) Transport planning. In M.C.J. Bliemer, C. Mulley and C. Moutou (eds.), *Handbook on Transport and Urban Planning in the Developed World*. Edward Elgar Publishing, London.

Willumsen, L.G. and Radovanać, M. (1988) Testing the practical value of the UMOT model. *International Journal of Transport Economics* **15**, 203–223.

Wilson, A.G. (1970) *Entropy in Urban and Regional Modelling*. Pion, London.

Wilson, A.G. (1974) *Urban and Regional Models in Geography and Planning*. Wiley, Chichester.

Wilson, A.G., Rees, P.H. and Leigh, C.M. (eds.) (1977) *Models of Cities and Regions: Theoretical and Empirical Developments*. Wiley, Chichester.

Winters, M., Davidson, G., Kao, D. and Teschke, K. (2011) Motivators and deterrents of bicycling: comparing influences on decisions to ride. *Transportation* **38**, 153–168.

Wittink, D., Krishnamurthi, L. and Nutter, J. (1982) Comparing derived importance weights accross attributes. *Journal of Consumer Research* **8**, 471–474.

Wonnacott, T.H. and Wonnacott, R.J. (1990) *Introductory Statistics for Business and Economics*. Wiley, New York.

Wood, W., Quinn, J.M. and Kashy, D.A. (2002) Habit in everyday life: thought, emotion, and action. *Journal of Personality and Social Psychology* **83**, 1281–1297.

Wooldridge, J. (2010) *Econometric Analysis of Cross-Section and Panel Data*. Second Edition. MIT Press, Cambridge, MA.

Wootton, H.J., Ness, M.P. and Burton, R.S. (1981) Improved direction signs and the benefits for road users. *Traffic Engineering and Control* **22**, 264–268.

Wootton, H.J. and Pick, G.W. (1967) A model for trips generated by households. *Journal of Transport Economics and Policy* **1**, 137–153.

Yamamoto, T. and Komori, R. (2010) Mode choice analysis with imprecise location information. *Transportation* **37**, 491–503.

Yáñez, M.F., Cherchi, E., Heydecker, B.G. and Ortúzar, J. de D. (2011) On the treatment of repeated observations in panel data: efficiency of mixed logit parameter estimates. *Networks and Spatial Economics* **11**, 393–418.

Yáñez, M.F., Cherchi, E. and Ortúzar, J. de D. (2009) Inertia and shock effects on mode choice panel data: implications of the Transantiago implementation. *12th International Conference on Travel Behaviour Research*, Jaipur, India.

Yáñez, M.F., Cherchi, E. and Ortúzar, J. de D. (2010a) Defining inter-alternative error structures for joint RP-SP modeling: some new evidence. *Transportation Research Record* **2175**, 65–73.

Yáñez, M.F., Mansilla, P. and Ortúzar, J. de D. (2010b) The Santiago Panel: measuring the effects of implementing Transantiago. *Transportation* **37**, 125–149.

Ye, X., Konduri, K., Pendyala, R.M., Sana, B. and Waddell, P. (2009) A methodology to match distributions of both household and person attributes in the generation of synthetic populations. *88th Annual TRB Meeting*, Washington, DC, USA.

Yen, J., Mahmasani, H.S. and Herman, R. (1998) A model of employee participation in telecommuting programs based on stated preference data. In J. de D. Ortúzar, D.A. Hensher and S.R. Jara-Díaz (eds.), *Travel Behaviour Research: Updating the State of Play*. Pergamon, Oxford.

Youn, H., Gastner, M.T. and Jeong, H. (2008) Price of anarchy in transportation networks: efficiency and optimality control. *Physical Review Letters* **101**, 128701.

Young, W. and Richardson, A.J. (1980) Residential location preference models: compensatory and non-compensatory approaches. *Proceedings 8th PTRC Summer Annual Meeting*, University of Warwick, England.

Yu, X. and Prevedouros, P.D. (2017) Risk assessment: method and case study for traffic projects. *Journal of Modern Transportation* **25**, 236–249.

Zahavi, Y. (1979) The UMOT project. *Report No. DoT-RSPA-DPB-20-79-3*, US Department of Transportation, Washington, DC.

Zhang, X. (ed.) (2022) *Cities for Driverless Vehicles: Planning the Future Built Environment with Shared Mobility*. ICE Publishing, London.

Zhao, Y. and Kockelman, K. (2002) Propagation of uncertainty through demand models: an exploratory analysis. *Annals of Regional Science* **36**, 145–163.

Zimowski, M., Tourangeau, R., Ghadialy, R and Pedlow, S. (1997) *Non-Response in Household Travel Surveys*. TMIP Report Prepared for the Federal Highway Administration, US Department of Transportation by NORC, Chicago.

Zmud, J., Ecola, L., Phleps, P. and Feige, I. (2013) *The Future of Mobility: Scenarios for the United States in 2030*. Institute for Mobility Research and Rand Corporation, Washington, DC.

Zöllig-Renner, C., Nicolai, T.W. and Nagel, K. (2015) Agent-based land use transport interaction modelling: state of the art. In M. Bierlaire, A. de Palma, R. Hurtubia and P. Waddell (eds.), *Integrated Transport and Land Use Modelling for Sustainable Cities*. EPFL Press/Routledge, London.

Index

Modelling Transport, Fifth Edition. Juan de Dios Ortúzar and Luis G. Willumsen.
© 2024 John Wiley & Sons Ltd. Published 2024 by John Wiley & Sons Ltd.
Companion website: www.wiley.com/go/ortuzar5e